Lecture Notes in Comp

Founding Editors

Gerhard Goos
Juris Hartmanis

Editorial Board Members

Elisa Bertino, *Purdue University, West Lafayette, IN, USA*
Wen Gao, *Peking University, Beijing, China*
Bernhard Steffen , *TU Dortmund University, Dortmund, Germany*
Moti Yung , *Columbia University, New York, NY, USA*

The series Lecture Notes in Computer Science (LNCS), including its subseries Lecture Notes in Artificial Intelligence (LNAI) and Lecture Notes in Bioinformatics (LNBI), has established itself as a medium for the publication of new developments in computer science and information technology research, teaching, and education.

LNCS enjoys close cooperation with the computer science R & D community, the series counts many renowned academics among its volume editors and paper authors, and collaborates with prestigious societies. Its mission is to serve this international community by providing an invaluable service, mainly focused on the publication of conference and workshop proceedings and postproceedings. LNCS commenced publication in 1973.

Aleš Leonardis · Elisa Ricci · Stefan Roth ·
Olga Russakovsky · Torsten Sattler · Gül Varol
Editors

Computer Vision – ECCV 2024

18th European Conference
Milan, Italy, September 29–October 4, 2024
Proceedings, Part LXXXII

 Springer

Editors
Aleš Leonardis
University of Birmingham
Birmingham, UK

Elisa Ricci
University of Trento
Trento, Italy

Stefan Roth
Technical University of Darmstadt
Darmstadt, Germany

Olga Russakovsky
Princeton University
Princeton, NJ, USA

Torsten Sattler
Czech Technical University in Prague
Prague, Czech Republic

Gül Varol
École des Ponts ParisTech
Marne-la-Vallée, France

ISSN 0302-9743   ISSN 1611-3349 (electronic)
Lecture Notes in Computer Science
ISBN 978-3-031-73006-1   ISBN 978-3-031-73007-8 (eBook)
https://doi.org/10.1007/978-3-031-73007-8

© The Editor(s) (if applicable) and The Author(s), under exclusive license
to Springer Nature Switzerland AG 2025

This work is subject to copyright. All rights are solely and exclusively licensed by the Publisher, whether the whole or part of the material is concerned, specifically the rights of translation, reprinting, reuse of illustrations, recitation, broadcasting, reproduction on microfilms or in any other physical way, and transmission or information storage and retrieval, electronic adaptation, computer software, or by similar or dissimilar methodology now known or hereafter developed.
The use of general descriptive names, registered names, trademarks, service marks, etc. in this publication does not imply, even in the absence of a specific statement, that such names are exempt from the relevant protective laws and regulations and therefore free for general use.
The publisher, the authors and the editors are safe to assume that the advice and information in this book are believed to be true and accurate at the date of publication. Neither the publisher nor the authors or the editors give a warranty, expressed or implied, with respect to the material contained herein or for any errors or omissions that may have been made. The publisher remains neutral with regard to jurisdictional claims in published maps and institutional affiliations.

This Springer imprint is published by the registered company Springer Nature Switzerland AG
The registered company address is: Gewerbestrasse 11, 6330 Cham, Switzerland

If disposing of this product, please recycle the paper.

Foreword

Welcome to the proceedings of the European Conference on Computer Vision (ECCV) 2024, the 18th edition of the conference, which has been held biennially since its founding in France in 1990. We are holding the event in Italy for the third time, in a convention centre where we can accommodate nearly 7000 in-person attendees.

The journey to ECCV 2024 started at CVPR 2019, where the general chairs met at a sketchy bar in Long Beach to discuss ideas on how to organize an amazing future ECCV. After that, Covid came and changed conferences probably forever, but we maintained our excitement to organize an awesome in-person conference in 2024, particularly having seen the wonderful event that was ECCV 2022 in Tel Aviv. And here we are, 5 years later, delighted that we finally get to welcome all of you to Milan!

First and foremost, ECCV is about the exchange of scientific ideas, and hence the most important organizational task is the selection of the programme. To our Program Chairs, Aleš Leonardis, Elisa Ricci, Gül Varol, Olga Russakovsky, Stefan Roth, and Torsten Sattler, our entire community owe a huge debt of gratitude. As you will read in their Preface to the proceedings, they dealt with over 8,500 submissions, and their diligence, thoroughness, and sheer effort has been inspirational. From the early stages of designing the call for papers, through recruiting and selecting area chairs and reviewers, to ensuring every paper has a fair and thorough assessment, their professionalism, commitment, and attention to detail has been exemplary. From all the choices that we made to organize ECCV, choosing this amazing team of Program Chairs has been, without a doubt, our very best decision.

The Program Chairs were advised and supported by an Ethics Review Committee, Chloé Bakalar, Kate Saenko, Remi Denton, and Yisong Yue, who provided invaluable input on papers where reviewers or Area Chairs had raised ethical concerns.

The Publication Chairs, Jovita Lukasik, Michael Möller, François Brémond, and Mahmoud Ali, did great work in assembling the camera-ready papers into these Springer volumes, dealing with numerous issues that arose promptly and professionally.

With the ever-expanding importance of workshops, tutorials, and demos at our major conferences comes a corresponding increase in the challenge of organizing over 70 workshops and 9 tutorials attached to the main conference. The Workshop and Tutorial Chairs, Alessio Del Bue, Jordi Pont-Tuset, Cristian Canton, and Tatiana Tommasi, and Demo Chairs, Hyung Chang and Marco Cristani, worked tirelessly to coordinate the very varied demands and structures of these different events, yielding an extremely rich auxiliary program which will enhance the conference experience for everyone who attends.

The Industry Chairs, Shaogang Gong and Cees Snoek, working with Limor Urfaly and Lior Gelfand, managed to attract numerous companies and organizations from all around the world to the industrial Expo, making ECCV 2024 an international event that not only showcases excellent foundational research, but also presents how such research can be transformed in real products. It is notable that together with the big companies, several start-ups chose ECCV to present their business ideas and activities.

As usual in the recent main conference, we also organized Speed Mentoring and Doctoral Consortium events. While the latter is an institutional occasion allowing senior PhDs to show their work, the former is especially important to answer to the many doubts and uncertainties young scholars have while undertaking a career in Computer Vision and Artificial Intelligence. We warmly thank our Social Activities Chairs, Raffaella Lanzarotti, Simone Bianco and Giovanni Farinella, and the Doctoral Consortium Chairs, Cigdem Beyan and Or Litany, for their dedication to organize such events, so important for our young colleagues, as well as all the mentors who were available to share their experience.

As the conference becomes larger and larger, the overall budget increases to a level where uncertainties in estimates of attendance can result in significant losses, perhaps enough to significantly damage the prospects of running the conference in the future. This risk imposed the need to require authors to commit to full conference registrations, with the undesirable consequent possibility to cause hardship to authors, particularly student authors, from organizations where funding attendance is difficult. To mitigate this effect, we allocated travel grants to students who might otherwise have had difficulty in attending. Our Diversity Chairs, David Fouhey and Rita Cucchiara, put tremendous effort into a wonderfully careful and thoughtful programme, balancing diversity in all its forms with a changing budget landscape, allowing us to support the participation of many people who might not otherwise have been able to attend.

This year's conference also marks the move to a unified, multi-year conference website, based on that used for other leading computer vision and AI conferences. This was an effort that we decided to take on to make the ECCV website more familiar to the community and support the organization of the future ECCV editions. Our thanks go to Lee Campbell of Eventhosts who managed the entire design and development of the new website and provided responsive support and updates throughout the conference organization timeline.

Our Finance Chairs, Gérard Medioni and Nicole Finn, provided invaluable advice and assistance on all aspects of the financial planning of the conference, in conjunction with the professional conference organizers AIM Group, where Lavinia Ricci led a fantastic team that managed thousands of details from space planning to food to registrations. In particular, Elder Bromley and Mara Carletti were a tremendous support to the entire conference organization. The conference centre, Mico DMC, also provided an excellent service, showing considerable flexibility in adapting the venue and processes to the particular demands of an academic-focused conference.

Finally, we cannot forget to thank our Social Media Chair, Kosta Derpanis, supported by Abby Stylianou and Jia-Bin Huang, who, with their presence on social media, spread ECCV news and responded to small and big questions in a timely manner, helping the community to promptly receive information and fix their problems.

October 2024

Andrew Fitzgibbon
Laura Leal-Taixé
Vittorio Murino

Preface

ECCV 2024 saw a remarkable 48% increase in submissions, with 8585 valid papers submitted, compared to 5804 in ECCV 2022, 5150 in 2020, and 2439 in 2018. Of these, 2387 papers (27.9%) were accepted for publication, with 200 (2.3% overall) selected for oral presentations.

A total of 173 complete submissions were desk-rejected for various reasons. Common issues included revealing author identities in the paper or supplementary material (for example: by including the author names, referring to specific grants in the acknowledgments, including links to GitHub accounts with visible author information, etc.), exceeding the page limit or otherwise significantly altering the submission template, or posting an arXiv preprint explicitly mentioning the work as an ECCV 2024 submission. The desk-rejected submissions also include 14 papers identified as dual submissions to other concurrent conferences, such as ICML, NeurIPS, and MICCAI. Seven papers were rejected among the accepted papers due to plagiarism issues, after the Program Chairs conducted a thorough plagiarism check with the iThenticate software followed by manual inspection.

The double-blind review process was managed using the CMT system, ensuring anonymity between authors and Area Chairs/reviewers. Each paper received at least three reviews, amounting to over 26,000 reviews in total. To maintain the quality and the fairness of the reviews, we recruited 469 Area Chairs (ACs) and more than 8200 reviewers. Of the recruited reviewers, 7293 ultimately participated.

Area Chairs were chosen for their technical expertise and reputation; many of them had served in similar roles at other top conferences. We also recruited additional Area Chairs through an open call for self-nominations. Among the selected ACs, around 18% were women. Geographical distribution was taken into account during the selection process, with 31% of ACs working in Europe, 38% in the Americas, 27% in Asia, and 4% in the rest of the world. Despite our best efforts to consider diversity factors in the selection process, the gender diversity and geographic diversity of ACs is still below what we were hoping for.

To ensure a smooth decision-making process, we grouped the ACs into triplets. Upon accepting their role, ACs were requested to enter into the CMT system domain conflicts and subject areas, as well as information about their DBLP, TPMS (Toronto Paper Matching System), and OpenReview profiles. These were used to determine whether different subject areas were sufficiently covered, to detect conflicts, and to automatically form AC triplets. An AC triplet was a group of three ACs who worked together throughout the process, which helped to facilitate discussion, calibrate the decision-making process, and provide support for first-time ACs. AC triplets were manually refined by Program Chairs to ensure no triplet contained more than two major time zones to enable easier coordination for virtual AC meetings. Paper assignments were made such that no AC within the triplet was conflicted with any of the papers. Each AC was responsible for overseeing 18 papers on average.

In each triplet, we identified a Lead AC who was an experienced member of the computer vision community willing to take on additional administrative duties. The primary responsibilities of Lead ACs included coordinating the AC triplet to ensure deadlines were met and offering guidance to less experienced ACs. They also served as the main contact person with the Program Chairs and helped in handling papers if an original AC was unable to complete their tasks.

Reviewers were invited from reviewer pools of previous conferences and also selected from the pool of authors. We asked experienced ACs to recommend more potential reviewers. We further recruited reviewers through self-nominations, where researchers volunteered to review for ECCV 2024 via an open call. The self-nominations were then filtered, selecting those reviewers with at least two papers published in related conferences such as ECCV, ICCV, CVPR, ICLR, NeurIPS, etc.

Similarly to ACs, reviewers were also asked to update their CMT, Toronto Paper Matching System (TPMS), and OpenReview profiles. Reviewers had the option to classify themselves as either senior researchers or junior researchers and students, with the number of assigned papers varying accordingly (a quota of 6 for senior researchers with 3 or more times reviewing experience, and a quota of 4 otherwise). On average, each reviewer was assigned about 4 papers.

Conflicts of interest among authors, ACs, and reviewers were managed automatically via conflict domains and co-authorship information from DBLP. Paper assignments to ACs and reviewers were determined using an algorithm that combined subject-area affinity scores from CMT, TPMS, and OpenReview, along with additional criteria such as ensuring that every paper had at least one non-student reviewer. For AC assignments, we additionally incorporated a newly built affinity score based on embedding similarity with the papers on the Google Scholar profile of the ACs. Once the assignments were made, ACs and reviewers were asked to promptly report any undetected conflicts of interest or other concerns that would preclude them from handling the assigned papers to ensure immediate reassignment, if necessary.

Given the widespread use of Large Language Models (LLMs), ECCV 2024, like previous conferences, implemented a special policy regulating their use. Authors were permitted to use any tools, including LLMs, in preparing their papers, but they remained fully responsible for any misrepresentation, factual inaccuracies, or plagiarism. Reviewers were also allowed to use LLMs to refine the wording of their reviews; however, they were held accountable for the accuracy of their content and were strictly prohibited from inputting submissions into an LLM. One challenge we ran into was that tools that detect LLM-generated content were not able to distinguish between text polished vs. text written from scratch by an LLM. Thus, it became quite difficult to enforce the policy in practice.

Overall, the challenges encountered during the review process were consistent with those faced at previous computer vision conferences. For instance, a small number of reviewers ultimately did not submit their reviews or provided brief, uninformative reviews, necessitating last-minute reassignment to emergency reviewers. As the community expands and the number of submissions rises, securing enough qualified reviewers and performing a quality check of the received reviews becomes increasingly difficult.

The review process included a rebuttal phase where authors were given a week to address any concerns raised by the reviewers. Each response was limited to a single PDF page using a predefined template. Rebuttals with formatting or anonymity violations were removed by the Program Chairs. ACs then facilitated discussions with reviewers to evaluate the merits of each submission. While the goal was to reach a consensus, the final decision was left to the ACs in the triplet to ensure fairness. As is common, the ACs within each triplet were tasked with organizing meetings to discuss papers, especially borderline cases. The entire process was conducted online, with no in-person meetings. For each paper, a primary and a secondary AC were assigned from the AC triplet. The primary AC was responsible for entering the decisions and meta-reviews into the CMT system, which were then reviewed and approved by the secondary ACs. Besides making accept/reject decisions, the ACs were also required to provide recommendations regarding oral/poster presentations and award nominations.

Reviewers and ACs were asked to flag papers with potential ethical concerns. An ethics review committee, consisting of four leading researchers from both industry and academia, was appointed. The role of the committee was to evaluate papers escalated by the AC for further investigation. In total, 31 papers were flagged to the committee. The committee requested that the authors of 15 papers address specific concerns in their camera-ready submissions. The authors all complied with the requests.

Following the recent IEEE decision to ban the use of the photograph of Lena Forsén, we automatically identified all submissions that included this image. Although the ban did not officially apply to ECCV, we nevertheless shared the context with the authors of the affected papers and suggested that they remove the photograph in the camera-ready submission. All authors complied.

To help the ACs to track the overall progress of the reviewing and decision-making processes, we created AC triplet activity dashboards for each AC triplet (via Google sheets), which were updated on a regular basis. During the review process, the activity dashboard enabled ACs to track the progress of incoming reviews for each paper and promptly identify papers that necessitated attention, e.g., papers potentially requiring the invitation of emergency reviewers, or papers with suspiciously short reviews. During the decision-making process, the dashboard was also used by the Program Chairs to identify papers with incomplete review information, e.g., for which reviewers did not enter their final recommendation, the AC did not enter their final decision or meta-review, the secondary AC did not confirm the recommendation, etc.

At the end of the decision-making progress, the Program Chairs carefully examined the very small number of cases where ACs overruled a unanimous reviewer recommendation for acceptance or rejection. In such instances, it was crucial for the primary and secondary ACs to explain in detail the motivation behind the decision. To ensure a fair evaluation, an additional Area Chair, not originally part of the decision-making triplet, was appointed to provide an independent opinion and advise the Program Chairs in determining whether to uphold or reverse the overturn.

Inevitably, after decisions were released, some authors were dissatisfied. The authors were given the possibility to appeal the final decision in case of procedural issues encountered during the review process by filling out a form to directly contact the Program Chairs. Valid reasons for appealing a decision included policy errors (e.g., reviewers or ACs enforcing a non-existent policy), clerical errors (e.g., the meta-review clearly indicated an intention to accept a paper, but it was mistakenly rejected), and significant misunderstandings by the reviewers or ACs. In total, we received 59 appeals and upheld 6 of them. One was due to a clear policy error by an AC, and the paper decision changed from reject to accept following a successful appeal. The other 5 papers were already conditionally accepted – but the AC/reviewers wrongly flagged them as papers for which the associated dataset was required to be released prior to final acceptance; this condition was removed following the appeal.

After the camera-ready deadline, additional checks were conducted, particularly for papers contributing a new dataset and for those flagged for potential ethical concerns. Specifically, during the submission phase, authors were required to specify if the primary contribution of their papers was the release of a new dataset, and for these papers, a valid dataset link had to be provided by the camera-ready deadline. A total of 435 submissions included a dataset as part of their contribution, and all complied with the requirement to release the dataset by the camera-ready deadline.

We sincerely thank all the ACs and reviewers for their invaluable contributions to the review process. Their dedication and expertise were crucial in maintaining the high standards of the conference. We would also like to thank our Technical Program Chair, Sascha Hornauer, who did tremendous work behind the scenes, and to the entire CMT team for their prompt support with any technical issues.

October 2024

Aleš Leonardis
Elisa Ricci
Stefan Roth
Olga Russakovsky
Torsten Sattler
Gül Varol

Organization

General Chairs

Andrew Fitzgibbon Graphcore, UK
Laura Leal-Taixé NVIDIA, Italy
Vittorio Murino University of Verona & University of Genoa, Italy

Program Chairs

Aleš Leonardis University of Birmingham, UK
Elisa Ricci University of Trento, Italy
Stefan Roth Technical University of Darmstadt, Germany
Olga Russakovsky Princeton University, USA
Torsten Sattler Czech Technical University in Prague, Czech Republic
Gül Varol Ecole des Ponts Paris Tech, France

Technical Program Chair

Sascha Hornauer Mines Paris – PSL, France

Industrial Liaison Chairs

Cees Snoek University of Amsterdam, Netherlands
Shaogang Gong Queen Mary University of London, UK

Publication Chairs

Francois Bremond Inria, France
Mahmoud Ali Inria, France
Jovita Lukasik University of Siegen, Germany
Michael Moeller University of Siegen, Germany

Poster Chairs

Aljoša Ošep — Carnegie Mellon University, USA
Zuzana Kukelova — Czech Technical University in Prague, Czech Republic

Diversity Chairs

David Fouhey — New York University, USA
Rita Cucchiara — Università di Modena e Reggio Emilia, Italy

Local Chairs

Raffaella Lanzarotti — Università degli Studi di Milano, Italy
Simone Bianco — University of Milano-Bicocca, Italy

Conference Ombuds

Georgia Gkioxari — California Institute of Technology, USA
Greg Mori — Borealis AI & Simon Fraser University, Canada

Ethics Review Committee

Chloé Bakalar — Meta, USA
Kate Saenko — Boston University, USA
Remi Denton — Google Research, USA
Yisong Yue — Caltech, Asari AI & Latitude AI, USA

Workshop and Tutorial Chairs

Alessio Del Bue — Istituto Italiano di Tecnologia, Italy
Cristian Canton — Meta AI, USA
Jordi Pont-Tuset — Google DeepMind, Switzerland
Tatiana Tommasi — Politecnico di Torino, Italy

Finance Chairs

Gerard Medioni Amazon, USA
Nicole Finn c to c events, USA

Demo Chairs

Hyung Chang University of Birmingham, UK
Marco Cristani University of Verona, Italy

Publicity and Social Media Chairs

Konstantinos Derpanis York University & Samsung AI Centre Toronto, Canada
Jia-Bin Huang University of Maryland College Park, USA
Abby Stylianou Saint Louis University, USA

Social Activities Chairs

Giovanni Maria Farinella University of Catania, Italy
Raffaella Lanzarotti Università degli Studi di Milano, Italy
Simone Bianco University of Milano-Bicocca, Italy

Doctoral Consortium Chairs

Cigdem Beyan University of Verona, Italy
Or Litany NVIDIA & Technion, USA

Web Developer

Lee Campbell Eventhosts, USA

Area Chairs

Ehsan Adeli	Stanford University, USA
Aishwarya Agrawal	University of Montreal & Mila & DeepMind, Canada
Naveed Akhtar	University of Western Australia, Australia
Yagiz Aksoy	Simon Fraser University, Canada
Karteek Alahari	Inria, France
Jose Alvarez	NVIDIA, USA
Djamila Aouada	Interdisciplinary Centre for Security, Reliability and Trust, University of Luxembourg, Luxembourg
Andre Araujo	Google, Brazil
Pablo Arbelaez	Universidad de los Andes, Colombia
Iro Armeni	Stanford University, USA
Yuki Asano	University of Amsterdam, Netherlands
Karl Åström	Lund University, Sweden
Mathieu Aubry	Ecole des Ponts ParisTech, France
Shai Bagon	Weizmann Institute of Science, Israel
Song Bai	ByteDance, Singapore
Xiang Bai	Huazhong University of Science and Technology, China
Yang Bai	Tencent, China
Lamberto Ballan	University of Padova, Italy
Linchao Bao	University of Birmingham, UK
Ronen Basri	Weizmann Institute of Science, Israel
Sara Beery	Massachusetts Institute of Technology, USA
Vasileios Belagiannis	University of Erlangen–Nuremberg, Germany
Serge Belongie	University of Copenhagen, Denmark
Sagie Benaim	Hebrew University of Jerusalem, Israel
Rodrigo Benenson	Google, Switzerland
Cigdem Beyan	University of Bergamo, Italy
Bharat Bhatnagar	Meta Reality Labs Research, Switzerland
Tolga Birdal	Imperial College London, UK
Matthew Blaschko	KU Leuven, Belgium
Federica Bogo	Meta Reality Labs Research, Switzerland
Timo Bolkart	Google, Switzerland
Katherine Bouman	California Institute of Technology, USA
Lubomir Bourdev	Primepoint, USA
Edmond Boyer	Inria, France
Yuri Boykov	University of Waterloo, Canada
Eric Brachmann	Niantic, Germany

Wieland Brendel	University of Tübingen, Germany
Gabriel Brostow	University College London, UK
Michael Brown	York University, Canada
Andrés Bruhn	University of Stuttgart, Germany
Andrei Bursuc	valeo.ai, France
Benjamin Busam	Technical University of Munich, Germany
Holger Caesar	TU Delft, Netherlands
Jianfei Cai	Monash University, Australia
Simone Calderara	University of Modena and Reggio Emilia, Italy
Octavia Camps	Northeastern University, Boston, USA
Joao Carreira	DeepMind, UK
Silvia Cascianelli	University of Modena and Reggio Emilia, Italy
Ayan Chakrabarti	Google Research, USA
Tat-Jen Cham	Nanyang Technological University, Singapore
Antoni Chan	City University of Hong Kong, China
Manmohan Chandraker	University of California, San Diego, USA
Xiaojun Chang	University of Technology Sydney, Australia
Devendra Singh Chaplot	Mistral AI, USA
Liang-Chieh Chen	TikTok, USA
Long Chen	Hong Kong University of Science and Technology, China
Mei Chen	Microsoft, USA
Shizhe Chen	Inria, France
Shuo Chen	RIKEN, Japan
Xilin Chen	Institute of Computing Technology, Chinese Academy of Sciences, China
Anoop Cherian	Mitsubishi Electric Research Labs, USA
Tat-Jun Chin	University of Adelaide, Australia
Minsu Cho	Pohang University of Science and Technology, Korea
Hisham Cholakkal	Mohamed bin Zayed University of Artificial Intelligence, United Arab Emirates
Yung-Yu Chuang	National Taiwan University, Taiwan
Ondrej Chum	Czech Technical University in Prague, Czech Republic
Marcella Cornia	University of Modena and Reggio Emilia, Italy
Marco Cristani	University of Verona, Italy
Cristian Canton	Meta AI, USA
Zhaopeng Cui	Zhejiang University, China
Angela Dai	Technical University of Munich, Germany
Jifeng Dai	Tsinghua University, China
Yuchao Dai	Northwestern Polytechnical University, China

Dima Damen	University of Bristol, UK
Kostas Daniilidis	University of Pennsylvania, USA
Antitza Dantcheva	Inria, France
Raoul de Charette	Inria, France
Shalini De Mello	NVIDIA Research, USA
Tali Dekel	Weizmann Institute of Science, Israel
Ilke Demir	Intel Corporation, USA
Joachim Denzler	Friedrich Schiller University Jena, Germany
Konstantinos Derpanis	York University, Canada
Jose Dolz	École de technologie supérieure Montreal, Canada
Jiangxin Dong	Nanjing University of Science and Technology, China
Hazel Doughty	Leiden University, Netherlands
Iddo Drori	Boston University and Columbia University, USA
Enrique Dunn	Stevens Institute of Technology, USA
Thibaut Durand	Borealis AI, Canada
Sayna Ebrahimi	Google, USA
Mohamed Elhoseiny	King Abdullah University of Science and Technology, Saudi Arabia
Bin Fan	University of Science and Technology Beijing, China
Yi Fang	New York University, USA
Giovanni Maria Farinella	University of Catania, Italy
Ryan Farrell	Brigham Young University, USA
Alireza Fathi	Google, USA
Christoph Feichtenhofer	Meta & FAIR, USA
Rogerio Feris	MIT-IBM Watson AI Lab & IBM Research, USA
Basura Fernando	Agency for Science, Technology and Research (A*STAR), Singapore
David Fouhey	New York University, USA
Katerina Fragkiadaki	Carnegie Mellon University, USA
Friedrich Fraundorfer	Graz University of Technology, Austria
Oren Freifeld	Ben-Gurion University, Israel
Huan Fu	University of Sydney, China
Yasutaka Furukawa	Simon Fraser University, Canada
Andrea Fusiello	University of Udine, Italy
Fabio Galasso	Sapienza University, Italy
Jürgen Gall	University of Bonn, Germany
Orazio Gallo	NVIDIA Research, USA
Chuang Gan	MIT-IBM Watson AI Lab, USA
Lianli Gao	University of Electronic Science and Technology of China, China

Shenghua Gao	University of Hong Kong, China
Sourav Garg	University of Adelaide, Australia
Efstratios Gavves	University of Amsterdam, Netherlands
Peter Gehler	Zalando, Germany
Theo Gevers	University of Amsterdam, Netherlands
Golnaz Ghiasi	Google DeepMind, USA
Rohit Girdhar	Carnegie Mellon University, USA
Xavier Giro-i-Nieto	Universitat Politecnica de Catalunya, Spain
Ross Girshick	Allen Institute for Artificial Intelligence, USA
Zan Gojcic	NVIDIA, Switzerland
Vladislav Golyanik	Max Planck Institute for Informatics, Germany
Stephen Gould	Australian National University, Australia
Liangyan Gui	University of Illinois Urbana-Champaign, USA
Fatma Guney	Koc University, Turkey
Yulan Guo	Sun Yat-sen University, China
Yunhui Guo	University of Texas at Dallas, USA
Mohit Gupta	University of Wisconsin-Madison, USA
Minh Ha Quang	RIKEN Center for Advanced Intelligence Project, Japan
Bumsub Ham	Yonsei University, Korea
Bohyung Han	Seoul National University, Korea
Kai Han	University of Hong Kong, China
Xiaoguang Han	Shenzhen Research Institute of Big Data & Chinese University of Hong Kong (Shenzhen), China
Tatsuya Harada	University of Tokyo & RIKEN, Japan
Tal Hassner	Weir AI, USA
Xuming He	ShanghaiTech University, China
Felix Heide	Princeton & Algolux, USA
Anthony Hoogs	Kitware, USA
Timothy Hospedales	Edinburgh University, UK
Mahdi Hosseini	Concordia University, Canada
Peng Hu	College of Computer Science, Sichuan University, China
Gang Hua	Wormpex AI Research, USA
Jia-Bin Huang	University of Maryland College Park, USA
Qixing Huang	University of Texas at Austin, USA
Sharon Xiaolei Huang	Pennsylvania State University, USA
Junhwa Hur	Google, USA
Nazli Ikizler-Cinbis	Hacettepe University, Turkey
Eddy Ilg	Saarland University, Germany
Ahmet Iscen	Google, France

Nathan Jacobs	Washington University in St. Louis, USA
Varun Jampani	Stability AI, USA
C. V. Jawahar	International Institute of Information Technology - Hyderabad, India
Laszlo Jeni	Carnegie Mellon University, USA
Jiaya Jia	Chinese University of Hong Kong, China
Qin Jin	Renmin University of China, China
Jungseock Joo	NVIDIA & University of California, Los Angeles, USA
Fredrik Kahl	Chalmers, Sweden
Yannis Kalantidis	NAVER LABS Europe, France
Vicky Kalogeiton	Ecole Polytechnique, IP Paris, France
Evangelos Kalogerakis	University of Massachusetts Amherst, USA
Hiroshi Kawasaki	Kyushu University, Japan
Margret Keuper	University of Mannheim & Max Planck Institute for Informatics, Germany
Salman Khan	Mohamed bin Zayed University of Artificial Intelligence, United Arab Emirates
Anna Khoreva	Bosch Center for Artificial Intelligence, Germany
Gunhee Kim	Seoul National University, Korea
Junmo Kim	Korea Advanced Institute of Science and Technology, Korea
Min H. Kim	Korea Advanced Institute of Science and Technology, Korea
Seon Joo Kim	Yonsei University, Korea
Tae-Kyun Kim	Korea Advanced Institute of Science and Technology & Imperial College London, Korea
Benjamin Kimia	Brown University, USA
Hedvig Kjellström	KTH Royal Institute of Technology, Sweden
Laurent Kneip	ShanghaiTech University, China
A. Sophia Koepke	University of Tübingen, Germany
Iasonas Kokkinos	University College London, UK
Wai-Kin Adams Kong	Nanyang Technological University, Singapore
Piotr Koniusz	Data61/CSIRO & Australian National University, Australia
Jana Kosecka	George Mason University, USA
Adriana Kovashka	University of Pittsburgh, USA
Hilde Kuehne	University of Bonn, Germany
Zuzana Kukelova	Czech Technical University in Prague, Czech Republic
Ajay Kumar	Hong Kong Polytechnic University, China
Kiriakos Kutulakos	University of Toronto, Canada

Suha Kwak	Pohang University of Science and Technology, Korea
Zorah Laehner	University of Bonn, Germany
Shang-Hong Lai	National Tsing Hua University, Taiwan
Iro Laina	University of Oxford, UK
Jean-Francois Lalonde	Université Laval, Canada
Loic Landrieu	École des Ponts ParisTech, France
Diane Larlus	Naver Labs Europe, France
Viktor Larsson	Lund University, Sweden
Stéphane Lathuilière	Telecom-Paris, France
Gim Hee Lee	National University of Singapore, Singapore
Yong Jae Lee	University of Wisconsin-Madison, USA
Bastian Leibe	RWTH Aachen University, Germany
Ales Leonardis	University of Birmingham, UK
Vincent Lepetit	Ecole des Ponts ParisTech, France
Fuxin Li	Oregon State University, USA
Hongdong Li	Australian National University, Australia
Xi Li	Zhejiang University, China
Yin Li	University of Wisconsin-Madison, USA
Yu Li	International Digital Economy Academy, Singapore
Xiaodan Liang	Sun Yat-sen University, China
Shengcai Liao	Inception Institute of Artificial Intelligence, United Arab Emirates
Jongwoo Lim	Seoul National University, Korea
Ser-Nam Lim	University of Central Florida, USA
Stephen Lin	Microsoft Research, China
Tsung-Yi Lin	NVIDIA Research, USA
Yen-Yu Lin	National Yang Ming Chiao Tung University, Taiwan
Yutian Lin	Wuhan University, China
Zhe Lin	Adobe Research, USA
Haibin Ling	Stony Brook University, USA
Or Litany	NVIDIA, USA
Jiaying Liu	Peking University, China
Jun Liu	Lancaster University, UK
Miaomiao Liu	Australian National University, Australia
Si Liu	Beihang University, China
Sifei Liu	NVIDIA, USA
Wei Liu	Tencent, China
Yanxi Liu	Pennsylvania State University, USA
Yaoyao Liu	Johns Hopkins University, USA

Yebin Liu	Tsinghua University, China
Zicheng Liu	Microsoft, USA
Ziwei Liu	Nanyang Technological University, Singapore
Ismini Lourentzou	University of Illinois, Urbana-Champaign, USA
Chen Change Loy	Nanyang Technological University, Singapore
Feng Lu	Beihang University, China
Le Lu	Alibaba Group, USA
Oisin Mac Aodha	University of Edinburgh, UK
Luca Magri	Politecnico di Milano, Italy
Michael Maire	University of Chicago, USA
Subhransu Maji	University of Massachusetts, Amherst, USA
Atsuto Maki	KTH Royal Institute of Technology, Sweden
Yasushi Makihara	Osaka University, Japan
Massimiliano Mancini	University of Trento, Italy
Kevis-Kokitsi Maninis	Google Research, Switzerland
Kenneth Marino	Google DeepMind, USA
Renaud Marlet	Ecole des Ponts ParisTech & Valeo, France
Niki Martinel	University of Udine, Italy
Jiri Matas	Czech Technical University in Pragu, Czech Republic
Yasuyuki Matsushita	Osaka University, Japan
Simone Melzi	University of Milano-Bicocca, Italy
Thomas Mensink	Google Research, Netherlands
Michele Merler	IBM Research, USA
Pascal Mettes	University of Amsterdam, Netherlands
Ajmal Mian	University of Western Australia, Australia
Krystian Mikolajczyk	Imperial College London, UK
Ishan Misra	Facebook AI Research, USA
Niloy Mitra	University College London, UK
Anurag Mittal	Indian Institute of Technology Madras, India
Davide Modolo	Amazon, USA
Davide Moltisanti	University of Bath, UK
Philippos Mordohai	Stevens Institute of Technology, USA
Francesc Moreno	Institut de Robotica i Informatica Industrial, Spain
Pietro Morerio	Istituto Italiano di Tecnologia, Italy
Greg Mori	Simon Fraser University & Borealis AI, Canada
Roozbeh Mottaghi	FAIR @ Meta, USA
Yadong Mu	Peking University, China
Vineeth N. Balasubramanian	Indian Institute of Technology, India
Hajime Nagahara	Osaka University, Japan
Vinay Namboodiri	University of Bath, UK
Liangliang Nan	Delft University of Technology, Netherlands

Michele Nappi	University of Salerno, Italy
Srinivasa Narasimhan	Carnegie Mellon University, USA
P. J. Narayanan	International Institute of Information Technology, Hyderabad, India
Nassir Navab	Technical University of Munich, Germany
Ram Nevatia	University of Southern California, USA
Natalia Neverova	Facebook AI Research, France
Juan Carlos Niebles	Salesforce & Stanford University, USA
Michael Niemeyer	Google, Germany
Simon Niklaus	Adobe Research, USA
Ko Nishino	Kyoto University, Japan
Nicoletta Noceti	MaLGa-DIBRIS & Università degli Studi di Genova, Italy
Matthew O'Toole	Carnegie Mellon University, USA
Jean-Marc Odobez	Idiap Research Institute, École polytechnique fédérale de Lausanne, Switzerland
Francesca Odone	Università di Genova, Italy
Takayuki Okatani	Tohoku University & RIKEN Center for Advanced Intelligence Project, Japan
Carl Olsson	Lund University, Sweden
Vicente Ordonez	Rice University, USA
Aljosa Osep	Technical University of Munich, Germany
Martin R. Oswald	University of Amsterdam, Netherlands
Andrew Owens	University of Michigan, USA
Tomas Pajdla	Czech Technical University in Prague, Czechia
Manohar Paluri	Meta, USA
Jinshan Pan	Nanjing University of Science and Technology, China
In Kyu Park	Inha University, Korea
Jaesik Park	Seoul National University, Korea
Despoina Paschalidou	Stanford, USA
Georgios Pavlakos	University of Texas at Austin, USA
Vladimir Pavlovic	Rutgers University, USA
Marcello Pelillo	University of Venice, Italy
David Picard	École des Ponts ParisTech, France
Hamed Pirsiavash	University of California, Davis, USA
Bryan Plummer	Boston University, USA
Thomas Pock	Graz University of Technology, Austria
Fabio Poiesi	Fondazione Bruno Kessler, Italy
Marc Pollefeys	ETH Zurich & Microsoft, Switzerland
Jean Ponce	Ecole normale supérieure - PSL, France
Gerard Pons-Moll	University of Tübingen, Germany

Jordi Pont-Tuset	Google, Switzerland
Fatih Porikli	Qualcomm R&D, USA
Brian Price	Adobe, USA
Guo-Jun Qi	Westlake University, USA
Long Quan	Hong Kong University of Science and Technology, China
Venkatesh Babu Radhakrishnan	Indian Institute of Science, India
Hossein Rahmani	Lancaster University, UK
Deva Ramanan	Carnegie Mellon University, USA
Vignesh Ramanathan	Facebook, USA
Vikram V. Ramaswamy	Princeton University, USA
Nalini Ratha	University at Buffalo, State University of New York, USA
Yogesh Rawat	University of Central Florida, USA
Hamid Rezatofighi	Monash University, Australia
Helge Rhodin	University of British Columbia, Canada
Stephan Richter	Apple, Germany
Anna Rohrbach	Technical University of Darmstadt, Germany
Marcus Rohrbach	Technical University of Darmstadt, Germany
Gemma Roig	Goethe University Frankfurt, Germany
Negar Rostamzadeh	Google, Canada
Paolo Rota	University of Trento, Italy
Stefan Roth	Technical University of Darmstadt, Germany
Carsten Rother	University of Heidelberg, Germany
Subhankar Roy	University of Trento, Italy
Michael Rubinstein	Google, USA
Christian Rupprecht	University of Oxford, UK
Olga Russakovsky	Princeton University, USA
Bryan Russell	Adobe Research, USA
Michael Ryoo	Stony Brook University & Google, USA
Reza Sabzevari	Delft University of Technology, Netherlands
Mathieu Salzmann	École polytechnique fédérale de Lausanne, Switzerland
Dimitris Samaras	Stony Brook University, USA
Aswin Sankaranarayanan	Carnegie Mellon University, USA
Sudeep Sarkar	University of South Florida, USA
Yoichi Sato	University of Tokyo, Japan
Manolis Savva	Simon Fraser University, Canada
Simone Schaub-Meyer	Technical University of Darmstadt, Germany
Bernt Schiele	Max Planck Institute for Informatics, Germany
Cordelia Schmid	Inria & Google, France
Nicu Sebe	University of Trento, Italy

Laura Sevilla-Lara	University of Edinburgh, UK
Fahad Shahbaz Khan	Mohamed bin Zayed University of Artificial Intelligence, United Arab Emirates
Qi Shan	Apple, USA
Shiguang Shan	Institute of Computing Technology, Chinese Academy of Sciences, China
Viktoriia Sharmanska	University of Sussex & Imperial College London, UK
Eli Shechtman	Adobe Research, USA
Evan Shelhamer	DeepMind, UK
Lu Sheng	Beihang University, China
Boxin Shi	Peking University, China
Jianbo Shi	University of Pennsylvania, USA
Mike Zheng Shou	National University of Singapore, Singapore
Abhinav Shrivastava	University of Maryland, USA
Aliaksandr Siarohin	Snap, USA
Kaleem Siddiqi	McGill University, Canada
Leonid Sigal	University of British Columbia, Canada
Oriane Siméoni	valeo.ai, France
Krishna Kumar Singh	Adobe Research, USA
Richa Singh	Indian Institute of Technology Jodhpur, India
Yale Song	Facebook AI Research, USA
Concetto Spampinato	University of Catania, Italy
Srinath Sridhar	Brown University, USA
Bjorn Stenger	Rakuten, Japan
Abby Stylianou	Saint Louis University, USA
Akihiro Sugimoto	National Institute of Informatics, Japan
Chen Sun	Brown University, USA
Deqing Sun	Google, USA
Jian Sun	Xi'an Jiaotong University, China
Min Sun	National Tsing Hua University, Taiwan
Qianru Sun	Singapore Management University, Singapore
Yu-Wing Tai	Dartmouth College, China
Ayellet Tal	Technion, Israel
Ping Tan	Hong Kong University of Science and Technology, China
Robby Tan	National University of Singapore, Singapore
Chi-Keung Tang	Hong Kong University of Science and Technology, China
Makarand Tapaswi	International Institute of Information Technology, Hyderabad & Wadhwani AI, India
Justus Thies	Technical University of Darmstadt, Germany

Radu Timofte	University of Wurzburg & ETH Zurich, Germany
Giorgos Tolias	Czech Technical University in Prague, Czechia
Federico Tombari	Google & Technical University of Munich, Switzerland
Tatiana Tommasi	Politecnico di Torino, Italy
James Tompkin	Brown University, USA
Matthew Trager	AWS AI Labs, USA
Anh Tran	VinAI, Vietnam
Du Tran	Google, USA
Shubham Tulsiani	Carnegie Mellon University, USA
Sergey Tulyakov	Snap, USA
Dimitrios Tzionas	University of Amsterdam, Netherlands
Maria Vakalopoulou	CentraleSupelec, France
Joost van de Weijer	Computer Vision Center, Spain
Laurens van der Maaten	Meta, USA
Jan van Gemert	Delft University of Technology, Netherlands
Sebastiano Vascon	Ca' Foscari University of Venice & European Centre for Living Technology, Italy
Nuno Vasconcelos	University of California, San Diego, USA
Mayank Vatsa	Indian Institute of Technology Jodhpur, India
Javier Vazquez-Corral	Autonomous University of Barcelona, Spain
Andrea Vedaldi	Oxford University, UK
Olga Veksler	University of Waterloo, Canada
Luisa Verdoliva	University Federico II of Naples, Italy
Rene Vidal	University of Pennsylvania, USA
Bo Wang	CtrsVision, USA
Heng Wang	TikTok, USA
Jingdong Wang	Baidu, China
Jingya Wang	ShanghaiTech University, China
Ruiping Wang	Institute of Computing Technology, Chinese Academy of Sciences, China
Shenlong Wang	University of Illinois, Urbana-Champaign, USA
Ting-Chun Wang	NVIDIA, USA
Wenguan Wang	Zhejiang University, China
Xiaolong Wang	University of California, San Diego, USA
Xiaoqian Wang	Purdue University, USA
Xiaoyu Wang	Hong Kong University of Science and Technology, USA
Xin Wang	Microsoft Research, USA
Xinchao Wang	National University of Singapore, Singapore
Yang Wang	Concordia University, Canada
Yiming Wang	Fondazione Bruno Kessler, Italy

Yu-Chiang Frank Wang	National Taiwan University, Taiwan
Yu-Xiong Wang	University of Illinois, Urbana-Champaign, USA
Zhangyang Wang	University of Texas at Austin, USA
Olivia Wiles	DeepMind, UK
Christian Wolf	Naver Labs Europe, France
Michael Wray	University of Bristol, UK
Jiajun Wu	Stanford University, USA
Jianxin Wu	Nanjing University, China
Shangzhe Wu	Stanford University, USA
Ying Wu	Northwestern University, USA
Yu Wu	Wuhan University, China
Jianwen Xie	Baidu Research, USA
Saining Xie	New York University, USA
Weidi Xie	Shanghai Jiao Tong University, China
Dan Xu	Hong Kong University of Science and Technology, China
Huijuan Xu	Pennsylvania State University, USA
Xiangyu Xu	Xi'an Jiaotong University, China
Yan Yan	Illinois Institute of Technology, USA
Jiaolong Yang	Microsoft Research, China
Kaiyu Yang	California Institute of Technology, USA
Linjie Yang	ByteDance AI Lab, USA
Ming-Hsuan Yang	University of California, Merced, USA
Yanchao Yang	University of Hong Kong, China
Yi Yang	Zhejiang University, China
Angela Yao	National University of Singapore, Singapore
Mang Ye	Wuhan University, China
Kwang Moo Yi	University of British Columbia, Canada
Xi Yin	Facebook, USA
Chang D. Yoo	Korea Advanced Institute of Science and Technology, Korea
Shaodi You	University of Amsterdam, Netherlands
Jingyi Yu	Shanghai Tech University, China
Ning Yu	Netflix Eyeline Studios, USA
Qi Yu	Rochester Institute of Technology, USA
Stella Yu	University of Michigan, USA
Yizhou Yu	University of Hong Kong, China
Junsong Yuan	University at Buffalo, State University of New York, USA
Lu Yuan	Microsoft, USA
Sangdoo Yun	NAVER AI Lab, Korea
Hongbin Zha	Peking University, China

Jing Zhang	Australian National University, Australia
Lei Zhang	Hong Kong Polytechnic University, China
Ning Zhang	Facebook AI, USA
Richard Zhang	Adobe, USA
Tianzhu Zhang	University of Science and Technology of China, China
Hengshuang Zhao	University of Hong Kong, China
Qi Zhao	University of Minnesota, USA
Sicheng Zhao	Tsinghua University, China
Liang Zheng	Australian National University, Australia
Wei-Shi Zheng	Sun Yat-sen University, China
Zhun Zhong	University of Nottingham, UK
Bolei Zhou	University of California, Los Angeles, USA
Kaiyang Zhou	Hong Kong Baptist University, China
Luping Zhou	University of Sydney, Australia
S. Kevin Zhou	University of Science and Technology of China, China
Lei Zhu	Hong Kong University of Science and Technology (Guangzhou), China
Linchao Zhu	Zhejiang University, China
Andrew Zisserman	University of Oxford, UK
Daniel Zoran	DeepMind, UK
Wangmeng Zuo	Harbin Institute of Technology, China

Technical Program Committee

Sathyanarayanan N. Aakur	Abulikemu Abuduweili	Abhinav Agarwalla
Andrea Abate	Hanno Ackermann	Shivam Aggarwal
Wim Abbeloos	Chandranath Adak	Sara Aghajanzadeh
Amos L. Abbott	Aikaterini Adam	Shashank Agnihotri
Rameen Abdal	Lukáš Adam	Harsh Agrawal
Salma Abdel Magid	Kamil Adamczewski	Antonio Agudo
Rabab Abdelfattah	Donald Adjeroh	Shai Aharon
Ahmed Abdelkader	Mohammed Adnan	Saba Ahmadi
Eslam Mohamed Abdelrahman	Mahmoud Afifi	Waqar Ahmed
	Triantafyllos Afouras	Byeongjoo Ahn
Ahmed Abdelreheem	Akshay Agarwal	Chanho Ahn
Akash Abdu Jyothi	Madhav Agarwal	Jaewoo Ahn
Kfir Aberman	Nakul Agarwal	Namhyuk Ahn
Shahira Abousamra	Sharat Agarwal	Seoyoung Ahn
Yazan Abu Farha	Shivang Agarwal	Nilesh A. Ahuja
Shady Abu-Hussein	Vatsal Agarwal	Yihao Ai

Abhishek Aich
Emanuele Aiello
Noam Aigerman
Kiyoharu Aizawa
Kenan Emir Ak
Reza Akbarian Bafghi
Sai Aparna Aketi
Derya Akkaynak
Ibrahim Alabdulmohsin
Yuval Alaluf
Md Ferdous Alam
Stephan Alaniz
Badour A. Sh AlBahar
Paul Albert
Isabela Albuquerque
Emanuel Aldea
Luca Alessandrini
Emma Alexander
Konstantinos P.
    Alexandridis
Stamatis Alexandropoulos
Saghir Alfasly
Mohsen Ali
Waqar Ali
Daniel Aliaga
Sadegh Aliakbarian
Garvita Allabadi
Thiemo Alldieck
Antonio Alliegro
Jurandy Almeida
Gal Alon
Hadi Alzayer
Ibtihel Amara
Giuseppe Amato
Sameer Ambekar
Niki Amini-Naieni
Shir Amir
Roberto Amoroso
Dongsheng An
Jie An
Junyi An
Liang An
Xiang An
Yuexuan An
Zhaochong An

Zhulin An
Saket Anand
Pavan Kumar Anasosalu
    Vasu
Codruta O. Ancuti
Cosmin Ancuti
Fernanda Andaló
Connor Anderson
Juan Andrade-Cetto
Bruno Andreis
Alexander Andreopoulos
Bjoern Andres
Shivangi Aneja
Dragomir Anguelov
Damith Kawshan
    Anhettigama
Aditya Annavajjala
Michel Antunes
Tejas Anvekar
Saeed Anwar
Evlampios Apostolidis
Srikar Appalaraju
Relja Arandjelović
Nikita Araslanov
Alexandre Araujo
Helder Araujo
Eric Arazo
Dawit Mureja Argaw
Anurag Arnab
Federica Arrigoni
Alexey Artemov
Aditya Arun
Nader Asadi
Kumar Ashutosh
Vishal Asnani
Mahmoud Assran
Amir Atapour-Abarghouei
Nikos Athanasiou
Ali Athar
ShahRukh Athar
Rowel O. Atienza
Benjamin Attal
Matan Atzmon
Nicolas Audebert
Romaric Audigier

Tristan
    Aumentado-Armstrong
Yannis Avrithis
Muhammad Awais
Md Rabiul Awal
Görkay Aydemir
Mehmet Aygün
Melika Ayoughi
Ali Ayub
Kumar Ayush
Mehdi Azabou
Basim Azam
Mohammad Farid
    Azampour
Hossein Azizpour
Vivek B. S.
Yunhao Ba
Mohammadreza Babaee
Deepika Bablani
Sudarshan Babu
Roman Bachmann
Inhwan Bae
Jeonghun Baek
Kyungjune Baek
Seungryul Baek
Piyush Nitin Bagad
Mohammadhossein Bahari
Gaetan Bahl
Sherwin Bahmani
Bing Bai
Haoyue Bai
Jiawang Bai
Jinbin Bai
Lei Bai
Qingyan Bai
Shutao Bai
Xuyang Bai
Yalong Bai
Yuanchao Bai
Yue Bai
Yunpeng Bai
Ziyi Bai
Daniele Baieri
Sungyong Baik
Ramon Baldrich

Coloma Ballester
Irene Ballester
Sonat Baltaci
Biplab Banerjee
Monami Banerjee
Sandipan Banerjee
Snehasis Banerjee
Soumya Banerjee
Sreya Banerjee
Jihwan Bang
Ankan Bansal
Nitin Bansal
Siddhant Bansal
Chong Bao
Hangbo Bao
Runxue Bao
Wentao Bao
Yiwei Bao
Zhipeng Bao
Zhongyun Bao
Amir Bar
Omer Bar Tal
Manel Baradad Jurjo
Lorenzo Baraldi
Lorenzo Baraldi
Tal Barami
Danny Barash
Daniel Barath
Adrian Barbu
Nick Barnes
Alina J. Barnett
German Barquero
Joao P. Barreto
Jonathan T. Barron
Rodrigo C. Barros
Luca Barsellotti
Myles Bartlett
Kristijan Bartol
Hritam Basak
Dina Bashkirova
Chaim Baskin
Lennart C. Bastian
Abhipsa Basu
Soumen Basu
Peyman Bateni

Anil Batra
David Bau
Zuria Bauer
Eduard Gabriel Bazavan
Federico Becattini
Nathan A. Beck
Jan Bednarik
Peter A. Beerel
Ardhendu Behera
Harkirat Singh Behl
Jens Behley
William Beksi
Soufiane Belharbi
Giovanni Bellitto
Ismail Ben Ayed
Abdessamad Ben Hamza
Yizhak Ben-Shabat
Guy Ben-Yosef
Robert Benavente
Assia Benbihi
Yasser Benigmim
Urs Bergmann
Amit H Bermano
Jesus Bermudez-Cameo
Florian Bernard
Stefano Berretti
Gabriele Berton
Petra Bevandić
Matthew Beveridge
Lalit Bhagat
Yash Bhalgat
Homanga Bharadhwaj
Skanda Bharadwaj
Ambika Saklani Bhardwaj
Goutam Bhat
Gaurav Bhatt
Neel P. Bhatt
Deblina Bhattacharjee
Anish Bhattacharya
Rajarshi Bhattacharya
Uttaran Bhattacharya
Apratim Bhattacharyya
Anand Bhattad
Binod Bhattarai
Snehal Bhayani

Brojeshwar Bhowmick
Aritra Bhowmik
Neelanjan Bhowmik
Amran Bhuiyan
Ankan Kumar Bhunia
Ayan Kumar Bhunia
Jing Bi
Qi Bi
Xiuli Bi
Michael Bi Mi
Hao Bian
Jia-Wang Bian
Shaojun Bian
Simone Bianco
Pia Bideau
Xiaoyu Bie
Roberto Bigazzi
Mario Bijelic
Bahri Batuhan Bilecen
Guillaume-Alexandre Bilodeau
Alexander Binder
Massimo Bini
Niccolò Bisagno
Carmen Bisogni
Anthony R. Bisulco
Sandika Biswas
Sanket Biswas
Ksenia Bittner
Kevin Blackburn-Matzen
Nathaniel Blanchard
Hunter Blanton
Emile Blettery
Hermann Blum
Deyu Bo
Janusz Bobulski
Marius Bock
Lea Bogensperger
Simion-Vlad Bogolin
Moritz Böhle
Piotr Bojanowski
Alexey Bokhovkin
Georg Bökman
Ladislau Boloni
Daniel Bolya

Lorenzo Bonicelli
Gianpaolo Bontempo
Vivek Boominathan
Adrian Bors
Damian Borth
Matteo Bortolon
Davide Boscaini
Laurie Nicholas Bose
Shirsha Bose
Mark Boss
Andrea Bottino
Larbi Boubchir
Alexandre Boulch
Terrance E. Boult
Amine Bourki
Walid Bousselham
Fadi Boutros
Nicolas C. Boutry
Thierry Bouwmans
Richard S. Bowen
Kevin W. Bowyer
Ivaylo Boyadzhiev
Aidan Boyd
Joseph Boyd
Aljaz Bozic
Behzad Bozorgtabar
Manuel Brack
Samarth Brahmbhatt
Nikolas Brasch
Cameron E. Braunstein
Toby P. Breckon
Mathieu Bredif
Romain Brégier
Carlo Bretti
Joel R. Brogan
Clifford Broni-Bediako
Andrew Brown
Ellis L. Brown
Thomas Brox
Qingwen Bu
Tong Bu
Shyamal Buch
Alvaro Budria
Ignas Budvytis
José M. Buenaposada

Kyle R. Buettner
Marcel C. Bühler
Nhat-Tan Bui
Adrian Bulat
Valay M. Bundele
Nathaniel J. Burgdorfer
Ryan D. Burgert
Evgeny Burnaev
Pietro Buzzega
Marco Buzzelli
Kwon Byung-Ki
Fabian Caba
Felipe Cadar
Martin Cadik
Bowen Cai
Changjiang Cai
Guanyu Cai
Haoming Cai
Hongrui Cai
Jiarui Cai
Jinyu Cai
Kaixin Cai
Lile Cai
Minjie Cai
Mu Cai
Pingping Cai
Rizhao Cai
Ruisi Cai
Ruojin Cai
Shen Cai
Shengqu Cai
Weidong Cai
Weiwei Cai
Wenjia Cai
Wenxiao Cai
Yancheng Cai
Yuanhao Cai
Yujun Cai
Yuxuan Cai
Zhaowei Cai
Zhipeng Cai
Zhixi Cai
Zhongang Cai
Zikui Cai
Salvatore Calcagno

Roberto Caldelli
Akin Caliskan
Lilian Calvet
Necati Cihan Camgoz
Tommaso Campari
Dylan Campbell
Guglielmo Camporese
Yigit Baran Can
Shaun Canavan
Gemma Canet Tarrés
Marco Cannici
Ang Cao
Anh-Quan Cao
Bing Cao
Bingyi Cao
Chenjie Cao
Dongliang Cao
Hu Cao
Jiahang Cao
Jiale Cao
Jie Cao
Jiezhang Cao
Jinkun Cao
Meng Cao
Mingdeng Cao
Peng Cao
Pu Cao
Qinglong Cao
Qingxing Cao
Shengcao Cao
Song Cao
Weiwei Cao
Xiangyong Cao
Xiao Cao
Xu Cao
Xu Cao
Yan-Pei Cao
Yang Cao
Yichao Cao
Ying Cao
Yuan Cao
Yuhang Cao
Yukang Cao
Yulong Cao
Yun-Hao Cao

Yunhao Cao
Yutong Cao
Zhangjie Cao
Zhiguo Cao
Zhiwen Cao
Ziang Cao
Giacomo Capitani
Luigi Capogrosso
Andrea Caraffa
Jaime S. Cardoso
Chris Careaga
Zachariah Carmichael
Giuseppe Cartella
Vincent Cartillier
Paola Cascante-Bonilla
Lucia Cascone
Pol Caselles Rico
Angela Castillo
Lluis Castrejon
Modesto
  Castrillón-Santana
Francisco M. Castro
Pedro Castro
Jacopo Cavazza
Oya Celiktutan
Luigi Celona
Jiazhong Cen
Fernando J. Cendra
Fabio Cermelli
Duygu Ceylan
Eunju Cha
Junbum Cha
Rohan Chabra
Rohan Chacko
Julia Chae
Nicolas Chahine
Junyi Chai
Lucy Chai
Tianrui Chai
Wenhao Chai
Zenghao Chai
Anish Chakrabarty
Goirik Chakrabarty
Anirban Chakraborty
Deep Chakraborty

Souradeep Chakraborty
Punarjay Chakravarty
Jacob Chalk
Loick Chambon
Chee Seng Chan
David Chan
Dorian Y. Chan
Kin Sun Chan
Matthew Chan
Adrien Chan-Hon-Tong
Sampath Chanda
Shivam Chandhok
Prashanth Chandran
Kenneth Chaney
Chirui Chang
Di Chang
Dongliang Chang
Hong Chang
Hyung Jin Chang
Ju Yong Chang
Matthew Chang
Ming-Ching Chang
MingFang Chang
Nadine Chang
Qi Chang
Seunggyu Chang
Wei-Yi Chang
Xiaojun Chang
Yakun Chang
Yi Chang
Zheng Chang
Sumohana S.
  Channappayya
Hanqing Chao
Pradyumna Chari
Korawat Charoenpitaks
Dibyadip Chatterjee
Moitreya Chatterjee
Pratik Chaudhari
Muawiz Chaudhary
Ritwick Chaudhry
Ruchika Chavhan
Sofian Chaybouti
Zhengping Che
Gullal Singh Cheema

Kunal Chelani
Rama Chellappa
Allison Y. Chen
Anpei Chen
Aozhu Chen
Bin Chen
Bin Chen
Bin Chen
Bohong Chen
Bor-Chun Chen
Bowei Chen
Brian Chen
Chang Wen Chen
Changan Chen
Changhao Chen
Changhuai Chen
Chaofan Chen
Chaofan Chen
Chaofeng Chen
Chen Chen
Cheng Chen
Chengkuan Chen
Chieh-Yun Chen
Chong Chen
Chun-Fu Richard Chen
Congliang Chen
Cuiqun Chen
Da Chen
Daoyuan Chen
Defang Chen
Dian Chen
Dian Chen
Ding-Jie Chen
Dingfan Chen
Dong Chen
Dongdong Chen
Fangyi Chen
Guang-Yong Chen
Guangyao Chen
Guangyi Chen
Guanying Chen
Guikun Chen
Guo Chen
Haibo Chen
Haiwei Chen

Hansheng Chen
Hanting Chen
Hanzhi Chen
Hao Chen
Haodong Chen
Haokun Chen
Haoran Chen
Haoxin Chen
Haoyu Chen
Haoyu Chen
Hong-You Chen
Honghua Chen
Honglin Chen
Hongming Chen
Huanran Chen
Hui Chen
Hwann-Tzong Chen
Jiacheng Chen
Jiajing Chen
Jianqiu Chen
Jiaqi Chen
Jiaxin Chen
Jiayi Chen
Jie Chen
Jie Chen
Jie Chen
Jieneng Chen
Jierun Chen
Jinghui Chen
Jingjing Chen
Jintai Chen
Jinyu Chen
Joya Chen
Jun Chen
Jun Chen
Jun Chen
Jun-Cheng Chen
Jun-Kun Chen
Junjie Chen
Kai Chen
Kai Chen
Kaifeng Chen
Kefan Chen
Kejiang Chen
Kuan-Wen Chen

Kunyao Chen
Li Chen
Liang Chen
Lianggangxu Chen
Liangyu Chen
Ling-Hao Chen
Linghao Chen
Liyan Chen
Min Chen
Min-Hung Chen
Minghao Chen
Minghui Chen
Nenglun Chen
Peihao Chen
Pengguang Chen
Ping Chen
Qi Chen
Qi Chen
Qi Chen
Qiang Chen
Qimin Chen
Qingcai Chen
Rongyu Chen
Rui Chen
Runjian Chen
Shang-Fu Chen
Shen Chen
Sherry X. Chen
Shi Chen
Shiming Chen
Shiyan Chen
Shizhe Chen
Shoufa Chen
Shu-Yu Chen
Shuai Chen
Si Chen
Siheng Chen
Simin Chen
Sixiang Chen
Sizhe Chen
Songcan Chen
Tao Chen
Tianshui Chen
Tianyi Chen
Tsai-Shien Chen

Wei Chen
Weidong Chen
Weihua Chen
Weijie Chen
Weikai Chen
Wuyang Chen
Xi Chen
Xi Chen
Xiang Chen
Xiangyu Chen
Xianyu Chen
Xiao Chen
Xiao Chen
Xiaohan Chen
Xiaokang Chen
Xiaotong Chen
Xin Chen
Xin Chen
Xin Chen
Xinghao Chen
Xingyu Chen
Xingyu Chen
Xinlei Chen
Xiuwei Chen
Xuanbai Chen
Xuanhong Chen
Xuesong Chen
Xuweiyi Chen
Xuxi Chen
Yanbei Chen
Yang Chen
Yaofo Chen
Yen-Chun Chen
Yi-Ting Chen
Yi-Wen Chen
Yilun Chen
Yinbo Chen
Yingcong Chen
Yingwen Chen
Yiting Chen
Yixin Chen
Yixin Chen
Yu Chen
Yuanhong Chen
Yuanqi Chen

Yuedong Chen	Ziyang Chen	Seunggeun Chi
Yueting Chen	An-Chieh Cheng	Zhixiang Chi
Yuhao Chen	Bowen Cheng	Naoya Chiba
Yuhao Chen	De Cheng	Julian Chibane
Yujin Chen	Erkang Cheng	Aditya Chinchure
Yukang Chen	Feng Cheng	Eleni Chiou
Yun-Chun Chen	Guangliang Cheng	Mia Chiquier
Yunjin Chen	Hao Cheng	Chiranjeev Chiranjeev
Yunlu Chen	Hao Cheng	Kashyap Chitta
Yuntao Chen	Harry Cheng	Hsu-kuang Chiu
Yushuo Chen	Ho Kei Cheng	Wei-Chen Chiu
Yuxiao Chen	Jingchun Cheng	Chee Kheng Chng
Zehui Chen	Jun Cheng	Shin-Fang Chng
Zerui Chen	Ka Leong Cheng	Donghyeon Cho
Zewen Chen	Kelvin B. Cheng	Gyusang Cho
Zeyuan Chen	Kevin Ho Man Cheng	Hanbyel Cho
Zeyuan Chen	Lechao Cheng	Hoonhee Cho
Zhang Chen	Shuo Cheng	Jae Won Cho
Zhao-Min Chen	Silin Cheng	Jang Hyun Cho
Zhaoliang Chen	Ta-Ying Cheng	Jungchan Cho
Zhaoxi Chen	Tianhang Cheng	Junhyeong Cho
Zhaoyu Chen	Weihao Cheng	Nam Ik Cho
Zhaozheng Chen	Wencan Cheng	Seokju Cho
Zhaozheng Chen	Wensheng Cheng	Seungju Cho
Zhe Chen	Xi Cheng	Suhwan Cho
Zhen Chen	Xize Cheng	Sunghyun Cho
Zhen Chen	Xuxin Cheng	Sungjun Cho
Zheng Chen	Yen-Chi Cheng	Sungmin Cho
Zhenghao Chen	Yi Cheng	Wonwoong Cho
Zhengyu Chen	Yihua Cheng	Nathaniel E. Chodosh
Zhenyu Chen	Yu Cheng	Jaesung Choe
Zhi Chen	Zesen Cheng	Junsuk Choe
Zhibo Chen	Zezhou Cheng	Changwoon Choi
Zhimin Chen	Zhanzhan Cheng	Chiho Choi
Zhineng Chen	Zhen Cheng	Ching Lam Choi
Zhiwei Chen	Zhi-Qi Cheng	Dongyoung Choi
Zhixiang Chen	Ziang Cheng	Dooseop Choi
Zhiyang Chen	Ziheng Cheng	Hongsuk Choi
Zhiyuan Chen	Soon Yau Cheong	Hyesong Choi
Zhuangzhuang Chen	Anoop Cherian	Jaewoong Choi
Zhuo Chen	Daniil Cherniavskii	Janghoon Choi
Zhuoxiao Chen	Ajad Chhatkuli	Jin Young Choi
Zihan Chen	Cheng Chi	Jinwoo Choi
Zijiao Chen	Haoang Chi	Jinyoung Choi
Ziwen Chen	Hyung-gun Chi	Jonghyun Choi

Jongwon Choi
Jooyoung Choi
Kanghyun Choi
Myungsub Choi
Seokeon Choi
Wonhyeok Choi
Bhavin Choksi
Jaegul Choo
Baptiste Chopin
Ayush Chopra
Gene Chou
Shih-Han Chou
Divya Choudhary
Siddharth Choudhary
Sunav Choudhary
Rohan Choudhury
Vasileios Choutas
Ka-Ho Chow
Pinaki Nath Chowdhury
Sanjoy Chowdhury
Sammy Christen
Fu-Jen Chu
Qi Chu
Ruihang Chu
Sanghyeok Chu
Wei-Ta Chu
Wen-Hsuan Chu
Wen-Hsuan Chu
Wei Qin Chuah
Andrei Chubarau
Ilya Chugunov
Sanghyuk Chun
Se Young Chun
Hyungjin Chung
Inseop Chung
Jaeyoung Chung
Jihoon Chung
Jiwan Chung
Joon Son Chung
Thomas A. Ciarfuglia
Umur A. Ciftci
Antonio E. Cinà
Ramazan Gokberk Cinbis
Anthony Cioppa
Javier Civera

Christopher A. Clark
James J. Clark
Ronald Clark
Brian S. Clipp
Mark Coates
Federico Cocchi
Felipe Codevilla
Cecília Coelho
Danny Cohen-Or
Forrester Cole
Julien Colin
John Collomosse
Marc Comino-Trinidad
Nicola Conci
Marcos V. Conde
Alexandru Paul Condurache
Runmin Cong
Tianshuo Cong
Wenyan Cong
Yuren Cong
Alessandro Conti
Andrea Conti
Enric Corona
John R. Corring
Jaime Corsetti
Jason J. Corso
Theo W. Costain
Guillaume Couairon
Robin Courant
Davide Cozzolino
Donato Crisostomi
Francesco Croce
Florinel Alin Croitoru
Ioana Croitoru
James L. Crowley
Steve Cruz
Gabriela Csurka
Alfredo Cuesta Infante
Aiyu Cui
Hainan Cui
Jiali Cui
Jiequan Cui
Justin Cui
Qiongjie Cui

Ruikai Cui
Yawen Cui
Ying Cui
Yuning Cui
Zhen Cui
Zhenyu Cui
Ziteng Cui
Benjamin J. Culpepper
Xiaodong Cun
Federico Cunico
Claudio Cusano
Sebastian Cygert
Guido Maria D'Amely di Melendugno
Moreno D'Incà
Feipeng Da
Mosam Dabhi
Rishabh Dabral
Konstantinos M. Dafnis
Matthew L. Daggitt
Anders B. Dahl
Bo Dai
Bo Dai
Haixing Dai
Mengyu Dai
Peng Dai
Peng Dai
Pingyang Dai
Qi Dai
Qiyu Dai
Wenrui Dai
Wenxun Dai
Zuozhuo Dai
Nicola Dall'Asen
Naser Damer
Bo Dang
Meghal Dani
Duolikun Danier
Donald G. Dansereau
Quan Dao
Son Duy Dao
Trung Tuan Dao
Mohamed Daoudi
Ahmad Darkhalil
Rangel Daroya

Abhijit Das
Abir Das
Anurag Das
Ayan Das
Nilotpal Das
Partha Das
Soumi Das
Srijan Das
Sukhendu Das
Swagatam Das
Ananyananda Dasari
Avijit Dasgupta
Gourav Datta
Achal Dave
Ishan Rajendrakumar Dave
Andrew Davison
Aram Davtyan
Axel Davy
Youssef Dawoud
Laura Daza
Francesca De Benetti
Daan de Geus
Alfredo S. De Goyeneche
Riccardo de Lutio
Maria De Marsico
Christophe De Vleeschouwer
Tonmoay Deb
Sayan Deb Sarkar
Dale Decatur
Thanos Delatolas
Fabien Delattre
Mauricio Delbracio
Ginger D. Delmas
Andong Deng
Bailin Deng
Bin Deng
Boyang Deng
Chaorui Deng
Cheng Deng
Congyue Deng
Jiacheng Deng
Jiajun Deng
Jiankang Deng
Jingjing Deng

Jinhong Deng
Kangle Deng
Kenan Deng
Lei Deng
Liang-Jian Deng
Shijian Deng
Weijian Deng
Xiaoming Deng
Xueqing Deng
Ye Deng
Yongjian Deng
Youming Deng
Yu Deng
Zhijie Deng
Zhongying Deng
Matteo Denitto
Mohammad Mahdi Derakhshani
Alakh Desai
Kevin Desai
Nishq P. Desai
Sai Vikas Desai
Jean-Emmanuel Deschaud
Frédéric Devernay
Rahul Dey
Arturo Deza
Helisa Dhamo
Amaya Dharmasiri
Ankit Dhiman
Vikas Dhiman
Yan Di
Zonglin Di
Ousmane A. Dia
Renwei Dian
Haiwen Diao
Xiaolei Diao
Gonçalo José Dias Pais
Abdallah Dib
Juan Carlos Dibene Simental
Sebastian Dille
Christian Diller
Mariella Dimiccoli
Anastasios Dimou
Or Dinari

Caiwen Ding
Changxing Ding
Choubo Ding
Fangqiang Ding
Guiguang Ding
Guodong Ding
Henghui Ding
Hui Ding
Jian Ding
Jian-Jiun Ding
Jianhao Ding
Li Ding
Liang Ding
Lihe Ding
Ming Ding
Runyu Ding
Shouhong Ding
Shuangrui Ding
Shuxiao Ding
Tianjiao Ding
Tianyu Ding
Wenhao Ding
Xiaohan Ding
Xinpeng Ding
Yaqing Ding
Yong Ding
Yuhang Ding
Yuqi Ding
Zheng Ding
Zhengming Ding
Zhipeng Ding
Zhuangzhuang Ding
Zihan Ding
Zihan Ding
Ziluo Ding
Anh-Dung Dinh
Markos Diomataris
Christos Diou
Alish Dipani
Alara Dirik
Ajay Divakaran
Abdelaziz Djelouah
Nemanja Djuric
Thanh-Toan Do
Tien Do

Anh-Dzung Doan
Bao Gia Doan
Khoa D. Doan
Carl Doersch
Andreea Dogaru
Nehal Doiphode
Csaba Domokos
Bowen Dong
Haoye Dong
Jiahua Dong
Jianfeng Dong
Jinpeng Dong
Junhao Dong
Junting Dong
Li Dong
Minjing Dong
Nanqing Dong
Peijie Dong
Qiaole Dong
Qihua Dong
Qiulei Dong
Runpei Dong
Shichao Dong
Shuting Dong
Sixun Dong
Siyan Dong
Tian Dong
Wei Dong
Wei Dong
Weisheng Dong
Xiaoyi Dong
Xiaoyu Dong
Xingping Dong
Yinpeng Dong
Zhenxing Dong
Jonathan C. Donnelly
Naren Doraiswamy
Gianfranco Doretto
Michael Dorkenwald
Muskan Dosi
Huanzhang Dou
Shuguang Dou
Yiming Dou
Zhiyang Frank Dou
Zi-Yi Dou

Arthur Douillard
Michail C. Doukas
Matthijs Douze
Amil Dravid
Vincent Drouard
Dawei Du
Jia-Run Du
Jinhao Du
Ruoyi Du
Weiyu Du
Xiaodan Du
Ye Du
Yingjun Du
Yong Du
Yuanqi Du
Yuntao Du
Zexing Du
Zhuoran Du
Radhika Dua
Haodong Duan
Haoran Duan
Huiyu Duan
Jiafei Duan
Jiali Duan
Peiqi Duan
Ye Duan
Yuchen Duan
Yueqi Duan
Zhihao Duan
Amanda Cardoso Duarte
Shiv Ram Dubey
Anastasia Dubrovina
Daniel Duckworth
Timothy Duff
Nicolas Dufour
Rahul Duggal
Shivam Duggal
Bardienus P. Duisterhof
Felix Dülmer
Matteo Dunnhofer
Chi Nhan Duong
Elona Dupont
Nikita Durasov
Zoran Duric
Mihai Dusmanu

Titir Dutta
Ujjal Kr Dutta
Isht Dwivedi
Sai Kumar Dwivedi
Sebastian Dziadzio
Thomas Eboli
Ramin Ebrahim Nakhli
Benjamin Eckart
Takeharu Eda
Johan Edstedt
Alexei A. Efros
Nikos Efthymiadis
Bernhard Egger
Max Ehrlich
Iván Eichhardt
Farshad Einabadi
Martin Eisemann
Peter Eisert
Mohamed El Banani
Mostafa El-Khamy
Alaa El-Nouby
Randa I. M. Elanwar
James Elder
Abdelrahman Eldesokey
Ismail Elezi
Ehsan Elhamifar
Moshe Eliasof
Sara Elkerdawy
Qing En
Ian Endres
Andreas Engelhardt
Francis Engelmann
Martin Engilberge
Kenji Enomoto
Chanho Eom
Guy Erez
Linus Ericsson
Maria C. Escobar
Marcos Escudero-Viñolo
Sedigheh Eslami
Valter Estevam
Ali Etemad
Martin Nicolas Everaert
Ivan Evtimov
Ralph Ewerth

Fevziye Irem Eyiokur
Cristobal Eyzaguirre
Gabriele Facciolo
Mohammad Fahes
Masud ANI Fahim
Fabrizio Falchi
Alex Falcon
Aoxiang Fan
Baojie Fan
Bin Fan
Chao Fan
David Fan
Haoqiang Fan
Hehe Fan
Heng Fan
Hongyi Fan
Huijie Fan
Junsong Fan
Lei Fan
Lei Fan
Lifeng Fan
Lue Fan
Mingyuan Fan
Qi Fan
Rui R. Fan
Xiran Fan
Yifei Fan
Yue Fan
Zejia Fan
Zhaoxin Fan
Zhenfeng Fan
Zhentao Fan
Zhipeng Fan
Zhiwen Fan
Zicong Fan
Sean Fanello
Chaowei Fang
Chuan Fang
Gongfan Fang
Guian Fang
Han Fang
Jiansheng Fang
Jiemin Fang
Jin Fang
Jun Fang
Kun Fang
Pengfei Fang
Rui Fang
Sheng Fang
Xiang Fang
Xiaolin Fang
Xiuwen Fang
Zhaoyuan Fang
Zhiyuan Fang
Eros Fanì
Matteo Farina
Farzan Farnia
Aiman Farooq
Azade Farshad
Mohsen Fayyaz
Arash Fayyazi
Ben Fei
Jingjing Fei
Xiaohan Fei
Zhengcong Fei
Michael Felsberg
Berthy T. Feng
Bin Feng
Brandon Y. Feng
Chao Feng
Chen Feng
Chengjian Feng
Chun-Mei Feng
Fan Feng
Guorui Feng
Hao Feng
Jianjiang Feng
Lan Feng
Lei Feng
Litong Feng
Mingtao Feng
Qianyu Feng
Ruicheng Feng
Ruili Feng
Ruoyu Feng
Ryan T. Feng
Shiwei Feng
Songhe Feng
Tuo Feng
Wei Feng
Weitao Feng
Weixi Feng
Xuelu Feng
Yanglin Feng
Yao Feng
Yue Feng
Zhanxiang Feng
Zhenhua Feng
Zunlei Feng
Clara Fernandez
Victoria Fernandez
 Abrevaya
Miguel-Ángel
 Fernández-Torres
Sira Ferradans
Claudio Ferrari
Ethan Fetaya
Mustansar Fiaz
Guénolé Fiche
Panagiotis P. Filntisis
Marco Fiorucci
Kai Fischer
Tobias Fischer
Tom Fischer
Volker Fischer
Robert B. Fisher
Alessandro Flaborea
Corneliu O. Florea
Georgios Floros
Alejandro Fontan
Tomaso Fontanini
Lin Geng Foo
Wolfgang Förstner
Niki M. Foteinopoulou
Simone Foti
Simone Foti
Victor Fragoso
Gianni Franchi
Jean-Sebastien Franco
Emanuele Frascaroli
Moti Freiman
Rafail Fridman
Sara Fridovich-Keil
Felix Friedrich
Stanislav Frolov

Andre Fu
Bin Fu
Changcheng Fu
Changqing Fu
Jianlong Fu
Jie Fu
Jingjing Fu
Keren Fu
Lan Fu
Lele Fu
Minghao Fu
Qichen Fu
Rao Fu
Taimeng Fu
Tianwen Fu
Yanwei Fu
Ying Fu
Yonggan Fu
Yun Fu
Yuqian Fu
Zehua Fu
Zhenqi Fu
Zipeng Fu
Wolfgang Fuhl
Yasuhisa Fujii
Yuki Fujimura
Katsuki Fujisawa
Kent Fujiwara
Takuya Funatomi
Christopher Funk
Antonino Furnari
Ryo Furukawa
Ryosuke Furuta
David Futschik
Akshay Gadi Patil
Siddhartha Gairola
Rinon Gal
Adrian Galdran
Silvio Galesso
Meirav Galun
Rofida Gamal
Weijie Gan
Yiming Gan
Yulu Gan
Vineet Gandhi

Kanchana Vaishnavi
 Gandikota
Aditya Ganeshan
Aditya Ganeshan
Swetava Ganguli
Sujoy Ganguly
Hanan Gani
Alireza Ganjdanesh
Harald Ganster
Roy Ganz
Angela F. Gao
Bin-Bin Gao
Changxin Gao
Chen Gao
Chongyang Gao
Daiheng Gao
Daoyi Gao
Difei Gao
Hang Gao
Hongchang Gao
Huiyu Gao
Junyu Gao
Junyu Gao
Kaifeng Gao
Katelyn Gao
Kuofeng Gao
Lin Gao
Ling Gao
Maolin Gao
Quankai Gao
Ruopeng Gao
Shanghua Gao
Shangqi Gao
Shangqian Gao
Shaobing Gao
Shenyuan Gao
Xiang Gao
Xiangjun Gao
Yang Gao
Yipeng Gao
Yuan Gao
Yue Gao
Yunhe Gao
Zhanning Gao
Zhi Gao

Zhihan Gao
Zhitong Gao
Zhongpai Gao
Ziteng Gao
Nicola Garau
Elena Garces
Noa Garcia
Nuno Cruz Garcia
Ricardo Garcia Pinel
Guillermo
 Garcia-Hernando
Saurabh Garg
Mathieu Garon
Risheek Garrepalli
Pablo Garrido
Quentin Garrido
Stefano Gasperini
Vincent Gaudilliere
Chandan Gautam
Shivam Gautam
Paul Gavrikov
Jonathan Z. Gazak
Chongjian Ge
Liming Ge
Runzhou Ge
Shiming Ge
Songwei Ge
Weifeng Ge
Wenhang Ge
Yanhao Ge
Yixiao Ge
Yunhao Ge
Yuying Ge
Zhiqi Ge
Zongyuan Ge
James Gee
Chen Geng
Daniel Geng
Xue Geng
Zhengyang Geng
Zichen Geng
Kyle Genova
Georgios Georgakis
Mariana-Iuliana
 Georgescu

Yiangos Georgiou
Markos Georgopoulos
Stamatios Georgoulis
Michal Geyer
Mina Ghadimi Atigh
Abhijay Ghildyal
Reza Ghoddoosian
Mohsen Gholami
Peyman Gholami
Enjie Ghorbel
Soumya Suvra Ghosal
Anindita Ghosh
Anurag Ghosh
Arnab Ghosh
Arthita Ghosh
Aurobrata Ghosh
Shreya Ghosh
Soham Ghosh
Sreyan Ghosh
Andrea Giachetti
Paris Giampouras
Alexander Gielisse
Andrew Gilbert
Guy Gilboa
Nate Gillman
Jhony H. Giraldo
Roger Girgis
Sharath Girish
Mario Valerio Giuffrida
Francesco Giuliari
Nikolaos Gkanatsios
Ioannis Gkioulekas
Alexi Gladstone
Derek Gloudemans
Aurele T. Gnanha
Hyojun Go
Akos Godo
Danilo D. Goede
Arushi Goel
Kratarth Goel
Rahul Goel
Shubham Goel
Vidit Goel
Tejas Gokhale
Vignesh Gokul

Aditya Golatkar
Micah Goldblum
S. Alireza Golestaneh
Gabriele Goletto
Lluis Gomez
Guillermo Gomez-Trenado
Alex Gomez-Villa
Nuno Gonçalves
Biao Gong
Boqing Gong
Chen Gong
Chengyue Gong
Dayoung Gong
Dong Gong
Jingyu Gong
Kaixiong Gong
Kehong Gong
Liyu Gong
Mingming Gong
Ran Gong
Rui Gong
Tao Gong
Xiaojin Gong
Xuan Gong
Xun Gong
Yifan Gong
Yu Gong
Yuanhao Gong
Yuning Gong
Yunye Gong
Cristina I. González
Adam Goodge
Nithin Gopalakrishnan
   Nair
Cameron Gordon
Dipam Goswami
Gaurav Goswami
Hanno Gottschalk
Chenhui Gou
Jianping Gou
Shreyank N. Gowda
Mohit Goyal
Julia Grabinski
Helmut Grabner
Patrick L. Grady

Alexandros Graikos
Eric Granger
Douglas R. Gray
Michael Alan Greenspan
Connor Greenwell
David Griffiths
Artur Grigorev
Alexey A. Gritsenko
Ivan Grubišić
Geonmo Gu
Jianyang Gu
Jiaqi Gu
Jiayuan Gu
Jinwei Gu
Peiyan Gu
Qiao Gu
Tianpei Gu
Xianfan Gu
Xiang Gu
Xiangming Gu
Xiao Gu
Xiuye Gu
Yanan Gu
Yiwen Gu
Yuchao Gu
Yuming Gu
Yun Gu
Zeqi Gu
Zhangxuan Gu
Banglei Guan
Junfeng Guan
Qingji Guan
Shanyan Guan
Tianrui Guan
Tongkun Guan
Ziqiao Guan
Ziyu Guan
Denis A. Gudovskiy
Antoine Guédon
Paul Guerrero
Ricardo Guerrero
Liangke Gui
Sadaf Gulshad
Manuel Günther
Chen Guo

Chenqi Guo
Chuan Guo
Guodong Guo
Haiyun Guo
Heng Guo
Hengtao Guo
Hongji Guo
Jia Guo
Jianwei Guo
Jianyuan Guo
Jiayi Guo
Jie Guo
Jie Guo
Jingcai Guo
Jingyuan Guo
Jinyang Guo
Lanqing Guo
Meng-Hao Guo
Mengxi China Guo
Minghao Guo
Pengfei Guo
Pengsheng Guo
Pinxue Guo
Qi Guo
Qin Guo
Qing Guo
Qingpei Guo
Saidi Guo
Shi Guo
Shuxuan Guo
Song Guo
Taian Guo
Tiantong Guo
Tianyu Guo
Wen Guo
Wenzhong Guo
Xianda Guo
Xiaobao Guo
Xiefan Guo
Yangyang Guo
Yanhui Guo
Yao Guo
Yichen Guo
Yong Guo
Yuan-Chen Guo
Yuliang Guo
Yuxiang Guo
Yuyu Guo
Zixian Guo
Ziyu Guo
Zujin Guo
Aarush Gupta
Akash Gupta
Akshita Gupta
Ankush Gupta
Anshul Gupta
Anubhav Gupta
Anup Kumar Gupta
Honey Gupta
Kunal Gupta
pravir singh gupta
Rohit Gupta
Saumya Gupta
Corina Gurau
Yeti Z. Gurbuz
Saket Gurukar
Siddharth Gururani
Vladimir Guzov
Matthew A. Gwilliam
Hyunho Ha
Ryo Hachiuma
Isma Hadji
Armin Hadzic
Daniel Haehn
Nicolai Haeni
Ronny Haensch
Meera Hahn
Oliver Hahn
Nguyen Hai
Yuval Haitman
Levente Hajder
Moayed Haji-Ali
Sina Hajimiri
Alexandros Haliassos
Peter M. Hall
Bumsub Ham
Ryuhei Hamaguchi
Abdullah J. Hamdi
Max Hamilton
Lars Hammarstrand
Hasan Abed Al Kader Hammoud
Beining Han
Boran Han
Dongyoon Han
Feng Han
Guangxing Han
Guangzeng Han
Hu Han
Jiaming Han
Jin Han
Jungong Han
Junlin Han
Kai Han
Kang Han
Kun Han
Ligong Han
Longfei Han
Mei Han
Mingfei Han
Muzhi Han
Sungwon Han
Tengda Han
Tengda Han
Wencheng Han
Xinran Han
Xintong Han
Yizeng Han
Yufei Han
Zhenjun Han
Zhi Han
Zhongyi Han
Zongbo Han
Zongyan Han
Ankur Handa
Tiankai Hang
Yucheng Hang
Asif Hanif
Param Hanji
Joëlle Hanna
Niklas Hanselmann
Nicklas A. Hansen
Fusheng Hao
Shaozhe Hao
Shengyu Hao

Shijie Hao
Tianxiang Hao
Yanbin Hao
Yu Hao
Zekun Hao
Takayuki Hara
Mehrtash Harandi
Sanjay Haresh
Haripriya Harikumar
Adam Harley
David M. Hart
Md Yousuf Harun
Md Rakibul Hasan
Connor Hashemi
Atsushi Hashimoto
Khurram Azeem Hashmi
Nozomi Hata
Ali Hatamizadeh
Ryuichiro Hataya
Timm Haucke
Joakim Bruslund Haurum
Stephen Hausler
Mohammad Havaei
Hideaki Hayashi
Zeeshan Hayder
Chengan He
Conghui He
Dailan He
Fan He
Fazhi He
Gaoqi He
Haoyu He
Hongliang He
Jiangpeng He
Jianzhong He
Jiawei He
Ju He
Jun-Yan He
Junfeng He
Lihuo He
Liqiang He
Nanjun He
Ruian He
Runze He
Ruozhen He

Shengfeng He
Shuting He
Songtao He
Tao He
Tianyu He
Tong He
Wei He
Xiang He
Xiangteng He
Xiangyu He
Xiaoxiao He
Xin He
Xingyi He
Xingzhe He
Xinlei He
Xinwei He
Yang He
Yang He
Yannan He
Yeting He
Yinan He
Ying He
Yisheng He
Yuhang He
Yuhang He
Zecheng He
Zewei He
Zexin He
Zhenyu He
Ziwen He
Ramya S. Hebbalaguppe
Peter Hedman
Deepti B. Hegde
Sindhu B. Hegde
Matthias Hein
Mattias Paul Heinrich
Aral Hekimoglu
Philipp Henzler
Byeongho Heo
Jae-Pil Heo
Miran Heo
Stephane Herbin
Alexander Hermans
Pedro Hermosilla Casajus
Jefferson E. Hernandez

Monica Hernandez
Charles Herrmann
Roei Herzig
Fabian Herzog
Georg Hess
Mauricio Hess-Flores
Robin Hesse
Chamin P. Hewa
    Koneputugodage
Masatoshi Hidaka
Richard E. L. Higgins
Mitchell K. Hill
Carlos Hinojosa
Tobias Hinz
Yusuke Hirota
Elad Hirsch
Shinsaku Hiura
Chih-Hui Ho
Man M. Ho
Tuan N. A. Hoang
Jennifer Hobbs
Jiun Tian Hoe
David T. Hoffmann
Derek Hoiem
Yannick Hold-Geoffroy
Lukas Höllein
Gregory I. Holste
Hiroto Honda
Cheeun Hong
Cheng-Yao Hong
Danfeng Hong
Deokki Hong
Fa-Ting Hong
Fangzhou Hong
Feng Hong
Guan Zhe Hong
Je Hyeong Hong
Ji Woo Hong
Jie Hong
Joanna Hong
Lanqing Hong
Lingyi Hong
Sangwoo Hong
Sungeun Hong
Sunghwan Hong

Susung Hong
Weixiang Hong
Xiaopeng Hong
Yan Hong
Yi Hong
Yuchen Hong
Julia Hornauer
Maxwell C. Horton
Eliahu Horwitz
Mir Rayat Imtiaz Hossain
Tonmoy Hossain
Mehdi Hosseinzadeh
Kazuhiro Hotta
Andrew Z. Hou
Fei Hou
Hao Hou
Ji Hou
Jian Hou
Jinghua Hou
Qibin Hou
Qiqi Hou
Ruibing Hou
Saihui Hou
Tingbo Hou
Xinhai Hou
Yuenan Hou
Yunzhong Hou
Zhi Hou
Lukas Hoyer
Petr Hruby
Jun-Wei Hsieh
Chih-Fan Hsu
Gee-Sern Hsu
Hung-Min Hsu
Yen-Chi Hsu
Anthony Hu
Benran Hu
Bo Hu
Dapeng Hu
Dongting Hu
Guosheng Hu
Haigen Hu
Hanjiang Hu
Hanzhe Hu
Hengtong Hu

Hezhen Hu
Jian Hu
Jian-Fang Hu
Jie Hu
Juan Hu
Junjie Hu
Kai Hu
Lianyu Hu
Lily Hu
MengShun Hu
Minghui Hu
Mu Hu
Panwen Hu
Panwen Hu
Ruizhen Hu
Shengshan Hu
Shiyu Hu
Shizhe Hu
Shu Hu
Tao Hu
Tao Hu
Vincent Tao Hu
Wenbo Hu
Xiaodan Hu
Xiaolin Hu
Xiaoling Hu
Xiaotao Hu
Xiaowei Hu
Xinting Hu
Xinting Hu
Xixi Hu
Xuefeng Hu
Yang Hu
Yaosi Hu
Yinlin Hu
Yueyu Hu
Zeyu Hu
Zhanhao Hu
zhengyu hu
Zhenzhen Hu
Zhiming Hu
Zhiming Hu
Zhongyun Hu
Zijian Hu
Andong Hua

Hang Hua
Miao Hua
Wei Hua
Mengdi Huai
Binbin Huang
Bo Huang
Buzhen Huang
Chao Huang
Chao-Tsung Huang
Chaoqin Huang
Ching-Chun Huang
Cong Huang
Danqing Huang
Di Huang
Feihu Huang
Haibin Huang
Haiyang Huang
Hao Huang
Hsin-Ping Huang
Huaibo Huang
Ian Y. Huang
Jiabo Huang
Jiahui Huang
Jiancheng Huang
Jiangyong Huang
Jiaxing Huang
Jin Huang
Jinfa Huang
Jing Huang
Jonathan Huang
Junkai Huang
Junwen Huang
Kejie Huang
Kuan-Chih Huang
Lei Huang
Libo Huang
Lin Huang
Linjiang Huang
Linyan Huang
linzhi Huang
Lun Huang
Luojie Huang
Mingzhen Huang
Qidong Huang
Qiusheng Huang

Runhui Huang
Ruqi Huang
Shaofei Huang
Sheng Huang
Sheng-Yu Huang
Shiyuan Huang
Shuaiyi Huang
Shuangping Huang
Siteng Huang
Siyu Huang
Siyuan Huang
Tao Huang
Thomas E. Huang
Tianyu Huang
Wenjian Huang
Wenke Huang
Xiaohu Huang
Xiaohua Huang
Xiaoke Huang
Xiaoshui Huang
Xiaoyang Huang
Xijie Huang
Xinyu Huang
Xuhua Huang
Yan Huang
Yaping Huang
Ye Huang
Yi Huang
Yi Huang
Yi Huang
Yi-Hua Huang
Yifei Huang
Yihao Huang
Ying Huang
Yizhan Huang
You Huang
Yue Huang
Yufei Huang
Yuge Huang
Yukun Huang
Yuyao Huang
Zaiyu Huang
Zehao Huang
Zeyi Huang
Zhe Huang

Zhewei Huang
Zhida Huang
Zhiqi Huang
Zhiyu Huang
Zhongzhan Huang
Ziling Huang
Zilong Huang
Ziqi Huang
Zixuan Huang
Ziyuan Huang
Jaesung Huh
Ka-Hei Hui
Le Hui
Tianrui Hui
Xiaofei Hui
Ahmed Imtiaz Humayun
Thomas Hummel
Jing Huo
Shuwei Huo
Yuankai Huo
Julio Hurtado
Rukhshanda Hussain
Mohamed Hussein
Noureldien Hussein
Chuong Minh Huynh
Tran Ngoc Huynh
Muhammad Huzaifa
Inwoo Hwang
Jaehui Hwang
Jenq-Neng Hwang
Sehyun Hwang
Seunghyun Hwang
Sukjun Hwang
Sunhee Hwang
Wonjun Hwang
Jeongseok Hyun
Sangeek Hyun
Ekaterina Iakovleva
Tomoki Ichikawa
A. S. M. Iftekhar
Masaaki Iiyama
Satoshi Ikehata
Sunghoon Im
Woobin Im
Nevrez Imamoglu

Abdullah Al Zubaer Imran
Tooba Imtiaz
Nakamasa Inoue
Naoto Inoue
Eldar Insafutdinov
Catalin Ionescu
Radu Tudor Ionescu
Nasar Iqbal
Umar Iqbal
Go Irie
Muhammad Zubair Irshad
Francesco Isgrò
Yasunori Ishii
Berivan Isik
Syed M. S. Islam
Mariko Isogawa
Vamsi Krishna K. Ithapu
Koichi Ito
Leonardo Iurada
Shun Iwase
Sergio Izquierdo
Sarah Jabbour
David Jacobs
David E. Jacobs
Yasamin Jafarian
Azin Jahedi
Achin Jain
Ayush Jain
Jitesh Jain
Kanishk Jain
Yash Jain
Nikita Jaipuria
Tomas Jakab
Muhammad Abdullah
  Jamal
Hadi Jamali-Rad
Stuart James
Nataraj Jammalamadaka
Donggon Jang
Sujin Jang
Young Kyun Jang
Youngjoon Jang
Steeven Janny
Paul Janson
Maximilian Jaritz

Ronnachai Jaroensri
Guillaume Jaume
Sajid Javed
Saqib Javed
Bhavin Jawade
Hirunima Jayasekara
Senthilnath Jayavelu
Guillaume Jeanneret
Pranav Jeevan
Rohit Jena
Tomas Jenicek
Simon Jenni
Hae-Gon Jeon
Jeimin Jeon
Seogkyu Jeon
Boseung Jeong
Jongheon Jeong
Yonghyun Jeong
Yoonwoo Jeong
Koteswar Rao Jerripothula
Artur Jesslen
Nikolay Jetchev
Ankit Jha
Sumit K. Jha
I-Hong Jhuo
Deyi Ji
Ge-Peng Ji
Hui Ji
Jianmin Ji
Jiayi Ji
Jingwei Ji
Naye Ji
Pengliang Ji
Shengpeng Ji
Wei Ji
Xiang Ji
Xiaopeng Ji
Xiaozhong Ji
Xinya Ji
Yu Ji
Yuanfeng Ji
Zhanghexuan Ji
Baoxiong Jia
Ding Jia
Jinrang Jia

Menglin Jia
Shan Jia
Shuai Jia
Tong Jia
Wenqi Jia
Xiaojun Jia
Xiaosong Jia
Xu Jia
Yiren Jian
Bo Jiang
Borui Jiang
Chen Jiang
Guangyuan Jiang
Hai Jiang
Haiyang Jiang
Haiyong Jiang
Hanwen Jiang
Hanxiao Jiang
Hao Jiang
Haobo Jiang
Huaizu Jiang
Huajie Jiang
Jianmin Jiang
Jianwen Jiang
Jiaxi Jiang
Jin Jiang
Jing Jiang
Kaixun Jiang
Kui Jiang
Li Jiang
Liming Jiang
Meirui Jiang
Ming Jiang
Ming Jiang
Peng Jiang
Peng-Tao Jiang
Runqing Jiang
Siyang Jiang
Tianjian Jiang
Tingting Jiang
Weisen Jiang
Wen Jiang
Xingyu Jiang
Xueying Jiang
Xuhao Jiang

Yanru Jiang
Yi Jiang
Yifan Jiang
Yingying Jiang
Yue Jiang
Yuming Jiang
Zeren Jiang
Zheheng Jiang
Zhenyu Jiang
Zhongyu Jiang
Zi-Hang Jiang
Zixuan Jiang
Ziyu Jiang
Zutao Jiang
Zutao Jiang
Jianbin Jiao
Jianbo Jiao
Licheng Jiao
Ruochen Jiao
Shuming Jiao
Wenpei Jiao
Zequn Jie
Dongkwon Jin
Gaojie Jin
Haian Jin
Haibo Jin
Kyong Hwan Jin
Lei Jin
Lianwen Jin
Linyi Jin
Peng Jin
Peng Jin
Sheng Jin
SouYoung Jin
Weiyang Jin
Xiao Jin
Xiaojie Jin
Xin Jin
Xin Jin
Yeying Jin
Yi Jin
Ying Jin
Zhenchao Jin
Chenchen Jing
Junpeng Jing

Mengmeng Jing
Taotao Jing
Jeonghee Jo
Yeonsik Jo
Younghyun Jo
Cameron A. Johnson
Faith M. Johnson
Maxwell Jones
Michael J. Jones
R. Kenny Jones
Ameya Joshi
Shantanu H. Joshi
Brendan Jou
Chen Ju
Wei Ju
Xuan Ju
Yan Ju
Felix Juefei-Xu
Florian Jug
Yohan Jun
Kim Jun-Seong
Masum Shah Junayed
Chanyong Jung
Claudio R. Jung
Hoin Jung
Hyunyoung Jung
Sangwon Jung
Steffen Jung
Sacha Jungerman
Hari Chandana K.
Prajwal K. R.
Berna Kabadayi
Anis Kacem
Anil Kag
Kumara Kahatapitiya
Bernhard Kainz
Ivana Kajic
Ioannis Kakogeorgiou
Niveditha Kalavakonda
Mahdi M. Kalayeh
Ajinkya Kale
Anmol Kalia
Sinan Kalkan
Jayateja Kalla
Tarun Kalluri

Uday Kamal
Sandesh Kamath
Chandra Kambhamettu
Meina Kan
Sai Srinivas Kancheti
Takuhiro Kaneko
Cuicui Kang
Dahyun Kang
Guoliang Kang
Hyolim Kang
Jaeyeon Kang
Juwon Kang
Li-Wei Kang
Mingon Kang
MinGuk Kang
Minjun Kang
Weitai Kang
Zhao Kang
Yash Mukund Kant
Yueying Kao
Saarthak Kapse
Aupendu Kar
Oğuzhan Fatih Kar
Ozgur Kara
Tamás Karácsony
Srikrishna Karanam
Neerav Karani
Mert Asim Karaoglu
Laurynas Karazija
Navid Kardan
Amirhossein Kardoost
Nour Karessli
Michelle Karg
Mohammad Reza Karimi
    Dastjerdi
Animesh Karnewar
Arjun M. Karpur
Shyamgopal Karthik
Korrawe Karunratanakul
Tejaswi Kasarla
Satyananda Kashyap
Yoni Kasten
Marc A. Kastner
Hirokatsu Kataoka
Isinsu Katircioglu

Kai Katsumata
Ilya Kaufman
Manuel Kaufmann
Chaitanya Kaul
Prannay Kaul
Prakhar Kaushik
Isaak Kavasidis
Ryo Kawahara
Yuki Kawana
Justin Kay
Evangelos Kazakos
Jingcheng Ke
Lei Ke
Tsung-Wei Ke
Wei Ke
Zhanghan Ke
Nikhil V. Keetha
Thomas Kehrenberg
Marilyn Keller
Rohit Keshari
Janis Keuper
Daniel Keysers
Hrant Khachatrian
Seyran Khademi
Wesley A. Khademi
Taras Khakhulin
Samir Khaki
Hasam Khalid
Umar Khalid
Amr Khalifa
Asif Hussain Khan
Faizan Farooq Khan
Muhammad Haris Khan
Naimul Khan
Qadeer Khan
Zeeshan Khan
Bishesh Khanal
Pulkit Khandelwal
Ishan Khatri
Muhammad Uzair Khattak
Vahid Reza Khazaie
Vaishnavi M. Khindkar
Rawal Khirodkar
Pirazh Khorramshahi
Sahil S. Khose

Kourosh Khoshelham
Sena Kiciroglu
Benjamin Kiefer
Kotaro Kikuchi
Mert Kilickaya
Benjamin Killeen
Beomyoung Kim
Boah Kim
Bumsoo Kim
Byeonghwi Kim
Byoungjip Kim
Changhoon Kim
Changick Kim
Chanho Kim
Chanyoung Kim
Dahun Kim
Dahyun Kim
Diana S. Kim
Dong-Jin Kim
Donggun Kim
Donghyun Kim
Dongkeun Kim
Dongwan Kim
Dongyoung Kim
Eun-Sol Kim
Eunji Kim
Euyoung Kim
Geeho Kim
Giseop Kim
Guisik Kim
Gwanghyun Kim
Hakyeong Kim
Hanjae Kim
Hanjung Kim
Heewon Kim
Howon Kim
Hyeongwoo Kim
Hyeongwoo Kim
Hyo Jin Kim
Hyung-Il Kim
Hyunwoo J. Kim
Insoo Kim
Jae Myung Kim
Jeongsol Kim
Jihwan Kim

Jinkyu Kim
Jinwoo Kim
Jongyoo Kim
Joonsoo Kim
Jung Uk Kim
Junho Kim
Junho Kim
Junsik Kim
Kangyeol Kim
Kunhee Kim
Kwang In Kim
Manjin Kim
Minchul Kim
Minji Kim
Minjung Kim
Namil Kim
Namyup Kim
Nayeong Kim
Sanghyun Kim
Seong Tae Kim
Seungbae Kim
Seungryong Kim
Seungwook Kim
Soo Ye Kim
Soohwan Kim
Sungnyun Kim
Sungyeon Kim
Sunnie S. Y. Kim
Tae Hyun Kim
Tae Hyung Kim
Taehoon Kim
Taehun Kim
Taehwan Kim
Taekyung Kim
Taeoh Kim
Taewoo Kim
Won Hwa Kim
Wonjae Kim
Woo Jae Kim
Young Min Kim
YoungBin Kim
Youngeun Kim
Youngseok Kim
Youngwook Kim
Akisato Kimura

Andreas Kirsch
Nikita Kister
Furkan Osman Kınlı
Marcus Klasson
Florian Kleber
Tzofi M. Klinghoffer
Jan P. Klopp
Florian Kluger
Hannah Kniesel
David M. Knigge
Byungsoo Ko
Dohwan Ko
Jongwoo Ko
Takumi Kobayashi
Muhammed Kocabas
Yeong Jun Koh
Kathlén Kohn
Subhadeep Koley
Nick Kolkin
Soheil Kolouri
Jacek Komorowski
Deying Kong
Fanjie Kong
Hanyang Kong
Hui Kong
Kyeongbo Kong
Lecheng Kong
Lingdong Kong
Linghe Kong
Lingshun Kong
Naejin Kong
Quan Kong
Shu Kong
Xianghao Kong
Xiangtao Kong
Xiangwei Kong
Xiaoyu Kong
Xin Kong
Youyong Kong
Yu Kong
Zhenglun Kong
Aishik Konwer
Nicholas C. Konz
Gwanhyeong Koo
Juil Koo

Julian F. P. Kooij
George Kopanas
Sanjeev J. Koppal
Bruno Korbar
Giorgos Kordopatis-Zilos
Dimitri Korsch
Adam Kortylewski
Divya Kothandaraman
Suraj Kothawade
Iuliia Kotseruba
Sasikanth Kotti
Alankar Kotwal
Shashank Kotyan
Alexandros Kouris
Petros Koutras
Rama Kovvuri
Dilip Krishnan
Praveen Krishnan
Ranganath Krishnan
Rohan M. Krishnan
Georg Krispel
Alexander Krull
Tianshu Kuai
Haowei Kuang
Zhengfei Kuang
Andrey Kuehlkamp
David Kügler
Arjan Kuijper
Anna Kukleva
Jonas Kulhanek
Peter Kulits
Akshay R. Kulkarni
Ashutosh C. Kulkarni
Kuldeep Kulkarni
Nilesh Kulkarni
Abhinav Kumar
Abhishek Kumar
Akash Kumar
Avinash Kumar
B. V. K. Vijaya Kumar
Chandan Kumar
Pulkit Kumar
Ratnesh Kumar
Sateesh Kumar
Satish Kumar

Suryansh Kumar
Yogesh Kumar
Nupur Kumari
Sudhakar Kumawat
Nilakshan Kunananthaseelan
Rohit Kundu
Souvik Kundu
Meng-Yu Jennifer Kuo
Weicheng Kuo
Shuhei Kurita
Yusuke Kurose
Takahiro Kushida
Uday Kusupati
Alina Kuznetsova
Jobin K. V.
Henry Kvinge
Ho Man Kwan
Hyeokjun Kweon
Donghyeon Kwon
Gihyun Kwon
Heeseung Kwon
Hyoukjun Kwon
Myung-Joon Kwon
Taein Kwon
YoungJoong Kwon
Cameron Kyle-Davidson
Christos Kyrkou
Jorma Laaksonen
Patrick Labatut
Yann Labbé
Manuel Ladron de Guevara
Florent Lafarge
Jean Lahoud
Bolin Lai
Farley Lai
Jian-Huang Lai
Shenqi Lai
Xin Lai
Yu-Kun Lai
Yung-Hsuan Lai
Zeqiang Lai
Zhengfeng Lai
Barath Lakshmanan
Rohit Lal

Rodney LaLonde
Hala Lamdouar
Meng Lan
Yushi Lan
Federico Landi
George V. Landon
Chunbo Lang
Jochen Lang
Nico Lang
Georg Langs
Raffaella Lanzarotti
Dong Lao
Yixing Lao
Yizhen Lao
Zakaria Laskar
Alexandros Lattas
Chun Pong Lau
Shlomi Laufer
Justin Lazarow
Svetlana Lazebnik
Duy Tho Le
Hieu Le
Hoang Le
Hoang Le
Thi-Thu-Huong Le
Trung Le
Trung-Nghia Le
Tung Thanh Le
Hoàng-Ân Lê
Herve Le Borgne
Guillaume Le Moing
Erik Learned-Miller
Tim Lebailly
Byeong-Uk Lee
Byung-Kwan Lee
Cheng-Han Lee
Chul Lee
Daeun Lee
Dogyoon Lee
Dong Hoon Lee
Eugene Eu Tzuan Lee
Eung-Joo Lee
Gyuseong Lee
Hsin-Ying Lee
Hwee Kuan Lee

Hyeongmin Lee
Hyungtae Lee
Jae Yong Lee
Jaeho Lee
Jaeseong Lee
Jaewon Lee
Jangho Lee
Jangwon Lee
Jihyun Lee
Jiyoung Lee
Jong-Seok Lee
Jongho Lee
Jongmin Lee
Joo-Ho Lee
Joon-Young Lee
Joonseok Lee
Jungbeom Lee
Jungho Lee
Jungwoo Lee
Junha Lee
Junhyun Lee
Junyong Lee
Kibok Lee
Kuan-Ying Lee
Kwonjoon Lee
Kwot Sin Lee
Kyungmin Lee
Kyungmoon Lee
Minhyeok Lee
Minsik Lee
Pilhyeon Lee
Saehyung Lee
Sangho Lee
Sanghyeok Lee
Sangmin Lee
Sehun Lee
Sehyung Lee
Seon-Ho Lee
Seong Hun Lee
Seongwon Lee
Seung Hyun Lee
Seung-Ik Lee
Seungho Lee
Seunghun Lee
Seungmin Lee

Seungyong Lee
Sohyun Lee
Suhyeon Lee
Sungho Lee
Sungmin Lee
Suyoung Lee
Taehyun Lee
Wooseok Lee
Yao-Chih Lee
Yi-Lun Lee
Yonghyeon Lee
Youngwan Lee
Leonidas Lefakis
Bowen Lei
Chenyang Lei
Chenyi Lei
Jiahui Lei
Na Lei
Qinqian Lei
Yinjie Lei
Thomas Leimkuehler
Abe Leite
Abdelhak Lemkhenter
Jiaxu Leng
Luziwei Leng
Zhiying Leng
Hendrik P. A. Lensch
Jan E. Lenssen
Ted Lentsch
Simon Lepage
Stefan Leutenegger
Filippo Leveni
Axel Levy
Ailin Li
Aixuan Li
Baiang Li
Baoxin Li
Bin Li
Bing Li
Bing Li
Bo Li
Bowen Li
Boying Li
Changlin Li
Changlin Li

Chao Li
Chenghong Li
Chenglin Li
Chenglong Li
Chengze Li
Chun-Guang Li
Daiqing Li
Dasong Li
Dian Li
Dong Li
Fangda Li
Feiran Li
Fenghai Li
Gen Li
Guanbin Li
Guangrui Li
Guihong Li
Guorong Li
Haifeng Li
Han Li
Hang Li
Hangyu Li
Hanhui Li
Hao Li
Hao Li
Haoang Li
Haoran Li
Haoxiang Li
Haoxin Li
He Li
Heng Li
Hengduo Li
Hongshan Li
Hongwei Bran Li
Hongxiang Li
Hongyang Li
Hongyu Li
Huafeng Li
Huan Li
Hui Li
Jiacheng Li
Jiahao Li
Jialu Li
Jiaman Li
Jiangmeng Li

Jiangtong Li
Jiangyuan Li
Jianing Li
Jianwei Li
Jianwu Li
Jiaqi Li
Jiaqi Li
Jiatong Li
Jiaxuan Li
Jiazhi Li
Jichang Li
Jie Li
Jin Li
Jinglun Li
Jingzhi Li
Jingzong Li
Jinlong Li
Jinlong Li
Jinpeng Li
Jinxing Li
Jun Li
Jun Li
Junbo Li
Juncheng Li
Junxuan Li
Junyi Li
Kai Li
Kaican Li
Kailin Li
Ke Li
Kehan Li
Keyu Li
Kun Li
Kunchang Li
Kunpeng Li
Lei Li
Lei Li
Li Li
Li Erran Li
Liang Li
Lin Li
Lincheng Li
Liulei Li
Liunian Harold Li
Lujun Li

Manyi Li
Maomao Li
Meng Li
Mengke Li
Mengtian Li
Mengtian Li
Ming Li
Ming Li
Minghan Li
Mingjie Li
Nannan Li
Nianyi Li
Peike Li
Peizhao Li
Peng Li
Pengpeng Li
Pengyu Li
Ping Li
Puhao Li
Qiang Li
Qing Li
Qingyong Li
Qiufu Li
Qizhang Li
Ren Li
Rong Li
Rongjie Li
Ru Li
Rui Li
Ruibo Li
Ruihui Li
Ruilong Li
Ruining Li
Ruixuan Li
Runze Li
Ruoteng Li
Shaohua Li
Shasha Li
Shigang Li
Shijie Li
Shikun Li
Shile Li
Shuai Li
Shuai Li
Shuang Li

Shuwei Li
Si Li
Siyao Li
Siyuan Li
Siyuan Li
Taihui Li
Tianye Li
Wanhua Li
Wanqing Li
Wei Li
Wei Li
Wei Li
Wei-Hong Li
Weihao Li
Weijia Li
Weiming Li
Wenbin Li
Wenbo Li
Wenhao Li
Wenjie Li
Wenshuo Li
Wentong Li
Wenxi Li
Xiang Li
Xiang Li
Xiang Li
Xiang Li
Xiang Li
Xiangtai Li
Xiangyang Li
Xianzhi Li
Xiao Li
Xiao Li
Xiaoguang Li
Xiaomeng Li
Xiaoming Li
Xiaoqi Li
Xiaoqiang Li
Xiaotian Li
Xiaoyu Li
Xin Li
Xin Li
Xin Li
Xinghui Li
Xingyi Li

Xingyu Li
Xinjie Li
Xinyu Li
Xiu Li
Xiujun Li
Xuan Li
Xuanlin Li
Xuelong Li
Xuelu Li
Xueqian Li
Ya-Li Li
Yanan Li
Yang Li
Yang Li
Yangyan Li
Yanjing Li
Yansheng Li
Yanwei Li
Yanyu Li
Yaohui Li
Yaowei Li
Yawei Li
Yi Li
Yi Li
Yicong Li
Yicong Li
Yifei Li
Yijin Li
Yijun Li
Yijun Li
Yikang Li
Yimeng Li
Yiming Li
Yiming Li
Yingwei Li
Yiting Li
Yixuan Li
Yize Li
Yizhuo Li
Yong Li
Yong-Lu Li
Yongjie Li
Yuanman Li
Yuanming Li
Yuelong Li

Yuexiang Li
Yuezun Li
Yuhang Li
Yuheng Li
Yulin Li
Yumeng Li
Yunfan Li
Yunheng Li
Yunqiang Li
Yunsheng Li
Yuyan Li
Yuyang Li
Zejian Li
Zekun Li
Zekun Li
Zhangheng Li
Zhangzikang Li
Zhaoshuo Li
Zhaowen Li
Zhe Li
Zhe Li
Zhen Li
Zhen Li
Zhen Li
Zheng Li
Zhengqin Li
Zhengyuan Li
Zhenyu Li
Zhichao Li
Zhihao Li
Zhihao Li
Zhiheng Li
Zhiqi Li
Zhixuan Li
Zhong Li
Zhuoling Li
Zhuowan Li
Zhuowei Li
Zhuoxiao Li
Zihan Li
Ziqiang Li
Wen Qiao Li
Dongze Lian
Long Lian
Qing Lian

Ruyi Lian
Zhouhui Lian
Jin Lianbao
Chao Liang
Chia-Kai Liang
Dingkang Liang
Feng Liang
Gongbo Liang
Hanxue Liang
Hao Liang
Hui Liang
Jiadong Liang
Jiajun Liang
Jian Liang
Jingyun Liang
Jinxiu S. Liang
Junwei Liang
Kaiqu Liang
Ke Liang
Kevin J. Liang
Luming Liang
Mingfu Liang
Pengpeng Liang
Siyuan Liang
Xiaoxiao Liang
Xinran Liang
Xiwen Liang
Yang Liang
Yixun Liang
Yongqing Liang
Youwei Liang
Yuanzhi Liang
Zhexin Liang
Zhihao Liang
Zhixuan Liang
Kang Liao
Liang Liao
Minghui Liao
Ting-Hsuan Liao
Wei Liao
Xin Liao
Yinghong Liao
Yue Liao
Zhibin Liao
Ziwei Liao

Benedetta Liberatori
Daniel J. Lichy
Maiko Lie
Qin Likun
Isaak Lim
Teck Yian Lim
Bannapol Limanond
Baijiong Lin
Beibei Lin
Cheng Lin
Chenhao Lin
Chia-Wen Lin
Chieh Hubert Lin
Chuang Lin
Chung-Ching Lin
Chunyu Lin
Ci-Siang Lin
Di Lin
Fanqing Lin
Feng Lin
Fudong Lin
Guangfeng Lin
Haotong Lin
Haozhe Lin
Hubert Lin
Hui Lin
Jason Lin
Jianxin Lin
Jiaqi Lin
Jiaying Lin
Jiehong Lin
Jierui Lin
Jintao Lin
Kai-En Lin
Ke Lin
Kevin Lin
Kevin Qinghong Lin
Kuan Heng Lin
Kun-Yu Lin
Kunyang Lin
Kwan-Yee Lin
Lijian Lin
Liqiang Lin
Liting Lin
Luojun Lin

Mingyuan Lin
Qiuxia Lin
Shaohui Lin
Shih-Yao Lin
Sihao Lin
Siyou Lin
Tiancheng Lin
Tsung-Yu Lin
Wanyu Lin
Wei Lin
Wei Lin
Wen-Yan Lin
Wenbin Lin
Xiangbo Lin
Xianhui Lin
Xiaofan Lin
Xiaofeng Lin
Xin Lin
Xudong Lin
Xue Lin
Xuxin Lin
Ya-Wei Eileen Lin
Yan-Bo Lin
Yancong Lin
Yi Lin
Yijie Lin
Yiming Lin
Yiqi Lin
Yiqun Lin
Yongliang Lin
Yu Lin
Yuanze Lin
Yuewei Lin
Zhi-Hao Lin
Zhiqiu Lin
ZhiWei Lin
Zinan Lin
Ziyi Lin
David B. Lindell
Philipp Lindenberger
Jingwang Ling
Jun Ling
Yongguo Ling
Zhan Ling
Alexander Liniger

Stefan P. Lionar
Phillip Lippe
Lahav O. Lipson
Joey Litalien
Ron Litman
Mattia Litrico
Dor Litvak
Aishan Liu
Ajian Liu
Akide L. Y. Liu
Andrew Liu
Ao Liu
Bang Liu
Benlin Liu
Bin Liu
Bin Liu
Bing Liu
Binghao Liu
Bingyuan Liu
Bo Liu
Bo Liu
Bo Liu
Boning Liu
Bowen Liu
Boxiao Liu
Chang Liu
Chang Liu
Chang Liu
Chang Liu
Chao Liu
Chengxin Liu
Chengxu Liu
Chih-Ting Liu
Chuanjian Liu
Chun-Hao Liu
Daizong Liu
Decheng Liu
Di Liu
Difan Liu
Dong Liu
Dongnan Liu
Fang Liu
Fang Liu
Fangyi Liu
Feng Liu

Fengbei Liu
Fenglin Liu
Fengqi Liu
Furui Liu
Fuxiao Liu
Haisong Liu
Han Liu
Hanwen Liu
Hanyuan Liu
Hao Liu
Haolin Liu
Haotian Liu
Haozhe Liu
Heshan Liu
Hong Liu
Hongbin Liu
Hongfu Liu
Hongyu Liu
Hsueh-Ti Derek Liu
Huidong Liu
Isabella Liu
Ji Liu
Jia-Wei Liu
Jiachen Liu
Jiaheng Liu
Jiahui Liu
Jiaming Liu
Jiancheng Liu
Jiang Liu
Jianmeng Liu
Jiashuo Liu
Jiawei Liu
Jiawei Liu
Jiayang Liu
Jiayi Liu
Jie Liu
Jie Liu
Jie Liu
Jihao Liu
Jing Liu
Jing Liu
Jing Liu
Jingyuan Liu
Jingyuan Liu
Jiuming Liu
Jiyuan Liu
Jun Liu
Kang-Jun Liu
Kangning Liu
Kenkun Liu
Kunhao Liu
Li Liu
Lijuan Liu
Lingbo Liu
Lingqiao Liu
Liu Liu
Liyang Liu
Meng Liu
Mengchen Liu
Miao Liu
Ming Liu
Minghao Liu
Minghua Liu
Mingxuan Liu
Mingyuan Liu
Nan Liu
Nian Liu
Ning Liu
Peidong Liu
Peirong Liu
Peiye Liu
Pengju Liu
Ping Liu
Qi Liu
Qiankun Liu
Qihao Liu
Qing Liu
Qingjie Liu
Richard Liu
Risheng Liu
Rui Liu
Ruicong Liu
Ruoshi Liu
Ruyu Liu
Shaohui Liu
Shaoteng Liu
Shaowei Liu
Sheng Liu
Shenglan Liu
Shikun Liu
Shilong Liu
Shuaicheng Liu
Shuaicheng Liu
Shuming Liu
Songhua Liu
Tao Liu
Tian Yu Liu
Tianci Liu
Tianshan Liu
Tongliang Liu
Tyng-Luh Liu
Wei Liu
Weifeng Liu
Weixiao Liu
Weiyu Liu
Wen Liu
Wenxi Liu
Wenyu Liu
Wenyu Liu
Wu Liu
Xian Liu
Xianglong Liu
Xianpeng Liu
Xiao Liu
Xiaohong Liu
Xiaoyu Liu
Xiaoyu Liu
Xihui Liu
Xin Liu
Xin Liu
Xinchen Liu
Xingtong Liu
Xingyu Liu
Xinhang Liu
Xinhui Liu
Xinpeng Liu
Xinwei Liu
Xinyu Liu
Xiulong Liu
Xiyao Liu
Xu Liu
Xubo Liu
Xudong Liu
Xueting Liu
Xueyi Liu

Yan Liu
Yanbin Liu
Yang Liu
Yang Liu
Yang Liu
Yang Liu
Yang Liu
Yanwei Liu
Yaojie Liu
Ye Liu
Yi Liu
Yihao Liu
Yingcheng Liu
Yingfei Liu
Yipeng Liu
Yipeng Liu
Yixin Liu
Yizhang Liu
Yong Liu
Yong Liu
Yonghuai Liu
Yongtuo Liu
Yu Liu
Yu-Lun Liu
Yu-Shen Liu
Yuan Liu
Yuang Liu
Yuanpei Liu
Yuanpeng Liu
Yuanwei Liu
Yuchen Liu
Yuchen Liu
Yuchi Liu
Yueh-Cheng Liu
Yufan Liu
Yuhao Liu
Yuliang Liu
Yun Liu
Yun Liu
Yun Liu
Yunfan Liu
Yunfei Liu
Yunze Liu
Yupei Liu
Yuqi Liu
Yuyang Liu
Yuyuan Liu
Zhaoqiang Liu
Zhe Liu
Zhe Liu
Zhen Liu
Zheng Liu
Zhenguang Liu
Zhi Liu
Zhihua Liu
Zhijian Liu
Zhili Liu
Zhuoran Liu
Ziquan Liu
Ziyi Liu
Zuxin Liu
Zuyan Liu
Josep Llados
Ling Lo
Shao-Yuan Lo
Liliana Lo Presti
Sylvain Lobry
Yaroslava Lochman
Fotios Logothetis
Suhas Lohit
Marios Loizou
Vishnu Suresh Lokhande
Cheng Long
Chengjiang Long
Fuchen Long
Guodong Long
Rujiao Long
Shangbang Long
Teng Long
Xiaoxiao Long
Zijun Long
Ivan Lopes
Vasco Lopes
Adrian Lopez-Rodriguez
Javier Lorenzo-Navarro
Yujing Lou
Brian C. Lovell
Weng Fei Low
Changsheng Lu
Chun-Shien Lu
Daohan Lu
Dongming Lu
Erika Lu
Fan Lu
Guangming Lu
Guo Lu
Hao Lu
Hao Lu
Hongtao Lu
Jiachen Lu
Jiaxin Lu
Jiwen Lu
Lewei Lu
Liying Lu
Quanfeng Lu
Shenyu Lu
Shun Lu
Tao Lu
Xiangyong Lu
Xiankai Lu
Xin Lu
Xuanchen Lu
Xuequan Lu
Yan Lu
Yang Lu
Yanye Lu
Yawen Lu
Yifan Lu
Yongchun Lu
Yongxi Lu
Yu Lu
Yu Lu
Yuzhe Lu
Zhichao Lu
Zhihe Lu
Zijia Lu
Tianyu Luan
Pauline Luc
Simon Lucey
Timo Lüddecke
Jonathon Luiten
Jovita Lukasik
Ao Luo
Cheng Luo
Chuanchen Luo

Donghao Luo
Fangzhou Luo
Gen Luo
Gongning Luo
Hongchen Luo
Jiahao Luo
Jiebo Luo
Jinqi Luo
Jinqi Luo
Jun Luo
Katie Z. Luo
Kunming Luo
Lei Luo
Mandi Luo
Mi Luo
Ruotian Luo
Sihui Luo
Tiange Luo
Wenhan Luo
Xiao Luo
Xiaotong Luo
Xiongbiao Luo
Xu Luo
Yadan Luo
Yawei Luo
Ye Luo
Yisi Luo
Yong Luo
You-Wei Luo
Yuanjing Luo
Zelun Luo
Zhengxiong Luo
Zhengyi Luo
Zhiming Luo
Zhipeng Luo
Zhongjin Luo
Zilin Luo
Ziyang Luo
Tung M. Luu
Diogo C. Luvizon
Jun Lv
Pei Lv
Yunqiu Lv
Zhaoyang Lv
Gengyu Lyu

Jiancheng Lyu
Jipeng Lyu
Junfeng Lyu
Mengyao Lyu
Mingzhi Lyu
Weimin Lyu
Xiaoyang Lyu
Xinyu Lyu
Yiwei Lyu
Youwei Lyu
Ailong Ma
Andy J. Ma
Benteng Ma
Bingpeng Ma
Chao Ma
Chuofan Ma
Cong Ma
Cuixia Ma
Fan Ma
Fangchang Ma
Fei Ma
Guozheng Ma
Haoyu Ma
Hengbo Ma
Huimin Ma
Jiahao Ma
Jianqi Ma
Jiawei Ma
Jiayi Ma
Kai Ma
Kede Ma
Lei Ma
Li Ma
Lin Ma
Liqian Ma
Lizhuang Ma
Mengmeng Ma
Ning Ma
Qianli Ma
Rui Ma
Shijie Ma
Shiqiang Ma
Shiqing Ma
Shuailei Ma
Sizhuo Ma

Tao Ma
Teli Ma
Wenxuan Ma
Wufei Ma
Xianzheng Ma
Xiaoxuan Ma
Xinyin Ma
Xinzhu Ma
Xu Ma
Yeyao Ma
Yifeng Ma
Yuexiao Ma
Yuexin Ma
Yunsheng Ma
Zhan Ma
Zhanyu Ma
Ziping Ma
Ziqiao Ma
Muhammad Maaz
Anish Madan
Neelu Madan
Spandan Madan
Sai Advaith Maddipatla
Rishi Madhok
Filippo Maggioli
Simone Magistri
Marcus Magnor
Sabarinath Mahadevan
Shweta Mahajan
Aniruddha Mahapatra
Sarthak Kumar Maharana
Behrooz Mahasseni
Upal Mahbub
Arif Mahmood
Kaleel Mahmood
Mohammed Mahmoud
Tanvir Mahmud
Jinjie Mai
Helena de Almeida Maia
Josef Maier
Shishira R. Maiya
Snehashis Majhi
Orchid Majumder
Sagnik Majumder
Ilya Makarov

Sina Malakouti
Hashmat Shadab Malik
Mateusz Malinowski
Utkarsh Mall
Srikanth Malla
Clement Mallet
Dimitrios Mallis
Abed Malti
Yunze Man
Oscar Mañas
Karttikeya Mangalam
Fabian Manhardt
Ioannis Maniadis Metaxas
Fahim Mannan
Rafal Mantiuk
Dongxing Mao
Jiageng Mao
Wei Mao
Weian Mao
Weixin Mao
Ye Mao
Yongsen Mao
Yunyao Mao
Yuxin Mao
Zhiyuan Mao
Emanuela Marasco
Matthew Marchellus
Alberto Marchisio
Diego Marcos
Alina E. Marcu
Riccardo Marin
Manuel J. Marín-Jiménez
Octave Mariotti
Dejan Markovic
Imad Eddine Marouf
Valerio Marsocci
Diego Martin Arroyo
Ricardo Martin-Brualla
Brais Martinez
Renato Martins
Damien Martins Gomes
Tetiana Martyniuk
Pierre Marza
David Masip
Carlo Masone

Timothée Masquelier
André G. Mateus
Minesh Mathew
Yusuke Matsui
Bruce A. Maxwell
Christoph Mayer
Prasanna Mayilvahanan
Amir Mazaheri
Amrita Mazumdar
Pratik Mazumder
Alessio Mazzucchelli
Amarachi B. Mbakwe
Scott McCloskey
Naga Venkata Kartheek
    Medathati
Henry Medeiros
Guofeng Mei
Haiyang Mei
Jie Mei
Jieru Mei
Kangfu Mei
Lingjie Mei
Xiaoguang Mei
Dennis Melamed
Luke Melas-Kyriazi
Iaroslav Melekhov
Yifang Men
Ricardo A. Mendoza-León
Depu Meng
Fanqing Meng
Jingke Meng
Lingchen Meng
Qier Meng
Qingjie Meng
Quan Meng
Yanda Meng
Zibo Meng
Otniel-Bogdan Mercea
Pablo Mesejo
Safa Messaoud
Nico Messikommer
Nando Metzger
Christopher Metzler
Vasileios Mezaris
Liang Mi

Zhenxing Mi
S. Mahdi H. Miangoleh
Bo Miao
Changtao Miao
Jiaxu Miao
Zichen Miao
Bjoern Michele
Christian Micheloni
Marko Mihajlovic
Zoltán Á. Milacski
Simone Milani
Leo Milecki
Roy Miles
Christen Millerdurai
Monica Millunzi
Chaerin Min
Cheol-Hui Min
Dongbo Min
Hyun-Seok Min
Jie Min
Juhong Min
Kyle Min
Yifei Min
Yuecong Min
Zhixiang Min
Matthias Minderer
Di Ming
Qi Ming
Xiang Ming
Riccardo Miotto
Aymen Mir
Pedro Miraldo
Parsa Mirdehghan
Seyed Ehsan Mirsadeghi
Muhammad Jehanzeb
    Mirza
Ashkan Mirzaei
Dmytro Mishkin
Anand Mishra
Ashish Mishra
Samarth Mishra
Shlok K. Mishra
Diganta Misra
Abhay Mittal
Gaurav Mittal

Surbhi Mittal
Trisha Mittal
Taiki Miyanishi
Daisuke Miyazaki
Hong Mo
Kaichun Mo
Sangwoo Mo
Sicheng Mo
Sicheng Mo
Zhipeng Mo
Michael Moeller
Peyman Moghadam
Hadi Mohaghegh
  Dolatabadi
Salman Mohamadi
Mirgahney H. Mohamed
Deen Dayal Mohan
Fnu Mohbat
Satyam Mohla
Tony C. W. Mok
Liliane Momeni
Pascal Monasse
Ajoy Mondal
Anindya Mondal
Mathew Monfort
Tom Monnier
Yusuke Monno
Eduardo F. Montesuma
Gyeongsik Moon
Taesup Moon
WonJun Moon
Dror Moran
Julie R. C. Mordacq
Deeptej S. More
Arthur Moreau
Davide Morelli
Luca Morelli
Pedro Morgado
Alexandre Morgand
Henrique Morimitsu
Matteo Moro
Lia Morra
Matteo Mosconi
Ali Mosleh

Sayed Mohammad
  Mostafavi Isfahani
Saman Motamed
Chong Mou
Dana Moukheiber
Pierre Moulon
Ramy A. Mounir
Théo Moutakanni
Fangzhou Mu
Jiteng Mu
Yao Mark Mu
Manasi Muglikar
Yasuhiro Mukaigawa
Amitangshu Mukherjee
Avideep Mukherjee
Prerana Mukherjee
Tanmoy Mukherjee
Anirban Mukhopadhyay
Soumik Mukhopadhyay
Yusuke Mukuta
Ravi Teja Mullapudi
Lea Müller
Norman Müller
Chaithanya Kumar
  Mummadi
Muhammad Akhtar Munir
Subrahmanyam Murala
Sanjeev Muralikrishnan
Ana C. Murillo
Nils Murrugarra-Llerena
Mohamed Adel Musallam
Damien Muselet
Josh David Myers-Dean
Byeonghu Na
Taeyoung Na
Muhammad Ferjad Naeem
Sauradip Nag
Pravin Nagar
Rajendra Nagar
Varun Nagaraja
Tushar Nagarajan
Seungjun Nah
Shu Nakamura
Gaku Nakano
Yuta Nakashima

Kiyohiro Nakayama
Mitsuru Nakazawa
Krishna Kanth Nakka
Yuesong Nan
Karthik Nandakumar
Paolo Napoletano
Syed S. Naqvi
Dinesh Reddy
  Narapureddy
Supreeth
  Narasimhaswamy
Kartik Narayan
Sriram Narayanan
Fabio Narducci
Erickson R. Nascimento
Muzammal Naseer
Kamal Nasrollahi
Lakshmanan Nataraj
Vishwesh Nath
Avisek Naug
Alexander Naumann
K. L. Navaneet
Pablo Navarrete Michelini
Shah Nawaz
Nazir Nayal
Niv Nayman
Amin Nejatbakhsh
Negar Nejatishahidin
Reyhaneh Neshatavar
Pedro C. Neto
Lukáš Neumann
Richard Newcombe
Alejandro Newell
Evonne Ng
Kam Woh Ng
Trung T. Ngo
Tuan Duc Ngo
Anh Nguyen
Anh Duy Nguyen
Cuong Cao Nguyen
Duc Anh Nguyen
Hoang Chuong Nguyen
Huy Hong Nguyen
Khai Nguyen
Khanh-Binh Nguyen

Khanh-Duy Nguyen
Khoi Nguyen
Khoi D. Nguyen
Kiet A. Nguyen
Ngoc Cuong Nguyen
Pha Nguyen
Phi Le Nguyen
Phong Ha Nguyen
Rang Nguyen
Tam V. Nguyen
Thao Nguyen
Thuan Hoang Nguyen
Toan Tien Nguyen
Trong-Tung Nguyen
Van Nguyen Nguyen
Van-Quang Nguyen
Thuong Nguyen Canh
Thien Trang Nguyen Vu
Haomiao Ni
Jiangqun Ni
Minheng Ni
Yao Ni
Zhen-Liang Ni
Zixuan Ni
Dong Nie
Hui Nie
Jiahao Nie
Lang Nie
Liqiang Nie
Qiang Nie
Ying Nie
Yinyu Nie
Yongwei Nie
Aditya Nigam
Kshitij N. Nikhal
Nick Nikzad
Jifeng Ning
Rui Ning
Xuefei Ning
Li Niu
Muyao Niu
Shuaicheng Niu
Wei Niu
Xuesong Niu
Yi Niu

Yulei Niu
Zhenxing Niu
Zhong-Han Niu
Shohei Nobuhara
Jongyoun Noh
Junhyug Noh
Nadhira Noor
Parsa Nooralinejad
Sotiris Nousias
Tiago Novello
Gal Novich
David Novotny
Slawomir Nowaczyk
Ewa M. Nowara
Evangelos Ntavelis
Valsamis Ntouskos
Leonardo Nunes
Oren Nuriel
Zhakshylyk Nurlanov
Simbarashe Nyatsanga
Lawrence O'Gorman
Anton Obukhov
Michael Oechsle
Ferda Ofli
Changjae Oh
Dongkeun Oh
Junghun Oh
Seoung Wug Oh
Youngtaek Oh
Hiroki Ohashi
Takehiko Ohkawa
Takeshi Oishi
Takahiro Okabe
Fumio Okura
Daniel Olmeda Reino
Suguru Onda
Trevine S. J. Oorloff
Michael Opitz
Roy Or-El
Jose Oramas
Jordi Orbay
Tribhuvanesh Orekondy
Evin Pınar Örnek
Alessandro Ortis
Magnus Oskarsson

Julian Ost
Daniil Ostashev
Mayu Otani
Naima Otberdout
Hatef Otroshi Shahreza
Yassine Ouali
Amine Ouasfi
Cheng Ouyang
Wanli Ouyang
Wenqi Ouyang
Xu Ouyang
Poojan B. Oza
Milind G. Padalkar
Johannes C. Paetzold
Gautam Pai
Anwesan Pal
Simone Palazzo
Avinash Paliwal
Cristina Palmero
Chengwei Pan
Fei Pan
Hao Pan
Jianhong Pan
Junting Pan
Liang Pan
Lili Pan
Linfei Pan
Liyuan Pan
Tai-Yu Pan
Xichen Pan
Xingjia Pan
Xinyu Pan
Yingwei Pan
Zhaoying Pan
Zhihong Pan
Zixuan Pan
Zizheng Pan
Rohit Pandey
Saurabh Pandey
Bo Pang
Guansong Pang
Lu Pang
Meng Pang
Tianyu Pang
Youwei Pang

Ziqi Pang
Omiros Pantazis
Juan J. Pantrigo
Hsing-Kuo Kenneth Pao
Marina Paolanti
Joao P. Papa
Samuele S. Papa
Dim P. Papadopoulos
Symeon Papadopoulos
George Papandreou
Toufiq Parag
Chethan Parameshwara
Foivos Paraperas
  Papantoniou
Shaifali Parashar
Alejandro Pardo
Jason R. Parham
Kranti K. Parida
Rishubh Parihar
Chunghyun Park
Daehee Park
Dongmin Park
Dongwon Park
Eunbyung Park
Eunhyeok Park
Eunil Park
Geon Yeong Park
Gyeong-Moon Park
Hyoungseob Park
Jae Sung Park
JaeYoo Park
Jin-Hwi Park
Jinhyung Park
Jinyoung Park
Jongwoo Park
JoonKyu Park
JungIn Park
Junheum Park
Kiru Park
Kwanyong Park
Seongsik Park
Seulki Park
Song Park
Sungho Park
Sungjune Park

Taesung Park
Yeachan Park
Gaurav Parmar
Paritosh Parmar
Maurizio Parton
Magdalini Paschali
Vito Paolo Pastore
Or Patashnik
Gaurav Patel
Maitreya Patel
Diego Patino
Suvam Patra
Viorica Patraucean
Badri Narayana Patro
Danda Pani Paudel
Angshuman Paul
Sneha Paul
Soumava Paul
Sudipta Paul
Sujoy Paul
Rémi Pautrat
Ioannis Pavlidis
Svetlana Pavlitska
Raju Pavuluri
Kim Steenstrup Pedersen
Marco Pedersoli
Adithya Pediredla
Pieter Peers
Jiju Peethambaran
Sen Pei
Wenjie Pei
Yuru Pei
Simone Alberto Peirone
Chantal Pellegrini
Latha Pemula
Abhirama Subramanyam
  V. B. Penamakuri
Adrian Penate-Sanchez
Baoyun Peng
Bo Peng
Can Peng
Cheng Peng
Chi-Han Peng
Chunlei Peng
Jie Peng

Jingliang Peng
Kebin Peng
Kunyu Peng
Liang Peng
Liangzu Peng
Pai Peng
Peixi Peng
Sida Peng
Songyou Peng
Wei Peng
Wen-Hsiao Peng
Xi Peng
Xiaojiang Peng
Yi-Xing Peng
Yuxin Peng
Zhiliang Peng
Ziqiao Peng
Matteo Pennisi
Or Perel
Gabriel Perez
Gustavo Perez
Juan C. Perez
Andres Felipe Perez
  Murcia
Eduardo Pérez-Pellitero
Neehar Peri
Skand Peri
Gabriel J. Perin
Federico Pernici
Chiara Pero
Elia Peruzzo
Marco Pesavento
Dmitry M. Petrov
Ilya A. Petrov
Mathis Petrovich
Vitali Petsiuk
Tomas Pevny
Shubham Milind Phal
Chau Pham
Hai X. Pham
Khoi Pham
Long Hoang Pham
Trong Thang Pham
Trung X. Pham
Tung Pham

Hoang Phan
Huy Phan
Minh Hieu Phan
Julien Philip
Stephen Phillips
Cheng Perng Phoo
Hao Phung
Shruti S. Phutke
Weiguo Pian
Yongri Piao
Luigi Piccinelli
A. J. Piergiovanni
Sara Pieri
Vipin Pillai
Wu Pingyu
Silvia L. Pintea
Francesco Pinto
Maura Pintor
Giovanni Pintore
Vittorio Pippi
Robinson Piramuthu
Fiora Pirri
Leonid Pishchulin
Francesca Pistilli
Francesco Pittaluga
Fabio Pizzati
Edward Pizzi
Benjamin Planche
Iuliia Pliushch
Chiara Plizzari
Ryan Po
GIovanni Poggi
Matteo Poggi
Kilian Pohl
Chandradeep Pokhariya
Ashwini Pokle
Matteo Polsinelli
Adrian Popescu
Teodora Popordanoska
Nikola Popović
Ronald Poppe
Samuele Poppi
Andrea Porfiri Dal Cin
Angelo Porrello
Pedro Porto Buarque de Gusmão
Rudra P. K. Poudel
Kossar Pourahmadi Meibodi
Hadi Pouransari
Ali Pourramezan Fard
Omid Poursaeed
Anish J. Prabhu
Mihir Prabhudesai
Aayush Prakash
Aditya Prakash
Shraman Pramanick
Mantini Pranav
B. H. Pawan Prasad
Meghshyam Prasad
Prateek Prasanna
Ekta Prashnani
Bardh Prenkaj
Derek S. Prijatelj
Véronique Prinet
Malte Prinzler
Victor Adrian Prisacariu
Federica Proietto Salanitri
Sergey Prokudin
Bill Psomas
Dongqi Pu
Mengyang Pu
Nan Pu
Shi Pu
Rita Pucci
Kuldeep Purohit
Senthil Purushwalkam
Waqas A. Qazi
Charles R. Qi
Chenyang Qi
Haozhi Qi
Jiaxin Qi
Lei Qi
Mengshi Qi
Peng Qi
Xianbiao Qi
Xiangyu Qi
Yuankai Qi
Zhangyang Qi
Guocheng Qian
Hangwei Qian
Jianing Qian
Qi Qian
Rui Qian
Shengsheng Qian
Shengyi Qian
Shenhan Qian
Wen Qian
Xuelin Qian
Yaguan Qian
Yijun Qian
Yiming Qian
Zhenxing Qian
Wenwen Qiang
Feng Qiao
Fengchun Qiao
Xiaotian Qiao
Yanyuan Qiao
Yi-Ling Qiao
Yu Qiao
Hangyu Qin
Haotong Qin
Jie Qin
Peiwu Qin
Siyang Qin
Wenda Qin
Xuebin Qin
Xugong Qin
Yang Qin
Yipeng Qin
Yongqiang Qin
Yuzhe Qin
Zequn Qin
Zeyu Qin
Zheng Qin
Zhenyue Qin
Ziheng Qin
Jiaxin Qing
Congpei Qiu
Haibo Qiu
Hang Qiu
Heqian Qiu
Jiayan Qiu
Jielin Qiu

Longtian Qiu
Mufan Qiu
Ri-Zhao Qiu
Weichao Qiu
Xuchong Qiu
Xuerui Qiu
Yuda Qiu
Yuheng Qiu
Zhongxi Qiu
Maan Qraitem
Chao Qu
Linhao Qu
Yanyun Qu
Kha Gia Quach
Ruijie Quan
Fabio Quattrini
Yvain Queau
Faisal Z. Qureshi
Rizwan Qureshi
Hamid R. Rabiee
Paolo Rabino
Ryan L. Rabinowitz
Petia Radeva
Bhaktipriya Radharapu
Krystian Radlak
Bodgan Raducanu
M. Usman Rafique
Francesco Ragusa
Sahar Rahimi Malakshan
Tanzila Rahman
Aashish Rai
Arushi Rai
Shyam Nandan Rai
Zobeir Raisi
Amit Raj
Kiran Raja
Sachin Raja
Deepu Rajan
Jathushan Rajasegaran
Gnana Praveen Rajasekhar
Ramanathan Rajendiran
Marie-Julie Rakotosaona
Gorthi Rama Krishna Sai Subrahmanyam

Sai Niranjan Ramachandran
Santhosh Kumar Ramakrishnan
Srikumar Ramalingam
Michaël Ramamonjisoa
Ravi Ramamoorthi
Shanmuganathan Raman
Mani Ramanagopal
Ashish Ramayee Asokan
Andrea Ramazzina
Jason Rambach
Sai Saketh Rambhatla
Sai Saketh Rambhatla
Clément Rambour
Francois Bernard Julien Rameau
Visvanathan Ramesh
Adín Ramírez Rivera
Haoxi Ran
Xuming Ran
Aakanksha Rana
Srinivas Rana
Kanchana N. Ranasinghe
Poorva G. Rane
Aneesh Rangnekar
Harsh Rangwani
Viresh Ranjan
Anyi Rao
Sukrut Rao
Yongming Rao
ZhiBo Rao
Carolina Raposo
Hanoona Abdul Rasheed
Amir Rasouli
Deevashwer Rathee
Christian Rathgeb
Avinash Ravichandran
Bharadwaj Ravichandran
Arijit Ray
Dripta S. Raychaudhuri
Sonia Raychaudhuri
Haziq Razali
Daniel Rebain
William T. Redman

Albert W. Reed
Aniket Rege
Christoph Reich
Christian Reimers
Simon Reiß
Konstantinos Rematas
Tal Remez
Davis Rempe
Bin Ren
Chao Ren
Chuan-Xian Ren
Dayong Ren
Dongwei Ren
Jiawei Ren
Jiaxiang Ren
Jing Ren
Mengwei Ren
Pengfei Ren
Pengzhen Ren
Qibing Ren
Shuhuai Ren
Sucheng Ren
Tianhe Ren
Weihong Ren
Wenqi Ren
Xuanchi Ren
Yanli Ren
Yihui Ren
Yixuan Ren
Yufan Ren
Zhenwen Ren
Zhihang Ren
Zhiyuan Ren
Zhongzheng Ren
Jose Restom
George Retsinas
Ambareesh Revanur
Ferdinand Rewicki
Manuel Rey Area
Md Alimoor Reza
Farnoush Rezaei Jafari
Hamed Rezazadegan Tavakoli
Rafael S. Rezende
Wonjong Rhee

Anthony D. Rhodes
Daniel Riccio
Alexander Richard
Christian Richardt
Luca Rigazio
Benjamin Risse
Dominik Rivoir
Luigi Riz
Mamshad Nayeem Rizve
Antonino M. Rizzo
Wes J. Robbins
Damien Robert
Jonathan Roberts
Joseph Robinson
Antonio Robles-Kelly
Mrigank Rochan
Chris Rockwell
Chris Rockwell
Ivan Rodin
Erik Rodner
Ranga Rodrigo
Andres C. Rodriguez
Cristian Rodriguez
Carlos Rodriguez-Pardo
Antonio J.
  Rodriguez-Sanchez
Barbara Roessle
Paul Roetzer
Alina Roitberg
Javier Romero
Meitar Ronen
Keran Rong
Xuejian Rong
Yu Rong
Marco Rosano
Bodo Rosenhahn
Gabriele Rosi
Candace Ross
Andreas Rössler
Giulio Rossolini
Mohammad Rostami
Edward Rosten
Daniel Roth
Karsten Roth
Mark S. Rothermel

Matthias Rottmann
Anastasios Roussos
Aniket Roy
Anirban Roy
Debaditya Roy
Shuvendu Roy
Sudipta Roy
Ahana Roy Choudhury
Amit Roy-Chowdhury
Aruni RoyChowdhury
Dávid Rozenberszki
Denys Rozumnyi
Lixiang Ru
Lingyan Ruan
Shulan Ruan
Viktor Rudnev
Daniel Rueckert
Nataniel Ruiz
Ewelina Rupnik
Evgenia Rusak
Chris Russell
Marc Rußwurm
Fiona Ryan
Dawid Damian Rymarczyk
DongHun Ryu
Sari Saba-Sadiya
Robert Sablatnig
Mohammad Sabokrou
Ragav Sachdeva
Ali Sadeghian
Arka Sadhu
Sadra Safadoust
Bardia Safaei
Ryusuke Sagawa
Avishkar Saha
Gobinda Saha
Oindrila Saha
Aditya Sahdev
Lakshmi Babu Saheer
Aadarsh Sahoo
Pritish Sahu
Aneeshan Sain
Nirat Saini
Saurabh Saini
Kuniaki Saito

Shunsuke Saito
Rahul Sajnani
Fumihiko Sakaue
Parikshit V. Sakurikar
Riccardo Salami
Soorena Salari
Mohammadreza Salehi
Leonard Salewski
Driton Salihu
Benjamin Salmon
Cristiano Saltori
Joel Saltz
Tim Salzmann
Sina Samangooei
Babak Samari
Nermin Samet
Fawaz Sammani
Leo Sampaio Ferraz
  Ribeiro
Shailaja Keyur Sampat
Alessio Sampieri
Jorge Sanchez
Pedro Sandoval-Segura
Nong Sang
Shengtian Sang
Patsorn Sangkloy
Depanshu Sani
Juan C. Sanmiguel
Hiroaki Santo
Joshua Santoso
Bikash Santra
Soubhik Sanyal
Hitesh Sapkota
Ayush Saraf
Nikolaos Sarafianos
István Sárándi
Kyle Sargent
Andranik Sargsyan
Josip Šarić
Mert Bulent Sariyildiz
Abhijit Sarkar
Anirban Sarkar
Chayan Sarkar
Michel Sarkis
Paul-Edouard Sarlin

Sara Sarto
Josua Sassen
Srikumar Sastry
Imari Sato
Takami Sato
Shin'ichi Satoh
Ravi Kumar Satzoda
Jack Saunders
Corentin Sautier
Mattia Savardi
Bogdan Savchynskyy
Mohamed Sayed
Marin Scalbert
Gianluca Scarpellini
Gerald Schaefer
Guilherme G. Schardong
David Schinagl
Phillip Schniter
Patrick Schramowski
Matthias Schubert
Peter Schüffler
Samuel Schulter
René Schuster
Klamer Schutte
Luca Scofano
Jesse Scott
Marcel Seelbach Benkner
Karthik Seemakurthy
Mattia Segù
Santi Seguí
Sinisa Segvic
Constantin Marc Seibold
Roman Seidel
Lorenzo Seidenari
Taiki Sekii
Yusuke Sekikawa
Matan Sela
Pratheba Selvaraju
Agniva Sengupta
Ahyun Seo
Jinhwan Seo
Junyoung Seo
Kwanggyoon Seo
Seonguk Seo
Seunghyeon Seo

Jinseok Seol
Hongje Seong
Ana F. Sequeira
Dario Serez
Dario Serez
David Serrano-Lozano
Pratinav Seth
Francesco Setti
Giorgos Sfikas
Mohammad Amin Shabani
Faisal Shafait
Anshul Shah
Chintan Shah
Jay Shah
Ketul Shah
Mubarak Shah
Viraj Shah
Mohamad Shahbazi
Muhammad Bilal B.
  Shaikh
Abdelrahman M. Shaker
Greg Shakhnarovich
Md Salman Shamil
Fahad Shamshad
Caifeng Shan
Dandan Shan
Hongming Shan
Xiaojun Shan
Chong Shang
Fanhua Shang
Jinghuan Shang
Lei Shang
Sifeng Shang
Wei Shang
Yuzhang Shang
Yuzhang Shang
Sukrit Shankar
Dian Shao
Mingwen Shao
Rui Shao
Ruizhi Shao
Shuai Shao
Shuwei Shao
Ron A. Shapira Weber
S. M. A. Sharif

Aashish Sharma
Avinash Sharma
Charu Sharma
Prafull Sharma
Prasen Kumar Sharma
Allam Shehata
Mark Sheinin
Sumit Shekhar
Oleksandr Shekhovtsov
Chuanfu Shen
Fei Shen
Fengyi Shen
Furao Shen
Hui-liang Shen
Jiajun Shen
Jianghao Shen
Jiangrong Shen
Jiayi Shen
Li Shen
Li-Yong Shen
Linlin Shen
Maying Shen
Qiu Shen
Qiuhong Shen
Shuai Shen
Shuhan Shen
Siqi Shen
Tianwei Shen
Tong Shen
Xiaolong Shen
Xiaoqian Shen
Yan Shen
Yanqing Shen
Yilin Shen
Ying Shen
Yiqing Shen
Yuan Shen
Yucong Shen
Yuhan Shen
Yunhang Shen
Zehong Shen
Zengming Shen
Zhijie Shen
Zhiqiang Shen
Hualian Sheng

Tao Sheng
Yichen Sheng
Zehua Sheng
Shivanand Venkanna Sheshappanavar
Ivaxi Sheth
Baoguang Shi
Botian Shi
Dachuan Shi
Daqian Shi
Haizhou Shi
Hengcan Shi
Jia Shi
Jing Shi
Jingang Shi
QingHongYa Shi
Ruoxi Shi
Tianyang Shi
Weishi Shi
Wu Shi
Wuxuan Shi
Xiaodan Shi
Xiaoshuang Shi
Xiaoyu Shi
Xingjian Shi
Xinyu Shi
Xuepeng Shi
Yichun Shi
Yujiao Shi
Zhenbo Shi
Zheng Shi
Zhensheng Shi
Zhenwei Shi
Zhihao Shi
Zifan Shi
Takashi Shibata
Meng-Li Shih
Yichang Shih
Dongseok Shim
Wataru Shimoda
Ilan Shimshoni
Changha Shin
Gyungin Shin
Hyungseob Shin
Inkyu Shin

Seungjoo Shin
Ukcheol Shin
Yooju Shin
Young Min Shin
Koichi Shinoda
Kaede Shiohara
Suprosanna Shit
Palaiahnakote Shivakumara
Sindi Shkodrani
Michal Shlapentokh-Rothman
Debaditya Shome
Hyounguk Shon
Sulabh Shrestha
Aman Shrivastava
Ayush Shrivastava
Gaurav Shrivastava
Aleksandar Shtedritski
Dong Wook Shu
Han Shu
Jun Shu
Xiangbo Shu
Xiujun Shu
Yang Shu
Bing Shuai
Hong-Han Shuai
Qing Shuai
Changjian Shui
Pushkar Shukla
Mustafa Shukor
Hubert P. H. Shum
Nina Shvetsova
Chenyang Si
Jianlou Si
Zilin Si
Mennatullah Siam
Sven Sickert
Désiré Sidibé
Ioannis Siglidis
Alberto Signoroni
Karan Sikka
Pedro Silva
Julio Silva-Rodríguez
Hyeonjun Sim

Jae-Young Sim
Chonghao Sima
Christian Simon
Martin Simon
Alessandro Simoni
Enis Simsar
Abhishek Singh
Apoorv Singh
Ashish Singh
Bharat Singh
Jasdeep Singh
Jaskirat Singh
Krishnakant Singh
Manish Kumar Singh
Mannat Singh
Nikhil Singh
Pravendra Singh
Rajat Vikram Singh
Simranjit Singh
Darshan Singh S.
Utkarsh Singhal
Dipika Singhania
Vasu Singla
Abhishek Kumar Sinha
Animesh Sinha
Sanjana Sinha
Saptarshi Sinha
Sudipta Sinha
Sophia A. Sirko-Galouchenko
Josef Sivic
Elena Sizikova
Geri Skenderi
Gregory Slabaugh
Habib Slim
Dmitriy Smirnov
James S. Smith
William Smith
Noah Snavely
Kihyuk Sohn
Bolivar E. Solarte
Mattia Soldan
Sobhan Soleymani
Samik Some
Nagabhushan Somraj

Jeany Son
Seung Woo Son
Byung Cheol Song
Chen Song
Guanglu Song
Jie Song
Jifei Song
Li Song
Liangchen Song
Lin Song
Luchuan Song
Mingli Song
Ran Song
Sibo Song
Sifan Song
Siyang Song
Weilian Song
Weinan Song
Wenfeng Song
Xiangchen Song
Xibin Song
Xinhang Song
Yafei Song
Yang Song
Yi-Yang Song
Yizhi Song
Yue Song
Zeen Song
Zhenbo Song
Zikai Song
Ekta Sood
Tomáš Souček
Rajiv Soundararajan
Albin Soutif-Cormerais
Jeremy Speth
Indro Spinelli
Jon Sporring
Manogna Sreenivas
Arvind Krishna Sridhar
Deepak Sridhar
Balaji Vasan Srinivasan
Pratul Srinivasan
Anuj Srivastava
Astitva Srivastava
Dhruv Srivastava

Koushik Srivatsan
Pierre-Luc St-Charles
Ioannis Stamos
Anastasis Stathopoulos
Colton Stearns
Jan Steinbrener
Jan-Martin O. Steitz
Sinisa Stekovic
Federico Stella
Michael Stengel
Alexandros Stergiou
Gleb Sterkin
Rainer Stiefelhagen
Noah Stier
Timo N. Stoffregen
Vladan Stojnić
Nick O. Stracke
Ombretta Strafforello
Julian Straub
Nicola Strisciuglio
Vitomir Struc
Yannick Strümpler
Joerg Stueckler
Chi Su
Hang Su
Hang Su
Kun Su
Rui Su
Shaolin Su
Sitong Su
Xingzhe Su
Xiu Su
Yao Su
Yiyang Su
Yongyi Su
Zhaoqi Su
Zhixun Su
Zhuo Su
Zhuo Su
Iago Suárez
Arulkumar Subramaniam
Sanjay Subramanian
A. Subramanyam
Swathikiran Sudhakaran
Yusuke Sugano

Masanori Suganuma
Yumin Suh
Mohammed Suhail
Xiuchao Sui
Yang Sui
Yao Sui
Heung-Il Suk
Pavel Suma
Baigui Sun
Baochen Sun
Bin Sun
Bo Sun
Changchang Sun
Che Sun
Cheng Sun
Chong Sun
Chunyi Sun
Gan Sun
Guofei Sun
Guoxing Sun
Haifeng Sun
Hanqing Sun
Haoliang Sun
He Sun
Heming Sun
Hongbin Sun
Huiming Sun
Jennifer J. Sun
Jian Sun
Jiande Sun
Jianhua Sun
Jiankai Sun
Jipeng Sun
Keqiang Sun
Lei Sun
Lichao Sun
Long Sun
Mingjie Sun
Peize Sun
Pengzhan Sun
Qiyue Sun
Shangquan Sun
Shanlin Sun
Shuyang Sun
Tao Sun

Tiancheng Sun
Wei Sun
Weiwei Sun
Weixuan Sun
Xianfang Sun
Xiaohang Sun
Xiaoshuai Sun
Xiaoxiao Sun
Ximeng Sun
Xuxiang Sun
Yanan Sun
Yasheng Sun
Yihong Sun
Ying Sun
Yixuan Sun
Yu Sun
Yuan Sun
Yuchong Sun
Zeren Sun
Zhanghao Sun
Zhaodong Sun
Zhaohui H. Sun
Zhicheng Sun
Zhicheng Sun
Haomiao Sun
Varun Sundar
Shobhita Sundaram
Minhyuk Sung
Kalyan Sunkavalli
Yucheng Suo
Indranil Sur
Saksham Suri
Naufal Suryanto
Vadim Sushko
David Suter
Roman Suvorov
Fnu Suya
Teppei Suzuki
Kunal Swami
Archana Swaminathan
Gurumurthy Swaminathan
Robin Swanson
Eran Swears
Alexander Swerdlow
Sirnam Swetha
Tabish A. Syed
Tanveer Syeda-Mahmood
Stanislaw K. Szymanowicz
Sethuraman T. V.
Calvin-Khang T. Ta
The-Anh Ta
Babak Taati
Samy Tafasca
Andrea Tagliasacchi
Haowei Tai
Yuan Tai
Francesco Taioli
Peng Taiying
Keita Takahashi
Naoya Takahashi
Jun Takamatsu
Nicolas Talabot
Hugues G. Talbot
Hossein Talebi
Davide Talon
Gary Tam
Toru Tamaki
Dipesh Tamboli
Andong Tan
Bin Tan
Cheng Tan
David Joseph New Tan
Fuwen Tan
Guang Tan
Jianchao Tan
Jing Tan
Jingru Tan
Lei Tan
Mingkui Tan
Mingxing Tan
Shuhan Tan
Shunquan Tan
Weimin Tan
Xin Tan
Zhentao Tan
Zhentao Tan
Masayuki Tanaka
Chen Tang
Chengzhou Tang
Chenwei Tang
Fan Tang
Feng Tang
Hao Tang
Haoran Tang
Jiajun Tang
Jiapeng Tang
Jiaxiang Tang
Jie Tang
Junshu Tang
Keke Tang
Luming Tang
Luyao Tang
Lv Tang
Ming Tang
Quan Tang
Shengji Tang
Sheyang Tang
Shitao Tang
Shixiang Tang
Tao Tang
Weixuan Tang
Xu Tang
Yang Tang
Yansong Tang
Yehui Tang
Yu-Ming Tang
Zheng Tang
Zhipeng Tang
Zitian Tang
Md Mehrab Tanjim
Julian Tanke
An Tao
Chaofan Tao
Chenxin Tao
Jiale Tao
Junli Tao
Keda Tao
Ming Tao
Ran Tao
Wenbing Tao
Xinhao Tao
Jean-Philippe G. Tarel
Laia Tarres
Laia Tarrés
Enzo Tartaglione

Keisuke Tateno
SaiKiran K. Tedla
Antonio Tejero-de-Pablos
Bugra Tekin
Purva Tendulkar
Minggui Teng
Ruwan Tennakoon
Andrew Beng Jin Teoh
Konstantinos Tertikas
Piotr Teterwak
Piotr Teterwak
Anh Thai
Kartik Thakral
Nupur Thakur
Sadbhawna Thakur
Balamurugan Thambiraja
Vikas Thamizharasan
Kevin Thandiackal
Sushil Thapa
Daksh Thapar
Jonas Theiner
Christian Theobalt
Spyridon Thermos
Fida Mohammad Thoker
Christopher L. Thomas
Diego Thomas
William Thong
Mamatha Thota
Mukund Varma Thottankara
Changyao Tian
Chunwei Tian
Jinyu Tian
Kai Tian
Lin Tian
Tai-Peng Tian
Xin Tian
Xinyu Tian
Yapeng Tian
Yu Tian
Yuan Tian
Yuesong Tian
Yunjie Tian
Yuxin Tian
Zhuotao Tian

Mert Tiftikci
Javier Tirado-Garín
Garvita Tiwari
Lokender Tiwari
Anastasia Tkach
Andrea Toaiari
Sinisa Todorovic
Pavel Tokmakov
Tri Ton
Adam Tonderski
Jinguang Tong
Peter Tong
Xin Tong
Zhan Tong
Francesco Tonini
Alessio Tonioni
Alessandro Torcinovich
Marwan Torki
Lorenzo Torresani
Fabio Tosi
Matteo Toso
Anh T. Tran
Hung Tran
Linh-Tam Tran
Minh-Triet Tran
Ngoc-Trung Tran
Phong Tran
Jonathan Tremblay
Alex Trevithick
Aditay Tripathi
Subarna Tripathi
Felix Tristram
Gabriele Trivigno
Emanuele Trucco
Prune Truong
Thanh-Dat Truong
Tomasz Trzcinski
Fu-Jen Tsai
Yu-Ju Tsai
Michael Tschannen
Tze Ho Elden Tse
Ethan Tseng
Yu-Chee Tseng
Shahar Tsiper
Hanzhang Tu

Rong-Cheng Tu
Yuanpeng Tu
Zhengzhong Tu
Zhigang Tu
Narek Tumanyan
Anil Osman Tur
Haithem Turki
Mehmet Ozgur Turkoglu
Daniyar Turmukhambetov
Victor G. Turrisi da Costa
Tinne Tuytelaars
Bartlomiej Twardowski
Radim Tylecek
Christos Tzelepis
Seiichi Uchida
Hideaki Uchiyama
Vishaal Udandarao
Mostofa Rafid Uddin
Kohei Uehara
Tatsumi Uezato
Nicolas Ugrinovic
Youngjung Uh
Norimichi Ukita
Amin Ullah
Markus Ulrich
Ardian Umam
Mesut Erhan Unal
Mathias Unberath
Devesh Upadhyay
Paul Upchurch
Shagun Uppal
Yoshitaka Ushiku
Anil Usumezbas
Yuzuko Utsumi
Roy Uziel
Anil Vadathya
Sharvaree Vadgama
Pratik Vaishnavi
Gregory Vaksman
Matias A. Valdenegro Toro
Lucas Valença
Eduardo Valle
Ernest Valveny
Laurens van der Maaten
Wouter Van Gansbeke

Nanne van Noord
Max W. F. van Spengler
Lorenzo Vaquero
Farshid Varno
Cristina Vasconcelos
Francisco Vasconcelos
Igor Vasiljevic
Florin-Alexandru Vasluianu
Subeesh Vasu
Arun Balajee Vasudevan
Vaibhav S. Vavilala
Kyle Vedder
Vijay Veerabadran
Ronny Xavier Velastegui Sandoval
Senem Velipasalar
Andreas Velten
Raviteja Vemulapalli
Deepika Vemuri
Edward Vendrow
Jonathan Ventura
Lucas Ventura
Jakob Verbeek
Dor Verbin
Eshan Verma
Manisha Verma
Monu Verma
Sahil Verma
Constantin Vertan
Eli Verwimp
Noranart Vesdapunt
Jordan J. Vice
Sara Vicente
Kavisha Vidanapathirana
Dat Viet Thanh Nguyen
Sudheendra Vijayanarasimhan
Sujal T. Vijayaraghavan
Deepak Vijaykeerthy
Elliot Vincent
Yael Vinker
Duc Minh Vo
Huy V. Vo
Khoa H. V. Vo

Romain Vo
Antonin Vobecky
Michele Volpi
Riccardo Volpi
Igor Vozniak
Nicholas Vretos
Vibashan V. S.
Ngoc-Son Vu
Tuan-Anh Vu
Khiem Vuong
Mårten Wadenbäck
Neal Wadhwa
Sophia J. Wagner
Muntasir Wahed
Nobuhiko Wakai
Devesh Walawalkar
Jacob Walker
Matthew Walmer
Matthew R. Walter
Bo Wan
Guancheng Wan
Jia Wan
Jin Wan
Jun Wan
Qiyang Wan
Renjie Wan
Wei Wan
Xingchen Wan
Yecong Wan
Zhexiong Wan
Ziyu Wan
Karan Wanchoo
Alex Jinpeng Wang
Angtian Wang
Baoyuan Wang
Benyou Wang
Biao Wang
Bin Wang
Bing Wang
Binghui Wang
Binglu Wang
Can Wang
Ce Wang
Changwei Wang
Chao Wang

Chaoyang Wang
Chen Wang
Chen Wang
Chen Wang
Chengrui Wang
Chien-Yao Wang
Chu Wang
Chuan Wang
Congli Wang
Dadong Wang
Di Wang
Dong Wang
Dong Wang
Dongdong Wang
Dongkai Wang
Dongqing Wang
Dongsheng Wang
X. Wang
Fan Wang
Fangfang Wang
Fangjinhua Wang
Fei Wang
Feng Wang
Feng Wang
Fu-Yun Wang
Gaoang Wang
Guangcong Wang
Guangming Wang
Guangrun Wang
Guangzhi Wang
Guanshuo Wang
Guo-Hua Wang
Guoqing Wang
Guoqing Wang
Haixin Wang
Haiyan Wang
Han Wang
Hanjing Wang
Hanyu Wang
Hao Wang
Hao Wang
Haobo Wang
Haochen Wang
Haochen Wang
Haohan Wang

Haoqi Wang
Haoran Wang
Haotao Wang
Haoxuan Wang
HaoYu Wang
Hengkang Wang
Hengli Wang
Hengyi Wang
Hesheng Wang
Hong Wang
Hongjun Wang
Hongxiao Wang
Hongyu Wang
Hongzhi Wang
Hua Wang
Huafeng Wang
Huan Wang
Huijie Wang
Huiyu Wang
Jiadong Wang
Jiahao Wang
Jiahao Wang
Jiahao Wang
Jiakai Wang
Jialiang Wang
Jiamian Wang
Jian Wang
Jiang Wang
Jiangliu Wang
Jianjia Wang
Jianyi Wang
Jianyuan Wang
Jiaqi Wang
Jiashun Wang
Jiayi Wang
Jiaze Wang
Jin Wang
Jinfeng Wang
Jingbo Wang
Jinghua Wang
Jingkang Wang
Jinglong Wang
Jinglu Wang
Jinpeng Wang
Jinqiao Wang

Jue Wang
Jun Wang
Jun Wang
Junjue Wang
Junke Wang
Junxiao Wang
Kai Wang
Kai Wang
Kai Wang
Kai Wang
Kaihong Wang
Kewei Wang
Keyan Wang
Kun Wang
Kup Wang
Lan Wang
Lanjun Wang
Le Wang
Lei Wang
Lei Wang
Lezi Wang
Liansheng Wang
Liao Wang
Lijuan Wang
Lijun Wang
Limin Wang
Lin Wang
Linwei Wang
Lishun Wang
Lixu Wang
Liyuan Wang
Lizhen Wang
Lizhi Wang
Longguang Wang
Luozhou Wang
Luting Wang
Mang Wang
Manning Wang
Mei Wang
Mengjiao Wang
Mengmeng Wang
Miaohui Wang
Min Wang
Naiyan Wang
Nannan Wang

Ning-Hsu Wang
Pei Wang
Peihao Wang
Peiqi Wang
Peng Wang
Pengfei Wang
Pengkun Wang
Pichao Wang
Pu Wang
Qi Wang
Qian Wang
Qiang Wang
Qiang Wang
Qiangchang Wang
Qianqian Wang
Qifei Wang
Qilong Wang
Qin Wang
Qing Wang
Qingzhong Wang
Qitong Wang
Qiufeng Wang
Ronggang Wang
Rui Wang
Rui Wang
Rui Wang
Ruibin Wang
Ruisheng Wang
Ruoyu Wang
Sai Wang
Sen Wang
Sen Wang
Shan Wang
Shaoru Wang
Sheng Wang
Sheng-Yu Wang
Shengze Wang
Shida Wang
Shijie Wang
Shipeng Wang
Shiping Wang
Shiyu Wang
Shizun Wang
Shuhui Wang
Shujun Wang

Shunli Wang
Shunxin Wang
Shuo Wang
Shuo Wang
Shuo Wang
Shuzhe Wang
Siqi Wang
Siwei Wang
Song Wang
Song Wang
Su Wang
Tan Wang
Tao Wang
Taoyue Wang
Teng Wang
Tengfei Wang
Tiancai Wang
Tianqi Wang
Tianyang Wang
Tianyu Wang
Tong Wang
Tsun-Hsuan Wang
Tuanfeng Wang
Tuanfeng Y. Wang
Wei Wang
Weihan Wang
Weikang Wang
Weimin Wang
Weiqiang Wang
Weixi Wang
Weiyao Wang
Weiyun Wang
Wen Wang
Wenbin Wang
Wenhao Wang
Wenjing Wang
Wenqian Wang
Wentao Wang
Wenxiao Wang
Wenxuan Wang
Wenzhe Wang
Xi Wang
Xi Wang
Xiang Wang
Xiao Wang

Xiao Wang
Xiao Wang
Xiaobing Wang
Xiaofeng Wang
Xiaohan Wang
Xiaosen Wang
Xiaosong Wang
Xiaoxing Wang
Xiaoyang Wang
Xijun Wang
Xijun Wang
Xinggang Wang
Xinghan Wang
Xinjiang Wang
Xinshao Wang
Xintong Wang
Xizi Wang
Xu Wang
Xuan Wang
Xuanhan Wang
Xue Wang
Xueping Wang
Xuyang Wang
Yali Wang
Yan Wang
Yan Wang
Yang Wang
Yangang Wang
Yangtao Wang
Yaohui Wang
Yaoming Wang
Yaxing Wang
Yaxiong Wang
Yi Wang
Yi Ru Wang
Yidong Wang
Yifan Wang
Yifeng Wang
Yifu Wang
Yikai Wang
Yilin Wang
Yilun Wang
Yin Wang
Yinggui Wang
Yingheng Wang

Yingqian Wang
Yipei Wang
Yiqun Wang
Yiran Wang
Yiwei Wang
Yixu Wang
Yizhi Wang
Yizhou Wang
Yizhou Wang
Yong Wang
Yu Wang
Yu-Shuen Wang
Yuan-Gen Wang
Yuchen Wang
Yude Wang
Yue Wang
Yuesong Wang
Yufei Wang
Yufu Wang
Yuguang Wang
Yuhan Wang
Yujia Wang
Yulin Wang
Yunke Wang
Yuting Wang
Yuxi Wang
YuXin Wang
Yuzheng Wang
Ze Wang
Zedong Wang
Zehan Wang
Zengmao Wang
Zeyu Wang
Zeyu Wang
Zhao Wang
Zhaokai Wang
Zhaowen Wang
Zhe Wang
Zhen Wang
Zhen Wang
Zhendong Wang
Zheng Wang
Zheng Wang
Zheng Wang
Zhengyi Wang

Zhennan Wang
Zhenting Wang
Zhenyi Wang
Zhenyu Wang
Zhenzhi Wang
Zhepeng Wang
Zhi Wang
Zhibo Wang
Zhihao Wang
Zhihui Wang
Zhijie Wang
Zhikang Wang
Zhixiang Wang
Zhiyong Wang
Zhongdao Wang
Zhonghao Wang
Zhouxia Wang
Zhu Wang
Zian Wang
Zifu Wang
Zihao Wang
Zijian Wang
Ziqiang Wang
Ziqin Wang
Ziqing Wang
Zirui Wang
Zirui Wang
Ziwei Wang
Ziyan Wang
Ziyang Wang
Ziyi Wang
Ziyun Wang
Frederik Warburg
Syed Talal Wasim
Daniel Watson
Jamie Watson
Ethan Weber
Silvan Weder
Jan Dirk Wegner
Chen Wei
Donglai Wei
Fangyin Wei
Fangyun Wei
Guoqiang Wei
Jia Wei

Jiacheng Wei
Kaixuan Wei
Kun Wei
Longhui Wei
Megan Wei
Mian Wei
Mingqiang Wei
Pengxu Wei
Ping Wei
Qiuhong Anna Wei
Shikui Wei
Tianyi Wei
Wei Wei
Wenqi Wei
Xian Wei
Xin Wei
Xing Wei
Xinyue Wei
Xiu-Shen Wei
Yi Wei
Yixuan Wei
Yunchao Wei
Yuxiang Wei
Yuxiang Wei
Zeming Wei
Zhipeng Wei
Zihao Wei
Zimian Wei
Jean-Baptiste Weibel
Luca Weihs
Martin Weinmann
Michael Weinmann
Bihan Wen
Bowen Wen
Chao Wen
Chenglu Wen
Chuan Wen
Congcong Wen
Jie Wen
Jing Wen
Qiang Wen
Rui Wen
Sijia Wen
Song Wen
Xiang Wen

Xin Wen
Yilin Wen
Youpeng Wen
Yuanbo Wen
Yuxin Wen
Chung-Yi Weng
Junwu Weng
Shuchen Weng
Wenming Weng
Yijia Weng
Zhenzhen Weng
Thomas Westfechtel
Christopher Johannes
  Wewer
Spencer Whitehead
Tobias Jan Wieczorek
Thaddäus Wiedemer
Julian Wiederer
Kevin Tirta Wijaya
Asiri Wijesinghe
Kimberly Wilber
Jeffrey R. Willette
Bryan M. Williams
Williem Williem
Christian Wilms
Benjamin Wilson
Richard Wilson
Felix Wimbauer
Vanessa Wirth
Scott Wisdom
Calden Wloka
Alex Wong
Chau-Wai Wong
Chi-Chong Wong
Ka Wai Wong
Kelvin Wong
Kok-Seng Wong
Kwan-Yee K. Wong
Yongkang Wong
Sangmin Woo
Simon S. Woo
Markus Worchel
Scott Workman
Marcel Worring
Safwan Wshah

Aming Wu
Bo Wu
Bojian Wu
Boxi Wu
Changguang Wu
Chaoyi Wu
Chen Henry Wu
Cheng-En Wu
Chenming Wu
Chenyan Wu
Chenyun Wu
Cho-Ying Wu
Chongruo Wu
Cong Wu
Dayan Wu
Di Wu
Dongming Wu
Fangzhao Wu
Fuxiang Wu
Gaojie Wu
Guanyao Wu
Guile Wu
Haiping Wu
Haiwei Wu
Haiyu Wu
Han Wu
Haoning Wu
Haoning Wu
Haotian Wu
Hefeng Wu
Huisi Wu
Jane Wu
Jay Zhangjie Wu
Jhih-Ciang Wu
Ji-Jia Wu
Jialian Wu
Jiaye Wu
Jimmy Wu
Jing Wu
Jing Wu
Jinjian Wu
Jiqing Wu
Jun Wu
Junfeng Wu
Junlin Wu

Junru Wu
Junyang Wu
Junyi Wu
Letian Wu
Lifang Wu
Lin Yuanbo Wu
Liwen Wu
Min Wu
Minye wu
Peng Wu
Penghao Wu
Qian Wu
Qiangqiang Wu
Qianyi Wu
Qingbo Wu
Rongliang Wu
Rui Wu
Rundi Wu
Shuang Wu
Shuzhe Wu
Tao Wu
Tao Wu
Tao Wu
Te-Lin Wu
Tianfu Wu
Tianhao Wu
Tianhao Wu
Ting-Wei Wu
Tong Wu
Tong Wu
Tsung-Han Wu
Tz-Ying Wu
Weibin Wu
Weijia Wu
Xian Wu
Xiao Wu
Xiaodong Wu
Xiaohe Wu
Xiaoqian Wu
Xiaoyang Wu
Xindi Wu
Xingjiao Wu
Xinxiao Wu
Xiuzhe Wu
Yang Wu

Yangzheng Wu
Yanze Wu
Yanzhao Wu
Yawen Wu
Yicheng Wu
Ying Nian Wu
Yingwen Wu
Yong Wu
Yuanwei Wu
Yue Wu
Yue Wu
Yuqun Wu
Yushu Wu
Yushuang Wu
Zhe Wu
Zheng Wu
Zhi-Fan Wu
Zhihao Wu
Zhijie Wu
Zhiliang Wu
Zhonghua Wu
Zijie Wu
Ziyi Wu
Zizhao Wu
Zongwei Wu
Zongyu Wu
Zongze Wu
Stefanie Wuhrer
Jamie M. Wynn
Monika Wysoczańska
Jianing Xi
Teng Xi
Bin Xia
Changqun Xia
Haifeng Xia
Jiaer Xia
Jiahao Xia
Kun Xia
Mingxuan Xia
Shihong Xia
Weihao Xia
Xiaobo Xia
Yan Xia
Ye Xia
Yifei Xia

Zhaoyang Xia
Zhihao Xia
Zhihua Xia
Zhuofan Xia
Zimin Xia
Chuhua Xian
Wenqi Xian
Yongqin Xian
Donglai Xiang
Jinhai Xiang
Liuyu Xiang
Tian-Zhu Xiang
Tiange Xiang
Wangmeng Xiang
Xiaoyu Xiang
Yuanbo Xiangli
Anqi Xiao
Aoran Xiao
Bei Xiao
Chunxia Xiao
Fanyi Xiao
Han Xiao
Jiancong Xiao
Jimin Xiao
Jing Xiao
Jing Xiao
Jun Xiao
Junbin Xiao
Junfei Xiao
Mingqing Xiao
Qingyang Xiao
Ruixuan Xiao
Taihong Xiao
Yang Xiao
Yanru Xiao
Yao Xiao
Yijun Xiao
Yuting Xiao
Zehao Xiao
Zeyu Xiao
Zihao Xiao
Binhui Xie
Chaohao Xie
Chi Xie
Christopher Xie

Chuanlong Xie
Fei Xie
Guo-Sen Xie
Haozhe Xie
Hongtao Xie
Jiahao Xie
Jiaxin Xie
Jin Xie
Jinheng Xie
Jiu-Cheng Xie
Jiyang Xie
Junyu Xie
Liuyue Xie
Ming-Kun Xie
Mingyang Xie
Qian Xie
Tingting Xie
Weicheng Xie
Xianghui Xie
Xiaohua Xie
Xudong Xie
Yichen Xie
Yiming Xie
You Xie
Yuan Xie
Yusheng Xie
Yutong Xie
Zeke Xie
Zhenda Xie
Zhenyu Xie
ZiYang Xie
Chaoyue Xing
Fuyong Xing
Jinbo Xing
Xiaoyan Xing
Xiaoying Xing
XiMing Xing
Xin Xing
Yazhou Xing
Yifan Xing
Yun Xing
Zhen Xing
Hongkai Xiong
Jingjing Xiong
Jinhui Xiong

Junwen Xiong
Peixi Xiong
Wei Xiong
Weihua Xiong
Yu Xiong
Yuanhao Xiong
Yuanjun Xiong
Yuwen Xiong
Zhexiao Xiong
Zhiwei Xiong
Yuliang Xiu
Alessio Xompero
An Xu
Angchi Xu
Baixin Xu
Bicheng Xu
Bo Xu
Chao Xu
Chenfeng Xu
Chenshu Xu
Chenxin Xu
Chi Xu
Dejia Xu
Dongli Xu
Feng Xu
Gangwei Xu
Haiming Xu
Haiyang Xu
Han Xu
Haofei Xu
Haohang Xu
Haoran Xu
Hongbin Xu
Hongmin Xu
Jiale Xu
Jianjin Xu
Jiaqi Xu
Jie Xu
Jilan Xu
Jinglin Xu
Jingyi Xu
Jun Xu
Kai Xu
Katherine Xu
Ke Xu

Kele Xu
Lan Xu
Lian Xu
Liang Xu
Linning Xu
Lumin Xu
Manjie Xu
Mengde Xu
Mengdi Xu
Mengmeng Frost Xu
Min Xu
Ming Xu
Mutian Xu
Peiran Xu
Peng Xu
Qi Xu
Qiang Xu
Qiangeng Xu
Qingshan Xu
Qingyang Xu
Qiuling Xu
Ran Xu
Renzhe Xu
Ruikang Xu
Runsen Xu
Runsheng Xu
Shichao Xu
Sirui Xu
Tongda Xu
Wanting Xu
Wei Xu
Weiwei Xu
Wenjia Xu
Wenju Xu
Wenqiang Xu
Xiang Xu
Xianghao Xu
Xiangyu Xu
Xiangyu Xu
Xiaogang Xu
Xiaohao Xu
Xin Xu
Xin Xu
Xin-Shun Xu
Xing Xu

Xinli Xu
Xinyu Xu
Xiuwei Xu
Xiyan Xu
Xudong Xu
Xuemiao Xu
Xun Xu
Yan Xu
Yan Xu
Yan Xu
Yangyang Xu
Yanwu Xu
Yating Xu
Yi Xu
Yi Xu
Yi Xu
Yihong Xu
YiKun Xu
Yinghao Xu
Yingyan Xu
Yinshuang Xu
Yiran Xu
Yixing Xu
Yongchao Xu
Yue Xu
Yufei Xu
Yunqiu Xu
Zexiang Xu
Zhan Xu
Zhe Xu
Zhengqin Xu
Zhenlin Xu
Zhiqiu Xu
Zhiyuan Xu
Zhongcong Xu
Zhuoer Xu
Zipeng Xu
Ziyue Xu
Zongyi Xu
Ziwei Xuan
Danna Xue
Fanglei Xue
Fei Xue
Feng Xue
Han Xue

Jianru Xue
Le Xue
Lixin Xue
Mingfu Xue
Nan Xue
Qinghan Xue
Shangjie Xue
Xiangyang Xue
Zihui Xue
Abhay Yadav
Amit Kumar Singh Yadav
Takuma Yagi
Tomas F Yago Vicente
I. Zeki Yalniz
Kota Yamaguchi
Shin'ya Yamaguchi
Burhaneddin Yaman
Toshihiko Yamasaki
Kohei Yamashita
Lee Juliette Yamin
Chaochao Yan
Hongyu Yan
Jiexi Yan
Kai Yan
Pei Yan
Qingan Yan
Qingsen Yan
Qingsong Yan
Rui Yan
Shaoqi Yan
Shi Yan
Siming Yan
Siming Yan
Siyuan Yan
Weilong Yan
Wending Yan
Xiangyi Yan
Xinchen Yan
Xingguang Yan
Xueting Yan
Yan Yan
Yichao Yan
Zhaoyi Yan
Zhiqiang Yan
Zhiyuan Yan

Zike Yan
Zizheng Yan
Keiji Yanai
Pinar Yanardag
Anqi Yang
Anqi Joyce Yang
Bangbang Yang
Baoyao Yang
Bin Yang
Binwei Yang
Bo Yang
Bo Yang
Boyu Yang
Changdi Yang
Chao Yang
Charig Yang
Cheng-Fu Yang
Cheng-Yen Yang
Chenhongyi Yang
Chuanguang Yang
De-Nian Yang
Dingcheng Yang
Dingkang Yang
Dong Yang
Erkun Yang
Fan Yang
Fan Yang
Fan Yang
Fan Yang
Fan Yang
Feng Yang
Fengting Yang
Fengxiang Yang
Fengyuan Yang
Fu-En Yang
Gang Yang
Gengshan Yang
Guandao Yang
Guanglei Yang
Haitao Yang
Hanqing Yang
Heran Yang
Honghui Yang
Huanrui Yang
Huiyuan Yang

Huizong Yang
Hunmin Yang
Jiange Yang
Jiaqi Yang
Jiawei Yang
Jiayu Yang
Jiazhi Yang
Jie Yang
Jie Yang
Jiewen Yang
Jihan Yang
Jing Yang
Jingkang Yang
Jinhui Yang
Jinlong Yang
Jinrong Yang
Jinyu Yang
Kaicheng Yang
Kailun Yang
Lan Yang
Le Yang
Lehan Yang
Lei Yang
Lei Yang
Lei Yang
Li Yang
Lihe Yang
Ling Yang
Lingxiao Yang
Linlin Yang
Lixin Yang
Longrong Yang
Lu Yang
Luwei Yang
Michael Ying Yang
Min Yang
Ming Yang
MingKun Yang
Mouxing Yang
Muli Yang
Peiyu Yang
Qi Yang
Qian Yang
Qiushi Yang
Ren Yang

Rui Yang
Ruihan Yang
Sejong Yang
Shan Yang
Shangrong Yang
Shiqi Yang
Shuai Yang
Shuai Yang
Shuang Yang
Shuo Yang
Shusheng Yang
Sibei Yang
Siwei Yang
Siyuan Yang
Siyuan Yang
Song Yang
Songlin Yang
Tianyu Yang
Tong Yang
Wankou Yang
Wenhan Yang
Wenhan Yang
Wenjie Yang
Wenqi Yang
William Yang
Xi Yang
Xi Yang
Xiangpeng Yang
Xiao Yang
Xiaofeng Yang
Xiaoshan Yang
Xin Jeremy Yang
Xingyi Yang
Xinlong Yang
Xitong Yang
Xiulong Yang
Xu Yang
Xuan Yang
Xue Yang
Xuelin Yang
Xun Yang
Yan Yang
Yan Yang
Yang Yang
Yaokun Yang

Yezhou Yang
Yiding Yang
Yijun Yang
Yijun Yang
Yin Yang
Yinfei Yang
Yixin Yang
Yongqi Yang
Yongqi Yang
Yue Yang
Yuewei Yang
Yuezhi Yang
Yujiu Yang
Yung-Hsu Yang
Yuwei Yang
Ze Yang
Ze Yang
Zetong Yang
Zhangsihao Yang
Zhaoyuan Yang
Zhen Yang
Zhenpei Yang
Zhibo Yang
Zhiwei Yang
Zhiwen Yang
Zhiyuan Yang
Zhuoqian Yang
Ziyan Yang
Ziyun Yang
Zongxin Yang
Zuhao Yang
Chengtang Yao
Cong Yao
Hantao Yao
Jiawen Yao
Lina Yao
Mingde Yao
Mingshuai Yao
Qingsong Yao
Shunyu Yao
Taiping Yao
Ting Yao
Xincheng Yao
Xinwei Yao
Xu Yao

Xufeng Yao
Yao Yao
Yazhou Yao
Yue Yao
Ziwei Yao
Sudhir Yarram
Rajeev Yasarla
Mohsen Yavartanoo
Botao Ye
Dengpan Ye
Fei Ye
Hanrong Ye
Jianglong Ye
Jiarong Ye
Jin Ye
Jingwen Ye
Jinwei Ye
Junjie Ye
Keren Ye
Maosheng Ye
Meng Ye
Meng Ye
Muchao Ye
Nanyang Ye
Peng Ye
Qi Ye
Qian Ye
Qinghao Ye
Qixiang Ye
Ruolin Ye
Shuquan Ye
Tian Ye
Vickie Ye
Wenqian Ye
Xinchen Ye
Yufei Ye
Moon Ye-Bin
Yousef Yeganeh
Chun-Hsiao Yeh
Raymond Yeh
Yu-Ying Yeh
Florence Yellin
Sriram Yenamandra
Tarun Yenamandra
Promod Yenigalla

Chandan Yeshwanth
Dong Yi
Hongwei Yi
Kai Yi
Ran Yi
Renjiao Yi
Xinyu Yi
Alper Yilmaz
Jonghwa Yim
Aoxiong Yin
Fei Yin
Fukun Yin
Jia-Li Yin
Ming Yin
Nan Yin
Ruihong Yin
Tianwei Yin
Wenzhe Yin
Xiaoqi Yin
Yingda Yin
Yu Yin
Yufeng Yin
Zhenfei Yin
Xianghua Ying
Xiaowen Ying
Naoto Yokoya
Chen YongCan
ByungIn Yoo
Innfarn Yoo
Jinsu Yoo
Sungjoo Yoo
Hee Suk Yoon
Jae Shin Yoon
Jihun Yoon
Sangwoong Yoon
Sejong Yoon
Sung Whan Yoon
Sung-Hoon Yoon
Sunjae Yoon
Youngho Yoon
Youngseok Yoon
Youngseok Yoon
Yuichi Yoshida
Ryota Yoshihashi
Yusuke Yoshiyasu

Chenyu You
Haoran You
Haoxuan You
Shan You
Yang You
Yingxuan You
Yurong You
Chan-Hyun Youn
Kim Youwang
Nikolaos-Antonios Ypsilantis
Baosheng Yu
Bei Yu
Bruce X. B. Yu
Chaohui Yu
Chunlin Yu
Cunjun Yu
Dahai Yu
En Yu
En Yu
Fenggen Yu
Gang Yu
Haibao Yu
Hanchao Yu
Hang Yu
Hao Yu
Hao Yu
Haojun Yu
Heng Yu
Hong-Xing Yu
Houjian Yu
Jianhui Yu
Jiashuo Yu
Jing Yu
Jiwen Yu
Jiyang Yu
Kaicheng Yu
Lei Yu
Lidong Yu
Lijun Yu
Mulin Yu
Peilin Yu
Qian Yu
Qihang Yu
Qing Yu
Rui Yu
Ruixuan Yu
Runpeng Yu
Shaozuo Yu
Shuzhi Yu
Sihyun Yu
Tan Yu
Tao Yu
Tianjiao Yu
Wei Yu
Weihao Yu
Wenwen Yu
Xi Yu
Xiaohan Yu
Xin Yu
Xin Yu
Xuehui Yu
Yingchen Yu
Yongsheng Yu
Yunlong Yu
Zehao Yu
Zhaofei Yu
Zhengdi Yu
Zhengdi Yu
Zhixuan Yu
Zhongzhi Yu
Zhuoran Yu
Zitong Yu
Chun Yuan
Chunfeng Yuan
Hangjie Yuan
Haobo Yuan
Jiakang Yuan
Jiangbo Yuan
Liangzhe Yuan
Maoxun Yuan
Shanxin Yuan
Shengming Yuan
Shuai Yuan
Shuaihang Yuan
Wentao Yuan
Xiaoding Yuan
Xiaoyun Yuan
Xin Yuan
Xin Yuan
Yixuan Yuan
Yu-Jie Yuan
Yuan Yuan
Yuan Yuan
Yuhui Yuan
Zheng Yuan
Zhuoning Yuan
Mehmet Kerim Yücel
Dongxu Yue
Haixiao Yue
Kaiyu Yue
Tao Yue
Xiangyu Yue
Zihao Yue
Zongsheng Yue
Heeseung Yun
Jooyeol Yun
Juseung Yun
Kimin Yun
Se-Young Yun
Sukwon Yun
Tian Yun
Raza Yunus
Ekim Yurtsever
Eloi Zablocki
Riccardo Zaccone
Martin Zach
Muhammad Zaigham Zaheer
Ilya Zakharkin
Egor Zakharov
Abhaysinh S. Zala
Pierluigi Zama Ramirez
Eduard Sebastian Zamfir
Amir Zamir
Luca Zancato
Yuan Zang
Yuhang Zang
Zelin Zang
Pietro Zanuttigh
Giacomo Zara
Samira Zare
Olga Zatsarynna
Denis Zavadski
Vitjan Zavrtanik

Jan Zdenek
Yanjie Ze
Bernhard Zeisl
John Zelek
Oliver Zendel
Ailing Zeng
Chong Zeng
Dan Zeng
Fangao Zeng
Haijin Zeng
Huimin Zeng
Jia Zeng
Jiabei Zeng
Kuo-Hao Zeng
Libing Zeng
Ling-An Zeng
Ming Zeng
Pengpeng Zeng
Runhao Zeng
Tieyong Zeng
Wei Zeng
Yan Zeng
Yanhong Zeng
Yawen Zeng
Yuyuan Zeng
Zilai Zeng
Ziyao Zeng
Kaiwen Zha
Ruyi Zha
Yaohua Zha
Bohan Zhai
Qiang Zhai
Runtian Zhai
Wei Zhai
YiKui Zhai
Yuanhao Zhai
Yunpeng Zhai
De-Chuan Zhan
Fangneng Zhan
Guanqi Zhan
Huangying Zhan
Huijing Zhan
Kun Zhan
Xueying Zhan
Aidong Zhang

Baochang Zhang
Baoheng Zhang
Baoming Zhang
Biao Zhang
Bingfeng Zhang
Binjie Zhang
Bo Zhang
Bo Zhang
Borui Zhang
Bowen Zhang
Can Zhang
Ce Zhang
Chang-Bin Zhang
Chao Zhang
Chao Zhang
Chen-Lin Zhang
Cheng Zhang
Cheng Zhang
Chenghao Zhang
Chenyangguang Zhang
Chi Zhang
Chongyang Zhang
Chris Zhang
Chuhan Zhang
Chunhui Zhang
Chuyu Zhang
Congyi Zhang
Daichi Zhang
Dan Zhang
Daoan Zhang
Daoqiang Zhang
David Junhao Zhang
Dexuan Zhang
Dingwen Zhang
Dingyuan Zhang
Dongsu Zhang
Fan Zhang
Fan Zhang
Fang-Lue Zhang
Feilong Zhang
Frederic Z. Zhang
Fuyang Zhang
Gang Zhang
Gengwei Zhang
Gengyu Zhang

Gengyuan Zhang
Gongjie Zhang
GuiXuan Zhang
Guofeng Zhang
Guozhen Zhang
Hang Zhang
Hang Zhang
Hanwang Zhang
Hao Zhang
Hao Zhang
Hao Zhang
Haokui Zhang
Haonan Zhang
Haotian Zhang
Hengrui Zhang
Hongguang Zhang
Hongrun Zhang
Hongyuan Zhang
Howard Zhang
Huaidong Zhang
Huaiwen Zhang
Hui Zhang
Hui Zhang
Jason Y. Zhang
Ji Zhang
Jiahui Zhang
Jiakai Zhang
Jiaming Zhang
Jian Zhang
Jianfu Zhang
Jiangning Zhang
Jianhua Zhang
Jianming Zhang
Jianpeng Zhang
Jianping Zhang
Jianrong Zhang
Jichao Zhang
Jie Zhang
Jie Zhang
Jie Zhang
Jimuyang Zhang
Jing Zhang
Jing Zhang
Jinghao Zhang
Jingyi Zhang

Jinlu Zhang
Jiqing Zhang
Jiyuan Zhang
Junbo Zhang
Junge Zhang
Junyi Zhang
Juyong Zhang
Kai Zhang
Kai Zhang
Kaidong Zhang
Kaihao Zhang
Kaipeng Zhang
Kaiyi Zhang
Ke Zhang
Ke Zhang
Kui Zhang
Le Zhang
Le Zhang
Lefei Zhang
Lei Zhang
Leo Yu Zhang
Li Zhang
Lianbo Zhang
Liang Zhang
Liangpei Zhang
Lin Zhang
Linfeng Zhang
Liqing Zhang
Lu Zhang
Malu Zhang
Manyuan Zhang
Mengmi Zhang
Mengqi Zhang
Mi Zhang
Min Zhang
Min-Ling Zhang
Mingda Zhang
Mingfang Zhang
Minghui Zhang
Mingjin Zhang
Mingyuan Zhang
Minjia Zhang
Ni Zhang
Pan Zhang
Peiyan Zhang
Pengze Zhang
Pingping Zhang
Qi Zhang
Qi Zhang
Qian Zhang
Qiang Zhang
Qijian Zhang
Qiming Zhang
Qing Zhang
Qing Zhang
Renrui Zhang
Rongyu Zhang
Ruida Zhang
Ruimao Zhang
Ruixin Zhang
Runze Zhang
Sanyi Zhang
Shan Zhang
Shanghang Zhang
Shaofeng Zhang
Sheng Zhang
Shengping Zhang
Shengyu Zhang
Shimian Zhang
Shiwei Zhang
Shizhou Zhang
Shu Zhang
Shuo Zhang
Siwei Zhang
Song-Hai Zhang
Tao Zhang
Tianyun Zhang
Ting Zhang
Tong Zhang
Weixia Zhang
Wendong Zhang
Wenlong Zhang
Wenqiang Zhang
Wentao Zhang
Wentian Zhang
Wenxiao Zhang
Wenxuan Zhang
Xi Zhang
Xiang Zhang
Xiang Zhang
Xianling Zhang
Xiao Zhang
Xiaohan Zhang
Xiaoming Zhang
Xiaoran Zhang
Xiaowei Zhang
Xiaoyun Zhang
Xikun Zhang
Xin Zhang
Xinfeng Zhang
Xingchen Zhang
Xingguang Zhang
Xingxuan Zhang
Xiong Zhang
Xiuming Zhang
Xu Zhang
Xuanyang Zhang
Xucong Zhang
Xuying Zhang
Yabin Zhang
Yabo Zhang
Yachao Zhang
Yahui Zhang
Yan Zhang
Yan Zhang
Yanan Zhang
Yang Zhang
Yanghao Zhang
Yawen Zhang
Yechao Zhang
Yi Zhang
Yi Zhang
Yi Zhang
Yi-Fan Zhang
Yifan Zhang
Yifei Zhang
Yifeng Zhang
Yihao Zhang
Yihua Zhang
Yimeng Zhang
Yiming Zhang
Yin Zhang
Yinan Zhang
Yinda Zhang
Ying Zhang

Yingliang Zhang
Yitian Zhang
Yixin Zhang
Yiyuan Zhang
Yongfei Zhang
Yonggang Zhang
Yonghua Zhang
Youjian Zhang
Youmin Zhang
Youshan Zhang
Yu Zhang
Yu Zhang
Yuan Zhang
Yuechen Zhang
Yuexi Zhang
Yufei Zhang
Yuhan Zhang
Yuhang Zhang
Yunchao Zhang
Yunhe Zhang
Yunhua Zhang
Yunpeng Zhang
Yunzhi Zhang
Yuxin Zhang
Yuyao Zhang
Zaixi Zhang
Zeliang Zhang
Zewei Zhang
Zeyu Zhang
Zhang Zhang
Zhao Zhang
Zhaoxiang Zhang
Zhen Zhang
Zheng Zhang
Zheng Zhang
Zhenyu Zhang
Zhenyu Zhang
Zheyuan Zhang
Zhicheng Zhang
Zhilu Zhang
Zhishuai Zhang
Zhitian Zhang
Zhiwei Zhang
Zhixing Zhang
Zhiyuan Zhang

Zhong Zhang
Zhongping Zhang
Zhongqun Zhang
Zicheng Zhang
Zicheng Zhang
Zihao Zhang
Ziming Zhang
Ziqi Zhang
Qilong Zhangli
Bin Zhao
Bingchen Zhao
Bingyin Zhao
Cairong Zhao
Can Zhao
Chen Zhao
Dong Zhao
Dongxu Zhao
Fang Zhao
Feng Zhao
Fuqiang Zhao
Gangming Zhao
Ganlong Zhao
Guiyu Zhao
Haimei Zhao
Hanbin Zhao
Handong Zhao
Jian Zhao
Jiaqi Zhao
Jie Zhao
Kai Zhao
Kaifeng Zhao
Lei Zhao
Liang Zhao
Lirui Zhao
Long Zhao
Luxi Zhao
Mingyang Zhao
Minyi Zhao
Na Zhao
Nanxuan Zhao
Pu Zhao
Qi Zhao
Qian Zhao
Qibin Zhao
Qingsong Zhao

Qingyu Zhao
Qinyu Zhao
Rongchang Zhao
Rui Zhao
Rui Zhao
Rui Zhao
Ruiqi Zhao
Shanshan Zhao
Shihao Zhao
Shiyu Zhao
Shizhen Zhao
Shuai Zhao
Siheng Zhao
Tianchen Zhao
TianHao Zhao
Tiesong Zhao
Wang Zhao
Wangbo Zhao
Weichao Zhao
Weiyue Zhao
Wenda Zhao
Wenliang Zhao
Xiangyun Zhao
Xiaoming Zhao
Xiaonan Zhao
Xiaoqi Zhao
Xin Zhao
Xin Zhao
Xingyu Zhao
Xu Zhao
Yajie Zhao
Yang Zhao
Yifan Zhao
Ying Zhao
Yiqun Zhao
Yiqun Zhao
Yizhou Zhao
Yizhou Zhao
Yucheng Zhao
Yue Zhao
Yunhan Zhao
Yuyang Zhao
Yuzhi Zhao
Zelin Zhao
Zengqun Zhao

Zhen Zhao
Zhenghao Zhao
Zhengyu Zhao
Zhou Zhao
Zibo Zhao
Zimeng Zhao
Ziwei Zhao
Ziwei Zhao
Zixiang Zhao
Jin Zhe
Anling Zheng
Ce Zheng
Chaoda Zheng
Chengwei Zheng
Chuanxia Zheng
Duo Zheng
Ervine Zheng
Guangcong Zheng
Haitian Zheng
Haiyong Zheng
Hao Zheng
Haotian Zheng
Haoxin Zheng
Huan Zheng
Huan Zheng
Jia Zheng
Jian Zheng
Jian-Qing Zheng
Jianqiao Zheng
Jianwei Zheng
Jin Zheng
Jingxiao Zheng
Kaiwen Zheng
Kecheng Zheng
Léon Zheng
Meng Zheng
Naishan Zheng
Peng Zheng
Qi Zheng
Qian Zheng
Rongkun Zheng
Shen Zheng
Shuai Zheng
Shuhong Zheng
Shunyuan Zheng

Siming Zheng
Tianhang Zheng
Wenting Zheng
Wenzhao Zheng
Xiaozheng Zheng
Xiawu Zheng
Xu Zheng
Yajing Zheng
Yalin Zheng
Yang Zheng
Ye Zheng
Yinglin Zheng
Yinqiang Zheng
Yu Zheng
Zangwei Zheng
Zehan Zheng
Zerong Zheng
Zhaoheng Zheng
Zhedong Zheng
Zhuo Zheng
Zilong Zheng
Shuaifeng Zhi
Tiancheng Zhi
Bineng Zhong
Fangcheng Zhong
Fangwei Zhong
Guoqiang Zhong
Nan Zhong
Yaoyao Zhong
Yijie Zhong
Yiqi Zhong
Yiran Zhong
Yiwu Zhong
Yunshan Zhong
Zichun Zhong
Ziming Zhong
Brady Zhou
Chong Zhou
Chu Zhou
Chunluan Zhou
Da-Wei Zhou
Dawei Zhou
Dewei Zhou
Dingfu Zhou
Donghao Zhou

Dongzhan Zhou
Fan Zhou
Hang Zhou
Hang Zhou
Hao Zhou
Hao Zhou
Haoyi Zhou
Hong-Yu Zhou
Honglu Zhou
Huayi Zhou
Jiahuan Zhou
Jiaming Zhou
Jian Zhou
Jianan Zhou
Jiantao Zhou
Jianxiong Zhou
JIngkai Zhou
Junsheng Zhou
Kailai Zhou
Kaiyang Zhou
Keyang Zhou
Kun Zhou
Lei Zhou
Mingyang Zhou
Mingyi Zhou
Mingyuan Zhou
Mo Zhou
Pan Zhou
Peng Zhou
Peng Zhou
Qianyi Zhou
Qianyu Zhou
Qihua Zhou
Qin Zhou
Qinqin Zhou
Qunjie Zhou
Sheng Zhou
Shenglong Zhou
Shijie Zhou
Shuchang Zhou
Sihang Zhou
Tao Zhou
Tianfei Zhou
Xiangdong Zhou
Xiaoqiang Zhou

Xin Zhou
Xingyi Zhou
Yan-Jie Zhou
Yang Zhou
Yang Zhou
Yanqi Zhou
Yi Zhou
Yi Zhou
Yichao Zhou
Yichen Zhou
Yin Zhou
Yiyi Zhou
Yu Zhou
Yucheng Zhou
Yufan Zhou
Yunsong Zhou
Yuqian Zhou
Yuxiao Zhou
Yuxuan Zhou
Zhenyu Zhou
Zijian Zhou
Zikun Zhou
Ziqi Zhou
Zixiang Zhou
Zongwei Zhou
Alex Z. Zhu
Benjin Zhu
Bin Zhu
Bin Zhu
Chenming Zhu
Chenyang Zhu
Deyao Zhu
Dongxiao Zhu
Fangrui Zhu
Fei Zhu
Feida Zhu
Fengqing Maggie Zhu
Guibo Zhu
Haidong Zhu
Hanwei Zhu
Hao Zhu
Hao Zhu
Heming Zhu
Jiachen Zhu
Jianke Zhu

Jiawen Zhu
Jiayin Zhu
Jinjing Zhu
Junyi Zhu
Kai Zhu
Ke Zhu
Lanyun Zhu
Lin Zhu
Linchao Zhu
Liyuan Zhu
Meilu Zhu
Muzhi Zhu
Qingtian Zhu
Ronghang Zhu
Rui Zhu
Rui Zhu
Rui-Jie Zhu
Ruizhao Zhu
Shengjie Zhu
Sijie Zhu
Siyu Zhu
Tyler Zhu
Wang Zhu
Weicheng Zhu
Wenwu Zhu
Xiangyu Zhu
Xiaofeng Zhu
Xiaoguang Zhu
Xiaosu Zhu
Xiaoyu Zhu
Xingkui Zhu
Xinxin Zhu
Xiyue Zhu
Yangguang Zhu
Yanjun Zhu
Yao Zhu
Ye Zhu
Feida Zhu
Fengqing Maggie Zhu
Guibo Zhu
Haidong Zhu
Hanwei Zhu
Hao Zhu
Hao Zhu
Heming Zhu

Jiachen Zhu
Jianke Zhu
Jiawen Zhu
Jiayin Zhu
Jinjing Zhu
Junyi Zhu
Kai Zhu
Ke Zhu
Lanyun Zhu
Lin Zhu
Linchao Zhu
Liyuan Zhu
Meilu Zhu
Muzhi Zhu
Qingtian Zhu
Ronghang Zhu
Rui Zhu
Rui Zhu
Rui-Jie Zhu
Ruizhao Zhu
Shengjie Zhu
Sijie Zhu
Siyu Zhu
Tyler Zhu
Wang Zhu
Weicheng Zhu
Wenwu Zhu
Xiangyu Zhu
Xiaofeng Zhu
Xiaoguang Zhu
Xiaosu Zhu
Xiaoyu Zhu
Xingkui zhu
Xinxin Zhu
Xiyue Zhu
Yangguang Zhu
Yanjun Zhu
Yao Zhu
Ye Zhu
Ye Zhu
Yichen Zhu
Yingying Zhu
Yousong Zhu
Yuansheng Zhu
Yurui Zhu

Zhen Zhu
Zhenwei Zhu
Zhenyao Zhu
Zhifan Zhu
Zhigang Zhu
Zhihong Zhu
Zihan Zhu
Zixin Zhu
Zunjie Zhu
Bingbing Zhuang
Jia-Xin Zhuang
Jiafan Zhuang
Peiye Zhuang
Wanyi Zhuang
Weiming Zhuang
Yihong Zhuang
Yixin Zhuang
Mingchen Zhuge
Tao Zhuo
Wei Zhuo
Yaoxin Zhuo
Bartosz Zieliński
Wojciech Zielonka
Filippo Ziliotto
Karel Zimmermann
Primo Zingaretti
Nikolaos Zioulis
Liu Ziyin
Mohammad Zohaib
Yongshuo Zong
Zhuofan Zong
Maria Zontak
Gaspard Zoss
Changqing Zou
Chuhang Zou
Danping Zou
Dongqing Zou
Haoming Zou
Longkun Zou
Shihao Zou
Xingxing Zou
Xueyan Zou
Yang Zou
Yuexian Zou
Yuli Zou
Yuliang Zou
Yunhao Zou
Zhiming Zou
Zihang Zou
Silvia Zuffi
Idil Esen Zulfikar
Maria A. Zuluaga
Ronglai Zuo
Xingxing Zuo
Xinxin Zuo
Yifan Zuo
Yiming Zuo
Reyer Zwiggelaar
Vlas Zyrianov

Contents – Part LXXXII

Exploring Conditional Multi-modal Prompts for Zero-Shot HOI Detection 1
  Ting Lei, Shaofeng Yin, Yuxin Peng, and Yang Liu

Training-Free Video Temporal Grounding Using Large-Scale Pre-trained Models 20
  Minghang Zheng, Xinhao Cai, Qingchao Chen, Yuxin Peng, and Yang Liu

Revisit Self-supervised Depth Estimation with Local Structure-from-Motion 38
  Shengjie Zhu and Xiaoming Liu

FAMOUS: High-Fidelity Monocular 3D Human Digitization Using View Synthesis 57
  Vishnu Mani Hema, Shubhra Aich, Christian Haene, Jean-Charles Bazin, and Fernando De la Torre

Efficient Learning of Event-Based Dense Representation Using Hierarchical Memories with Adaptive Update 74
  Uday Kamal and Saibal Mukhopadhyay

SNP: Structured Neuron-Level Pruning to Preserve Attention Scores 90
  Kyunghwan Shim, Jaewoong Yun, and Shinkook Choi

Multi-Granularity Sparse Relationship Matrix Prediction Network for End-to-End Scene Graph Generation 105
  Lei Wang, Zejian Yuan, and Badong Chen

Flash-Splat: 3D Reflection Removal with Flash Cues and Gaussian Splats 122
  Mingyang Xie, Haoming Cai, Sachin Shah, Yiran Xu, Brandon Y. Feng, Jia-Bin Huang, and Christopher A. Metzler

PALM: Predicting Actions through Language Models 140
  Sanghwan Kim, Daoji Huang, Yongqin Xian, Otmar Hilliges, Luc Van Gool, and Xi Wang

Motion Keyframe Interpolation for Any Human Skeleton via Temporally Consistent Point Cloud Sampling and Reconstruction 159
  Clinton Mo, Kun Hu, Chengjiang Long, Dong Yuan, and Zhiyong Wang

SwiftBrush V2: Make Your One-Step Diffusion Model Better Than Its Teacher 176
Trung Dao, Thuan Hoang Nguyen, Thanh Le, Duc Vu, Khoi Nguyen, Cuong Pham, and Anh Tran

Learning to Localize Actions in Instructional Videos with LLM-Based Multi-pathway Text-Video Alignment 193
Yuxiao Chen, Kai Li, Wentao Bao, Deep Patel, Yu Kong, Martin Renqiang Min, and Dimitris N. Metaxas

Improving Hyperbolic Representations via Gromov-Wasserstein Regularization 211
Yifei Yang, Wonjun Lee, Dongmian Zou, and Gilad Lerman

VSViG: Real-Time Video-Based Seizure Detection via Skeleton-Based Spatiotemporal ViG 228
Yankun Xu, Junzhe Wang, Yun-Hsuan Chen, Jie Yang, Wenjie Ming, Shuang Wang, and Mohamad Sawan

DiffSurf: A Transformer-Based Diffusion Model for Generating and Reconstructing 3D Surfaces in Pose 246
Yusuke Yoshiyasu and Leyuan Sun

Exploiting Supervised Poison Vulnerability to Strengthen Self-supervised Defense 265
Jeremy Styborski, Mingzhi Lyu, Yi Huang, and Adams Kong

Dense Hand-Object (HO) GraspNet with Full Grasping Taxonomy and Dynamics 284
Woojin Cho, Jihyun Lee, Minjae Yi, Minje Kim, Taeyun Woo, Donghwan Kim, Taewook Ha, Hyokeun Lee, Je-Hwan Ryu, Woontack Woo, and Tae-Kyun Kim

Human Pose Recognition via Occlusion-Preserving Abstract Images 304
Saad Manzur and Wayne Hayes

DA-BEV: Unsupervised Domain Adaptation for Bird's Eye View Perception 322
Kai Jiang, Jiaxing Huang, Weiying Xie, Jie Lei, Yunsong Li, Ling Shao, and Shijian Lu

SlimFlow: Training Smaller One-Step Diffusion Models with Rectified Flow 342
Yuanzhi Zhu, Xingchao Liu, and Qiang Liu

PhysGen: Rigid-Body Physics-Grounded Image-to-Video Generation 360
    Shaowei Liu, Zhongzheng Ren, Saurabh Gupta, and Shenlong Wang

Depth-Aware Blind Image Decomposition for Real-World Adverse
Weather Recovery 379
    Chao Wang, Zhedong Zheng, Ruijie Quan, and Yi Yang

DreamSampler: Unifying Diffusion Sampling and Score Distillation
for Image Manipulation 398
    Jeongsol Kim, Geon Yeong Park, and Jong Chul Ye

Reshaping the Online Data Buffering and Organizing Mechanism
for Continual Test-Time Adaptation 415
    Zhilin Zhu, Xiaopeng Hong, Zhiheng Ma, Weijun Zhuang, Yaohui Ma,
    Yong Dai, and Yaowei Wang

Personalized Privacy Protection Mask Against Unauthorized Facial
Recognition 434
    Ka-Ho Chow, Sihao Hu, Tiansheng Huang, and Ling Liu

PosterLlama: Bridging Design Ability of Language Model
to Content-Aware Layout Generation 451
    Jaejung Seol, Seojun Kim, and Jaejun Yoo

PreciseControl: Enhancing Text-to-Image Diffusion Models
with Fine-Grained Attribute Control 469
    Rishubh Parihar, V. S. Sachidanand, Sabariswaran Mani,
    Tejan Karmali, and R. Venkatesh Babu

Author Index 489

Exploring Conditional Multi-modal Prompts for Zero-Shot HOI Detection

Ting Lei[1], Shaofeng Yin[1], Yuxin Peng[1], and Yang Liu[1,2](✉)

[1] Wangxuan Institute of Computer Technology, Peking University, Beijing, China
{ting_lei,pengyuxin,yangliu}@pku.edu.cn, yin_shaofeng@stu.pku.edu.cn
[2] State Key Laboratory of General Artificial Intelligence,
Peking University, Beijing, China

Abstract. Zero-shot Human-Object Interaction (HOI) detection has emerged as a frontier topic due to its capability to detect HOIs beyond a predefined set of categories. This task entails not only identifying the interactiveness of human-object pairs and localizing them but also recognizing both seen and unseen interaction categories. In this paper, we introduce a novel framework for zero-shot HOI detection using Conditional Multi-Modal Prompts, namely CMMP. This approach enhances the generalization of large foundation models, such as CLIP, when fine-tuned for HOI detection. Unlike traditional prompt-learning methods, we propose learning decoupled vision and language prompts for interactiveness-aware visual feature extraction and generalizable interaction classification, respectively. Specifically, we integrate prior knowledge of different granularity into conditional vision prompts, including an input-conditioned instance prior and a global spatial pattern prior. The former encourages the image encoder to treat instances belonging to seen or potentially unseen HOI concepts equally while the latter provides representative plausible spatial configuration of the human and object under interaction. Besides, we employ language-aware prompt learning with a consistency constraint to preserve the knowledge of the large foundation model to enable better generalization in the text branch. Extensive experiments demonstrate the efficacy of our detector with conditional multi-modal prompts, outperforming previous state-of-the-art on unseen classes of various zero-shot settings. The code and models are available at https://github.com/ltttpku/CMMP.

Keywords: Human-object interaction detection · Zero-shot learning

1 Introduction

Human-object interaction (HOI) detection has been introduced by [8] and plays an important role in understanding high-level human-centric scenes. Given an

---

Supplementary Information The online version contains supplementary material available at https://doi.org/10.1007/978-3-031-73007-8_1.

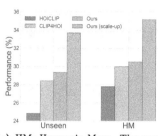 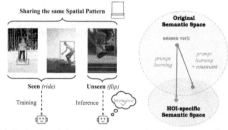

(a) HM: Harmonic Mean. The averaged performance on unseen classes and harmonic mean across all zero-shot settings of HICO-DET.

(b) Left: Spatial cues help recognize the interactiveness of unseen HOI concepts; Right: The training trajectory of prompts.

Fig. 1. (a) Previous detectors struggle with the delicate balance between seen and unseen classes, resulting in a low harmonic mean (HM) and poor performance on unseen classes. In contrast, our method effectively addresses this balance issue, leading to significant improvement and establishing a new state-of-the-art benchmark for unseen classes. (b) Our model uses visual spatial cues during feature extraction to help recognize the interactiveness of unseen HOI concepts and utilize constraint prompt learning for better generalizability on unseen classes.

image, HOI detection aims to localize human and object pairs and recognize their interactions, *i.e.* a set of <human, object, action> triplets. Traditionally, human-object interaction detectors can be categorized as one- or two-stage. One-stage methods leverage multi-stream networks [22,41] or encoder-decoder architectures [4,14,23,31,35,58,62] to predict HOI triplets from a holistic image context in an end-to-end manner. Two-stage methods [20,21,25,26,28,43,49,50,54,61] first localize humans and objects separately using off-the-shelf detectors (*e.g.*, DETR [2]), followed by utilizing the region features from the localized areas to predict the interaction class.

Despite recent advances, most previous works lack generalizability to unseen HOIs. Although some zero-shot HOI detectors [10,12,23,40,42] have been proposed in recent years, some of them [10,12] can only detect HOIs of unseen compositions, but fail to incorporate language priors and can't generalize to unseen verbs. Besides, many previous zero-shot detectors [30,31] encounter challenges in achieving a nuanced balance between seen and unseen classes, leading to a low harmonic mean and poor performance on unseen classes, as illustrated in Fig. 1a.

Given the combinatorial nature of HOIs, constructing a HOI dataset with all possible HOIs is prohibitively expensive. This motivates us to investigate a HOI detector that can be applied to a wide range of previously unseen interactions with powerful generalizability. Zero-shot HOI detection has the following two challenges: 1) how to extract interactiveness-aware features for human-object pairs to determine whether they interact with each other when confronted with unseen HOI concepts, and 2) how to recognize the unseen interaction types accurately.

To address the aforementioned issues, *we propose CMMP, which divides HOI detection into two subtasks: interactiveness-aware visual feature extraction and generalizable interaction classification. The design aids in reducing their dependence on one another and error propagation between them.* Inspired by the rapid advancements in large vision-language foundation models and their remarkable zero-shot capability, our objective in this study is to devise a method capable of seamlessly adapting these models to the HOI task using efficient multi-modal prompt learning techniques.

For the first subtask, we introduce conditional vision prompts tailored to guide the extraction of interactiveness- and spatial-aware visual features. This guidance enables the model to generalize its capacity to determine the interactivity of human-object pairs across previously unseen classes. Specifically, we propose conditional vision prompts incorporating two priors: an input-conditioned instance prior and a global spatial pattern prior. The input-conditioned instance prior encompasses both the spatial configuration and semantics of all detected instances in the input image. This encourages the model to treat both seen and potentially unseen interactive instances equally. As depicted on the left side of Fig. 1b, the global spatial pattern prior provides representative plausible spatial configuration of the human and object engaged in interaction, serving as a bridge to discriminate interactivity between seen and unseen HOIs. Overall, the input-conditioned instance prior offers fine-grained details about individual instances, while the global spatial pattern prior provides a broader contextual understanding of interactions. As a result, these two conditions contain complementary information that guides spatial-aware visual feature learning for the HOI detection task. Subsequently, we employ the attention mechanism [38] to fuse the above knowledge embedded in conditional vision prompts into the image encoder from the early spatial-aware and fine-grained feature maps, where valuable information lies for the HOI detection task. The conditional vision prompts assist the model in extending its capacity to determine the interactivity of human-object pairs from seen categories to unseen ones.

For the subtask of interaction classification, we propose language prompts that are unaware of spatial information. The language prompts provide a unified context for both seen and unseen HOIs, allowing the model to leverage knowledge learned from seen classes to classify HOIs that include unseen verbs. Besides, we use human-designed prompts as a regularizer to keep the learned text prompts from diverging too much. This constraint preserves the origin semantic space learned by large foundation models, as shown on the right of Fig. 1b, and thus may be better for potential real-world scenario applications where arbitrary novel actions may occur.

We propose decoupled vision and language prompts for the above two subtasks to prevent mutual inhibition, respectively. These conditional multi-modal prompts serve as a hub to build a connection between seen and unseen categories. We evaluate our detector with conditional multi-modal prompts under various zero-shot settings.Experimental results demonstrate that our method achieves an effective balance between seen and unseen classes, achieving the highest harmonic mean of performance and the best results on unseen classes, as shown in Fig. 1a.

Our contributions can be summarized as follows: *(1)* To the best of our knowledge, we first propose a multi-modal prompt learning method for large foundation models in zero-shot human-object interaction detection to improve visual-language feature alignment and zero-shot knowledge transfer. *(2)* Through careful prompt design, we reveal the inherent capacity of the large foundation model for precise discrimination of fine-grained HOIs, enhancing the generalization ability of our CMMP. *(3)* Our model sets a new state-of-the-art for HOI detection on unseen classes in various zero-shot settings, significantly outperforming all previous methods.

2 Related Work

2.1 Human-Object Interaction Detection

With the development of large-scale datasets [3,8,16,22] and deep learning-based methods [19,36,37,39,45,53], HOI learning has been rapidly progressing in two main streams: one- and two-stage approaches. One-stage HOI detectors [4,14,15,18,23,32,35,44,57] usually formulate HOI detection task as a set prediction problem originating from DETR [2] and perform object detection and interaction prediction in a parallel [4,14,35] or sequentially [23]. In contrast, two-stage methods [17,20,26,49–51,54] usually utilize pre-trained detectors [2,9,34] to detect human and object proposals and exhaustively enumerate all possible human-object pairs in the first stage. Then they design an independent module to predict the multi-label interactions of each human-object pair in the second stage. Despite their improved performance, most previous models rely heavily on full annotations with predefined HOI categories and thus are costly to scale further. Moreover, they lack the generalization capability to deal with unseen HOI categories as shown in Fig. 1a. In contrast to them, our work targets zero-shot HOI detection with the help of an off-the-shelf object detector and vision-language model in a two-stage manner.

2.2 Zero-Shot Human-Object Interaction Detection

Zero-shot HOI detection aims at detecting interactions unseen in the training set, which is essential for developing practical HOI detection systems that can function effectively in real-world scenarios. ConsNet [28] converts HOI categories and their components into a graph and distributes knowledge among its nodes. VCL [10] recombines object representations and human representations to compose unseen HOI samples. FCL [12] proposes to generate fake object representations for human-object recombination. ATL [11] exploits additional object datasets for HOI detection to discover novel HOI categories. However, lacking the help of semantics, the above methods aren't capable of detecting HOIs including unseen actions.

To incorporate language priors in zero-shot HOI detection, prevailing approaches [23,30,31,42,47] propose to incorporate knowledge from CLIP [33] to achieve zero-shot HOI detection. The natural generalizability of language aids models in recognizing HOIs, even those with unseen actions. RLIP [47] proposes a Relational Language-Image Pre-training strategy for HOI detection. EoID [42] distills the distribution of action probability from CLIP to the HOI model and designs an interactive score module combined with a two-stage bipartite matching algorithm to achieve interaction distinguishment. GEN-VLKT [23] utilizes CLIP text embeddings for prompted HOI labels to initialize the classifier and employs CLIP visual features to guide the learning of interactive representations. HOICLIP [31] adopts the one-stage design following GEN-VLKT [23] and proposes query-based knowledge retrieval for efficient knowledge transfer for HOI detection with CLIP. Besides, it exploits zero-shot CLIP knowledge as a training-free enhancement during inference. Despite the progressing generalizability, previous one-stage methods utilize each query or pair thereof to localize the human-object pair jointly, often leading to overfitting of the decoder to seen categories.

The most relevant work to ours is CLIP4HOI [30], which integrates CLIP into a previously established two-stage method [50], thereby achieving a disentangled two-stage paradigm for zero-shot HOI detection. Different from CLIP4HOI, our method employs conditional multi-modal prompts to directly transform the feature space of large foundation models. This transformation moves from understanding image-level and first-order semantics to comprehending instance-level and second-order semantic information within images, resulting in better generalizability to unseen HOI concepts.

2.3 Prompt Learning

Recently, the development of large vision-language models (VLMs), *e.g.*, CLIP [33], emerges and finds its applications in various downstream tasks [7, 27, 29, 46, 52, 55, 56]. Inspired by prompt learning in language tasks, CoOp [60] first proposes to use context tokens as language prompts in the image classification task. Co-CoOp [59] proposes to explicitly condition language prompts on image instances. Recently, other approaches for adapting V-L models through prompting have been proposed. MaPLe [13] proposes a coupling function to explicitly condition vision prompts on their language counterparts, to provide more flexibility to align the vision-language representations. However, existing methods primarily focus on prompt learning for image classification, which may not be suitable for HOI detection. DetPro [5] and PromptDet [6] propose a novel prompt learning method based on first-order individual instance detection. However, they lack the understanding of second-order pair-wise relationships in images, which is crucial in the HOI detection task. [23] and [40] first propose applying static template prompts or learnable language prompts in the HOI detection task, respectively. However, they ignore the fact that HOI detection involves considering regional spatial information, making it distinct from image classification. Therefore, how to design tailored *spatial-aware* prompts specifically designed for the HOI detection task is critical. Given that interactiveness-aware visual feature extraction and interaction classification are distinct subtasks, we propose to employ decoupled multi-modal prompts for these two subtasks to reduce error propagation between them.

3 Method

3.1 Overview

HOI detection aims to detect all interactive human-object pairs and predict the interactive relationship for them. Formally, we define the interaction as a quadruple (b_h, b_o, a, o): b_h, b_o represent the bounding box of humans and objects and $a \in \mathbb{A}, o \in \mathbb{O}$ represent the human action and object category, where $\mathbb{A} = \{1, 2, ..., A\}$ and $\mathbb{O} = \{1, 2, ..., O\}$ denote the human action and object set, respectively. Then given an image $\mathbf{I}$, our goal is to predict all quadruples that exist in $\mathbf{I}$. To avoid struggling with multi-task learning [48] and the risk of overfitting to the joint positional distribution of human-object pairs for seen HOIs [30], we follow the previous two-stage design for HOI detection [30,50]: human-object detection and interaction classification. In the first stage, we use an off-the-shelf object detector D, e.g., DETR [2], and apply appropriate filtering strategies to extract all instances and exhaustively enumerate the detected instances to compose human-object pairs. Then in the second stage, we first encode the image $\mathbf{I}$ using a pretrained image encoder $\mathrm{E_I}$, i.e., $f_I = \mathrm{E_I}(\mathbf{I}) \in \mathbb{R}^{H \times W \times C}$. We define the union region b_u as the smallest rectangular region that contains b_h, b_o. Then following the multi-branch architecture of previous HOI detection works [10,12], we utilize b_h, b_o, and b_u to extract features for the human region, the object region, and the interaction region from the feature map f_I via ROI-align [9], respectively.

As illustrated in Fig. 2, CMMP tackles the zero-shot HOI detection task by dividing it into two subtasks: interactiveness-aware feature extraction and interaction classification. To propagate knowledge from seen HOI categories to unseen HOI categories and eliminate the dependence between the two subtasks, we propose decoupled vision prompts P_V and language prompts P_L for the image encoder $\mathrm{E_I}$ and text encoder $\mathrm{E_T}$, respectively. Specifically, in the image branch, we incorporate instance-level visual prior C_{ins} derived from the input image and global spatial patterns C_{GSP} obtained from the dataset to construct conditional vision prompts P_V. The instance-level visual prior emphasizes the unique characteristics of each detected instance in the input image, encompassing their spatial configurations and semantics. It enables the encoder to treat both seen and potential interactive instances with equal significance. In contrast, the global spatial pattern captures broader relationships and patterns among objects or entities within the scene, creating representative plausible spatial configurations of the human and object under interaction. The input-conditioned instance prior furnishes knowledge of individual instances, while the global spatial pattern prior offers a wider contextual comprehension of interactions. Consequently, these two conditions provide complementary guidance for spatial-aware visual feature learning in the HOI detection task. These prompts are then incorporated into the image encoder to refine its capabilities, transitioning from image-level individual instance comprehension to understanding region-level pair-wise relations. The conditional vision prompts can alert the image encoder $\mathrm{E_I}$ to all potential interactive instances within the image. For the text branch, we

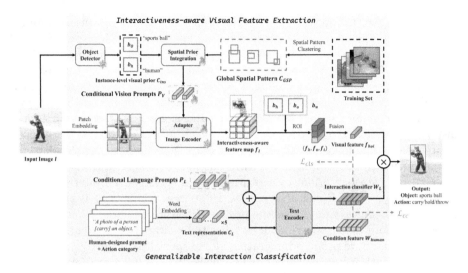

Fig. 2. The overall framework of CMMP. The proposed method splits zero-shot HOI detection into two subtasks: interactiveness-aware visual feature extraction and generalizable interaction classification. We propose decoupled vision and text prompts for each subtask to eliminate the dependence between them and break error-propagation in-between. The conditional vision prompts (P_V) are used to inject spatial- and interactiveness-aware knowledge into the image encoder and are explicitly constrained by instance-level visual prior (C_{ins}) and global spatial pattern (C_{GSP}). The conditional language prompts (P_L) are constrained by the human-designed prompts (C_L) through a regularization loss. (Best viewed in color.)

feed-forward the learnable prompts P_L alongside human-designed prompts C_L to E_T. This operation yields the weights of interaction classifier W_L, facilitating the computation of interaction scores for the provided human-object pair. Besides, to improve the generalization of large foundation models when fine-tuned on downstream tasks, we incorporate language priors from the pretrained vision-language model by enforcing a consistency constraint between W_L and the condition embeddings W_{human}.

3.2 Interactiveness-Aware Visual Feature Extraction

Given that the image encoder of the large foundation model we adopt [33] is originally trained via Contrastive Language-Image Pre-Training on large-scale image-text pairs, its inherent capability might be limited to grasping image-level first-order semantics. To endow the image encoder with the capability of distinguishing the interactiveness of all human-object pairs within a given image, we propose to integrate prior knowledge of varying granularity into conditional vision prompts to empower the image encoder to comprehend region-level second-order semantics specifically tailored for the HOI task, as illustrated in Fig. 2.

To encourage the model to treat the seen and potentially unseen interactive instances equally, we utilize instance-level information as input-conditioned prior knowledge into the conditional vision prompts. Given an input image $\mathbf{I}$, we first follow [17], utilizing a pretrained object detector to obtain all the instance-level prior knowledge C_{ins} including (1) bounding boxes b, which captures the spatial information of detected objects. The spatial configuration might provide cues for understanding interactiveness and is good at transferring as it is instance-agnostic. (2) confidence scores s, which reflect the quality and uncertainty of the candidate instances. (3) semantic embeddings e of the detected instances, which are obtained by CLIP language encoder and enable P_V to leverage the category priors to capture which objects can be interacted with. Formally, $C_{ins} = \text{MLP}(\text{concat}(b, s, e)) \in \mathbb{R}^{N_{ins}*d_{ins}}$, where N_{ins} is the number of detected instances and d_{ins} is the projected feature dimension.

Furthermore, to encourage each instance to be aware of its potential interactive counterpart, we propose integrating global spatial patterns C_{GSP} from the training set with instance-level prior knowledge C_{ins} using a spatial prior integration module SPI. Specifically, for each annotated interactive human-object pair (b_h^i, b_o^i), we first compute its unary and pairwise spatial features sp_i used by Zhang et al. [49]. These features include the normalized unary box center, width, and height, as well as pairwise metrics like intersection-over-union, relative area, and direction. More details are provided in the supplementary material. Subsequently, we employ the K-means algorithm to identify the clustering centers of $\{sp_i\}_{i=1}^{N_{hoi}}$ and utilize them as representative spatial patterns of interactive human-object pairs, which are denoted as C_{GSP}. The global spatial interactive pattern provides a category-agnostic representative plausible spatial configuration between human and object during interaction, serving as a bridge to discern the interactivity between seen and unseen HOI concepts. Different from the input-conditioned instance prior that offers fine-grained details about individual instances, the global spatial pattern prior provides a broader contextual understanding of interactions, thus offering supplementary prior knowledge and enhancing the understanding of the interactions. Overall, the construction of $P_V \in \mathbb{R}^{N_{ins}*d_{ins}}$ can be formulated as:

$$P_V = \text{SPI}(C_{ins}, C_{GSP}), \tag{1}$$

where the spatial prior integration module SPI is implemented via transformer decoder layers, C_{ins} is treated as query, and C_{GSP} is treated as key and value. Then, we incorporate P_V into the image encoder E_I through cross-attention mechanism [38]. Formally, we denote $X_i \in \mathbb{R}^{hw \times d}$ as the feature map of i-th block of E_I. We first project the feature dimension of X_i to d_{ins} ($d_{ins} << d$) through MLP:

$$X_i' = \text{MLP}(X_i), \tag{2}$$

where X_i' shares the same feature dimension with P_V. Then we inject context knowledge P_V into X_i through the cross-attention (Attn) mechanism:

$$X_i = X_i + \text{MLP}(\text{Attn}(X_i', P_V, P_V)), \tag{3}$$

where X_i' is treated as query and P_V is treated as key and value. The purpose of conditional vision prompts P_V is to utilize its spatial information to enhance the image feature map X_i, allowing X_i to acquire valuable spatial information from P_V.

3.3 Generalizable Interaction Classification

To learn task-specific representations while also retaining generalized CLIP knowledge, we employ language-aware prompt learning with a consistency constraint in the text branch, as shown in Fig. 2. The constraint ensures that the learned prototypes of seen and unseen classes leave a reasonable separation margin among each other and do not diverge too far apart. Specifically, for each action class $a \in \mathbb{A}$, we first format it using the human-designed prompt "A photo of a person [verb-ing] an object". We denote $P_L = [P_L^1, P_L^2, ..., P_L^S]$ as the learnable context words, where S denotes the number of learnable prompts. The context words P_L are shared among all classes and thus serve as a bridge between the semantics of seen and unseen categories. The final representation of class a can be obtained by concatenation of learnable context words P_L and the word embedding of the above sentence C_L^a. Then the prototype of the class a can be obtained by the text encoder $\mathrm{E_T}$:

$$W_L^a = \mathrm{E_T}(concat(P_L, C_L^a)), a \in \mathbb{A} \qquad (4)$$

The prototypes should be the representative features belonging to the corresponding category. Given a sample, the similarity with a prototype could represent how likely it belongs to the category. After performing l_2-normalization on all prototypes W_L^a, the interaction classifier W_L is then constructed from prototypes of all target classes' embeddings:

$$W_L = \mathrm{concat}(W_L^1, W_L^2, ..., W_L^A) \qquad (5)$$

To further utilize the feature space learned by the text encoder of VLMs and improve generalization for unseen classes, we propose to use human-designed prompts to guide the feature space of the learnable language prompts. The constraint ensures that prototypes of seen and unseen classes leave a reasonable separation margin among each other and do not diverge too far apart. We apply a regularization loss to reduce the discrepancy between the feature representation of P_L and that of the human-designed language prompts C_L. Specifically, we encourage the soft prompt P_L^i to be encoded close to its corresponding human-designed prompt C_L^i through a conditional constraint loss ($\mathcal{L}_{cc}$), which can be formulated as:

$$\mathcal{L}_{cc} = -\sum_{i=1}^{A} \log \frac{\exp(\cos(W_L^i, W_{hum}^i))}{\sum_{j=1}^{A} \exp(\cos(W_L^i, W_{hum}^j))}, \qquad (6)$$

where $W_{hum} = \mathrm{E_T}(C_L)$ is the encoded features of human-designed prompts C_L and W_L is the feature representation of P_L.

3.4 Training CMMP

Based on the interactiveness-aware feature map f_I and the extracted bounding boxes b_h, b_o, and b_u, we first apply ROI-Pooling to extract features for different regions:

$$f_{hum}, f_{obj}, f_{inter} = \text{ROI}(f_I, b_h), \text{ROI}(f_I, b_o), \text{ROI}(f_I, b_u) \tag{7}$$

The interaction classifier W_L is composed of prototypes of all target classes as described in Sect. 3.3. We then calculate the action prediction s_{ho} for the corresponding human-object pair as follows:

$$s_{ho} = (\alpha_{hum} f_{hum} + \alpha_{obj} f_{obj} + \alpha_{inter} f_{inter}) W_L^T \tag{8}$$

where α_{hum}, α_{obj}, and α_{inter} are set to learnable parameters. We incorporate the object confidence scores into the final scores of each human-object pair. We denote σ as the sigmoid function. The final score s_{ho}^{final} is computed as:

$$s_{ho}^{final} = \sigma(s_{ho}) \cdot (s_h)^\lambda \cdot (s_o)^\lambda, \tag{9}$$

where s_h and s_o are confidence scores given by object detector D, and $\lambda > 1$ is a constant that is used to suppress overconfident objects during inference. The whole model is trained on focal loss [24] $\mathcal{L}_{cls}$ for action classification and language regularization loss $\mathcal{L}_{cc}$ at the same time. We use λ_{cc} as the hyper-parameter weight. The overall objective function is formulated as:

$$\mathcal{L} = \mathcal{L}_{cls} + \lambda_{cc} \mathcal{L}_{cc} \tag{10}$$

4 Experiments

4.1 Experiment Setting

Dataset. HICO-DET [3] is a dataset for detecting human-object interactions in images and has 47,776 images (38,118 in train set and 9,658 in test set) and is annotated with <human, verb, object> triplets. 600 HOI categories in HICO-DET are composed of 80 object classes and 117 verb classes, including no interaction labels.

Zero-Shot Setups. To validate our model's zero-shot performance, we evaluate our model on four zero-shot settings on HICO-DET: 1) Unseen Composition (UC), where the training data contains all categories of object and verb but misses some HOI triplet categories. 2) Rare First Unseen Combination (RF-UC) [12], which prioritizes rare HOI categories when selecting held-out HOI categories. 3) Non-rare First Unseen Combination (NF-UC) [12], which prioritizes non-rare HOI categories instead. Therefore, the training set of the NF-UC setting contains much fewer samples and thus is more challenging. 4) Unseen Verb (UV) [23], which is set to discover novel categories of actions and reflects a unique characteristic of zero-shot HOI detection.

Evaluation Metric. Following the common evaluation protocol, we use the mean average precision (mAP) to examine the model performance. A detected human-object pair is considered as a true positive if 1) both the predicted human and object boxes have the Interaction-over-Union (IOU) ratio greater than 0.5 with regards to the ground-truth boxes. 2) the predicted HOI categories are accurate.

4.2 Implementation Details

We follow the standard protocol of existing zero-shot two-stage HOI detectors [1, 10] to fine-tune DETR on all the instance-level annotations of the training set of HICO-DET before training CMMP. For the conditional language prompts P_L, we set the length of context words S to be 16. The weight λ_{cc} for the consistency loss is set to 1.0 during training. We utilize ViT-B/16 as our backbone if not otherwise stated. See more details in the supplementary material.

4.3 Compare with the State-of-the-Art Methods

We evaluate the performance of our model and compare it with existing zero-shot HOI detectors under UC, RF-UC, NF-UC, UO, and UV settings of the HICO-DET [3] dataset.

As shown in Table 1, our CMMP has demonstrated exceptional performance by outperforming all previous detectors by a significant margin on the unseen classes. Furthermore, our CMMP performs comparably to the previous detectors on the seen classes, resulting in an overall outstanding performance. To be specific, compared to the previous state-of-the-art methods, our CMMP achieves a relative mAP gain of 6.82%, 3.44%, 2.07%, 6.20%, and 0.81% on unseen classes on five zero-shot settings, respectively. As shown in the last line of each type in Table 1, scaling our CMMP by utilizing the ViT-L/14 backbone to match the FLOPs of CLIP4HOI results in superior performance across all splits. The performance gap demonstrates our model's ability to excel in both spatial relation extraction for visual features and prototype learning for interaction classification. Notably, since unseen classes under the NF-UC setting are sometimes more common and semantically straightforward, both our model and previous models [23, 42] may perform better on the unseen split than on the seen split. We also observe that our model performs better on unseen than seen splits in the UO setting, unlike related works. This is likely because CLIP already understands common objects (e.g., bicycles, cars) in the unseen splits. Our consistency constraint preserves CLIP's knowledge, reducing overfitting compared to CLIP4HOI.

Furthermore, previous methods exhibit severe performance degradation between seen and unseen classes, indicating a lack of generalizability. Our model, on the other hand, could alleviate the problem to a large extent and has a high potential for generalization to previously unseen HOIs, confirming the effectiveness of our multi-modal prompts with constraints.

As shown in Table 2, we further compare our CMMP with other methods under the fully supervised setting on HICO-DET and V-COCO datasets. We observe that CMMP improves our baseline model by 5.44 mAP on the full split of HICO-DET, showing the effectiveness of our method design. When scaled to match the FLOPs of CLIP4HOI, our model achieves state-of-the-art performance across all splits of HICO-DET.

Table 1. Performance comparison for zero-shot HOI detection. UC, UO, and UV denote unseen composition, unseen object, and unseen verb settings, respectively. RF- and NF- denote rare first and non-rare first. #TP/#AP denotes the number of Trainable/All Parameters. † denotes the scaled-up version utilizing the ViT-L/14 backbone. HM denotes harmonic mean.

Method	Type	#TP/#AP	FLOPs	Unseen↑	Seen↑	Full↑	HM↑
HOICLIP [31]	UC	–	–	23.15	31.65	29.93	26.74
CLIP4HOI [30]	UC	71.2M/262.4M	186G	27.71	33.25	32.11	30.23
CMMP (Ours)	UC	2.3M/193.4M	114G	29.60	32.39	31.84	30.93
CMMP†(Ours)	UC	5.4M/433.2M	168G	**34.46**	**37.15**	**36.56**	**35.75**
GEN-VLKT [23]	RF-UC	–	–	21.36	32.91	30.56	25.91
EoID [42]	RF-UC	–	–	22.04	31.39	29.52	25.90
HOICLIP [31]	RF-UC	–	–	25.53	34.85	32.99	29.47
CLIP4HOI [30]	RF-UC	71.2M/262.4M	186G	28.47	35.48	34.08	31.59
CMMP (Ours)	RF-UC	2.3M/193.4M	114G	29.45	32.87	32.18	31.07
CMMP†(Ours)	RF-UC	5.4M/433.2M	168G	**35.98**	**37.42**	**37.13**	**36.69**
GEN-VLKT [23]	NF-UC	–	–	25.05	23.38	23.71	24.19
EoID [42]	NF-UC	–	–	26.77	26.66	26.69	26.71
HOICLIP [31]	NF-UC	–	–	26.39	28.10	27.75	27.22
CLIP4HOI [30]	NF-UC	71.2M/262.4M	186G	31.44	28.26	28.90	29.77
CMMP (Ours)	NF-UC	2.3M/193.4M	114G	32.09	29.71	30.18	30.85
CMMP†(Ours)	NF-UC	5.4M/433.2M	168G	**33.52**	**35.53**	**35.13**	**34.50**
CLIP4HOI [30]	UO	71.2M/262.4M	186G	31.79	32.73	32.58	32.25
CMMP (Ours)	UO	2.3M/193.4M	114G	33.76	31.15	31.59	32.40
CMMP†(Ours)	UO	5.4M/433.2M	168G	**39.67**	**36.15**	**36.74**	**37.83**
GEN-VLKT [23]	UV	–	–	20.96	30.23	28.74	24.76
EoID [42]	UV	–	–	22.71	30.73	29.61	26.12
HOICLIP [31]	UV	–	–	24.30	32.19	31.09	27.69
CLIP4HOI [30]	UV	71.2M/262.4M	186G	26.02	31.14	30.42	28.35
CMMP (Ours)	UV	2.3M/193.4M	114G	26.23	32.75	31.84	29.13
CMMP†(Ours)	UV	5.4M/433.2M	168G	**30.84**	**37.28**	**36.38**	**33.75**

Table 2. Performance comparison under fully supervised settings of HICO-DET and V-COCO. †: scaled-up version.

Method	FLOPs	HICO-DET			V-COCO
		Rare	Non-rare	Full	AP_{role}^{S2}
Baseline	–	26.64	28.15	27.80	56.2
CLIP4HOI [30]	186G	33.95	35.74	35.33	**66.3**
CMMP	114G	32.26	33.53	33.24	61.2
CMMP†	168G	**37.75**	**38.25**	**38.14**	64.0

Table 3. Ablation on network modules under the Unseen Verb setting. CLP: Conditional Language Prompts. CVP: Conditional Vision Prompts. GSP: Global Spatial Pattern.

Setting	Unseen	Seen	Full
w/o CLP	19.40	32.13	30.35
w/o CVP	20.56	26.03	25.27
w/o Instance-level Prior	25.07	31.27	30.40
w/o GSP	24.92	31.66	30.71
CMMP (Ours)	**26.23**	**32.75**	**31.84**

Table 4. Ablation on the consistency constraint of the language prompts under the Unseen Verb setting.

λ_{cc}	Unseen	Seen	Full
0	24.49	32.48	31.36
1.0	**26.23**	**32.75**	**31.84**
2.0	24.45	32.24	31.15

4.4 Ablation Study

Network Modules. As shown in Table 3, we study the effectiveness of different modules of CMMP under the unseen verb setting of HICO-DET. We observe the following behaviors related to the use of different modules in the HOID task: *(1)* As shown in lines 1–2 of Table 3, in the absence of conditional prompts from one modality, the model's performance is hindered by the inherent limitations of another frozen modality, particularly when dealing with unseen classes. Specifically, the model without conditional language/vision prompts exhibits a 6.83%, and 5.67% mAP drop on unseen classes, respectively. Moreover, the model without conditional vision prompts experiences a clear performance decline in seen classes, compared to the one without conditional language prompts. This is primarily because the image encoder of the foundational model is initially tailored for comprehending image-level, first-order semantics, making it challenging to directly adapt to tasks requiring region-level, second-order relationship understanding. *(2)* In the absence of any prior knowledge incorporated into the image branch, the model's performance noticeably declines, especially on the unseen classes, as shown in lines 3–4 of Table 3. However, when these conditions are combined, our model achieves its optimal performance on both seen and unseen classes, underscoring the complementary nature of these priors in enhancing the generalizability of HOI detection.

Constraints for Language Prompts. The role of the consistency loss is to serve as a regularization term, allowing the text prompt P_L to learn contextual information through learnable context U_L while preventing excessive deviation from the CLIP text feature space, avoiding a decrease in generalization performance. We conduct experiments using various weights for the consistency loss, as presented in the Table 4. We observe that: *(1)* When changing the weight λ_{cc}, the changes in model performance are mainly shown in the unseen categories. This indicates that the regularization loss primarily affects the model's generalization ability. *(2)* When λ_{cc} is set to 0, the lack of constraint in the text prompt might cause the textual features to deviate from the CLIP feature space, and decrease the performance on unseen categories. *(3)* As λ_{cc} is increased to 1.0, the performance on unseen categories improves, demonstrating an enhancement in model generalization. However, further increasing λ_{cc} could potentially result in U_L becoming useless, constraining the model's capacity and leading to a decrease in the final performance.

4.5 Qualitative Results

As shown in Fig. 3, we present several qualitative results of successful HOI detections. The visualized HOIs contain unseen verbs, *e.g.*, the verbs "wear" and "swing" which don't appear in the training set in the unseen verb setting. Our model successfully detects a human-wearing-tie triplet and a human-swing-baseball-bat triplet as shown in Fig. 3a and Fig. 3c, which shows the powerful generalizability of our detector.

(a) wearing a tie   (b) blocking a frisbee   (c) swing a baseball bat   (d) ride a bike

Fig. 3. Visualization of successfully detected HOIs in the unseen verb setting. Each detected human-object pair is connected by a red line, with the corresponding interaction score overlaid above the human box. All the images contain unseen HOIs made up of unseen verbs and seen objects. (Color figure online)

5 Conclusion

We propose CMMP, a novel technique adapting large foundation models for the challenging task of zero-shot HOI detection via conditional multi-modal prompts. Our model separates zero-shot HOI detection into two subtasks: extracting spatial-aware visual features and interaction classification, and dealing with them using decoupled multi-modal prompts to break error-propagation in-between. By carefully designing prompts, we harness the inherent capabilities of large foundation models to precisely discern fine-grained HOIs, thereby enhancing CMMP's generalization ability. Experimental results across five zero-shot settings show that CMMP outperforms all previous methods by a large margin, establishing a new state-of-the-art for zero-shot HOI detection.

Acknowledgements. This work was supported by grants from the National Natural Science Foundation of China (62372014, 61925201, 62132001, U22B2048).

References

1. Bansal, A., Rambhatla, S.S., Shrivastava, A., Chellappa, R.: Detecting human-object interactions via functional generalization. In: Proceedings of the AAAI Conference on Artificial Intelligence, vol. 34, pp. 10460–10469 (2020)
2. Carion, N., Massa, F., Synnaeve, G., Usunier, N., Kirillov, A., Zagoruyko, S.: End-to-end object detection with transformers. In: Vedaldi, A., Bischof, H., Brox, T., Frahm, J.-M. (eds.) ECCV 2020. LNCS, vol. 12346, pp. 213–229. Springer, Cham (2020). https://doi.org/10.1007/978-3-030-58452-8_13
3. Chao, Y.W., Liu, Y., Liu, X., Zeng, H., Deng, J.: Learning to detect human-object interactions. In: 2018 IEEE Winter Conference on Applications of Computer Vision (WACV), pp. 381–389. IEEE (2018)
4. Chen, M., Liao, Y., Liu, S., Chen, Z., Wang, F., Qian, C.: Reformulating hoi detection as adaptive set prediction. In: Proceedings of the IEEE/CVF Conference on Computer Vision and Pattern Recognition, pp. 9004–9013 (2021)
5. Du, Y., Wei, F., Zhang, Z., Shi, M., Gao, Y., Li, G.: Learning to prompt for open-vocabulary object detection with vision-language model. In: Proceedings of the IEEE/CVF Conference on Computer Vision and Pattern Recognition (CVPR), pp. 14084–14093 (2022)
6. Feng, C., et al.: PromptDet: towards open-vocabulary detection using uncurated images. In: Proceedings of the European Conference on Computer Vision (2022)
7. Gao, P., et al.: CLIP-adapter: better vision-language models with feature adapters. arXiv preprint arXiv:2110.04544 (2021)
8. Gupta, S., Malik, J.: Visual semantic role labeling. arXiv preprint arXiv:1505.04474 (2015)
9. He, K., Gkioxari, G., Dollár, P., Girshick, R.: Mask R-CNN. In: Proceedings of the IEEE International Conference on Computer Vision, pp. 2961–2969 (2017)
10. Hou, Z., Peng, X., Qiao, Yu., Tao, D.: Visual compositional learning for human-object interaction detection. In: Vedaldi, A., Bischof, H., Brox, T., Frahm, J.-M. (eds.) ECCV 2020. LNCS, vol. 12360, pp. 584–600. Springer, Cham (2020). https://doi.org/10.1007/978-3-030-58555-6_35
11. Hou, Z., Yu, B., Qiao, Y., Peng, X., Tao, D.: Affordance transfer learning for human-object interaction detection. In: Proceedings of the IEEE/CVF Conference on Computer Vision and Pattern Recognition, pp. 495–504 (2021)
12. Hou, Z., Yu, B., Qiao, Y., Peng, X., Tao, D.: Detecting human-object interaction via fabricated compositional learning. In: Proceedings of the IEEE/CVF Conference on Computer Vision and Pattern Recognition, pp. 14646–14655 (2021)

13. khattak, M.U., Rasheed, H., Maaz, M., Khan, S., Khan, F.S.: MaPLe: multi-modal prompt learning. In: The IEEE/CVF Conference on Computer Vision and Pattern Recognition (2023)
14. Kim, B., Lee, J., Kang, J., Kim, E.S., Kim, H.J.: HOTR: end-to-end human-object interaction detection with transformers. In: Proceedings of the IEEE/CVF Conference on Computer Vision and Pattern Recognition, pp. 74–83 (2021)
15. Kim, S., Jung, D., Cho, M.: Relational context learning for human-object interaction detection. In: Proceedings of the IEEE/CVF Conference on Computer Vision and Pattern Recognition, pp. 2925–2934 (2023)
16. Kuznetsova, A., et al.: The open images dataset V4. Int. J. Comput. Vision **128**(7), 1956–1981 (2020)
17. Lei, T., Caba, F., Chen, Q., Jin, H., Peng, Y., Liu, Y.: Efficient adaptive human-object interaction detection with concept-guided memory. In: Proceedings of the IEEE/CVF International Conference on Computer Vision, pp. 6480–6490 (2023)
18. Lei, T., Yin, S., Liu, Y.: Exploring the potential of large foundation models for open-vocabulary hoi detection. In: Proceedings of the IEEE/CVF Conference on Computer Vision and Pattern Recognition, pp. 16657–16667 (2024)
19. Li, Y.L., et al.: Detailed 2D-3D joint representation for human-object interaction. In: Proceedings of the IEEE/CVF Conference on Computer Vision and Pattern Recognition, pp. 10166–10175 (2020)
20. Li, Y.L., Liu, X., Wu, X., Li, Y., Lu, C.: HOI analysis: integrating and decomposing human-object interaction. Adv. Neural. Inf. Process. Syst. **33**, 5011–5022 (2020)
21. Li, Y.L., et al.: Transferable interactiveness knowledge for human-object interaction detection. In: Proceedings of the IEEE/CVF Conference on Computer Vision and Pattern Recognition, pp. 3585–3594 (2019)
22. Liao, Y., Liu, S., Wang, F., Chen, Y., Qian, C., Feng, J.: PPDM: parallel point detection and matching for real-time human-object interaction detection. In: Proceedings of the IEEE/CVF Conference on Computer Vision and Pattern Recognition, pp. 482–490 (2020)
23. Liao, Y., Zhang, A., Lu, M., Wang, Y., Li, X., Liu, S.: GEN-VLKT: simplify association and enhance interaction understanding for HOI detection. In: Proceedings of the IEEE/CVF Conference on Computer Vision and Pattern Recognition, pp. 20123–20132 (2022)
24. Lin, T.Y., Goyal, P., Girshick, R., He, K., Dollár, P.: Focal loss for dense object detection. In: Proceedings of the IEEE International Conference on Computer Vision, pp. 2980–2988 (2017)
25. Liu, X., Li, Y.L., Wu, X., Tai, Y.W., Lu, C., Tang, C.K.: Interactiveness field in human-object interactions. In: Proceedings of the IEEE/CVF Conference on Computer Vision and Pattern Recognition, pp. 20113–20122 (2022)
26. Liu, Y., Chen, Q., Zisserman, A.: Amplifying key cues for human-object-interaction detection. In: Vedaldi, A., Bischof, H., Brox, T., Frahm, J.-M. (eds) ECCV 2020. LNCS, vol. 12359, pp. 248–265. Springer, Cham (2020). https://doi.org/10.1007/978-3-030-58568-6_15
27. Liu, Y., Zhang, J., Chen, Q., Peng, Y.: Confidence-aware pseudo-label learning for weakly supervised visual grounding. In: Proceedings of the IEEE/CVF International Conference on Computer Vision, pp. 2828–2838 (2023)
28. Liu, Y., Yuan, J., Chen, C.W.: ConsNet: learning consistency graph for zero-shot human-object interaction detection. In: Proceedings of the 28th ACM International Conference on Multimedia, pp. 4235–4243 (2020)

29. Luo, D., Huang, J., Gong, S., Jin, H., Liu, Y.: Zero-shot video moment retrieval from frozen vision-language models. In: Proceedings of the IEEE/CVF Winter Conference on Applications of Computer Vision, pp. 5464–5473 (2024)
30. Mao, Y., Deng, J., Zhou, W., Li, L., Fang, Y., Li, H.: CLIP4HOI: towards adapting clip for practical zero-shot hoi detection. Adv. Neural. Inf. Process. Syst. **36**, 45895–45906 (2023)
31. Ning, S., Qiu, L., Liu, Y., He, X.: HOICLIP: efficient knowledge transfer for hoi detection with vision-language models. In: Proceedings of the IEEE/CVF Conference on Computer Vision and Pattern Recognition, pp. 23507–23517 (2023)
32. Park, J., Park, J.W., Lee, J.S.: ViPLO: vision transformer based pose-conditioned self-loop graph for human-object interaction detection. In: Proceedings of the IEEE/CVF Conference on Computer Vision and Pattern Recognition, pp. 17152–17162 (2023)
33. Radford, A., et al.: Learning transferable visual models from natural language supervision. In: International Conference on Machine Learning, pp. 8748–8763. PMLR (2021)
34. Ren, S., He, K., Girshick, R., Sun, J.: Faster R-CNN: towards real-time object detection with region proposal networks. Adv. Neural Inf. Process. Syst. **28** (2015)
35. Tamura, M., Ohashi, H., Yoshinaga, T.: QPIC: query-based pairwise human-object interaction detection with image-wide contextual information. In: Proceedings of the IEEE/CVF Conference on Computer Vision and Pattern Recognition, pp. 10410–10419 (2021)
36. Tian, Y., Fu, Y., Zhang, J.: Transformer-based under-sampled single-pixel imaging. Chin. J. Electron. **32**(5), 1151–1159 (2023)
37. Ulutan, O., Iftekhar, A., Manjunath, B.S.: VSGNet: spatial attention network for detecting human object interactions using graph convolutions. In: Proceedings of the IEEE/CVF Conference on Computer Vision and Pattern Recognition, pp. 13617–13626 (2020)
38. Vaswani, A., et al.: Attention is all you need. Adv. Neural Inf. Process. Syst. **30** (2017)
39. Wang, G., Li, Z., Chen, Q., Liu, Y.: OED: towards one-stage end-to-end dynamic scene graph generation. In: Proceedings of the IEEE/CVF Conference on Computer Vision and Pattern Recognition, pp. 27938–27947 (2024)
40. Wang, S., Duan, Y., Ding, H., Tan, Y.P., Yap, K.H., Yuan, J.: Learning transferable human-object interaction detector with natural language supervision. In: Proceedings of the IEEE/CVF Conference on Computer Vision and Pattern Recognition, pp. 939–948 (2022)
41. Wang, T., Yang, T., Danelljan, M., Khan, F.S., Zhang, X., Sun, J.: Learning human-object interaction detection using interaction points. In: Proceedings of the IEEE/CVF Conference on Computer Vision and Pattern Recognition, pp. 4116–4125 (2020)
42. Wu, M., et al.: End-to-end zero-shot hoi detection via vision and language knowledge distillation. arXiv preprint arXiv:2204.03541 (2022)
43. Wu, X., Li, Y.L., Liu, X., Zhang, J., Wu, Y., Lu, C.: Mining cross-person cues for body-part interactiveness learning in hoi detection. In: Avidan, S., Brostow, G., Cissé, M., Farinella, G.M., Hassner, T. (eds.) ECCV 2022. LNCS, vol. 13664, pp. 121–136. Springer, Cham (2022). https://doi.org/10.1007/978-3-031-19772-7_8
44. Xie, C., Zeng, F., Hu, Y., Liang, S., Wei, Y.: Category query learning for human-object interaction classification. In: Proceedings of the IEEE/CVF Conference on Computer Vision and Pattern Recognition, pp. 15275–15284 (2023)

45. Xu, Z., Chen, Q., Peng, Y., Liu, Y.: Semantic-aware human object interaction image generation. In: Forty-first International Conference on Machine Learning
46. Yang, D., Liu, Y.: Active object detection with knowledge aggregation and distillation from large models. In: Proceedings of the IEEE/CVF Conference on Computer Vision and Pattern Recognition, pp. 16624–16633 (2024)
47. Yuan, H., et al.: RLIP: relational language-image pre-training for human-object interaction detection. In: Advances in Neural Information Processing Systems (NeurIPS) (2022)
48. Zhang, A., et al.: Mining the benefits of two-stage and one-stage hoi detection. Adv. Neural. Inf. Process. Syst. **34**, 17209–17220 (2021)
49. Zhang, F.Z., Campbell, D., Gould, S.: Spatially conditioned graphs for detecting human-object interactions. In: Proceedings of the IEEE/CVF International Conference on Computer Vision, pp. 13319–13327 (2021)
50. Zhang, F.Z., Campbell, D., Gould, S.: Efficient two-stage detection of human-object interactions with a novel unary-pairwise transformer. In: Proceedings of the IEEE/CVF Conference on Computer Vision and Pattern Recognition, pp. 20104–20112 (2022)
51. Zhang, F.Z., Yuan, Y., Campbell, D., Zhong, Z., Gould, S.: Exploring predicate visual context in detecting human-object interactions. In: Proceedings of the IEEE/CVF International Conference on Computer Vision (ICCV), pp. 10411–10421 (2023)
52. Zhang, R., et al.: Tip-adapter: training-free CLIP-adapter for better vision-language modeling. arXiv preprint arXiv:2111.03930 (2021)
53. Zhang, T., Fu, Y., Zhang, J.: Deep guided attention network for joint denoising and demosaicing in real image. Chin. J. Electron. **33**(1), 303–312 (2024)
54. Zhang, Y., Pan, Y., Yao, T., Huang, R., Mei, T., Chen, C.W.: Exploring structure-aware transformer over interaction proposals for human-object interaction detection. In: Proceedings of the IEEE/CVF Conference on Computer Vision and Pattern Recognition, pp. 19548–19557 (2022)
55. Zheng, M., Cai, X., Chen, Q., Peng, Y., Liu, Y.: Zero-shot video temporal grounding using large-scale pre-trained models. In: Proceedings of the European Conference on Computer Vision (2024)
56. Zheng, M., Gong, S., Jin, H., Peng, Y., Liu, Y.: Generating structured pseudo labels for noise-resistant zero-shot video sentence localization. In: Proceedings of the 61st Annual Meeting of the Association for Computational Linguistics (Volume 1: Long Papers), pp. 14197–14209 (2023)
57. Zheng, S., Xu, B., Jin, Q.: Open-category human-object interaction pre-training via language modeling framework. In: Proceedings of the IEEE/CVF Conference on Computer Vision and Pattern Recognition, pp. 19392–19402 (2023)
58. Zhong, X., Ding, C., Li, Z., Huang, S.: Towards hard-positive query mining for DETR-based human-object interaction detection. In: Avidan, S., Brostow, G., Cissé, M., Farinella, G.M., Hassner, T. (eds.) ECCV 2022, Part XXVII. LNCS, vol. 13687, pp. 444–460. Springer, Cham (2022). https://doi.org/10.1007/978-3-031-19812-0_26
59. Zhou, K., Yang, J., Loy, C.C., Liu, Z.: Conditional prompt learning for vision-language models. In: Proceedings of the IEEE/CVF Conference on Computer Vision and Pattern Recognition, pp. 16816–16825 (2022)
60. Zhou, K., Yang, J., Loy, C.C., Liu, Z.: Learning to prompt for vision-language models. Int. J. Comput. Vision **130**(9), 2337–2348 (2022)

61. Zhou, P., Chi, M.: Relation parsing neural network for human-object interaction detection. In: Proceedings of the IEEE/CVF International Conference on Computer Vision, pp. 843–851 (2019)
62. Zou, C., et al.: End-to-end human object interaction detection with hoi transformer. In: Proceedings of the IEEE/CVF conference on computer vision and pattern recognition, pp. 11825–11834 (2021)

Training-Free Video Temporal Grounding Using Large-Scale Pre-trained Models

Minghang Zheng[1], Xinhao Cai[1], Qingchao Chen[2], Yuxin Peng[1], and Yang Liu[1,3(✉)]

[1] Wangxuan Institute of Computer Technology, Peking University, Beijing, China
{minghang,qingchao.chen,pengyuxin,yangliu}@pku.edu.cn
[2] National Institute of Health Data Science, Peking University, Beijing, China
xinhao.cai@stu.pku.edu.cn
[3] State Key Laboratory of General Artificial Intelligence, Peking University, Beijing, China

Abstract. Video temporal grounding aims to identify video segments within untrimmed videos that are most relevant to a given natural language query. Existing video temporal localization models rely on specific datasets for training, with high data collection costs, but exhibit poor generalization capability under the across-dataset and out-of-distribution (OOD) settings. In this paper, we propose a **T**raining-**F**ree **V**ideo **T**emporal **G**rounding (TFVTG) approach that leverages the ability of pre-trained large models. A naive baseline is to enumerate proposals in the video and use the pre-trained visual language models (VLMs) to select the best proposal according to the vision-language alignment. However, most existing VLMs are trained on image-text pairs or trimmed video clip-text pairs, making it struggle to (1) grasp the relationship and distinguish the temporal boundaries of multiple events within the same video; (2) comprehend and be sensitive to the dynamic transition of events (the transition from one event to another) in the video. To address these issues, firstly, we propose leveraging large language models (LLMs) to analyze multiple sub-events contained in the query text and analyze the temporal order and relationships between these events. Secondly, we split a sub-event into dynamic transition and static status parts and propose the dynamic and static scoring functions using VLMs to better evaluate the relevance between the event and the description. Finally, for each sub-event description provided by LLMs, we use VLMs to locate the top-k proposals that are most relevant to the description and leverage the order and relationships between sub-events provided by LLMs to filter and integrate these proposals. Our method achieves the best performance on zero-shot video temporal grounding on Charades-STA and ActivityNet Captions datasets without any training and demonstrates better generalization capabilities in cross-dataset and OOD settings. Code is available at https://github.com/minghangz/TFVTG.

Keywords: Video Temporal Grounding · Zero-shot Learning · Large Language Model · Vision Language Model

---

Supplementary Information The online version contains supplementary material available at https://doi.org/10.1007/978-3-031-73007-8_2.

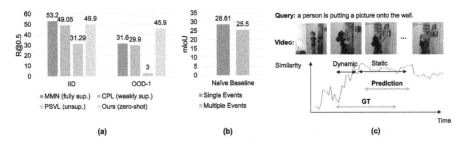

Fig. 1. (a) Evaluation results of existing methods and our method under the IID and OOD setting on the Charades-STA dataset. (b) Evaluation results of the naive baseline on the ActivityNet Datasets when the query describes single or multiple events. (c) The query-frame similarity obtained from the BLIP-2 Q-Former. The naive baseline based on BLIP-2 tends to predict the static parts of the video and ignores the dynamic transitions.

1 Introduction

Video temporal grounding [7] aims to localize the most semantically relevant segment in untrimmed videos according to free-form natural language queries. It has broad application potential in video surveillance [4], video summarization [29], and other fields. Existing video temporal grounding methods [7,11,14,27,40,56–58,62,63] mainly train models on manually annotated data to understand the alignment between video segments and natural language queries. These methods have achieved remarkable improvements recently on specific datasets, such as ActivityNet Captions [16] and Charades-STA [7]. However, collecting high-quality video temporal grounding datasets is time-consuming and labor-intensive, which hinders the large-scale real-world application of these methods. In addition, they heavily rely on the distribution of the training dataset and the performance degrades significantly in out-of-distribution (OOD) or cross-dataset settings according to previous research [3, 19, 50, 53]. As shown in Fig. 1(a), we report the OOD performance[1] on the Charades-STA dataset of recent fully supervised method MMN [44], weakly supervised method CPL [61], and unsupervised method PSVL [33]. As we can see, their performance all has a significant drop. This is because these models are trained on small-scale video temporal grounding datasets that exhibit certain biases, leading to poor generalization of the models. Therefore, we aim to design a *training-free video temporal grounding* approach that *does not rely on specific video temporal grounding datasets*, so that it can be better generalized to real application scenarios.

Recently, large-scale pre-trained models [1,21,26,34,36,42,43,55] have shown strong generalization ability in zero-shot retrieval [28,52], VQA [9,31], detection [17,18,47] et al. This inspires us to transfer the powerful generalization ability of them to video temporal grounding tasks. A naive baseline is to enumerate proposals in the video and use the pre-trained visual language models

---

[1] We follow [50] and plot the performance under the OOD-1 setting in the figure.

(VLMs) [21,34,42,43] to assess the alignment between these proposals and the query and select the proposal with the highest score. However, this approach has the following drawbacks. *Firstly, it can be challenging for VLMs to understand the temporal boundaries of multiple events in untrimmed video.* Most of the existing image-text or video-text pre-trained VLMs are trained to align single images (e.g. CLIP [34], BLIP [21]) or trimmed video clips (e.g. InterVideo [43]) with texts. However, in the video temporal grounding, the model needs to understand multiple sequential events and their temporal relationships, such as 'She sprays it with a spray bottle and continues brushing her hair'. The skill is seldom emphasized in the image or trimmed video pretraining, and as shown in Fig. 1(b), the naive baseline has a poorer performance when the query describes multiple events. *Secondly, we find that directly selecting the most relevant proposal using VLMs often leads to overlooking the dynamic transitions at the beginning of an event.* As shown in Fig. 1(c), we show the similarity between video frames and the query text using the BLIP-2 Q-Former [20]. It can be found that in the naive baseline's prediction, the beginning of the event where a person gradually picks up the picture and approaches the wall is ignored. We think this is because these VLMs are trained directly by visual-textual contrastive learning. In such a training paradigm, the model's primary objective is to associate the most discriminative visual cues with their corresponding text descriptions, rather than emphasizing the need to focus on dynamic transitions and ensure the completeness of localized event boundaries.

To address the above problems, we propose to combine the ability of large language models (LLMs) [1,26,36] to understand and reason about queries and the ability of visual language models (VLMs) to align vision and text in a training-free manner. Specifically, for the challenge of understanding videos and queries that contain multiple events, we propose to prompt the LLMs to analyze the multiple events that may be contained in the query and give a text description of each single event as sub-queries, the order in which each event occurs, and the relationship between events (sequential or simultaneous). For the sub-query of each single event, we can use the VLMs to locate its possible occurrence in the video. To better use the VLMs to locate the video clip corresponding to the single event query and solve the problem of ignoring the dynamic transitions in the video while enhancing localization completeness, we propose to consider both dynamic transition and static status after transition explicitly when measuring the similarity between the proposal and the text query. For example in Fig. 1(c), for the query 'a person put a picture on the wall', the dynamic transition part is where the person gradually raises the picture and approaches the wall, while the static status part is where the person has already hung the picture on the wall and is looking at the camera. A good proposal should begin with a dynamic segment, exhibiting a notable increase in video-text similarity and followed by a static post-status segment characterized by a high average video-text similarity within the static segment, while displaying a low average similarity outside of it. To evaluate each proposal, we compute a matching score by summing the dynamic score, which measures the rate of similarity change within its dynamic segment, and the static score, which evaluates the comparative similarity within and outside its static post-status segment. Then, we select the top proposals with the highest matching scores as the localization results of the single event description. Finally, the localization results of each single event are combined with the LLM's judgment of the relationship and order of events to filter and integrate the final predictions.

Our contributions are summarized as follows. (1) We propose a training-free pipeline for video temporal grounding (TFVTG) using pre-trained large language models (LLMs) and vison-language models (VLMs). We use the LLMs to split the original query into sub-events and reason the temporal order and relationship between them, use VLMs to localize each sub-event, and filter and integrate the localization results based on the temporal order and relationships. (2) To help VLM better understand the dynamic transitions in the video, we divide the events into dynamic and static parts and model them separately. For the dynamic part, measures the rate of similarity change, and for the static part, we measure the comparative similarity within and outside. (3) Our method achieves the best performance on zero-shot video temporal grounding on both the Charades-STA [7] and ActivityNet Captions [16] datasets and has a greater advantage in cross-dataset and OOD settings.

2 Related Work

2.1 Video Temporal Localization

Fully supervised video temporal grounding methods [5,7,11,14,25,27,32,44,50, 57] typically train models using manually annotated queries and start and end timestamps. For example, MMN [44] trains models to distinguish matched and unmatched video-sentence pairs collected from within videos and across videos; VTimeLLM [10] is the first to fine-tune large language models using video temporal grounding data. However, fully supervised methods are often influenced by annotation bias, leading to poor generalization. Weakly supervised learning [13,60,61] and unsupervised learning [23,33,51,59] are often used to mitigate the high dependence on human annotation. For instance, PSVL [33] and SPL [59] train models by generating pseudo-labels within videos. Nevertheless, even without using manually annotated data, biases present in the training videos can still affect the generalization of these methods. Therefore, in this work, we focus on training-free video temporal grounding, aiming for stronger generalization and applicability in real-world scenarios. Recently, Luo et al. [28] and VTG-GPT [46] made the first attempt to training-free video temporal grounding. Luo et al. measure the similarity between video proposals and the text query using VLM while VTG-GPT measures the similarity between the video frame captions and the text query using LLM. They then make a prediction based on the similarity. However, they overlooked the order and relationship between the possible multiple events within the query, as well as the issue of VLM's insensitivity to dynamic transitions in the video due to their training scheme. In contrast, we propose to infer multiple sub-events contained in the query and their order and relationship through LLM, model dynamic changes in the video explicitly to assist VLM in better localizing each sub-event, and filter and integrate the localization results based on the order and relationship between events inferred by LLM.

2.2 Bias in Video Temporal Localization Datasets

The mainstream video temporal grounding datasets [7,16] suffer from certain biases, which affect the generalization ability of models trained on these datasets. Many studies [3,19,50,53] have investigated biases in video temporal grounding datasets. For example, [50,53] study biases in the location of target segments. When there are significant changes in the distribution of locations, existing methods exhibit noticeable performance degradation. [19] explores biases in query texts and proposes the ActivityNet-CG and Charades-CG datasets to evaluate the generalization ability of existing methods in the novel combinations of phrases and novel words. [3] studies the cross-dataset generalization ability of existing models and finds that models trained on specific datasets perform poorly when testing on another dataset. Some works [3,19,49,50] have focused on improving model generalization to specific biases. However, they are difficult to generalize to address various types of biases in video temporal grounding. Therefore, we attempt to study training-free video temporal grounding to overcome reliance on specific datasets and achieve better generalization across various scenarios.

2.3 Large-Scale Pretrained Models in Video Understanding

In recent years, with the development of large-scale corpus [2,35], model architecture [37,38], and computational resources, large language models (LLMs) [1,26,36] have achieved rapid development, demonstrating powerful capabilities in text generation, chat, problem-solving, reasoning, et al. On the other hand, visual language models (VLMs) [21,34,43,48] have also shown strong generalization abilities in tasks such as multimodal alignment and retrieval. Some existing methods attempt to combine VLMs with LLMs in video tasks. For example, ChatVideo [41] utilizes pre-trained models such as trajectory detection, video captioning, and speech recognition to convert videos into text, which serves as input to LLMs for video understanding. Video-LLaMA [55], and VTimeLLM [10] project video features into the token embedding space of LLMs through fine-tuning to enable LLMs to understand videos. However, these methods perform poorly on video temporal grounding tasks as shown in Table 1. Even VTimeLLM, which specifically fine-tunes LLM using video temporal grounding data, still exhibits a significant performance gap compared to the traditional video temporal grounding method. We think that this may be because fine-tuning VLMs and LLMs on video temporal grounding datasets degrades their generalization, and encoding videos solely as input token sequences for LLMs makes it difficult for LLMs to accurately understand the time boundaries of various events. Therefore, we propose a training-free pipeline to combine the capabilities of LLMs and VLMs for video temporal grounding tasks. We leverage the strengths of both:

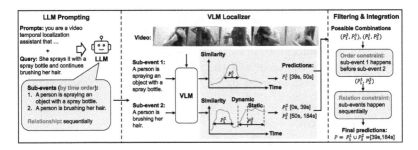

Fig. 2. The pipeline of the proposed method. Firstly, the LLM prompting leverages the large language model (LLM) to analyze sub-events contained in the query and reason the order and the temporal relationship of these sub-events. Then, the VLM localizer uses the vision language models to localize the sub-event in the video. The VLM localizer calculates the similarity between the video frames and the sub-event descriptions, enumerates event proposals in the video, and explicitly considers both dynamic transition and static status post-transition when measuring the similarity between the proposal and the text query, thus selecting proposals as the localization results. Finally, we filter and integrate the results of the VLM localizer based on the order and relationship of sub-events inferred by LLM to make the final prediction.

using LLMs to reason the sub-events contained in queries, their occurrence order, and relationships, using VLMs to measure the vision-text similarity and localize each sub-event, and integrating the final predictions based on the inferred sub-event order and relationships from LLM.

3 Method

The overview of our model design is illustrated in Fig. 2. Our method consists of three parts: Firstly, since the query may describe multiple events that happened sequentially or simultaneously in the video, we propose the LLM prompting to leverage LLM for analyzing sub-events contained in the query and reason the order and the temporal relationship of these sub-events. Then, we propose the VLM localizer that uses the VLM to localize the sub-event in the video. The VLM localizer first calculates the similarity between the video frames and the sub-event descriptions. To solve the problem that VLMs are not sensitive enough to the dynamic transitions in the video, we propose explicitly considering both dynamic transitions and static states following these transitions when evaluating the similarity between the proposal and the text query. A good proposal should commence with a dynamic segment showing a significant rise in video-text similarity, followed by a static post-transition segment characterized by a high average video-text similarity within the static segment, while maintaining a low average similarity outside of it. To assess each proposal, we enumerate the plausible dynamic and static segments of every event proposal and calculate a matching score by aggregating the dynamic and static scores. The dynamic score measures the rate of similarity change within its dynamic segment, and the static score evaluates the comparative similarity within and outside its static post-status segment. The VLM localizer returns the top-k proposals with the highest sum of dynamic and static scores as the localization results. Finally, we filter and integrate the results of the VLM localizer based on the order and relationship of sub-events inferred by LLM to make the final prediction.

3.1 LLM Prompting

The large language model (LLM) demonstrates powerful capabilities in instruction following, context understanding, and reasoning. Therefore, we propose to prompt LLM to analyze query texts, identifying multiple potential events therein, and analyzing the order and relationship of these events. Specifically, we request the large language model to provide the following information:

– Reasoning: Analyze the user's query and infer the sub-events it may contain.
– Order of sub-events: Provide each sub-events in chronological order.
– Relationships between sub-events: Consider three types of relationships: Single event, simultaneously (i.e. the sub-events occur simultaneously), and sequentially (i.e. the sub-events occur sequentially).
– Textual descriptions: Generate text descriptions for each sub-event.

The complete prompt will be provided in the supplementary materials.

3.2 Grounding with Vison Language Model

Inspired by the powerful multimodal alignment capabilities of VLM, we propose to use VLM as a localizer to locate each sub-event in the video.

Specifically, we choose BLIP-2 Q-Former [20] following [22,55] as the VLM localizer. For a sub-event description c and each video frame $v_1, ..., v_N$, we use VLM to extract their text features $f^c \in \mathbb{R}^D$ and vision features $F^v = [f_1^v, ..., f_N^v] \in \mathbb{R}^{N \times D}$, where N is the number of video frames and D is the feature dimension. As the text and vision space are well aligned, we can directly use the cosine similarity of the text and vision features to measure the relevance of the sub-event description and the video frame:

$$S = \frac{f^c F^{v\intercal}}{\|f^c\|\|F^v\|} \in \mathbb{R}^N \quad (1)$$

We find that directly enumerating proposals within the video and selecting the proposal with the highest average similarity often leads the model to only predict static status in the event while ignoring the dynamic transition at the beginning of an event. In 56.1% of the testing data in the Charades-STA dataset, the naive baseline predicts a start timestamp that is after the ground truth start timestamp. For example, for the query "a person sits down" the model tends to predict segments where the person is already seated on the chair rather than the process of the person gradually from standing up to sitting down. A good event should commence with a dynamic segment showing a significant rise in video-text similarity, followed by a static post-transition segment characterized by a high average video-text similarity within the static segment, while maintaining a low average similarity outside of it. To assess each proposal, we enumerate the plausible dynamic and static segments of every event proposal and calculate a matching score by aggregating the dynamic and static scores. The dynamic score measures the rate of similarity change within its dynamic segment, and the static score evaluates the comparative similarity within and outside its static post-status segment:

Dynamic Scoring. Considering the continuity of the video, in the dynamic part of the target segment, the video content transitions from another event to the target event described by the query, so the relevance between the video frames and the query should quickly increase. Our dynamic scoring aims to provide quantitative scores for segments where the relevance between the video and the query increases rapidly. Firstly, to eliminate the influence of video jitter, we apply a Gaussian filter to the similarity S: $\hat{S} = G(S)$, where $G(\cdot)$ is the Gaussian filter. We expect that the dynamic section contains as many parts as possible where the video-text correlation significantly increases. Therefore, we calculate the difference in similarity $\hat{S}$: $D_i = \hat{S}_i - \hat{S}_{i-1}$. Given a proposal starts from the i-th frame and ends with the k-th frame, we require that all the differential values within the proposal exceed a certain threshold δ. If this condition is met, we sum up the differential values in the proposal to obtain the dynamic score for that proposal:

$$S_{i,k}^{dynamic} = \begin{cases} \sum_{l=i}^{k} D_l, & D_l > \delta, \forall l \in [i,k] \\ 0, & otherwise \end{cases} \quad (2)$$

Static Scoring. Inspired by SPL [59], in the static part, we require that the most relevant event for a given query should satisfy the requirement that videos within the event have high relevance to the query and videos outside the event have low relevance to the query. Therefore, given a proposal starts from the k-th frame and ends with the j-th frame, we calculate the average similarity within the proposal and the average similarity outside the proposal, and use the difference between them as the static scores:

$$S_{k,j}^{static} = \frac{1}{j-k} \sum_{l \in [k,j]} S_l - \frac{1}{N-(j-k)} \sum_{l \notin [k,j]} S_l \quad (3)$$

where N is the number of frames.

To localize the target video clip using dynamic and static scoring, we enumerate event proposals in the video. For each proposal (i, j), due to the varying lengths of transition segments in different events, we enumerate all feasible timestamps k where the transition just finished and divide the proposal into two parts: the dynamic part (i,k) and static part (k,j). We then calculate the sum of the dynamic score and static score, and select the maximum value as the score for this proposal:

$$S_{i,j}^{final} = \max_{k=i}^{j}(S_{i,k}^{dynamic} + S_{k,j}^{static}) \qquad (4)$$

Finally, we select the top-k proposal with the highest final score S^{final} as the localization results of the VLM localizer.

3.3 Prediction Filtering and Integration

The VLM locator returns the top-k predictions for each sub-event description. To obtain the final prediction, we propose to filter and integrate these predictions from the VLP localizer based on the order of occurrence of sub-events and their relationships derived from LLM. Firstly, for each sub-event, we enumerate one of its top-k predictions, and there are total k^m combinations, where m is the number of sub-events. For a combination $P_1, P_2, ..., P_m$, we filter these combinations by the order constraint: if LLM determines that P_i from s^i to e^i should occur before P_j from s^j to e^j, but the start time of P_i is later than the end time of P_j (i.e. $s^i > e^j$), then this combination will be discarded. After filtering, we can calculate the sum of scores S^{final} returned by the VLM localizer for each combination, selecting the combination with the highest score, and merging these proposals based on the relation constraint: If LLM determines that these sub-events should occur simultaneously, we take the intersection of these proposals as the final prediction; otherwise, we take the union of these proposals as the final prediction:

$$P^{final} = \begin{cases} P_1 \cap P_2 \cap ... \cap P_m, & \text{relation is 'simultaneously'} \\ P_1 \cup P_2 \cup ... \cup P_m, & \text{relation is 'sequentially'} \end{cases} \qquad (5)$$

4 Experiments

To comprehensively validate the effectiveness of our method, we conduct experiments on the Charades-STA [7] and ActivityNet Captions [16] datasets. We compare our method with existing methods under the IID, OOD, and cross-dataset settings respectively to demonstrate the generalization capability of our method. We also conduct ablation studies to evaluate the effectiveness of each module.

4.1 Experimental Setup

Dataset: We conduct experiment on two benchmark ActivityNet Captions [16] and Charades-STA [7]. *ActivityNet Captions* contains 20K videos, which is originally collected for video captioning. There are 37,417/17,505/17,031 video-query pairs in the train /val_1/val_2 split. We follow previous works [28,44] and report the performance on the val_2 split. *Charades-STA* is built based on the Charades dataset. There are 12,408/3,720 video-query pairs in the train/test split. We report the performance on the test split.

Evaluation Metrics: We follow the evaluation metrics 'R@m' and 'mIoU' in the previous work [28,44], where m is the predefined temporal Intersection over Union (IoU) threshold. The metric 'R@m' represents the percentage of predictions that have the IoU larger than m. The metric 'mIoU' represents the average IoU of all the predictions.

Implementation Details: We follow [22,55] to use the BLIP-2 Q-former [20] as the vision-language model. For the large language model, we use the GPT-4 Turbo API. For the VLM localizer, we downsample the input video to 3 FPS and use VLM to calculate the similarity between video frames and text. The localizer returns $k = 3$ predictions for each sub-event. For the hyperparameter, we set δ to 5×10^{-4} across all the datasets.

4.2 Comparison with the SOTAs

Comparison Under the IID Setting. In Table 1, we compare the performance of our method with existing methods under the IID setting. We use the official splits of Charades-STA and ActivityNet Captions. It can be observed that our method achieves the best performance in the zero-shot setting. For example, on the Charades-STA dataset, our method surpasses the second-ranked VTG-GPT [46] by 6.29% on the R@0.5 metric. Our method also outperforms unsupervised training methods by 9.27% on the R@0.5 metric on the Charades-STA dataset. The methods that utilize both LLM and VLM, such as VideoLLaMA [55] which aligns visual features to the input token space of LLM, have a poor performance on video temporal grounding. Although VTimeLLM [10] and GroundingGPT [24] further finetune LLM using data from ActivityNet-Captions or Charades-STA, the performance remains suboptimal. Our method combines the advantages of LLM in text understanding and reasoning with the advantages of VLM in visual-text alignment, resulting in superior performance.

Comparison Under the OOD Setting. To study the generalization capability of our method, we conducted experiments under OOD settings. We consider three OOD settings: novel location, novel text, and cross-dataset.

For the novel location, we follow DCM [50] by inserting a segment of random-generated video at the beginning of test videos, as shown in Table 2. In Table 3, we also test the performance on the Charades-CD [53] dataset, which alters the distribution of target location by resplitting the training and test data. As we can see, our method outperforms recent fully supervised methods in this setting without training, proving its superior generalization capability.

For the novel text, we follow VISA [19] and test on the Charades-CG [19] dataset as shown in Table 3. 'Novel-composition' indicates the text contains novel combinations of training vocabulary, while 'novel-word' indicates the text contains novel words. Our method achieves the best performance under the novel-

Table 1. Evaluation Results on the Charades-STA and ActivityNet Captions Datasets.

Method	Setting	VLM	LLM	Charades-STA				ActivityNet Captions			
				R@0.3	R@0.5	R@0.7	mIoU	R@0.3	R@0.5	R@0.7	mIoU
2D-TAN [57]	fully	✗	✗	–	39.81	23.25	–	58.75	44.05	27.38	–
EMB [11]				**72.50**	58.33	39.25	**53.09**	**64.13**	44.81	26.07	**45.59**
MGSL-Net [25]				–	63.98	41.03	–	–	51.87	31.42	–
EaTR [14]				–	**68.47**	**44.92**	–	–	**58.18**	**37.64**	–
CRM [12]	weakly	✗	✗	53.66	34.76	16.37	–	55.26	32.19	–	–
CNM [60]				60.39	35.43	15.45	–	55.68	33.33	–	–
CPL [61]				66.40	49.24	22.39	–	55.73	31.37	–	–
Huang et al. [13]				**69.16**	**52.18**	**23.94**	**45.20**	**58.07**	**36.91**	–	**41.02**
Gao et al. [8]	unsup.[a]	✓	✗	46.69	20.14	8.27	–	46.15	26.38	11.64	–
PSVL [33]				46.47	31.29	14.17	31.24	44.74	30.08	14.74	29.62
PZVMR [39]				46.83	33.21	18.51	32.62	45.73	31.26	**17.84**	30.35
Kim et al. [15]				52.95	37.24	19.33	36.05	47.61	**32.59**	15.42	31.85
SPL [59]				**60.73**	**40.70**	**19.62**	**40.47**	**50.24**	27.24	15.03	**35.44**
GroundingGPT [24]	fully[b]	✓	✓	–	29.6	11.9	–	–	–	–	–
VTimeLLM-13B [10]				**55.3**	**34.3**	**14.7**	**34.6**	**44.8**	**29.5**	**14.2**	**31.4**
VideoChat-7B [22]	zero-shot	✓	✓	9.0	3.3	1.3	6.5	8.8	3.7	1.5	7.2
VideoLLaMA-7B [55]		✓	✓	10.4	3.8	0.9	7.1	6.9	2.1	0.8	6.5
VideoChatGPT-7B [30]		✓	✓	20.0	7.7	1.7	13.7	26.4	13.6	6.1	18.9
Luo et al. [28]		✓	✗	56.77	42.93	20.13	37.92	48.28	27.90	11.57	32.37
VTG-GPT [46]		✓	✓	59.48	43.68	**25.94**	39.81	47.13	28.25	12.84	30.49
Ours w/o LLM		✓	✗	65.46	48.01	22.07	43.37	48.84	26.64	13.10	33.61
Ours		✓	✓	**67.04**	**49.97**	24.32	**44.51**	**49.34**	**27.02**	**13.39**	**34.10**

[a] Some of them claim to be under the zero-shot setting. However, they still require video data for training. We follow [28] to classify them as unsupervised methods.
[b] They use the data in the ActivityNet Captions or Charades-STA to finetune LLMs.

word setting. Under the novel-composition setting, although not as competitive as methods specifically designed for this scenario, such as DeCo and VISA, our method still outperforms other fully supervised approaches.

For the cross-dataset setting, we follow Debias-TLL [3] where the models are trained on ActivityNet Captions and tested on the Charades-STA, as shown in Table 4. Notably, almost all fully supervised methods experience significant performance degradation when applied across datasets, while our method remains unaffected as it does not rely on training data distribution. These experiments demonstrate that our method has superior generalization capabilities, making it more suitable for practical application requirements.

4.3 Ablation Studies

To validate the effectiveness of each module, we conduct ablation studies on the Charades-STA dataset.

Effectiviness of Each Component. Table 5 shows the ablation studies on the effectiveness of the proposed modules. When disenabling the VLM localizer, we use the naive baseline. When disenabling Filtering&Integration, we simply take the top-1 predictions of the localizer for each sub-event and consider their union box as the final prediction. (1) From the 2nd row of the table, it can be observed

Table 2. Results under OOD setting on the Charades and ActivityNet Dataset.

Method	Setting	Charades-STA						ActivityNet-Captions					
		OOD-1			OOD-2			OOD-1			OOD-2		
		R@0.5	R@0.7	mIoU	R@0.5	R@0.7	mIoU	R@0.5	R@0.7	mIoU	R@0.5	R@0.7	mIoU
LGI [32]	fully	42.1	18.6	41.2	35.8	13.5	37.1	16.3	6.2	22.2	11.0	3.9	17.3
2D-TAN [57]		27.1	13.1	25.7	21.1	8.8	22.5	16.4	6.6	23.2	11.5	3.9	19.4
MMN [44]		31.6	13.4	33.4	27.0	9.3	30.3	20.3	7.1	26.2	14.1	**5.2**	20.6
VDI [27]		25.9	11.9	26.7	20.8	8.7	22.0	**20.9**	7.1	**27.6**	14.3	**5.2**	**23.7**
DCM [50]		44.4	**19.7**	42.3	38.5	15.4	**39.0**	18.2	**7.9**	24.4	12.9	4.8	20.7
CNM [60]	weakly	9.9	1.7	21.6	6.1	0.5	16.6	**6.1**	**0.4**	21.0	**2.5**	**0.1**	16.8
CPL [61]		29.9	8.5	32.2	24.9	6.3	30.5	4.7	**0.4**	21.1	2.1	**0.2**	17.7
PSVL [33]	unsup.	3.0	0.7	8.2	2.2	0.4	6.8	–	–	–	–	–	–
PZVMR [39]		–	8.6	25.1	–	6.5	28.5	–	4.4	28.3	–	2.6	19.1
Luo et al. [28]	zero-shot	40.3	18.2	38.2	38.9	17.0	37.8	18.4	6.8	21.1	**18.6**	7.4	20.6
Ours		**45.9**	**20.8**	**43.0**	**43.8**	**20.0**	**42.6**	20.4	**11.2**	**31.7**	18.5	**10.0**	**30.3**

Table 3. Results under OOD setting on the Charades-CD and Charades-CG Dataset.

Method	Setting	Charades-CD			Charades-CG					
		test-ood			novel-composition			novel-word		
		R@0.3	R@0.5	R@0.7	R@0.5	R@0.7	mIoU	R@0.5	R@0.7	mIoU
2D-TAN [57]	fully	43.45	30.77	11.75	30.91	12.23	29.75	29.36	13.21	28.47
TSP-PRL [45]		31.93	19.37	6.20	16.30	2.04	13.52	14.83	2.61	14.03
SCDM [54]		**52.38**	**41.60**	**22.22**	27.73	12.25	30.84	–	–	–
VISA [19]		–	–	–	45.41	**22.71**	**42.03**	42.35	20.88	40.18
DeCo [49]		–	–	–	**47.39**	21.06	40.70	–	–	–
WSSL [6]	weakly	35.86	23.67	8.27	3.61	1.21	8.26	2.79	0.73	**7.92**
CPL [61]		–	–	–	39.11	15.60	35.53	45.90	22.88	–
SPL [59]	unsup.	62.96	38.25	15.53	–	–	–	–	–	–
Luo et al. [28]	zero-shot	–	–	–	40.27	16.27	–	45.04	21.44	–
Ours		**65.07**	**49.24**	**23.05**	**43.84**	**18.68**	**40.19**	**56.26**	**28.49**	**46.90**

that using LLM alone without Filtering&Integration may hurt some metrics. This is because the descriptions of sub-events usually only capture a part of the semantics of the original query, and the localizing results are inaccurate without Filtering&Integration. (2) From the 3rd row of the table, it can be seen that when both LLM prompting and Filtering&Integration are used, the model outperforms the naive baseline by 1.46% in mIoU. (3) From the 4th row of the table, our proposed VLM localizer shows a significant performance improvement, with a 5.69% increase in R@0.5 compared to the naive baseline. (4) In the 6th row, when all three modules are enabled, performance is further improved, demonstrating the effectiveness of our method.

Effectiviness of VLM. Table 6 shows the effectiveness of our two scoring functions in the VLM Localizer. (1) From the second row of the table, it can be

Table 4. Cross-dataset performance when training on ActivityNet captions and evaluate on Charades-STA.

Method	R@1 R@0.5	R@1 R@0.7	R@5 R@0.5	R@5 R@0.7
SCDM [54]	15.91	6.19	54.04	30.39
2D-TAN [57]	15.81	6.30	59.06	31.53
Debias-TLL [3]	21.45	10.38	62.34	32.90
Ours	**49.97**	**24.32**	**83.5**	**42.2**

Table 5. Ablations on each component.

	LLM prompting	VLM localizer	Filtering & Integration	R@0.5	R@0.7	mIoU
1	✗	✗	✗	42.32	18.91	31.61
2	✓			43.17	18.56	32.14
3	✓		✓	44.12	19.21	33.07
4		✓		48.01	22.07	43.37
5	✓	✓		48.41	21.94	42.76
6	✓	✓	✓	**49.97**	**24.32**	**44.51**

Table 6. Ablations on VLM localizer.

Dynamic Scoring	Static Scoring	R@0.5	R@0.7	mIoU
✗	✗	42.32	18.91	31.61
✓		47.63	20.13	41.68
	✓	45.48	22.02	41.81
✓	✓	**48.01**	**22.07**	**43.37**

Table 7. Ablations on LLM prompting.

Order Constraint	Relation Constraint	R@0.5	R@0.7	mIoU
✗	✗	42.32	18.91	31.61
✓		43.01	19.03	31.73
	✓	43.97	19.11	32.76
✓	✓	**44.12**	**19.21**	**33.07**

observed that the dynamic scoring significantly improves performance, with an increase of 10.07% in mIoU. Since most VLMs are trained on image-text or trimmed video clip-text data, they are not sensitive enough to the dynamic transitions between different events in the same video. Dynamic scoring models the dynamic transitions implicitly, thus demonstrating better performance. (2) From the third row of the table, when using static scoring, the mIoU of the naive baseline improves by 10.2%. Static scoring compared with the naive baseline not only requires high visual-textual relevance within the event but also requires low visual-textual relevance outside the event, thereby avoiding the model's focus solely on the most discriminative video segments. Combining both approaches further enhances model performance, demonstrating the effectiveness of our VLM localizer.

In Table 8, we report the performance of different VLMs, including the image-text pre-trained model (CLIP [34] and BLIP-2 [20]) and video-text pre-trained model (InterVideo [43], ViCLIP [42]). It can be observed that BLIP-2

Table 8. Ablations on the VLMs.

VLMs	Type	R@0.5	R@0.7	mIoU
CLIP [34]	Image	42.68	18.92	38.89
BLIP-2 [20]		**48.01**	**22.07**	**43.37**
InterVideo [43]	Video	44.60	20.51	40.72
ViCLIP [42]		44.01	20.48	40.25

Table 9. Ablations on the LLMs.

LLMs	R@0.5	R@0.7	mIoU
None	48.01	22.07	43.37
Gemini-1.0-Pro [36]	48.97	22.76	44.12
GPT-3.5 Turbo	49.23	23.11	**44.69**
GPT-4 Turbo	**49.97**	**24.32**	44.51

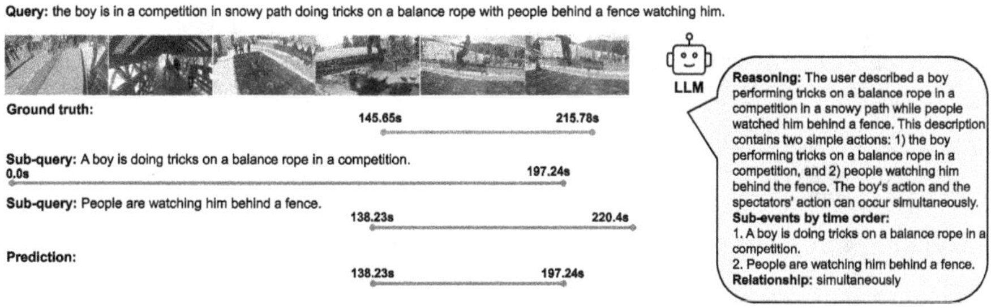

Fig. 3. Qualitative results on the ActivityNet Captions dataset.

exhibits the best performance, even surpassing models trained on video-text data. We attribute this to the fact that the pretraining data for image-text is much larger than that for video-text (e.g. LAION400M [35] v.s. WebVid10M [2]), thus BLIP-2 demonstrates better generalization capability. Additionally, our designed dynamic scoring helps BLIP-2 better understand the dynamic transition in the videos. Notably, in Table 1, VideoLLaMA [55] and VideoChat [22] utilize the frozen BLIP-2 Q-Former. VTG-GPT employs MiniGPT [64], which is also based on the frozen BLIP-2. Therefore, our comparison with them is fair. Additionally, Luo et al. [28] use InterVideo as the VLM, and as shown in Table 8, our performance using InterVideo still surpasses them.

Effectiviness of LLM. (1) We utilize the order and relationships of sub-events provided by LLM to filter and integrate the predictions of the VLM localizer. Table 7 verifies the effectiveness of Filtering&Integration. It can be observed that both order and relation constraints improve the performance, with the most significant improvement when both are used simultaneously. (2) In Table 9, we also report the performance of different LLMs, including Gemini-1.0-Pro, GPT-3.5 Turbo, and GPT-4 Turbo. The results indicate that more powerful LLM (e.g. GPT-4) can lead to better performance.

4.4 Qualitative Results

Figure 3 presents a visualization. It can be seen that the LLM successfully splits the query into two sub-events and analyzes that both of these sub-events should occur simultaneously in the target query. Our VLM localizer successfully localizes these two sub-events, and their intersection forms the final prediction. More qualitative results can be found in the supplementary materials.

5 Conclusion

In this work, we study the problem of training-free video temporal grounding. We leverage the ability of LLM and VLM, requiring no specific video temporal localization dataset. We propose leveraging LLM to analyze multiple sub-events contained in the query and analyze the temporal order and relationships between these events. Then, we explicitly model the dynamic transition and static status in the video and use the VLM to localize the sub-events and leverage the order and relationships provided by LLMs to integrate the predictions. Our method achieves the best performance on zero-shot video temporal grounding on Charades-STA and ActivityNet Captions datasets without any training and demonstrates better generalization in cross-dataset and OOD settings.

Limitations. LLM are not always reliable in reasoning the order and relationships between sub-events, which can negatively impact the performance. How to validate the reliability of outputs from LLM can be studied in the future.

Acknowledgement. This work was supported by grants from the National Natural Science Foundation of China (62372014, 61925201, 62132001, U22B2048).

References

1. Achiam, J., et al.: GPT-4 technical report. arXiv preprint arXiv:2303.08774 (2023)
2. Bain, M., Nagrani, A., Varol, G., Zisserman, A.: Frozen in time: a joint video and image encoder for end-to-end retrieval. In: Proceedings of the IEEE/CVF International Conference on Computer Vision, pp. 1728–1738 (2021)
3. Bao, P., Mu, Y.: Learning sample importance for cross-scenario video temporal grounding. arXiv preprint arXiv:2201.02848 (2022)
4. Collins, R.T., et al.: A system for video surveillance and monitoring. VSAM Final Rep. **2000**(1–68), 1 (2000)
5. Croitoru, I., et al.: Moment detection in long tutorial videos. In: Proceedings of the IEEE/CVF International Conference on Computer Vision, pp. 2594–2604 (2023)
6. Duan, X., Huang, W., Gan, C., Wang, J., Zhu, W., Huang, J.: Weakly supervised dense event captioning in videos. Adv. Neural Inf. Process. Syst. **31** (2018)
7. Gao, J., Sun, C., Yang, Z., Nevatia, R.: TALL: temporal activity localization via language query. In: Proceedings of the IEEE International Conference on Computer Vision, pp. 5267–5275 (2017)
8. Gao, J., Xu, C.: Learning video moment retrieval without a single annotated video. IEEE Trans. Circuits Syst. Video Technol. **32**(3), 1646–1657 (2021)
9. Guo, J., et al.: From images to textual prompts: zero-shot visual question answering with frozen large language models. In: Proceedings of the IEEE/CVF Conference on Computer Vision and Pattern Recognition, pp. 10867–10877 (2023)
10. Huang, B., Wang, X., Chen, H., Song, Z., Zhu, W.: VTimellm: empower LLM to grasp video moments. arXiv preprint arXiv:2311.18445 (2023)
11. Huang, J., Jin, H., Gong, S., Liu, Y.: Video activity localisation with uncertainties in temporal boundary. In: European Conference on Computer Vision. pp. 724–740. Springer (2022)
12. Huang, J., Liu, Y., Gong, S., Jin, H.: Cross-sentence temporal and semantic relations in video activity localisation. In: Proceedings of the IEEE/CVF International Conference on Computer Vision, pp. 7199–7208 (2021)
13. Huang, Y., Yang, L., Sato, Y.: Weakly supervised temporal sentence grounding with uncertainty-guided self-training. In: Proceedings of the IEEE/CVF Conference on Computer Vision and Pattern Recognition, pp. 18908–18918 (2023)

14. Jang, J., Park, J., Kim, J., Kwon, H., Sohn, K.: Knowing where to focus: event-aware transformer for video grounding. In: Proceedings of the IEEE/CVF International Conference on Computer Vision, pp. 13846–13856 (2023)
15. Kim, D., Park, J., Lee, J., Park, S., Sohn, K.: Language-free training for zero-shot video grounding. In: Proceedings of the IEEE/CVF Winter Conference on Applications of Computer Vision, pp. 2539–2548 (2023)
16. Krishna, R., Hata, K., Ren, F., Fei-Fei, L., Niebles, J.C.: Dense-captioning events in videos. In: 2017 IEEE International Conference on Computer Vision (ICCV) (2017)
17. Lei, T., Yin, S., Liu, Y.: Exploring the potential of large foundation models for open-vocabulary hoi detection. In: Proceedings of the IEEE/CVF Conference on Computer Vision and Pattern Recognition, pp. 16657–16667 (2024)
18. Lei, T., Yin, S., Peng, Y., Liu, Y.: Exploring conditional multi-modal prompts for zero-shot hoi detection. In: Proceedings of the European Conference on Computer Vision (ECCV) (2024)
19. Li, J., et al.: Compositional temporal grounding with structured variational cross-graph correspondence learning. In: Proceedings of the IEEE/CVF Conference on Computer Vision and Pattern Recognition, pp. 3032–3041 (2022)
20. Li, J., Li, D., Savarese, S., Hoi, S.: BLIP-2: bootstrapping language-image pre-training with frozen image encoders and large language models. arXiv preprint arXiv:2301.12597 (2023)
21. Li, J., Li, D., Xiong, C., Hoi, S.: BLIP: bootstrapping language-image pre-training for unified vision-language understanding and generation. arXiv preprint arXiv:2201.12086 (2022)
22. Li, K., et al.: VideoChat: chat-centric video understanding. arXiv preprint arXiv:2305.06355 (2023)
23. Li, Z., Wang, P., Wang, Z., Zhan, D.C.: FlowGANAnomaly: flow-based anomaly network intrusion detection with adversarial learning. Chin. J. Electron. **33**(1), 58–71 (2024)
24. Li, Z., et al.: LEGO: language enhanced multi-modal grounding model. arXiv preprint arXiv:2401.06071 (2024)
25. Liu, D., Qu, X., Di, X., Cheng, Y., Xu, Z., Zhou, P.: Memory-guided semantic learning network for temporal sentence grounding. arXiv preprint arXiv:2201.00454 (2022)
26. Liu, H., Li, C., Wu, Q., Lee, Y.J.: Visual instruction tuning. Adv. Neural Inf. Process. Syst. **36** (2024)
27. Luo, D., Huang, J., Gong, S., Jin, H., Liu, Y.: Towards generalisable video moment retrieval: visual-dynamic injection to image-text pre-training. In: Proceedings of the IEEE/CVF Conference on Computer Vision and Pattern Recognition, pp. 23045–23055 (2023)
28. Luo, D., Huang, J., Gong, S., Jin, H., Liu, Y.: Zero-shot video moment retrieval from frozen vision-language models. In: Proceedings of the IEEE/CVF Winter Conference on Applications of Computer Vision, pp. 5464–5473 (2024)
29. Ma, Y.F., Lu, L., Zhang, H.J., Li, M.: A user attention model for video summarization. In: Proceedings of the Tenth ACM International Conference on Multimedia, MULTIMEDIA 2002, pp. 533–542. Association for Computing Machinery, New York (2002). https://doi.org/10.1145/641007.641116
30. Maaz, M., Rasheed, H., Khan, S., Khan, F.S.: Video-ChatGPT: towards detailed video understanding via large vision and language models. arXiv preprint arXiv:2306.05424 (2023)

31. Mo, W., Liu, Y.: Bridging the gap between 2D and 3D visual question answering: a fusion approach for 3D VQA. In: Proceedings of the AAAI Conference on Artificial Intelligence, vol. 38, pp. 4261–4268 (2024)
32. Mun, J., Cho, M., Han, B.: Local-global video-text interactions for temporal grounding. In: Proceedings of the IEEE/CVF Conference on Computer Vision and Pattern Recognition, pp. 10810–10819 (2020)
33. Nam, J., Ahn, D., Kang, D., Ha, S.J., Choi, J.: Zero-shot natural language video localization. In: Proceedings of the IEEE/CVF International Conference on Computer Vision, pp. 1470–1479 (2021)
34. Radford, A., et al.: Learning transferable visual models from natural language supervision. In: International Conference on Machine Learning, pp. 8748–8763. PMLR (2021)
35. Schuhmann, C., et al.: LAION-400M: open dataset of clip-filtered 400 million image-text pairs. arXiv preprint arXiv:2111.02114 (2021)
36. Team, G., et al.: Gemini: a family of highly capable multimodal models. arXiv preprint arXiv:2312.11805 (2023)
37. Tian, Y., Fu, Y., Zhang, J.: Transformer-based under-sampled single-pixel imaging. Chin. J. Electron. **32**(5), 1151–1159 (2023)
38. Vaswani, A., et al.: Attention is all you need. Adv. Neural Inf. Process. Syst. **30** (2017)
39. Wang, G., Wu, X., Liu, Z., Yan, J.: Prompt-based zero-shot video moment retrieval. In: Proceedings of the 30th ACM International Conference on Multimedia, pp. 413–421 (2022)
40. Wang, H., Zha, Z.J., Li, L., Liu, D., Luo, J.: Structured multi-level interaction network for video moment localization via language query. In: Proceedings of the IEEE/CVF Conference on Computer Vision and Pattern Recognition (CVPR), pp. 7026–7035 (2021)
41. Wang, J., et al.: ChatVideo: a tracklet-centric multimodal and versatile video understanding system. arXiv preprint arXiv:2304.14407 (2023)
42. Wang, Y., et al.: InternVid: a large-scale video-text dataset for multimodal understanding and generation. arXiv preprint arXiv:2307.06942 (2023)
43. Wang, Y., et al.: InternVideo: general video foundation models via generative and discriminative learning. arXiv preprint arXiv:2212.03191 (2022)
44. Wang, Z., Wang, L., Wu, T., Li, T., Wu, G.: Negative sample matters: a renaissance of metric learning for temporal grounding. In: AAAI, pp. 2613–2623. AAAI Press (2022)
45. Wu, J., Li, G., Liu, S., Lin, L.: Tree-structured policy based progressive reinforcement learning for temporally language grounding in video. In: Proceedings of the AAAI Conference on Artificial Intelligence, vol. 34, pp. 12386–12393 (2020)
46. Xu, Y., Sun, Y., Xie, Z., Zhai, B., Du, S.: VTG-GPT: tuning-free zero-shot video temporal grounding with GPT. Appl. Sci. **14**(5), 1894 (2024)
47. Yang, D., Liu, Y.: Active object detection with knowledge aggregation and distillation from large models. In: Proceedings of the IEEE/CVF Conference on Computer Vision and Pattern Recognition, pp. 16624–16633 (2024)
48. Yang, D., Xu, Z., Mo, W., Chen, Q., Huang, S., Liu, Y.: 3D vision and language pretraining with large-scale synthetic data. IJCAI (2024)
49. Yang, L., Kong, Q., Yang, H.K., Kehl, W., Sato, Y., Kobori, N.: Deco: Decomposition and reconstruction for compositional temporal grounding via coarse-to-fine contrastive ranking. In: Proceedings of the IEEE/CVF Conference on Computer Vision and Pattern Recognition, pp. 23130–23140 (2023)

50. Yang, X., Feng, F., Ji, W., Wang, M., Chua, T.S.: Deconfounded video moment retrieval with causal intervention. In: Proceedings of the 44th International ACM SIGIR Conference on Research and Development in Information Retrieval, pp. 1–10 (2021)
51. Ye, Z., He, X., Peng, Y.: Unsupervised cross-media hashing learning via knowledge graph. Chin. J. Electron. **31**(6), 1081–1091 (2022)
52. Yelamarthi, S.K., Reddy, S.K., Mishra, A., Mittal, A.: A zero-shot framework for sketch based image retrieval. In: Proceedings of the European Conference on Computer Vision (ECCV), pp. 300–317 (2018)
53. Yuan, Y., Lan, X., Wang, X., Chen, L., Wang, Z., Zhu, W.: A closer look at temporal sentence grounding in videos: dataset and metric. In: Proceedings of the 2nd International Workshop on Human-Centric Multimedia Analysis, pp. 13–21 (2021)
54. Yuan, Y., Ma, L., Wang, J., Liu, W., Zhu, W.: Semantic conditioned dynamic modulation for temporal sentence grounding in videos. Adv. Neural Inf. Process. Syst. **32** (2019)
55. Zhang, H., Li, X., Bing, L.: Video-LLaMA: an instruction-tuned audio-visual language model for video understanding. arXiv preprint arXiv:2306.02858 (2023)
56. Zhang, M., et al.: Multi-stage aggregated transformer network for temporal language localization in videos. In: Proceedings of the IEEE/CVF Conference on Computer Vision and Pattern Recognition, pp. 12669–12678 (2021)
57. Zhang, S., Peng, H., Fu, J., Luo, J.: Learning 2D temporal adjacent networks for moment localization with natural language. In: Proceedings of the AAAI Conference on Artificial Intelligence, vol. 34, pp. 12870–12877 (2020)
58. Zhao, Y., Zhao, Z., Zhang, Z., Lin, Z.: Cascaded prediction network via segment tree for temporal video grounding. In: Proceedings of the IEEE/CVF Conference on Computer Vision and Pattern Recognition (CVPR), pp. 4197–4206 (2021)
59. Zheng, M., Gong, S., Jin, H., Peng, Y., Liu, Y.: Generating structured pseudo labels for noise-resistant zero-shot video sentence localization. In: Proceedings of the 61st Annual Meeting of the Association for Computational Linguistics (Volume 1: Long Papers), pp. 14197–14209 (2023)
60. Zheng, M., Huang, Y., Chen, Q., Liu, Y.: Weakly supervised video moment localization with contrastive negative sample mining. In: Proceedings of the AAAI Conference on Artificial Intelligence (2022)
61. Zheng, M., Huang, Y., Chen, Q., Peng, Y., Liu, Y.: Weakly supervised temporal sentence grounding with gaussian-based contrastive proposal learning. In: Proceedings of the IEEE/CVF Conference on Computer Vision and Pattern Recognition (CVPR) (2022)
62. Zheng, M., Li, S., Chen, Q., Peng, Y., Liu, Y.: Phrase-level temporal relationship mining for temporal sentence localization. In: Proceedings of the AAAI Conference on Artificial Intelligence (2023)
63. Zhou, H., Zhang, C., Luo, Y., Chen, Y., Hu, C.: Embracing uncertainty: decoupling and de-bias for robust temporal grounding. In: 2021 IEEE/CVF Conference on Computer Vision and Pattern Recognition (CVPR), pp. 8441–8450. IEEE Computer Society, Los Alamitos (2021)
64. Zhu, D., Chen, J., Shen, X., Li, X., Elhoseiny, M.: MiniGPT-4: enhancing vision-language understanding with advanced large language models. arXiv preprint arXiv:2304.10592 (2023)

Revisit Self-supervised Depth Estimation with Local Structure-from-Motion

Shengjie Zhu[✉] and Xiaoming Liu

Department of Computer Science and Engineering, Michigan State University,
East Lansing, MI 48824, USA
zhusheng@msu.edu, liuxm@cse.msu.edu

Abstract. Both self-supervised depth estimation and Structure-from-Motion (SfM) recover scene depth from RGB videos. Despite sharing a similar objective, the two approaches are disconnected. Prior works of self-supervision backpropagate losses defined within immediate neighboring frames. Instead of learning-through-loss, this work proposes an alternative scheme by performing local SfM. First, with calibrated RGB or RGB-D images, we employ a depth and correspondence estimator to infer depthmaps and pair-wise correspondence maps. Then, a novel bundle-RANSAC-adjustment algorithm jointly optimizes camera poses and one depth adjustment for each depthmap. Finally, we fix camera poses and employ a NeRF, however, without a neural network, for dense triangulation and geometric verification. Poses, depth adjustments, and triangulated sparse depths are our outputs. For the first time, we show self-supervision within 5 frames already benefits SoTA supervised depth and correspondence models. Despite self-supervision, our pose algorithm has certified global optimality, outperforming optimization-based, learning-based, and NeRF-based prior arts. The project page is held in the link.

Keywords: Self-supervision · Depth · Pose · Structure-from-Motion

1 Introduction

Monocular depth estimation [17,29] infers depthmap from a single image. It is an essential vision task with applications in AR/VR [41], autonomous driving [19], and 3D reconstruction [8]. Most methods [5,43,45,69] supervise the model with groundtruth collected from stereo cameras [71] or LiDAR [19]. Recently, self-supervised depth [20,21,76] has drawn significant attention due to its potential to scale up depth learning from massive unlabeled RGB videos.

Classic SfM methods [1,12,47,48,52,65] also reconstruct scene depth from unlabled RGB videos. Despite its relevance, SfM is rarely applied to self-supervised depth learning. We outline two potential reasons. First, SfM is an

---

Supplementary Information The online version contains supplementary material available at https://doi.org/10.1007/978-3-031-73007-8_3.

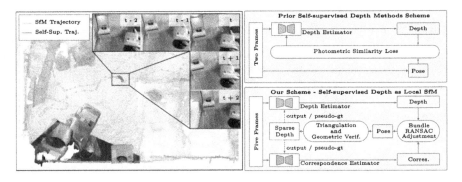

Fig. 1. Revisit Self-supervision with Local SfM. The work proposes alternating the learning-through-loss with a local SfM pipeline for self-supervised depth estimation. We summarize our differences. On self-supervision: (1) Instead of using naive two-view camera poses, we propose a Bundle-RANSAC-Adjustment pose optimization algorithm with multi-view constraints. (2) Instead of backpropagating through a loss, we produce a sparse point cloud with explicit triangulation and geometric verification. The point cloud serves as either output or pseudo-groundtruth for self-supervision. On SfM: (1) Our local SfM is adapted to use estimated monocular depthmaps and automatically resolve their scale inconsistency between pairs of images. (2) We maintain accuracy under significant sparse view variations, e.g., red trajectories. We generalize SfM to as few as 5 frames, similar to the number of images used to define self-supervision loss. (Color figure online)

off-the-shelf algorithm unrelated to the depth estimator. Scale ambiguity renders SfM poses and depths at different scales compared to depth models. Second, self-supervision has a well-defined training scheme to work with universal unlabeled videos. It backpropagates through photometric loss computed within immediate neighboring frames, e.g., red trajectory in Fig. 1. In contrast, SfM is more selective to input videos. It requires images of diverse view variations (green trajectory in Fig. 1), being inaccurate and unstable when applied to a small frame window.

This work connects self-supervision with SfM. We replace the self-supervision loss with a complete SfM pipeline that maintains robustness to a local window. Shown in Fig. 2, with N frames as input, our algorithm outputs $N-1$ camera poses, $N-1$ depth adjustments, and the sparse triangulated point cloud. In initialization, N monocular depthmaps and $N \times (N-1)$ pairwise correspondence maps are inferred. Next, we propose a Bundle-RANSAC-Adjustment pose estimation algorithm that retains accuracy for second-long videos. The algorithm utilizes the 3D priors from mono-depthmap to compensate for the deficient camera views. Correspondingly, we optimize $N-1$ depth adjustments to alleviate the depth scale ambiguity by temporally aligning to the root frame depth.

The Bundle-RANSAC-Adjustment extends two-view RANSAC with multi-view bundle-adjustment (BA). The algorithm has quadratic complexity, designed for parallel GPU computation. We RANdomly SAmple and hypothesize a set

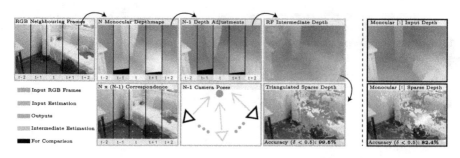

Fig. 2. Local Structure-from-Motion. With N neighboring frames, we extract monocular depthmaps and pairwise dense correspondence maps with methods, *e.g.*, ZoeDepth [5] and PDC-Net [57]. Next, skipping the root frame, we optimize the rest $N-1$ camera poses and depth adjustments. The depth adjustments render input depthmaps **temporally consistent**. Fixing poses and adjustments, we use the Radiance Field (RF) for triangulation and output a geometrically verified sparse root depthmap. Our local SfM applies **self-supervision** with only 5 RGB frames. Yet, our sparse output already outperforms the input supervised depth with SoTA performance.

of normalized poses. In **C**onsensus checking, we apply BA to evaluate a robust inlier-counting scoring function over multi-view images. Camera scales and depth adjustments are determined during BA to maximize the scoring function.

Next, we freeze the optimized poses and employ a Radiance Field (RF), *i.e.*, a NeRFF [39] without a neural network, for triangulation. We optimize RF to achieve multi-view depthmap and correspondence consistency within a shared 3D frustum volume. For outputs, we apply geometric verification to extract multi-view consistent point cloud, *i.e.*, a sparse root depthmap.

Figure 1 contrasts our method with prior self-supervised depth and SfM methods. To our best knowledge, there has not been prior work showing geometry-based self-supervised depth benefits supervised models. However, self-supervision is supposed to augment supervised models with unlabeled data. In Fig. 2, our unique pipeline gives the **first** evident results, that self-supervision with **as few as** 5 frames already benefits supervised models.

On top of depths, our multi-view RANSAC pose has certified global optimality under a robust scoring function. It outperforms prior arts in optimization-based [48,77], learning-based [55,61], and NeRF-based [58] pose algorithms.

Beyond pose and depth, our method has diverse applications. The depth adjustments from our method provide empirically consistent depthmaps, important for AR image compositing. When given RGB-D inputs, our method enables self-supervised correspondence estimation. Our accurate pose estimation gives improved projective correspondence than the SoTA supervised correspondence input. An example is in Fig. 9. We summarize our contributions as:

– We propose a novel local SfM algorithm with Bundle-RANSAC-Adjustment.
– We show the **first** evident result that self-supervised depth with **as few as** 5 frames already benefit SoTA supervised models.

- We achieve SoTA sparse-view pose estimation performance.
- We enable self-supervised temporally consistent depthmaps.
- We enable self-supervised correspondence estimation with 5 RGB-D frames.

2 Related Works

Structure-from-Motion. SfM is a comprehensive task [48,65]. A typical pipeline is, correspondence extraction [7,36,59], two-view initialization [3,30,32], triangulation [31,42], and local & global bundle-adjustment [48,65]. Classic methods require diverse view variations for accurate reconstruction. Our method compensates SfM on scarse camera views via introducing deep depth estimator. Further, we suggest SfM itself is a self-supervised learning pipeline, as in Fig. 1. Finally, our SfM is not up-to-scale and shares the metric space as the input depthmap.

Sparse Multi-view Pose Estimation. Estimating poses from sparse frames is crucial for self-supervision [11,20,46,70,74], video depth estimation [22, 55,60,77], and sparse-view NeRF [14,27,34,40,58]. Camera poses are estimated either by learning [11,20,22,55], optimization [73,77] or together with NeRF [34,58]. We propose an additional multi-view RANSAC pipeline with improved accuracy.

Self-supervised Depth and Correspondence Estimation. Multiple works improve self-supervised depth in different ways, including learning loss [20, 44,63], architecture [23,75], camera pose [6,10,38,73], joint with semantics segmentation [76], and using large-scale data [53,66]. Recently, [53] shows self-supervision only performs on-par with supervised models under substantially more data. [66] shows the benefit of self-supervision via exploiting nongeometry monocular semantic consistency. Our method shows the first evident results where self-supervision benefits supervised models with only 5 consecutive frames.

Consistent Depth Estimation. AR applications necessitate temporally consistent depthmaps, *i.e.*, depthmaps from different temporal frames reside in the same 3D space. Recent works [37,72] align depthmap according to the poses and points from the off-the-shelf COLMAP algorithm. Our method seamlessly integrates SfM with monocular depthmaps, outputting consistent depth and poses.

Test Time Refinement (TTR). TTR aims to improve self-supervised/ supervised depth estimators in testing time with RGB video [9,10,28,50,64]. Methods [25,56] rely on off-the-shelf algorithms for pseudo depth and pose labels. Recently, [25] first shows TTR improves supervised models. TTR is our downstream application, which details strategies for utilizing noisy pseudo-labels.

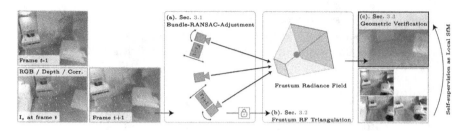

Fig. 3. Algorithm Overview. After extracting monodepths and correspondence maps from inputs: (a) We apply Bundle-RANSAC-Adjustment to optimize $N-1$ camera poses $\mathcal{P}$ and $N-1$ depth adjustments $\mathcal{R}$. (b) We fix poses and depth adjustments and optimize a frustum Radiance Field (RF) for triangulation. (c) We apply geometric verification to extract multi-view consistent 3D points via rendering with RF. We further detail step (a) in Figs. 4, 5, and 6, and steps (b) and (c) in Fig. 7.

3 Methodology

Our method runs sequentially. From N calibrated images $\mathcal{I}$, we extract N monocular depthmaps $\mathcal{D}$ and $N \times (N-1)$ pair-wise dense correspondence $\mathcal{C}$. We split the N images into one root frame $\mathbf{I}_o$ in the center of the N-frame window where $o = \lfloor \frac{N+1}{2} \rfloor$, and $N-1$ support frames $\mathbf{I}_i$, where $i \in \mathbb{N}^+ = [1, N]\{o\}$. In Sect. 3.1, after setting the root frame as identity pose, we use Bundle-RANSAC-Adjustment to optimize $N-1$ poses $\mathcal{P}$ and $N-1$ depth adjustments $\mathcal{R}$. Next, in Section 3.2, we apply triangulation by optimizing a frustum Radiance Field (RF) $\mathbf{V}$, *i.e.*, a NeRF without network. Finally, in Sect. 3.3, we apply geometric verification by rendering multi-view consistent 3D points from RF. An overview is in Fig. 3.

3.1 Bundle-RANSAC-Adjustment Pose Estimation

We generalize two-view RANSAC with multi-view constraints through Bundle-Adjustment. Section 3.1.1 describes our pipeline. In Sect. 3.1.2, we propose Hough transform to accelerate computation. We discuss the time complexity in Sect. 3.1.3.

3.1.1 Optimization Pipeline

RANdom **SA**mple. We use five-point algorithm [32] as the minimal solver. We execute it between root and each support frame, extracting a pool of $(N-1) \times K$ normalized poses (*i.e.*, pose of unit translation), $\overline{\mathcal{Q}} = \{\overline{\mathbf{P}}_i^k \mid i \in \mathbb{N}^+, k \in [1, K]\}$, where $\overline{\mathbf{P}}_i^k \in \mathbb{R}^{3 \times 4}$. The K is the number of normalized poses extracted per frame. We term a set of $N-1$ normalized poses as a group $\overline{\mathcal{P}} \in \mathbb{R}^{(N-1) \times 3 \times 4}$. Two-view RANSAC enumerates over single normalized pose $\overline{\mathbf{P}}$. Our multi-view algorithm hence enumerates over normalized pose group $\overline{\mathcal{P}}$. We initialize the optimal group $\overline{\mathcal{P}}^*$ as the top candidate from K poses of $\overline{\mathcal{Q}}$ for each frame. See examples in Fig. 4.

Bundle-Adjustment Consensus. While computing consensus counts, the camera scales $\mathcal{S}$ and depth adjustments $\mathcal{R}$ are automatically determined with bundle-adjustment to maximize a robust scoring function:

$$\rho_i = \phi(\overline{\mathcal{P}}) = \max_{\mathcal{S},\mathcal{R}} f(\mathcal{S}, \mathcal{R} \mid \overline{\mathcal{P}}, \mathcal{D}, \mathcal{C}). \tag{1}$$

Search for Optimal Group. Our multi-view RANSAC has a significantly larger solution space than two-view RANSAC. With N view inputs, we determine the optimal group out of K^{N-1} combinations. Hence, we iteratively search for the optimal group with a greedy strategy. For each epoch, we ablate $(N-1)(K-1)$ additional pose groups:

$$\overline{\mathcal{P}}_i^k = \overline{\mathcal{P}}_i^* \, \{\overline{\mathbf{P}}_i^*\} \cup \{\overline{\mathbf{P}}_i^k\}, \tag{2}$$

where $i \in \mathbb{N}^+$ and $k \in [1, K]$. Combine Eq. (2) and Fig. 4, taking frame i as an example, we replace the optimal pose $\overline{\mathbf{P}}_i^*$ by its $K-1$ other candidates $\overline{\mathbf{P}}_i^k$, generating $K-1$ groups. For N frames, we have $(N-1)(K-1)+1$ groups. We apply bundle-adjustment to each group to evaluate Eq. (1). As shown in Fig. 3 and Fig. 4, we select the normalized pose together with its optimized scales and depth adjustments that maximize the scores as the output,

$$\mathcal{P}_i^* = b(\overline{\mathcal{P}}_i^*, \mathcal{S}_i^*), \; \mathcal{R}_i^* = \mathcal{R}_i^k, \; \text{where} \; k = \arg\max\{\rho_i^k\}, \; \overline{\mathcal{P}}_i^* = \overline{\mathcal{P}}_i^k, \; \mathcal{S}_i^* = \mathcal{S}_i^k, \tag{3}$$

where $b(\cdot)$ combines normalized poses with scales. Figure 2 third column plots an adjusted temporal consistent depthmap after applying $\mathcal{R}^*$. In Fig. 4, the algorithm terminates when the maximum score stops increasing.

Scoring Function. Similar to other RANSAC methods, we adopt robust inlier-counting based scoring functions. Expand Eq. (1) for a specific group $\overline{\mathcal{P}}$:

$$\phi(\overline{\mathcal{P}}) = \sum_{i, i \neq j} \sum_j f_{i,j}(s_i, s_j, r_i, r_j \mid \overline{\mathbf{P}}_i, \overline{\mathbf{P}}_j, \mathbf{D}_i, \mathbf{D}_j, \mathbf{C}_{i,j}), \tag{4}$$

where i, j are frame index. We set per-frame camera scale, depth, depth adjustment, and correspondence as $s \in \mathcal{S}$, $\mathbf{D} \in \mathcal{D}$, $r \in \mathcal{R}$, and $\mathbf{C} \in \mathcal{C}$. The scoring function $f_{i,j}(\cdot)$ has various forms. First, we describe a 2D scoring function:

$$f_{i,j}^{2D}(\cdot) = \sum_m \mathbf{1}\left(\|\pi(s_i, s_j, r_i \mid \overline{\mathbf{P}}_i, \overline{\mathbf{P}}_j, d_i^m) - \mathbf{c}_{i,j}^m\|_2 < \lambda^{2D}\right), \tag{5}$$

where $m \in [1, M]$ indexes sampled pixels per frame pair. $f_{i,j}^{2D}(\cdot)$ measures the inlier count between depth projected correspondence and input correspondence. $\pi(\cdot)$ is projection process. Intrinsic is skipped. d and $\mathbf{c}$ are depth and correspondence sampled from $\mathbf{D}$ and $\mathbf{C}$. An example is in Fig. 5. The $\mathbf{1}(\cdot)$ is the

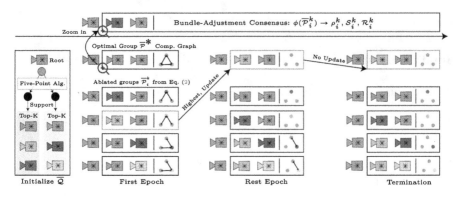

Fig. 4. Pose Optimization Pipeline. We show a sample execution when $N=3$ and $K=3$. We initialize normalized pose candidates pool $\overline{\mathcal{Q}}$. Optimal group $\overline{\mathcal{P}}^*$ is set to top candidates within $\overline{\mathcal{Q}}$. In each epoch, Eq. (2) ablates pose group $\overline{\mathcal{P}}_i^k$. Each group is scored with Eq. (1) via BA with Hough Transform, detailed in Sect. 3.1.2. The optimal group with the highest score is updated with Eq. (3). Termination occurs when the maximum score stabilizes. We maintain quadratic complexity by avoiding repetitive computation after the first epoch, shown with the Comp. Graph, detailed in Sect. 3.1.3.

indicator function. The projected pixel is an inlier if it resides within the circle of radius λ^{2D} and center at correspondence $\mathbf{c}_{i,j}^m$ (denoted as $\mathbf{p}_j$ in Fig. 5). $\mathbf{c}_{i,j}^m$ is sampled from correspondence map $\mathbf{C}_{i,j}$. Second, we introduce a 3D scoring function:

$$f_{i,j}^{3D}(\cdot) = \sum_m \mathbf{1}\left(\|\pi^{-1}(s_i \mid \overline{\mathbf{P}}_i, r_i, d_i^m) - \pi^{-1}(s_j \mid \overline{\mathbf{P}}_j, r_j, d_j^m)\|_2 < \lambda^{3D}\right). \quad (6)$$

Depth pair d_i and d_j is determined by correspondence. Unlike the 2D one, the 3D function fixes depth adjustment r. Function $\pi^{-1}(\cdot)$ back-projects 3D point.

3.1.2 Hough Transform Acceleration

Maximizing Eq. (1) for each pose group is computationally prohibitive, as shown in Fig. 4. We propose Hough Transform for acceleration. We use Eq. (5), the 2D function $f^{2D}(\cdot)$ as an example for illustration. See our motivation in Fig. 5.

Hough Transform. The relative pose between $\overline{\mathbf{P}}_i$ and $\overline{\mathbf{P}}_j$ is defined as:

$$\mathbf{P}_{i,j} = \mathbf{P}_j \mathbf{P}_i^{-1} = [\mathbf{R}_{i,j} s_{i,j} \overline{\mathbf{t}}_{i,j}] = [\mathbf{R}_j \mathbf{R}_i^{-1} - s_i \mathbf{R}_j \mathbf{R}_i^{-1} \overline{\mathbf{t}}_i + s_j \overline{\mathbf{t}}_j], \quad (7)$$

where $\mathbf{R}, \overline{\mathbf{t}}$, and s are rotation, normalized translation and pose scale. From Eq. (7) and Fig. 5, $\overline{\mathbf{t}}_{i,j}$ is controlled by the scale s_i and s_j, and thus we have:

$$\lim_{s_i \to +\inf} \overline{\mathbf{t}}_{i,j} = -\mathbf{R}_j \mathbf{R}_i^{-1} \overline{\mathbf{t}}_i, \quad \lim_{s_j \to +\inf} \overline{\mathbf{t}}_{i,j} = \overline{\mathbf{t}}_j. \quad (8)$$

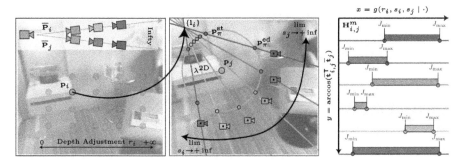

Fig. 5. Hough Transform between Two Normalized Poses. With fixed normalized poses, there exists three variables, scales s_i & s_j of $\overline{\mathbf{P}}_i$ & $\overline{\mathbf{P}}_j$ and adjustment r_i. Pixel $\mathbf{p}_i$ and $\mathbf{p}_j$ are corresponded. Ablating pose scales maps pixel $\mathbf{p}_i$ to a set of epipolar lines $\{\mathbf{l}_i\}$, however, bounded by Red and Green at infinite scales. We have three observations. First, with fixed normalized poses, epipolar lines $\mathbf{l}_i$ have limited possibilities. Second, scale s and depth adjustment d are equivalent, both adjusting projection on epipolar line. Third, per epipolar line, to be an inlier, the projection has to reside within the line-circle intersection, between $\mathbf{p}_\pi^{st}$ and $\mathbf{p}_\pi^{ed}$. The observations motivate us to discretize the solution space to a 2D matrix, i.e., Hough Transform. Right figure plots an example transformation $\mathbf{H}_{i,j}^m$ from frame i to j on the mth pixel $\mathbf{p}_i$. (Color figure online)

For a pixel $\mathbf{p}_i$ on frame i, its corresponding epipolar line $\mathbf{l}_i$ on frame j is:

$$\mathbf{l}_i = \mathbf{K}^{-\mathsf{T}}[\overline{\mathbf{t}}_{i,j}]_\times \mathbf{R}_{i,j}\mathbf{K}^{-1}\mathbf{p}_i. \tag{9}$$

Equation (8) and Eq. (9) suggest the epipolar line has limited possibilities. Operation $[\cdot]_\times$ is the cross product in matrix form. Further, as the depth reprojected pixel $\mathbf{p}_\pi$ of $\mathbf{p}_i$ always locate on the epipolar line $\mathbf{l}_i$ [24], we have:

$$\mathbf{l}_i^\mathsf{T}\mathbf{p}_\pi = 0, \quad \mathbf{p}_\pi = \pi(s_i, s_j, r_i \mid \overline{\mathbf{P}}_i, \overline{\mathbf{P}}_j, d_i). \tag{10}$$

To be an inlier of the scoring function $f^{2D}(\cdot)$, we have:

$$\|\mathbf{p}_\pi - \mathbf{p}_j\|_2 \leq \lambda^{2D}. \tag{11}$$

Combining Eq. (10), Eq. (11) and Fig. 5, to be an inlier, the projected pixel $\mathbf{p}_\pi$ has to reside within the line segment, with two end-points computed by the line-circle intersection. The circle centers at corresponded pixel $\mathbf{p}_j$ on frame j with a radius λ^{2D}. We denote the two end-points $\mathbf{p}_\pi^{st}$ and $\mathbf{p}_\pi^{ed}$. We add their calculation in Supp. Function $J(\cdot)$ follows [77] Supp. Equation (4), which maps a projected pixel $\mathbf{p}_\pi$ and adjusted depth $r_i d_i$ to camera scale $s_{i,j}$ as: $s_{i,j} = J(\overline{\mathbf{P}}_{i,j}, r_i d_i, \mathbf{p}_\pi)$.

Corollary 1. *A pixel is an inlier iff:*

$$J(\overline{\mathbf{P}}_{i,j}, r_i d_i, \mathbf{p}_\pi^{st}) \leq s_{i,j} \leq J(\overline{\mathbf{P}}_{i,j}, r_i d_i, \mathbf{p}_\pi^{ed}). \tag{12}$$

Corollary 2. *Scale and depth are equivalent as;*

$$s_{i,j} = J(\overline{\mathbf{P}}_{i,j}, r_i d_i, \mathbf{p}_\pi) = r_i \cdot J(\overline{\mathbf{P}}_{i,j}, d_i, \mathbf{p}_\pi). \tag{13}$$

See Fig. 5 and proof in Supp. Combine Eqs. (12) and (13),

$$J(\overline{\mathbf{P}}_{i,j}, d_i, \mathbf{p}_\pi^{\text{st}}) \leq \frac{s_{i,j}}{r_i} \leq J(\overline{\mathbf{P}}_{i,j}, d_i, \mathbf{p}_\pi^{\text{ed}}). \tag{14}$$

Set $g(\cdot)$ maps the variables under optimization to intermediate term $\frac{s_{i,j}}{r_i}$:

$$J(\overline{\mathbf{P}}_{i,j}, d_i, \mathbf{p}_\pi^{\text{st}}) \leq g(r_i, s_i, s_j \mid \overline{\mathbf{P}}_i, \overline{\mathbf{P}}_j) \leq J(\overline{\mathbf{P}}_{i,j}, d_i, \mathbf{p}_\pi^{\text{ed}}). \tag{15}$$

The ith pixel is an inlier if and only if its projection satisfies Eq. (15). Note, the value space of function $g(\cdot)$ is mapped to a 2D space $\mathbf{H}$ after Hough Transform:

$$x = g(r_i, s_i, s_j \mid \overline{\mathbf{P}}_i, \overline{\mathbf{P}}_j), \ y = \arccos(\overline{\mathbf{t}}_{i,j}^\top \overline{\mathbf{t}}_j), \tag{16}$$

where x and y are transformed coordinates. From Eq. (16), x is a synthesized translation magnitude and y is angular variable. We then set $x \in [0, x_{\max}]$, and $y \in [0, \theta_{\max}]$, where $\theta_{\max} = \arccos(-\overline{\mathbf{t}}_j^\top \mathbf{R}_{i,j} \overline{\mathbf{t}}_i)$. Finally, the value of $\mathbf{H}$ is:

$$\forall y \in [0, \theta_{\max}], \ \mathbf{H}(x \mid y) = 1, \text{if } x \in [J_{\min}, J_{\max}], \tag{17}$$

where $J_{\min}$ and $J_{\max}$ are the two bounds from Eq. (15). The transformation over the scoring function $f_{i,j}^{2D}$ with all M sampled pixels between frame $\mathbf{I}_i$ and $\mathbf{I}_j$:

$$\mathbf{H}_{i,j} = \sum_m \mathbf{H}_{i,j}^m, \ f_{i,j}^{2D}(s_i, s_j, r_i \mid \overline{\mathbf{P}}_i, \overline{\mathbf{P}}_j) = \mathbf{H}_{i,j}(x, y), \tag{18}$$

where x and y are functions of s_i, s_j, r_i. Equation (1) becomes:

$$\phi(\overline{\mathcal{P}}) = \max_{\mathcal{S}, \mathcal{R}} \sum_i \sum_{j, j \neq i} \mathbf{H}_{i,j}(x(\mathcal{S}, \mathcal{R}), y(\mathcal{S}, \mathcal{R})). \tag{19}$$

In our implementation, we discretize $\mathbf{H}_{i,j}$ to a 2D matrix.

Accelerate Bundle-Adjustment Consensus. The BA determines $N-1$ camera scales and $N-1$ depth adjustments to maximize the scoring function $\phi(\cdot)$ in Eq. (19). With Hough transform, BA maximizes the summarized intensity via **indexing** $N \times (N-1)$ Hough transform matrices $\mathbf{H}$. It avoids BA repetitively enumerating all sampled pixels. Figure 6 shows an example optimization process.

Certified Global Optimality of robust inlier-counts scoring function Eq. (5) and Eq. (6) are achieved after optimization. See Fig. 8 for more analysis.

Optimization with RGB-D. With GT depthmap, the algorithm switches to the 3D scoring function $f_{i,j}^{3D}(\cdot)$. The depth adjustment is fixed to 1 and the 2D line-circle intersection becomes 3D line-sphere intersection. See Supp. for details.

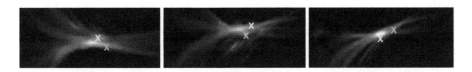

Fig. 6. Visualize Hough Transform Matrix $\mathbf{H}_i^j$ from Eq. (18). Area with higher intensity suggests more inlier counts. Given normalized pose group, for N views, there exists $N \times (N-1)$ matrices $\mathbf{H}_i^j$, constraining $N-1$ scale and $N-1$ adjustments. We plot the start and end points after optimizing Eq. (19) in the figure.

3.1.3 Computational Complexity

Naive Time Complexity. From Eq. (2) and Fig. 4, in each epoch, we evaluate $(N-1)(K-1)$ pose groups with Hough Transform Acceleration. Suppose each group takes T iterations to optimize Eq. (19), the time complexity is:

$$\mathcal{O}((N-1)(K-1) \cdot N(N-1) \cdot (M+T)), \tag{20}$$

where each group computes $N(N-1)$ Hough matrices $\mathbf{H}$. Each matrix enumerates M sampled pixels, see Eq. (18). Maximizing Eq. (19) becomes indexing $\mathbf{H}$, hence has a constant time complexity T, where $T << M$.

Counting Unique Hough Matrices. Most computation is spent on Hough matrices. In Fig. 4, each connection in the computation graph suggests two unique Hough matrices. We minimize time complexity by only computing **unique** Hough matrices. In Fig. 4 first epoch, the initial optimal group $\overline{\mathcal{P}}^*$ has $N(N-1)$ matrices. Each ablated group only differs by one pose, hence introducing $2(N-1)(N-1)(K-1)$ matrices. The first-epoch complexity is then:

$$\mathcal{O}^H(N(N-1)M + 2(N-1)^2(K-1)M) + \mathcal{O}^{\mathrm{BA}}(N(N-1)(K-1)T). \tag{21}$$

Only the Hough transform is accelerated. As $T << M$, the complexity of BA is neglectable. After the first epoch, $\overline{\mathcal{P}}^*$ only updates one pose per epoch, hence introducing $2(N-2)(K-1)$ matrices. The complexity for the rest epochs is,

$$\mathcal{O}^H(2(N-2)(K-1)M) + \mathcal{O}^{\mathrm{BA}}(N(N-1)(K-1)T). \tag{22}$$

While Eq. (22) has linear complexity, our method only updates one pose per epoch. Updating poses in all frames like other SfM methods is still quadratic.

3.2 Frustum Radiance Field Triangulation

Frustum Radiance Field. Now, we fix the optimized pose $\mathcal{P}^*$. Then we employ a frustum radiance field $\mathbf{V}$ of size $H \times W \times D$ for dense triangulation. Field $\mathbf{V}$ is defined over the root frame $\mathbf{I}_o$ and shares similarity with the categorical depthmap [4,17]. We follow [58,62] in rendering the depth d. The RGB estimation is skipped as unrelated. A 3D ray originated from pixel $\mathbf{p}_i$ at frame i is

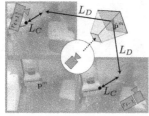 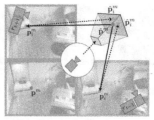

(a) Triangulation (b) Geometric Verification

Fig. 7. Triangulation optimizes frustum RF for multiview consistency *w.r.t.* depth and correspondence. **Geometric Verification** inferences RF for sparse multiview consistent 3D points. For simplicity, in (a), we only plot L_c defined from the root frame.

discretized into a set of 3D points and depth labels. With slight abuse of notation, we denote $\{\hat{\mathbf{p}}_{i,t} = \mathbf{o} + d_t \mathbf{r} \mid t \in [1, T]\}$, where $\hat{\mathbf{p}}$ is a 3D point, d_t is depth label and $\mathbf{r}$ is ray direction. Set integration interval $\delta_t = d_{t+1} - d_t$, depth d is:

$$d(\mathbf{p}_i) = \sum_t \alpha_t d_t, \ \alpha_t = T_t(1 - \exp(-\sigma_t \delta_t)), \ T_t = \exp(-\sum_{t' \in [1,t]} \sigma_{t'} \delta_{t'}). \quad (23)$$

We set the camera origin of frame i as $\mathbf{o}$. Instead of regressing occupancy δ with MLP [58,62], we directly interpolate the radiance field $\mathbf{V}$:

$$\delta_t = \mathbf{V}(u, v, w), \text{where } [u\ v\ w]^\mathsf{T} = \pi(\mathbf{E}, \hat{\mathbf{p}}_{i,t}). \quad (24)$$

Matrix $\mathbf{E}$ is the identity matrix. Function $\pi(\cdot)$ is projection function. Compared to using the MLP, frustum radiance field $\mathbf{V}$ is more computationally efficient [16].

Triangulation. Classic triangulation method [48] operates on a single 3D point. The RF provides additional constraints where all optimized points share a canonical 3D volume. In Fig. 7, we supervise $\mathbf{V}$ for multi-view consistency between dense depthmap $\mathcal{D}$ and correspondence map $\mathcal{C}$. On depth:

$$L_D = \frac{1}{NM} \sum_i \sum_m \|\pi(\mathbf{P}_i, \hat{\mathbf{p}}^m) - d_i^m\|_1. \quad (25)$$

Here, $\hat{\mathbf{p}}^m$ is rendered from the root frame, following depth computed with Eq. (23). To apply correspondence consistency, we have:

$$L_C = \frac{1}{N(N-1)M} \sum_i \sum_{j, j \neq i} \sum_m \|\pi(\mathbf{P}_j, \hat{\mathbf{p}}_i^m) - \mathbf{q}_{i,j}^m\|_1, \quad (26)$$

where $\hat{\mathbf{p}}_i^m = \pi^{-1}(\mathbf{P}_i, \mathbf{p}_i^m, d(\mathbf{p}_i^m))$, $\mathbf{p}_i^m = \pi(\mathbf{P}_i, \hat{\mathbf{p}}^m)$. With slight abuse of notation, function $\pi(\cdot)$ returns depth for L_D, and location for L_C. We **always** first render from the root frame and subsequently project to N frames. From there, we project to other supported frames again, forming $N(N-1)$ pairs.

Table 1. Self-Supervised Depth Estimation. We apply self-supervision with 5 frames via executing the local SfM. We output improved sparse depthmaps over SoTA supervised inputs. The evaluation is conducted over the root frame.

Dataset	Method	Density	$\delta_{0.5}$	δ_1	SI_{log}	A.Rel	S.Rel	RMS	RMS_{log}
ScanNet [13]	ZoeDepth [5]	9.1%	0.877	0.963	6.655	0.056	0.016	0.154	0.075
	↳ Ours		**0.902**	**0.976**	**5.901**	**0.050**	**0.014**	**0.149**	**0.070**
	ZeroDepth [35]	5.6%	0.641	0.834	12.860	0.124	0.086	0.337	0.152
	↳ Ours		**0.686**	**0.877**	**9.463**	**0.106**	**0.067**	**0.295**	**0.133**
	Metric3D [68]	2.6%	0.804	0.946	6.708	0.067	0.020	0.150	0.084
	↳ Ours		**0.854**	**0.968**	**4.170**	**0.055**	**0.014**	**0.125**	**0.068**
KITTI360 [33]	ZoeDepth [5]	4.0%	0.677	0.899	14.154	0.103	0.490	3.521	0.153
	↳ Ours		**0.719**	**0.910**	**13.220**	**0.094**	**0.474**	**3.499**	**0.145**
	ZeroDepth [35]	4.5%	0.584	0.844	16.468	0.132	0.819	3.486	0.183
	↳ Ours		**0.654**	**0.877**	**13.881**	**0.115**	**0.772**	**3.395**	**0.164**
	Metric3D [68]	3.2%	0.846	0.958	9.226	0.072	0.508	2.194	0.104
	↳ Ours		**0.860**	**0.963**	**8.896**	**0.068**	**0.487**	**2.139**	**0.101**

Table 2. Consistent Depth Estimation. We measure the numerical improvement by aligning the support frame depthmaps to the root frame with our depth adjustment scalars. The evaluation is conducted on support frames on ScanNet [13].

Method	$\delta_{0.5}$	δ_1	SI_{log}	A.Rel	S.Rel	RMS	RMS_{log}
ZoeDepth [5]	0.658	0.894	9.242	0.104	0.039	0.255	0.128
↳ Ours	**0.793**	**0.942**	9.242	**0.079**	**0.024**	**0.203**	**0.105**
ZeroDepth [35]	0.351	0.589	20.145	0.254	0.223	0.565	0.287
↳ Ours	**0.490**	**0.725**	20.145	**0.199**	**0.156**	**0.457**	**0.237**
Metric3D [68]	0.533	0.753	12.425	0.216	0.339	0.495	0.228
↳ Ours	**0.664**	**0.838**	12.425	**0.137**	**0.126**	**0.345**	**0.175**

3.3 Geometric Verification

With the RF optimized, we apply geometric verification to acquire sparse multi-view consistent 3D points, as in Fig. 7:

$$\mathcal{C} = \{\sum_{i, i \neq o} c_i^m \geq n^c\}, \; c_i^m = 1 \text{ if } \sum_{i, i \neq o} \|\hat{\mathbf{p}}_i^m - \hat{\mathbf{p}}^m\|_2 \leq \lambda^c. \tag{27}$$

We follow the same rendering process as training, where $\hat{\mathbf{p}}_i^m$ is computed with Eq. (26). First, we render 3D points from the root frame, project them to other views, and render 3D points from there again. A point is valid if a minimum of n^c views are consistent with the root.

4 Experiments

4.1 Self-supervised Depth Estimation

We benchmark whether self-supervision benefits supervised depth in unseen test data. For the correspondence estimator, we use PDC-Net [57]. For depth

Table 3. Self-Supervised Correspondence Estimation. We improve correspondence with RGB-D inputs, using metrics from [57]. The entry train and test are training and testing datasets of correspondence estimators. [Key: M = MegaDepth, S = ScanNet]

Method	Train	Test	PCK-1	PCK-3	PCK-5	AEPE
PDC-Net [57]			0.119	0.511	0.743	4.612
↳ LightedDepth [77]	M	S	0.061	0.341	0.563	6.590
↳ Ours			**0.178**	**0.658**	**0.866**	**2.898**
RoMa [15]			0.144	0.583	0.815	3.333
↳ LightedDepth [77]	S	S	0.066	0.359	0.588	5.974
↳ Ours			**0.183**	**0.638**	**0.844**	**3.067**

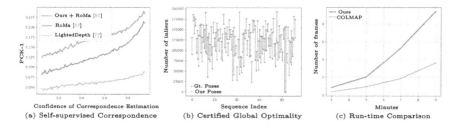

(a) Self-supervised Correspondence (b) Certified Global Optimality (c) Run-time Comparison

Fig. 8. Ablation Studies on the ScanNet.

estimators, we adopt recently published in-the-wild depth estimator, including ZoeDepth [5], ZeroDepth [35], and Metric3D [68]. We evaluate with ScanNet [13] and KITTI360 [33] where all models perform zero-shot prediction.

Test Data. In dense correspondence estimation, methods [57,58,78] output confidence score per correspondence. We follow [57,58] to set a minimum threshold of 0.95. We run on ScanNet test split and it returns 92 sequences with sufficient correspondence. We form our test split by sampling 5 neighboring frames per valid sequence. Similarly, we run on KITTI360 data and randomly select 100×5 test split, *i.e.*, 100 sequences with 5 frames each. We consider it a comprehensive experiment. Similar to SPARF [58], our triangulation trains a NeRF-like structure. For reference, SPARF experiment on DTU dataset [26] includes only 15 sequences each with 3 images. In comparison, we include around 100 sequences.

Evaluation Protocols. We evaluate on **root** frame. We remove the scale ambiguity in the local SfM system to correctly reflect depth improvement. Specifically, we adjust all 5 depthmaps by an identical scalar computed between estimated root and GT depthmap, *i.e.*, the median scaling [21]. This eliminates scale ambiguity in the root frame while preserving it in support frames.

Results. In Table 1, our point cloud has a density of 2.6%–9.1%, which amounts to **10−30k** points on a 480×640 image. On accuracy, we have **unanimous** improvement over all supervised models of both datasets. Especially, we outperform strong baselines of ZoeDepth on ScanNet and Metric3D on KITTI360.

Fig. 9. Self-supervised Correspondence Estimation enabled by our method with RGB-D inputs. The correspondence error is marked by the radius of the circle.

4.2 Consistent Depth Estimation

We evaluate on ScanNet. We follow Sect. 4.1 data split but evaluate the **support** frames. Temporal consistent depth is essential for AR applications [37]. Table 2 reflects the performance gain by aligning support frames to root with adjustments, which are jointly estimated with camera poses, see Fig. 2 and Fig. 3.

4.3 Self-supervised Correspondence Estimation

Real-world image correspondence label is expensive, *e.g.* KITTI provides only 200 optical flow labels. Existing datasets, such as MegaDepth and ScanNet, require large-scale 3D reconstruction with manual verification. Hence, correspondence estimators can not fine-tune on general RGB-D datasets like NYUv2 [51] or KITTI [18]. But our method enables self-supervised correspondence estimation on RGB-D data when using 3D scoring function Eq. (6). The camera poses are optimized with the point cloud specified by depthmap and correspondence. The accurate pose in turn improves projective correspondence. In Table 3, with 5 RGB-D frames, our method improves projective correspondence over inputs. We use the same test split as Sect. 4.1. The evaluation accumulates correspondence of each frame pair. Figure 8a shows our improvement is **unanimous** over both confident and unconfident estimation. A visual example is in Fig. 9.

4.4 Sparse-View Pose Estimation

Comparison with Optimization-Based and Learning-Based Poses. Previous studies either evaluate two-view pose [22,55], or SLAM-like odometry [61]. For more comparison, following Sect. 4.1 ScanNet split, we keep root frame and gradually add neighboring frames. In Table 4, LightedDepth [77] and ours both use PDC-Net [57] correspondence and ZoeDepth [5] mono-depth. COLMAP [48] uses PDC-Net correspondence. In evaluation, we follow [58] in aligning to GT poses. In Table 4, our **zero-shot** pose accuracy significantly outperforms all prior arts, including [22,55,61] with ScanNet [13] or ScanNet++ [67] in their training set. See Supp. Table 4 for complete comparison from 3 to 9 frames. In Fig. 8, we attribute our superiority to certified global optimality over robust measurements.

Table 4. Sparse-view Pose Comparison with optimization-based and learning-based methods. We only compare against COLMAP on its success sequences. Please see the complete comparison from 3 to 9 frames in Supp. Table 1. Our method performs **zero-shot** testing on ScanNet while outperforming DeepV2D [55], DRO [22] with ScaNet [13] in training set. DUSt3R [61] trains on a similar dataset ScanNet++ [67].

Frames	Method	Zero-shot	Suc. (%)	PCK-3	C3D-3	Rot.	Trans.
5	COLMAP [48]	✓	36.7	0.584	0.863	0.577	1.296
	Ours	✓	100.0	**0.727**	**0.904**	**0.422**	**1.062**
	DeepV2D [55] - ScanNet	✗		0.526	0.805	0.945	1.496
	DeepV2D [55] - NYUv2	✓		0.530	0.771	1.041	1.568
	DeepV2D [55] - KITTI	✓		0.125	0.387	4.908	4.231
	LightedDepth [77]	✓		0.651	0.832	0.469	1.550
	DRO [22] - ScanNet	✗	100.0	0.656	0.853	0.385	1.200
	DRO [22] - KITTI	✓		0.003	0.211	3.610	5.469
	DUSt3R [61] w.o. Intrinsic	✓		0.364	0.705	0.487	2.074
	DUSt3R [61] w.t. Intrinsic	✓		0.594	0.824	0.570	1.759
	Ours	✓		**0.799**	**0.900**	**0.368**	**1.120**

Table 5. Sparse-view Pose Comparison with NeRF-based methods following [58].

Method	Frames	LLFF [49]		Replica [54]	
		Rot.	Trans.	Rot.	Trans.
BARF [34]		2.04	11.6	3.35	16.96
RegBARF [34,40]		1.52	5.0	3.66	20.87
DistBARF [2,34]	3	5.59	26.5	2.36	7.73
SCNeRF [27]		1.93	11.4	0.65	4.12
SPARF [58]		0.53	2.8	**0.15**	**0.76**
Ours		**0.46**	**1.9**	0.52	4.09

Comparison with NeRF-Based Poses. Sparse view NeRF methods optimize NeRF jointly with camera poses, mandating a sophisticated and time-consuming optimization scheme. *E.g.*, SPARF [58], takes one day to optimize the pose and NeRF. Typically, their poses are initialized with COLMAP. Our method provides an alternative initialization with superior performance. In Table 5, our initialization achieves better or on-par pose performance than SoTA [58] while only taking ~3 minutes (Fig. 7). Our lower performance on Replica dataset might be due to ZoeDepth not being trained on synthetic data. Our work suggests the straightforward "first-pose-then-NeRF" scheme also applies to short videos.

Certified Global Optimality. In Fig. 8b, our Bundle-RANSAC-Adjustment **always** finds more inliers than groundtruth poses. To our best knowledge, we are the **first** work that extends RANSAC to a multi-view system.

Run-Time. In Fig. 8c, we run approximately 3× slower than COLMAP. But both have quadratic complexity. With 3/5/7/9 frames, we take 0.8/2.0/5.3/9.4 minutes on RTX 2080 Ti GPU, while COLMAP uses 0.3/0.9/1.8/3.6 minutes on

Intel Xeon 4216 CPU. COLMAP runs sequentially. But our method is highly parallelized. Our core operation Hougn Transform scales up with more GPUs.

5 Conclusion

By revisiting self-supervision with local SfM, we first show self-supervised depth benefits SoTA supervised model with only 5 frames. We have SoTA sparse-view pose accuracy, applicable to NeRF rendering. We have diverse applications including self-supervised correspondence and consistent depth estimation.

Limitation. The NeRF-like triangulation constrains our method from applying to large-scale self-supervised learning. Its efficiency requires improvement.

References

1. Agarwal, S., et al.: Building Rome in a day. Commun. ACM **54**, 105–112 (2011)
2. Barron, J.T., Mildenhall, B., Verbin, D., Srinivasan, P.P., Hedman, P.: Mip-NeRF 360: unbounded anti-aliased neural radiance fields. In: CVPR (2022)
3. Beder, C., Steffen, R.: Determining an initial image pair for fixing the scale of a 3D reconstruction from an image sequence. In: Joint Pattern Recognition Symposium (2006)
4. Bhat, S.F., Alhashim, I., Wonka, P.: AdaBins: depth estimation using adaptive bins. In: CVPR (2021)
5. Bhat, S.F., Birkl, R., Wofk, D., Wonka, P., Müller, M.: ZoeDepth: zero-shot transfer by combining relative and metric depth. arXiv preprint arXiv:2302.12288 (2023)
6. Bian, J., et al.: Unsupervised scale-consistent depth and ego-motion learning from monocular video. In: NeurIPS (2019)
7. Brown, M., Hua, G., Winder, S.: Discriminative learning of local image descriptors. PAMI **33**, 43–57 (2010)
8. Butime, J., Gutierrez, I., Corzo, L.G., Espronceda, C.F.: 3D reconstruction methods, a survey. In: VISAPP (2006)
9. Casser, V., Pirk, S., Mahjourian, R., Angelova, A.: Depth prediction without the sensors: leveraging structure for unsupervised learning from monocular videos. In: AAAI (2019)
10. Chen, Y., Schmid, C., Sminchisescu, C.: Self-supervised learning with geometric constraints in monocular video: connecting flow, depth, and camera. In: ICCV (2019)
11. Clark, R., Bloesch, M., Czarnowski, J., Leutenegger, S., Davison, A.J.: LS-Net: learning to solve nonlinear least squares for monocular stereo. ECCV (2018)
12. Crandall, D., Owens, A., Snavely, N., Huttenlocher, D.: Discrete-continuous optimization for large-scale structure from motion. In: CVPR (2011)
13. Dai, A., Chang, A., Savva, M., Halber, M., Funkhouser, T., Nießner, M.: ScanNet: Richly-annotated 3D reconstructions of indoor scenes. In: PAMI (2017)
14. Deng, K., Liu, A., Zhu, J.Y., Ramanan, D.: Depth-supervised NeRF: fewer views and faster training for free. In: CVPR (2022)
15. Edstedt, J., Sun, Q., Bökman, G., Wadenbäck, M., Felsberg, M.: RoMa: revisiting robust losses for dense feature matching. arXiv preprint arXiv:2305.15404 (2023)

16. Fridovich-Keil, S., Yu, A., Tancik, M., Chen, Q., Recht, B., Kanazawa, A.: Plenoxels: radiance fields without neural networks. In: CVPR (2022)
17. Fu, H., Gong, M., Wang, C., Batmanghelich, K., Tao, D.: Deep ordinal regression network for monocular depth estimation. In: CVPR (2018)
18. Geiger, A., Lenz, P., Stiller, C., Urtasun, R.: Vision meets robotics: the KITTI dataset. IJRR **32**, 1231–1237 (2013)
19. Geiger, A., Lenz, P., Urtasun, R.: Are we ready for autonomous driving? The KITTI vision benchmark suite. In: CVPR (2012)
20. Godard, C., Mac Aodha, O., Firman, M., Brostow, G.J.: Digging into self-supervised monocular depth estimation. In: ICCV (2019)
21. Gordon, A., Li, H., Jonschkowski, R., Angelova, A.: Depth from videos in the wild: unsupervised monocular depth learning from unknown cameras. In: ICCV (2019)
22. Gu, X., Yuan, W., Dai, Z., Zhu, S., Tang, C., Dong, Z., Tan, P.: DRO: deep recurrent optimizer for video to depth. IEEE Robot. Autom. Lett. **8**, 2844–2851 (2023)
23. Guizilini, V., Ambruş, R., Chen, D., Zakharov, S., Gaidon, A.: Multi-frame self-supervised depth with transformers. In: CVPR (2022)
24. Hartley, R., Zisserman, A.: Multiple View Geometry in Computer Vision. Cambridge University Press, Cambridge (2003)
25. Izquierdo, S., Civera, J.: SfM-TTR: using structure from motion for test-time refinement of single-view depth networks. In: CVPR (2023)
26. Jensen, R., Dahl, A., Vogiatzis, G., Tola, E., Aanæs, H.: Large scale multi-view stereopsis evaluation. In: CVPR (2014)
27. Jeong, Y., Ahn, S., Choy, C., Anandkumar, A., Cho, M., Park, J.: Self-calibrating neural radiance fields. In: CVPR (2021)
28. Kuznietsov, Y., Proesmans, M., Van Gool, L.: CoMoDA: continuous monocular depth adaptation using past experiences. In: WACV (2021)
29. Lee, J.H., Han, M.K., Ko, D.W., Suh, I.H.: From big to small: multi-scale local planar guidance for monocular depth estimation. arXiv preprint arXiv:1907.10326 (2019)
30. Lepetit, V., Moreno-Noguer, F., Fua, P.: Ep n p: an accurate o (n) solution to the p n p problem. IJCV (2009)
31. Li, H.: A practical algorithm for l triangulation with outliers. In: CVPR (2007)
32. Li, H., Hartley, R.: Five-point motion estimation made easy. In: ICPR (2006)
33. Liao, Y., Xie, J., Geiger, A.: KITTI-360: a novel dataset and benchmarks for urban scene understanding in 2D and 3D. IEEE Trans. Pattern Anal. Mach. Intell. **45**, 3292–3310 (2022)
34. Lin, C.H., Ma, W.C., Torralba, A., Lucey, S.: BARF: bundle-adjusting neural radiance fields. In: ICCV (2021)
35. Liu, R., Wu, R., Van Hoorick, B., Tokmakov, P., Zakharov, S., Vondrick, C.: Zero-1-to-3: zero-shot one image to 3D object. arXiv preprint arXiv:2303.11328 (2023)
36. Lowe, D.G.: Distinctive image features from scale-invariant keypoints. IJCV **60**, 91–110 (2004)
37. Luo, X., Huang, J.B., Szeliski, R., Matzen, K., Kopf, J.: Consistent video depth estimation. ToG (2020)
38. Mahjourian, R., Wicke, M., Angelova, A.: Unsupervised learning of depth and ego-motion from monocular video using 3D geometric constraints. In: CVPR (2018)
39. Mildenhall, B., Srinivasan, P.P., Tancik, M., Barron, J.T., Ramamoorthi, R., Ng, R.: NeRF: representing scenes as neural radiance fields for view synthesis. Commun. ACM **65**, 99–106 (2021)

40. Niemeyer, M., Barron, J.T., Mildenhall, B., Sajjadi, M.S., Geiger, A., Radwan, N.: RegNeRF: regularizing neural radiance fields for view synthesis from sparse inputs. In: CVPR (2022)
41. Niu, L., et al.: Making images real again: a comprehensive survey on deep image composition. arXiv preprint arXiv:2106.14490 (2021)
42. Olsson, C., Eriksson, A., Hartley, R.: Outlier removal using duality. In: CVPR (2010)
43. Piccinelli, L., Sakaridis, C., Yu, F.: IDISC: internal discretization for monocular depth estimation. In: CVPR (2023)
44. Pillai, S., Ambruș, R., Gaidon, A.: SuperDepth: self-supervised, super-resolved monocular depth estimation. In: ICRA (2019)
45. Ranftl, R., Lasinger, K., Hafner, D., Schindler, K., Koltun, V.: Towards robust monocular depth estimation: mixing datasets for zero-shot cross-dataset transfer. PAMI **44**, 1623–1637 (2020)
46. Ranjan, A., et al.: Competitive collaboration: joint unsupervised learning of depth, camera motion, optical flow and motion segmentation. In: CVPR (2019)
47. Sarlin, P.E., Lindenberger, P., Larsson, V., Pollefeys, M.: Pixel-perfect structure-from-motion with feature metric refinement. PAMI (2023)
48. Schonberger, J.L., Frahm, J.M.: Structure-from-motion revisited. In: CVPR (2016)
49. Shafiei, M., Bi, S., Li, Z., Liaudanskas, A., Ortiz-Cayon, R., Ramamoorthi, R.: Learning neural transmittance for efficient rendering of reflectance fields (2021)
50. Shu, C., Yu, K., Duan, Z., Yang, K.: Feature-metric loss for self-supervised learning of depth and egomotion. In: Vedaldi, A., Bischof, H., Brox, T., Frahm, J.-M. (eds.) ECCV 2020. LNCS, vol. 12364, pp. 572–588. Springer, Cham (2020). https://doi.org/10.1007/978-3-030-58529-7_34
51. Silberman, N., Hoiem, D., Kohli, P., Fergus, R.: Indoor segmentation and support inference from RGBD images. In: Fitzgibbon, A., Lazebnik, S., Perona, P., Sato, Y., Schmid, C. (eds.) ECCV 2012. LNCS, vol. 7576, pp. 746–760. Springer, Heidelberg (2012). https://doi.org/10.1007/978-3-642-33715-4_54
52. Snavely, N., Seitz, S.M., Szeliski, R.: Photo tourism: exploring photo collections in 3D. In: Siggraph (2006)
53. Spencer, J., Russell, C., Hadfield, S., Bowden, R.: Kick back & relax: learning to reconstruct the world by watching SlowTV. In: ICCV (2023)
54. Straub, J., et al.: The Replica dataset: a digital replica of indoor spaces. arXiv preprint arXiv:1906.05797 (2019)
55. Teed, Z., Deng, J.: DeepV2D: video to depth with differentiable structure from motion. In: ICLR (2020)
56. Tiwari, L., Ji, P., Tran, Q.-H., Zhuang, B., Anand, S., Chandraker, M.: Pseudo RGB-D for self-improving monocular SLAM and depth prediction. In: Vedaldi, A., Bischof, H., Brox, T., Frahm, J.-M. (eds.) ECCV 2020. LNCS, vol. 12356, pp. 437–455. Springer, Cham (2020). https://doi.org/10.1007/978-3-030-58621-8_26
57. Truong, P., Danelljan, M., Timofte, R., Van Gool, L.: PDC-Net+: enhanced probabilistic dense correspondence network. PAMI **45**, 10247–10266 (2023)
58. Truong, P., Rakotosaona, M.J., Manhardt, F., Tombari, F.: SpaRF: neural radiance fields from sparse and noisy poses. In: CVPR (2023)
59. Tuytelaars, T., Mikolajczyk, K., et al.: Local invariant feature detectors: a survey. Found. Trends® Comput. Graph. Vision (2008)
60. Ummenhofer, B., et al.: DeMoN: depth and motion network for learning monocular stereo. In: CVPR (2017)
61. Wang, S., Leroy, V., Cabon, Y., Chidlovskii, B., Revaud, J.: DUSt3R: geometric 3D vision made easy. arXiv preprint arXiv:2312.14132 (2023)

62. Wang, Z., Wu, S., Xie, W., Chen, M., Prisacariu, V.A.: NeRF–: neural radiance fields without known camera parameters. arXiv preprint arXiv:2102.07064 (2021)
63. Watson, J., Firman, M., Brostow, G.J., Turmukhambetov, D.: Self-supervised monocular depth hints. In: ICCV (2019)
64. Watson, J., Mac Aodha, O., Prisacariu, V., Brostow, G., Firman, M.: The temporal opportunist: self-supervised multi-frame monocular depth. In: CVPR (2021)
65. Wu, C.: Towards linear-time incremental structure from motion. In: 3DV (2013)
66. Yang, L., Kang, B., Huang, Z., Xu, X., Feng, J., Zhao, H.: Depth anything: unleashing the power of large-scale unlabeled data. arXiv preprint arXiv:2401.10891 (2024)
67. Yeshwanth, C., Liu, Y.C., Nießner, M., Dai, A.: ScanNet++: a high-fidelity dataset of 3D indoor scenes. In: ICCV (2023)
68. Yin, W., Zhang, C., Chen, H., Cai, Z., Yu, G., Wang, K., Chen, X., Shen, C.: Metric3D: towards zero-shot metric 3D prediction from a single image. In: ICCV (2023)
69. Yuan, W., Gu, X., Dai, Z., Zhu, S., Tan, P.: New CRFs: neural window fully-connected CRFs for monocular depth estimation. In: CVPR (2022)
70. Zhan, H., Garg, R., Weerasekera, C.S., Li, K., Agarwal, H., Reid, I.: Unsupervised learning of monocular depth estimation and visual odometry with deep feature reconstruction. In: CVPR (2018)
71. Zhang, Z.: Microsoft kinect sensor and its effect. IEEE Multimed. **19**(2), 4–10 (2012)
72. Zhang, Z., Cole, F., Tucker, R., Freeman, W.T., Dekel, T.: Consistent depth of moving objects in video. TOG **40**, 1–12 (2021)
73. Zhao, W., Liu, S., Shu, Y., Liu, Y.J.: Towards better generalization: joint depth-pose learning without PoseNet. In: CVPR (2020)
74. Zhou, T., Brown, M., Snavely, N., Lowe, D.G.: Unsupervised learning of depth and ego-motion from video. In: CVPR (2017)
75. Zhou, Z., Dong, Q.: Two-in-one depth: bridging the gap between monocular and binocular self-supervised depth estimation. In: ICCV (2023)
76. Zhu, S., Brazil, G., Liu, X.: The edge of depth: explicit constraints between segmentation and depth. In: CVPR (2020)
77. Zhu, S., Liu, X.: LightedDepth: video depth estimation in light of limited inference view angles. In: CVPR (2023)
78. Zhu, S., Liu, X.: PMatch: paired masked image modeling for dense geometric matching. In: CVPR (2023)

FAMOUS: High-Fidelity Monocular 3D Human Digitization Using View Synthesis

Vishnu Mani Hema[1](✉) ⓘ, Shubhra Aich[1] ⓘ, Christian Haene[2] ⓘ,
Jean-Charles Bazin[2] ⓘ, and Fernando De la Torre[1] ⓘ

[1] Carnegie Mellon University, Pittsburgh, USA
vmanihem@alumni.cmu.edu
[2] Pittsburgh, USA

Abstract. The advancement in deep implicit modeling and articulated models has significantly enhanced the process of digitizing human figures in 3D from just a single image. While state-of-the-art methods have greatly improved geometric precision, the challenge of accurately inferring texture remains, particularly in obscured areas such as the back of a person in frontal-view images. This limitation in texture prediction largely stems from the scarcity of large-scale and diverse 3D datasets, whereas their 2D counterparts are abundant and easily accessible. To address this issue, our paper proposes leveraging extensive 2D fashion datasets to enhance both texture and shape prediction in 3D human digitization. We incorporate 2D priors from the fashion dataset to learn the occluded back view, refined with our proposed domain alignment strategy. We then fuse this information with the input image to obtain a fully textured mesh of the given person. Through extensive experimentation on standard 3D human benchmarks, we demonstrate the superior performance of our approach in terms of both texture and geometry. Code and dataset is available at https://github.com/humansensinglab/FAMOUS.

Keywords: Human digitization · 2D prior · Fashion dataset

1 Introduction

High-fidelity 3D human digitization (e.g., [34]) is crucial for a wide range of virtual reality (VR) and augmented reality (AR) applications. This technology is extensively used in industries like gaming and entertainment [3], animation [7], visual effects, virtual fashion experiences [6], fitness programs [2], and virtual

---

C. Haene and J.-C. Bazin—Independent Researcher.

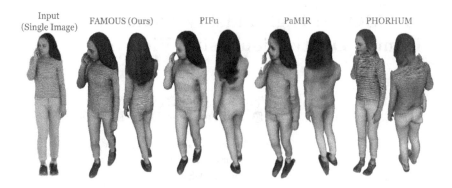

Fig. 1. Visual comparison between monocular 3D human digitization methods. Given a single image input, we demonstrate the renderings of textured mesh generated by our approach (FAMOUS) and SOTA pipelines (*i.e.* predicting both the shape and texture): PIFu [33], PaMIR [52], and PHORHUM [10]. Our approach presents significantly improved results in terms of texture and geometry.

training simulations. It also has significant applications in healthcare, anthropology [5], and forensics [1,36] to name a few. This advanced 3D digitization enables the creation of highly realistic avatars, enhancing the immersive experience in virtual environments by providing detailed and lifelike interactions.

Creating such high-precision avatars requires generating both texture and shape in meticulous detail. Traditionally, this involves collecting data using a 3D scanner and/or a multi-camera setup [16]. This data is then processed to create a mesh through algorithms, followed by texture mapping, rigging, and further refinements often involving human artists. However, this semi-manual and multi-step process is not easily scalable for contemporary VR, MR, and AR platforms, which cater to millions of users globally, like the rapidly evolving digital environments of the metaverse.

Creating an automated method that can handle millions of users is crucial for bringing virtual reality into the mainstream. Recent advancements in deep learning have spurred several initiatives towards automation and scalability [33,34,43,44]. However, while the geometric accuracy of these state-of-the-art methods is impressive, the texture quality of unseen areas is still lacking, especially in moderately textured clothing (see Fig. 1). Some methods even overlook this aspect entirely [43,44].

Our approach focuses on achieving both realistic texture quality and geometric accuracy, see Fig. 1. Additionally, we address practical concerns like bandwidth limitations in large-scale VR platforms. To tackle this, we propose an approach for 3D human digitization using just a single image per user. Considering the massive user base, ranging from millions to potentially billions, even uploading two images per user would significantly increase bandwidth demands.

PIFuHD [34] is the first attempt regarding full-resolution monocular 3D human reconstruction that builds on top of the pioneering work of PIFu [33] based on implicit function (IF) models. Incorporation of articulated models from the SMPL family [26, 28] to guide the implicit function further improves the geometric reconstruction quality for challenging poses in ICON [44]. However, this guided reconstruction tends to overfit the articulated model and thus fails to produce realistic results for typical fashion poses (details in the supplementary Section B). Note that among these methods, only PIFu [33] predicts both shape and texture (in a coarse resolution), whereas PIFu-HD, ICON, and ECON focus on the reconstruction task only (i.e. no texture). PHORHUM [10] is one of the most recent entry dealing with both human shape and texture based on the albedo surface color and shading information but still fails to extract semantically accurate texture for the occluded region. DIFU [39] and 2K2K [17] also, effectively reconstructs geometry from high resolution image but DIFU fails to obtain high fidelity textures and 2K2K doesn't focus on texture generation.

We believe that a key obstacle in achieving satisfactory texture quality in existing literature is the limited diversity of textures in the relatively small pool of available 3D samples [4, 46]. Acquiring detailed 3D scans is a time-consuming and costly process. Conversely, there is an abundance of varied cloth textures in 2D images, for example accessible in large-scale 2D fashion datasets [25] and online. This contrast points to an opportunity for leveraging these extensive 2D resources to enhance texture quality in 3D models. Drawing from this insight, our paper leverages extensive 2D fashion datasets to enhance the texture quality of 3D models. As will be shown in the experiments section, this approach not only improves texture fidelity but also boosts the geometric precision of the models.

Our approach FAMOUS integrates the rich textural data from 2D datasets into our 3D modeling process through a technique of view synthesis, utilizing pretrained hallucinator [30]. We then iteratively refine the hallucinator, focusing on disentangled factors as outlined in our methodology section. This self-supervised process, which we term "disentangled domain alignment", effectively aligning to the limited variety found in 3D datasets. Our extensive experiments demonstrate that by merging abundant 2D data with a smaller set of 3D scans in this manner, we can produce 3D models of superior fidelity, both for texture and geometry, compared to existing state-of-the-art methods. To our knowledge, this is the first instance of using 2D datasets in conjunction with limited 3D dataset through the lens of hallucinators, using domain alignment strategy focused on disentanglement factors.

Overall, below is the summary of our contributions:

- We propose FAMOUS, a framework that generates a high-fidelity textured mesh given a single RGB image. We harness the rich context available in the form of 2D fashion datasets through the lens of a hallucinator resulting in improved 3D shape and texture, especially for the occluded views.
- To this end, we also contribute a large-scale 2D fashion dataset, a derivative from Deep Fashion HD [25] but with full body front back image pairs with their COCO-skeleton keypoints annotations. We hope this new dataset will complement future research in the community.

- We introduce a self-supervised disentangled domain alignment approach to iteratively enable the hallucinator generalize to the distribution of the target 3D dataset.
- Finally, we will release our complete codebase along with the dataset splits to facilitate more open-source research in this subdomain.

2 Related Work

2.1 3D Human Digitization

Multi-view Reconstruction. Earlier work [42,54] in visual hull-based surface reconstruction required images to be captured from multiple viewpoints. The reconstructed avatar needed further postprocessing based on multiview silhouettes to compensate for the additional complexity induced by the topology of the cloth and self-occlusion, in particular. However, the visual hull based approaches fall short on fine-grained reconstruction when the number of input views is limited. On this account, deep volumetric stereo [20] attempts to predict the volumetric occupancy that can capture dynamic clothed humans from highly sparse views. A similar approach is used in [15] that uses a trained autoencoder to generate a deep prior to enable high-end volumetric captures. However, the inherent requirement for a multi-camera setup for these systems poses additional constraints regarding scalability to mass users.

Single-View Reconstruction. These approaches can be categorized into explicit shape-based methods (i.e. voxel grid techniques, template meshes), implicit function-based approaches, and hybrid methods combining both.

1) Explicit Shape Based Approaches. The introduction of SMPL [26] made the estimation of pose and shape in 3D human reconstruction tractable. SMPL works based on a linear combination of a small set of body shapes and pose parameters, which allows it to generate to a wide range of realistic body shapes and poses. Octopus [9] takes SMPL further by introducing SMPL-D which learns the SMPL body parameter and additional 3D vertex offsets that model clothing, hair, and details. The primary limitation of SMPL and SMPL-D is their fixed topology of the template mesh, which means that these template models are difficult to fit to arbitrary topological changes such as the disappearance of body parts and clothes with substantially different topologies. Other methods use nonparametric depth map [38] or point cloud [47]. However, scaling up these approaches to model the loose or diverse clothing is nontrivial.

2) Implicit Function Based Approaches. This kind of models offer several advantages, such as topology-agnosticism and the ability to represent arbitrary 3D clothed human shapes. PIFu [33] introduces pixel-aligned implicit human shape reconstruction – the first purely implicit function based occupancy and texture predictors for 3D human digitization. PIFuHD [34] improves the geometric details with multi-resolution network architectures and estimated normal maps. However, both these methods tend to overfit to the body poses in training distribution (i.e. fashion poses) [44] GeoPIFu [18], Self-Portraits [23], PINA [14], and S3 [45] overcome this limitation by introducing geometric priors to regularize the deep implicit representation.

3) Hybrid Appraoches. PaMIR [52], DeepMultiCap [51], JIFF [12], ARCH [21], ARCH++ [19] and ICON [44] attempt to combine both the explicit, parametric body models with continuous, implicit representations. Despite the promising performance of these approaches emphasizing primarily on the modeling innovations, the fundamental limitation regarding the scarcity of detailed 3D scans remains somewhat unexplored. This is evident in case the texture pattern of cloth topology is far from the quite limited training distribution of the 3D scans (Figs. 1 and 7).

2.2 Pose-Guided Person Image Synthesis

These approaches synthesize novel views of a person from a reference (source) image and a target pose. Controllable GAN based ones [27,35] extract person attributes from various segmentation maps. Dense deformations and flow-based methods [8,22] generate aligned features and estimate appearance flow between references and desired targets. GFLA [31] obtains a global flow field and occlusion mask to match the patches from the source image to the target pose. Some other methods [22,24] use 3D geometric details that fit SMPL mesh onto 2D images and query the source appearance using the 3D context. PISE [49] and CASD [53] parse maps to guide the view synthesis in the target pose. CoCosNet [50] applies cross-attention to extract dense correspondences but are limited in lower resolution due to their quadratic memory footprint. Recently, NTED [30] proposes an efficient attention mechanism based on semantic attributes to achieve promising texture quality but lacks 3D consistency.

3 Our Approach: FAMOUS

FAMOUS aims to infer high-fidelity 3D shape and texture of a clothed human from a single monocular image. Figure 2 illustrates our complete framework comprising of three steps: (1) Distributionally Aligned Hallucinator (DAH) for generation of the back-view image. (2) Geometry prediction, and (3) texture prediction from the reference image and generated back image.

3.1 Distributionally Aligned Hallucinator (DAH)

A pivotal stage in our process involves generating a back image from the frontal view to provide an additional viewpoint for constructing a 3D model. While this task can be achieved easily using methods trained on 3D scans [33,52], these approaches are typically limited by small training datasets and struggle to generalize well, particularly in term of texture. In this paper, we propose leveraging large-scale 2D datasets, the source (e.g., Fashion dataset) alongside a limited

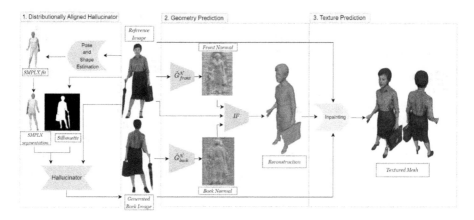

Fig. 2. Method overview. Given a single frontal image of a subject, FAMOUS generates a high-quality textured mesh. Our distributionally aligned hallucinator (Sect. 3.1) first predicts the back view based on the reference image, SMPL-X segmentation map and silhouette. Next, the normal map generator $\{(\hat{G}_f^\mathcal{N}, \hat{G}_b^\mathcal{N})\}$ takes the reference image and generated back image as input and output the respective normal maps. The reference image and the normal estimates are leveraged by the implicit function network (IF) for the geometry prediction (Sect. 3.2). Finally, our learnable texture prediction module refines the 3D texture aggregated from the input reference image and predicted back view (Sect. 3.3).

amount of 3D data, the target (e.g., Render-People dataset [4]) to train a robust hallucinator. This hallucinator is designed to generate a back view from a frontal image. Drawing inspiration from recent research [30], we pretrained the hallucinator by augmenting an existing large-scale 2D dataset to account for partial occlusion of the samples (semantic changes) and different body poses (pose changes). Following this augmentation, we conduct simultaneous training and fine-tuning on the target 3D dataset (RenderPeople [4]).

It's worth noting that in our current implementation, we utilize the state-of-the-art sparse attention-based StyleGAN hallucinator [30] rather than the most recent diffusion-based approach [11] due to the substantial memory requirements of the latter.

Problems with Existing Approaches. As mentioned above, SOTA methods for monocular 3D human digitization primarily revolve around training models using 3D scans with limited texture variations. Hence, these approaches fail to achieve good generalization in terms of texture. To overcome this bottleneck, a scalable solution is to leverage the abundant and easily accessible fashion 2D datasets. Directly incorporating the hallucinators [11,30] train on source fashion datasets into our pipeline does not yield satisfactory results for the target 3D dataset. That is, if we train our models in 2D Fashion dataset (e.g., DeepFashion [25]), and test it in target 3D dataset (e.g., Render people [4]), the results are not satisfactory. The same occurs when we fine-tune a pretrained model. The

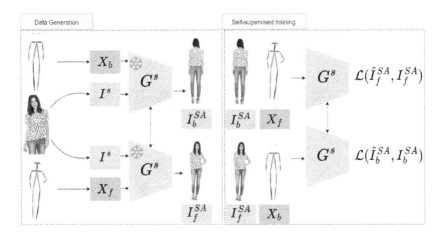

Fig. 3. Semantic Alignment (SA): Given a partial image I^s and full body COCO skeleton (X_f, X_b) from our 2D fashion dataset (i.e. source), we generate the front/back pairs (I_f^{SA}, I_b^{SA}) similar to our full body target 3D dataset (thus aligning the *semantics*) with a frozen hallucinator (G^s). Next, we finetune the hallucinator on these full body pseudo pairs from one another. The hallucinator is initialized with pretrained weights from [30], indicated by superscript s. The weight sharing in each stage is indicated by the bidirectional dotted arrows. The default loss function ($\mathcal{L}$) proposed in [30] is used.

domain gap between these two datasets leads to performance degradation of the hallucinator on the target data distribution. Domain adaptation methods such as [40,41] fall short due to the insufficient support [13] of our source and target dataset in terms of pose and texture variations, respectively (see supplementary Section A for examples and more detailed explanation). To address this issues, we propose the Disentangled Domain Alignment (DDA) approach.

Disentangled Domain Alignment (DDA). An image can be represented by its style, semantics, pose, and view [29]. Our hallucinator [30] can learn to disentangle these factors from an image to perform a task like view synthesis. We find that the domain gap between the two datasets in the context of the hallucinator depends on these disentanglement factors, so we introduce a factorized alignment approach rooted in this concept. We refer to this approach as DDA. The key steps of our DDA scheme are outlined as follows.

Semantic Alignment (SA): The goal of SA is to shift the data distribution learned by the hallucinator from the source to the target distribution based on the semantics. The main difference, in terms of semantics, is the obvious lack of a sufficient number of full body image pairs covering the front/back view in the source dataset. More specifically, only about 8.5% of the source fashion dataset [25] contains full body pairs and 10% of which encompasses back views. This is prevalent for the fashion datasets since most of these image pairs

Fig. 4. Dataset: The sets of images randomly sampled from the source dataset (top) with their corresponding SA (mid) and PA (bottom) pairs.

only highlight a single cloth (i.e. upper/lower half). To address this limitation, we first generate full body (i.e. target semantics) image pairs for each sample in the source dataset. For this step, we sample image and guidance with target semantics (i.e. full body COCO key points) from the source dataset and generate full body image pairs with a pretrained hallucinator. Then, we finetune the hallucinator with these generated image pairs in a self-supervised manner, as shown in Fig. 3. We find this pretraining stage improves the texture prediction, due to the alignment of the hallucinator weights more towards the full body semantic distribution, as will be shown in the experiments section (Table 2).

To this end, we also release these full body image pairs and their COCO-skeleton keypoints annotations to facilitate future research. A few samples of the generated image pairs are shown in Fig. 4. Note that the front/back images do not necessarily mean front/back canonical fashion poses.

Pose Alignment (PA): Following the semantic alignment, we perform pose alignment to further shift the data distribution learned by our hallucinator based on the target pose distribution. We obtain the pseudo pairs $\{(I_f^{\mathcal{PA}}, I_b^{\mathcal{PA}})\}$ from the SA pairs and guidance (with target pose distribution) from the target dataset, as shown in Fig. 5. The sampling for the guidance is based on a discriminator (checkpointed after vanilla hallucinator training [30]) score thresholding to prevent noisy reconstruction of the PA pairs (See supplementary Section A for details). This somewhat conservative threshold allows us to select a subset with a higher degree of realism that eventually prevents degeneracies during further finetuning. Finally, we finetune the semantically aligned hallucinator from the last step with these pose-aligned pseudo pairs. Note that we also switch the guidance structure (used to represent pose) from COCO keypoints (2D) to SMPL-X

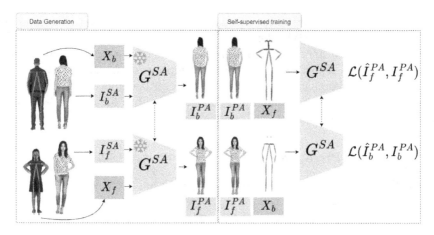

Fig. 5. Pose Alignment (PA): Given the subset of the pseudo pairs (I_f^{SA}, I_b^{SA}) after SA, we generate the front/back pairs following the pose from the target distribution (I_f^{PA}, I_b^{PA}) to finetune the semantically aligned hallucinator further. The hallucinator is initialized with weights obtained after SA.

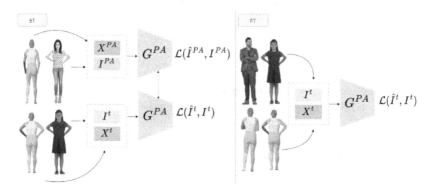

Fig. 6. The updated hallucinator weights after PA is transferred to this stage (indicated by superscript of G). We replace the guidance X from COCO key points to SMPL-X segmentation map for better expressivity. **Simultaneous training:** The pseudo pairs after PA, I^{PA}, and the 2D renderings of our target 3D dataset I^t are trained together to align the style (i.e. cloth texture and topology) from both datasets or domains. **Finetuning:** The hallucinator is finetuned directly on the 2D renderings of our target 3D dataset.

segmentation map and silhouette [28], which we empirically find in reconstructing the occluded regions more accurately due to its better expressiveness.

Simultaneous Training (ST)/Finetuning (FT): At this stage, we obtain the pose-aligned pseudo pairs $\{(I_f^{PA}, I_b^{PA})\}$. This updated subset represents the source style distribution (i.e. texture/clothing patterns) in the target semantic and pose distribution. For ST, this adapted subset $\{I^{PA}\}$ along with the 2D renderings $\{I^t\}$ from our target 3D dataset are jointly used to finetune our PA hallucinator for the style alignment (Fig. 6). In FT, the 2D renderings $\{I^t\}$ are directly used to finetune our PA hallucinator. In the experiments section, Table 2 shows the ablations of the relative improvements after incorporating these individual alignment stages in the order prescribed above. During training, we use the GT COCO keypoints, and SMPL-X fits, but for testing, we obtain the SMPL-X fits using off the shelf pose and shape estimation model [48].

3.2 Geometry Prediction

This section describes the process for shape generation (using implicit functions) and texture composition.

Background on Implicit Functions. For any point $\mathbf{x} \in \mathbb{R}^3$, the implicit function (IF) methods learn the conditional probability for occupancy $p(\mathbf{x}|\Pi_\mathbf{x})$, where $\Pi_\mathbf{x}$ is the 2D projection of $\mathbf{x}$. PIFu [33], the precursor of PIFuHD [34], estimates the probability as $p(\mathbf{x}|\Pi_\mathbf{x}) = f(\mathcal{F}(\Pi_\mathbf{x}), \Delta)$, where $\mathcal{F}$ is the pixel aligned feature, and Δ is the depth estimate of x in camera space.

PIFuHD enhances the vanilla PIFu framework with a (low/high) dual-resolution formulation:

$$p_c(\mathbf{x}|\Pi_\mathbf{x}) = f^\theta(f^\lambda(\mathcal{F}_c(\Pi_\mathbf{x}), \Delta)) \tag{1}$$

where p_c is the occupancy field at a coarse-resolution, and F_c is the corresponding pixel-aligned feature. Also, the implicit function is split into a composition as $f^\theta(f^\lambda)$, where f^λ jointly contributes to both low/high-resolution occupancy predictions, and f^θ is employed for low-resolution prediction only. Next, the high-resolution model predicts the final occupancy as:

$$p(\mathbf{x}|\Pi_\mathbf{x}) = f(\mathcal{F}(\Pi_\mathbf{x}), f^\lambda(\mathcal{F}_c(\Pi_\mathbf{x}), \Delta)) \tag{2}$$

where $\mathcal{F}$ is a separate fine resolution 2D pixel-aligned feature and $f^\lambda(.)$ is the same joint embedding from Eq. 1.

Occupancy Prediction for Shape Estimation. Our distributionally aligned hallucinator (DAH), as detailed in Sect. 3.1, is designed to be versatile and compatible with various shape inference methods. These include purely implicit function (IF) approaches such as PIFu and PIFuHD, as well as hybrid methods that combine implicit functions with explicit articulated model fitting, like SMPL/SMPL-X. Notably, IF approaches generally excel in producing high-quality reconstructions for typical fashion poses, as highlighted in the literature, while hybrid methods tend to shine in handling challenging and acrobatic poses. However, our primary objective is to create lifelike high-quality avatars for AR/VR/MR applications. Once the avatar is initially constructed from a standard pose, further animation of the 3D model becomes feasible, even for complex movements. Consequently, in this paper, we adopt the purely IF-based network, PIFuHD, as the baseline for enhancing DAH.

Examining Eqs. 1 and 2, it is evident that PIFuHD directly estimates depth ($\Delta(\ldots)$). Additionally, pixel aligned features $\mathcal{F}_c$ and $\mathcal{F}$ are leveraged for the occupancy prediction of query 3D points. Our observations indicate that relying solely on the features derived from the back normal map predicted (from front image) by PIFuHD falls short in estimating fine details in the occluded region. To address this limitation, apart from the improved learning of cloth textures from the 2D fashion datasets, we reroute our DAH-predicted back view as input to the back normal estimator (as illustrated in Fig. 2). This modification ultimately results in superior performance for the occluded back region (Please refer to the supplementary Section B for the visualization).

3.3 Texture Prediction

To obtain texture on the reconstructed 3D mesh described earlier, we employ a texture refinement network similar to Tex-PIFu [33], but with some modifications that enhance texture quality. Firstly, we project the front (reference) and back (synthesized) views onto their corresponding pixel-aligned query point. This differs from the approach used in PIFu [33], where masking was not necessary due to the absence of additional synthesized views in the monocular case. Subsequently, our completion network, based on UNet [32] and MLP, refines the texture at the per-vertex level using an $\mathcal{L}_1$ loss. To achieve more accurate high-frequency texture prediction, the MLP employed for vertex-level feature refinement is equipped with SIREN activation [37]. For further insights and ablations, please consult the supplementary materials Section C.

4 Experiments

Datasets: We utilized the RenderPeople dataset [4] for our evaluation, comprising 950 3D scans. Out of these, 865 scans were allocated for training, while the remaining 85 were reserved for evaluation purposes. To ensure reproducibility, we will make these data splits available alongside our codebase. The 2D renderings derived from these scans were generated at a yaw interval of 10°. Additionally, we employed PyMAFX [48] to obtain the SMPLX fits and their corresponding segmentation maps. For our 2D fashion dataset, we turned to DeepFashion [25]. We specifically used the high-resolution pairs from the original training split in DeepFashion for training, following the approach outlined in our vanilla hallucinator [30].

Training and Implementation: Our framework is implemented in PyTorch. The disentangled domain alignement scheme is implemented on top of the vanilla 2D hallucinator codebase [30]. Moreover, the occupancy and texture completion subsystems are built on the official releases containing partial implementations only [34]. Unlike these official implementations, we will make our complete source code publicly available to facilitate further research into this direction. For DDA

Table 1. Quantitative comparison of texture and shape on RenderPeople [4] test set.

Methods	Texture				Shape		
	FID (↓)	IS (↑)	LPIPS (↓)	SSIM (↑)	Chamfer (↓)	P2S (↓)	Nml MSE (↓)
PIFu [33]	67.6852	3.5190	0.1551	0.9101	1.6631	1.4427	0.0554
PAMIR [52]	70.8481	3.6684	0.1542	0.8963	2.6157	2.0091	0.0738
PIFuHD [34]	-	-	-	-	1.5709	1.4025	0.0526
ICON [44]	-	-	-	-	2.4936	2.0389	0.0680
ECON [43]	-	-	-	-	2.7133	2.4414	0.0742
FAMOUS (Ours)	55.0711	3.8655	0.1425	0.9140	1.4102	1.2718	0.0371

and occupancy prediction, we follow the same hyperparameter settings from the respective literature [30,33,34]. For texture completion, however, we use the Adam optimizer with the learning rate of $1e^{-4}$ to train for 20 epochs. For the 3D dataset, we generate 2D renderings at 10° intervals. All the models are trained on a dedicated 8-GPU NVIDIA RTX A4000 server.

4.1 Evaluation

SOTA Comparison: Table 1 shows the comparison with the SOTA methods both in terms of texture and shape. The texture scores are not reported for the methods focusing only on shape (i.e. PIFuHD [34], ICON [44], and ECON [43]).

In terms of texture quality, our method, FAMOUS, outperforms its counterparts in 3 out of 4 metrics, achieving approximately 19% and 6% relative improvement in FID and IS, respectively, over the second-best alternative. For SSIM, PIFu [33] exhibits similar performance compared to ours. It is important to note that SSIM places greater emphasis on smoother predictions with reduced high-frequency variations, and this aligns with the characteristics of PIFu's predictions, as evidenced in Fig. 7. In the figure, the sharp, localized variations in both normals and textures from PIFu appear as a subdued version of the ground truth, whereas FAMOUS preserves the intricate details that are captured by deep perceptual metrics, such as FID and IS.

Regarding shape reconstruction, our method outperforms the SOTA methods across the board, with PIFuHD [34] as the second-best contender (Table 1). Note that the recent shape-only approaches, i.e. ICON [44] and ECON [43], (with the publicly available codebases open-sourced by the respective authors), perform worse than PIFuHD. Based on further qualitative investigation, we find these SMPLX model based methods excel others for the more difficult acrobatic poses. Such extreme poses are not present in Fashion images, and PIFuHD is preferred for those cases [43]. On a related note, RenderPeople [4], our target 3D dataset, contains mostly casual poses – somewhat similar to the canonical case, and so, more appropriate for our evaluation than the extreme cases. Nonetheless, we include them for completeness. Please refer to the supplementary Section B for additional visualizations in this regard.

Table 2. Ablation study of the hallucinator on Render People.

Experiments	FID (↓)	LPIPS (↓)	SSIM (↑)	IS (↑)
FT	77.1910	0.1007	0.8645	2.7750
FT+SA	73.0653	0.0935	0.8678	2.7222
FT+SA+PA	67.7944	0.0824	**0.8929**	2.6282
ST	71.0900	0.0935	0.8678	2.7228
ST+SA	77.2460	0.0927	0.8683	2.5015
ST+SA+PA	**66.8911**	**0.0822**	0.8899	**3.2810**

FT ≡ Finetuning; ST ≡ Simultaneous Training
SA ≡ Semantic Alignment; PA ≡ Pose Alignment

4.2 Ablation Studies

Disentangled Domain Alignment (DDA): Table 2 shows progressive improvements for the back (occluded) view prediction with our DDA alignments on the pretrained hallucinator (Sect. 3.1). DDA improves the prediction quality for both finetuning (FT) on the 2D renderings of our 3D dataset and simultaneously training (ST) with both **aligned** 2D fashion images (source) and the 2D renderings (target).

For finetuning, one thing to note is that IS keeps degrading as we keep doing more alignment for finetuning (FT → FT+SA → FT+SA+PA) whereas the other 3 metrics gradually improve. This is because the distribution of the generated samples in each stage of alignment gets closer to the distribution of 2D renderings on the test set (gradually lower FID). However, the 3D dataset lacks sufficient number of samples with diverse textures, and so, finetuning with just the corresponding 2D renderings lead to an apparent loss of realism reflected by IS degradation.

For simultaneous training (ST), FID and IS degrade while LPIPS and SSIM improve slightly after incorporating the semantic alignment (SA). This is because the SA images from the source (2D fashion dataset) contains predictions with variable amount of realism – some of which are failed cases. Simultaneously tuning the hallucinator with this unfiltered set of generated samples alongside 2D renderings lead to poor convergence which is the reason behind the high perceptual losses for ST+SA in Table 2. However, the additional discriminator scoring in the beginning of pose alignment helps prevent such convergence issues, thus leading to significant improvements in the final stage (ST+SA vs. ST+SA+PA). Also, simultaneously training with the aligned 2D fashion images and the 2D renderings equivalent to Style Alignment (TA) as described in Sect. 3.1. Thus, the improvement of (ST+SA+PA) over (FT+SA+PA) in Table 2 shows the effectiveness of our complete disentangled alignment. Lastly, SSIM for FT is slightly better than ST due to the potential overfitting of the sufficient statistics on the 2D renderings in case of FT where the support of ST [13] will likely generalize better.

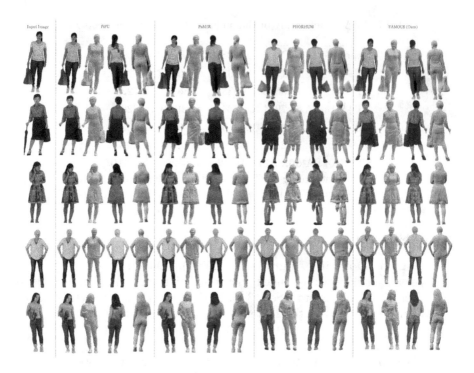

Fig. 7. Qualitative comparison of the SOTA pipelines predicting both shape and texture with ours on the RenderPeople test set for the images shown on the left. From left to right: Input image, PIFu [33], PaMIR [52], PHORHUM [10], and FAMOUS (ours). For each of these methods, we provide 4 visualizations – front orthographic view (2D rendering), front surface normals, back orthographic view (2D rendering), and back surface normals in order. FAMOUS visually outperforms these SOTA approaches in predicting both the occluded texture and surface delicacies.

5 Conclusion and Future Work

This paper introduces a complete framework for 3D human digitization from one image, emphasizing the generation of high-fidelity textures. In contrast to SOTA methods, we propose a new approach that harnesses the wealth of 2D datasets to predict arbitrary texture patterns, addressing the scarcity of similar 3D datasets. To achieve this, we leverage recent advancements in 2D hallucinators, incorporating a gradual domain alignment strategy based on disentangled factors. This integration of information from 2D datasets leads to a significant increase in texture prediction accuracy while also improving the accuracy of shape inference. To the best of our knowledge, this represents the first work to primarily focus on enhancing the quality of arbitrary textures for 3D human digitization by utilizing large-scale 2D datasets.

In terms of limitations and future directions, both 3D texture and shape predictions face challenges when dealing with rare clothing types that are not present in either the 2D/3D datasets. Additionally, texture prediction may encounter difficulties with rare or highly acrobatic poses that fall outside the scope of the training distribution.

References

1. Forensic Science Professor Brings Her Innovative VR Tech to Roger Williams University. https://www.rwu.edu/news/news-archive/forensic-science-professor-brings-her-innovative-vr-tech-rwu
2. IGOODI Official. https://www.linkedin.com/pulse/virtual-fitness-training-future-realistic-avatars-igoodi-official
3. Readyplayer. https://readyplayer.me
4. RenderPeople. https://renderpeople.com/3d-people
5. Virtual Fieldwork: Using Virtual Reality (VR) in Anthropology. https://www.unf.edu/dhi/projects/current/virtual-fieldwork-using-virtual-reality-in-anthropology.html
6. Walmart Virtual Try-On. https://www.walmart.com/cp/virtual-try-on/4879497
7. Weta FX. https://www.wetafx.co.nz/films
8. AlBahar, B., Lu, J., Yang, J., Shu, Z., Shechtman, E., Huang, J.B.: Pose with style: detail-preserving pose-guided image synthesis with conditional StyleGAN. ACM ToG **40**, 1–11 (2021)
9. Alldieck, T., Magnor, M., Bhatnagar, B.L., Theobalt, C., Pons-Moll, G.: Learning to reconstruct people in clothing from a single RGB camera. In: CVPR (2019)
10. Alldieck, T., Zanfir, M., Sminchisescu, C.: Photorealistic monocular 3D reconstruction of humans wearing clothing. In: CVPR (2022)
11. Bhunia, A.K., et al.: Person image synthesis via denoising diffusion model. In: CVPR (2023)
12. Cao, Y., Chen, G., Han, K., Yang, W., Wong, K.Y.K.: JIFF: jointly-aligned implicit face function for high quality single view clothed human reconstruction. In: CVPR (2022)
13. Cover, T.M., Thomas, J.A.: Elements of Information Theory. Wiley-Interscience, USA (2006)
14. Dong, Z., Guo, C., Song, J., Chen, X., Geiger, A., Hilliges, O.: PINA: learning a personalized implicit neural avatar from a single RGB-D video sequence. In: CVPR (2022)
15. Gilbert, A., Volino, M., Collomosse, J., Hilton, A.: Volumetric performance capture from minimal camera viewpoints. In: ECCV (2018)
16. Guo, K., et al.: The relightables: volumetric performance capture of humans with realistic relighting. ACM ToG **38**, 1–19 (2019)
17. Han, S.H., Park, M.G., Yoon, J.H., Kang, J.M., Park, Y.J., Jeon, H.G.: High-fidelity 3D human digitization from single 2k resolution images. In: Proceedings of the IEEE/CVF Conference on Computer Vision and Pattern Recognition (CVPR) (2023)
18. He, T., Collomosse, J., Jin, H., Soatto, S.: Geo-PIFu: geometry and pixel aligned implicit functions for single-view human reconstruction. In: NeurIPS (2020)
19. He, T., Xu, Y., Saito, S., Soatto, S., Tung, T.: ARCH++: animation-ready clothed human reconstruction revisited. In: ICCV (2021)
20. Huang, Z., et al.: Deep volumetric video from very sparse multi-view performance capture. In: ECCV (2018)

21. Huang, Z., Xu, Y., Lassner, C., Li, H., Tung, T.: ARCH: animatable reconstruction of clothed humans. In: CVPR (2020)
22. Li, Y., Huang, C., Loy, C.C.: Dense intrinsic appearance flow for human pose transfer. In: CVPR (2019)
23. Li, Z., Yu, T., Zheng, Z., Liu, Y.: Robust and accurate 3D portraits in seconds. IEEE TPAMI (2022)
24. Liu, W., Piao, Z., Min, J., Luo, W., Ma, L., Gao, S.: Liquid warping GAN: a unified framework for human motion imitation, appearance transfer and novel view synthesis. In: ICCV (2019)
25. Liu, Z., Luo, P., Qiu, S., Wang, X., Tang, X.: DeepFashion: powering robust clothes recognition and retrieval with rich annotations. In: CVPR (2016)
26. Loper, M., Mahmood, N., Romero, J., Pons-Moll, G., Black, M.J.: SMPL: a skinned multi-person linear model. ACM Trans. Graph. **34**(6) (2015)
27. Men, Y., Mao, Y., Jiang, Y., Ma, W.Y., Lian, Z.: Controllable person image synthesis with attribute-decomposed GAN. In: CVPR (2020)
28. Pavlakos, G., et al.: Expressive body capture: 3D hands, face, and body from a single image. In: CVPR (2019)
29. Prabhudesai, M., Lal, S., Patil, D., Tung, H.Y., Harley, A.W., Fragkiadaki, K.: Disentangling 3D prototypical networks for few-shot concept learning. In: ICLR (2021)
30. Ren, Y., Fan, X., Li, G., Liu, S., Li, T.H.: Neural texture extraction and distribution for controllable person image synthesis. In: CVPR (2022)
31. Ren, Y., Yu, X., Chen, J., Li, T.H., Li, G.: Deep image spatial transformation for person image generation. In: CVPR (2020)
32. Ronneberger, O., Fischer, P., Brox, T.: U-net: convolutional networks for biomedical image segmentation. In: Navab, N., Hornegger, J., Wells, W.M., Frangi, A.F. (eds.) MICCAI 2015. LNCS, vol. 9351, pp. 234–241. Springer, Cham (2015). https://doi.org/10.1007/978-3-319-24574-4_28
33. Saito, S., Huang, Z., Natsume, R., Morishima, S., Kanazawa, A., Li, H.: PIFu: pixel-aligned implicit function for high-resolution clothed human digitization. In: ICCV (2019)
34. Saito, S., Simon, T., Saragih, J., Joo, H.: PIFuHD: multi-level pixel-aligned implicit function for high-resolution 3D human digitization. In: CVPR (2020)
35. Sarkar, K., Golyanik, V., Liu, L., Theobalt, C.: Style and pose control for image synthesis of humans from a single monocular view (2021)
36. Sieberth, T., Dobay, A., Affolter, R., Ebert, L.C.: Applying virtual reality in forensics - a virtual scene walkthrough. Forensic Sci. Med. Pathol. **15**, 41–47 (2019)
37. Sitzmann, V., Martel, J., Bergman, A., Lindell, D., Wetzstein, G.: Implicit neural representations with periodic activation functions. In: NeurIPS (2020)
38. Smith, D., Loper, M., Hu, X., Mavroidis, P., Romero, J.: FACSIMILE: fast and accurate scans from an image in less than a second. In: ICCV (2019)
39. Song, D.Y., , Lee, H., Seo, J., Cho, D.: DiFu: depth-guided implicit function for clothed human reconstruction (2023)
40. Sun, B., Saenko, K.: Deep CORAL: correlation alignment for deep domain adaptation. In: Hua, G., Jégou, H. (eds.) ECCV 2016. LNCS, vol. 9915, pp. 443–450. Springer, Cham (2016). https://doi.org/10.1007/978-3-319-49409-8_35
41. Tzeng, E., Hoffman, J., Zhang, N., Saenko, K., Darrell, T.: Deep domain confusion: maximizing for domain invariance. CoRR abs/1412.3474 (2014)
42. Vlasic, D., Baran, I., Matusik, W., Popović, J.: Articulated mesh animation from multi-view silhouettes. ACM Trans. Graph (2008)

43. Xiu, Y., Yang, J., Cao, X., Tzionas, D., Black, M.J.: ECON: explicit clothed humans optimized via normal integration. In: CVPR (2023)
44. Xiu, Y., Yang, J., Tzionas, D., Black, M.J.: ICON: implicit clothed humans obtained from normals. In: CVPR (2022)
45. Yang, Z., et al.: S3: neural shape, skeleton, and skinning fields for 3D human modeling. In: CVPR (2021)
46. Yu, T., Zheng, Z., Guo, K., Liu, P., Dai, Q., Liu, Y.: Function4D: real-time human volumetric capture from very sparse consumer RGBD sensors. In: CVPR (2021)
47. Zakharkin, I., Mazur, K., Grigorev, A., Lempitsky, V.: Point-based modeling of human clothing. In: ICCV (2021)
48. Zhang, H., et al.: PyMAF-X: towards well-aligned full-body model regression from monocular images. IEEE TPAMI **45**, 12287–12303 (2023)
49. Zhang, J., Li, K., Lai, Y.K., Yang, J.: PISE: person image synthesis and editing with decoupled GAN. In: CVPR (2021)
50. Zhang, P., Zhang, B., Chen, D., Yuan, L., Wen, F.: Cross-domain correspondence learning for exemplar-based image translation. In: CVPR (2020)
51. Zheng, Y., et al.: DeepMultiCap: performance capture of multiple characters using sparse multiview cameras. In: ICCV (2021)
52. Zheng, Z., Yu, T., Liu, Y., Dai, Q.: PaMIR: parametric model-conditioned implicit representation for image-based human reconstruction. IEEE TPAMI **44**(6) (2022)
53. Zhou, X., Yin, M., Chen, X., Sun, L., Gao, C., Li, Q.: Cross attention based style distribution for controllable person image synthesis. In: Avidan, S., Brostow, G., Cissé, M., Farinella, G.M., Hassner, T. (eds.) ECCV 2022. LNCS, vol. 13675, pp. 161–178. Springer, Cham (2022). https://doi.org/10.1007/978-3-031-19784-0_10
54. Zuo, X., Du, C., Wang, S., Zheng, J., Yang, R.: Interactive visual hull refinement for specular and transparent object surface reconstruction. In: ICCV (2015)

Efficient Learning of Event-Based Dense Representation Using Hierarchical Memories with Adaptive Update

Uday Kamal[✉] and Saibal Mukhopadhyay

Georgia Institute of Technology, Atlanta, GA, USA
{ukamal6,smukhopadhyay6}@gatech.edu

Abstract. Leveraging the high temporal resolution of an event-based camera requires highly efficient event-by-event processing. However, dense prediction tasks require explicit pixel-level association, which is challenging for event-based processing frameworks. Existing works aggregate the events into a static frame-like representation at the cost of a much slower processing rate and high compute cost. To address this challenge, this work introduces an event-based spatiotemporal representation learning framework for efficiently solving dense prediction tasks. We uniquely handle the sparse, asynchronous events using an unstructured, set-based approach and project them into a hierarchically organized multi-level latent memory space that preserves the pixel-level structure. Low-level event streams are dynamically encoded into these latent structures through an explicit attention-based spatial association. Unlike existing works that update these memory stacks at a fixed rate, we introduce a data-adaptive update rate that recurrently keeps track of the past memory states and learns to update the corresponding memory stacks only when it has substantial new information, thereby improving the overall compute latency. Our method consistently achieves competitive performance across different event-based dense prediction tasks while ensuring much lower latency compared to the existing methods.

Keywords: Event-Based Vision · Memory-Augmented Representation Learning · Efficient Processing · Event-based dense prediction

1 Introduction

Event-based cameras [5, 10, 35, 35], which are bioinspired and aim to mimic the brains efficient data processing, represent a departure from the traditional frame-based cameras. They feature pixels that independently record asynchronous events, capturing brightness changes with high temporal precision (microsecond order) and offering substantial dynamic range (>120 dB). Their low power usage and swift response are particularly beneficial in scenarios with rapid motion or sudden light shifts, such as drones or autonomous vehicles [9, 15, 16, 36, 42].

Nonetheless, these advantages necessitate the development of specialized processing architectures to fully harness their capabilities.

Existing learning-based approaches for event data have primarily relied on grouping events into synchronized stacks before feeding them into convolutional neural networks (CNNs) [4,7,13]. However, this strategy not only undermines the high temporal resolution and asynchronicity inherent in event data, but also offers a limited temporal context to the foregoing recurrent neural networks (RNNs) [14,19]. Recent endeavors in event-based processing techniques for object recognition-such as time-surface representations [23], 3D event-clouds [39], and graph-based methods [25]-have mostly prioritized sparse processing for computational efficiency. Yet, they fall short in performance when compared to their dense-representation counterparts [32].

Efficient processing of event-triggered asynchronous data necessitates algorithms that construct and continuously refine an internal representation, correlating new and historical events across both space and time. Although recurrent modules per pixel could trace past events [7], this approach significantly increases processing and memory requirements with higher spatial resolution. Recently proposed work [21] has shown impressive compute efficiency due to its associative memory augmented recurrent representation learning that preserves sparsity and enables event-by-event spatiotemporal correlation. However, their method has only been evaluated on simple classification tasks and it is nontrivial to scale it on dense tasks due to the lack of explicit pixel-level association to the memory states and its reliance on a single-stage memory. A hierarchical latent memory-based method has been proposed [17] that encodes the event representation into a stack of memory states with a multi-rate update mechanism. Despite achieving superior performance across diverse dense tasks, this method still relies on a hand-tuned update rate for the memory stacks limiting its adaptability and optimality in real-time applications. Moreover, unlike EventFormer [21], they do not consider the inherent recurrent nature of their memory states that can further avoid redundant updates, and improve latency. While it is desired to change this update rate adaptively depending on the input scenario, it is not trivial to learn them directly since it is still required to keep track of the states in the absence of an update. At the same time, computing the association between two subsequent memory levels for each time step can be compute-intensive, especially for earlier memory levels with higher spatial resolution. Essentially, we need a recurrent mechanism that efficiently builds up representation with incoming events and triggers an update only when it has accumulated sufficient new information.

To this end, we present a novel architecture for efficient processing of event-based dense tasks. Our architecture introduces recurrent processing of memory-augmented representation learning by leveraging a multi-stage memory structure that enables gradual feature enhancement. Here, the lowest memory level encodes the local dynamic information by projecting the unstructured, sparse events into a spatially structured, dense latent space. Subsequent memory levels learn to extract higher-level static and global contexts. Instead of updating the whole compute stack at each time step, our method uses an adaptive memory update mechanism that can learn the optimal update rates from data, thereby enhancing the models responsiveness to the dynamic scenes and its overall efficiency. Moreover, rather than using the entire set of memory states to perform an association between two subsequent levels, we introduce a global pooling operation to compress the memory states, recurrently process them and adaptively skip updates in the absence of any substantial new information. Effectively, this operation saves computation and avoids redundant updates due to any low-level information gain. In other words, we avoid updating if the changes in the memory states between events are minor in the global context. To the best of our knowledge, we are the first to introduce an adaptive component to event-based memory models, which crucially improves the computational load required for its real-world application.

We evaluate our method on several event-based prediction tasks and datasets that have traditionally leveraged dense information, including semantic segmentation, object detection, and depth estimation. Our method demonstrates superior reduction in latency (50%–70%) while maintaining competitive task performance, compared to state-of-the-art baselines.

To summarize, our contributions are:

– We introduce an end-to-end learning framework with an adaptive update rate for multi-stage memory architecture.
– We propose a global context-based association between two contiguous memory levels to efficiently process their recurrent updates.
– We experimentally show that our method achieves a significant reduction in latency while maintaining competitive performance on different event-based dense perception tasks.

2 Related Works

We discuss the relevant prior works from two different directions: building an effective representation for event camera data and extracting useful features from this representation for the downstream tasks.

Event Representation: Building an effective representation for event-based data is one of the key challenges for its efficient processing mostly due to its highly sparse and asynchronous nature. Both manually designed and data-adaptive representation space have been explored in several prior works that consider aggregating event information into a dense, frame-like structure. Histogram-based methods, such as hierarchy of event-based time surface (HOTS) [23], and histograms of averaged time surface (HATS) [34], encode the timestamp information of the events by computing a time-ordered representation. However, their hand-crafted nature of event encoding can result in sub-optimal representation. To this end, data-adaptive methods have been proposed that rely on end-to-end learnable representation. Event spike tensor (EST) [13] introduced learnable kernels for histogram-based methods, further improving performance. Voxel Grid-based representation has also been considered where AED [26] and ASTMNet [24] introduced data-adaptive kernel and sampling mechanism. However, they need to keep track of past events for temporal correlation and perform redundant computations every time a new event occurs. Memory-augmented representation can overcome this challenge by internally tracking past events in its memory states. Matrix-LSTM [7] proposed a pixel-wise, LSTM-based memory model that keeps track of past events at each pixel location independently. However, such representation does not consider any spatial correlation and, therefore requires a compute-expensive feature extractor. EventFormer [21] addresses this challenge by proposing a set-based spatiotemporal association while building the memory states. However, scaling EventFormer for dense prediction is challenging due to the absence of any explicit correspondence between the memory states and the pixel location.

Event Feature Extraction: Extracting useful features from the event representation is crucial for the downstream task. Popular feature extraction architectures like CNNs, which are originally designed for frame-based data, have been extensively explored in prior works [4,28]. These methods process discretized events aggregated into a dense representation, discarding their inherent sparsity and temporal correlation. Several new directions have been explored to leverage the sparse and asynchronous nature of the event data, including sparse asynchronous CNN architecture [29] and graph neural networks [25,32]. Despite their superior compute efficiency, graph-based methods have shown limited performance compared to the dense, synchronous methods on complex tasks [32]. MLP and Transformer-based architectures have also been explored in this direction. These methods consider events as streams of space-time point clouds and leverage simple MLP-based PointNet [33,38] or Transformer [31] architectures to extract useful features. Although these methods have shown superior performance on simple classification tasks, it is challenging to scale them up for complex, dense tasks while maintaining event-by-event processing. To address this challenge, a hierarchical latent memory-based solution has been proposed [17] that encodes raw events into a structured memory space and leverages it for downstream dense tasks. Despite performing well on several dense tasks, this method relies on a hand-tuned, fixed update rate without considering the relative information gain with respect to its past update, which perform suboptimally and cause redundant computation.

3 Method

3.1 Preliminaries

An event-based camera contains independent sensors at every pixel location that can asynchronously respond to the brightness change [11]. Specifically, within a time interval τ, events produced at pixel locations $(x,y)_k$ can be mathematically expressed as, $\mathcal{E}_\tau = \{(x_k, y_k, t_k, p_k) \mid t_k \in \tau, k \in \mathcal{N}\}$ where t_k is the event timestamps, $p_k \in \{-1, +1\}$ corresponds to relative change in brightness, and

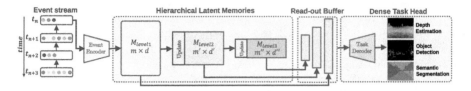

Fig. 1. Overview of the overall architecture. It consists of an event encoder that directly maps the raw events into the lowest level of latent memory (M_{level1}) at each time step. Subsequent memory levels are updated adaptively by utilizing the previous memory states. The readout buffer extracts and stores the memory states to be used by the task-specific decoder at the end.

$\mathcal{N}$ denotes the total number of events. We want to learn a parametric mapping $\mathcal{F} : \mathcal{E}_\tau \rightarrow \{\mathcal{M}^i_{\mathcal{E}_\tau}\}$ to encode the unstructured list of events into a stack of hierarchically structured memory representation $\{\mathcal{M}^i_{\mathcal{E}_\tau}\}$ where i denotes the level of hierarchy that can be directly integrated with off-the-shelf task heads (decoders) to solve event-based dense tasks. We also want $\mathcal{F}$ to be capable of updating its memory representation adaptively (i.e. only when there is sufficient new information) to avoid redundant computation.

3.2 Overview of the Framework

We leverage a hierarchy of latent memory stacks to extract high-level contextual information from the low-level event streams. Specifically, spatially unstructured event streams are first projected into the lowest level of structured latent memory space, which is followed by multiple higher memory levels. Figure 1 shows a high-level overview of the method. Since very low-level events most often do not contribute much to the higher level of representation, we propose a learnable memory update mechanism that allows higher levels of memories to select between preserving or updating its past states, thereby resulting in an adaptive-rate network architecture and saving computation. We introduce a compressed-space memory association to compare the global context of subsequent memory states and leverage a recurrent module to keep track of the associations. The updated memory states are extracted and stored in its corresponding readout buffer. A task-specific decoder at the final layers exploits these stored buffer states at each time step and computes prediction.

Event Representation and Encoding: We consider window-based pixel association of the events [17]. Specifically, the lowest level memory $M_{level1} \in \mathbb{R}^{m \times D}$ contains m rows where each row stores D-dimensional representation from a $w \times w$ sized window (with a stride of w) at a spatial location (i, j). At any time step t_n, with N number of events triggered within the time-interval of $[t_{n-1}, t_n]$, we compute the relative position $(\hat{x}_k, \hat{y}_k)$ within the window w_{ij}, and relative timestamp $\hat{t}_k$ (within the interval of $[t_{n-1}, t_n]$) for each event. This low-level event information is then projected into a D-dimensional embedding vector π_k using the following transformations:

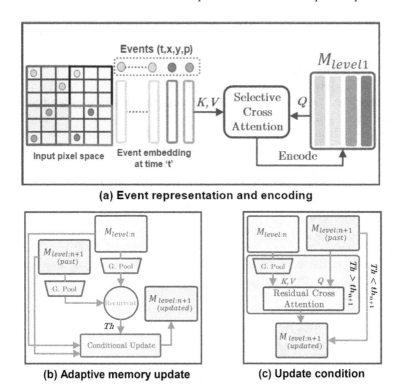

Fig. 2. Details of the different components of the architecture. (a) Embeddings computed from raw events are spatially encoded into the lowest memory level using cross-attention, (b) Global states from two subsequent memory levels are sequentially processed to generate score (Th) for the adaptive update, (c) memory states are either updated using residual cross attention or retained (therefore bypassing expensive operation) depending on the Th value.

$$\pi_k = LN([F_{pos}(\hat{x}_k, \hat{y}_k), \ F_t(\hat{t}_k), \ F_p(p_k)]), \tag{1}$$

where F_{pos}, F_t, F_p each denotes an MLP with two hidden layers for event position, timestamp, and polarity, respectively; LN denotes Layer Normalization [3]; and [,] denotes concatenation operation. This transformed event representation is next encoded to the memory cell using cross-attention, as shown in Fig. 2(a). Specifically, we compute cross attention between a memory cell s_{ij} in M_{level1} and all the n-event embeddings $\pi \in \mathbb{R}^{n \times D}$ that fall within the spatial window w_{ij}, therefore selectively updating only those memory cells that receive any events:

$$q_{ij} = \mathcal{Q}(s_{ij}), \quad K = \mathcal{K}(\pi), \quad V = \mathcal{V}(\pi)$$

$$s_{ij}^{new} = \text{softmax}\left(\frac{q_{ij}K^T}{\sqrt{D}}V\right) \tag{2}$$

where $\mathcal{Q}, \mathcal{K}, \mathcal{V}$ are the MLP-based mapping for computing query, key, and value, respectively.

Adaptive Update of the Memory: The hierarchical memory structure requires the update to the next memory level to be conditioned on the previous memory states. We achieve this by computing a window-based multi-head cross-attention (WCA) [17,27] between two subsequent memory levels. We compute the *Query* from the current memory level (M_{n+1}) while the *Key* and *Value* from a downsampled version of the previous memory level (M_n):

$$\hat{M}_{n+1} = WCA(M_{n+1}, \text{Down}(M_n) + M_{n+1}$$
$$M_{n+1}^{updated} = \text{MLP}(\text{LN}(\hat{M}_{n+1})) + \hat{M}_{n+1} \quad (3)$$

where MLP denotes two conv 1×1 layers with GELU activation [18], and Down is a spatially downsampling operation with strided convolution. We avoid redundant updates to the memories by introducing a learnable update score, Th. Specifically, we consider the global representation of two subsequent memory levels and recurrently compute the update score for each time step:

$$M_{n+1}^G = \text{G.Pool}(M_{n+1}), \quad M_n^G = \text{G.Pool}(M_n)$$
$$\hat{Th} = \mathcal{R}(M_{n+1}^G, M_n^G), \quad Th = \text{sigmoid}(\text{MLP}(\hat{Th})) \quad (4)$$

where G.Pool denotes the global pooling operation $\text{G.Pool}: M_{\{.\}} \in \mathbb{R}^{m \times d} \to M_{\{.\}}^G \in \mathbb{R}^{1 \times d}$, $\mathcal{R}$ denotes the recurrent operation (a GRU unit), and MLP here denotes a linear mapping to a scalar value. Subsequent memory update occurs only when this Th exceeds a certain threshold (th). We also reset the hidden state of $\mathcal{R}$ following each update.

Readout Buffers: Readout buffers [17] extract the memory states and utilize a point-wise convolution layer to further refine them to be utilized by the task head at the end at every time step. The operation is defined as:

$$ro_n = c_{ro}(LN(M_n))) \quad (5)$$

where ro_n, and M_n are the refined buffer state, and memory states at level n, respectively; c_{ro} is a conv 1×1 layer with Group Normalization [40] and SiLU activation [8].

4 Experiments

Base Training and Inference Setup: We consider three different event-based dense prediction tasks for experiments: segmentation, object detection, and depth estimation. To evaluate our performance, we use both the algorithmic performance (task-specific metrics) and the inference latency. For our network configuration, we use three memory levels with $D_i \in \{32, 64, 128\}$ (with increasing memory dimension for higher levels) and window stride size $s_i \in \{4, 8, 16\}$.

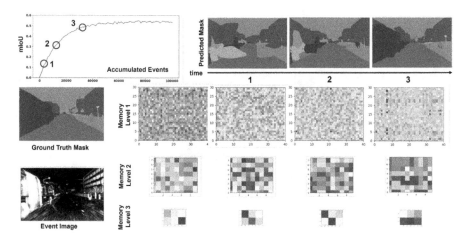

Fig. 3. Temporal evolution of the memory states across different levels. The prediction performance gradually improves as we get more and more events while the memory states tend to form distinct patterns.

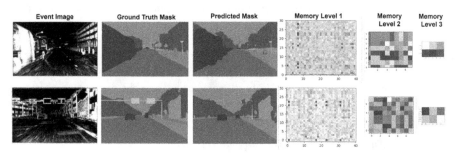

Fig. 4. Visualization of memory states for different inputs where we observe unique patterns but similar trends (sparse to dense) across different memory levels.

Unless otherwise specified, we use 4 as the number of heads for all cross-attention operations, and the 0.5 as the threshold (th) for adaptive update. For training, we use Adam [22] as our optimizer with a batch size of 4 and an initial learning rate (LR) of 5.0e−4 with a cosine LR scheduler. For fair comparison with other methods, we use the same hardware setup as [17] with NVIDIA Tesla V100 GPU to report latency.

4.1 Qualitative Analysis

Temporal Evolution of Multi-scale Memory: We first visualize the temporal evolution of the latent memory states of our proposed method trained on DSEC-Semantic dataset [15] as shown in Fig. 3) where we plot the first channel of each memory level. We observe that all memory states start from an initial state but as we get more and more events, the states get more refined and start

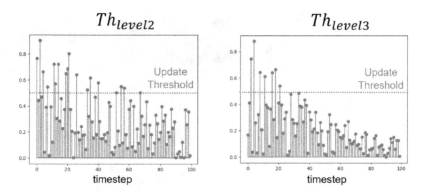

Fig. 5. Learned memory update score across different levels where the score gradually decays over time with a higher decay rate for higher memory levels.

to form a distinct pattern. Specifically, the lower level memory tends to form a sparse representation while the higher levels tend to form a more dense representation gradually. This can be explained by the fact that the higher resolution of the lower-level memory captures more coarse-grained features which get more refined throughout the stacks. This is evident from the gradual improvement in both quantitative (mIoU) and qualitative results (predicted mask) as we get more and more events. We also visualize the converged memory pattern for two different input samples in Fig. 4 where we observe visibly distinguishable patterns in the state representation with a similar sparse to dense pattern formation trend across the memory stacks.

Adaptive Multi-scale Update Across Hierarchies: We analyze the temporal adaptation of the learned update scores, Th_{level_n}, across the memory stacks of our proposed method trained on the DSEC-Semantic dataset as shown in Fig. 5). We observe that an update to the higher level memory is not necessary throughout the whole observation period where most of the updates occur during the initial period of observation. Our method adaptively learns to have fewer updates to the higher-level memory compared to the lower ones. The intuition behind this is that for the static dense prediction task, the higher level memory captures more fine-grained, global information that requires less frequent updates as we get more and more events. The proposed recurrent formulation ensures keeping track of the skipped updates so that they can influence future ones. We also show in Sect. 4.3 that the proposed adaptive update mechanism ensures faster convergence to the memory states for higher-level memory, thereby reducing redundant computation and improving the overall latency.

4.2 Quantitative Results

The section describes the quantitative experimental results of our method compared to the existing methods on several event-based dense prediction tasks including semantic segmentation, object detection and depth estimation.

Table 1. Results of (a) semantic segmentation task on DSEC-Semantic dataset, and (b) object detection on GEN1 dataset. Latency values are measured on the TeslaV100 GPU system.

Method	Decoder	mIoU	Latency (ms)
EV-SegNet [1]	UNet	51.8	-
ESS [37]	UNet	53.3	16.2
HMNet-B3 [17]	UPerNet	53.9	9.7
HMNet-L3 [17]	UPerNet	57.1	13.9
Ours	UPerNet	49.9	4.5

(a)

Method	Decoder	mAP	Latency (ms)
MatrixLSTM [7]	YOLOv3	31.0	-
NGA [20]	YOLOv3	35.9	-
RED [30]	SSD	40.0	11.6
Asynet [6]	YOLO	12.9	-
AEGNN [32]	YOLO	16.3	-
AED [26]	YOLOX	45.4	13.1
ASTMNet [24]	SSD	46.7	35.6
HMNet-B3 [17]	YOLOX	45.2	7.0
HMNet-L3 [17]	YOLOX	47.1	7.9
Ours	YOLOX	44.8	3.2

(b)

Semantic Segmentation: We conduct event-based semantic segmentation experiments on the DSEC-Semantic dataset [15] with 640 × 480 pixel resolution event-camera data recorded under real-world street scenes. The pixel-wise semantic segmentation annotation for this dataset are recorded from RGB images at 20 Hz with a total 8,082 and 2,809 frames available for training and testing, respectively. We follow the same setup in [17] with 11 object classes for our experiments and UPerNet [41] as the task head. We train this model for 50k iterations with our base training setup. Table 1b shows the result. ESS [37] adopts additional finetuning on RGB-based data while our method relies on event data only. HMNetB3/L3 achieves the best-performing mIoU but they incur higher latency mostly due to the redundant memory updates at a higher level. Our method achieves a balance between computation and task-specific performance with a competitive mIoU score while outperforming all the existing methods in terms of latency, with a 50%/70% improvement over the SoTA methods HMNetB3/L3.

Object Detection: We conduct event-based object detection experiments on GEN1 dataset [30] with 304 × 240 pixel resolution event-camera data recorded under a real-world driving scenario. This dataset includes 2,358 event sequences each with 60 sec time duration. These sequences are further divided into 1,459, 429, and 470 for training, validation, and testing, respectively. The bounding box annotations are available at 1 Hz to 4 Hz, depending on the sequence. There are two object level annotations: pedestrian and car. We adopt the same YOLOX [12] detector head as used in [17], and train the model end to end for 300k iterations. Table 1a shows the comparative results with the existing methods. Our method outperforms most of the recurrent baselines (MatrixLSTM, and RED) both in terms of mAP and latency. The best-performing recurrent method ASTM-Net requires significant compute latency mostly due to its long accumulation time. Our method achieves competitive performance against these methods while reducing 60% latency of the existing SoTA method (HMNet-L3).

Table 2. Results of depth estimation task on MVSEC dataset. Latency values are measured on the TeslaV100 GPU system.

Method	Decoder	outdoor day1			outdoor nigh1			Latency (ms)
		REL ↓	RMS↓	RMSlog↓	REL↓	RMS ↓	RMSlog↓	
E2Depth [19]	UNet	0.346	8.564	0.421	0.591	11.210	0.646	–
RAMNet [14]	UNet	0.303	8.526	0.424	0.583	13.340	0.830	9.0
HMNet-B3 [17]	UNet	0.270	7.101	0.332	0.323	8.935	0.462	5.0
HMNet-L3 [17]	UNet	0.254	6.890	0.319	0.323	9.008	0.482	6.9
Ours	UNet	0.292	7.985	0.386	0.358	9.441	0.498	2.3

Depth Estimation: We conduct event-based depth estimation experiments on real MVSEC dataset [42] with the street-scene recording of 346 × 260 pixel resolution DAVIS event-camera data. A LiDAR sensor is used to record the ground truth depth map at 20 Hz. We follow the same training and testing pipeline in [17] with the outdoor day2 sequences for training and outdoor day1 and outdoor night1 sequences for evaluation. We adopt the pre-training pipeline proposed in [14] and the UNet-based decoder in [17]. Table 2 shows the comparative results with the existing methods. Although RAMNet leverages a fusion of event and image space, our method can outperform it on both day and night sequences while achieving 74% latency improvement. While HMNet-B3/L3 with fixed update achieves the best result, our method with dynamic update can still perform competitively while reducing the latency by 50%/67%.

4.3 Ablation Experiments

One of our key contributions is the adaptive update to the memory states. A straightforward alternative is having a fixed, uniform update rate across all the memory levels. To better understand the effect of adaptation, we compare the performance of the model with adaptive updates to that of the model without it (i.e. replacing it with the uniform update mechanism). We conduct several ablation experiments for one of the vision tasks – specifically, segmentation with the DSEC-semantic dataset. Our method also relies on a threshold parameter (th) to determine the update decision, which can make our method sensitive to this parameter. Therefore, we also conduct a sensitivity analysis of this parameter.

Adaptive vs. Uniform Memory Update: We investigate the impact of the proposed adaptive memory update mechanism in Fig. 6. We compute the mean

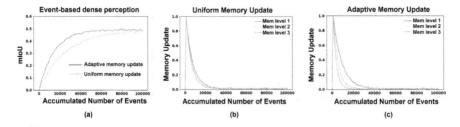

Fig. 6. Comparison between adaptive and uniform memory update on DSEC-Semantic dataset: (a) event-based dense prediction results, and visualization of memory update across different levels for (b) uniform and (c) adaptive memory update, respectively. Here *Memory Update* denotes the mean L_1 norm between two successive memory states normalized for each level.

Table 3. Ablation results on (a) adaptive vs uniform update (b) sensitivity analysis on the update threshold (th) on DSEC-Semantic dataset

Method	mIoU	Latency (ms)
Ours (Uniform update)	49.7	5.8
Ours (Adaptive update)	49.9	4.5

(a)

Method	Update threshold	mIoU	Latency (ms)
Ours (Adaptive update)	0.5	49.9	4.5
	0.3	49.8	4.8
	0.7	46.8	4.1

(b)

L_1 norm of the difference between two successive memory states at each level and normalize them to quantitatively analyze the change in memory states that occurred with the incoming events. As a baseline, we compare it against a uniform update rate across all hierarchies. We observe that adaptive memory update enables higher levels of memory to converge at a much faster rate, thereby reducing redundant computation (Fig. 6(c)). On the other hand, there is not much difference in convergence rates for the uniform update (Fig. 6(b)). Updating higher levels of memory at the same rate as the lowest one can introduce not only redundant computation but also noisy updates, which can slow down the convergence rate and degrade performance, as we can observe in Fig. 6(a)) and Table 3a. The Hierarchical memory model with a uniform update slightly degrades the performance while increasing the overall latency by 28%.

Sensitivity Analysis of Update Threshold: We conduct a sensitivity analysis of the update threshold for our proposed adaptive memory update score Th_{level_n} where we vary the threshold by ± 0.2. Table 3b shows the result. We observe that the overall latency has a profound impact on the threshold since it directly controls the number of updates to the memory (lower update count and improved latency with a higher threshold value). However, increasing the threshold too much can degrade performance as shown in the result. A higher threshold value may prevent updates necessary for learning an effective representation, therefore harming the performance.

5 Future Directions

One of the limitations of our proposed method is its sensitivity to the threshold parameter for memory update as shown in Sect. 4.3. Determining the optimal threshold can be non-trivial, especially if higher memory levels were to be used in the model. One potential solution to this could be learning this threshold in an end-to-end manner [2].

Our proposed adaptive update mechanism currently relies primarily on a single modality (event-camera data). Multimodal fusion between events and frames has been explored in recent works [14,17]. However, it is non-trivial to directly apply our method in a multimodal setting since different modality needs to incorporate different update rates. An interesting future direction could be learning modality-conditioned adaptive updates while fusing different modes together in a compute-efficient manner.

6 Conclusion

In this work, we introduce a novel adaptive update-based hierarchical memory network for low latency event-based dense perception. The proposed method processes the sparse, unstructured raw events and directly encodes them into a compact latent memory space while preserving sparsity. To avoid redundant memory operations and improve latency, we propose a recurrent update mechanism that learns to adaptively update the memory states only when there is sufficient new information. Results on different event-based dense prediction tasks demonstrate the effectiveness of the proposed method. Future works can explore novel adaptive update mechanisms to enable efficient multimodal fusion such as fusion between events and frames.

Acknowledgements. This work is supported by the Defense Advanced Research Projects Agency (DARPA) under Grant Numbers GR00019153. The views and conclusions contained in this document are those of the authors and should not be interpreted as representing the official policies, either expressed or implied, of the Department of Defence, DARPA, or the U.S. Government.

The corresponding author would like to thank Saurabh Das for his feedback during the early stage of this project and acknowledge Eley Ng for her insightful discussion while preparing this manuscript.

References

1. Alonso, I., Murillo, A.C.: EV-SegNet: semantic segmentation for event-based cameras. In: Proceedings of the IEEE/CVF Conference on Computer Vision and Pattern Recognition Workshops (2019)
2. Azarian, K., Bhalgat, Y., Lee, J., Blankevoort, T.: Learned threshold pruning. arXiv preprint arXiv:2003.00075 (2020)
3. Ba, J.L., Kiros, J.R., Hinton, G.E.: Layer normalization. arXiv preprint arXiv:1607.06450 (2016)

4. Baldwin, R., Liu, R., Almatrafi, M.M., Asari, V.K., Hirakawa, K.: Time-ordered recent event (TORE) volumes for event cameras. IEEE Trans. Pattern Anal. Mach. Intell. **45**, 2519–2532 (2022)
5. Brandli, C., Berner, R., Yang, M., Liu, S.C., Delbruck, T.: A 240 × 180 130 db 3 μs latency global shutter spatiotemporal vision sensor. IEEE J. Solid-State Circuits **49**(10), 2333–2341 (2014). https://doi.org/10.1109/JSSC.2014.2342715
6. Cannici, M., Ciccone, M., Romanoni, A., Matteucci, M.: Asynchronous convolutional networks for object detection in neuromorphic cameras. In: Proceedings of the IEEE/CVF Conference on Computer Vision and Pattern Recognition Workshops (2019)
7. Cannici, M., Ciccone, M., Romanoni, A., Matteucci, M.: A differentiable recurrent surface for asynchronous event-based data. In: Vedaldi, A., Bischof, H., Brox, T., Frahm, J.-M. (eds.) ECCV 2020. LNCS, vol. 12365, pp. 136–152. Springer, Cham (2020). https://doi.org/10.1007/978-3-030-58565-5_9
8. Elfwing, S., Uchibe, E., Doya, K.: Sigmoid-weighted linear units for neural network function approximation in reinforcement learning. Neural Netw. **107**, 3–11 (2018)
9. Falanga, D., Kleber, K., Scaramuzza, D.: Dynamic obstacle avoidance for quadrotors with event cameras. Sci. Robot. **5**(40), eaaz9712 (2020)
10. Finateu, T., et al.: 5.10 a 1280 × 720 back-illuminated stacked temporal contrast event-based vision sensor with 4.86 μm pixels, 1.066geps readout, programmable event-rate controller and compressive data-formatting pipeline. In: 2020 IEEE International Solid- State Circuits Conference - (ISSCC), pp. 112–114 (2020). https://doi.org/10.1109/ISSCC19947.2020.9063149
11. Gallego, G., et al.: Event-based vision: a survey. IEEE Trans. Pattern Anal. Mach. Intell. **44**(1), 154–180 (2020)
12. Ge, Z., Liu, S., Wang, F., Li, Z., Sun, J.: YOLOx: Exceeding YOLO series in 2021. arXiv preprint arXiv:2107.08430 (2021)
13. Gehrig, D., Loquercio, A., Derpanis, K.G., Scaramuzza, D.: End-to-end learning of representations for asynchronous event-based data. In: Proceedings of the IEEE/CVF International Conference on Computer Vision, pp. 5633–5643 (2019)
14. Gehrig, D., Rüegg, M., Gehrig, M., Hidalgo-Carrió, J., Scaramuzza, D.: Combining events and frames using recurrent asynchronous multimodal networks for monocular depth prediction. IEEE Robot. Autom. Lett. **6**(2), 2822–2829 (2021)
15. Gehrig, M., Aarents, W., Gehrig, D., Scaramuzza, D.: DSEC: a stereo event camera dataset for driving scenarios. IEEE Robot. Autom. Lett. **6**(3), 4947–4954 (2021)
16. Hagenaars, J.J., Paredes-Vallés, F., Bohté, S.M., De Croon, G.C.: Evolved neuromorphic control for high speed divergence-based landings of MAVs. IEEE Robot. Autom. Lett. **5**(4), 6239–6246 (2020)
17. Hamaguchi, R., Furukawa, Y., Onishi, M., Sakurada, K.: Hierarchical neural memory network for low latency event processing. In: CVPR, pp. 22867–22876 (2023)
18. Hendrycks, D., Gimpel, K.: Gaussian error linear units (GELUs). arXiv preprint arXiv:1606.08415 (2016)
19. Hidalgo-Carrió, J., Gehrig, D., Scaramuzza, D.: Learning monocular dense depth from events. In: 2020 International Conference on 3D Vision (3DV), pp. 534–542. IEEE (2020)
20. Hu, Y., Delbruck, T., Liu, S.-C.: Learning to exploit multiple vision modalities by using grafted networks. In: Vedaldi, A., Bischof, H., Brox, T., Frahm, J.-M. (eds.) ECCV 2020. LNCS, vol. 12361, pp. 85–101. Springer, Cham (2020). https://doi.org/10.1007/978-3-030-58517-4_6

21. Kamal, U., Dash, S., Mukhopadhyay, S.: Associative memory augmented asynchronous spatiotemporal representation learning for event-based perception. In: The Eleventh International Conference on Learning Representations (2022)
22. Kingma, D.P., Ba, J.: Adam: a method for stochastic optimization. arXiv preprint arXiv:1412.6980 (2014)
23. Lagorce, X., Orchard, G., Galluppi, F., Shi, B.E., Benosman, R.B.: HOTS: a hierarchy of event-based time-surfaces for pattern recognition. IEEE Trans. Pattern Anal. Mach. Intell. **39**(7), 1346–1359 (2016)
24. Li, J., Li, J., Zhu, L., Xiang, X., Huang, T., Tian, Y.: Asynchronous spatiotemporal memory network for continuous event-based object detection. IEEE Trans. Image Process. **31**, 2975–2987 (2022)
25. Li, Y., et al.: Graph-based asynchronous event processing for rapid object recognition. In: Proceedings of the IEEE/CVF International Conference on Computer Vision, pp. 934–943 (2021)
26. Liu, B., Xu, C., Yang, W., Yu, H., Yu, L.: Motion robust high-speed light-weighted object detection with event camera. IEEE Trans. Instrum. Meas. **72**, 1–13 (2023)
27. Liu, Z., et al.: Swin transformer: hierarchical vision transformer using shifted windows. In: Proceedings of the IEEE/CVF International Conference on Computer Vision, pp. 10012–10022 (2021)
28. Maqueda, A.I., Loquercio, A., Gallego, G., García, N., Scaramuzza, D.: Event-based vision meets deep learning on steering prediction for self-driving cars. In: Proceedings of the IEEE Conference on Computer Vision and Pattern Recognition, pp. 5419–5427 (2018)
29. Messikommer, N., Gehrig, D., Loquercio, A., Scaramuzza, D.: Event-based asynchronous sparse convolutional networks. In: Vedaldi, A., Bischof, H., Brox, T., Frahm, J.-M. (eds.) ECCV 2020. LNCS, vol. 12353, pp. 415–431. Springer, Cham (2020). https://doi.org/10.1007/978-3-030-58598-3_25
30. Perot, E., De Tournemire, P., Nitti, D., Masci, J., Sironi, A.: Learning to detect objects with a 1 megapixel event camera. Adv. Neural. Inf. Process. Syst. **33**, 16639–16652 (2020)
31. Sabater, A., Montesano, L., Murillo, A.C.: Event transformer. A sparse-aware solution for efficient event data processing. In: Proceedings of the IEEE/CVF Conference on Computer Vision and Pattern Recognition, pp. 2677–2686 (2022)
32. Schaefer, S., Gehrig, D., Scaramuzza, D.: AEGNN: asynchronous event-based graph neural networks. In: Proceedings of the IEEE/CVF Conference on Computer Vision and Pattern Recognition, pp. 12371–12381 (2022)
33. Sekikawa, Y., Hara, K., Saito, H.: EventNet: asynchronous recursive event processing. In: Proceedings of the IEEE/CVF Conference on Computer Vision and Pattern Recognition, pp. 3887–3896 (2019)
34. Sironi, A., Brambilla, M., Bourdis, N., Lagorce, X., Benosman, R.: HATS: histograms of averaged time surfaces for robust event-based object classification. In: Proceedings of the IEEE Conference on Computer Vision and Pattern Recognition, pp. 1731–1740 (2018)
35. Son, B., et al.: 4.1 a 640 × 480 dynamic vision sensor with a 9 μm pixel and 300meps address-event representation. In: 2017 IEEE International Solid-State Circuits Conference (ISSCC), pp. 66–67 (2017). https://doi.org/10.1109/ISSCC.2017.7870263
36. Sun, S., Cioffi, G., De Visser, C., Scaramuzza, D.: Autonomous quadrotor flight despite rotor failure with onboard vision sensors: frames vs. events. IEEE Robot. Autom. Lett. **6**(2), 580–587 (2021)

37. Sun, Z., Messikommer, N., Gehrig, D., Scaramuzza, D.: ESS: learning event-based semantic segmentation from still images. In: Avidan, S., Brostow, G., Cissé, M., Farinella, G.M., Hassner, T. (eds.) ECCV 2022. LNCS, vol. 13694, pp. 341–357. Springer, Cham (2022)
38. Vemprala, S., Mian, S., Kapoor, A.: Representation learning for event-based visuomotor policies. In: Ranzato, M., Beygelzimer, A., Dauphin, Y., Liang, P., Vaughan, J.W. (eds.) Advances in Neural Information Processing Systems, vol. 34, pp. 4712–4724. Curran Associates, Inc. (2021)
39. Wang, Q., Zhang, Y., Yuan, J., Lu, Y.: Space-time event clouds for gesture recognition: from RGB cameras to event cameras. In: 2019 IEEE Winter Conference on Applications of Computer Vision (WACV), pp. 1826–1835. IEEE (2019)
40. Wu, Y., He, K.: Group normalization. In: Proceedings of the European Conference on Computer Vision (ECCV), pp. 3–19 (2018)
41. Xiao, T., Liu, Y., Zhou, B., Jiang, Y., Sun, J.: Unified perceptual parsing for scene understanding. In: Proceedings of the European Conference on Computer Vision (ECCV), pp. 418–434 (2018)
42. Zhu, A.Z., Thakur, D., Özaslan, T., Pfrommer, B., Kumar, V., Daniilidis, K.: The multivehicle stereo event camera dataset: an event camera dataset for 3D perception. IEEE Robot. Autom. Lett. **3**(3), 2032–2039 (2018)

SNP: Structured Neuron-Level Pruning to Preserve Attention Scores

Kyunghwan Shim, Jaewoong Yun, and Shinkook Choi(✉)

Nota Inc., Seoul, Korea
{kyunghwan.shim,jwyun,shinkook.choi}@nota.ai

Abstract. Multi-head self-attention (MSA) is a key component of Vision Transformers (ViTs), which have achieved great success in various vision tasks. However, their high computational cost and memory footprint hinder their deployment on resource-constrained devices. Conventional pruning approaches can only compress and accelerate the MSA module using head pruning, although the head is not an atomic unit. To address this issue, we propose a novel graph-aware neuron-level pruning method, Structured Neuron-level Pruning (SNP). SNP prunes neurons with less informative attention scores and eliminates redundancy among heads. Specifically, it prunes graphically connected query and key layers having the least informative attention scores while preserving the overall attention scores. Value layers, which can be pruned independently, are pruned to eliminate inter-head redundancy. Our proposed method effectively compresses and accelerates Transformer-based models for both edge devices and server processors. For instance, the DeiT-Small with SNP runs 3.1 times faster than the original model and achieves performance that is 21.94% faster and 1.12% higher than the DeiT-Tiny. Additionally, SNP accelerates the efficiently designed Transformer model, EfficientFormer, by 1.74 times on the Jetson Nano with acceptable performance degradation. Source code is at https://github.com/Nota-NetsPresso/SNP.

Keywords: Network pruning · Network compression · Compact models

1 Introduction

Vision Transformers (ViTs) [7,17,24] have outperformed or matched the performance of state-ofthe-art convolutional neural networks (CNNs) [10,22,23,28] on various computer vision tasks. The success of ViTs is attributed to the Multi-head Self-Attention (MSA) module, which captures intricate relationships in data. However, the Transformer architecture entails substantial computational resources, posing challenges for practical applications on edge devices with constrained storage and computational capabilities. To address this issue, we leverage the graphical components of the MSA module to reduce the dimension of

Supplementary Information The online version contains supplementary material available at https://doi.org/10.1007/978-3-031-73007-8_6.

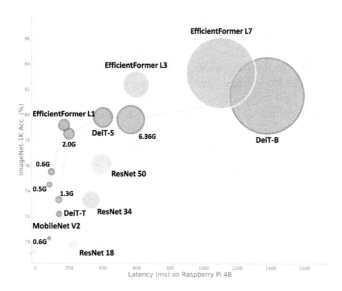

Fig. 1. Comparison of model size, speed, and performance. ImageNet-1K classification results. Latency is profiled by Rasbperry Pi 4B. The connected lines represent the compressed models paired with the original model. The size of each circle indicates the number of parameters in respective model. The number adjacent to each compressed model indicates its compressed FLOPs.

interconnected layers, aimed at effectively reducing their computing budgets while achieving hardware-agnostic speedups.

From the perspective of structured pruning, which aims to reduce the number of dimensions in convolutional or linear layers, the MSA module contains two prunable objectives: heads and neurons. Head pruning, which removes the number of heads, is relatively intuitive to implement due to the reduced complexity of graphical elements compared to neuron-level pruning. In contrast, neuron-level pruning reduces the dimension of individual layers in each head of the MSA module, necessitating a comprehensive understanding of the graphical connectivity of the MSA module. A recent work [31] implements neuron-level pruning by zeroing out individual filters without considering the model's graphical connectivity. This indiscriminate zeroing can negatively affect both accuracy and throughput performance [26].

In this paper, we introduce a novel graph-aware neuron-level pruning method called Structured Neuron-level Pruning (SNP) to accelerate and compress ViTs effectively. We propose two pruning criteria based on the function of each layer within the MSA module. SNP prunes filter pairs of query and key layers containing fewer contributions to attention scores. Moreover, SNP aims to reduce redundancy across heads by eliminating the redundant filter from the value layer.

Furthermore, by removing identical filter indices for all the graphically connected layers, SNP could accelerate various Transformer models on various devices without additional libraries, as shown in Fig. 1.

In summary, the major contributions of this paper are:

- We propose a novel graph-aware neuron-level pruning method, SNP, for Transformer models. SNP is the first method to use the graphical characteristics of the MSA module to measure the importance score of neurons.
- To the best of our knowledge, this is the first work to accelerate Transformer models on any device using neuron-level pruning only.
- SNP achieves significant acceleration while maintaining the original performance on several models. Compressed DeiT-Small outperforms DeiT-Tiny by 1.12% in accuracy, with similar FLOPs, and reduces inference time on various edge devices. Furthermore, the proposed method accelerates the efficiently designed Transformer model, EfficientFormer [16], more than two times with acceptable performance degradation.

2 Related Work

2.1 Compressing Vision Transformers

The ViTs [7,17,24] achieve high performance in numerous vision tasks without specialized image processing modules such as convolutions. The key concept of ViTs is to segment images into patch sequences, convert these patches to token embeddings, and then process them through Transformer encoders [25]. ViTs consist only of Transformer blocks, making them likely to be over-parameterized. For this reason, recent works have aimed to reduce computational cost [3,24] and be memory efficient [18,26]. DeiT [24] proposes lightweight ViT architectures through knowledge distillation [13]. ToMe [3] proposes accelerating ViTs by directly combining similar tokens, without the need for training. Liu et al. [18] propose post-training quantization method using nuclear norm based mixed-precision scheme that does not require fine-tuning for the Vision Transformer.

2.2 Pruning Vision Transformers

Unstructured and Structured Pruning. Pruning methods can be broadly categorized into two types, unstructured and structured pruning. Unstructured pruning sets individual weights or parameters to zero, resulting in irregular sparse matrices [9,14]. Compressed models using unstructured pruning tend to maintain relatively high performance for a given pruning ratio. However, they necessitate additional libraries, such as cuSPARSE [5], Automatic SParsity [20], or SparseDNN [27] to accelerate sparse matrix computations.

Structured pruning, on the other hand, involves the removal of entire groups of units, such as filters or attention heads. This can be implemented using "masking" (zeroing out) [11,12,32,33], or by "pruning" [8,15]. Structured pruning by masking [11,12,32,33] simply sets the group of units to zero, which requires additional libraries to accelerate the model, as unstructured pruning. "Pruning" [8,15], on the other hand, requires a comprehensive understanding of the network's graphical connectivity, including element-wise operations that enforce the same input shape. By considering the graphical connectivity and pruning identical filter indices for inter-connected layers, structured pruning can achieve acceleration on any devices.

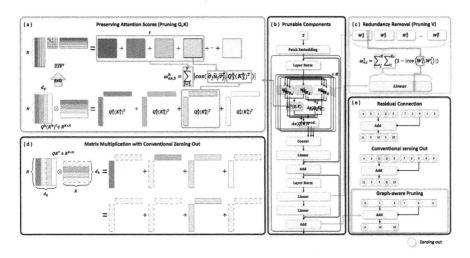

Fig. 2. Proposed SNP methods, on each prunable component of the Transformer block. **(a)** SNP for query and key layers to preserve attention scores. **(b)** prunable components of the Transformer block. **(c)** SNP for value and other layers, including FFN and patch embedding. **(d)** conventional zeroing out in the matrix multiplication operator. **(e)** conventional zeroing out and graph-aware pruning in the residual connection.

Head and Neuron-Level Pruning. Structured pruning for the MSA module has two pruning objectives: head and neuron. Head pruning [29,30] reduces the number of heads, while neuron-level pruning [31] reduces the dimension of each query, key, and value layer in each head. Recent studies for pruning ViTs have focused on head pruning. X-Pruner [30] proposes a novel head pruning method for ViTs that introduces explainability-aware masks and measure the head importance, resulting in superior model compression. WDPruning [29] propose a method to control the number of attention heads and blocks via threshold on learnable parameters.

UVC [31] utilizes knowledge distillation alongside several pruning techniques, such as head pruning, block pruning, and neuron-level pruning. However, neuron-level pruning of UVC is carried out in a masking (zeroing out) manner, converting the weight matrix into a sparse matrix. For this reason, UVC necessitates additional libraries or hardware for accelerating sparse matrices, otherwise, the compressed model with UVC cannot achieve latency gain from the neuron-level pruning.

3 Methodology

3.1 Preliminaries

MSA module takes a single input $X \in \mathbb{R}^{N \times d}$, where N denotes the input vector length, and d represents the hidden size. It comprises H heads, each consisting of three linear layers: query, key, and value. Each layer denoted as $W^h_{\{q,k,v\}} \in$

$\mathbb{R}^{d \times d_{\{q,k,v\}}}$, where h represents h-th head and $d_{\{q,k,v\}}$ indicates the hidden size of each query, key, and value layer. $\{Q, K, V\}^h$ signifies the output of each query, key, and value layer in h-th head, with a shape of $\mathbb{R}^{N \times d_{\{q,k,v\}}}$.

The self-attention operation for h-th head can be expressed as follows:

$$\text{As}^h(X) = Q^h \cdot (K^h)^T \tag{1}$$

$$\text{Att}^h(X) = \text{Softmax}(\frac{\text{As}^h(X)}{\sqrt{d_q}}) \cdot V^h \tag{2}$$

$$\text{MSA}(X) = \text{Concat}(\text{Att}^1(X), ..., \text{Att}^H(X)) \tag{3}$$

here, As, Att, and MSA represent the functions to calculate attention scores, attention module, and MSA module respectively.

Matrix multiplication, unlike a residual connection, yields zero when either of the inputs is masked, regardless of the value of the other input. Therefore, applying a conventional zeroing-out approach to neuron-level pruning in the MSA module can lead to unintended results, as shown in Fig. 2(d). To address this issue, we introduce two graph-aware neuron-level pruning criteria to compress and expedite the MSA module.

3.2 Preserving Attention Scores

Attention scores, Eq. (1), of the MSA module learn long-range dependencies between image features by focusing on different parts of the image when processing different features. These attention scores can be recognized as a series of outer product of Q_i^h and $(K_i^h)^T$ as follows:

$$\begin{aligned}\text{As}^h(X) &= Q^h \cdot (K^h)^T \\ &= \sum_{i=1}^{d_q} Q_i^h \cdot (K_i^h)^T \end{aligned} \tag{4}$$

from this perspective, neuron-level pruning, reducing the dimension of query d_q and key d_k, inevitably distorts the attention scores. For this reason, preserving attention scores is essential to maintain the high performance of the original model.

To alleviate the distortion, we maintain the graphically connected query-key filter pair $(Q_i^h$ and $K_i^h)$, constituting a filter-by-attention score $(Q_i^h \cdot (K_i^h)^T \in \mathbb{R}^{N \times N})$, that retains the most significant aspects of the overall attention scores. To identify the most informative filter pair, we initially employ singular value decomposition (SVD) to decompose the original model's attention scores.

$$\begin{aligned}\text{As}^h(X) &= \tilde{U} \cdot \tilde{S} \cdot \tilde{V}^T \\ &= \sum_{j=1}^{N} \tilde{\sigma}_j \cdot \tilde{u}_j \cdot \tilde{v}_j^T \end{aligned} \tag{5}$$

where $\tilde{U}$ and $\tilde{V}$ are the left and right singular vector matrices, respectively, and $\tilde{S}$ is the diagonal matrix of singular values, with $\tilde{\sigma}_1 \geq \tilde{\sigma}_2 \geq ... \geq \tilde{\sigma}_N$.

SVD is a technique for extracting the most informative components of a matrix, those with large singular values, while discarding the less informative components, those with small singular values. To retain the spatial relationships captured by the attention mechanism, we prune the filter pair Q_i^h and K_i^h with least correlation with the most informative components of the attention scores.

To measure the correlation, we adopt the cosine similarity between the attention scores of the i-th filter and the j-th rank matrix, as normalizing the attention scores reduces the impact of magnitude-based approaches. Consequently, the importance score $w_{as,i}^h$ is defined as follows:

$$w_{as,i}^h = \sum_{j=1}^{r} |\cos((Q_i^h \cdot (K_i^h)^T), (\tilde{\sigma}_j \cdot \tilde{u}_j \cdot \tilde{v}_j^T))| \\ = \sum_{j=1}^{r} |\frac{(Q_i^h \cdot (K_i^h)^T) \cdot (\tilde{\sigma}_j \cdot \tilde{u}_j \cdot \tilde{v}_j^T)}{||(Q_i^h \cdot (K_i^h)^T)|| \cdot ||(\tilde{\sigma}_j \cdot \tilde{u}_j \cdot \tilde{v}_j^T)||}| \quad (6)$$

where r is a hyperparameter dictating the quantity of ranks, within the range of $1 \leq r \leq N$, to be compared with the i-th attention scores, while the remaining $N - r$ singular values $(\tilde{\sigma}_{r+1}, ..., \tilde{\sigma}_N)$ are discarded.

Optimizing r for each attention module can contribute to preserving informative filters and sustaining higher performance. For this reason, we adopt Variational Bayesian Matrix Factorization (VBMF) [21] to determine the optimal rank r for the attention scores. VBMF inherently addresses the challenge of determining the number of latent factors (ranks) in matrix factorization using its probabilistic approach.

Importance score $w_{as,i}^h$, defined in Eq. (6), represents the importance of the i-th filter for both query and key layers. A larger $w_{as,i}^h$ indicates that the filter has a greater impact on the main component of the attention scores, while a lower $w_{as,i}^h$ indicates that the filter is less important and can be removed without significantly affecting the attention scores.

3.3 Inter-head Redundancy Removal

In the preceding section, we outlined the approach to preserving attention scores even with reduced embedding dimensions in the query and key layers. Here, we introduce a pruning method for the value and other layers, such as FFN or patch embedding layer.

Previous works [2,19] have revealed that a significant proportion of attention heads can be removed without causing significant performance deterioration. To remove this inter-head redundancy through neuron-level pruning, we propose to measure the distance between all the value layers of MSA module, irrespective of the heads. The importance score $w_{v,i}^h$ of the i-th filter in the value layer of the h-th head is as follows:

$$\omega_{v,i}^h = \sum_{j=1}^{H} \sum_{l=1}^{d_v} (1 - |\cos(W_i^h, W_l^j)|) \tag{7}$$

The importance score of the value layer, denoted as $\omega_{v,i}^h$, indicates the redundancy of filter i in the value layer compared to all value layers of the heads within the corresponding MSA layer. Consequently, filters with the lowest importance scores will be pruned in the initial stages of the pruning process.

3.4 Accelerating Transformers

As described in Sect. 3.2, SNP removes the least correlated filter pairs (Q_i^h, K_i^h) to preserve attention scores and enhance the efficiency of the MSA module across various devices.

Most MSA modules, including MSA modules in DeiTs, are designed to compute all heads in parallel. For achieving this parallel computation, SNP ensures that the number of filter dimensions are consistent across heads, even though the filter indices for each head are independently selected.

The green highlighted boxes in Fig. 2(b) represent layers connected by a single residual connection at the last add layer of the Transformer block. Unlike the matrix multiplication operator, the output of the residual connection becomes zero when all the interconnected layers return zero for specific filter indices. For this reason, the residual connection with masking always exceeds the performance of actual pruning, which restricts the set of possible pruning patterns [26].

To accelerate the residual connection layers, all interconnected layers should be pruned identically. To achieve this, we sum up all the calculated importance scores for interconnected layers based on their filter indices, as shown in Fig. 2(e). Subsequently, we prune the least important filter indices of all the connected layers.

4 Experimental Results

To ensure a fair comparison with existing methods, we apply SNP to prune the DeiT [24] architectures trained on the ImageNet-1K [6] dataset. Additionally, we conduct experiments on the efficient Transformer model, EfficientFormer-L1, to confirm the robustness of SNP. Furthermore, a series of ablation studies are conducted to gain a comprehensive understanding of our methodology.

4.1 Implementation Details

The overall pruning and fine-tuning process are executed on the pre-trained DeiT[1] and EfficientFormer-L1[2] released from the official implementation on ImageNet-1K. Throughout the fine-tuning phase of the pruned model, we maintain consistent settings across all models, except for the batch size and learning

---
[1] https://github.com/facebookresearch/deit.
[2] https://github.com/snap-research/EfficientFormer.

Table 1. Performance comparison of various pruning methods on ImageNet-1K. The "Pruning" column represents the pruning methods that corresponding methods use to compress the MSA module. Each of the methods contains one of follows: head pruning (HP), block pruning (BP), neuron-level pruning (NP), neuron-level sparsity (NS).

Model	Method	Pruning	Top-1 (%)	Top-5 (%)	GFLOPs	Params (M)
DeiT-T	Original [24]	–	72.20	91.10	1.3	5.7
	SSViTE [4]	HP	70.12	–	0.9	4.2
	WDPruning [29]	HP+BP	70.34	89.82	0.7	3.5
	X-Pruner [30]	HP	71.10	90.11	0.6	–
	SNP	**NP**	**70.29**	**90.01**	**0.6**	**3.0**
	UVC [31]	HP+NS+BP	70.60	–	0.5	–
DeiT-S	Original [24]	–	79.85	95.00	4.6	22.1
	SSViTE [4]	HP	79.22	–	3.1	14.6
	WDPruning [29]	HP+BP	78.38	94.05	2.6	13.3
	X-Pruner [30]	HP	78.93	94.24	2.4	–
	UVC [31]	HP+NS+BP	78.82	–	2.3	–
	SNP	**NP**	**78.52**	**94.37**	**2.0**	**10.0**
	SNP	**NP**	**73.32**	**91.66**	**1.3**	**6.4**
DeiT-B	Original [24]	–	81.80	95.59	17.6	86.6
	SSViTE [4]	HP	82.22	–	11.8	56.8
	WDPruning [29]	HP+BP	80.76	95.36	9.9	55.3
	X-Pruner [30]	HP	81.02	95.38	8.5	–
	UVC [31]	HP+NS+BP	80.57	–	8.0	–
	SNP	**NP**	**79.63**	**94.37**	**6.4**	**31.6**

rate. The batch size is set to 256, and to prevent weight explosion, we adjust the learning rate of the compressed model to 1/10 or 1/100 of the original model.

To evaluate the reduced latency using SNP, we have configured four testing scenarios: one on a CPU and another on GPU for both edge devices and server processors. We employ a standard PyTorch model for profiling on the server processors (Intel Xeon Silver 4210R and NVIDIA GeForce RTX 3090). Profiling on the Raspberry Pi 4B and Jetson Nano is conducted using the ONNX and TensorRT formats, respectively. All latencies are measured using a single image as an input, except for the GPU of the server processor (RTX 3090), where it is set to 64 images.

4.2 Quantitative Results

Comparison with Other Methods. Despite the constraints outlined in Section 3.4, SNP not only maintains accuracy comparable to existing methods but also significantly reduces inference time across diverse hardware and data types on the ImageNet-1K dataset, as shown in Table 1 and Table 2.

Table 2. Inference speed and Top-1 accuracy of the compressed model across different devices. Performance evaluation involves accuracy on ImageNet-1K and inference time for the compressed DeiTs and EfficientFormer-L1. Latency is benchmarked with 200 warm-up runs and averaged over 1000 runs. In latency measurement, a single image is used as an input, except for the RTX 3090, where 64 images are employed in a single batch.

Model	Top-1 (%)	GFLOPs	Edge devices (ms)		Server processors (ms)	
			Raspberry Pi 4B (.onnx)	Jetson Nano (.trt)	Xeon Silver 4210R (.pt)	RTX 3090 (.pt)
DeiT-Tiny	72.20	1.3	139.13	41.03	34.74	18.65
+ SNP (Ours)	70.29	0.6	81.63 (1.70×)	26.67 (1.54×)	25.25 (1.38×)	17.82 (1.05×)
DeiT-Small	79.80	4.6	401.27	99.32	53.37	46.13
+ SNP (Ours)	78.52	2.0	199.15 (2.01×)	45.51 (2.18×)	38.57 (1.38×)	32.91 (1.40×)
+ SNP (Ours)	73.32	1.3	136.68 (2.94×)	32.03 (3.10×)	33.46 (1.60×)	26.98 (1.71×)
DeiT-Base	81.80	17.6	1377.71	293.29	122.03	151.35
+ SNP (Ours)	79.63	6.4	565.68 (2.44×)	132.55 (2.21×)	64.65 (1.89×)	72.96 (2.07×)
EfficientFormer-L1	79.20	1.3	169.13	30.95	43.75	26.19
+ SNP (Ours)	75.53	0.6	95.12 (1.78×)	19.78 (1.56×)	38.25 (1.14×)	17.24 (1.52×)
+ SNP (Ours)	74.51	0.5	82.60 (2.05×)	17.76 (1.74×)	35.15 (1.24×)	16.01 (1.64×)

In a recent study [26], unconstrained masking generally outperforms post-training accuracy of pruned models by an average of 2.1% on ImageNet-1K. Compared to unconstrained head masking approaches like SSVITE [4] and X-Pruner [30], SNP achieves significantly higher compression rates for all DeiTs FLOPs (30.64% and 10.64%) with minimal performance degradation (0.83% and 0.87%), much less than the 2.1% average mentioned.

Furthermore, compared to other pruning approaches like WDPruning [29] and UVC [31], which use a combination of pruning techniques to compress DeiTs, SNP achieves comparable accuracy solely through neuron-level pruning. Notably, DeiT-Small with SNP outperforms WDPruning by 0.14%, with the removal of 3.3 million parameters and a reduction of 0.6 GFLOPs. Compared to UVC, SNP exhibits negligible performance degradation, averaging 0.51% across all DeiTs while using 7.65% fewer FLOPs.

Large Compressed vs. Small Hand-Crafted. DeiT-Small with SNP, a large pruned model, outperforms the smaller, hand-crafted DeiT-Tiny in both accuracy and latency, achieving a notable 1.12% improvement in top-1 accuracy while maintaining similar FLOPs. Additionally, DeiT-Small with SNP exhibits enhanced speed compared to the original DeiT-Tiny across edge devices and CPU-based server processors. Notably, its speed increases up to 21.94% compared to the original DeiT-Tiny running on Jetson Nano, a GPU-based edge device. This substantial performance gap underscores the superiority of the compressed model (DeiT-Small with SNP) over the smaller hand-crafted designed model (DeiT-Tiny) in both overall performance and speed.

Accelerating Transformer-Based Models. As depicted in Table 2, SNP achieves impressive acceleration of DeiTs by a factor of 1.44× to 2.44× on edge

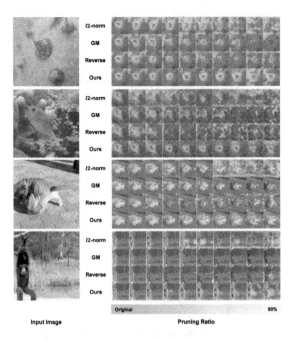

Fig. 3. Attention maps with varying pruning criteria and compression ratios. All query and key layers are locally pruned based on the specified pruning ratio without fine-tuning. The importance scores of $l2$-norm and GM on query and key layers are combined by filter index and pruned simultaneously. "Reverse" represents the reverse order of SNP.

devices and 1.05× to 2.07× on server processors. This acceleration is notable, with negligible average performance degradation of 1.79%, specifically 1.91%, 1.28%, and 2.17% for DeiT-Tiny, DeiT-Small, and DeiT-Base, respectively.

Compared to WDPruning, which employs both head and block pruning, SNP surpasses in terms of latency for DeiT-Small and DeiT-Base on the RTX 3090 GPU. SNP accelerates the original DeiTs by 1.38× and 2.07×, respectively, while WDPruning achieves a comparatively modest acceleration of 1.18× for both models.

The superiority of SNP becomes more evident when the target of neuron-level pruning, linear or convolutional layers, contribute a larger proportion of the original model's computation time. We find that SNP further accelerates original model 1.05× to 2.07× as the DeiT model's size grows on the RTX 3090 GPU. In contrast, WDPruning, involving the removal of entire layers and associated operators in both head and block, consistently shows acceleration rates of 1.18×.

To ascertain the robustness of SNP across diverse Transformer models, especially on the efficiently designed model, we conduct additional experiments on EfficientFormer-L1. SNP accelerates the model by 1.78× and 2.05× faster on Raspberry Pi 4B and 1.56× and 1.74× faster on Jetson Nano. Notably, the compressed EfficientFormer-L1 achieves acceptable accuracy of 75.53% and 74.51%.

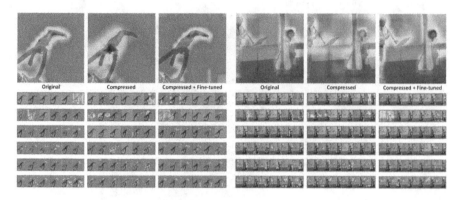

Fig. 4. Attention maps from the original, compressed, and fine-tuned DeiT-Tiny with SNP. The attention maps in the first row are visualized using the attention rollout [1]. Each red box contains three attention maps from each head of the MSA module, ordered accordingly.

4.3 Qualitative Results

In the subsequent sections, attention rollout [1] is employed to randomly selected images from the test datasets to visualize the attention maps of the DeiTs and assess the efficacy of SNP. This evaluation unfolds across two key facets:

- **Preserving the attention scores:** Visualize attention maps to verify the effectiveness of the proposed importance score in preserving attention scores.
- **Restoring the attention scores after SNP:** Visualize attention maps to find the effectiveness of SNP in restoring attention scores.

Preserving the Attention Scores. To visualize SNP effectiveness, we apply four pruning criteria to compress the query and key filter pairs in all MSA modules of DeiT-Tiny: SNP, $l2$-norm, geometric median (GM) [11], and "Reverse".

The "Reverse" criterion prioritizes pruning the most important filter pairs first, keeping the least important pairs until the end. However, both $l2$-norm and GM pruning criteria ignore the MSA module's graphical connectivity, evaluating importance scores independently for query and key layers. To handle identical filter indices during pruning, we aggregate scores based on filter indices, removing the least important indices from both layers.

In Fig. 3, attention maps for the original DeiT-Tiny and locally pruned models are presented across various pruning ratios (10% to 90% with 10% intervals). Our proposed method effectively maintains the original attention map even after pruning over 80% of the filters, whereas other methods show fragmented attention maps at much lower pruning ratios (typically 30% or less). These results highlight the potential of our neuron-level pruning criteria, utilizing SVD to preserve attention scores, in reducing the size and speeding up the execution of MSA modules without compromising accuracy.

Table 3. Performance of SNP without fine-tuning across different data quantities at various pruning ratios. To determine the proper number of images for calculating the importance score, we conduct local pruning on the query and key layers at pruning ratios of 10%, 30%, 50%, 70%, and 90%. All performance metrics are assessed without fine-tuning. The latency is measured on Raspberry Pi 4B.

Number of images	Performance by pruning ratio					
	Original	10%	30%	50%	70%	90%
1	72.20	71.15	67.96	57.60	38.68	11.17
4	72.20	70.82	67.86	57.65	38.05	11.86
16	72.20	70.72	67.42	58.60	39.46	9.11
64	72.20	70.83	67.55	59.50	42.13	12.83
256	72.20	70.75	68.14	59.58	42.70	14.48
Latency (ms)	139.13	133.60	129.29	119.41	117.48	112.32
Params (M)	5.7	5.59	5.41	5.23	5.06	4.88

Restoring the Attention Scores After SNP. Fig. 4 illustrates the overall and per-head attention maps of the original, compressed, and fine-tuned DeiT-Tiny, respectively. The first row shows the overall attention maps of the respective models. Notably, the compressed model maintains a well-preserved overall attention map, despite pruning all layers, including values and FFN, resulting in a 53% reduction in FLOPs and a 46% reduction in parameters. Especially, we can observe that attention map is well-restored after the fine-tuning process.

The twelve red boxes below the overall attention map depict per-head attention maps for twelve layers of each original, compressed, and fine-tuned models respectively. As shown in Fig. 4, it is evident that the attention maps for each head are effectively preserved and restored in each of the compressed and fine-tuned models.

4.4 Ablation Studies

Importance Scores by the Data Quantity. Since the attention scores depend on the input, as indicated in Eq. (1), the proposed importance scores for query and key filter pairs (Eq. (6)) may exhibit sensitivity to the distribution of the input image X. To validate the method's robustness, we compute importance scores using various image quantities, pruning query and key layers at different ratios, without fine-tuning process.

As depicted in Table 3, SNP demonstrates a slight advantage in preserving performance with an increasing number of images, outperforming models compressed with fewer images as the compression ratio rises. However, this improvement comes at the cost of increased computation time for SNP calculations. Considering these factors into account, we opt to use 64 images, which yield the second-best performance among the given number of images. This decision strikes a balance between achieving satisfactory performance and maintaining computational efficiency.

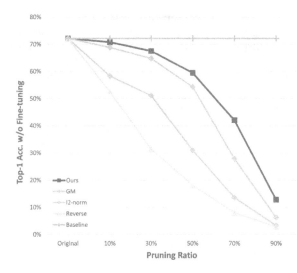

Fig. 5. Top-1 accuracy of compressed DeiT-Tiny on ImageNet using several pruning criteria without fine-tuning. Query and key layers are locally pruned using various pruning criteria: SNP, GM, $l2$-norm, reverse order of SNP ("Reverse"), and original DeiT-Tiny ("Baseline").

Performance Comparison Across Pruning Ratios. To assess the robustness of SNP, we examine the performance of DeiT-Tiny under various pruning criteria and different pruning ratios, as described in Sect. 4.3.

As illustrated in Fig. 5, SNP consistently outperforms other pruning criteria across all pruning ratios. In contrast, models compressed with "Reverse" criteria exhibit the lowest performance at all pruning ratios, underscoring the robustness of the proposed approach.

Both Fig. 3 and Fig. 5 confirm that SNP successfully preserves attention scores in both quantitative and qualitative aspects. Even with an 80% pruning ratio and without fine-tuning, SNP maintains original attention scores and outperforms other pruning criteria.

5 Conclusion

In this paper, we propose a novel graph-aware neuron-level pruning method, SNP, designed to compress and accelerate Transformer-based models. SNP proposes two pruning criteria for preserving attention scores and eliminating inter-head redundancy. Using SNP, a large compressed model outperforms small, hand-crafted designed models in both performance and latency on edge devices. Moreover, the compressed models exhibit astonishing results in latency on various devices, with negligible performance degradation.

As this work is a first attempt to accelerate MSA modules using neuron-level pruning alone, many challenges remain. One is incorporating other pruning methods such as head or block pruning for a more efficient Transformer model. Another challenge is applying SNP to other vision tasks, including image generation, which requires high computational costs on both training and inference.

We believe that these works encourage the adoption of model pruning as a tool, for both improving the applicability of ViTs in resource-constrained environments and reducing the training costs of large models by integrating with training process.

Acknowledgements. This work was supported by Artificial intelligence industrial convergence cluster development project funded by the Ministry of Science and ICT(MSIT, Korea)&Gwangju Metropolitan City.

References

1. Abnar, S., Zuidema, W.: Quantifying attention flow in transformers. arXiv preprint arXiv:2005.00928 (2020)
2. Bian, Y., Huang, J., Cai, X., Yuan, J., Church, K.: On attention redundancy: a comprehensive study. In: Proceedings of the 2021 Conference of the North American Chapter of the Association for Computational Linguistics: Human Language Technologies, pp. 930–945 (2021)
3. Bolya, D., Fu, C.Y., Dai, X., Zhang, P., Feichtenhofer, C., Hoffman, J.: Token merging: your VIT but faster. arXiv preprint arXiv:2210.09461 (2022)
4. Chen, T., Cheng, Y., Gan, Z., Yuan, L., Zhang, L., Wang, Z.: Chasing sparsity in vision transformers: an end-to-end exploration. Adv. Neural. Inf. Process. Syst. **34**, 19974–19988 (2021)
5. Demouth, J.: Sparse matrix-matrix multiplication on the GPU. Technical report, NVIDIA (2012)
6. Deng, J., Dong, W., Socher, R., Li, L.J., Li, K., Fei-Fei, L.: ImageNet: a large-scale hierarchical image database. In: 2009 IEEE Conference on Computer Vision and Pattern Recognition, pp. 248–255. IEEE (2009)
7. Dosovitskiy, A., et al.: An image is worth 16×16 words: transformers for image recognition at scale. arXiv preprint arXiv:2010.11929 (2020)
8. Fang, G., Ma, X., Song, M., Mi, M.B., Wang, X.: DepGraph: towards any structural pruning. In: Proceedings of the IEEE/CVF Conference on Computer Vision and Pattern Recognition, pp. 16091–16101 (2023)
9. Han, S., Mao, H., Dally, W.J.: Deep compression: compressing deep neural networks with pruning, trained quantization and Huffman coding. arXiv preprint arXiv:1510.00149 (2015)
10. He, K., Zhang, X., Ren, S., Sun, J.: Deep residual learning for image recognition. In: Proceedings of the IEEE Conference on Computer Vision and Pattern Recognition, pp. 770–778 (2016)
11. He, Y., Liu, P., Wang, Z., Hu, Z., Yang, Y.: Filter pruning via geometric median for deep convolutional neural networks acceleration. In: Proceedings of the IEEE/CVF Conference on Computer Vision and Pattern Recognition, pp. 4340–4349 (2019)
12. He, Y., Zhang, X., Sun, J.: Channel pruning for accelerating very deep neural networks. In: Proceedings of the IEEE International Conference on Computer Vision, pp. 1389–1397 (2017)
13. Hinton, G., Vinyals, O., Dean, J.: Distilling the knowledge in a neural network. arXiv preprint arXiv:1503.02531 (2015)

14. Lee, J., Park, S., Mo, S., Ahn, S., Shin, J.: Layer-adaptive sparsity for the magnitude-based pruning. arXiv preprint arXiv:2010.07611 (2020)
15. Li, H., Kadav, A., Durdanovic, I., Samet, H., Graf, H.P.: Pruning filters for efficient convnets. arXiv preprint arXiv:1608.08710 (2016)
16. Li, Y., et al.: EfficientFormer: vision transformers at MobileNet speed. Adv. Neural. Inf. Process. Syst. **35**, 12934–12949 (2022)
17. Liu, Z., et al.: Swin transformer: hierarchical vision transformer using shifted windows. In: Proceedings of the IEEE/CVF International Conference on Computer Vision, pp. 10012–10022 (2021)
18. Liu, Z., Wang, Y., Han, K., Zhang, W., Ma, S., Gao, W.: Post-training quantization for vision transformer. Adv. Neural. Inf. Process. Syst. **34**, 28092–28103 (2021)
19. Michel, P., Levy, O., Neubig, G.: Are sixteen heads really better than one? Adv. Neural. Inf. Process. Syst. **32** (2019)
20. Mishra, A., et al.: Accelerating sparse deep neural networks. arXiv preprint arXiv:2104.08378 (2021)
21. Nakajima, S., Sugiyama, M., Babacan, S.D., Tomioka, R.: Global analytic solution of fully-observed variational Bayesian matrix factorization. J. Mach. Learn. Res. **14**(1), 1–37 (2013)
22. Radosavovic, I., Kosaraju, R.P., Girshick, R., He, K., Dollár, P.: Designing network design spaces. In: Proceedings of the IEEE/CVF Conference on Computer Vision and Pattern Recognition, pp. 10428–10436 (2020)
23. Tan, M., Le, Q.: EfficientNet: rethinking model scaling for convolutional neural networks. In: International Conference on Machine Learning, pp. 6105–6114. PMLR (2019)
24. Touvron, H., Cord, M., Douze, M., Massa, F., Sablayrolles, A., Jégou, H.: Training data-efficient image transformers & distillation through attention. In: International Conference on Machine Learning, pp. 10347–10357. PMLR (2021)
25. Vaswani, A., et al.: Attention is all you need. Adv. Neural. Inf. Process. Syst. **30** (2017)
26. Wan, A., et al.: UPSCALE: unconstrained channel pruning. In: International Conference on Machine Learning, pp. 35384–35412. PMLR (2023)
27. Wang, Z.: SparseDNN: fast sparse deep learning inference on CPUs. arXiv preprint arXiv:2101.07948 (2021)
28. Xie, Q., Luong, M.T., Hovy, E., Le, Q.V.: Self-training with noisy student improves ImageNet classification. In: Proceedings of the IEEE/CVF Conference on Computer Vision and Pattern Recognition, pp. 10687–10698 (2020)
29. Yu, F., Huang, K., Wang, M., Cheng, Y., Chu, W., Cui, L.: Width & depth pruning for vision transformers. In: Proceedings of the AAAI Conference on Artificial Intelligence, vol. 36, pp. 3143–3151 (2022)
30. Yu, L., Xiang, W.: X-pruner: explainable pruning for vision transformers. In: Proceedings of the IEEE/CVF Conference on Computer Vision and Pattern Recognition, pp. 24355–24363 (2023)
31. Yu, S., et al.: Unified visual transformer compression. arXiv preprint arXiv:2203.08243 (2022)
32. Yu, S., Mazaheri, A., Jannesari, A.: Topology-aware network pruning using multi-stage graph embedding and reinforcement learning. In: International Conference on Machine Learning, pp. 25656–25667. PMLR (2022)
33. Yu, X., Serra, T., Ramalingam, S., Zhe, S.: The combinatorial brain surgeon: pruning weights that cancel one another in neural networks. In: International Conference on Machine Learning, pp. 25668–25683. PMLR (2022)

Multi-Granularity Sparse Relationship Matrix Prediction Network for End-to-End Scene Graph Generation

Lei Wang, Zejian Yuan, and Badong Chen(✉)

Institute of Artificial Intelligence and Robotics, Xi'an Jiaotong University,
Xi'an 710049, China
leiwangmail@stu.xjtu.edu.cn, {yuan.ze.jian,chenbd}@mail.xjtu.edu.cn

Abstract. Current end-to-end Scene Graph Generation (SGG) relies solely on visual representations to separately detect sparse relations and entities in an image. This leads to the issue where the predictions of entities do not contribute to the prediction of relations, necessitating post-processing to assign corresponding subjects and objects to the predicted relations. In this paper, we introduce a sparse relationship matrix that bridges entity detection and relation detection. Our approach not only eliminates the need for relation matching, but also leverages the semantics and positional information of predicted entities to enhance relation prediction. Specifically, a multi-granularity sparse relationship matrix prediction network is proposed, which utilizes three gated pooling modules focusing on filtering negative samples at different granularities, thereby obtaining a sparse relationship matrix containing entity pairs most likely to form relations. Finally, a set of sparse, most probable subject-object pairs can be constructed and used for relation decoding. Experimental results on multiple datasets demonstrate that our method achieves a new state-of-the-art overall performance. Our code is available at https://github.com/wanglei0618/Mg-RMPN.

Keywords: Scene Graph Generation · End-to-End · Sparse Relationship Matrix · Multi-Granularity

1 Introduction

Scene Graph Generation (SGG) is a fundamental visual comprehension task that captures semantic information by detecting relation triplets <subject entity, predicate, object entity> in an image. This structured representation can facilitate many downstream tasks, such as image captioning [4,41], visual question answering [9,28], image retrieval [11,25] and image generation [10,19]. The end-to-end SGG methods [6,16,30] can directly generate sparse relations from an

---

Supplementary Information The online version contains supplementary material available at https://doi.org/10.1007/978-3-031-73007-8_7.

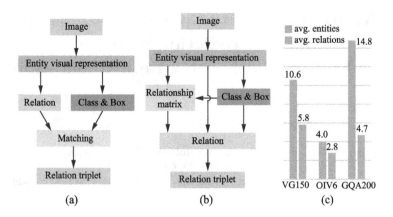

Fig. 1. (a) End-to-end SGG pipeline. (b) Our proposed end-to-end pipeline based on multi-granularity sparse relationship matrix prediction network. (c) The average number of entities and relations in the VG150, OIv6, and GQA200 datasets demonstrates the sparsity of relations in an image.

image based on a fixed number of relation queries, avoiding dense prediction that contains a lot of background relations in two-stage SGG. This leads to faster inference speed, and because fewer background relations are predicted, it exhibits superior predictive performance for rare samples of tail classes.

Current end-to-end approaches [6,16,22] still lag behind two-stage methods in performance, which is attributed to their reliance solely on visual representations to separately detect entities and relations, as shown in Fig. 1 (a). This leads to two drawbacks: (1) It requires post-processing to match the corresponding subjects and objects for the relations, and the model's performance is constrained by the capabilities of matching. (2) Due to the inability to determine the subject and object in advance, entity predictions cannot help with relation prediction, and relation prediction can only rely on visual features.

In this paper, we introduce a sparse relationship matrix that bridges entity detection and relation detection. Our approach not only eliminates the need for relation matching, but also leverages the semantics and positional information of predicted entities to enhance relation prediction. Figure 1 (b) shows our pipeline based on the "pair then relation" [30] framework, which first constructs a relationship matrix based on entity predictions to represent the possibility of forming a relation between two entities and then selects the sparsest, most likely subject-object pairs for relation prediction.

Compared to all possible pairs formed by the entity proposals, the subject-object pairs with relations are extremely sparse. Figure 1 (c) illustrates the average entities and relations in the datasets. Taking VG150 [13] as an example, when generating $N_e = 100$ entities, there are a total of $(100(100 − 1)) = 9900$ possible combinations. But the average relations per image is only 5.8, which means that more than 99.9% pairs are negative samples. Therefore, directly learning this extremely sparse relationship matrix is very challenging.

In this work, we propose a Multi-granularity sparse Relationship Matrix Prediction Network (Mg-RMPN) to partition the negative samples into different granularities, achieving layer-by-layer filtering of dense negative samples. Specifically, Mg-RMPN employs three Gated Pooling Modules (GPM) with identical structures but independent parameters, each of which focuses on filtering negative samples at different granularities. The GPMs are used to generate relation matrices at different granularities and construct the corresponding ground truth matrices for supervised learning. The experimental results from various datasets reveal that our approach not only accurately identifies numerous head-class samples, but also excels in predicting sparse tail-class samples, achieving state-of-the-art overall performance.

The contributions are summarized as follows: (1) An end-to-end SGG framework based on the relationship matrix is proposed that bridges entity detection and relation detection. (2) An Mg-RMPN is proposed to achieve sparse relationship matrix prediction based on multi-granularity negative sample learning. (3) Experimental results have confirmed that our method achieves state-of-the-art overall performance in both end-to-end and two-stage methods.

2 Related Works

2.1 Scene Graph Generation

Early work [5,27,32,36] in SGG is based on the two-stage pipeline, first employing a Faster-RCNN [24] object detector to generate entity proposals, and the class of each entity is predicted. Then, entities are paired to form relations, and all possible pairs are classified. Its drawback lies that dense relation pairs include numerous background relations, which dilute the sparse tail-class samples, thus aggravating the imbalance issue. Therefore, some recent efforts [8,23] have focused on addressing the imbalanced predictions, developing a series of rebalancing strategies such as logit adjustment [2,26], loss reweighting [18,33,35], and data resampling [17,31,37]. Unlike these methods, we introduce a relationship matrix to generate sparse subject-object pairs for relation prediction. The sparse relations generated in this way have a higher signal-to-noise ratio, which can alleviate the imbalance problem. The results of the experiments validate that our method maintains outstanding performance for head classes and also demonstrates high performance for tail classes.

2.2 End-to-End Scene Graph Generation

End-to-end SGG, inspired by DETR [1], generates sparse relations directly from an image based on queries. For instance, [6] introduced the Relation Transformer to directly generate a set of relations from visual features. However, the generated relations did not match the corresponding subjects and objects, hence [16] proposed a graph assembling module to address the relation matching. [30] further introduced a "pair then relation" framework that predetermines subject-object pairs, circumventing the relation matching. However, these approaches suffer from the issue of relying solely on visual representations to detect relations and entities separately. Therefore, this paper introduces Mg-RMPN to predict a sparse relationship matrix to link entity detection and relation detection, not only eliminating the need for relation matching but also utilizing the predicted entity semantics and positional information to assist in relation prediction.

3 Method

3.1 Problem Definition

SGG aims to generate a scene graph $\mathcal{G} = \{\mathcal{E}, \mathcal{R}\}$ from an input image $\mathcal{I}$. The scene graph consists of a set of n entities $\mathcal{E} = \{e_i\}_{i=1}^n$ and a set of m relations $\mathcal{R} = \{r_k\}_{k=1}^m$ between entities. The set of entities $\mathcal{E}$ can be further decomposed into a set of bounding boxes $\mathcal{B} = \{\mathbf{b}_i\}_{i=1}^n$ and a set of class labels $\mathcal{C} = \{c_i\}_{i=1}^n$ The generation of a scene graph $\mathcal{G}$ can be formulated as the joint probability distribution:

$$\Pr(\mathcal{G}|\mathcal{I}) = \Pr(\mathcal{B}, \mathcal{C}|\mathcal{I})\Pr(\mathcal{R}|\mathcal{I}, \mathcal{B}, \mathcal{C}), \tag{1}$$

where $\Pr(\mathcal{B}, \mathcal{C}|\mathcal{I})$ represents the entity representation obtained from the object detector, including the class labels and bounding boxes. $\Pr(\mathcal{R}|\mathcal{I}, \mathcal{B}, \mathcal{C})$ represents the relation prediction based on pairs of entities by the relation decoder. The process from $\Pr(\mathcal{B}, \mathcal{C}|\mathcal{I})$ to $\Pr(\mathcal{R}|\mathcal{I}, \mathcal{B}, \mathcal{C})$ requires our Mg-RMPN to predict the pairs of entities most likely to form a relation from all possible pairs of entities.

3.2 Overall Architecture

As shown in Fig. 2 (a), our method comprises three modules: (1) entity detection, (2) relationship matrix prediction, and (3) relation prediction. The entity detection is responsible for generating a set of entities based on the object detector, including their visual representations, class labels, and bounding boxes. The relationship matrix prediction is responsible for producing a sparse relationship matrix that represents the relevance between two entities, and then obtaining the indices of the subject and object entities that are most likely to form a relation from all possible combinations. Finally, the relation of these sparse subject-object pairs is predicted through the relation decoder.

3.3 Object Detector

This paper employs a deformable DETR [42] as the object detector, which contains a transformer encoder-decoder architecture with a set of entity queries. Here, the entity queries interact with encoder features through multi-scale deformable attention. Given an image $\mathcal{I}$, the object detector produces N_e entity representations $\mathbf{Q}_e = \{\mathbf{q}_i^e\}_{i=1}^{N_e} \in \mathbb{R}^{N_e \times d}$ from a set of learnable entity queries, where d is the embedding dimensions. Based on $\mathbf{Q}_e$, the object detector then follows with a classification head and a regression head to predict the entity's class predictions $\mathbf{C} = \{\hat{c}_i\}_{i=1}^n \in \mathbb{R}^{N_e \times C_e}$ and bounding boxes $\mathbf{B} = \{\hat{\mathbf{b}}_i\}_{i=1}^n \in \mathbb{R}^{N_e \times 4}$, respectively, where C_e is the number of entity classes.

3.4 Multi-Granularity Relationship Matrix Prediction Network

Given N_e entities, Multi-granularity sparse Relationship Matrix Prediction Network (Mg-RMPN) generates a $N_e \times N_e$ adjacency relationship matrix $\mathbf{M}_r$. Each node represents the probability of forming a relationship between two entities. Due to the sparse entities and relations in an image, $\mathbf{M}_r$ contains a large number of pairs without relations, that is, negative samples.

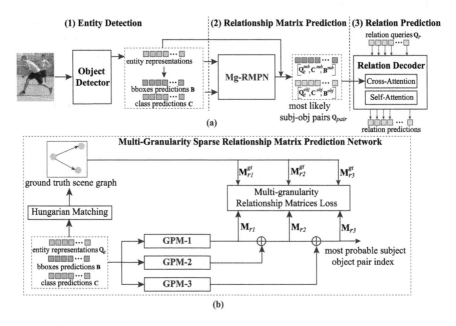

Fig. 2. (a) Overall pipeline of our end-to-end SGG method. (b) Multi-granularity sparse Relationship Matrix Prediction Network (Mg-RMPN). GPM is Gated Pooling Module.

Therefore, Mg-RMPN defined three $\mathbf{M}_r$ based on multi-granularity negative samples, where the three $\mathbf{M}_r$ are learned through three structurally identical but parameter-independent Gated Pooling Modules (GPM), as shown in Fig. 2 (b). With the outputs of the GPMs, Mg-RMPN is able to generate the final sparse relationship matrix, enabling the identification of most probable subject-object pairs to predict relations.

Gated Pooling Module. Unlike [30], which only uses visual features, our GPM determines the correlation of two entities based on their visual representation, linguistic prior and location information. Given a set of visual representations $\mathbf{Q}_e$ of the entities in an image, GPM first uses two fully connected networks $\mathcal{F}_{sub}$ and $\mathcal{F}_{obj}$ to project it to the visual representations of the subject and object, $\mathbf{Q}_{sub}$ and $\mathbf{Q}_{obj}$, respectively, as follows:

$$\mathbf{Q}_{sub} = \mathcal{F}_{sub}(\mathbf{Q}_e) = \{\mathbf{q}_i^{sub}\}_{i=1}^{N_e}, \quad \mathbf{Q}_{obj} = \mathcal{F}_{obj}(\mathbf{Q}_e) = \{\mathbf{q}_j^{obj}\}_{j=1}^{N_e}. \tag{2}$$

Subsequently, the visual similarity between the subject i and the object j can be calculated as $v_{ij}^{sim} = \mathbf{q}_i^{sub} \cdot (\mathbf{q}_j^{obj})^{\mathrm{T}}$.

Given a set of class prediction distributions $\mathbf{C}$ for entities, GPM uses the prior linguistic function $\mathcal{F}_{prior}$ from [26] to predict the correlation between two entities, $s_{ij}^{prior} = \mathcal{F}_{prior}(\hat{\mathbf{c}}_i, \hat{\mathbf{c}}_j)$. $\mathcal{F}_{prior}$ aims to capture the linguistic inductive bias that exists between two entities to enhance the prediction of the relational matrices. Similarly to [26], $\mathcal{F}_{prior}$ obtains a reasonable initialization based on the distribution of relations generated by entity pairs in the training dataset.

Based on the positions $\mathbf{B}$ of the entities, GPM uses the function $\mathcal{F}_{overlap}$ to obtain the relative position information between two entities $\mathbf{b}_{ij}^{overlap} = \mathcal{F}_{overlap}(\hat{\mathbf{b}}_i, \hat{\mathbf{b}}_j) \in \mathbb{R}^8$, where 8 elements indicate difference in x and y coordinates, width ratio, height ratio, IOU, center distance, area ratio, and aspect ratio. Additionally, we can also obtain the union bounding box $\hat{\mathbf{b}}_{ij}^u \in \mathbb{R}^9$ for the entity pairs, where 9 elements indicate the union bounding box coordinates (x_1, y_1, x_2, y_2), center $(\frac{x_1+x_2}{2}, \frac{y_1+y_2}{2})$, size $(x_2 - x_1, y_2 - y_1)$ and area $(x_2 - x_1)(y_2 - y_1)$, respectively. Then, we can obtain the relevance representation of the subject i and object j as follows:

$$\mathbf{e}_{ij} = [v_{ij}^{sim}, s_{ij}^{prior}, \mathbf{b}_{ij}^{overlap}, \hat{\mathbf{b}}_{ij}^u], \qquad (3)$$

where $[\cdot, \cdot]$ denotes the concatenation operation. Subsequently, the feature representation of the entity pair can be expressed as

$$\mathbf{f}_{ij} = [\mathbf{q}_i^{sub}, \mathbf{q}_j^{obj}, f_s(s_{ij}^{prior}), pos(\mathbf{b}_{ij}^{overlap}), pos(\hat{\mathbf{b}}_{ij}^u)], \qquad (4)$$

where f_s is a fully-connected layer for language prior encoding, and pos is a fully-connected layer for positional encoding. Then, GPM uses a gated pooling layer f_{gate} to obtain the probability of relation between subject i and object j as follows:

$$\hat{p}_{ij} = \mathrm{Sigmoid}(\sum f_{gate}(\mathbf{f}_{ij}) \odot \mathbf{e}_{ij}), \qquad (5)$$

where $\odot$ represents element-wise multiplication. Finally, we can obtain a $N_e \times N_e$ relationship matrix $\mathbf{M}_r = \mathrm{GPM}(\mathbf{Q}_e, \mathbf{C}, \mathbf{B})$. Note that we set the diagonal elements of $\mathbf{M}_r$ to zero, removing the cases where an entity serves as both subject and object simultaneously.

Multi-Granularity Sparse Relationship Matrix Prediction. Based on the output of three GPMs, we can construct three relationship matrices based on the granularity of negative samples and build the corresponding ground truth matrices (see Sect. 3.6) for supervised learning, enabling each GPM to focus on identifying negative samples of specific granularity.

Since the N_e entity proposals are typically much greater than the number of entities contained in an image, resulting in a large number of background entities among all possible pairs of entities. Since background entities do not form relationships, we divide the negative samples in the relationship matrix into three levels of granularity: (1) high-confidence negative samples composed of background entities (background-background), (2) medium-confidence negative samples constructed from background and foreground entities (entity-background or background-entity), and (3) low-confidence negative samples composed of foreground entities (entity-entity).

Then we can define three relationship matrices based on the output of the GPMs as follows.

$$\begin{aligned} \mathbf{M}_{r1} &= \text{GPM-1}(\mathbf{Q}_e, \mathbf{C}, \mathbf{B}) \\ \mathbf{M}_{r2} &= \mathbf{M}_{r1} + \text{GPM-2}(\mathbf{Q}_e, \mathbf{C}, \mathbf{B}) \\ \mathbf{M}_{r3} &= \mathbf{M}_{r2} + \text{GPM-3}(\mathbf{Q}_e, \mathbf{C}, \mathbf{B}) \end{aligned} \quad (6)$$

In Eq. 6, each GPM can focus on learning the filtering of negative samples at different granularities, as follows: (1) GPM-1 focuses on learning from pairs between background and filters out a large number of high-confidence negative samples. (2) GPM-2 focuses on learning from pairs between background and entities and filters out medium-confidence negative samples. (3) GPM-3 focuses on learning from no relation pairs between entities and only needs to filter out entity pairs that do not form relationships from the remaining entity pairs.

Based on the final sparse relationship matrix $\mathbf{M}_{r3}$, we can select the top-N_r subject-object pairs most likely to form relations and obtain their indices in the entity visual representations $\mathbf{Q}_e$, the entity's class predictions $\mathbf{C}$, and the bounding boxes $\mathbf{B}$.

3.5 Relation Decoder

With the $\mathbf{Q}_e$, $\mathbf{C}$ and $\mathbf{B}$ of subjects and objects, we can obtain the subject-object pair representation $\mathbf{Q}_{pair}$, which integrates visual, semantic, and positional encodings as follows:

$$\mathbf{Q}_{pair} = [\mathbf{Q}_e^{sub}, emb(\mathbf{C}^{sub}), pos(\mathbf{B}^{sub}), \mathbf{Q}_e^{obj}, emb(\mathbf{C}^{obj}), pos(\mathbf{B}^{obj})]. \quad (7)$$

where emb is a pre-trained Glove language model to acquire the word embedding.

In this paper, we adopt the relation decoder in [30], which consists of transformer decoders in the style of DETR, to predict relations. In the relation decoder, we initialize a relation query $\mathbf{Q}_r \in \mathbb{R}^{N_r \times d}$ as the query input, and the subject-object pair $\mathbf{Q}_{pair}$ is projected as the key and value of cross-attention. Subsequently, the relation representation after the Relation Decoder (RD) can be expressed as $\tilde{\mathbf{Q}}_r = \text{RD}(\mathbf{Q}_r, \mathbf{Q}_{pair})$. Then, the relation prediction for the entity pairs can be expressed as $\hat{\mathbf{r}} = W_{cls}\tilde{\mathbf{Q}}_r$, where W_{cls} is the liner relation classifier.

3.6 Training

Our end-to-end SGG is divided into three subtasks: (1) an entity detection task based on the object detector, (2) a sparse relationship matrix prediction task based on the Mg-RMPN, and (3) a relation prediction task based on the relation decoder. During training, each subtask generates the corresponding supervisory information and losses, as follows.

Entity Detection Loss. We uses the end-to-end deformable DETR [42] as object detector, which uses the set prediction loss proposed in DETR [1] by assigning the ground truth entities to the predictions. A cost function is applied to compute the matching cost between a prediction and a ground-truth entity. With the cost matrix, the entity prediction-ground truth assignment is computed with the Hungarian Matching [14]. Giver a set of N_e entity proposals $\{e_i\}_{i=1}^{N_e}$ from object detector, the set prediction loss for can be presented as:

$$\mathcal{L}_e = \Sigma_{i=1}^{N_e}[\mathcal{L}_{cls}^e + 1_{e_i \neq \phi}\mathcal{L}_{box}^e], \tag{8}$$

where $\mathcal{L}_{cls}^e$ denotes the cross-entropy loss for label classification and $e_i \neq \phi$ means that <background> is not assigned to the ith entity prediction. $\mathcal{L}_{box}^e$ consists of L_1 loss and generalized IoU loss for the box regression.

Multi-Granularity Relationship Matrices Loss. To predict the final sparse relationship matrix, we need to define the supervisory information and loss functions during the training phase.

Multi-granularity Supervision Matrices. In the training phase, we assign labels to the entity proposals $\{e_i\}_{i=1}^{N_e}$ by Hungarian matching and obtain the ground truth for each node in $\mathbf{M}_r$ regarding the subject, object, and whether they form a relation. In response to the sparsity of $\mathbf{M}_r$, we propose Mg-RMPN to achieve hierarchical identification of negative samples at different granularities.

Consequently, based on the granularity of the negative samples that each GPM focuses on in Sect. 3.4, the corresponding ground truth can be obtained. (1) $\mathbf{M}_{r1}^{gt}$ filters out all pairs composed of background entities as follows:

$$\mathbf{M}_{r1}^{gt} = \{p_{ij} \mid p_{ij} = \begin{cases} 0 & \text{if } (e_i = \phi \text{ and } e_j = \phi) \text{ or } i = j \\ 1 & \text{otherwise} \end{cases}, 1 \leq i \leq N_e, 1 \leq j \leq N_e\}, \tag{9}$$

where $e_i = \phi$ ($e_j = \phi$) means that <background> is assigned to the ith (jth) entity prediction, $i = j$ means removing self-connected entity pairs. (2) $\mathbf{M}_{r2}^{gt}$ further filters out all pairs composed of backgrounds and entities as follows:

$$\mathbf{M}_{r2}^{gt} = \{p_{ij} \mid p_{ij} = \begin{cases} 0 & \text{if } (e_i = \phi \text{ or } e_j = \phi) \text{ or } i = j \\ 1 & \text{otherwise} \end{cases}, 1 \leq i \leq N_e, 1 \leq j \leq N_e\}. \tag{10}$$

(3) $\mathbf{M}_{r3}^{gt}$ filters out all non-relationship pairs, and $\mathbf{M}_{r3}^{gt}$ is also the final ground truth relationship matrix of the scene graph, expressed as follows

$$\mathbf{M}_{r3}^{gt} = \{p_{ij} \mid p_{ij} = \begin{cases} 0 & \text{otherwise} \\ 1 & \text{if } r_{ij} = 1 \end{cases}, 1 \leq i \leq N_e, 1 \leq j \leq N_e\}, \tag{11}$$

where $r_{ij} = 1$ implies that there is a relation between subject i and object j. Here, the relationships of the negative sample subsets in $\mathbf{M}_{r1}^{gt}$, $\mathbf{M}_{r2}^{gt}$ and $\mathbf{M}_{r3}^{gt}$ are $\mathbf{M}_{r1|p_{ij}=0}^{gt} \subseteq \mathbf{M}_{r2|p_{ij}=0}^{gt} \subseteq \mathbf{M}_{r3|p_{ij}=0}^{gt}$.

Multi-granularity Relationship Matrices Loss. Considering that the large number of negative samples in $\mathbf{M}_r$ have different confidences, we modify the binary cross-entropy function using focal loss [20], which applies weighted discrimination on well-classified samples, forcing the model to focus on wrongly classified samples. Then, the multi-granularity learning loss is

$$\mathcal{L}_{Mg} = \mathcal{L}_{RM}^1(\mathbf{M}_{r1}, \mathbf{M}_{r1}^{gt}) + \mathcal{L}_{RM}^2(\mathbf{M}_{r2}, \mathbf{M}_{r2}^{gt}) + \mathcal{L}_{RM}^3(\mathbf{M}_{r3}, \mathbf{M}_{r3}^{gt}), \quad (12)$$

where the relationship matrices loss $\mathcal{L}_{RM}^k (k=1,2,3)$ is defined as

$$\begin{aligned}\mathcal{L}_{RM}^k(\mathbf{M}_r, \mathbf{M}_r^{gt}) = &-\sum_i \sum_j \{\alpha_k (1-\hat{p}_{ij})^\gamma p_{ij} \log(\hat{p}_{ij}) \\ &+ (1-\alpha_k)\hat{p}_{ij}^\gamma (1-p_{ij})\log(1-\hat{p}_{ij})\}/N_{pos} \\ &+ \|\mathbf{M}_r\|_1 + \|\mathbf{M}_r\|_2,\end{aligned} \quad (13)$$

where the first part is the focal loss based on binary cross-entropy, and the second part is the L1 and L2 regularization of $\mathbf{M}_r$, $\alpha_k(k=1,2,3)$ is hyperparameters that adjusts the weights of positive and negative samples, γ is a hyperparameter for focal loss, and N_{pos} is the number of positive samples in each mini-batch.

Relation Prediction Loss. In this paper, the relation decoder also considers relation prediction as a set prediction based on a query. Therefore, during the training phase, we also use Hungarian Matching to assign labels for relation prediction. Due to the imbalanced relation classes in SGG, we use Seesaw loss [29] to dynamically adjust the gradients of samples of different classes. The relation loss is as follows:

$$\mathcal{L}_r = -\sum_{i=1}^{C_r} y_i \log(\hat{\sigma}_i), \quad \hat{\sigma}_i = \frac{e^{\hat{r}_i}}{\sum_{i \neq j}^{C_r} \mathcal{S}_{ij} e^{\hat{r}_j} + e^{\hat{r}_i}}, \quad (14)$$

where C_r is the number of relation classes, $y_i \in \{0,1\}, 1 \leqslant i \leqslant C_r$ is the one-hot ground truth label, $\mathcal{S}_{ij}$ is a tunable balancing factor between different classes, detail can be found in [29].

In summary, the total loss of our method is

$$\mathcal{L} = \mathcal{L}_e + \lambda_1 \mathcal{L}_{Mg} + \lambda_2 \mathcal{L}_r, \quad (15)$$

where λ_1 and λ_2 are hyperparameters used to respectively adjust multi-granularity learning loss $\mathcal{L}_{Mg}$ and relation prediction loss $\mathcal{L}_r$.

4 Experiments

4.1 Experimental Settings

Datasets: We evaluate our proposed method on the following datasets: (1) **Visual Genome (VG150)** [13] is the most widely used dataset in SGG, consisting of the most frequent 150 object classes and 50 predicate classes. (2) **Open**

Images V6 (OIv6) [15] is a large-scale dataset proposed by Google. We follow the data processing and evaluation protocols in [17,40]. OIv6 consists of 601 object classes and 30 predicate classes. (2) **Generalized Question Answering (GQA200)** [9] is another vision and language benchmark with more than 3.8M relation annotations. We follow the data processing in [7], which consists of Top-200 object classes and Top-100 predicate classes.

Task and Evaluation Metrics: In this work, we focus on the Scene Graph Detection task, which detects all objects in an image and predicts their bounding boxes, labels, and relations. The initial evaluation metric Recall@K (R@K) was found to be dominated by head classes. Therefore, the mean Recall@K (mR@K) across all relations classes is proposed to evaluate unbiased SGG. However, focusing solely on mR@K and neglecting R@K can result in a tail bias. Therefore, we adopts the harmonic mean F@K of R@K and mR@K as an overall metric. For OIv6, the weighted evaluation metrics (wmAP$_{rel}$, wmAP$_{phr}$, score$_{wtd}$) are used for a more class-balanced evaluation. **Implementation Details:** We utilized the pre-trained deformable DETR [42] as the object detector. We set the loss hyperparameters λ_1 and λ_2 to 0.5 and 3 respectively. For the multi-granularity relationship matrices loss, we set $\alpha_k (k=1,2,3)$ to 0.75, 0.9, and 0.99, respectively, to control the weight of positive samples, with the focal loss parameter γ set to 2. The number of entities predicted by the detector N_e is 100, the number of most probable subject-object pairs selected N_r is 100, and the size of embedding dimemsions d is 256. Our model is implemented on 8 NVIDIA 3090 GPUs with learning rate $\times 10^{-4}$ and batch size 16 for 24 epochs. More implementation details are shown in the Supplementary Material.

4.2 Comparison with State of the Arts

VG150: Table 1 shows the comparison of our method on VG150, with methods divided into two-stage (above) and end-to-end (below). Due to the imbalance issue in SGG, the two-stage methods (Motifs, VCTree, GPS-Net, RelDN, PE-Net) perform better in R@K but show very poor performance in mR@K. Thus, many rebalancing techniques (VCTree+TDE, BGNN, SHA+GCL) have been developed. Although they improve mR@K, they inevitably impair R@K. For example, SHA+GCL, despite achieving the best mR@K, its recall performance significantly deteriorates, with R@50/100 being only 14.9/18.2.

Although the end-to-end methods are inferior to the two-stage methods on R@K, it demonstrates better potential on mR@K due to its advantage in sparse relation prediction, which does not dilute the scarce tail-class samples. Furthermore, compared to the rebalancing techniques of the two-stage methods, the end-to-end approach does not overly harm R@K and can achieve superior overall performance F@K, such as SGTR and Pair-Net.

Our Mg-RMPN leads comprehensively in end-to-end methods. Compared with two-stage methods, Mg-RMPN achieves competitive performance in R@K and significantly excels in mR@K, achieving the best overall performance F@K. For example, Mg-RMPN achieved 19.3/22.8 on F@50/100, surpassing the end-to-end and two-stage optimal methods Pair-Net and PE-Net, respectively.

Table 1. Performance comparison at scene graph detection task on VG150. ∗ denotes the results from [16].

Method	R@20	R@50	R@100	mR@20	mR@50	mR@100	F@20	F@50	F@100
Motifs [36]	**25.5**	**32.8**	**37.2**	5.0	6.8	7.9	8.4	11.3	13.0
VCTree [27]	24.5	31.9	36.2	5.4	7.4	8.7	8.8	12.0	14.0
VCTree+TDE [26]	14.0	19.4	23.2	6.9	9.3	11.1	9.2	12.6	15.0
GPS-Net [21]	22.3	28.9	33.2	6.9	8.7	9.8	**10.5**	13.4	15.1
BGNN [17]	–	31.0	35.8	–	10.7	12.6	–	15.9	18.6
RelDN [39]	–	31.4	35.9	–	6.0	7.3	–	10.1	12.1
SHA+GCL [7]	–	14.9	18.2	**14.2**	**17.9**	**20.9**	–	16.3	19.5
PE-Net [40]	–	30.7	35.2	–	12.4	14.5	–	**17.7**	**20.5**
FCSGG [22]	16.1	21.3	25.1	2.7	3.6	4.2	4.6	6.2	7.2
RelTR [6]	21.2	27.5	–	6.8	10.8	–	10.3	15.5	–
AS-Net∗ [3]	–	18.7	21.1	–	6.1	7.2	–	9.2	10.7
HOTR∗ [12]	–	23.5	27.7	–	9.4	12.0	–	13.4	16.7
SGTR [16]	–	24.6	28.4	–	12.0	15.2	–	16.1	19.8
Pair-Net [30]	18.8	24.9	29.3	8.9	12.4	15.4	12.1	16.6	20.2
Mg-RMPN(DETR)	21.0	27.1	31.3	9.9	13.5	16.2	13.5	18.0	21.3
Mg-RMPN(Ours)	22.5	29.1	33.5	10.3	14.4	17.3	14.1	19.3	22.8

We also compared with Pair-Net that is also based on relationship matrix learning. The experimental results show that our method is comprehensively superior to it, which demonstrates the effectiveness of our proposed Mg-RMPN. Furthermore, we also present the results of Mg-RMPN based on the DETR object detector to eliminate the influence of the detector. The results show that our Mg-RMPN(DETR) still achieves the best overall performance among both one-stage and two-stage methods.

Open Image V6: To validate the generalizability of Mg-RMPN, we performed experiments on Open Images V6 in Table 2 and compared with two-stage (above) and end-to-end (below) methods. Compared to classic SGG benchmarks, our method achieved the optimal results among all methods with 45.5, 77.8, 57.4 respectively in mR@50, R@50, F@50. Compared to open image benchmarks, we obtained competitive 35.5 and 36.4 in wmAP$_{rel}$ and wmAP$_{phr}$, and achieved the best result 43.6 in the overall metric score$_{wtd}$.

GQA200: We also conducted experiments on the more challenging GQA200, as shown in Table 3. Compared to two-stage methods (VTransE, Motifs, VCTree, SHA+GCL), our method's overall performance F@K is only slightly inferior to SHA+GCL on F@100, which sacrificed a significant amount of R@K to enhance mR@K by using rebalancing techniques. Compared to the end-to-end method Pair-Net, which is also based on relationship matrices, our method is comprehensively superior.

Table 2. Performance comparison on OIv6. ∗ denotes the results from [16,17]. †denote results reproduced with the authors' code.

Method	mR@50	R@50	F@50	wmAP$_{rel}$	wmAP$_{phr}$	score$_{wtd}$
Motifs∗ [36]	32.7	71.6	44.9	29.9	31.6	38.9
VCTree∗ [27]	33.9	74.1	46.5	**34.2**	33.1	40.2
RelDN∗ [39]	34.0	73.1	46.4	32.2	33.4	40.8
G-RCNN∗ [34]	34.0	74.5	46.7	33.2	**34.2**	41.8
GPS-Net∗ [21]	35.3	74.8	48.0	32.9	34.0	41.7
BGNN∗ [17]	40.5	75.0	52.6	33.5	**34.2**	42.1
RelTR [6]	–	71.7	–	37.2	37.5	43.0
AS-Ne∗ [3]	35.2	55.3	43.0	25.9	27.5	32.4
HOTR∗ [12]	40.1	52.7	45.5	19.4	21.5	26.7
SGTR∗ [16]	42.6	59.9	49.8	37.0	**38.7**	42.3
Pair-Net†[30]	44.5	77.4	56.5	31.8	32.4	40.3
Mg-RMPN(Ours)	**45.5**	**77.8**	**57.4**	35.5	36.4	**43.6**

Table 3. Performance comparison at scene graph detection task on GQA200. ∗ denotes the results from [7]. †denote results reproduced with the authors' code.

Method	R@50	R@100	mR@50	mR@100	F@50	F@100
VTransE∗ [38]	27.2	30.7	5.8	6.6	9.6	10.9
Motifs∗ [36]	**28.9**	**33.1**	6.4	7.7	10.5	12.5
VCTree∗ [27]	28.3	31.9	6.5	7.4	10.6	12.0
SHA+GCL∗ [7]	14.8	17.9	**17.8**	**20.1**	16.2	**18.9**
Pair-Net†[30]	20.2	23.4	10.6	12.6	13.9	16.4
Mg-RMPN (Ours)	23.2	25.7	12.8	14.5	**16.5**	18.5

4.3 Visualization of Multi-Granularity Relationship Matrices

Figure 3 presents a visualization of the multi-granularity relationship matrices learned by our Mg-RMPN, where (d), (e), (f) are the sparse relationship matrices learned after being filtered by GPM-1, GPM-2, GPM-3, respectively, and (a), (b), (c) are the corresponding ground truth relationship matrices defined in Sect. 3.6. The visualization results show that the dense negative samples in the relationship matrix are filtered layer by layer according to different granularities. The consistency of the generated multi-granularity relationship matrices with their ground truth indicates that the three GPMs of Mg-RMPN indeed focus on filtering negative samples of different granularities.

Figures 3 (h) and (g), respectively, present a zoomed-in view of the final relationship matrix and its ground truth. We can observe that the predicted relationship matrix successfully captures the true relationships, for example: "man wearing shirt", "man holding racket", "man wearing short" and "pole

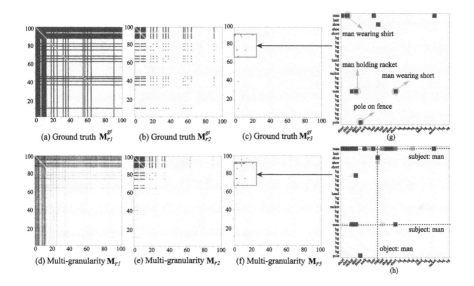

Fig. 3. Visualization of Multi-Granularity Relationship Matrices

on fence". Additionally, the relationship matrix also predicts some relationships that do not in the ground truth, as illustrated by the dashed lines in the rows and columns, which represent the relationships with the corresponding entity "man" as the subject and object, respectively. Due to the abundant presence of "man" in SGG, the relationship matrix tends to favor predicting relations that include "man", which aligns with the actual language induction bias. Therefore, the results of visualization indicate that our Mg-RMPN can accurately predict the sparse relationship matrix.

4.4 Ablation Studies

Table 4. Ablation studies of Mg-RMPN components on VG150.

GPM-1	GPM-2	GPM-3	R@20/50/100	mR@20/50/100	F@20/50/100
✓	✗	✗	6.5/10.7/15.2	2.9/4.8/6.6	4.0/6.6/9.2
✗	✓	✗	12.5/17.5/21.6	5.2/7.4/9.3	7.3/10.4/13.0
✓	✓	✗	12.8/18.1/22.3	5.8/8.5/10.7	8.0/11.6/14.5
✗	✗	✓	21.2/27.6/32.1	9.2/13.1/15.8	12.8/17.8/21.2
✓	✗	✓	21.7/28.2/32.6	9.8/13.8/16.5	13.5/18.5/21.9
✗	✓	✓	21.8/28.2/32.5	9.7/13.7/16.7	13.4/18.4/22.1
✓	✓	✓	**22.5/29.1/33.5**	**10.3/14.4/17.3**	**14.1/19.3/22.8**

Component Analysis of Mg-RMPN. To assess the effectiveness of each GPM in Mg-RMPN, we performed ablation experiments in Table 4. In Mg-RMPN, each GPM focuses on negative sample filtering of different granularities, and the three GPMs collaborate to complete the learning of the sparse relationship matrix. GPM-1 and GPM-2 are used to help GPM-3 filter negative samples. Using GPM-1 and GPM-2 individually cannot make full use of the ground truth, hence their results are very poor. The results obtained using only GPM-3 are also not good enough. Further increasing the assistance of GPMs for negative sample filtering can significantly improve the performance of the model. Mg-RMPN fully utilized three GPMs to achieve optimal results, confirming that learning negative samples based on different granularities can effectively improve the model's learning ability for sparse relationship matrices.

Table 5. Ablation studies of different input information for Mg-RMPN on VG150.

Visual	Semantic	Position	R@20/50/100	mR@20/50/100	F@20/50/100
✓	✗	✗	21.2/27.6/32.1	9.8/13.6/16.4	13.4/18.2/21.7
✓	✗	✓	22.2/28.7/33.1	10.1/14.1/16.9	13.9/18.9/22.4
✓	✓	✗	21.2/27.6/32.2	9.8/13.8/16.8	13.4/18.4/22.1
✓	✓	✓	**22.5/29.1/33.5**	**10.3/14.4/17.3**	**14.1/19.3/22.8**

Different Input Information for Mg-RMPN. This paper bridges entity detection and relation detection through a relationship matrix, making full use of the semantic and positional information of entities and solving the problem that end-to-end SGG relies solely on visual information to predict relations. Table 5 presents ablation studies of Mg-RMPN based on different input information. Compared to models that only utilize visual information, separately incorporating the semantic and positional information of entities can enhance the performance of the model. Furthermore, the improvement in introducing semantic and positional information simultaneously is significantly greater than in introducing a single modality of information, indicating that semantic and positional information are two complementary information. Hence, the semantic and positional information incorporated in our approach plays a crucial role in predicting relations. More hyperparameter analysis can be found in Supplementary Material.

5 Conclusion

In this paper, our proposed end-to-end method bridges entity detection and relation detection through a sparse relationship matrix, which not only eliminates the need for post-processing of relation matching but also leverages the semantics and positional information of predicted entities to enhance relation prediction. To predict the sparse relationship matrix, we propose a multi-granularity sparse relationship matrix prediction network, which utilizes three gated pooling modules focusing on filtering negative samples at different granularities. Finally, a set of sparse, most likely subject-object pairs can be constructed and used for relation decoding. The experimental results demonstrate that our method achieves an optimal overall performance in both the end-to-end and two-stage methods.

Acknowledgements. This work was supported by the National Natural Science Foundation of China (U21A20485, 61976170, 62088102) and the National Key R&D Program of China (2023YFB4704900).

References

1. Carion, N., Massa, F., Synnaeve, G., Usunier, N., Kirillov, A., Zagoruyko, S.: End-to-end object detection with transformers. In: Vedaldi, A., Bischof, H., Brox, T., Frahm, J.-M. (eds.) ECCV 2020. LNCS, vol. 12346, pp. 213–229. Springer, Cham (2020). https://doi.org/10.1007/978-3-030-58452-8_13
2. Chen, C., Zhan, Y., Yu, B., Liu, L., Luo, Y., Du, B.: Resistance training using prior bias: toward unbiased scene graph generation. In: Proceedings of the AAAI Conference on Artificial Intelligence, vol. 36, no. 1, pp. 212–220 (2022)
3. Chen, M., Liao, Y., Liu, S., Chen, Z., Wang, F., Qian, C.: Reformulating hoi detection as adaptive set prediction. In: Proceedings of the IEEE/CVF Conference on Computer Vision and Pattern Recognition, pp. 9004–9013 (2021)
4. Chen, S., Jin, Q., Wang, P., Wu, Q.: Say as you wish: fine-grained control of image caption generation with abstract scene graphs. In: Proceedings of the IEEE/CVF Conference on Computer Vision and Pattern Recognition, pp. 9962–9971 (2020)
5. Chen, T., Yu, W., Chen, R., Lin, L.: Knowledge-embedded routing network for scene graph generation. In: Proceedings of the IEEE/CVF Conference on Computer Vision and Pattern Recognition, pp. 6163–6171 (2019)
6. Cong, Y., Yang, M.Y., Rosenhahn, B.: RelTR: relation transformer for scene graph generation. IEEE Trans. Pattern Anal. Mach. Intell. (2023)
7. Dong, X., Gan, T., Song, X., Wu, J., Cheng, Y., Nie, L.: Stacked hybrid-attention and group collaborative learning for unbiased scene graph generation. In: Proceedings of the IEEE/CVF Conference on Computer Vision and Pattern Recognition, pp. 19427–19436 (2022)
8. Gao, L., et al.: Informative scene graph generation via debiasing. arXiv preprint arXiv:2308.05286 (2023)
9. Hudson, D.A., Manning, C.D.: GQA: a new dataset for real-world visual reasoning and compositional question answering. In: Proceedings of the IEEE/CVF Conference on Computer Vision and Pattern Recognition, pp. 6700–6709 (2019)
10. Johnson, J., Gupta, A., Fei-Fei, L.: Image generation from scene graphs. In: Proceedings of the IEEE Conference on Computer Vision and Pattern Recognition, pp. 1219–1228 (2018)
11. Johnson, J., et al.: Image retrieval using scene graphs. In: Proceedings of the IEEE Conference on Computer Vision and Pattern Recognition, pp. 3668–3678 (2015)
12. Kim, B., Lee, J., Kang, J., Kim, E.S., Kim, H.J.: HOTR: end-to-end human-object interaction detection with transformers. In: Proceedings of the IEEE/CVF Conference on Computer Vision and Pattern Recognition, pp. 74–83 (2021)

13. Krishna, R., et al.: Visual genome: connecting language and vision using crowdsourced dense image annotations. Int. J. Comput. Vis. **123**(1), 32–73 (2017)
14. Kuhn, H.W.: The hungarian method for the assignment problem. Nav. Res. Logist. Q. **2**(1–2), 83–97 (1955)
15. Kuznetsova, A., et al.: The open images dataset v4: unified image classification, object detection, and visual relationship detection at scale. Int. J. Comput. Vis. **128**(7), 1956–1981 (2020)
16. Li, R., Zhang, S., He, X.: SGTR: end-to-end scene graph generation with transformer. In: proceedings of the IEEE/CVF Conference on Computer Vision and Pattern Recognition, pp. 19486–19496 (2022)
17. Li, R., Zhang, S., Wan, B., He, X.: Bipartite graph network with adaptive message passing for unbiased scene graph generation. In: Proceedings of the IEEE/CVF Conference on Computer Vision and Pattern Recognition, pp. 11109–11119 (2021)
18. Li, W., Zhang, H., Bai, Q., Zhao, G., Jiang, N., Yuan, X.: PPDL: predicate probability distribution based loss for unbiased scene graph generation. In: Proceedings of the IEEE/CVF Conference on Computer Vision and Pattern Recognition, pp. 19447–19456 (2022)
19. Li, Y., Ma, T., Bai, Y., Duan, N., Wei, S., Wang, X.: PasteGAN: a semi-parametric method to generate image from scene graph. Adv. Neural Inf. Process. Syst. **32** (2019)
20. Lin, T.Y., Goyal, P., Girshick, R., He, K., Dollár, P.: Focal loss for dense object detection. In: Proceedings of the IEEE International Conference on Computer Vision, pp. 2980–2988 (2017)
21. Lin, X., Ding, C., Zeng, J., Tao, D.: GPS-Net: graph property sensing network for scene graph generation. In: Proceedings of the IEEE/CVF Conference on Computer Vision and Pattern Recognition, pp. 3746–3753 (2020)
22. Liu, H., Yan, N., Mortazavi, M., Bhanu, B.: Fully convolutional scene graph generation. In: Proceedings of the IEEE/CVF Conference on Computer Vision and Pattern Recognition, pp. 11546–11556 (2021)
23. Lyu, X., et al.: Generalized unbiased scene graph generation. arXiv preprint arXiv:2308.04802 (2023)
24. Ren, S., He, K., Girshick, R., Sun, J.: Faster R-CNN: towards real-time object detection with region proposal networks. Adv. Neural Inf. Process. Syst. **28** (2015)
25. Schroeder, B., Tripathi, S.: Structured query-based image retrieval using scene graphs. In: Proceedings of the IEEE/CVF Conference on Computer Vision and Pattern Recognition Workshops, pp. 178–179 (2020)
26. Tang, K., Niu, Y., Huang, J., Shi, J., Zhang, H.: Unbiased scene graph generation from biased training. In: Proceedings of the IEEE/CVF Conference on Computer Vision and Pattern Recognition, pp. 3716–3725 (2020)
27. Tang, K., Zhang, H., Wu, B., Luo, W., Liu, W.: Learning to compose dynamic tree structures for visual contexts. In: Proceedings of the IEEE/CVF Conference on Computer Vision and Pattern Recognition, pp. 6619–6628 (2019)
28. Teney, D., Liu, L., van Den Hengel, A.: Graph-structured representations for visual question answering. In: Proceedings of the IEEE Conference on Computer Vision and Pattern Recognition, pp. 1–9 (2017)
29. Wang, J., et al.: Seesaw loss for long-tailed instance segmentation. In: Proceedings of the IEEE/CVF Conference on Computer Vision and Pattern Recognition, pp. 9695–9704 (2021)
30. Wang, J., Wen, Z., Li, X., Guo, Z., Yang, J., Liu, Z.: Pair then relation: pair-net for panoptic scene graph generation. arXiv preprint arXiv:2307.08699 (2023)

31. Wang, L., Yuan, Z., Chen, B.: Learning to generate an unbiased scene graph by using attribute-guided predicate features. In: Proceedings of the AAAI Conference on Artificial Intelligence, vol. 37, pp. 2581–2589 (2023)
32. Xu, D., Zhu, Y., Choy, C.B., Fei-Fei, L.: Scene graph generation by iterative message passing. In: Proceedings of the IEEE Conference on Computer Vision and Pattern Recognition, pp. 5410–5419 (2017)
33. Yang, G., Zhang, J., Zhang, Y., Wu, B., Yang, Y.: Probabilistic modeling of semantic ambiguity for scene graph generation. In: Proceedings of the IEEE/CVF Conference on Computer Vision and Pattern Recognition, pp. 12527–12536 (2021)
34. Yang, J., Lu, J., Lee, S., Batra, D., Parikh, D.: Graph R-CNN for scene graph generation. In: Proceedings of the European Conference on Computer Vision (ECCV), pp. 670–685 (2018)
35. Yu, J., Chai, Y., Wang, Y., Hu, Y., Wu, Q.: Cogtree: cognition tree loss for unbiased scene graph generation. In: Zhou, Z.H. (ed.) Proceedings of the Thirtieth International Joint Conference on Artificial Intelligence, IJCAI-21, pp. 1274–1280. International Joint Conferences on Artificial Intelligence Organization, August 2021
36. Zellers, R., Yatskar, M., Thomson, S., Choi, Y.: Neural motifs: scene graph parsing with global context. In: Proceedings of the IEEE Conference on Computer Vision and Pattern Recognition, pp. 5831–5840 (2018)
37. Zhang, A. et al.: Fine-grained scene graph generation with data transfer. In: Avidan, S., Brostow, G., Cissé, M., Farinella, G.M., Hassner, T. (eds.) Computer Vision – ECCV 2022. ECCV 2022. LNCS, vol. 13687, pp. 409–424. Springer, Cham (2022). https://doi.org/10.1007/978-3-031-19812-0_24
38. Zhang, H., Kyaw, Z., Chang, S.F., Chua, T.S.: Visual translation embedding network for visual relation detection. In: Proceedings of the IEEE Conference on Computer Vision and Pattern Recognition, pp. 5532–5540 (2017)
39. Zhang, J., Shih, K.J., Elgammal, A., Tao, A., Catanzaro, B.: Graphical contrastive losses for scene graph parsing. In: Proceedings of the IEEE/CVF Conference on Computer Vision and Pattern Recognition, pp. 11535–11543 (2019)
40. Zheng, C., Lyu, X., Gao, L., Dai, B., Song, J.: Prototype-based embedding network for scene graph generation. In: Proceedings of the IEEE/CVF Conference on Computer Vision and Pattern Recognition, pp. 22783–22792 (2023)
41. Zhong, Y., Wang, L., Chen, J., Yu, D., Li, Y.: Comprehensive image captioning via scene graph decomposition. In: Vedaldi, A., Bischof, H., Brox, T., Frahm, J.-M. (eds.) ECCV 2020. LNCS, vol. 12359, pp. 211–229. Springer, Cham (2020). https://doi.org/10.1007/978-3-030-58568-6_13
42. Zhu, X., Su, W., Lu, L., Li, B., Wang, X., Dai, J.: Deformable detr: deformable transformers for end-to-end object detection. arXiv preprint arXiv:2010.04159 (2020)

Flash-Splat: 3D Reflection Removal with Flash Cues and Gaussian Splats

Mingyang Xie[1](✉), Haoming Cai[1], Sachin Shah[1], Yiran Xu[1], Brandon Y. Feng[2], Jia-Bin Huang[1], and Christopher A. Metzler[1]

[1] University of Maryland, College Park, USA
mingyang@umd.edu
[2] Massachusetts Institute of Technology, Cambridge, USA
https://flash-splat.github.io/

Abstract. We introduce a simple yet effective approach for separating transmitted and reflected light. Our key insight is that the powerful novel view synthesis capabilities provided by modern inverse rendering methods (e.g., 3D Gaussian splatting) allow one to perform flash/no-flash reflection separation using *unpaired measurements*—this relaxation dramatically simplifies image acquisition over conventional paired flash/no-flash reflection separation methods. Through extensive real-world experiments, we demonstrate our method, Flash-Splat, accurately reconstructs both transmitted and reflected scenes in 3D. Our method outperforms existing 3D reflection separation methods, which do not leverage illumination control, by a large margin.

1 Introduction

We are often surrounded by scenes with transparent surfaces, most notably glass, which introduce specular reflections. When viewing such scenes, we see a superimposition of transmitted and reflected light. This work focuses on the unsupervised separation of a transmitted 3D scene and a reflected 3D scene.

Reflection removal and separation have received considerable attention in the computational photography community. In addition to enhancing image quality and appeal, effective reflection separation methods can improve the robustness of downstream computer vision systems used in various applications, including robot navigation, classification, and 3D surface reconstruction. Separating transmitted and reflected 3D scenes is vital for various virtual reality tasks, such as 3D object extraction or editing.

Unfortunately, separating transmitted and reflected light from the sum of their intensities is a highly under-determined problem. To address this challenge,

---

M. Xie and H. Cai—Equal Contribution.

Supplementary Information The online version contains supplementary material available at https://doi.org/10.1007/978-3-031-73007-8_8.

© The Author(s), under exclusive license to Springer Nature Switzerland AG 2025
A. Leonardis et al. (Eds.): ECCV 2024, LNCS 15140, pp. 122–139, 2025.
https://doi.org/10.1007/978-3-031-73007-8_8

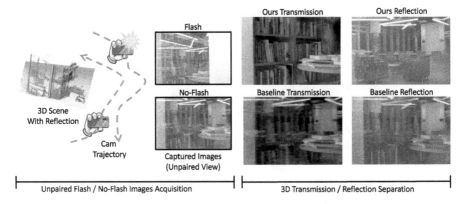

Fig. 1. Left: We separate the 3D transmitted and reflected scenes by capturing some views with camera flash and some views with no flash. **Right**: Our proposed Flash-Splat method achieves much better separation than the state-of-the-art unsupervised 3D separation method NeRFReN [13].

prior works have relied on various assumptions to perform single-image reflection removal. For instance, they have assumed the reflection is out-of-focus [2,56] or there is a noticeable double reflection caused by two sides of the glass [42]. However, these assumptions are not always true in real life. Other works have leveraged videos or multi-view images for reflection removal [1,11,12,15,16,33, 54]. Their advantages over single-image methods are (1) they can get "lucky" where some views have weaker reflections than others, and (2) they can utilize multi-view consistency to regularize the separation. However, these methods still struggle to overcome the fundamental ill-posed problem, especially under strong reflection. For example, in Fig. 1 the state-of-the-art unsupervised 3D reflection separation method, NeRFReN [13], fails to separate reflected and transmitted light from a collection of images captured under similar illumination conditions.

Introducing illumination control, i.e., flash/no-flash photography [24,52,53], can make the reflection separation problem significantly easier. Intuitively, the camera flash increases the intensities of the transmitted scene while leaving the reflected scene largely intact. Therefore, we can recover a reflection-free transmission scene by comparing images captured with and without flash. The core limitation is that it requires paired (tightly-aligned) flash/no-flash image captures—the camera cannot move between the captures. This paired measurement requirement represents a major barrier to effective in-the-wild reflection separation.

In this paper, we perform flash-based reflection separation without paired measurements by leveraging the powerful novel view synthesis capabilities of recently developed inverse differentiable rendering methods. Specifically, during acquisition, a user captures roughly half of the views with flash on and the other half with flash off. Then, by extending the powerful Gaussian Splatting [20] technique, we can construct 2D "pseudo-paired" flash/no-flash images, where one image in the pseudo flash/no-flash pair is captured, and the other one is synthesized with our 3D inverse rendering framework; we can also construct a 3D "pseudo-pair" of flash/no-flash 3D representations, where one 3D representation is reconstructed from only the flash images, and the other is reconstructed from only the no-flash images. The difference between the 2D pseudo-pair and the difference between the 3D pseudo-pair both serve as strong priors for the transmitted 3D scene, which significantly reduce the ill-posedness of the separation problem. As a byproduct of our 3D inverse differentiable rendering framework, our method, Flash-Splat, is also capable of performing novel view synthesis and depth estimation for each transmitted and reflected scene. We validate our proposed approach in real-world experiments and demonstrate its state-of-the-art performance.

Our contributions are:

- We propose a robust strategy, Flash-Splat, for 3D transmission-reflection separation and scene reconstruction, using flash illumination as a physical cue without requiring paired flash/no-flash captures.
- We introduce novel modifications to make 3D Gaussian Splatting illumination-aware, enhancing the quality of each separated 3D scene.
- We show that Flash-Splat excels in separating reflection and transmission, even when baseline methods fail, over real-world scenes.
- We demonstrate that Flash-Splat can perform high-quality novel view synthesis and depth estimation for both the transmitted and reflected 3D scenes.

2 Related Work

Reflection Removal. Existing reflection removal methods can generally be divided into three categories: single-frame, multi-frame and polarization-based. Single-frame approaches [2,7,8,14,17–19,21,26–29,31,32,42,44–47,50,51,55–58,60] only take a single image and remove the reflection. Multi-frame approaches [1,6,9,11,12,15,16,31,33,54] use multiple input frames as cue and produce multi-view consistent results. Polarization-based approaches [22,23,25,30,34,37] leverage the fact that the transmission is unpolarized while the reflection component varies when rotating the polarization filter. However, none of those methods aim to recover a 3D representation of the transmitted or the reflected scene.

3D Neural Scene Representations. To get more accurate 3D reconstruction for decomposition, we consider differentiable 3D neural representations. Neural Radiance Fields (NeRFs) [4,5,10,36] has received vast attentions in the past few years, for their accurate and consistent novel view synthesis results. Another line of works focuses on accurate 3D geometry, so they considers Signed Distance Function (SDF) [48,49] for better surface accuracy. Recently, 3D Gaussian Splatting (3DGS) [20,59] emerges for its fast training and inference speed.

Reflection Removal by Inverse Rendering. Previous methods consider solving reflection removal using inverse 3D rendering. ReflectionsIBR [43] as a pioneer proposes to separate each frame into a transmission and reflection layer combined with a binary reflection mask, and tries to reconstruct the scene using

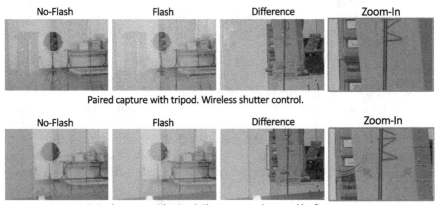

Fig. 2. Flash/No-Flash For Reflection Removal. The difference between paired flash and no-flash images is equivalent to taking a photo with flash in a dark environment, which gives us a reflection-free image (top). This is because flash increases the transmission brightness, but not the reflection brightness. Notice pairs must be tightly aligned for this method to work. Even tiny vibrations such as pressing the shutter button even when using a tripod produce artifacts (bottom).

an image-based rendering. Recently, NeRFReN [13] uses a NeRF to achieve better 3D reconstruction accuracy. NeuS-HSR [38], instead of focusing on reflection separation, uses Signed Distance Functions to achieve better surface reconstruction quality. Distinct from these 3D methods, our proposed method dramatically extends these approaches by incorporating variable illumination.

3 Method

3.1 Paired 2D Flash/No-Flash

Capturing a pair of flash and no-flash images of a scene from the same camera viewpoint allows one to reconstruct a reflection-free image.

Reflections exist because ambient light illuminates objects in the reflected scene and reflects off the glass onto the camera sensor. The captured composite scene with no flash $\mathbf{I}_N$ can be modeled as

$$\mathbf{I}_N = \mathbf{T}_N + \beta \circ \mathbf{R} \qquad (1)$$

where $\mathbf{T}_N$ is the transmission scene with no flash, $\mathbf{R}$ is the reflection scene and β is the reflective fraction factor (in the extreme case where the reflection is caused by a mirror, then β could be interpreted as the mask of the mirror in the scene). Now consider the case where the scene is captured with a flash co-located on the camera. If we assume that the scene behind glass is diffuse and that the camera flash is uniform, the camera flash will increase the intensity of all pixels proportionally. Therefore, we can formulate the flash image $\mathbf{I}_F$ as following:

$$\mathbf{I}_F = (1+\alpha)\mathbf{T}_N + (\beta + \beta_F) \circ \mathbf{R}, \tag{2}$$

where α and β_F represent the intensity increase of the transmitted and reflected scene due to flash, respectively. Assuming the direct reflection of the flash is outside the camera's field of view (i.e., the specular surface is not orthogonal to the camera view), the flash will have little effect on the brightness of the reflected scene. In common cases like glass, the impact of secondary reflections is also usually very low. Therefore, we may approximate β_F as close to zero,

$$\mathbf{I}_F \approx (1+\alpha)\mathbf{T}_N + \beta \circ \mathbf{R}. \tag{3}$$

As such, one technique used among photographers is subtracting a no-flash image $\mathbf{I}_N$ [24] from a flash image $\mathbf{I}_F$,

$$\mathbf{I}_F - \mathbf{I}_N \approx \alpha \mathbf{T}_N. \tag{4}$$

The difference is effectively a reflection-free transmitted scene scaled by some constant. Figure 2 demonstrates the impressive performance of this simple method.

Unfortunately, this process only works when we capture **paired** flash and no-flash images at the *same* location and orientation. Any small movement between the image pair causes the approach to break down. As illustrated in the bottom row of Fig. 2, even with a tripod, the slight motion caused by touching the exposure button (as opposed to using remote triggering) can introduce significant errors in the conventional flash/no-flash reflection separation process.

3.2 Unpaired 3D Flash/No-Flash

In this work, we extend the flash/no-flash idea to 3D and thus remove the requirement of capturing paired images, which makes flash-based reflection removal significantly easier and more practical. Instead of directly capturing paired multi-view images of a scene, we propose to first capture an arbitrary sequence of multi-view flash images of the scene, and then capture another sequence of multi-view no-flash images of the scene. These two sequences should be captured such that they approximately cover a similar range of perspectives.

Our 3D Flash/No-flash formulation is defined as follows. Following previous notations, we consider four 3D representations in total: transmission with flash $\mathbf{T}_F$, transmission without flash $\mathbf{T}_N$, reflection $\mathbf{R}$, and the reflective fraction factor β. To render a target pixel in a captured image, we blend the overlapping regions of the transmitted and reflected scenes. We then have,

$$\begin{aligned} \mathbf{I}_N &= \mathbf{T}_N + \beta \circ \mathbf{R} \\ \mathbf{I}_F &= \mathbf{T}_F + \beta \circ \mathbf{R} \end{aligned} \tag{5}$$

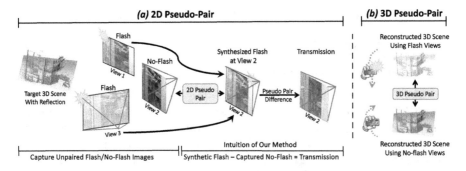

Fig. 3. Our Intuition: Construct 2D and 3D "pseudo-pairs" as Cues for Reflection Removal. Flash-Splat does not require *paired* flash/no-flash data. During the data capture stage, we collect *unpaired* flash/no-flash images from different views. In (a), if we captured a no-flash image at View 2, we can learn a 3D representation of the captured flash images at other views, and then synthesize a novel view of the flash image at View 2. As such, we have created a **2D pseudo-pair** of flash and no-flash images at View 2. If we then take the difference between the pseudo-pair as in Fig. 2, we get the transmission component of View 2 that is free of reflection. In (b), we reconstruct a 3D scene with flash using only the flash images (top); we also reconstruct a 3D scene without flash using only the no-flash images (bottom). As such, we have created a **3D pseudo-pair** of flash/no-flash scenes.

for flash (F) images and no-flash (N) images. Even though we are only capturing unpaired flash/no-flash views now, we can still associate them by creating 2 types of "pseudo-pairs" to aid reflection separation.

Firstly, we can construct **2D "pseudo-pairs"** via novel view synthesis of the missing flash/no-flash counterpart, as shown in Fig. 3a. Consider a specific view where only the flash image is taken. Utilizing inverse rendering techniques, we are able to synthesize a no-flash image at this exact same view by using the no-flash images taken at neighboring views. This synthesized no-flash image and the captured flash image form a 2D pseudo-pair. The difference image between this 2D pseudo-pair should be reflection-free, just like the difference between the 2D paired flash/no-flash images, as indicated in Eq. (4).

Secondly, we can construct a **3D "pseudo-pair"** by elevating the problem to the 3D space, as shown in Fig. 3b. More specifically, we can reconstruct a 3D scene with flash and another without flash, using only the views captured with flash and only the views captured without flash, respectively. We name these two reconstructed scenes as $\mathbf{I}_F^{Rec}$ and $\mathbf{I}_N^{Rec}$, to differentiate them with the ground truth 3D flash/no-flash scenes $\mathbf{I}_F$ and $\mathbf{I}_N$. $\mathbf{I}_F^{Rec}$ and $\mathbf{I}_N^{Rec}$ form a 3D pseudo-pair, as they are the same scene except that the transmitted part of $\mathbf{I}_F^{Rec}$ is brighter due to the flash. A 3D pseudo-pair difference can be used as a cue for the transmitted scene. Nevertheless, as $\mathbf{I}_F^{Rec}$ and $\mathbf{I}_N^{Rec}$ are separately reconstructed from 2 unpaired sets of data, they will be misaligned, thus the word "pseudo".

As such, we obtain important "flash cues" from the 2D and 3D pseudo-pairs, and use them as the high-level intuitions for our proposed method.

3.3 Proposed Pipeline for 3D Reflection Separation

In this subsection, we first explain how to incorporate flash cues to guide our reconstruction, then describe our overall optimization framework, and finally discuss how to adapt the loss functions to accommodate the RAW input images.

Regularizing Reflection Separation Using Flash Cues. While our high-level intuition is to construct pseudo-pairs as cues for reflection-free images, in this work, we want to reconstruct both the transmitted scene and the reflected scene. Therefore, we do not explicitly calculate the difference between the pseudo-pairs, but rather, use it as a regularization term to guide separation optimization. As shown in Eq. (3), $\mathbf{T}_F$ is expressed as $\mathbf{T}_N$ multiplied by a scalar $(1+\alpha)$, which enforces a linear relationship between them. Therefore, we choose to enforce the linearity between $\mathbf{T}_F$ and $\mathbf{T}_N$, which is equivalent to enforcing the constraint that the flash/no-flash difference is reflection-free.

While in the ideal case, $\mathbf{T}_F$ and $\mathbf{T}_N$ should form a strictly linear relationship, in reality, the camera flash might not be perfectly uniform; there is also a chance the secondary reflection of the flash does hit the camera sensor. As a result, this relationship between $\mathbf{T}_F$ and $\mathbf{T}_N$ should be close to linear, but might not perfectly linear. Therefore, we chose not to use this hard constraint, but use the Pearson Coefficient, which measures the linearity between $\mathbf{T}_F$ and $\mathbf{T}_N$:

$$\mathcal{L}_{linearity} = -\frac{\mathrm{cov}\left(\mathbf{T}_N, \mathbf{T}_F\right)}{\sqrt{\mathrm{var}\left(\mathbf{T}_N\right)\mathrm{var}\left(\mathbf{T}_F\right)}} \tag{6}$$

By minimizing this loss term, we encourage the 3D Gaussians to learn a reflection-free transmission, therefore reducing the ill-posedness of the separation.

Notably, while the analysis above holds true for both the 3D pseudo-pair (see Fig. 3b) and the 2D pseudo-pairs (see Fig. 3a), we only apply this linearity regularization to the 2D pseudo-pairs, as it is more straightforward to measure the linearity of images than 3D representations.

Initializing 3D Representations Using Flash Cues. Now we show how to utilize the 3D pseudo-pair to aid reflection separation. As illustrated in Fig. 2b, the 3D pseudo-pair, namely $\mathbf{I}_F^{Rec}$ and $\mathbf{I}_N^{Rec}$, are 3D representations of the target scene reconstructed from the flash views and no-flash views, respectively. Their difference should be the reflection-free 3D transmitted scene. However, given the highly ill-posed nature of the 3D scene reconstruction problem, it is very likely that the contents in $\mathbf{I}_F^{Rec}$ and $\mathbf{I}_N^{Rec}$ do not correspond with each other. As such, this difference between $\mathbf{I}_F^{Rec}$ and $\mathbf{I}_N^{Rec}$ should be viewed as a very rough estimate of the transmitted scene. Therefore, we decide to only use it to initialize the 3D representations $\mathbf{T}_F$, $\mathbf{T}_N$, $\mathbf{R}$, and β for better convergence.

We use 3DGS [20] as the 3D representation architecture, which is normally initialized from sparse point clouds. We first use structure from motion, e.g., [41], to obtain the sparse point clouds of the 3D pseudo-pair $\mathbf{I}_F^{Rec}$ and $\mathbf{I}_N^{Rec}$. Then we roughly align them via linear transformation to compensate for the difference in the camera coordinate systems. Afterwards, we compare these two sets of point

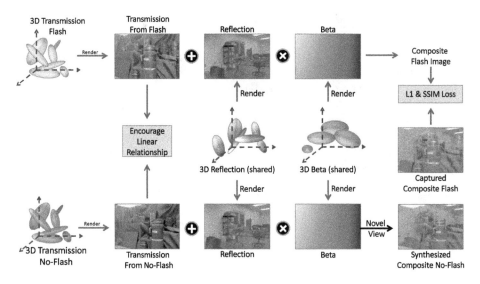

Fig. 4. Method Overview. We use 4 3DGSs [20] as our 3D representations for the transmitted scene with flash $\mathbf{T}_F$, the transmitted scene with no flash $\mathbf{T}_N$, the reflected scene $\mathbf{R}$ and the reflective fraction map β. Based on the Flash/No-flash technique, $\mathbf{R}$ and β are shared between the flash image and the no-flash image, while the relationship of $\mathbf{T}_F$ and $\mathbf{T}_N$ is close to linear. We initialize these 4 3DGSs using cues from the 3D pseudo-pair (see Fig. 3b and Sect. 3.3). In each iteration of optimization, our method operates on a single view. This figure, for instance, shows a view where we captured a flash image. There is NO no-flash image captured at this view. As shown in the top row, we use $\mathbf{T}_F$, $\mathbf{R}$, and β to render a flash image of this particular view and calculate losses with the captured ground truth flash image. Additionally, based on the cues from 2D pseudo-pairs, we calculate the Pearson linearity loss between $\mathbf{T}_F$ and $\mathbf{T}_N$ to encourage the linearity between them (see Fig. 3a and Sect. 3.3). We then back-propagate the gradients and update the weights of the 4 3DGSs.

clouds: for points in regions with increased intensities, we classify them as "transmitted points"; for points in regions with unchanged intensities, we classify them as "reflected points". Finally, we initialize the 3DGSs for $\mathbf{T}_F$, $\mathbf{T}_N$, and β from the "transmitted points", and the 3DGS for $\mathbf{R}$ from the "reflected points".

Note that this way of initialization relies on the 3D representation's compatibility with point clouds. It does not work with implicit neural 3D representations like NeRF [36]. When using NeRF as our 3D representations, we just randomly initialize the neural network and only rely on the previously discussed linearity regularization using 2D pseudo-pairs, which would still achieve better reflection removal performance than baselines, as will be shown in Sect. 6.

Overall Optimization Framework. As shown in Fig. 4, in each iteration of optimization, Flash-Splat operates on a single view. If we captured a flash image at this view (meaning that NO no-flash image was captured at this view), we follow Eq. (5) and use our 3D representations $\mathbf{T}_F$, $\mathbf{R}$, and β to render a flash image at this same view. Then we calculate the loss between the rendered flash image and the captured ground truth flash image (more on this in the next paragraph). Additionally, we also calculate the Pearson linearity loss between images rendered from $\mathbf{T}_F$ and $\mathbf{T}_N$ at this view (the 2D pseudo-pair). We then back-propagate the gradients and update the weights of the 4 3D representations $\mathbf{T}_F$, $\mathbf{T}_N$, $\mathbf{R}$, and β. In the next iteration, we perform similar computations with flash and no-flash swapped. By doing such alternative optimization, we are using the loss with the captured ground truth images to supervise the novel view synthesis ability of our 3D representations, while using the Pearson linearity loss to implicitly enforce the flash/no-flash prior.

Gamma Corrected Loss Function. Our complete loss function is:

$$\mathcal{L} = \lambda_1 \mathcal{L}_1 + \lambda_2 \mathcal{L}_{DSSIM} + \lambda_3 \mathcal{L}_{linearity} + \lambda_4 \mathcal{L}_{depth}, \tag{7}$$

where $\mathcal{L}_1$ and DSSIM [3] are computed between the captured RGB images and the rendered images; $\mathcal{L}_{depth}$ is a depth smoothness regularization adopted from NeRFReN [13]; λ_{1-4} are weightings for these 4 loss terms, respectively.

It is imperative to understand that for the difference of the flash/no-flash discrepancy to work, Eqs. 1 through 6 must be operational within the RAW image space. Nevertheless, our objective is for the $\mathcal{L}_1$ loss and DSSIM loss to be applied to tone-mapped images, as [35] demonstrates that learning within the RAW space predisposes the model to bias in favor of brighter pixels while neglecting the darker ones. Furthermore, it is our intention for the 3DGS model to output tone-mapped images, a decision driven by empirical observations indicating a potential underperformance when learning is conducted in the RAW domain. Consequently, we modify our loss function to incorporate $\mathcal{L}_1$ and DSSIM calculations as follows:

$$\mathcal{L} = \mathcal{L}\left(\gamma\left(\mathbf{I}^{raw}\right), \gamma\left(\gamma^{-1}\left(\mathbf{T}\right) + \beta\gamma^{-1}\left(\mathbf{R}\right)\right)\right) \tag{8}$$

where $\mathcal{L}$ can be either $\mathcal{L}_1$ or DSSIM loss, and we chose $\gamma(\mathbf{x}) = \mathbf{x}^{0.22}$ as gamma correction to tone-map RAW images.

4 Experimental Details

4.1 Dataset Collection

We captured a flash/no-flash dataset using a Canon Rebel R7 camera. Our camera settings included a fixed 0.25-s exposure time, 200 ISO sensitivity, an f-number of 5.6, and fixed white balancing. For each scene, we collected 30 images equally split into two categories: 15 with the built-in flash and 15 without. The typical distance between a flash view and the closest no-flash view is 5–10 cm, with 1–6 m separating the camera from the transmission scene objects. For a fair comparison with the baselines, we captured paired flash/no-flash data for a few scenes using a tripod, such that our method and the baselines use exactly the same views, with the only data difference being our method uses half flash and

Fig. 5. The Office scene. Top, middle, and bottom rows are the captured images, separated transmissions, and separated reflections, respectively. Our reflection separation approach is far more effective than NeRFReN [13] and Dong et al. [7].

half no-flash views. Note that our method does not see any paired flash/no-flash views. We process the raw images through a standard image signal processor consisting of white balancing, tone-mapping and gamma correction. We run COLMAP [41] on each method's corresponding input images to obtain camera poses and sparse point cloud estimations.

4.2 Baselines

We choose 2 baselines: NeRFReN [13], an unsupervised multi-view method based on 3D inverse rendering, and Dong et al. [7], a supervised deep learning approach. For fair comparison, we run both baselines twice, once on all flash images, and once on all no-flash images. This is to show that our method does not perform better because we use flash, but rather, because we use flash/no-flash cues.

4.3 Architecture and Optimization Details

Flash-Splat was implemented in PyTorch and run on an NVIDIA A6000 GPU. Our 4 3DGSs ($\mathbf{T}_F$, $\mathbf{T}_N$, $\mathbf{R}$, and β) have no shared parameters and are optimized with the same hyperparameters. Our implementation of 3DGS follows FSGS [59], a variant of 3DGS that supports feedforward settings (the original 3DGS [20] is not intended for feedforward scenes). Each 3DGS is initialized with 350K Gaussians and grow up to 500K. We optimize all 3DGSs for 5000 iterations on our flash/no-flash images sized 1200 × 800 pixels. Total running time is about 10 minutes per scene. As an alternative to 3DGS, we also explored using NeRF for 3D representations, but empirically found that it inferior to 3DGS; see Sect. 6.2.

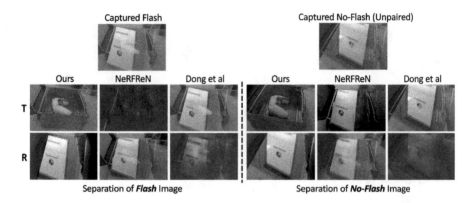

Fig. 6. The Game Controller Scene. Top, middle, and bottom rows are the captured images, separated transmissions, and separated reflections, respectively. Our reflection separation approach is far more effective.

Fig. 7. The Lens Stage Scene. Top, middle, and bottom rows are the captured images, separated transmissions, and separated reflections, respectively. Our reflection separation approach is far more effective.

5 Results

We compare our method's transmission-reflection separation ability against the baseline state-of-the-art methods. We also demonstrate novel view synthesis and depth estimation capabilities through our 3D inverse rendering framework.

Transmission-Reflection Separation. Our method, Flash-Splat, outperforms both baselines in transmission-reflection separation on our real-world scenes. In fact, the baselines fail completely to produce a reasonable separation. Figures 5, 6, 7 show results on indoor scenes. Figure 8 show results on an outdoor scene. Scene descriptions and results on 3 more scenes are included in the supplement.

Novel View Synthesis (NVS). Flash-Splat can perform novel view synthesis as it is based on inverse rendering. We compare our synthesis quality for scenes

Fig. 8. The Outdoor Scene. The captured images, separated transmissions, and separated reflections are shown in the top, middle, and bottom rows, respectively. We also include two additional reference images in the top row as references. "Paired F/nF Diff" denotes the difference between flash and no-flash paired images. "Glass Removed" shows true transmission by directly removing the glass.

Fig. 9. Novel View Synthesis (NVS). Our method does not compromise NVS quality, even when compared to dedicated NVS methods like NeRF and 3DGS.

with reflections against NeRF [36], NeRFReN [13], and 3DGS (our implementation of 3DGS still follows FSGS [59], as explained in Sect. 4.3). Figure 9 shows Flash-Splat does not compromise NVS performance, even when compared to dedicated NVS methods like NeRF and 3DGS. Rendered videos of our separated transmitted and reflected 3D scenes are in our project webpage.

Depth Estimation. Benefitting from the 3DGS base representation, Flash-Splat can also perform depth estimation on both transmitted and reflected scenes. Flash-Splat outperforms NeRFReN for both components (see Fig. 10).

6 Ablation Studies

6.1 With and Without Flash Cues

To demonstrate the importance of flash cues, we design a "flashless" variant of our proposed framework where we remove the flash cues from the 2D and 3D pseudo-pairs. This "flashless" framework is still a 3DGS-based approach, but does not utilize flash/no-flash photography at all. Its detailed architecture is illustrated in Figure 19 in the supplement. For a fair comparison, this flashless

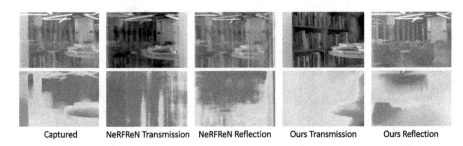

Captured NeRFReN Transmission NeRFReN Reflection Ours Transmission Ours Reflection

Fig. 10. Depth Estimation. The captured image's depth estimated by MiDaS [39,40] is shown in the leftmost column, which cannot differentiate between transmitted and reflected scenes. Our depths are much better than NeRFReN's [13].

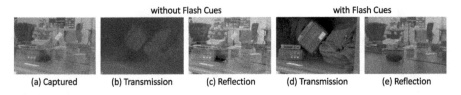

(a) Captured (b) Transmission (c) Reflection (d) Transmission (e) Reflection

Fig. 11. With and Without the Flash Cues. (b, c) shows the reflection separation result if we do not utilize the flash cues in our proposed framework, details of which are discussed in Sect. 6.1. (d, e) shows the result using the full version our proposed framework—the separated reflection and transmission are significantly better. This highlights the importance of our proposed flash cues.

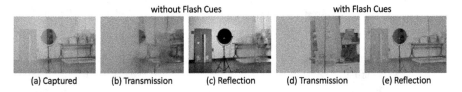

(a) Captured (b) Transmission (c) Reflection (d) Transmission (e) Reflection

Fig. 12. Equip NeRFReN [13] With Flash Cues from 2D Pseudo-Pairs. (b,c) show separation results from the original NeRFReN; (d,e) show separation results from NeRFReN incorporated our proposed flash cues. Since NeRFReN is not compatible with the point clouds initialization using the 3D pseudo pair, we only equip NeRFReN with the Pearson linearity regularization from the 2D pseudo pairs. Our 2D pseudo pairs regularization alone can achieve better reflection separation performance than the original NeRFReN.

framework is using the same number of total views as our proposed framework. Figure 11 shows that this flashless framework achieves significantly worse separation performance than our proposed Flash-Splat framework, which highlights the indispensability of the flash cues.

6.2 Replacing 3DGS with NeRF

To investigate the flash cues' impact on other 3D representations, we design another framework named Flash-NeRF, where we replace the 3DGS [20] representation with NeRF [36] and keep everything else the same. Since NeRF is not compatible with our pseudo pair point clouds initialization, we only utilize the Pearson linearity regularization from the 2D pseudo pairs as the flash cue. The Flash-NeRF framework can be seen as NeRFReN [13] plus our 2D pseudo pairs regularization. Figure 12 shows that our 2D pseudo pairs regularization significantly enhances NeRFReN's reflection separation performance.

Nevertheless, while this Flash-NeRF framework obtains an almost perfect transmitted scene, we can still notice obvious artifacts and floaters in the reflected scene. We find that using 3DGS as 3D representation can notably mitigate this issue, and thus decide to use 3DGS in our proposed framework.

6.3 Discussions and Limitations

Reflections in COLMAP: By Law of Reflection, reflected objects are equivalently "virtual objects" superimposed into the transmitted scene. We empirically verified reflections do not decrease COLMAP's view matching performance.

View Coverage: Although Flash-Splat does not need paired flash/no-flash images, it does require the flash and no-flash scenes to cover a similar range of perspectives. For instance, Flash-Splat would not work if we take flash images of the scene from the left and no-flash images from the right, as we would not be able to accurately synthesize pseudo-pairs in this case.

Curved Reflection: Flash-Splat currently cannot deal with scenes with curved reflective surfaces, e.g., curved glasses, since curved reflective surfaces will cause severe deformation of the reflected scene. We believe this is an important yet under-explored corner case for reflection removal, which we leave for future work.

Double Reflection: While double reflections can be a powerful cue for removing reflections when dealing with thick or double-pane glass [42], Flash-Splat currently does not handle scenes with obvious double reflections. Incorporating double-reflection–based cues is an interesting research direction.

Flash Strength: If the flash is too weak to illuminate the transmission scene, Flash-Splat would not have the necessary flash cues to remove the reflection.

Dynamic Scene: Because our 3D representation is static, objects moving between captured images result in blurry reconstructions.

7 Conclusion

We present a novel approach for transmission-reflection separation of 3D scenes through flash cues, significantly mitigating the ill-posedness of the task. By synthesizing "pseudo-paired" flash/no-flash images within a 3D inverse rendering framework based on Gaussian Splatting, we demonstrate superior reflection separation capabilities, particularly under challenging conditions where traditional methods falter. We validate our method on a new real-world dataset, showcasing its effectiveness and robustness. Our method not only unlocks practical reflection-removal but also enables novel view synthesis and depth estimation separately for the transmitted and reflected 3D scene.

Acknowledgements. This work was supported in part by AFOSR Young Investigator Program Award no. FA9550-22-1-0208, ONR Award no. N00014-23-1-2752, NSF CAREER Award no. 2339616, the Joint Directed Energy Transition Office, and a gift from Dolby Labs. We thank Kevin Zhang and Yi-Ting Chen for helpful discussions.

References

1. Alayrac, J.B., Carreira, J., Zisserman, A.: The visual centrifuge: model-free layered video representations. In: Proceedings of the IEEE/CVF Conference on Computer Vision and Pattern Recognition, pp. 2457–2466 (2019)
2. Arvanitopoulos, N., Achanta, R., Susstrunk, S.: Single image reflection suppression. In: Proceedings of the IEEE Conference on Computer Vision and Pattern Recognition, pp. 4498–4506 (2017)
3. Baker, A.H., Pinard, A., Hammerling, D.M.: On a structural similarity index approach for floating-point data. IEEE Trans. Vis. Comput. Graph. 1–13 (2023)
4. Barron, J.T., Mildenhall, B., Verbin, D., Srinivasan, P.P., Hedman, P.: Mip-NeRF 360: unbounded anti-aliased neural radiance fields. In: Proceedings of the IEEE/CVF Conference on Computer Vision and Pattern Recognition, pp. 5470–5479 (2022)
5. Chen, A., Xu, Z., Geiger, A., Yu, J., Su, H.: TensoRF: tensorial radiance fields. In: Avidan, S., Brostow, G., Cissé, M., Farinella, G.M., Hassner, T. (eds.) Computer Vision – ECCV 2022. ECCV 2022. LNCS, vol.13692, pp. 333–350. Springer, Cham (2022). https://doi.org/10.1007/978-3-031-19824-3_20
6. Chugunov, I., Shustin, D., Yan, R., Lei, C., Heide, F.: Neural spline fields for burst image fusion and layer separation. CVPR (2024)
7. Dong, Z., Xu, K., Yang, Y., Bao, H., Xu, W., Lau, R.W.: Location-aware single image reflection removal. In: Proceedings of the IEEE/CVF International Conference on Computer Vision, pp. 5017–5026 (2021)
8. Fan, Q., Yang, J., Hua, G., Chen, B., Wipf, D.: A generic deep architecture for single image reflection removal and image smoothing. In: Proceedings of the IEEE International Conference on Computer Vision, pp. 3238–3247 (2017)
9. Farid, H., Adelson, E.H.: Separating reflections and lighting using independent components analysis. In: Proceedings. 1999 IEEE Computer Society Conference on Computer Vision and Pattern Recognition (Cat. No PR00149), vol. 1, pp. 262–267. IEEE (1999)
10. Fridovich-Keil, S., Meanti, G., Warburg, F.R., Recht, B., Kanazawa, A.: K-planes: explicit radiance fields in space, time, and appearance. In: Proceedings of the IEEE/CVF Conference on Computer Vision and Pattern Recognition, pp. 12479–12488 (2023)
11. Gandelsman, Y., Shocher, A., Irani, M.: Double-DIP: unsupervised image decomposition via coupled deep-image-priors. In: Proceedings of the IEEE/CVF Conference on Computer Vision and Pattern Recognition, pp. 11026–11035 (2019)

12. Guo, X., Cao, X., Ma, Y.: Robust separation of reflection from multiple images. In: Proceedings of the IEEE Conference on Computer Vision and Pattern Recognition, pp. 2187–2194 (2014)
13. Guo, Y.C., Kang, D., Bao, L., He, Y., Zhang, S.H.: NeRFReN: neural radiance fields with reflections. In: Proceedings of the IEEE/CVF Conference on Computer Vision and Pattern Recognition, pp. 18409–18418 (2022)
14. Hariharan, B., Arbeláez, P., Girshick, R., Malik, J.: Hypercolumns for object segmentation and fine-grained localization. In: Proceedings of the IEEE Conference on Computer Vision and Pattern Recognition, pp. 447–456 (2015)
15. Hong, Y., Zheng, Q., Zhao, L., Jiang, X., Kot, A.C., Shi, B.: Panoramic image reflection removal. In: Proceedings of the IEEE/CVF Conference on Computer Vision and Pattern Recognition, pp. 7762–7771 (2021)
16. Hong, Y., Zheng, Q., Zhao, L., Jiang, X., Kot, A.C., Shi, B.: PAR2 NET: end-to-end panoramic image reflection removal. IEEE Trans. Pattern Anal. Mach. Intell. (2023)
17. Hu, Q., Guo, X.: Trash or treasure? An interactive dual-stream strategy for single image reflection separation. Adv. Neural. Inf. Process. Syst. **34**, 24683–24694 (2021)
18. Hu, Q., Guo, X.: Single image reflection separation via component synergy. In: Proceedings of the IEEE/CVF International Conference on Computer Vision, pp. 13138–13147 (2023)
19. Kee, E., Pikielny, A., Blackburn-Matzen, K., Levoy, M.: Removing reflections from raw photos. ArXiv **abs/2404.14414** (2024)
20. Kerbl, B., Kopanas, G., Leimkühler, T., Drettakis, G.: 3D gaussian splatting for real-time radiance field rendering. ACM Trans. Graph. **42**(4) (2023)
21. Kim, S., Huo, Y., Yoon, S.E.: Single image reflection removal with physically-based rendering. arXiv preprint arXiv:1904.11934 (2019)
22. Kong, N., Tai, Y.W., Shin, J.S.: A physically-based approach to reflection separation: from physical modeling to constrained optimization. IEEE Trans. Pattern Anal. Mach. Intell. **36**(2), 209–221 (2013)
23. Kong, N., Tai, Y.W., Shin, S.Y.: High-quality reflection separation using polarized images. IEEE Trans. Image Process. **20**(12), 3393–3405 (2011)
24. Lei, C., Chen, Q.: Robust reflection removal with reflection-free flash-only cues. In: Proceedings of the IEEE/CVF Conference on Computer Vision and Pattern Recognition, pp. 14811–14820 (2021)
25. Lei, C., Huang, X., Zhang, M., Yan, Q., Sun, W., Chen, Q.: Polarized reflection removal with perfect alignment in the wild. In: Proceedings of the IEEE/CVF Conference on Computer Vision and Pattern Recognition, pp. 1750–1758 (2020)
26. Levin, A., Weiss, Y.: User assisted separation of reflections from a single image using a sparsity prior. IEEE Trans. Pattern Anal. Mach. Intell. **29**(9), 1647–1654 (2007)
27. Levin, A., Zomet, A., Weiss, Y.: Learning to perceive transparency from the statistics of natural scenes. Adv. Neural Inf. Process. Syst. **15** (2002)
28. Levin, A., Zomet, A., Weiss, Y.: Separating reflections from a single image using local features. In: Proceedings of the 2004 IEEE Computer Society Conference on Computer Vision and Pattern Recognition, 2004. CVPR 2004, vol. 1, p. I. IEEE (2004)
29. Li, C., Yang, Y., He, K., Lin, S., Hopcroft, J.E.: Single image reflection removal through cascaded refinement. In: Proceedings of the IEEE/CVF Conference on Computer Vision and Pattern Recognition, pp. 3565–3574 (2020)

30. Li, R., Qiu, S., Zang, G., Heidrich, W.: Reflection separation via multi-bounce polarization state tracing. In: Vedaldi, A., Bischof, H., Brox, T., Frahm, J.-M. (eds.) ECCV 2020. LNCS, vol. 12358, pp. 781–796. Springer, Cham (2020). https://doi.org/10.1007/978-3-030-58601-0_46
31. Li, Y., Brown, M.S.: Exploiting reflection change for automatic reflection removal. In: Proceedings of the IEEE International Conference on Computer Vision, pp. 2432–2439 (2013)
32. Li, Y., Liu, M., Yi, Y., Li, Q., Ren, D., Zuo, W.: Two-stage single image reflection removal with reflection-aware guidance. Appl. Intell. 1–16 (2023)
33. Liu, Y.L., Lai, W.S., Yang, M.H., Chuang, Y.Y., Huang, J.B.: Learning to see through obstructions. In: Proceedings of the IEEE/CVF Conference on Computer Vision and Pattern Recognition, pp. 14215–14224 (2020)
34. Lyu, Y., Cui, Z., Li, S., Pollefeys, M., Shi, B.: Reflection separation using a pair of unpolarized and polarized images. Adv. Neural Inf. Process. Syst. **32** (2019)
35. Mildenhall, B., Hedman, P., Martin-Brualla, R., Srinivasan, P.P., Barron, J.T.: Nerf in the dark: high dynamic range view synthesis from noisy raw images. In: Proceedings of the IEEE/CVF Conference on Computer Vision and Pattern Recognition, pp. 16190–16199 (2022)
36. Mildenhall, B., Srinivasan, P.P., Tancik, M., Barron, J.T., Ramamoorthi, R., Ng, R.: NeRF: representing scenes as neural radiance fields for view synthesis. In: Vedaldi, A., Bischof, H., Brox, T., Frahm, J.-M. (eds.) ECCV 2020. LNCS, vol. 12346, pp. 405–421. Springer, Cham (2020). https://doi.org/10.1007/978-3-030-58452-8_24
37. Nayar, S.K., Fang, X.S., Boult, T.: Separation of reflection components using color and polarization. Int. J. Comput. Vis. **21**(3), 163–186 (1997)
38. Qiu, J., Jiang, P.T., Zhu, Y., Yin, Z.X., Cheng, M.M., Ren, B.: Looking through the glass: neural surface reconstruction against high specular reflections. In: Proceedings of the IEEE/CVF Conference on Computer Vision and Pattern Recognition, pp. 20823–20833 (2023)
39. Ranftl, R., Bochkovskiy, A., Koltun, V.: Vision transformers for dense prediction. ArXiv preprint (2021)
40. Ranftl, R., Lasinger, K., Hafner, D., Schindler, K., Koltun, V.: Towards robust monocular depth estimation: mixing datasets for zero-shot cross-dataset transfer. IEEE Trans. Pattern Anal. Mach. Intell. (TPAMI) (2020)
41. Schönberger, J.L., Zheng, E., Frahm, J.-M., Pollefeys, M.: Pixelwise view selection for unstructured multi-view stereo. In: Leibe, B., Matas, J., Sebe, N., Welling, M. (eds.) ECCV 2016. LNCS, vol. 9907, pp. 501–518. Springer, Cham (2016). https://doi.org/10.1007/978-3-319-46487-9_31
42. Shih, Y., Krishnan, D., Durand, F., Freeman, W.T.: Reflection removal using ghosting cues. In: Proceedings of the IEEE Conference on Computer Vision and Pattern Recognition, pp. 3193–3201 (2015)
43. Sinha, S.N., Kopf, J., Goesele, M., Scharstein, D., Szeliski, R.: Image-based rendering for scenes with reflections. ACM Trans. Graph. (TOG) **31**(4), 1–10 (2012)
44. Wan, R., Shi, B., Duan, L.Y., Tan, A.H., Gao, W., Kot, A.C.: Region-aware reflection removal with unified content and gradient priors. IEEE Trans. Image Process. **27**(6), 2927–2941 (2018)
45. Wan, R., Shi, B., Duan, L.Y., Tan, A.H., Kot, A.C.: CRRN: multi-scale guided concurrent reflection removal network. In: Proceedings of the IEEE Conference on Computer Vision and Pattern Recognition, pp. 4777–4785 (2018)

46. Wan, R., Shi, B., Li, H., Duan, L.Y., Kot, A.C.: Reflection scene separation from a single image. In: Proceedings of the IEEE/CVF Conference on Computer Vision and Pattern Recognition, pp. 2398–2406 (2020)
47. Wan, R., Shi, B., Li, H., Duan, L.Y., Kot, A.C.: Face image reflection removal. Int. J. Comput. Vis. **129**, 385–399 (2021)
48. Wang, P., Liu, L., Liu, Y., Theobalt, C., Komura, T., Wang, W.: Neus: learning neural implicit surfaces by volume rendering for multi-view reconstruction. NeurIPS (2021)
49. Wang, Y., Han, Q., Habermann, M., Daniilidis, K., Theobalt, C., Liu, L.: Neus2: fast learning of neural implicit surfaces for multi-view reconstruction. In: Proceedings of the IEEE/CVF International Conference on Computer Vision, pp. 3295–3306 (2023)
50. Wei, K., Yang, J., Fu, Y., Wipf, D., Huang, H.: Single image reflection removal exploiting misaligned training data and network enhancements. In: Proceedings of the IEEE/CVF Conference on Computer Vision and Pattern Recognition, pp. 8178–8187 (2019)
51. Wen, Q., Tan, Y., Qin, J., Liu, W., Han, G., He, S.: Single image reflection removal beyond linearity. In: Proceedings of the IEEE/CVF Conference on Computer Vision and Pattern Recognition, pp. 3771–3779 (2019)
52. Xia, Z., Gharbi, M., Perazzi, F., Sunkavalli, K., Chakrabarti, A.: Deep denoising of flash and no-flash pairs for photography in low-light environments. In: 2021 IEEE/CVF Conference on Computer Vision and Pattern Recognition (CVPR), pp. 2063–2072 (2020)
53. Xia, Z., Lawrence, J., Achar, S.: A dark flash normal camera. In: 2021 IEEE/CVF International Conference on Computer Vision (ICCV), pp. 2410–2419 (2020)
54. Xue, T., Rubinstein, M., Liu, C., Freeman, W.T.: A computational approach for obstruction-free photography. ACM Trans. Graph. (TOG) **34**(4), 1–11 (2015)
55. Yang, J., Gong, D., Liu, L., Shi, Q.: Seeing deeply and bidirectionally: a deep learning approach for single image reflection removal. In: Proceedings of the European Conference on Computer Vision (ECCV), pp. 654–669 (2018)
56. Yang, Y., Ma, W., Zheng, Y., Cai, J.F., Xu, W.: Fast single image reflection suppression via convex optimization. In: Proceedings of the IEEE/CVF Conference on Computer Vision and Pattern Recognition, pp. 8141–8149 (2019)
57. Zhang, X., Ng, R., Chen, Q.: Single image reflection separation with perceptual losses. In: Proceedings of the IEEE Conference on Computer Vision and Pattern Recognition, pp. 4786–4794 (2018)
58. Zheng, Q., Shi, B., Chen, J., Jiang, X., Duan, L.Y., Kot, A.C.: Single image reflection removal with absorption effect. In: Proceedings of the IEEE/CVF Conference on Computer Vision and Pattern Recognition, pp. 13395–13404 (2021)
59. Zhu, Z., Fan, Z., Jiang, Y., Wang, Z.: FSGS: real-time few-shot view synthesis using gaussian splatting. arXiv preprint arXiv:2312.00451 (2023)
60. Zou, Z., Lei, S., Shi, T., Shi, Z., Ye, J.: Deep adversarial decomposition: a unified framework for separating superimposed images. In: Proceedings of the IEEE/CVF Conference on Computer Vision and Pattern Recognition, pp. 12806–12816 (2020)

PALM: Predicting Actions through Language Models

Sanghwan Kim[1]($\boxtimes$), Daoji Huang[1], Yongqin Xian[2], Otmar Hilliges[1], Luc Van Gool[1,3,4], and Xi Wang[1]

[1] ETH Zürich, Zürich, Switzerland
kshwan0227ch@gmail.com
[2] Google, Mountain View, USA
[3] KU Leuven, Leuven, Belgium
[4] INSAIT, Sofia, Bulgaria

Abstract. Understanding human activity is a crucial yet intricate task in egocentric vision, a field that focuses on capturing visual perspectives from the camera wearer's viewpoint. Traditional methods heavily rely on representation learning that is trained on a large amount of video data. However, a major challenge arises from the difficulty of obtaining effective video representation. This difficulty stems from the complex and variable nature of human activities, which contrasts with the limited availability of data. In this study, we introduce PALM, an approach that tackles the task of long-term action anticipation, which aims to forecast forthcoming sequences of actions over an extended period. Our method PALM incorporates an action recognition model to track previous action sequences and a vision-language model to articulate relevant environmental details. By leveraging the context provided by these past events, we devise a prompting strategy for action anticipation using large language models (LLMs). Moreover, we implement maximal marginal relevance for example selection to facilitate in-context learning of the LLMs. Our experimental results demonstrate that PALM surpasses the state-of-the-art methods in the task of long-term action anticipation on the Ego4D benchmark. We further validate PALM on two additional benchmarks, affirming its capacity for generalization across intricate activities with different sets of taxonomies.

Keywords: egocentric vision · video understanding · action anticipation · large language model

1 Introduction

As wearable technology, such as smart glasses and body-worn cameras, becomes more prevalent, the availability of egocentric videos has grown significantly.

Supplementary Information The online version contains supplementary material available at https://doi.org/10.1007/978-3-031-73007-8_9.

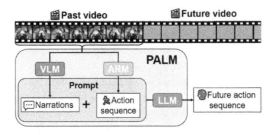

Fig. 1. PALM for long-term action anticipation. Given a video input, PALM uses a vision-language model (VLM) and an action recognition model (ARM) to compile a text prompt that describes the past video content. A large-language model (LLM) then takes the prompt and predicts future actions.

These videos capture activities from the camera wearer's viewpoint, providing a unique perspective that features individual behaviors in various contexts. Egocentric vision holds immense promise for understanding how humans execute specific tasks and activities. It enables numerous applications such as real-time systems that anticipate and respond to user needs [50,51,58], monitoring patient activities [37,39,61], and providing personalized instructions and procedures [27,42,43].

Existing works on understanding human activities mostly rely on learning representations from large-scale video datasets [15,46,49,56]. However, these approaches face two critical challenges. First, the complexity and variability of human activities make it challenging to obtain comprehensive representations, particularly for longer videos. Second, reliance on large datasets limits the model's capability to generalize to less common, long-tail classes and unseen scenarios. In this context, there is a pressing need for a more versatile approach to human activity understanding, adept at handling the complicated nature of human activities without overly relying on large-scale video data. Therefore, recent works have started to exploit the procedure knowledge and generalization ability of large language models (LLMs) [1,21,32,44,45].

Most prior works based on LLMs have been largely confined to generating instructions and plans given a high-level task description (e.g., "prepare the dinner table") and have not been tailored for intricate human activity understanding from videos. Indeed, there is a significant gap in the literature when it comes to the use of LLMs for grounded activity understanding, especially in challenges involving understanding long-term action sequences. In this work, we efficiently boost the versatile ability of LLMs in long-term action anticipation [12,18,20] by providing sufficient visual context in discrete text.

As shown in Fig. 1, we introduce a simple yet effective framework called PALM that combines descriptions of past action sequences and corresponding narrations to bridge the gap between visual and textual domains. In detail, PALM takes an input video with annotated action periods and leverages a vision-language model and an action recognition model to describe past events. We then design a prompt strategy for large language models (LLMs) to anticipate future actions given the context of these past events. To facilitate the predictive capabilities of LLMs, we use maximal marginal relevance (MMR) [7,59] to retrieve similar and diverse examples from the training set, enabling in-context learning for the LLMs. Through this prompting strategy, our system can tackle intricate tasks in understanding human activities, benefiting from existing foundation models. Experimental results demonstrate that our method PALM significantly improves performance in long-term action anticipation, as evaluated using the Ego4D benchmark [18]. Further evaluations on two additional benchmarks, EPIC-KITCHEN [12] and EGTEA [33], reveal the generalization ability of PALM across different contexts and datasets. Our contributions can be summarized as follows:

- We propose an effective textural representation of past events, consisting of both action sequences and accompanying narrations that provide additional context and environmental details.
- We design a novel prompting strategy that enables efficient in-context learning for large language models.
- PALM outperforms existing state-of-the-art methods in the task of long-term action anticipation, which is part of the Ego4D benchmark.

2 Related Work

2.1 Foundation Model

Foundation models serve as the cornerstone for various downstream tasks across different data modalities [5,64]. Here we consider two types of foundation models: vision-language models (VLMs) to generate the textual narrations of video and large language models (LLMs) to predict future action given visual context.

Vision-Language Model. VLMs utilize complementary characteristics of image and text datasets to perform multimodal tasks such as image captioning, visual question-answering, and visual reasoning [31,35,47]. Various works [10,57,65] attempt to unify vision-language tasks with a single model, overcoming the limited utility of previous models. In this paper, the image or video captioning ability of VLMs is utilized to directly transform visual information into narrative text descriptions. We empirically found that BLIP family [11,29,30] provides concise and informative captions. For instance, BLIP [30] proposes a novel encoder-decoder architecture specialized for multimodal pertaining, and BLIP-2 [29] significantly reduces the trainable parameters by using a querying transformer to bridge the modality gap between frozen vision models and language models. InstructBLIP [11] further performs instruction tuning on BLIP-2, enabling general-purpose models with a unified natural language interface.

Large Language Model. Language foundation models [3,6,48,53,54,62] have had a significant impact on the NLP field, demonstrating surprising ability to generalize across unseen tasks. With their extensive training data and large parameter size, LLMs have demonstrated the ability to learn from examples provided in input prompts, a concept known as in-context learning [6]. LLMs have also shown success in the tasks of decision-making and planning. For example, SayCan [24] combines a language model with learned low-level skills for grounded planning, while Socratic Models [60] formulate action prediction as text completion problems from captioned images. In this work, we explore how such procedure knowledge can be exploited in the complex action anticipation task and showcase the effectiveness of in-context learning for this purpose.

2.2 Long-Term Action Anticipation

Given an input video, the long-term action anticipation (LTA) task aims to predict the next actions in the future. In this work, we explore various action recognition models to estimate accurate action sequences and extensively compare our method with previous action anticipation models.

Action Recognition. SlowFast [16] proposes to fuse two channels of input that consist of a low frame and a high frame rate video, using 3D convolutional kernels. Similar to Slowfast, StillFast [49] adopts a specific architecture with two branches to effectively process high-resolution still images and low-resolution video frames. On the other hand, MViT [15] employs several channel-resolution scale stages to obtain a multiscale pyramid of features, based on vision transformers. Compared to previous works solely optimized on video inputs, EgoVLP [34] and EgoVLPv2 [46] pre-train video and text encoders with contrastive loss specialized for the egocentric domain. Each encoder provides useful feature representations for downstream video tasks such as natural language queries, video question-answering, and video summarization.

Action Anticipation. Baselines provided by Ego4D [18] employ the SlowFast [16] video encoder to extract features from sampled clips, followed by transformer layers and classification heads to predict the verb-noun pair of future action sequence. HierVL [2] utilizes a hierarchical contrastive training objective that encourages text-visual alignment at both the clip and video levels. ICVAE [36] proposes using a predicted scenario, which summarizes the activity depicted in the video, as a condition for action prediction. A few concurrent approaches [9,55,63] leverage the generalization abilities and abundant knowledge of LLMs to predict future actions. Compared to these LLM-based methods, PALM differentiates itself by utilizing vision-language model for effective extraction of visual context and employing in-context learning in combination with MMR.

3 Method

We introduce PALM, a comprehensive framework designed for video understanding and action anticipation. PALM comprises three key modules, as illustrated

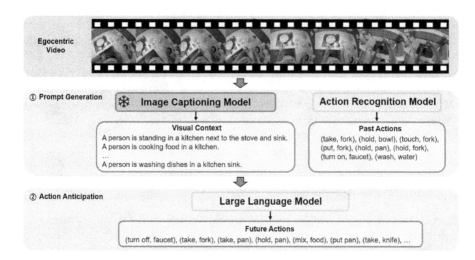

Fig. 2. Overview of PALM. Given an input video, an image captioning model and an action recognition model convert visual information of past activities into text format (visual context and past actions). Then, LLM utilizes visual context and past actions to predict the sequence of future actions. The image captioning model is frozen and in-context learning is adapted for LLM prompting.

in Fig. 2: an image captioning module (Sect. 3.2), an action recognition module (Sect. 3.3), and an action anticipation module based on LLMs (Sect. 3.4).

Given an untrimmed input video, the action recognition module generates labels for *past actions* in the form of verb and noun pairs (eight past actions in Fig. 2). The action recognition model is trained to produce valid verb and noun pairs as predefined by each dataset. Simultaneously, the image captioning module generates captions from the input video, providing additional visual information relevant to the events, termed *visual context*. The prompt is then constructed using the visual context and past actions, passed to LLMs to predict future actions. Additionally, we incorporate exemplars from the training set to facilitate in-context learning for LLMs. We demonstrate the effectiveness of our approach on the long-term action anticipation (LTA) task, achieving state-of-the-art (SOTA) performance on Ego4D benchmarks.

3.1 Task Description

The LTA task involves predicting a chronological sequence of future actions given an untrimmed video V. The input video V comprises multiple action segments with variable lengths, where each action segment corresponds to a single action (verb and noun pair). The segment boundaries for each action in the videos are provided as part of the input. Formally, after observing the video until time t (*e.g.*, $V_{:t}$), the LTA model sequentially predicts Z future actions that the camera-wearer is likely to perform:

$$\{\hat{n}_z, \hat{v}_z\}_{z=1}^{Z} \tag{1}$$

```
Question Dictionary = {
    Location: Where is the person?
    Detection: What objects are in this image?
    Action: What is the person doing in this image?
    Prediction: What will the person do later in this image?
    Interaction: What is the person interacting with in this image?
    Intention: What does the person want to do in this image?
}
```

Fig. 3. List of questions asking for information related to key concepts of egocentric image.

where $\hat{n}_z \in \mathcal{N}$ and $\hat{v}_z \in \mathcal{V}$, given the predefined noun and verb taxonomy $\mathcal{N}$ and $\mathcal{V}$. The input video $V_{:t}$ contains P action segments.

3.2 Image Captioning Module

The past action labels do not capture all the visual information and context presented in the video (*e.g..*, the location of the person or the background objects). Our idea is to use an image captioning model to capture the action context, providing more informative prompts related to the action performer for LLMs.

To enhance the descriptive power of captioning models, we formulate the image captioning task as a question-answering problem. Figure 3 displays a set of questions related to six key concepts that the models should aim to answer given an input image: location, detection, action, prediction, interaction, and intention. Given a question Q, the conditional text is formed as "Question: $<Q>$?, Answer:" where the answer part will be filled by the captioning model. After extensive experiments, we select intention-based question for conditional generation and VideoBLIP [25] as our image captioning model (see supplementary).

By posing these questions to the captioning models, we anticipate extracting useful descriptions beneficial for LLMs to predict future actions. In practice, the conditional caption generation yields consistent and accurate descriptions of the action context for later prediction. For image captioning, we use the middle frame of each clip segment to generate the captions, based on the assumption that the middle frame is highly likely to be the moment when the action is happening.

3.3 Action Recognition Module

The accurate past action descriptions are crucial, as they provide essential contextual information about past events in a given video input. Our model recognizes actions from the observed video clips, taking inspiration from the baseline model used in the Ego4D paper [18]. Visual features are extracted from a video encoder given n consecutive action segments ($n = 4$ in Fig. 4). We adopt the off-the-shelf video encoder of EgoVLP [34], a vision-language model pre-trained on Ego4D. Next, transformer layers are used to combine multiple visual features

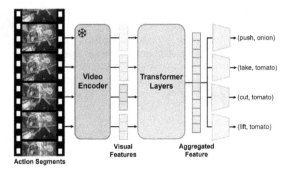

Fig. 4. Action recognition network. A frozen video encoder generates visual features from $n = 4$ continuous action segments, followed by transformer layers and classification heads.

into an aggregated feature, followed by n classification heads to recognize the verbs and nouns independently. The video encoder is frozen, and the rest of the model is trained with cross-entropy loss summed over n classification heads. We also try to exploit a text encoder by formulating the action recognition task as matching a proper action label with an input video, which does not improve the accuracy (See supplementary).

We empirically find that recognizing actions in isolation falls short in effectiveness compared to recognizing n actions within a continuous sequence. This is because actions in neighboring segments are often highly correlated, reflecting the natural continuity embedded in human activities. Thus, we use a sliding input video window that includes $n = 4$ segments with the window striding of one. The top-1 action is determined by identifying the most common action from the classifications across different windows. See supplementary for more details and ablation results.

3.4 Action Anticipation Module

In this section, we discuss the details of action anticipation with LLMs, from how the prompt is formatted to the details of inference and post-processing of raw outputs. Llama 2 (7B) [54], an open-source LLM, is employed in our framework.

Prompt Design. Following common practice, we design the prompt with a few examples to enable in-context learning (ICL), maximizing references given to LLMs. Figure 5 shows the template of our prompts, which consists of an instruction, a few examples, and a query. The instruction includes the task description and defines the input and output for LLMs. The examples for ICL, selected from the training set, include past and future actions (ground truth) and corresponding textual descriptions generated by the captioning model, which we call narrations. The query includes recognized past actions and generated narrations, and future actions are left empty for LLMs to complete.

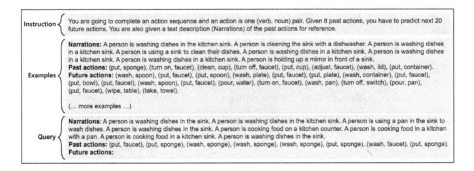

Fig. 5. Prompt structure for LLMs. The prompt is composed of an instruction, examples retrieved from the training set, and the query related to the prediction video. Each video description consists of narrations, past actions, and future actions.

Example Selection. Incorporating some examples that demonstrate proper generation will greatly improve the performance of LLMs by providing additional context and references [6,40]. Considering the diversity of egocentric videos, especially within the Ego4D dataset, randomly selected examples are unlikely to generalize well on various scenarios within the test set.

Therefore, we utilize a maximal marginal relevance (MMR) [7,59] based exemplar selection strategy. The main idea behind MMR is to select exemplars that are relevant to the query while exhibiting sufficient variety to provide robust generalization across the test set. To be specific, we iteratively select a set of examples $p_i \in T$ from the training set D that are semantically close to the query prompt q but also diverse enough to provide additional information.

$$p_j = \arg\max_{p_j \in D/T} \left[\lambda S(q, p_j) - (1-\lambda) \max_{p_i \in T} S(p_i, p_j) \right], \quad (2)$$

where S measures the semantic similarity between exemplars and the parameter λ strikes a balance between similarity and diversity. One exemplar includes past and future action sequences and corresponding captions in natural language. We use MPNet [52] to extract text embeddings and calculate cosine similarity to measure semantic similarity. We compare various exemplar selection strategies in Sect. 4.3.

Fine-Tuning LLM. To facilitate LLMs to generate proper action sequence, we fine-tune the LLMs based on the training set of each egocentric dataset. As fine-tuning whole parameters of the LLMs can be computationally inefficient and may lead to catastrophic forgetting [26], parameter-efficient fine-tuning is leveraged. Specifically, we utilize LoRA [19] with 8-bit quantization using a batch size of 8 and a learning rate of 5×10^{-4} for 10 epochs. The training set for fine-tuning consists of input prompts in the format of Fig. 5 with different exemplars and ground truth future actions as targets to be generated by the LLMs.

Inference of LLM and Post-processing. Following the evaluation protocol of Ego4d [18], we generate multiple predictions for each test video. To reduce the variability in predictions, we prompt LLMs multiple times using the same prompt structure, as shown in Fig. 5, but with different exemplars. After inference, we map the raw output of LLMs to the label space of predefined verbs and nouns. One should note that the output of LLMs may not always fall within the given taxonomy, despite all input prompts containing verbs and nouns from the predefined domain. For simplicity, we adopt a rule-based mapping (see details in supplementary).

4 Experiments

In this section, we evaluate PALM on the long-term action anticipation (LTA) task, shedding light on the various design aspects. We conduct a comprehensive evaluation using the Ego4D dataset, which stands as the largest repository of egocentric videos. We report the final performance metrics on the Ego4D [18], EPIC-KITCHEN-55 [12], and EGTEA Gaze+ [33] benchmarks.

4.1 Datasets

Ego4D [18]. With 243 h of egocentric video content encompassing 53 distinct scenarios, Ego4D offers a rich variety of data with a predefined annotation of 117 verbs and 521 nouns for the LTA task. Ego4D is available in two versions, v1 and v2, where Ego4D v1 is a subset of Ego4D v2 with fewer videos and a smaller verb and noun set. We use both versions of the dataset to enable extensive comparison with previous methods and adhere to the identical training, validation, and test splits as specified in the benchmark.

EPIC-KITCHENS-55 (EK-55) [12]. This dataset comprises 55 h of egocentric videos centered around cooking scenarios, featuring 125 annotated verbs and 352 nouns. We follow the same train and test splits as EGO-TOPO [38].

EGTEA Gaze+ (EGTEA) [33]. With approximately 26 h of egocentric cooking videos, EGTEA includes 19 verbs and 53 nouns. Similar to EK-55, we employ the same train and test splits as described in EGO-TOPO [38].

4.2 Comparison to the SOTA

Baselines. We include three baselines:

- **Last action** outputs the last action of past action sequence as Z future actions.
- **Repeat action** repeats the given past P actions to fill Z future actions.
- **Retrieve action** retrieves the most semantically similar example in the training set (as in the example selection step) and uses their future actions as the prediction.

Table 1. Comparison to the SOTA on the Ego4D test set. PALM outperforms all baselines on the Ego4D LTA benchmark in respect of edit distance (ED ↓).

Method	Version	Verb ED	Noun ED	Action ED
Slowfast [18]	v1	0.7389	0.7800	0.9432
VCLIP [14]	v1	0.7389	0.7688	0.9412
ICVAE [36]	v1	0.7410	0.7396	0.9304
HierVL [2]	v1	0.7239	0.7350	0.9276
AntGPT [63]	v1	0.6584	0.6546	0.8814
PALM (ours)	v1	**0.6559**	**0.6401**	**0.8613**
Last action	v2	0.8329	0.7258	0.9438
Repeat action	v2	0.7491	0.7073	0.9155
Retrieve action	v2	0.7417	0.7041	0.9127
Slowfast [18]	v2	0.7169	0.7359	0.9253
VideoLLM [9]	v2	0.721	0.725	0.921
AntGPT [63]	v2	0.6503	0.6498	0.8770
PALM (ours)	v2	**0.6471**	**0.6117**	**0.8503**

We also compare to three previous SOTA methods:

- **Slowfast** [18] is the baseline LTA model given by Ego4d.
- **VideoLLM** [9] converts multimodal data into a unified token sequence and utilizes fine-tuned LLMs
- **AntGPT** [63] leveraged action recognition model and fine-tuned LLMs to predict future actions.

Evaluation. We conform to the evaluation procedures set forth by the Ego4D [18] and calculate the edit distance (ED) using the Damerau-Levenshtein distance metric [13,28]. ED is computed separately for verbs, nouns, and the entire action sequences. Given $P = 8$ action segments of the input video, we report the minimum edit distance among $K = 5$ predicted sequences, each with a length of $Z = 20$ actions.

Results. Table 1 shows that PALM improves existing SOTA methods and achieves the best performance on the test set of the Ego4D v1 and v2 benchmark. In general, LLM-based approaches (VideoLLM, AntGPT, and PALM) predict more accurate future actions compared to other approaches based on representation learning. Notably, repeat and retrieve action baselines in Ego4D v2 perform better than Slowfast with respect to noun and action ED. These observations clearly indicate the advantages of leveraging foundation models and the limitations of feature learning methods on the LTA task. Moreover, PALM outperforms other LLM-based methods, which demonstrates the effectiveness of our approach to provide ample visual context to LLMs, equipped with action recognition and image captioning models.

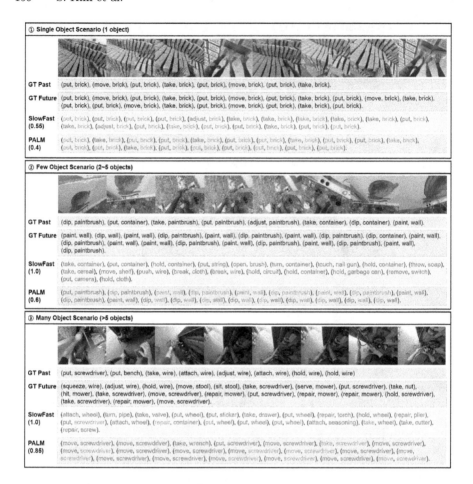

Fig. 6. Qualitative results of predictions on Ego4D. We show the comparison between the predictions from the Ego4D baseline using **SlowFast** [18] and the predictions from our model **PALM** respectively. For each of the three examples shown here, we include the middle frames of the previous eight action segments with the corresponding ground truth past actions **GT Past** and future actions **GT Future** alongside the predicted future actions. The green words represent correct predictions while the red words denote incorrect predictions. The edit distance between the predicted actions and ground truth actions is shown in parenthesis.

In Fig. 6, we perform a qualitative comparison between the Ego4D baseline using SlowFast [18] and PALM. The first example exemplifies a scenario where the person mainly interacts with a single object, in this case, a brick. Both methods can accurately predict the future noun "brick". In the second scenario, the person is drawing a painting on a wall, interacting with multiple objects such as a paintbrush, a container, and the wall. PALM predicts the repeated sequence of actions, unlike SlowFast which incorrectly anticipates unrelated actions such as putting a string, touching a nail gun, or throwing a soap, none of which are present in the input video. This also demonstrates the importance of previous actions in the anticipation of future actions. The final scenario portrays the person adjusting a wire in a repair shop, a task involving numerous items. While SlowFast and PALM recognize central objects such as a screwdriver, they both fail to predict a coherent sequence of repair actions. We hypothesize that a language model with expert knowledge can produce more meaningful anticipations, particularly in the context of repairing mechanics, where it is crucial to execute actions in a precise order. More qualitative results are shown in supplementary.

4.3 Ablations

We perform various ablation studies regarding action recognition, example selection methods, LLM models, and prompt contents. We evaluate on the validation set of Ego4D v2.

Table 2. Ablation on action recognition module. We compare different pre-trained video encoders by reporting the Top-1 accuracy (Acc. ↑) and edit distance (ED ↓) for verbs, nouns, and actions. For comparative purposes, we also include the ED when ground truth actions (GT action) are used in the prompts. The best Acc. and ED are highlighted in bold except those from GT action.

Method	Verb Acc.	Noun Acc.	Action Acc.	Verb ED	Noun ED	Action ED
MViT [15]	19.87	2.55	0.51	0.7291	0.8597	0.9630
SlowFast [16]	19.42	14.65	3.12	0.7252	0.7472	0.9278
StillFast [49]	19.12	19.41	4.06	0.7356	0.7369	0.9241
EgoVLPv2 [46]	33.84	40.63	16.31	0.7182	0.6651	0.8937
EgoVLP [34]	**40.32**	**45.53**	**20.63**	**0.7111**	**0.6465**	**0.8819**
GT action	–	–	–	0.6663	0.5887	0.8319

Action Recognition. The action recognition module is an essential part of our framework PALM as we utilize LLMs for predicting future actions and the accuracy of these predictions is highly influenced by the sequence of previous actions. We evaluate various video encoders trained on egocentric videos, including MViT [15], SlowFast [16], StillFast [49], EgoVLP [34], and EgoVLPv2 [46]. It is important to note that we employ a frozen video encoder without any fine-tuning, as depicted in Fig. 4.

Table 2 presents the Top-1 Accuracy (%) and the edit distance of each video encoder. EgoVLP emerges as the top performer, achieving the best accuracy and edit distance among all encoders. EgoVLP utilizes a custom contrastive loss to align egocentric video frames with narrations, effectively establishing a connection between visual and textual embeddings. However, EgoVLPv2 utilizes task-specific loss for question-answering tasks built on the EgoVLP setting, which results in slightly less informative video features for action recognition. Unlike EgoVLP and EgoVLPv2, other video encoders are solely trained on video frames, making it challenging to draw robust connections between their video features and textual information (verbs and nouns). Based on these results, we select EgoVLP as our video encoder.

It is expected that better accuracy in action recognition leads to a reduced edit distance, as accurate estimation of past actions is crucial for predicting future actions. Notably, the use of ground truth (GT) actions significantly outperforms all video encoders, which highlights the enhancement of action recognition as the foremost challenge within our framework. We believe that fine-tuning the video encoder on specific egocentric datasets would likely enhance recognition accuracy.

Table 3. Ablation on example selection strategies exhibits that MMR outperforms other selection methods.

Method	Verb ED	Noun ED	Action ED
Random	0.7362	0.6747	0.9002
Similarity	0.7204	0.6661	0.8923
MMR	**0.7179**	**0.6647**	**0.8908**

Example Selection. We compare different example selection strategies for in-context learning (ICL), including random, similarity, and maximal marginal relevance (MMR) [59]. The random method uniformly selects examples from the training set, whereas the similarity method selects the most semantically similar example to the query. MMR considers both the similarity and diversity of examples, striving for a balance between them. Table 3 demonstrates that MMR outperforms the other selection methods.

Table 4. Ablation on LLMs shows that larger LLMs perform better on the LTA task.

Method	Verb ED	Noun ED	Action ED
GPT-Neo (125M)	0.7657	0.6867	0.9119
GPT-Neo (1.3B)	0.7314	0.6759	0.8970
GPT-Neo (2.7B)	0.7178	0.6747	0.8943
Llama 2 (7B)	0.7179	0.6647	0.8908
Llama 2 (13B)	**0.6958**	**0.6634**	**0.8867**

LLM Ablation. We ablate the impact of LLMs by comparing different open-source models, primarily GPT-Neo [4] and Llama 2 [54] families. Table 4 indicates that larger language models are more capable of predicting future actions with their improved capacity for generalization across novel scenarios. This observation aligns with the performance of LLMs on general NLP benchmarks. Taking into account the computational resources at our disposal, we select Llama 2 with 7B parameters.

Table 5. Ablation on the content of the prompt. We use abbreviations for each content in the input prompt: $\mathcal{A}$ - action sequence, $\mathcal{C}$ - captions

Prompt Content	Verb ED	Noun ED	Action ED
$\mathcal{A}$	0.7203	0.6688	0.8947
$\mathcal{C}$	**0.7088**	0.6901	0.9064
$\mathcal{A} + \mathcal{C}$	0.7179	**0.6647**	**0.8908**

Prompt Contents. We examine different content included in the input prompt to validate the contribution of each module to final performance. PALM utilizes recognized past actions and generated image captions to compose prompts. The action sequence describes the temporal flow of activity in (verb, noun) format, while captions narrate visual information relevant to human. In Table 5, the action sequence with captions ($\mathcal{A} + \mathcal{C}$) achieves the lowest action ED, which thereby is used in our framework. Notably, caption-only prompt ($\mathcal{C}$) reaches the lowest verb ED, showing that narrations are capable of capturing dynamic motions of human activity, while action-only prompt ($\mathcal{A}$) effectively detects key objects interacted with human supported by relatively lower noun ED. This indicates that action sequence and narrations provide a beneficial complement, thereby supporting our decision to incorporate both in the prompts.

4.4 Evaluation on EK-55 and EGTEA

Evaluation. For EK-55 [12] and EGTEA [33], we adhere to the experimental setup established by EGO-TOPO [38] and employ mean average precision (mAP) for multi-label classification as the evaluation metric. The initial $R\%$ ($R = [25\%, 50\%, 75\%]$) of each video serves as input, with the objective of predicting the set of actions occurring in the remaining $(100 - R)\%$ of the video. Notably, actions here are confined to *verb* predictions only without *noun*, following previous conventions. We calculate the mAP on the validation set averaged over all R, examining various performance aspects, such as all target actions (ALL), frequently occurring actions (FREQ), and rarely occurring actions (RARE).

Results. As shown in Table 6, our method achieves SOTA results in the ALL and FREQ section while exhibiting comparable performance in the RARE section.

Table 6. Comparison to the SOTA on EK-55 and EGTEA. Performance is evaluated using the mAP metric (↑).

Method	EK-55			EGTEA		
	ALL	FREQ	RARE	ALL	FREQ	RARE
I3D [8]	32.7	53.3	23.0	72.1	79.3	53.3
ActionVLAD [17]	29.8	53.5	18.6	73.3	79.0	58.6
Timeception [22]	35.6	55.9	26.1	74.1	79.7	59.7
VideoGraph [23]	22.5	49.4	14.0	67.7	77.1	47.2
EGO-TOPO [38]	38.0	56.9	29.2	73.5	80.7	54.7
Anticipatr [41]	39.1	58.1	29.1	76.8	83.3	55.1
AntGPT [63]	40.1	58.8	**31.9**	80.2	84.8	72.9
PALM (ours)	**40.4**	**59.3**	30.3	**80.7**	**85.0**	**73.5**

The results demonstrate the effectiveness of PALM by capturing additional visual information with image captioning models. This aligns with the ablation study regarding prompt contents in Sect. 4.3 where captions contribute better on verb predictions. We observe that verb in RARE section are seldom represented in the captions, which results in relatively lower mAP in RARE section. Thus, we believe fine-tuning image captioning model or employing better video captioning model will further improve the performance accordingly.

5 Conclusion

We present PALM, a framework that leverages vision-language and large language models for long-term action anticipation. PALM achieves the best performance on the Ego4D benchmark, demonstrating the effectiveness of performing the LTA task in the language domain and the benefits of using foundation models. Our model also shows competitive results on two other benchmarks, EK-55 and EGTEA. As a result, we are optimistic that our work will spur further research in using commonsense knowledge extracted from LLMs for video understanding.

References

1. Ahn, M., et al.: Do as i can, not as i say: grounding language in robotic affordances. arXiv preprint arXiv:2204.01691 (2022)
2. Ashutosh, K., Girdhar, R., Torresani, L., Grauman, K.: HierVL: learning hierarchical video-language embeddings. In: Proceedings of the IEEE/CVF Conference on Computer Vision and Pattern Recognition, pp. 23066–23078 (2023)
3. Black, S., Gao, L., Wang, P., Leahy, C., Biderman, S.: GPT-NEO: large scale autoregressive language modeling with mesh-tensorflow. If you use this software, please cite it using these metadata, vol. 58 (2021)

4. Black, S., Gao, L., Wang, P., Leahy, C., Biderman, S.: GPT-NEO: large scale autoregressive language modeling with mesh-tensorflow (2021). https://doi.org/10.5281/zenodo.5297715, If you use this software, please cite it using these metadata
5. Bommasani, R., et al.: On the opportunities and risks of foundation models. arXiv preprint arXiv:2108.07258 (2021)
6. Brown, T., et al.: Language models are few-shot learners. Adv. Neural. Inf. Process. Syst. **33**, 1877–1901 (2020)
7. Carbonell, J., Goldstein, J.: The use of MMR, diversity-based reranking for reordering documents and producing summaries. In: Proceedings of the 21st Annual International ACM SIGIR Conference on Research and Development in Information Retrieval, pp. 335–336 (1998)
8. Carreira, J., Zisserman, A.: Quo vadis, action recognition? A new model and the kinetics dataset. In: proceedings of the IEEE Conference on Computer Vision and Pattern Recognition, pp. 6299–6308 (2017)
9. Chen, G., et al.: Videollm: modeling video sequence with large language models. arXiv preprint arXiv:2305.13292 (2023)
10. Cho, J., Lei, J., Tan, H., Bansal, M.: Unifying vision-and-language tasks via text generation. In: International Conference on Machine Learning, pp. 1931–1942. PMLR (2021)
11. Dai, W., et al.: Instructblip: towards general-purpose vision-language models with instruction tuning (2023)
12. Damen, D., et al.: The epic-kitchens dataset: collection, challenges and baselines. IEEE Trans. Pattern Anal. Mach. Intell. **43**(11), 4125–4141 (2020)
13. Damerau, F.J.: A technique for computer detection and correction of spelling errors. Commun. ACM **7**(3), 171–176 (1964)
14. Das, S., Ryoo, M.S.: Video+ clip baseline for ego4d long-term action anticipation. arXiv preprint arXiv:2207.00579 (2022)
15. Fan, H., et al.: Multiscale vision transformers. In: Proceedings of the IEEE/CVF International Conference on Computer Vision, pp. 6824–6835 (2021)
16. Feichtenhofer, C., Fan, H., Malik, J., He, K.: Slowfast networks for video recognition. In: Proceedings of the IEEE/CVF International Conference on Computer Vision, pp. 6202–6211 (2019)
17. Girdhar, R., Ramanan, D., Gupta, A., Sivic, J., Russell, B.: Actionvlad: learning spatio-temporal aggregation for action classification. In: Proceedings of the IEEE Conference on Computer Vision and Pattern Recognition, pp. 971–980 (2017)
18. Grauman, K., et al.: Ego4D: around the world in 3,000 hours of egocentric video. In: Proceedings of the IEEE/CVF Conference on Computer Vision and Pattern Recognition, pp. 18995–19012 (2022)
19. Hu, E.J., et al.: Lora: low-rank adaptation of large language models. In: International Conference on Learning Representations (2021)
20. Huang, D., Hilliges, O., Van Gool, L., Wang, X.: Palm: predicting actions through language models@ ego4d long-term action anticipation challenge 2023. arXiv preprint arXiv:2306.16545 (2023)
21. Huang, W., Abbeel, P., Pathak, D., Mordatch, I.: Language models as zero-shot planners: extracting actionable knowledge for embodied agents. In: International Conference on Machine Learning, pp. 9118–9147. PMLR (2022)
22. Hussein, N., Gavves, E., Smeulders, A.W.: Timeception for complex action recognition. In: Proceedings of the IEEE/CVF Conference on Computer Vision and Pattern Recognition, pp. 254–263 (2019)
23. Hussein, N., Gavves, E., Smeulders, A.W.: Videograph: recognizing minutes-long human activities in videos. arXiv preprint arXiv:1905.05143 (2019)

24. Ichter, B., et al.: Do as i can, not as i say: grounding language in robotic affordances. In: 6th Annual Conference on Robot Learning (2022)
25. Keunwoo Peter, Y.: Videoblip is a large vision-language model based on blip-2 that can generate texts conditioned on videos (2021). https://github.com/yukw777/VideoBLIP, If you use this software, please cite it using these metadata
26. Kirkpatrick, J., et al.: Overcoming catastrophic forgetting in neural networks. Proc. Nat. Acad. Sci. **114**(13), 3521–3526 (2017)
27. Leo, M., Medioni, G., Trivedi, M., Kanade, T., Farinella, G.M.: Computer vision for assistive technologies. Comput. Vis. Image Underst. **154**, 1–15 (2017)
28. Levenshtein, V.I., et al.: Binary codes capable of correcting deletions, insertions, and reversals. In: Soviet Physics Doklady, vol. 10, pp. 707–710. Soviet Union (1966)
29. Li, J., Li, D., Savarese, S., Hoi, S.: Blip-2: bootstrapping language-image pretraining with frozen image encoders and large language models. arXiv preprint arXiv:2301.12597 (2023)
30. Li, J., Li, D., Xiong, C., Hoi, S.: Blip: bootstrapping language-image pre-training for unified vision-language understanding and generation. In: International Conference on Machine Learning, pp. 12888–12900. PMLR (2022)
31. Li, L.H., Yatskar, M., Yin, D., Hsieh, C.J., Chang, K.W.: Visualbert: a simple and performant baseline for vision and language. arXiv preprint arXiv:1908.03557 (2019)
32. Li, S., et al.: Pre-trained language models for interactive decision-making. Adv. Neural. Inf. Process. Syst. **35**, 31199–31212 (2022)
33. Li, Y., Liu, M., Rehg, J.M.: In the eye of beholder: joint learning of gaze and actions in first person video. In: Proceedings of the European Conference on Computer Vision (ECCV), pp. 619–635 (2018)
34. Lin, K.Q., et al.: Egocentric video-language pretraining. Adv. Neural Inf. Process. Syst. **35**, 7575–7586 (2022)
35. Lu, J., Batra, D., Parikh, D., Lee, S.: Vilbert: pretraining task-agnostic visiolinguistic representations for vision-and-language tasks. Adv. Neural Inf. Process. Syst. **32** (2019)
36. Mascaró, E.V., Ahn, H., Lee, D.: Intention-conditioned long-term human egocentric action anticipation. In: Proceedings of the IEEE/CVF Winter Conference on Applications of Computer Vision, pp. 6048–6057 (2023)
37. Meditskos, G., Plans, P.M., Stavropoulos, T.G., Benois-Pineau, J., Buso, V., Kompatsiaris, I.: Multi-modal activity recognition from egocentric vision, semantic enrichment and lifelogging applications for the care of dementia. J. Vis. Commun. Image Represent. **51**, 169–190 (2018)
38. Nagarajan, T., Li, Y., Feichtenhofer, C., Grauman, K.: Ego-topo: environment affordances from egocentric video. In: Proceedings of the IEEE/CVF Conference on Computer Vision and Pattern Recognition, pp. 163–172 (2020)
39. Nakazawa, A., Honda, M.: First-person camera system to evaluate tender dementia-care skill. In: Proceedings of the IEEE/CVF International Conference on Computer Vision Workshops (2019)
40. Naveed, H., et al.: A comprehensive overview of large language models. arXiv preprint arXiv:2307.06435 (2023)
41. Nawhal, M., Jyothi, A.A., Mori, G.: Rethinking learning approaches for long-term action anticipation. In: Avidan, S., Brostow, G., Cissé, M., Farinella, G.M., Hassner, T. (eds.) Computer Vision – ECCV 2022. ECCV 2022. LNCS, vol. 13694, pp. 558–576. Springer, Cham (2022). https://doi.org/10.1007/978-3-031-19830-4_32

42. Neumann, L., Zisserman, A., Vedaldi, A.: Future event prediction: if and when. In: Proceedings of the IEEE/CVF Conference on Computer Vision and Pattern Recognition Workshops (2019)
43. OhnBar, E., Kitani, K., Asakawa, C.: Personalized dynamics models for adaptive assistive navigation systems. In: Conference on Robot Learning, pp. 16–39. PMLR (2018)
44. Pasca, R.G., et al.: Summarize the past to predict the future: natural language descriptions of context boost multimodal object interaction anticipation. In: Proceedings of the IEEE/CVF Conference on Computer Vision and Pattern Recognition, pp. 18286–18296 (2024)
45. Patel, D., Eghbalzadeh, H., Kamra, N., Iuzzolino, M.L., Jain, U., Desai, R.: Pretrained language models as visual planners for human assistance. arXiv preprint arXiv:2304.09179 (2023)
46. Pramanick, S., et al.: EgoVLPv2: egocentric video-language pre-training with fusion in the backbone. In: Proceedings of the IEEE/CVF International Conference on Computer Vision, pp. 5285–5297 (2023)
47. Radford, A., et al.: Learning transferable visual models from natural language supervision. In: International Conference on Machine Learning, pp. 8748–8763. PMLR (2021)
48. Radford, A., Wu, J., Child, R., Luan, D., Amodei, D., Sutskever, I., et al.: Language models are unsupervised multitask learners (2019)
49. Ragusa, F., Farinella, G.M., Furnari, A.: Stillfast: an end-to-end approach for short-term object interaction anticipation. In: Proceedings of the IEEE/CVF Conference on Computer Vision and Pattern Recognition, pp. 3635–3644 (2023)
50. Rodin, I., Furnari, A., Mavroeidis, D., Farinella, G.M.: Predicting the future from first person (egocentric) vision: a survey. Comput. Vis. Image Underst. **211**, 103252 (2021)
51. Ryoo, M.S., Fuchs, T.J., Xia, L., Aggarwal, J.K., Matthies, L.: Robot-centric activity prediction from first-person videos: what will they do to me? In: Proceedings of the Tenth Annual ACM/IEEE International Conference on Human-Robot Interaction, pp. 295–302 (2015)
52. Song, K., Tan, X., Qin, T., Lu, J., Liu, T.Y.: MPNet: masked and permuted pretraining for language understanding. Adv. Neural. Inf. Process. Syst. **33**, 16857–16867 (2020)
53. Touvron, H., et al.: LLAMA: open and efficient foundation language models. arXiv preprint arXiv:2302.13971 (2023)
54. Touvron, H., et al.: Llama 2: open foundation and fine-tuned chat models. arXiv preprint arXiv:2307.09288 (2023)
55. Wang, S., Zhao, Q., Do, M.Q., Agarwal, N., Lee, K., Sun, C.: Vamos: versatile action models for video understanding. arXiv preprint arXiv:2311.13627 (2023)
56. Wang, Y., et al.: Internvideo: general video foundation models via generative and discriminative learning. arXiv preprint arXiv:2212.03191 (2022)
57. Wang, Z., Yu, J., Yu, A.W., Dai, Z., Tsvetkov, Y., Cao, Y.: Simvlm: simple visual language model pretraining with weak supervision. In: International Conference on Learning Representations (2021)
58. Yao, Y., Xu, M., Choi, C., Crandall, D.J., Atkins, E.M., Dariush, B.: Egocentric vision-based future vehicle localization for intelligent driving assistance systems. In: 2019 International Conference on Robotics and Automation (ICRA), pp. 9711–9717. IEEE (2019)

59. Ye, X., Iyer, S., Celikyilmaz, A., Stoyanov, V., Durrett, G., Pasunuru, R.: Complementary explanations for effective in-context learning. arXiv preprint arXiv:2211.13892 (2022)
60. Zeng, A., et al.: Socratic models: composing zero-shot multimodal reasoning with language. In: The Eleventh International Conference on Learning Representations (2022)
61. Zhan, K., Faux, S., Ramos, F.: Multi-scale conditional random fields for first-person activity recognition. In: 2014 IEEE International Conference on Pervasive Computing and Communications (PerCom), pp. 51–59. IEEE (2014)
62. Zhang, S., et al.: OPT: open pre-trained transformer language models. arXiv preprint arXiv:2205.01068 (2022)
63. Zhao, Q., et al.: AntGPT: can large language models help long-term action anticipation from videos? arXiv preprint arXiv:2307.16368 (2023)
64. Zhou, C., et al.: A comprehensive survey on pretrained foundation models: a history from bert to chatgpt. arXiv preprint arXiv:2302.09419 (2023)
65. Zhou, L., Palangi, H., Zhang, L., Hu, H., Corso, J., Gao, J.: Unified vision-language pre-training for image captioning and vqa. In: Proceedings of the AAAI Conference on Artificial Intelligence, vol. 34, pp. 13041–13049 (2020)

Motion Keyframe Interpolation for Any Human Skeleton via Temporally Consistent Point Cloud Sampling and Reconstruction

Clinton Mo[1], Kun Hu[1(✉)], Chengjiang Long[2], Dong Yuan[1], and Zhiyong Wang[1]

[1] The University of Sydney, Darlington, NSW 2008, Australia
clmo6615@uni.sydney.edu.au, {kun.hu,dong.yuan,zhiyong.wang}@sydney.edu.au
[2] Meta Reality Labs, Burlingame, CA, USA
clong1@meta.com

Abstract. In the character animation field, modern supervised keyframe interpolation models have demonstrated exceptional performance in constructing natural human motions from sparse pose definitions. As supervised models, large motion datasets are necessary to facilitate the learning process; however, since motion is represented with fixed hierarchical skeletons, such datasets are incompatible for skeletons outside the datasets' native configurations. Consequently, the expected availability of a motion dataset for desired skeletons severely hinders the feasibility of learned interpolation in practice. To combat this limitation, we propose Point Cloud-based Motion Representation Learning (PC-MRL), an *unsupervised* approach to enabling cross-compatibility between skeletons for motion interpolation learning. PC-MRL consists of a skeleton obfuscation strategy using *temporal point cloud sampling*, and an *unsupervised skeleton reconstruction* method from point clouds. We devise a temporal point-wise K-nearest neighbors loss for unsupervised learning. Moreover, we propose First-frame Offset Quaternion (FOQ) and Rest Pose Augmentation (RPA) strategies to overcome necessary limitations of our unsupervised point cloud-to-skeletal motion process. Comprehensive experiments demonstrate the effectiveness of PC-MRL in motion interpolation for desired skeletons without supervision from native datasets.

Keywords: 3D point clouds · Human body motion · Dataset creation

1 Introduction

3D character animation workflows largely rely on the concepts of keyframing with interpolation, often referred to as the pose-to-pose principle [13]. By defining key poses at correct timings, algorithms can be employed to generate intermediate poses, and thereby eliminating the need for defining each frame

---

Supplementary Information The online version contains supplementary material available at https://doi.org/10.1007/978-3-031-73007-8_10.

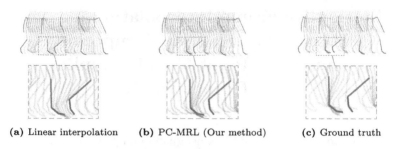

Fig. 1. Motion interpolation of walking keyframes conducted by our approach.

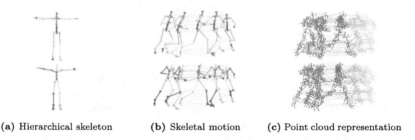

Fig. 2. Our method leverages point cloud representations (c) to obfuscate the hierarchical dependencies of skeletal motion. Joint colours correspond to ordinal indices in original motion data representation. Point clouds are coloured by temporal index.

individually [5,42]. Although this keyframe-based workflow is less costly than frame-by-frame production, human motion is often complex, and natural depictions demand a significant number of keyframes, especially for realistic motions that adhere to physical properties and constraints. Although motion capture (MoCap) is frequently employed as an alternative for achieving realistic human motions, MoCap often nonetheless necessitates subsequent keyframing to correct unwanted motion elements and address other imperfections.

In recent years, the interpolation process has been widely studied, particularly with the advent of machine learning methods for sequential data. Compared to conventional methods such as linear interpolation (LERP), machine learning methods have demonstrated the capability to derive visually natural motions from sparser keyframe sets [12,18,34,37,40], which we can observe in Fig. 1. Data-driven interpolation approaches require large motion capture datasets to learn the necessary motion features for effective synthesis of transitions between keyframes. In the field of 3D motions, this aspect significantly restricts the feasibility of learned interpolation methods in animation practice, as motion/pose data are tied to specific skeletal configurations and are not cross-compatible. That is, the dataset's native skeleton, i.e. *source skeleton*, is structurally different and incompatible with the desired skeleton, i.e. *target skeleton*.

Therefore, we propose a novel *point cloud-based motion representation learning* (PC-MRL) approach, in which skeletal hierarchies are obfuscated by sampling into the point cloud medium, and reconstructed into skeletal motion data using an *unsupervised* neural network. Unlike raw pose and motion data, the point cloud format is a non-hierarchic representation, and effectively obscures any skeletal configuration as displayed in Fig. 2. PC-MRL samples a set of points around the rest position of skeletons using a combination of uniform and normal distributions, and geometrically parents each point to an associated bone in order to effectively represent the associated skeletons' motion data in a temporally consistent manner. Subsequently, for the reconstruction of motion data into a given desired skeleton, we propose an *unsupervised* learning scheme, featuring a K-nearest neighbors (KNN) loss between cross-skeleton point clouds to optimize a transformer neural network. The network processes the point cloud-based motion representations and generates the corresponding target skeletal motion representations. By obscuring both source and predicted target motion data into the point cloud space, our strategy allows for a direct comparison, utilizing KNN to minimize the visual difference between the two point clouds.

The unsupervised nature of our target skeleton reconstruction scheme creates two main disparities between the expected and predicted motion features. Firstly, the exact roll rotation axis of bones cannot be represented in the point cloud format. To this end, we introduce *First-frame Offset Quaternion* (FOQ) representations to incorporate relative roll values, which are obtainable from *temporally consistent* point clouds. This approach is agnostic to absolute roll rotations, providing a standardisation mechanism for the resulting motion data within the context of motion interpolation learning. Second, KNN-based objectives learn skeletal representations in a geometrically optimal manner, which does not always reflect the desired skeletal behaviour. This disparity is commonly observed with smaller skeletal features, such as shoulder and hip bones. As such, we augment training motions using Rest Pose Augmentation (RPA) to increase the range of skeletal behaviours that our interpolation model learns. Extensive experiments demonstrate the effectiveness of our proposed method, achieving performance targets near the level of direct dataset supervision.

In summary, this paper presents the following key contributions:

1. We propose a novel method for achieving unsupervised human motion reconstruction from point cloud data, enabling skeleton-agnostic motion interpolation learning for the first time.
2. To train motion interpolation models, we formulate first-frame offset quaternions to represent bone rotations with relative roll data, as well as a rest position augmentation strategy to address skeletal configuration variability.
3. We perform comprehensive experiments to demonstrate our method's effectiveness towards learning motion interpolation, despite the absence of directly compatible datasets in the training process.

2 Related Works

We explore existing research on motion interpolation approaches, and *motion data modelling* methods in general. In addition, our proposed approach bears resemblance to *unsupervised approaches* for the motion re-targeting research problem, and as such, we will also analyse existing methods in this field.

2.1 Data-Driven Motion Modelling Methods

Contemporary research in motion modelling predominantly explore learned data-driven methods, using neural networks to capture, model, and establish correlations between the motion features of various skeletal configurations [29,35,41,51,55]. Motion features alone have demonstrated considerable efficacy in cross-skeleton imitation using a variety of strategies, including task-driven objectives [19,45], diffusion-based generation [25,49,52], and latent feature consistency [1–3,7,46]. Latent consistency in particular learns inter-skeleton correlations without the need for paired datasets, which reduces its barrier of entry.

Likewise, motion interpolation strategies have evolved through the use of learning based approaches. The inherent numerical precision of motion data has traditionally posed a challenge for the adoption of the learning approaches, due to their approximate nature [21,27]. This challenge is further amplified by the temporally sparse distribution of keyframes. Recently, a number of recurrent neural network based methods [17,18,43,53] and transformer-based networks [12,34,37] have demonstrated encouraging interpolation performance for the transition between keyframes. Additionally, various other motion modelling methods are capable for keyframe based interpolation, albeit with limited intermediate frame representation capabilities [19,25,45].

A major constraint inherent in all data-driven motion modelling methods is their dependency on datasets featuring specific skeletal configurations. To address this, skeleton-free training strategies have been explored for elementary tasks involving 3D characters. These include vertex-based pose transfer [7,8,31,47], and point cloud-based human shape reconstruction [23,48]. Notably, in these methods, 3D mesh and point cloud coordinates serve as general 3D data representations. Concurrently, other research has significantly minimized the required volume of training data, bringing it down to single example sequences. This reduction has been achieved through the use of patch matching techniques [28] and imitation learning within physics-based simulations [30,39]. Reinforcement learning has facilitated the generation of goal-driven motion without pre-existing datasets [32,38,50]. However, these simulation-based methods have typically been either exceedingly complex to implement in practice, or yield results that are too inflexible and/or prone to errors for animation workflows.

2.2 Motion Re-targeting Without Supervision

Conventional approaches in motion re-targeting have been a cornerstone of 3D animation for decades, primarily dedicated to converting motion capture data into 3D skeletal motions [4,15]. Typically, raw motion capture data is optically recorded, yielding primarily spatial information. In contrast, motion data is largely rotational in nature [5], often represented in $SO(3)$ space. Therefore, a principal objective of motion re-targeting solutions has been to effectively bridge these two distinct types of representations. In [14], extensive manually adjustable constraints were introduced for flexible re-targeting between skeletons of identical topology (i.e., same hierarchy, different bone lengths and rest poses). Key

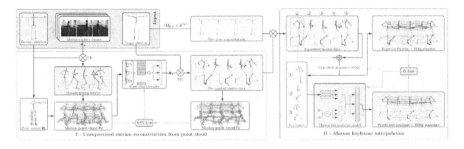

Fig. 3. Illustration of point cloud-based motion representation learning (PC-MRL) pipeline for keyframe interpolation to any human skeleton. Our unsupervised point cloud-to-pose decoding approach **(I)** first adapts motion features from existing datasets to new skeletons. We then employ the adapted motion data to supervise motion interpolation models **(II)**, using FOQ and rest pose augmentations to address remaining discrepancies between the adapted and expected target skeleton motion patterns.

practices in this approach, including end-effector matching, joint contacts, and inter-skeleton point correlations, are still relevant to current animation pipelines. Point correlations using Inverse Kinematics (IK) have also been proposed as an intermediate representation to facilitate the adaptation between topologically distinct skeletons [9,36], albeit with hierarchical pairing limitations.

In response, morphologically-independent motion control systems were proposed using IK joint chains and joint-wise constraints [20,26]. Such systems often expect highly standardised motion skeleton features, and are generally impractical for re-targeting motions produced for different control schemes. Additionally, the reliance on Euler angles for extensive classical re-targeting constraints poses a variety of issues regarding rotational freedom [11], as well as compatibility with animation workflows [33], which predominantly use quaternion-based systems.

Despite these advancements, the challenge of arbitrary skeleton re-targeting from motion datasets still remains largely unresolved, limiting any widespread application of data-driven pipelines. Our paper aims to effectively address this pervasive limitation.

3 Methodology

As illustrated in Fig. 3, our proposed method consists of two main components: *point cloud-to-motion reconstruction* and *motion keyframe interpolation*.

3.1 Quaternion-Based Motion Representations

We first declare a set of notations for representing skeletons. A skeleton S is defined by a tree of bones $\mathcal{B}$. A bone $\mathbf{v}_n \in \mathcal{B}$ is defined by its bone parent $\mathbf{v}_m \in \mathcal{B}$, and a *head position* $\mathbf{H}_n \in \mathbb{R}^3$ which denotes the bone's base offset from its parent head $\mathbf{H}_m$. In addition, a *tail position* $\mathbf{T}_n \in \mathbb{R}^3$ helps to define the end point of the bone's line segment representation, starting from its head position, of $\mathbf{v}_n$. The root bone $\mathbf{v}_0$ has no parent, and thus no head position $\mathbf{H}_0$. The set of $\mathbf{H}_n$ and $\mathbf{T}_n$ for each bone $\mathbf{v}_n \in \mathcal{B}$ make up the *rest position* of the skeleton.

Next, we define the properties of motion data. At any given frame t, a skeleton's *pose position* can be represented as global *3D positions* $\mathbf{p}_{n,t} \in \mathbb{R}^3$ and global *quaternion rotations* $\mathbf{q}_{n,t} \in \mathbb{R}^4$ for each bone $\mathbf{v}_n \in \mathcal{B}$. For a given motion sequence M, only the root bone is given a 3D position in pose position, i.e. $\mathbf{p}_{0,t}$, as all other pose positions are derived using Forwards Kinematics (FK [11,24]):

$$\mathbf{p}_{n,t} = \mathbf{p}_{m,t} + QR(\mathbf{H}_n, \mathbf{q}_{m,t}), \tag{1}$$

where QR is the quaternion rotation function, and $\mathbf{v}_m$ is the parent joint. In summary, a motion sequence M of $|M|$ frames is defined by:

$$M = \{\mathbf{p}_{0,t} \cup \{\mathbf{q}_{n,t} \mid \mathbf{v}_n \in \mathcal{B}\} \mid 0 \leq t < |M|\} \tag{2}$$

3.2 Point Cloud Obfuscation and Skeleton Reconstruction

Our approach is based on the hypothesis that 3D motion sequences can be visually represented by a more universal format: 3D point cloud sequences. By geometrically sampling the volume around a skeleton, point cloud data becomes a non-hierarchical representation of its pose(s), obfuscating the original skeletal configuration. Due to this decoupling between the skeleton configuration and motion representation, a successful motion data reconstruction from point cloud data would be the first step to enabling learnable cross-skeleton motion features. Formally, for a given source skeleton S_A with an associated motion sequence M_A, and any humanoid skeletal configuration S_B, the point cloud obfuscation and reconstruction module is to learn a function that can adequately perform $F_{S_A \to S_B}(M_A) = \hat{M}_B$, where $\hat{M}_B$ is a visually similar motion adaptation of M_A on skeleton S_B, which is an estimation of M_B.

Motion Data Obfuscation with Point Clouds. To sample the point cloud $\mathcal{X}$, we first generate a set of points $u \in \mathcal{X}$ along the line segments defined by the global head and tail positions for each bone $\mathbf{v}_n$. In the bone-local space to $\mathbf{v}_n$, this would mean sampling u by a uniformly distributed factor $\alpha \sim \mathcal{U}_{[0,1]}$. Specifically, $\alpha = 0$ aligns with the head position of the bone, and $\alpha = 1$ indicates the tail position. By further sampling u via a normal distribution, our sampling strategy can capture the surrounding volume around the skeleton. Formally, to sample u from its associated bone $\mathbf{v}_n$ within a standard deviation of σ, we have:

$$\begin{aligned} \mathbf{H}_x &\sim \mathcal{N}(\alpha \mathbf{T}_n, \sigma), \\ \mathbf{p}_{u,t} &= \mathbf{p}_{n,t} + QR(\mathbf{H}_x, \mathbf{q}_{n,t}) \end{aligned} \tag{3}$$

Notably, this sampling process as in Eq. 3 is associated with the skeletal configuration only, by which the sampled points are temporally consistent for further processing with the motion sequence. Specifically, each point maintains a constant position from its associated bone in the bone-local space. The lateral distance of each point from their parent bone segment allows point clouds to react to the bone's rotational roll axis.

In a pure point cloud representation, it is possible to remove the relationships between bones. However, to clearly distinguish symmetrical body features in different skeletons, we introduce and embed body group associations for the points during our reconstruction process. Specifically, every bone is categorised by one of these body groups: *[spine, left arm, right arm, left leg, right leg]*. Each sampled point is assigned q group attribute in line with the associated bone. The attribute is represented by a one-hot feature vector $\mathbf{g}_u$. To this end, we characterize $u = \{\mathbf{p}_{u,t}, \mathbf{g}_u\} \in \mathcal{X}$.

Skeleton Reconstruction. Given a sampled point cloud representation $\mathcal{X}_A$ from S_A and M_A as input, we aim to train a motion adaptation function $F_{S_A \to S_B}$, which is tasked with producing a visually similar skeletal motion $\hat{M}_B$ for S_B, using a temporal and set-based neural network. As shown in Fig. 3, to construct $F_{S_A \to S_B}$, we employ a Point Transformer [54] to learn embeddings for the unordered point cloud sets frame-wisely, and a standard temporal transformer to process the sequence and decode $\mathcal{X}_A$ into $\hat{M}_B$.

To optimize $F_{S_A \to S_B}$, our objective function maximises the similarity between the sampled point clouds of both M_A and $\hat{M}_B$. $\mathcal{X}_A$ can be viewed as the ground truth to $\mathcal{X}_B$, which enables an unsupervised strategy for $F_{S_A \to S_B}$. In detail, we first sample a point cloud $\mathcal{X}_B$ from S_B based on $\hat{M}_B$ derived by $F_{S_A \to S_B}$. Next, a **temporally consistent KNN-Loss objective** is devised to minimise the ℓ_2 distance between any given point of $\mathcal{X}_B$ from its K nearest neighbouring points in $\mathcal{X}_A$. The K nearest neighbours are determined based on inter-point distance throughout the entire motion sequence, instead of a frame-wise setting, as shown in Fig. 4. The points with differing body groups are excluded from the neighbour search. Mathematically, we have:

$$\delta(u_A, u_B) = \begin{cases} \sum_t ||\mathbf{p}_{u_A,t} - \mathbf{p}_{u_B,t}||_2, & \text{if } \mathbf{g}_{u_A} = \mathbf{g}_{u_B}, \\ \inf, & \text{otherwise,} \end{cases} \quad (4)$$

$$\mathcal{L}_{\text{KNN}}(\mathcal{X}_A, \mathcal{X}_B) = \sum_{u_A \in \mathcal{X}_A, u_B \in N_k(u_A, \mathcal{X}_B)} \delta(u_A, u_B), \quad (5)$$

where $N_k(u_A, \mathcal{X}_B)$ is the set containing points that are the k-nearest neighbors of u_A in $\mathcal{X}_B$, based on $\delta(u_A, u_B)$ distance.

In addition to the KNN-Loss, we introduce an optional end-effector loss in the skeleton space for the hand, foot, and head bones, which provides direct positional guidance for matched end-effector pairs E_{AB}:

$$\mathcal{L}_{\text{end}} = \sum_{(\mathbf{v}_A, \mathbf{v}_B) \in E_{AB}} \frac{1}{|M_A|} \sum_t ||\mathbf{p}_{\mathbf{v}_A,t} - \mathbf{p}_{\mathbf{v}_B,t}||_2. \quad (6)$$

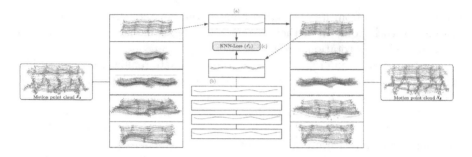

Fig. 4. Visualisation of our temporally consistent KNN objective, processed independently per body group. For each point from $\mathcal{X}_A$ (a), the total ℓ_2 distance throughout the sequence is measured between each point (b) from $\mathcal{X}_B$. The ℓ_2 distances from the K-nearest points are summed to produce KNN-Loss (c).

3.3 First-Frame Offset Quaternions

Point clouds lack absolute roll axis information for each bone. As such, we introduce a First-frame Offset Quaternion (FOQ) transform to incorporate the available *relative roll* data for motion modelling. We formulate all quaternions of a motion sequence relative to their values from the initial frame. In detail, for any quaternion sequence $Q = \{\mathbf{q}_0, \mathbf{q}_1, \mathbf{q}_2, ...\}$, we can simply obtain its FOQ transform as $\{\mathbf{q}_0 \times \mathbf{q}_0^{-1}, \mathbf{q}_1 \times \mathbf{q}_0^{-1}, \mathbf{q}_2 \times \mathbf{q}_0^{-1}, ...\}$, where $\mathbf{q}_0^{-1}$ is the quaternion conjugate of $\mathbf{q}_0$. An FOQ sequence can easily be converted back to a quaternion sequence as long as any $\mathbf{q}_t \in Q$ is known.

Crucially, FOQ is a **roll-invariant** rotation representation, that is, FOQ remains identical regardless of a bone's absolute roll position in the original data. Trivially, to adjust the rotational roll of a bone $\mathbf{v}_n$, we perform the quaternion multiplication for each frame: $\mathbf{q}_{n,t} \times \mathbf{r}$, where $\mathbf{r}$ is a roll quaternion. As a quaternion on the roll axis of $\mathbf{v}_n$, $\mathbf{r}$ is constrained within $\mathbf{r} = \alpha + \beta(x\mathbf{i} + y\mathbf{j} + z\mathbf{k})$, where x,y,z are the 3D XYZ values of $\mathbf{T}_n$, and α, β are scalars that control the roll magnitude. Note that $\mathbf{r}$ is constant throughout the sequence. Therefore, we can deduce the roll-invariant nature of any constant $\mathbf{r}$:

$$(\mathbf{q}_{n,t} \times \mathbf{r}) \times (\mathbf{q}_{0,t} \times \mathbf{r})^{-1} = \mathbf{q}_{n,t} \times \mathbf{r} \times \mathbf{r}^{-1} \times \mathbf{q}_{0,t}^{-1}$$
$$= \mathbf{q}_{n,t} \mathbf{q}_{0,t}^{-1}$$

3.4 Motion Interpolation Transformer and Rest Pose Augmentation

To leverage our point cloud-based representation learning for cross-skeleton interpolation, a transformer model - CITL [34] is adapted, allowing for an efficient interpolation pipeline with existing datasets to alternative skeletons. We introduce two key modifications to CITL and its training strategy, focusing on enhancing motion quality. First, quaternion predictions are substituted with FOQ predictions. This is inspired by the RNN-based method $TG_{complete}$ [18], which predicts quaternion offsets from the preceding frame rather than raw quaternions, thereby achieving greater generalisability and superior quality in interpolation.

Algorithm 1 IK target alignment with maximum real quaternion component

1: **Input:** $\mathbf{p}_{IK}$ = 3D IK target unit vector
2: $\qquad \mathbf{T}$ = resting tail position of associated bone
3: **Output:** $\mathbf{q}_{max}$ = output quaternion, to be applied on original bone quaternion
4: $\eta \leftarrow \frac{\mathbf{T}}{|\mathbf{T}|}$ $\qquad\qquad\qquad\qquad\qquad\qquad$ ▷ 3D unit coordinates $[\eta_x, \eta_y, \eta_z]$
5: $\omega \leftarrow \frac{\mathbf{p}_{IK}}{|\mathbf{p}_{IK}|}$ $\qquad\qquad\qquad\qquad\qquad$ ▷ 3D unit coordinates $[\omega_x, \omega_y, \omega_z]$
6: $\mathbf{q}_{min} \leftarrow \frac{0 + (\eta_x + \omega_x)\mathbf{i} + (\eta_y + \omega_y)\mathbf{j} + (\eta_z + \omega_z)\mathbf{k}}{2}$
7: $\mathbf{q}_{max} \leftarrow \mathbf{q}_{min} \times (0 + \eta_x \mathbf{i} + \eta_y \mathbf{j} + \eta_z \mathbf{k})$

For our model's CITL-based architecture, we substitute the original sinusoidal positional encoding scheme with learned relative positions [44]. While the original implementation relies on the continual nature of sinusoidal functions, we observe that relative positional embeddings with *zero initialisation* converges upon similarly continuous behaviour, and significantly lowers the complexity of positional attention. The lessened focus on positional relations allows the model to synthesise deeper behaviours within the pose space, improving overall generalisability. As a side effect, the model also accepts arbitrary length inputs beyond the training data length.

Notably, the intended rest pose of a skeletal configuration is often geometrically sub-optimal for point cloud matching, and can manifest in multiple forms. To address this, we devise a **rest position augmentation strategy**, leveraging a per-bone Inverse Kinematics (IK) system. RPA broadens the range of intended rest poses that the learned interpolation model can accommodate. In detail, during each phase of the Forward Kinematics (FK) process, we augment the bone tails with a random additional offset. We re-align the augmented global tail position as closely as possible to its original location using a quaternion rotation with the maximum possible real component. Algorithm 1 describes the method by which this quaternion can be derived. Since the real component describes, in cosine, the amount of rotation around an axis, this minimizes numerical disturbance from RPA while altering the visual representation to the desired extent.

4 Experimental Results

4.1 Datasets

To thoroughly evaluate the effectiveness of the proposed method, we conducted extensive experiments on the following widely used motion capture datasets:

- **LaFAN1** [18] contains long, high quality motions in controllable video game character styles. The dataset was recorded on a large motion capture stage, and manually cleaned to production standards. The native skeleton of LaFAN1 contains 6 spinal bones, and 4 bones for each limb.
- **Human3.6M** [6,22] comprises of 3.6 million motion frames. The motions captured in Human3.6M tend to be less dynamic, as they were recorded on a

relatively compact 4 × 3 m stage. Its native skeleton includes 5 spinal bones, 4 bones for each limb, and an additional finger bone for each hand.
- **CMU Mocap** [10] is a diverse motion dataset encompassing 113 categories. It features a mix of static and dynamic motions, captured on a 3 × 8 m stage. The skeleton is identical to that of Human3.6M, with the exception of the pelvis, which is divided into two hip bones and a lower back bone.

For PC-MRL, we designated the CMU MoCap skeleton as S_B. Its native dataset is exclusively utilized for testing purposes, while LaFAN1 and Human3.6M are used simultaneously for training.[1]

4.2 Implementation Details

In our experiments, point cloud obfuscation adopted a 256-point setting for the sampling process with $\sigma = 0.05$ and each point was characterized by a 64-element vector. To maintain consistency, we ensure that all bone offsets and motion positions are defined in **metre** units. For skeletal reconstruction, 4 Point Transformer [54] layers was utilized to construct the neural network. Within each layer, the feature size was doubled, and the size of the point set was downsampled by a factor of 4 through a farthest-first traversal approach, eventually resulting in a singular feature vector. In the final step, two linear layers with ReLU activations were utilized to transform the resulted feature vector in a frame-wise manner, producing the output skeletal representation. During training, we set $K = 8$ for the KNN loss to optimize the network. Additionally, the use of unit quaternion values in practice relies on modulus division for constraint, a process that can potentially lead to exploding gradients. To address this, we introduced an objective function $\mathcal{L}_q$ that measures the ℓ_2 norms' difference between the raw quaternion output and the expected unit quaternion. All experiments were trained and evaluated on a single NVIDIA GeForce RTX 3090 GPU.

4.3 Motion Interpolation Comparison Against Supervised Methods

To demonstrate the efficacy of our method, we produce and compare our method against the conventional LERP method, as well as three state-of-the-art interpolation models trained on the original CMU MoCap dataset. Specifically, we measure the performance of the RNN-based TG$_{complete}$ model [18], the BERT-based motion interpolation adaptation [12], and a transformer-based encoder-decoder approach [34]. We additionally provide results on a variant of our method trained without our RPA strategy. The experimental configuration for each state-of-the-art model is identical to their original implementations. To measure interpolation accuracy, we employ the standard ℓ_2 positional distance (L2P), ℓ_2 quaternion difference (L2Q), and NPSS [16] for visual similarity evaluation [18].

---

[1] The authors Clinton Mo, Kun Hu and Zhiyong Wang are signatories of the dataset licenses and produced all experimental results of the paper. Meta Inc. was not granted access to the datasets.

Table 1. Performance comparisons for various motion categories, measured in L2P, L2Q, and NPSS. All methods *except PC-MRL* are trained on the original CMU MoCap dataset. Each motion contains 128 frames with keyframes placed every 5, 15, or 30 frames. The top 2 results for each test are highlighted in **purple** and **blue** respectively.

Motion category	Method	L2P↓ 5	15	30	L2Q↓ 5	15	30	NPSS↓ 5	15	30
Basketball	LERP	0.0787	0.4009	0.8033	0.1588	0.5834	0.9611	**0.1315**	0.6120	1.8202
	TG$_{complete}$ [18]	0.1050	0.4419	0.7104	0.1859	0.6439	0.9303	0.1600	0.6579	**1.2136**
	BERT [12]	0.0889	0.3737	0.6865	0.1585	0.5560	0.8992	0.1689	0.6177	1.4587
	CITL [34]	**0.0625**	**0.2973**	**0.5708**	**0.1410**	**0.4932**	**0.8011**	**0.1047**	**0.4098**	**1.0959**
	PC-MRL w/o RPA	0.0767	0.3534	0.6715	**0.1505**	**0.5431**	**0.9049**	0.1364	0.5596	1.3471
	PC-MRL (Ours)	**0.0752**	**0.3443**	**0.6662**	0.1561	0.5495	0.9088	0.1397	**0.5572**	1.3015
Golf	LERP	**0.0189**	0.1229	0.3187	**0.0455**	**0.1870**	0.3987	**0.0587**	0.3979	1.0312
	TG$_{complete}$ [18]	0.0473	0.2616	0.4546	0.0693	0.3191	0.5297	0.0954	0.5062	1.1534
	BERT [12]	0.0276	0.1159	0.2967	0.0612	0.2082	0.4258	0.1099	0.3906	0.9255
	CITL [34]	**0.0201**	**0.1043**	**0.2589**	**0.0476**	**0.1766**	**0.3552**	**0.0636**	**0.2581**	**0.7799**
	PC-MRL w/out RPA	0.0274	0.1304	0.3622	0.0555	0.2065	0.4923	0.0911	0.3759	1.2083
	PC-MRL (Ours)	0.0273	**0.1130**	**0.2706**	0.0598	0.2012	**0.3844**	0.1049	**0.3716**	**0.8150**
Swimming	LERP	**0.0731**	**0.3680**	**0.7374**	**0.1326**	**0.5327**	**0.9476**	0.2147	0.9222	1.6972
	TG$_{complete}$ [18]	0.1292	0.5796	0.9141	0.1845	0.7420	1.1199	0.2660	0.9716	1.7698
	BERT [12]	0.1088	0.3896	0.7516	0.1819	0.5518	0.9622	0.2877	0.9074	1.6682
	CITL [34]	0.0905	0.3895	0.7475	0.1484	0.5535	0.9533	0.2084	**0.8317**	1.6122
	PC-MRL w/o RPA	**0.0722**	0.3754	0.7459	**0.1345**	0.5442	0.9594	**0.1944**	0.8731	1.6608
	PC-MRL (Ours)	0.0734	**0.3595**	**0.7139**	0.1355	**0.5331**	**0.9248**	**0.1956**	**0.8591**	**1.6370**
Walking & Running Locomotion	LERP	0.0455	0.2505	0.5527	0.1120	0.3997	0.6233	0.0655	0.2682	0.9071
	TG$_{complete}$ [18]	0.0540	0.1786	0.3103	0.1147	0.2866	**0.3866**	0.0895	0.2616	**0.4265**
	BERT [12]	0.0552	0.1998	0.3544	0.1111	0.3101	0.4762	0.0901	0.2760	0.6458
	CITL [34]	**0.0328**	**0.1034**	**0.2427**	**0.0942**	**0.2102**	**0.3384**	**0.0526**	**0.1568**	**0.4080**
	PC-MRL w/o RPA	0.0397	0.1657	0.3417	**0.1058**	0.2879	0.4722	**0.0640**	0.2288	0.6021
	PC-MRL (Ours)	**0.0377**	**0.1392**	**0.2848**	0.1064	**0.2778**	0.4281	0.0658	**0.2129**	0.5098

The results listed in Table 1 clearly demonstrate our method's ability to approach the accuracy levels exhibited by directly supervised state-of-the-art methods. PC-MRL consistently outperforms both the LERP standard and RNN-based TG$_{complete}$ model, particularly in longer keyframe interval scenarios where the quality and depth of learned motion features is most critical and impactful. Likewise, the inclusion of RPA also tends to improve the long interval performance of PC-MRL, particularly when precise movements (i.e. golf) or deeper motion features (i.e. locomotion) are expected. Given this, we can observe that RPA definitively improves the overall consistency of the PC-MRL method in scenarios that may be unseen or less effectively represented by our point cloud-to-motion system.

Native supervision, i.e. with CITL, unsurprisingly produces the most accurate interpolation model, with a notable exception of swimming motions, where our PC-MRL supervision produces even stronger results. We believe the abundance of low crawl motions, a similar motion to swimming in LaFAN1, provides more meaningful supervision over the original CMU dataset, which conversely has few similar motions once swimming motions are filtered out. On the other end of the spectrum, i.e. for high data scenarios such as walking and running locomotion from Fig. 5, all models including our PC-MRL approach observe the strongest improvements for learned methods over the naive LERP method.

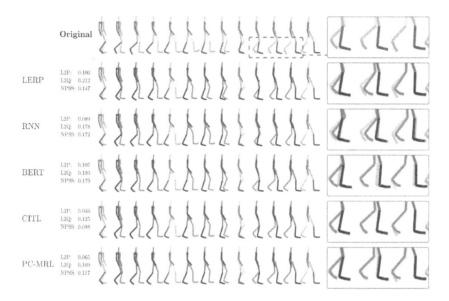

Fig. 5. Walking keyframe interpolation examples produced by methods from Table 1. Close-up layered comparisons against the original motion are provided on the right.

Both cases strongly support our method's capability to supervise highly intricate motion patterns, despite the incompleteness of the point cloud representation and reliance on relative rotations and geometric optimisation assumptions.

4.4 Cross-Skeleton Motion Re-targeting Experiments

We further conduct an experiment on motion re-targeting to directly benchmark our method against the sole existing state-of-the-art method for unpaired motion re-targeting with unrestricted skeletons, i.e. *primal skeletons* (PS) [1]. For direct motion re-targeting evaluation, we utilized the KNN-Loss and end-effector loss as metrics, owing to their effective measurement of visual similarity and non-sensitivity towards absolute roll axis correctness. Though absolute rolls

Table 2. Average unpaired LaFAN1 to CMU motion re-targeting performance of the primal skeleton method and our point cloud method. $\mathcal{L}_{\text{KNN}}$ is measured using 1024-point clouds at $K = 8$. Independently sampled point clouds from the original LaFAN1 skeleton are provided as optimal $\mathcal{L}_{\text{KNN}}$ reference. All values are scaled by ×100.

Model	$\mathcal{L}_{\text{KNN}}(\mathcal{X}_A, \mathcal{X}_B)$ ↓					$\mathcal{L}_{\text{end}}$ ↓
	$\sigma = 0$	$\sigma = 0.01$	$\sigma = 0.05$	$\sigma = 0.1$	$\sigma = 0.2$	
Original motion	1.0099	2.0234	5.7399	9.6636	16.9223	0
Primal skeleton [1]	8.1109	8.0353	9.4206	12.8291	20.6332	9.6990
PC-MRL (Ours)	**3.8714**	**4.1231**	**6.7277**	**10.3078**	**17.5261**	**5.6258**

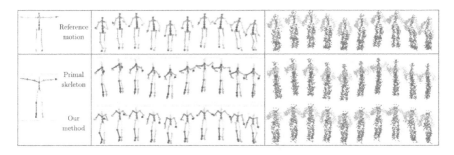

Fig. 6. Jumping motion re-targeting examples produced on the CMU skeleton by the primal skeleton method (middle row) and our method (bottom row). Sampled point cloud representations of each motion sequence are provided on the right.

are generally necessary for a complete motion re-targeting solution, this aspect is out of scope for our project as our main goal is motion interpolation.

Table 2 indicates the superior performance of our method over the state-of-the-art PS method in terms of visual similarity. At low σ values, the higher relative $\mathcal{L}_{KNN}$ of the primal skeleton method underscores its geometric deviations from the original motion. In contrast, our method demonstrates significantly better visual adherence to the original geometry. As σ increases, our method approaches near-perfect alignment with the original motion. Figure 6 visually indicates this, showing our method follows the original motion closely, while the latent consistency-based primal skeleton method struggles to generalise and decode the skeleton's internal structures without directly supervised objectives, such as the end-effector positions provided during training. In addition, Fig 7 demonstrates that, like existing re-targeting approaches, our method is able to produce adequate results on disproportionate skeletons.

Since our approach is largely focused on enabling non-native motion interpolation supervision, we additionally compare our PC-MRL method against a

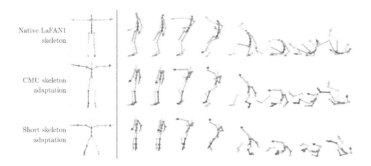

Fig. 7. Re-targeted falling motion from LaFAN1 (top) on the CMU skeleton (middle) and a short custom skeleton (bottom), performed by our point cloud-to-motion model.

Table 3. Average L2P, L2Q, and NPSS performance of CITL trained on native and PS-generated datasets, compared to our method.

Method	L2P↓			L2Q↓			NPSS↓		
	5	15	30	5	15	30	5	15	30
CITL - PS	0.2043	0.6332	1.1524	0.2318	0.7428	1.1442	0.1122	0.4276	1.4622
CITL - Native	0.0495	0.2306	0.4898	0.0979	0.3604	0.6609	0.1149	0.4684	1.1591
PC-MRL	0.0494	0.2337	0.4960	0.1030	0.3771	0.6813	0.1370	0.5433	1.1760

CITL model supervised by PS in Table 3. Due to the suboptimal re-targeting performance of PS, such interpolation models exhibit expectedly poor accuracy.

5 Limitations

Due to the inherent limitations of the point cloud representation, certain constraints on the applicability of our point cloud-to-motion method are inevitable. Primarily, the reliance of our point cloud re-targeting method on the relative nature of First-frame Offset Quaternions (FOQ) for most motion learning tasks necessitates the presence of at least *one known pose*. In the absence of a known pose, relative quaternions are incapable of reconstructing absolute rotational values, which are essential for motion applications. Secondly, due to our method of sampling point clouds with a consistent deviation from each bone segment, the resultant representation inevitably obscures finer details. This includes elements like fingers and facial controls, to an extent that is beyond rectification.

Our proposed objective function relies entirely on geometric segments, necessitating that all bones possess a non-zero length. This stipulation is deemed a reasonable prerequisite for generic human skeletons, as each bone is typically responsible for manipulating a discernible segment of the human body. Due to technical considerations, the skeletal configurations in each of our datasets included at least one bone of zero length, which we ignored in our experiments in favour of global quaternion rotations of all bones with non-zero length.

As a result of these limitations, we refrained from claiming state-of-the-art performance for general motion re-targeting tasks.

6 Conclusion

This paper introduces a novel learning-based motion interpolation method designed to enable cross-skeleton compatibility with existing motion datasets. Our proposed method PC-MRL employs a process of point cloud obfuscation and skeleton reconstruction. The point cloud space represents human motions in a non-hierarchic and skeleton-agnostic manner. It enables a KNN-based objective to be optimised without dataset supervision, guiding a neural network to generate high-quality motion features with any target skeleton. We address the rotational information loss of our point cloud format by presenting an offset quaternion strategy compatible with concurrent transformer-based models. Through extensive experiments, we have demonstrated the efficacy of PC-MRL in performing motion interpolation without relying on native motion data. Moreover, PC-MRL has achieved superior visual similarity metrics in the domain of motion re-targeting. We concluded our work by discussing the limitations of PC-MRL.

References

1. Aberman, K., Li, P., Lischinski, D., Sorkine-Hornung, O., Cohen-Or, D., Chen, B.: Skeleton-aware networks for deep motion retargeting **39**(4), 62–1 (2020)
2. Aberman, K., Wu, R., Lischinski, D., Chen, B., Cohen-Or, D.: Learning character-agnostic motion for motion retargeting in 2d **38**(4), 1–14 (2019)
3. Annabi, L., Ma, Z., et al.: Unsupervised motion retargeting for human-robot imitation. In: Companion of the 2024 ACM/IEEE International Conference on Human-Robot Interaction. ACM (2024)
4. Bodenheimer, B., Rose, C., Rosenthal, S., Pella, J.: The process of motion capture: dealing with the data. In: Thalmann, D., van de Panne, M. (eds.) Computer Animation and Simulation '97, LNCS, pp. 3–18. Eurographics. Springer, Vienna (1997). https://doi.org/10.1007/978-3-7091-6874-5_1
5. Burtnyk, N., Wein, M.: Computer-generated key-frame animation. J. SMPTE **80**(3), 149–153 (1971)
6. Catalin Ionescu, Fuxin Li, C.S.: Latent structured models for human pose estimation. In: International Conference on Computer Vision (2011)
7. Chen, H., Tang, H., Timofte, R., Gool, L.V., Zhao, G.: Lart: neural correspondence learning with latent regularization transformer for 3d motion transfer. Adv. Neural Inf. Process. Syst. **36** (2024)
8. Chen, J., Li, C., Lee, G.H.: Weakly-supervised 3D pose transfer with keypoints. In: ICCV, pp. 15156–15165 (2023)
9. Choi, K.J., Ko, H.S.: Online motion retargetting. J. Vis. Comput. Animat. **11**(5), 223–235 (2000)
10. CMU: Carnegie-mellon university motion capture database (2003). http://mocap.cs.cmu.edu/
11. Dam, E.B., Koch, M., Lillholm, M.: Quaternions, interpolation and animation, vol. 2. Citeseer (1998)
12. Duan, Y., Lin, Y., Zou, Z., Yuan, Y., Qian, Z., Zhang, B.: A unified framework for real time motion completion. In: AAAI, vol. 36, pp. 4459–4467 (2022)
13. Frank, T., Johnston, O.: Disney Animation: The Illusion of Life. Abbeville Publishing Group, New York (1981)
14. Gleicher, M.: Retargetting motion to new characters. In: Computer Graphics and Interactive Techniques, pp. 33–42 (1998)
15. Gleicher, M.: Animation from observation: motion capture and motion editing **33**(4), 51–54 (1999)
16. Gopalakrishnan, A., Mali, A., Kifer, D., Giles, L., Ororbia, A.G.: A neural temporal model for human motion prediction. In: CVPR, pp. 12116–12125 (2019)
17. Harvey, F.G., Pal, C.: Recurrent transition networks for character locomotion. In: SIGGRAPH Asia 2018 Technical Briefs, pp. 1–4 (2018)

18. Harvey, F.G., Yurick, M., Nowrouzezahrai, D., Pal, C.: Robust motion in-betweening **39**(4), 60–1 (2020)
19. He, C., Saito, J., Zachary, J., Rushmeier, H., Zhou, Y.: NeMF: neural motion fields for kinematic animation **35**, 4244–4256 (2022)
20. Hecker, C., Raabe, B., Enslow, R.W., DeWeese, J., Maynard, J., van Prooijen, K.: Real-time motion retargeting to highly varied user-created morphologies **27**(3), 1–11 (2008)
21. Hernandez, A., Gall, J., Moreno-Noguer, F.: Human motion prediction via spatio-temporal inpainting. In: ICCV, pp. 7134–7143 (2019)
22. Ionescu, C., Papava, D., Olaru, V., Sminchisescu, C.: Human3.6M: large scale datasets and predictive methods for 3D human sensing in natural environments. IEEE Trans. Pattern Anal. Mach. Intell. **36**(7), 1325–1339 (2014)
23. Jiang, H., Cai, J., Zheng, J.: Skeleton-aware 3D human shape reconstruction from point clouds. In: ICCV, pp. 5431–5441 (2019)
24. Kantor, I.L., Solodovnikov, A.S.: Hypercomplex Numbers: An Elementary Introduction to Algebras. In: Shenitzer, A. (ed.) Springer (1989). ISBN 9780387969800. lCCN 89006160
25. Karunratanakul, K., Preechakul, K., Suwajanakorn, S., Tang, S.: Guided motion diffusion for controllable human motion synthesis. In: ICCV, pp. 2151–2162 (2023)
26. Kulpa, R., Multon, F., Arnaldi, B.: Morphology-independent representation of motions for interactive human-like animation. In: Eurographics (2005)
27. Lehrmann, A.M., Gehler, P.V., Nowozin, S.: Efficient nonlinear Markov models for human motion. In: CVPR, pp. 1314–1321 (2014)
28. Li, P., Aberman, K., Zhang, Z., Hanocka, R., Sorkine-Hornung, O.: Ganimator: Neural motion synthesis from a single sequence **41**(4), 1–12 (2022)
29. Li, W., et al.: Modular design automation of the morphologies, controllers, and vision systems for intelligent robots: a survey. Vis. Intell. **1**(1), 2 (2023)
30. Li, Z., Peng, X.B., Abbeel, P., Levine, S., Berseth, G., Sreenath, K.: Reinforcement learning for versatile, dynamic, and robust bipedal locomotion control. arXiv preprint arXiv:2401.16889 (2024)
31. Liao, Z., Yang, J., Saito, J., Pons-Moll, G., Zhou, Y.: Skeleton-free pose transfer for stylized 3D characters. In: Avidan, S., Brostow, G., Cissé, M., Farinella, G.M., Hassner, T. (eds.) Computer Vision – ECCV 2022. ECCV 2022. LNCS, vol. 13662, pp. 640–656. Springer, Cham (2022). https://doi.org/10.1007/978-3-031-20086-1_37
32. Makoviychuk, V., et al.: Isaac gym: high performance gpu-based physics simulation for robot learning. arXiv preprint arXiv:2108.10470 (2021)
33. Merry, B., Marais, P., Gain, J.: Animation space: a truly linear framework for character animation. ACM Trans. Graph. (TOG) **25**(4), 1400–1423 (2006)
34. Mo, C.A., Hu, K., Long, C., Wang, Z.: Continuous intermediate token learning with implicit motion manifold for keyframe based motion interpolation. In: CVPR, pp. 13894–13903 (2023)
35. Mo, C.A., Hu, K., Mei, S., Chen, Z., Wang, Z.: Keyframe extraction from motion capture sequences with graph based deep reinforcement learning. In: ACM International Conference on Multimedia, pp. 5194–5202 (2021)
36. Monzani, J.S., Baerlocher, P., Boulic, R., Thalmann, D.: Using an intermediate skeleton and inverse kinematics for motion retargeting. In: Computer Graphics Forum, vol. 19, pp. 11–19. Wiley Online Library (2000)
37. Oreshkin, B.N., Valkanas, A., Harvey, F.G., Ménard, L.S., Bocquelet, F., Coates, M.J.: Motion inbetweening via deep δ-interpolator (2022)

38. Peng, X.B., Abbeel, P., Levine, S., van de Panne, M.: Deepmimic: example-guided deep reinforcement learning of physics-based character skills **37**(4), 143:1–143:14 (2018)
39. Peng, X.B., Ma, Z., Abbeel, P., Levine, S., Kanazawa, A.: Amp: adversarial motion priors for stylized physics-based character control **40**(4) (2021)
40. Qin, J., Zheng, Y., Zhou, K.: Motion in-betweening via two-stage transformers **41**(6), 1–16 (2022)
41. Reda, D., Won, J., Ye, Y., van de Panne, M., Winkler, A.: Physics-based motion retargeting from sparse inputs. ACM Comput. Graph. Interact. Tech. **6**(3), 1–19 (2023)
42. Reeves, W.T.: Inbetweening for computer animation utilizing moving point constraints **15**(3), 263–269 (1981)
43. Ren, T., et al.: Diverse motion in-betweening from sparse keyframes with dual posture stitching. IEEE Trans. Vis. Comput. Graph. (2024)
44. Shaw, P., Uszkoreit, J., Vaswani, A.: Self-attention with relative position representations. In: NAACL. Association for Computational Linguistics (2018)
45. Tiwari, G., Antić, D., Lenssen, J.E., Sarafianos, N., Tung, T., Pons-Moll, G.: Pose-NDF: modeling human pose manifolds with neural distance fields. In: Avidan, S., Brostow, G., Cissé, M., Farinella, G.M., Hassner, T. (eds.) Computer Vision – ECCV 2022. ECCV 2022. LNCS, vol. 13665, pp. 572–589. Springer, Cham (2022). https://doi.org/10.1007/978-3-031-20065-6_33
46. Villegas, R., Yang, J., Ceylan, D., Lee, H.: Neural kinematic networks for unsupervised motion retargetting. In: CVPR, pp. 8639–8648 (2018)
47. Wang, J., et al.: Zero-shot pose transfer for unrigged stylized 3D characters. In: CVPR, pp. 8704–8714 (2023)
48. Wang, K., Xie, J., Zhang, G., Liu, L., Yang, J.: Sequential 3D human pose and shape estimation from point clouds. In: CVPR, pp. 7275–7284 (2020)
49. Yuan, Y., Song, J., Iqbal, U., Vahdat, A., Kautz, J.: Physdiff: physics-guided human motion diffusion model. In: ICCV, pp. 16010–16021 (2023)
50. Yuan, Y., Wei, S.E., Simon, T., Kitani, K., Saragih, J.: Simpoe: simulated character control for 3D human pose estimation. In: CVPR, pp. 7159–7169 (2021)
51. Zhang, J., et al.: Skinned motion retargeting with residual perception of motion semantics & geometry. In: CVPR, pp. 13864–13872 (2023)
52. Zhang, M., Li, H., Cai, Z., Ren, J., Yang, L., Liu, Z.: Finemogen: fine-grained spatio-temporal motion generation and editing. Adv. Neural Inf. Process. Syst. **36** (2024)
53. Zhang, X., van de Panne, M.: Data-driven autocompletion for keyframe animation. In: ACM SIGGRAPH Conference on Motion, Interaction and Games, pp. 1–11 (2018)
54. Zhao, H., Jiang, L., Jia, J., Torr, P.H., Koltun, V.: Point transformer. In: ICCV, pp. 16259–16268 (2021)
55. Zhu, W., Yang, Z., Di, Z., Wu, W., Wang, Y., Loy, C.C.: Mocanet: motion retargeting in-the-wild via canonicalization networks. In: AAAI, vol. 36, pp. 3617–3625 (2022)

SwiftBrush V2: Make Your One-Step Diffusion Model Better Than Its Teacher

Trung Dao[1](✉), Thuan Hoang Nguyen[1], Thanh Le[1], Duc Vu[1], Khoi Nguyen[1], Cuong Pham[1,2], and Anh Tran[1]

[1] VinAI Research, Ho Chi Minh City, Vietnam
trung.dt880@gmail.com
[2] Posts and Telecommunications Institute of Technology, Ho Chi Minh City, Vietnam

Abstract. In this paper, we aim to enhance the performance of Swift-Brush, a prominent one-step text-to-image diffusion model, to be competitive with its multi-step Stable Diffusion counterpart. Initially, we explore the quality-diversity trade-off between SwiftBrush and SD Turbo: the former excels in image diversity, while the latter excels in image quality. This observation motivates our proposed modifications in the training methodology, including better weight initialization and efficient LoRA training. Moreover, our introduction of a novel clamped CLIP loss enhances image-text alignment and results in improved image quality. Remarkably, by combining the weights of models trained with efficient LoRA and full training, we achieve a new state-of-the-art one-step diffusion model, achieving an FID of 8.14 and surpassing all GAN-based and multi-step Stable Diffusion models.

Keywords: One-step Diffusion models · Text-to-image synthesis

1 Introduction

Text-to-image generation has experienced tremendous growth in recent years, allowing users to create high-quality images from simple descriptions. State-of-the-art models [2,27,33,35] could surpass humans in art competition [36] or produce synthetic images nearly indistinguishable from real ones [6]. Among popular text-to-image networks, Stable Diffusion (SD) models [34,35] are the most widely used due to their open-source accessibility. However, most SD models are designed as multi-step diffusion models, which require multiple forwarding steps to produce an output image. Such a slow and computationally expensive mechanism hinders the use of these models in real-time or on-device applications.

Recently, many works have tried to reduce the denoising steps required in text-to-image diffusion models. Notably, few recent studies have successfully

---

T. H. Nguyen, T. Le and D. Vu—Equal contribution.

Supplementary Information The online version contains supplementary material available at https://doi.org/10.1007/978-3-031-73007-8_11.

© The Author(s), under exclusive license to Springer Nature Switzerland AG 2025
A. Leonardis et al. (Eds.): ECCV 2024, LNCS 15140, pp. 176–192, 2025.
https://doi.org/10.1007/978-3-031-73007-8_11

Fig. 1. Our one-step diffusion model achieves an impressive FID of 8.14, generating high-quality and diverse results with a single UNet forwarding. The example images generated from the "A laughing cute grey rabbit with white stripe on the head, piles of gold coins in background, colorful, Disney Picture render, photorealistic" (first two rows) and "Portrait of a woman looking at the camera" (last two rows) prompts demonstrate our model's ability to create fast, visually appealing, and varied outputs.

developed **one-step diffusion models**, thus significantly speed up the image generation. While early attempts [10,26] produce blurry and malformed photos, subsequent methods produce sharp and high-quality outputs. These methods mainly distill knowledge from a pre-trained multi-step SD model (referred to as the *teacher model*) to a one-step *student model*. InstaFlow [24] employs Rectified Flows [23] in a multi-stage and computation-expensive training procedure. DMD [47] combines a reconstruction and a distribution matching loss as the training objectives, requiring massive pre-generated images from the teacher. SD Turbo [39] incorporates adversarial training alongside a score distillation loss, achieving photorealistic generation. However, it heavily relies on a large-scale image-text pair training dataset and, as later discussed, has a poor diversity. SwiftBrush [28] utilizes Variational Score Distillation (VSD) to transfer knowledge from the teacher network to the one-step student through a LoRA [12] intermediate teacher model. Notably, training SwiftBrush is simple, fast, and image-free, making it an intriguing method.

Despite these promising achievements, one-step text-to-image diffusion models still fall short of multi-step models in terms of the FID metric. On the standard COCO 2014 benchmark [22], SDv2.1 can achieve the lowest FID-30K score of 9.64 with classifier-free guidance (*cfg*) scale of 2, while the best-reported score from the one-step models of equivalent parameters scale is 11.49 [47]. The gap is expected since directly predicting a clean image from noise in a single step is much more challenging than via a multi-step scheme. Hence, one may believe that one-step text-to-image models could only approach or reach a similar performance as the teacher model but never exceeding it.

In this paper, we challenge this belief by seeking a one-step model that can surpass its multi-step teacher model quantitatively and qualitatively. Our solution drew inspiration from SwiftBrush, with its image-free training, enabling an effective, scalable, and flexible distillation. We examine its current state and compare it with SD Turbo. The comparison in Sect. 4.1 highlights a quality-diversity trade-off: SwiftBrush offers more diverse outputs due to its image-free and dynamic training, while SD Turbo yields high-quality but mode-collapse-like outputs due to its adversarial training. This insight drives us to initialize SwiftBrush training with SD Turbo, enhancing one-step student models significantly. Moreover, our extra clamped CLIP loss combined with SwiftBrush's flexible training mechanism empowers the student model to surpass the teacher. Lastly, we train the student on a larger text prompt dataset for better knowledge transfer between teacher and student models.

Given limited resources and the goal of offering effective and affordable model training solutions, we restrict the training on A100 40GB GPUs with affordable GPU hours. Such a restricted condition prevents us from employing efficiently the clamped CLIP loss and fully finetuning the student model. Hence, we propose two training schemes, one supporting full student model training without the extra loss and one employing LoRA-based model training associated with the mentioned auxiliary loss. Both training schemes produce high-quality output models that surpass all previous one-step diffusion-based approaches in most metrics. Especially when merging these two models using a simple weight linear interpolation, we obtain an one-step model with FID-30K of **8.77** on the COCO 2014 benchmark. Such a student model is the first to surpass its multi-step teacher model, breaking the common belief. It even exceeds all GAN-based text-to-image approaches [17,38] while also offering near real-time speed. With an extra regularization [47] on minimal real data, our merged model gets enhanced further to achieve the **FID score of 8.14**, setting a new standard for efficient and high-quality text-to-image models.

In summary, our contributions include (1) an analysis of representative existing diffusion-based text-to-image models to reveal the quality-diversity trade-off, (2) a simple but effective integration of SwiftBrush and SD Turbo to combine the advantages of both, (3) an extra clamped CLIP loss proposed to boost the image-text alignment of the student network and surpass the teacher model, (4) two resource-efficient training strategies to utilize the mentioned proposals, and (5) a fused one-step student model that is superior to its multi-step text-to-image teacher in all metrics and sets a new standard in this field.

2 Related Work

2.1 Text-to-Image Generation

Text-to-image generation involves synthesizing high-quality images based on input text prompts. This task has evolved over decades, transitioning from constrained domains like CUB [41] and COCO [22] to general domains such as LAION-5B [40]. This evolution is driven by the emergence of large vision-language models (VLMs) like CLIP [32] and ALIGN [15]. Leveraging these models and datasets, various approaches have been introduced, including autoregressive models like DALL-E [33], CogView [9], and Parti [48]; mask-based transformers such as MUSE [4] and MaskGIT [5]; GAN-based models such as StyleGAN-T [38], GigaGAN [17]; and diffusion models like GLIDE [29], Imagen [37], Stable Diffusion (SD) [34], DALL-E2 [33], DALL-E3 [2], and eDiff-I [1]. Among these, diffusion models are popular due to their ability to generate high-quality images. However, they typically require many-step sampling to generate high-quality images, limiting their real-time and on-device applications.

2.2 Accelerating Text-to-Image Diffusion Models

Efforts to accelerate diffusion model sampling include faster samplers and distillation techniques. Early methods reduce sampling steps to as few as 4–8 steps by incorporating Latent Consistency Models to distill latent diffusion models [7,26]. Recent studies have achieved one-step text-to-image generation by training a student model distilled from a pretrained multi-step diffusion model, employing various techniques such as Rectified Flows [24], reconstruction and distribution matching losses [47], and adversarial objectives [21,39,45,50]. However, the output images often exhibit blurriness and artifacts, and one-step methods still underperform compared to multi-step models while requiring large-scale text-image pairs for training.

Differentiating itself from the rest, SwiftBrush [28] proposed a one-step distillation technique that required only training on prompt inputs. The method gradually transfers knowledge from the teacher to the one-step student through an intermediate LoRA multi-step teacher. SwiftBrush's image-free training procedure offers a simple way to scale training data and extend the student model capability via auxiliary losses, which are not constrained by limited-size imagery training data. Therefore, while SwiftBrush also falls short in quality compared to the teacher model, we find its high potential for further development to produce a one-step student that even beat the multi-step teacher at its own game.

3 Background

Diffusion Models are generative models that transform a noise distribution into a target data distribution by simulating the diffusion process. This transformation involves gradually adding noise $\epsilon \sim \mathcal{N}(0, I)$ to clean image $\mathbf{x}_0$ over a series of T steps (forward process) and then learning to reverse this process (reverse process). The forward process can be formulated as:

$$\mathbf{x}_t = \alpha_t \mathbf{x}_0 + \sigma_t \epsilon \quad \forall t \in \overline{0, T} \tag{1}$$

where $\mathbf{x}_t$ is the data at time step t and $\{(\alpha_t, \sigma_t)\}_{t=1}^{T}$ is the noise schedule such that $(\alpha_T, \sigma_T) = (0, 1)$ and $(\alpha_0, \sigma_0) = (1, 0)$. On the other hand, the reverse process aims to reconstruct the original data from noise. Training involves minimizing the difference between predicted output from model ϵ_ϕ parameterized by ϕ and the actual added noise:

$$\min_{\phi} \mathbb{E}_{t \sim \mathcal{U}(0,T), \epsilon \sim \mathcal{N}(0,I)} \|\epsilon_\phi(\mathbf{x}_t, t) - \epsilon\|_2^2 \tag{2}$$

Text-to-Image Diffusion Models generate images from textual descriptions by integrating text embeddings within the inference process, aiming to align textual descriptions with visual outputs. A key challenge is the lack of mechanisms to ensure textual relevance and image fidelity. To deal with this, large-scale text-to-image diffusion models usually utilize classifier-free guidance which enhances text-image alignment without a separate classifier. This is achieved by interpolating the model's output with and without the conditioning text, controlled by a guidance scale γ. The model's final output with guidance is given by:

$$\hat{\epsilon}_\phi(\mathbf{x}_t, t, \mathbf{y}) = \epsilon_\phi(\mathbf{x}_t, t, \mathbf{y}) + \gamma \cdot (\epsilon_\phi(\mathbf{x}_t, t, \mathbf{y}) - \epsilon_\phi(\mathbf{x}_t, t)), \tag{3}$$

where $\epsilon_\phi(\mathbf{x}_t, t, \mathbf{y})$ is the predicted output conditioned on text embedding $\mathbf{y}$ and $\epsilon_\phi(\mathbf{x}_t, t)$ is the unconditional prediction, i.e., using null text. This formula showcases how classifier-free guidance directly influences the generation process, leading to more precise and relevant image outputs.

SwiftBrush [28] is a one-step text-to-image generative model, employing an image-free distillation technique inspired by text-to-3D synthesis methods. At the heart of this approach is re-purposing Variational Score Distillation (VSD) [42], a novel loss that tackles the challenges of over-smoothing and diversity reduction often seen in early text-to-3D work [31]. Specifically, SwiftBrush uses two teachers, one frozen teacher ϵ_ϕ and one LoRA [12] teacher ϵ_ψ, to guide the one-step student f_θ. Here, the LoRA teacher is to bridge the gap between the frozen teacher and the student. On one hand, the loss for training the student model is formalized as:

$$\nabla_\theta \mathcal{L}_{VSD} = \mathbb{E}_{t,\mathbf{y},\mathbf{z}} \left[w(t)(\hat{\epsilon}_\phi(\mathbf{x}_t, t, \mathbf{y}) - \hat{\epsilon}_\psi(\mathbf{x}_t, t, \mathbf{y})) \frac{\partial f_\theta(\mathbf{z}, \mathbf{y})}{\partial \theta} \right], \tag{4}$$

where $\mathbf{z} \sim \mathcal{N}(0, I)$ is the input noise to the student network, $\mathbf{x}_t = \alpha_t \hat{\mathbf{x}}_0 + \sigma_t \epsilon$ is the noise-added version at a time step t of the student's output $\hat{\mathbf{x}}_0 = f_\theta(\mathbf{z}, y)$, and $w(t)$ is the weighting of the loss. On the other hand, the LoRA teacher is trained using diffusion loss $\|\mathbb{E}_{t,\epsilon,y}[\epsilon_\psi(\mathbf{x}_t, t, \mathbf{y}) - \epsilon]\|_2^2$. SwiftBrush alternates between the training of the student model and the LoRA teacher until convergence.

4 Proposed Methods

In this section, we begin by conducting an in-depth analysis of the quality-diversity trade-off in representative diffusion-based text-to-image models (Sect. 4.1). Subsequently, we discuss our strategy for incorporating the strengths of SwiftBrush and SD Turbo (Sect. 4.2). Lastly, we explore various approaches to enhance the distillation process and post-training procedures (Sect. 4.3 to 4.5). An overview of our methodologies is presented in Fig. 2.

Table 1. Comparison between the multi-step teacher (SDv2.1), SD Turbo, and Swift-Brush on the zero-shot MS COCO-2014 30K benchmark. The best scores are in **bold**, while the better scores among one-step models are underlined.

Model Name	NFE	FID↓	CLIP↑	Precision↑	Recall↑
SD 2.1 ($cfg = 2$)	25	**9.64**	0.31	0.57	0.53
SD 2.1 ($cfg = 4.5$)	25	12.26	**0.33**	0.61	0.41
SD 2.1 ($cfg = 7.5$)	25	15.93	**0.33**	0.59	0.36
SD Turbo	1	16.10	**0.33**	**0.65**	0.35
SwiftBrush	1	<u>15.46</u>	0.30	0.47	<u>0.46</u>

4.1 Quality-Diversity Trade-Off in Existing Models

We first analyze the properties of the teacher model, SDv2.1, and existing one-step diffusion-based text-to-image models. For the teacher model, we assess its performance across different guidance scales. Here, we select SwiftBrush and SD Turbo due to their quality and distinct training procedures. SwiftBrush relies solely on score distillation from the teacher in its image-free training, while SD Turbo trains on real images with adversarial and distillation loss. We conduct our analysis on the COCO 2014 benchmark and report relevant metrics in Table 1.

When assessing the multi-step teacher's performance, the classifier-free guidance scale (cfg) plays a crucial role. A low cfg (e.g., $cfg = 2$) yields a low FID score of 9.64, driven by high output diversity (recall = 0.53). However, this setting results in weak alignment between images and prompts (CLIP score = 0.30) and lower image quality (low precision). Conversely, a large cfg (e.g., $cfg = 7.5$) markedly improves text-image alignment (CLIP score = 0.33) but restricts diversity (recall = 0.36), resulting in a poor FID score of 15.93. Moderate cfg values (e.g., $cfg = 4.5$) strike a better balance, offering the highest precision score.

When evaluating the one-step students, we notice distinct behaviors. SD Turbo, benefiting from adversarial training on real images, yields highly naturalistic outputs with an exceptionally high precision score, surpassing even those of the multi-step teacher. However, this results in poor diversity, reflected in a low recall of 0.35. Conversely, SwiftBrush adopts an image-free training approach, allowing flexible combinations of random-noise latents and input prompts. Such a relaxed supervision enables the student model to generate more diverse outputs but at the expense of quality (Table 1). We further verify this finding by a qualitative evaluation, illustrated in Fig. 1. When given identical input prompts, SD Turbo generates realistic yet similar outputs. In contrast, SwiftBrush produces a wider range of outcomes, albeit with greatly distorted artifacts. Regardless, both one-step models exhibit FID scores around 15–16, significantly higher than the teacher's best score. Observing a quality-diversity trade-off in existing one-step diffusion model such as SD Turbo and SwiftBrush, we aim to combine these two to leverage the strengths of both.

4.2 SwiftBrush and SD Turbo Integration

In this section, we explore strategies for effectively merging SwiftBrush and SD Turbo to enhance the quality-diversity trade-off. A direct approach is unifying their training procedure, i.e., combining adversarial training from SD Turbo and Variational Score Distillation from SwiftBrush. However, this simplistic approach proves challenging due to computational demands and potential failure. While SwiftBrush's image-free procedure is easier to implement, reproducing SD Turbo's training process is complex and resource-intensive. The presence of the discriminator complicates the training, necessitating significant VRAM and dataset requirements. Additionally, SD Turbo's intense supervision may constrain SwiftBrush's loose guidance, limiting output diversity.

Based on our discussions, we opt not to utilize SD Turbo's adversarial training. Instead, we leverage its pretrained weights to initialize the student network within SwiftBrush's training framework. This straightforward approach proves highly effective. As can be seen in the comparison between the second and the third row in Table 2, the resulting model has improved FID and recall. By employing SD Turbo's pretrained weights, we provide a solid foundation for the training model to maintain high-quality outputs, while SwiftBrush's image-free training process gradually enhances generation diversity.

4.3 In-Training Improvements

Besides data efficiency and diversity promotion, SwiftBrush's image-free training still has room for improvement. First, it allows an easy means to scale up training data by collecting more prompt inputs. This task is simple, given the abundance of textual datasets and the availability of large language models, unlike the costly and labor-intensive task of collecting image-text pair data commonly required. Second, by not forcing the output of the student model to be the same as that of the teacher, SwiftBrush allows the student to go even beyond the quality and capability of the teacher. We can advocate it to happen by adding extra auxiliary loss functions in SwiftBrush training. In this section, we will discuss the implementation of those ideas for improvement.

Implications of Dataset Size. SwiftBrush's image-free approach allows for scalable training datasets without limitations. To explore the dataset's impact on SwiftBrush performance, we conducted supplementary experiments by augmenting the dataset with an additional 2M prompts from the LAION dataset [40] to the original 1.5M deduplicated prompts from the JourneyDB dataset [30]. Analysis (Table 2) reveals improved performance with the expanded dataset. Specifically, this leads to a significant improvement in terms of FID and precision, suggesting a positive correlation between dataset size and the quality of the generated outputs. However, a slight degradation in recall was observed, indicating a potential trade-off between image diversity and overall quality. Furthermore, despite an increase in CLIP score compared to the previous version, there remains room for improvements in terms of text alignment.

Table 2. The impact of SD Turbo initialization and dataset size on Swift-Brush's performance, compared to SD Turbo on the zero-shot COCO-2014 benchmark.

Model Name	Training size	FID↓	CLIP↑	Precison↑	Recall↑
SD Turbo	–	16.10	**0.33**	**0.65**	0.35
SwiftBrush	1.3M	15.46	0.30	0.47	0.46
SwiftBrush (Turbo init.)	1.3M	14.59	0.30	0.43	**0.55**
SwiftBrush (Turbo init.)	3.3M	**11.27**	0.31	0.48	0.54

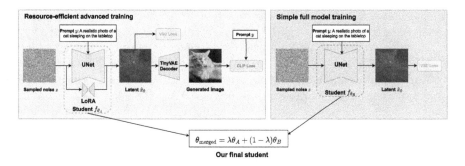

Fig. 2. SwiftBrush v2 overview: two versions of the student model: a fully finetuned model trained with the Variational Score Distillation (VSD) loss, and a LoRA finetuned model trained with both VSD and CLIP loss. The final model is obtained by merging the two student models, leveraging the strengths of both training schemes.

Tackling the Text-alignment Problem. To refine the coherence between textual prompts and visual outputs, we integrate an additional CLIP loss within the distillation process. However, naively employing such loss between the student model's predictions and the original textual prompts poses challenges , as overoptimizing for the CLIP score potentially degrade image quality. We observed issues such as blurriness, increased color saturation, and the emergence of textual artifacts within the generated images.

To address this, we propose clamping the CLIP value during training with ReLU activation. This aims to balance text alignment with preserving image quality, ensuring the model maintains visual integrity. Additionally, we introduce dynamic scheduling to control the influence of CLIP loss, gradually reducing its weight to zero by the end of distillation. This balanced approach integrates visual-textual alignment and image fidelity effectively. Our clamped CLIP loss is formulated as:

$$\mathcal{L}_{\text{CLIP}} = \max\left(0, \tau - \langle \mathcal{E}_{\text{image}}\left(\mathcal{D}\left(f_\theta(\mathbf{z}, \mathbf{y})\right)\right), \mathcal{E}_{\text{text}}(\mathbf{y})\rangle\right), \tag{5}$$

where $\mathcal{E}_{\text{image}}$ and $\mathcal{E}_{\text{text}}$ represent the CLIP image and text encoders, respectively. $\mathcal{D}$ is the VAE decoder used to map the latent back to the image. The term τ introduces a threshold on the desired cosine similarity $\langle \cdot, \cdot \rangle$ between the image and text embeddings, preventing the model from overemphasizing textual alignment at the expense of image quality.

4.4 Resource-Efficient Training Schemes

While our CLIP loss is highly beneficial, it comes with memory and computation costs. Particularly, the CLIP image encoder can only work on image space, requiring decoding the predicted latent to image via the image decoder $\mathcal{D}$ as can be seen in Eq. (5). We find incorporating CLIP loss into SwiftBrush's full-model distillation significantly slows down training speed, particularly on GPUs with moderate VRAM. This urges us to design a resource-efficient training scheme to fully exploit the proposed CLIP loss in constrained setting.

It is possible to significantly reduce memory requirements during fine-tuning with the LoRA framework [12], where only a set of small-rank parameters are trained. Also, to compute the CLIP loss, the predicted latent goes through a large VAE decoder, increasing training length and memory consumption. To address this, we integrate TinyVAE [3], a compact variant of Stable Diffusion's VAE. TinyVAE sacrifices some fine detail in images but preserves overall structure and object identity comparable to the original VAE. This approach maintains training efficiency close to those of the original fully fine-tuned model, as shown in Table 6 and Sect. 5.3.

4.5 Post-training Improvements

Recent literature [20] has shown a growing interest in model fusion techniques, aiming to integrate models performing distinct subtasks into a unified multitask model [16,46] or combining fine-tuned iterations to create an enhanced version [8, 14,43]. Our research focuses on the latter, particularly in the context of one-step text-to-image diffusion models. These models, although designed for the same task, differ in their training objectives, providing each with unique advantages. By merging these models, we aim to create a new model that captures the strength of each model without increasing model size or inference costs. Given two one-step diffusion models with weights θ_A and θ_B and an interpolation weight λ, we merge them using a simple linear interpolation of the weights:

$$\theta_{\text{merged}} = \lambda \theta_A + (1 - \lambda)\theta_B. \tag{6}$$

We empirically demonstrate the benefit of such interpolation scheme with SD Turbo, known for its precision and strong text alignment, and the original SwiftBrush, which excels in diversity. In our empirical analysis (refer to Fig. 3), we observe that by interpolating from one model to the other, all evaluated metrics (except for the CLIP score) show improvement at some optimal point. This indicates that the fused model could potentially outperform the original models. These findings underscore the potential of model fusion techniques in enhancing model efficacy, as evidenced by the metric analysis.

As discussed in Sect. 4.4, our proposal suggests two training schemes. We can either train the student model with LoRA and TinyVAE utilizing VSD and CLIP losses or fully finetune the student model employing only VSD loss. These two training schemes lead to two resulting one-step models with different behaviors, making them ideal ingredients for merging. By merging these models, we obtain the final model output of our proposed SwiftBrush v2 framework.

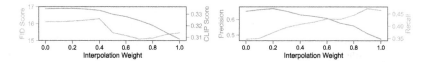

Fig. 3. The effect of weight interpolation upon FID, CLIP score, precision, and recall calculated on the zero-shot MS COCO-2014 benchmark. 0.0 indicates SD Turbo, and 1.0 indicates the original SwiftBrush.

5 Experiments

5.1 Experimental Setup

Evaluation Metrics. Our text-to-image model is evaluated using the "zero-shot" setting, i.e., trained on some datasets and tested on another dataset, undergoing comprehensive evaluation across three key aspects: image quality, diversity, and textual fidelity. We use the Fréchet Inception Distance (FID) [11] on resized 256 × 256 images as our primary metric for evaluating image quality, consistent with prior text-to-image research [18]. In addition, we employ precision [19] as a complementary metric to FID. For evaluating diversity, we rely on the recall metric [19]. Textual alignment is measured using the CLIP [32] score and the Human Preference Score v2 [44] (HPSv2).

Datasets. We utilize two training datasets: (1) 1.3M prompts from JourneyDB [30], and (2) an expanded dataset incorporating 2M prompts from LAION [40], totaling 3.3M prompts. Additionally, a human-feedback dataset, comprising 200K pairs from LAION-Aesthetic-6.25+ [40], can be optionally used for further image regularization [47], which is around 5% of the total training data. We use the MS COCO-2014 validation set as the standard zero-shot text-to-image benchmark, consistent with established practices in the field [10,18,24,26,38,39]. Samples are generated from the first 30k prompts, with the entire dataset serving as the reference for obtaining metrics. For HPSv2, we adopt the evaluation protocol from [44].

Training Details. Our method is built on top of SwiftBrush [28] with our proposed modifications. We conduct all our training on four NVIDIA A100 40GB GPUs, with training durations of one or three days depending on the dataset (JourneyDB alone or combined with LAION prompts). The batch size is 16 per GPU, and we use learning rates of $1e^{-6}$ and $1e^{-3}$ for the student and LoRA teacher, respectively, with the AdamW optimizer [25]. Our approach utilizes Stable Diffusion 2.1 [34] with $cfg = 4.5$ as the frozen teacher and LoRA teacher initialized with rank $r = 64$ and scaling $\gamma = 128$. As for the LoRA student, we set $r = 256$ and $\gamma = 512$ to enhance its learning capacity. In addition, we introduce the clamped CLIP loss with a margin of $\tau = 0.35$, starting with a weight of 0.1 and gradually reducing to zero. We use ViT-B/32 [13] as the backbone for CLIP image and text feature extraction. Finally, we merge two final models with $\lambda = 0.5$, further details are available in the Appendix.

Table 3. Quantitative comparisons between our method and others on zero-shot MS COCO-2014 benchmark. For multi-step SD models, we report each with the *cfg* that returns the best FID, e.g., *cfg* = 3 for SDv1.5 and *cfg* = 2 for SDv2.1. We also report performance of the teacher model (SDv2.1 with *cfg* = 4.5). [†] denotes reported numbers, [‡] denotes our rerun based on provided GitHubs. '-' denotes unreported data. Ours* indicates training with additional image regularization.

Method	NFEs	FID↓	CLIP↑	Precision↑	Recall↑
StyleGAN-T [38][†]	1	13.90	–	–	–
GigaGAN [17][†]	1	9.09	–	–	–
SDv1.5 [34] (*cfg* = 3)[†]	25	8.78	0.30	0.59	**0.53**
SDv2.1 [34] (*cfg* = 2)[‡]	25	9.64	0.31	0.57	**0.53**
SDv2.1 [34] (*cfg* = 4.5)[‡]	25	12.26	**0.33**	0.61	0.41
SD Turbo [39][‡]	1	16.10	**0.33**	**0.65**	0.35
UFOGen [45][†]	1	12.78	–	–	–
MD-UFOGen [50][†]	1	11.67	–	–	–
HiPA [49][†]	1	13.91	0.31	–	–
InstaFlow-0.9B [24][‡]	1	13.33	0.30	0.53	0.45
DMD [47][†]	1	11.49	0.32	–	–
SwiftBrush	1	15.46	0.30	0.47	0.46
Ours	1	8.77	0.32	0.55	**0.53**
Ours*	1	**8.14**	0.32	0.57	0.52

5.2 Comparison with Prior Approaches

Quantitative Results. Table 3 presents a comprehensive quantitative comparison between our approach and prior text-to-image models. This encompasses GAN-based models (group 1), multi-step diffusion models (group 2), and a variety of distillation techniques (group 3), both with and without image supervision. Our approach outperforms all competitors, notably achieving superior results even without direct image regularization. Remarkably, our distilled student models exceed their teacher model, SDv2.1, in FID scores by a significant margin while maintaining equivalent model size and inference times comparable to SD Turbo or SwiftBrush. Our model effectively addresses previous text alignment issues observed in SwiftBrush, being close to the CLIP scores of SD Turbo and multi-step models. Precision metrics show high-realism akin to the reference dataset, enhanced solely with a text-driven training dataset. Notably, ours exhibits significant recall improvements due to its image-free nature. Image-based regularization further improves student quality with a small reduction in recall.

In terms of HPSv2 (Table 4), our approach achieves competitive scores compared to the multi-step teacher model SDv2.1 and other distillation methods. Particularly, our model with image regularization achieves the highest HPSv2 scores for photos and remains close to the top performers in other categories.

Table 4. HPSv2 comparisons between our method and previous work. † denotes reported numbers, ‡ denotes our rerun based on provided GitHubs. Ours* indicates training with additional image regularization. **Bold** indicates the best, while <u>underline</u> indicates the second best.

Method	Anime	Photo	Concept Art	Painting
SDv2.1†	<u>27.48</u>	26.89	<u>26.86</u>	**27.46**
SD Turbo‡	**27.98**	<u>27.59</u>	**27.16**	<u>27.19</u>
InstaFlow†	25.98	26.32	25.79	25.93
BOOT†	25.29	25.16	24.40	24.61
SwiftBrush†	26.91	27.21	26.32	26.37
Ours	27.13	27.56	26.69	26.76
Ours*	27.25	**27.62**	<u>26.86</u>	26.77

Qualitative Results. We provide a qualitative comparison in Fig. 5. Our model produces higher quality compared to its teachers and one-step counterparts. In the first row, despite correctly illustrating the jumping action, other models yield deformed shapes of cats. This issue also reappears in the third row. In contrast, our model realistically represents the action and the cat's figure. SDv2.1 encounters out-of-frame issues in the fourth row, while others generate facial deformities in the last two rows. Meanwhile, our model produces artifact-free faces that are aligned with the text description regarding hairstyle and expression.

Among the counterparts, SD Turbo shows the best image quality but also the worst diversity. We reaffirm its weakness when comparing it to our method in Fig. 6. SD Turbo tends to generate similar images, as evidenced by the similar details and colors for different samples in each prompt. In contrast, our model can generate diverse views and colors in the first prompt as well as different environments in the second prompt.

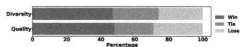

Fig. 4. User survey. We asked participants to compare the quality and diversity of images generated by our method and its teacher model across 20 random text prompts.

We also surveyed 250 participants to compare our distilled model with its multi-step teacher, SDv2.1. Participants evaluated images generated by each model for 20 random prompts, and the results (Fig. 4) show that our method consistently matched or exceeded the teacher in quality and diversity.

5.3 Ablation Studies

Effect of each proposed component is summarized in Table 5. We compare two student training schemes: (1) full model training and (2) efficient model training. Initializing the model from SD Turbo [39] significantly improves the FID in both schemes. Adding prompts from LAION [40] leads to notable FID

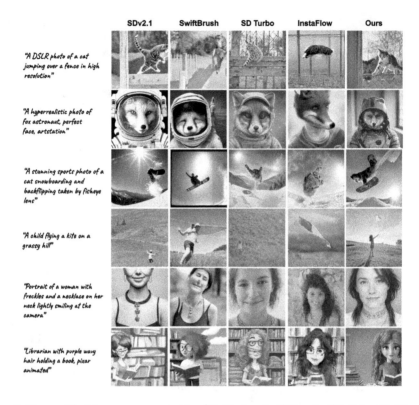

Fig. 5. Exemplified images generated by SD Turbo, SwiftBrush, SDv2.1 with 50 sampling steps, InstaFlow-0.9B and Ours. Images in the same row are sampled from the same text prompt, while images in the same column are from the same model.

Fig. 6. Exemplified images generated by SD Turbo and Ours to demonstrate our method's high diversity.

improvements in both cases, reaching 11.27 in (1) and 11.02 in (2). Also, efficient training allows for the use of the Clamped CLIP loss, resulting in a considerable FID improvement. Notably, combining models from both schemes further reduces the FID to 8.77, surpassing their SDv2.1 teacher. Finally, with extra image regularization, we achieve a further boost, reaching an FID of 8.14.

Detailed study in clamped CLIP loss is shown in Table 6. Naively applying CLIP loss [32] worsens performance, increasing the FID by 5 points. In contrast,

Table 5. Ablation of our methods upon zero-shot MS COCO-2014 30K. **Bold** indicates the best, while underline indicates the second best.

Label	SD Turbo	CLIP	LAION	FID↓	CLIP↑	Precision↑	Recall↑
SwiftBrush				15.46	0.30	0.47	0.46
		Fully training					
	✓			14.59	0.29	0.43	**0.55**
A	✓		✓	11.27	0.31	0.48	<u>0.54</u>
		Efficient training					
	✓			13.21	0.32	0.61	0.38
	✓	✓		11.70	**0.33**	**0.63**	0.42
B	✓	✓	✓	11.02	0.32	0.51	0.52
Ours	Merge A and B			<u>8.77</u>	0.32	0.55	0.53
Ours*	Merge A and B w/ regularization			**8.14**	<u>0.32</u>	<u>0.57</u>	0.52

Table 6. Ablation of our enhanced CLIP loss and resource-efficient training scheme.

CLIP	Clamped	Scheduler	Efficient	FID↓	CLIP↑	GPU days
				14.59	0.297	**4.1**
✓				21.32	**0.337**	7.8
✓	✓			13.19	0.319	7.8
✓	✓	✓	✓	**11.70**	0.330	<u>4.3</u>

our proposed clamped CLIP loss (Eq. 5) is highly effective, reducing the FID to 13.19. Moreover, employing a weight scheduler for the clamped CLIP loss further enhances performance, resulting in an FID of 11.70.

6 Discussion and Conclusion

Limitations: Despite the promising results, our distilled model still inherits some limitations from the teacher model, such as compositional problems. To address these limitations, future work could explore the integration of auxiliary losses that focus on cross-attention mechanisms during the distillation process.

Societal Impact: Our advancements improve high-quality image synthesis speed and accessibility. However, misuse of our advancements could spread misinformation and manipulate public perception. Thus, responsible use and safeguards are crucial to ensure that the benefits outweigh the risks.

Conclusion: This paper proposes a novel method to enhance SwiftBrush, a one-step text-to-image diffusion models. We address the quality-diversity trade-off by initializing the SwiftBrush student model with SD Turbo's pretrained weights and incorporating efficient in-training techniques as well as margin CLIP loss and large-scale dataset training. Also, by weight merging and optional image regularization, we achieve an outstanding FID score of 8.14, surpassing existing approaches in both GAN-based and one-step diffusion-based text-to-image generation while maintaining near real-time inference speed.

References

1. Balaji, Y., et al.: eDiff-I: text-to-image diffusion models with an ensemble of expert denoisers. arXiv preprint arXiv:2211.01324 (2022)
2. Betker, J., et al.: Improving image generation with better captions. Comput. Sci. **2**(3), 8 (2023). https://cdn.openai.com/papers/dall-e-3.pdf
3. Bohan, O.B.: Madebyollin/taesd, March 2024. https://github.com/madebyollin/taesd
4. Chang, H., et al.: Muse: text-to-image generation via masked generative transformers. arXiv preprint arXiv:2301.00704 (2023)
5. Chang, H., Zhang, H., Jiang, L., Liu, C., Freeman, W.T.: Maskgit: masked generative image transformer. In: Proceedings of the IEEE/CVF Conference on Computer Vision and Pattern Recognition, pp. 11315–11325 (2022)
6. Check, F.: Images appearing to show Donald Trump arrest created by AI. Reuters, March 2023. https://www.reuters.com/article/idUSL1N35T2TU
7. Chen, J., et al.: PIXART-δ: fast and controllable image generation with latent consistency models (2024)
8. Choshen, L., Venezian, E., Slonim, N., Katz, Y.: Fusing finetuned models for better pretraining (2022)
9. Ding, M., et al.: Cogview: mastering text-to-image generation via transformers. Adv. Neural. Inf. Process. Syst. **34**, 19822–19835 (2021)
10. Gu, J., Zhai, S., Zhang, Y., Liu, L., Susskind, J.: BOOT: data-free distillation of denoising diffusion models with bootstrapping. arXiv preprint arXiv:2306.05544 (2023)
11. Heusel, M., Ramsauer, H., Unterthiner, T., Nessler, B., Hochreiter, S.: GANs trained by a two time-scale update rule converge to a local nash equilibrium. Adv. Neural Inf. Process. Syst. **30** (2017)
12. Hu, E.J., et al.: Lora: low-rank adaptation of large language models. In: International Conference on Learning Representations (ICLR) (2021)
13. Ilharco, G., et al.: Openclip, July 2021. https://doi.org/10.5281/zenodo.5143773, if you use this software, please cite it as below
14. Izmailov, P., Podoprikhin, D., Garipov, T., Vetrov, D., Wilson, A.G.: Averaging weights leads to wider optima and better generalization. arXiv preprint arXiv:1803.05407 (2018)
15. Jia, C., et al.: Scaling up visual and vision-language representation learning with noisy text supervision. In: International Conference on Machine Learning, pp. 4904–4916. PMLR (2021)
16. Jin, X., Ren, X., Preotiuc-Pietro, D., Cheng, P.: Dataless knowledge fusion by merging weights of language models. arXiv preprint arXiv:2212.09849 (2022)
17. Kang, M., et al.: Scaling up GANs for text-to-image synthesis. In: Proceedings of the IEEE/CVF Conference on Computer Vision and Pattern Recognition, pp. 10124–10134 (2023)
18. Kang, M., et al.: Scaling up GANs for text-to-image synthesis. In: Proceedings of the IEEE Conference on Computer Vision and Pattern Recognition (CVPR) (2023)

19. Kynkäänniemi, T., Karras, T., Laine, S., Lehtinen, J., Aila, T.: Improved precision and recall metric for assessing generative models. Adv. Neural Inf. Process. Syst. **32** (2019)
20. Li, W., Peng, Y., Zhang, M., Ding, L., Hu, H., Shen, L.: Deep model fusion: a survey. arXiv preprint arXiv:2309.15698 (2023)
21. Lin, S., Wang, A., Yang, X.: SDXL-lightning: progressive adversarial diffusion distillation (2024)
22. Lin, T.-Y., et al.: Microsoft COCO: common objects in context. In: Fleet, D., Pajdla, T., Schiele, B., Tuytelaars, T. (eds.) ECCV 2014. LNCS, vol. 8693, pp. 740–755. Springer, Cham (2014). https://doi.org/10.1007/978-3-319-10602-1_48
23. Liu, X., Gong, C., Liu, Q.: Flow straight and fast: learning to generate and transfer data with rectified flow. In: International Conference on Learning Representations (2023)
24. Liu, X., Zhang, X., Ma, J., Peng, J., Liu, Q.: InstaFlow: one step is enough for high-quality diffusion-based text-to-image generation. arXiv preprint arXiv:2309.06380 (2023)
25. Loshchilov, I., Hutter, F.: Decoupled weight decay regularization. arXiv preprint arXiv:1711.05101 (2017)
26. Luo, S., Tan, Y., Huang, L., Li, J., Zhao, H.: Latent consistency models: synthesizing high-resolution images with few-step inference. arXiv preprint arXiv:2310.04378 (2023)
27. Midjourney: Midjourney. https://www.midjourney.com
28. Nguyen, T.H., Tran, A.: Swiftbrush: one-step text-to-image diffusion model with variational score distillation. In: Proceedings of the IEEE/CVF Conference on Computer Vision and Pattern Recognition (2024)
29. Nichol, A., et al.: Glide: towards photorealistic image generation and editing with text-guided diffusion models. In: International Conference on Machine Learning (2021)
30. Pan, J., et al.: JourneyDB: a benchmark for generative image understanding. Adv. Neural Inform. Process. Syst. (2023)
31. Poole, B., Jain, A., Barron, J.T., Mildenhall, B.: Dreamfusion: text-to-3d using 2d diffusion. In: International Conference on Learning Representations (2022)
32. Radford, A., et al.: Learning transferable visual models from natural language supervision. In: International Conference on Learning Representations, pp. 8748–8763. PMLR (2021)
33. Ramesh, A., Dhariwal, P., Nichol, A., Chu, C., Chen, M.: Hierarchical text-conditional image generation with clip latents. arXiv preprint arXiv:2204.06125 (2022). **1**(2), 3
34. Rombach, R., Blattmann, A., Lorenz, D., Esser, P., Ommer, B.: High-resolution image synthesis with latent diffusion models. In: 2022 IEEE/CVF Conference on Computer Vision and Pattern Recognition (CVPR), pp. 10674–10685 (2021)
35. Rombach, R., Blattmann, A., Lorenz, D., Esser, P., Ommer, B.: High-resolution image synthesis with latent diffusion models. In: Proceedings of the IEEE/CVF Conference on Computer Vision and Pattern Recognition, pp. 10684–10695 (2022)
36. Roose, K.: AI-Generated Art Won a Prize. Artists Aren't Happy. N.Y. Times, September 2022. https://www.nytimes.com/2022/09/02/technology/ai-artificial-intelligence-artists.html
37. Saharia, C., et al.: Photorealistic text-to-image diffusion models with deep language understanding. Adv. Neural. Inf. Process. Syst. **35**, 36479–36494 (2022)

38. Sauer, A., Karras, T., Laine, S., Geiger, A., Aila, T.: Stylegan-t: unlocking the power of gans for fast large-scale text-to-image synthesis. arXiv preprint arXiv:2301.09515 (2023)
39. Sauer, A., Lorenz, D., Blattmann, A., Rombach, R.: Adversarial diffusion distillation. arXiv preprint arXiv:2311.17042 (2023)
40. Schuhmann, C., et al.: LAION-5B: an open large-scale dataset for training next generation image-text models. Adv. Neural Inform. Process. Syst. (2022)
41. Wah, C., Branson, S., Welinder, P., Perona, P., Belongie, S.: The caltech-ucsd birds-200-2011 dataset (2011)
42. Wang, Z., et al.: ProlificDreamer: high-fidelity and diverse text-to-3D generation with variational score distillation. Adv. Neural Inform. Process. Syst. (2023)
43. Wortsman, M., et al.: Model soups: averaging weights of multiple fine-tuned models improves accuracy without increasing inference time. In: International Conference on Machine Learning, pp. 23965–23998. PMLR (2022)
44. Wu, X., et al.: Human Preference Score v2: a solid benchmark for evaluating human preferences of text-to-image synthesis. arXiv preprint arXiv:2306.09341 (2023)
45. Xu, Y., Zhao, Y., Xiao, Z., Hou, T.: Ufogen: you forward once large scale text-to-image generation via diffusion GANs (2023)
46. Yadav, P., Tam, D., Choshen, L., Raffel, C., Bansal, M.: Resolving interference when merging models. arXiv preprint arXiv:2306.01708 (2023)
47. Yin, T., et al.: One-step diffusion with distribution matching distillation. In: Proceedings of the IEEE/CVF Conference on Computer Vision and Pattern Recognition (2024)
48. Yu, J., et al.: Scaling autoregressive models for content-rich text-to-image generation. arXiv preprint arXiv:2206.10789 (2022)
49. Zhang, Y., Hooi, B.: Hipa: enabling one-step text-to-image diffusion models via high-frequency-promoting adaptation. arXiv preprint arXiv:2311.18158 (2023)
50. Zhao, Y., Xu, Y., Xiao, Z., Hou, T.: Mobilediffusion: subsecond text-to-image generation on mobile devices (2023)

Learning to Localize Actions in Instructional Videos with LLM-Based Multi-pathway Text-Video Alignment

Yuxiao Chen[1](✉), Kai Li[2], Wentao Bao[4], Deep Patel[3], Yu Kong[4], Martin Renqiang Min[3], and Dimitris N. Metaxas[1]

[1] Rutgers University, Piscataway, USA
{yc984,dnm}@cs.rutgers.edu
[2] Meta, Menlo Park, USA
[3] NEC Labs America-Princeton, Princeton, USA
{dpatel,renqiang}@nec-labs.com
[4] Michigan State University, East Lansing, USA
{baowenta,yukong}@msu.edu

Abstract. Learning to localize temporal boundaries of procedure steps in instructional videos is challenging due to the limited availability of annotated large-scale training videos. Recent works focus on learning the cross-modal alignment between video segments and ASR-transcripted narration texts through contrastive learning. However, these methods fail to account for the alignment noise, *i.e.*, irrelevant narrations to the instructional task in videos and unreliable timestamps in narrations. To address these challenges, this work proposes a novel training framework. Motivated by the strong capabilities of Large Language Models (LLMs) in procedure understanding and text summarization, we first apply an LLM to filter out task-irrelevant information and summarize task-related procedure steps (LLM-steps) from narrations. To further generate reliable pseudo-matching between the LLM-steps and the video for training, we propose the Multi-Pathway Text-Video Alignment (MPTVA) strategy. The key idea is to measure alignment between LLM-steps and videos via multiple pathways, including: (1) step-narration-video alignment using narration timestamps, (2) direct step-to-video alignment based on their long-term semantic similarity, and (3) direct step-to-video alignment focusing on short-term fine-grained semantic similarity learned from general video domains. The results from different pathways are fused to generate reliable pseudo step-video matching. We conducted extensive experiments across various tasks and problem settings to evaluate our proposed method. Our approach surpasses state-of-the-art methods in three downstream tasks: procedure step grounding, step localization, and narration grounding by 5.9%, 3.1%, and 2.8%.

Supplementary Information The online version contains supplementary material available at https://doi.org/10.1007/978-3-031-73007-8_12.

1 Introduction

Instructional videos are a type of educational content presented in video format. They demonstrate the steps required to perform specific tasks, such as cooking, repairing cars, and applying makeup [15,16,32,40]. These videos are invaluable resources for individuals seeking to learn new skills or knowledge in various domains. Instructional videos are in long duration since they aim to provide comprehensive visual guidance and instructions on task execution. Therefore, developing an automated system that can temporally localize procedural steps within instructional videos is crucial to facilitate the accessibility and utility of these educational resources.

Previous works addressed the task of localizing procedure steps either in a fully supervised [4,5,23,40] or weakly supervised [7,16,28] manners. These methods train models with annotated temporal boundaries or sequence orders of procedure steps. Although these methods have achieved notable performance, they require massive training videos with annotations, which are expensive and labor-intensive to collect. Recent methods [14,18,30,37] have explored training models using narrated instructional videos. Specifically, they train models to learn the cross-modal alignment between the text narrations and video segments through contrastive learning [31], where the temporally overlapped video segments and narrations are treated as positive pairs.

However, this approach is suboptimal since narrations are *noisily* aligned with video contents [18]. The noise mainly originates from two aspects. Firstly, some narrations may describe the information that is irrelevant to the task of the video. For instance, as illustrated in Fig. 1, narrations highlighted in red boxes discuss the demonstrator's interest in a food or background sounds, which are not related to the task of making a salad. Furthermore, the temporal correspondence between video frames and narrations derived from narrations' timestamps may not be consistently reliable since demonstrators may introduce steps before or

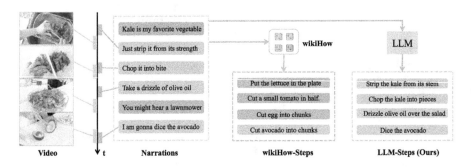

Fig. 1. Illustration of different text information in narrated instruction videos. The orange and green bars along the time (t) axis denote the temporal boundaries of procedure steps and timestamps of narrations, respectively. Sentences highlighted in green indicate task-relevant information, while those in red are task-irrelevant. (Color figure online)

after executing them. As illustrated in Fig. 1, the demonstrator introduces the procedure "take a drizzle of olive oil" before actually performing it.

Previous methods [25,30,50] propose to mitigate the first problem by leveraging the procedure steps defined by the human-constructed knowledge base wikiHow [22], denoted as wikiHow-steps, to supplement or replace narrations. However, wikiHow-steps may not always align with the actions demonstrated in the videos, as a task can be executed in ways that are not outlined in the knowledge base. For example, as shown in Fig. 1, the wikiHow-step "cutting tomato" is absent in the videos, while some steps performed in the videos, such as "take a drizzle of olive oil", are not included in the knowledge bases. Additionally, previous stuies [18,30] employs self-training technologies to tackle the issue of temporal misalignment. They first train models with supervision signals derived from narration timestamps, and then use the trained models to filter out unmatched narrations and establish new pseudo-alignment between narrations and videos. However, the effectiveness of these methods may be constrained, since the models employed to filter noises are trained on noisy narrations.

To address the above challenges, we propose a novel framework for training procedure steps localization models on narrated instructional videos. Figure 2 gives an overview of our proposed methods. Initially, we utilize a Large Language Model (LLM) [3,35,41,42] to filter out task-irrelevant information and summarize the procedure steps from narrations. LLMs have demonstrated strong capabilities in understanding procedures [1,12,49], thus serving as a knowledge base for instructional tasks. Additionally, LLMs possess robust text summarization capabilities [33,43]. Therefore, by using LLM, we can extract *video-specific* task-related procedure steps, denoted as LLM-steps. Compared with wikiHow-steps, which is designed to provide a general introduction to a task, our video-specific LLM-steps can align more closely with the video contents, as shown in Fig. 1.

Furthermore, we propose the Multi-Pathway Text-Video Alignment (MPTVA) strategy to generate reliable pseudo-matching between videos and LLM-steps. Our key insight is to measure the alignment between LLM-steps and video segments using different pathways, each of which captures their relationships from different aspects and provides complementary information. Consequently, combining those information can effectively filter out noise. Specifically, we leverage three distinct alignment pathways. The first pathway is designed to leverage the temporal correspondence derived from the timestamps of narrations. It first identifies narrations semantically similar to the LLM-steps and subsequently matches each LLM-step with video segments that fall within the timestamps of their semantically similar narrations. To mitigate the impact of timestamp noise, we further directly align LLM-steps with videos using a video-text alignment model pre-trained on long-duration instructional video data [30]. This pathway measures the alignment based on long-term global text-video semantic similarity. Additionally, we directly align LLM-steps with video segments utilizing a video-text foundation model [45] pre-trained on short-duration video-text datasets from various domains. This pathway not only captures fine-grained, short-term relationships but also incorporates broader knowledge learned from various video domains. We combine the alignment scores from the three pathways using mean-pooling, and then generate pseudo-labels from the obtained alignment score. It effectively leverages the mutual agreement among different pathways, which has been demonstrated to be effective in filtering noise for pseudo-labeling [18].

We conducted extensive experiments across various tasks and problem settings to evaluate our proposed method. Our approach significantly outperforms state-of-the-art methods in three downstream tasks: procedure step grounding, action step localization, and narration grounding. Notably, our method exhibits improvements over previous state-of-the-art methods by 5.9%, 2.8%, and 3.1% for procedure step grounding, narration grounding, and action step localization. We also draw the following key observations: (1) Models trained with our LLM-steps significantly outperform those trained with wikiHow-steps and a combination of wikiHow-steps and narrations by 10.7% and 6%, demonstrating the superior effectiveness of our LLM-steps. (2) When employing the supervision generated by our proposed MPTVA, models trained with narrations outperform those trained with wikiHow-steps. This underscores the limitations of using wikiHow-steps for training (3) Our ablation study reveals that applying all three alignment pathways improves the performance by 4.2% and 5.4% compared to the best model utilizing a single pathway, and by 3% and 4.5% compared to the best model utilizing two pathways. This indicates that the complementary information contributed by each pathway results in more reliable pseudo-labels.

2 Related Work

Step Localization in Instructional Videos. Previous studies on step localization in instructional videos can be classified into fully-supervised [4,5,23,40] and weakly-supervised [7,10,16,28] methods. Fully-supervised methods train models to directly predict the annotated boundaries of procedure steps. However, the main limitation of these approaches is the high cost to collect temporal annotations. On the other hand, weakly supervised methods only require the order or occurrence information of procedure steps for training, thus reducing the cost of annotation. However, collecting such annotations remains costly, as it requires annotators to watch entire videos. Recent work [14,18,30,37] attempts to eliminate labeling efforts by using narrated instructional videos. Their main focus is on how to leverage the noisy supervision signals provided by narrations. For example, some works [14,37] proposed learning an optimal sequence-to-sequence alignment between narrations and video segments by utilizing DTW [13,36]. However, these methods are not suited for the downstream tasks, such as step grounding, where the order of steps is unknown. Our method is more related to VINA [30] and TAN [18]. TAN proposed filtering out noisy alignments through mutual agreement between two models with distinct architectures. VINA extracted task-specific procedure steps from the WikiHow knowledge base to obtain auxiliary clean text descriptions. It aslo explored generating pseudo-alignments between procedure steps and videos using pre-trained long-term video-text alignment models. In contrast, our method leverages LLMs to extract video-specific task procedure steps, which align more closely with video contents than task-specific procedures used by VINA. Furthermore, our proposed MPTVA leverages a broader range of information which is extracted from various alignment paths and models pre-trained on diverse datasets, to generate pseudo-labels.

Learning from Noisily-Aligned Visual-Text Data. Recent studies have achieved remarkable progress in extracting knowledge from large-scale, weakly aligned visual-text datasets. In the field of images [19,34], it has been demonstrated that features with strong generalization abilities to various downstream tasks can be learned from large-scale weakly aligned image-text datasets through contrastive learning. In the video domain, this method has been extended to enhance video understanding. Various pretraining tasks [2,6,20,45,47] have been proposed to capture dynamic information in videos, and extensive video-text datasets are constructed to train the model. In the domain of instructional video analysis, researchers [25,29,31,46,50] mainly focus on learning feature representations from the HowTo100M dataset [32]. They mainly utilize narrations and video frames that are temporally overlapped as sources for contrastive learning or masked-based modeling. In this work, we leverage knowledge learned by these models from various data sources and pertaining tasks to measure the correspondence between video content and LLM-steps. We then combine the results to generate pseudo-alignment between LLM-steps and videos.

Training Data Generation Using LLMs. In order to reduce the cost of data collection, recent studies have explored leveraging LLMs to automatically generate training data. In the NLP domain [17,39,48], LLMs are employed to generate annotations for various tasks such as relevance detection, topic classification, and sentiment analysis. In the image-text area, LLMs are employed to generate question-answer text pairs from image captions for instruction tuning [9,11,26,27]. In video datasets, LLMs are primarily used to enrich the text descriptions of videos. For instance, MAXI [24] leverages LLMs to generate textual descriptions of videos from their category names for training. InternVideo [45] utilizes LLMs to generate the action-level descriptions of videos from their frame-level captions. HowtoCap [38] leveraged LLMs to enhance the detail in narrations of instructional videos. By contrast, our approach apply the procedure understanding and text summarization capabilities of LLMs to filter out irrelevant information and summarize the task-related procedural steps from narrations.

3 Method

3.1 Problem Introduction

In this work, we have a set of narrated instruction videos for training. A video $\mathcal{V}$ consists of a sequence of non-overlapping segments $\{v_1, v_2, ..., v_T\}$, where v_i

is the i-th segment and T is the number of video segments. The narration sentences $\mathcal{N} = \{n_1, n_2, ..., n_K\}$ of the video, where K is the number of narrations, are transcribed from the audio by using the ASR system. Each narration n_i is associated with the start and end timestamps within the video. We aim to use these videos and narrations to train a model Φ that takes a video and a sequence of sentences as inputs, and produces an alignment measurement between each video segment and sentence. The trained Φ is deployed to identify the temporal regions of procedural steps within the videos.

A straightforward way to solve the problem is to assume $\mathcal{N}$ are matched with the video segments within their timestamps, while unmatched with others. Specifically, Φ takes $\mathcal{V}$ and $\mathcal{N}$ as inputs to predict a similarity matrix between $\mathcal{V}$ and $\mathcal{N}$:

$$\hat{\mathbf{A}} = \phi(\mathcal{N}, \mathcal{V}) \in \mathbb{R}^{K \times T} \tag{1}$$

The model Φ is trained to maximize the similarity between matched video segments and narrations, while decrease those between unmatched ones. This is achieved by minimizing the MIL-NCE loss [31]:

$$\mathcal{L}(\mathbf{Y}^{NV}, \hat{\mathbf{A}}) = -\frac{1}{K} \sum_{k=1}^{K} \log \left(\frac{\sum_{t=1}^{T} Y_{k,t}^{NV} \exp(\hat{A}_{k,t}/\eta)}{\sum_{t=1}^{T} \exp(\hat{A}_{k,t}/\eta)} \right), \tag{2}$$

where $\mathbf{Y}^{NV}$ is the *pseudo matching matrix* between $\mathcal{N}$ and $\mathcal{V}$ constructed by the timestamps of narrations. $Y_{k,t}^{NV}$ is the element at the k-th row and t-th column. It is equal to one if the v_t is within the timestamps of n_k, and zero otherwise. η is the softmax temperature parameter.

However, the solution is sub-optimal, since n_i might not be relevant to any visual content of V, and the matching relationship encoded in $\mathbf{Y}^{NV}$ is unreliable, due to the temporal misalignment problem of timestamps.

3.2 Proposed Framework

Figure 2 gives an overview of our proposed methods. We first apply an LLM to process narrations $\mathcal{N}$, filtering out task-unrelated information and summarizing task-relevant procedure steps $\mathcal{S}$. Subsequently, we extract the pseudo matching matrix $\mathbf{Y}^{SV}$ between $\mathcal{S}$ and $\mathcal{V}$ by exploring multiple alignment pathways. The pseudo-alignments $\mathbf{Y}^{SV}$ are used as supervision to train the model Φ to learn the alignment between $\mathcal{S}$ and $\mathcal{V}$.

Extracting Procedure Steps Using LLM. Recent studies demonstrate that LLMs possess a strong capability for text summarization [33,43] and understanding task procedures [1,12,49]. Motivated by this, we propose using LLMs to filter video-irrelevant information and extract task-related procedural steps from narrations. To this end, we design the following prompt $\mathbf{P}$, which instructs the LLM model to (1) summarize the procedure steps relevant to the task shown in the video and (2) filter out irrelevant information, such as colloquial sentences:

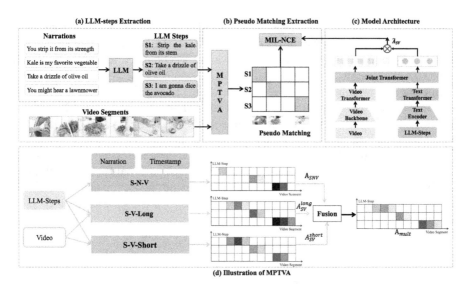

Fig. 2. Overview of our proposed method. (a): We first use an LLM to summarize task-relevant LLM-steps from narrations. (b): We then extract the pseudo-matching between LLM-steps and video segments using our proposed MPTVA. (c): The extracted pseudo-alignments are used as supervision to train the model to minimize the MIL-NCE loss. (d): The illustration of the proposed MPTVA strategy for pseudo-label generation.

"I will give you the text narrations extracted from an instruction video. Your task is to summarize the procedure steps that are relevant to the task of the video from the inputs. Please filter colloquial sentences in the speech."

This prompt, along with the narrations $\mathcal{N}$, is concatenated and then fed into an LLM model, resulting in the LLM-generated procedural steps $\mathcal{S}$ (denoted LLM-steps). The process is formulated as follows:

$$\mathcal{S} = LLM([\mathbf{P}; \mathcal{N}]). \tag{3}$$

We denote $\mathcal{S} = \{s_1, s_2 ..., s_L\}$ where s_i is the i-th out of L LLM-steps.

Figure 1 shows examples of obtained LLM-steps. We can see that (1) $\mathcal{S}$ are more descriptive than $\mathcal{N}$, and (2) most of the information unrelated to the video task contained in $\mathcal{N}$, such as "You might hear a lawnmower", have been filtered out in $\mathcal{S}$. In addition, we observe $\mathcal{S}$ has retained most of the procedure information related to the task, and can align more contents of the videos than the wikiHow-steps. which are widely in the previous work [25,30,50]. Please refer to the supplementary material for more examples.

Multi-pathway Text-Video Alignment. The obtained LLM-steps are not accompanied by annotations indicating temporal alignment with videos. To tackle this challenge, we propose the Multi-Pathway Text-Video Alignment (MPTVA) strategy to generate reliable pseudo-alignment between LLM-steps and video segments for training. The key idea is to measure the alignment between $\mathcal{S}$ and $\mathcal{V}$ in different pathways, each of which provides complementary information to others. The results from different pathways are then fused to extract the pseudo-matching relationship.

Specifically, we measure their alignment using three different pathways. These include: (1) aligning $\mathcal{S}$ with narrations $\mathcal{N}$ and then to $\mathcal{V}$ (S-N-V), (2) directly aligning S to $\mathcal{V}$ using the long-term video-text alignment models pre-trained in the instructional video domain (S-V-long), and (3) direct alignment using a video-text foundation model pre-trained on short-term video-text datasets from various domains (S-V-short). The three pathways capture the relationship between $\mathcal{S}$ and $\mathcal{V}$ based on different information, including the timestamps of narrations, long-term global text-video semantic similarity, and short-term fine-grained text-video semantic similarity.

The S-N-V path first identifies semantically similar narrations of $\mathcal{S}$, and then associates $\mathcal{S}$ with the video frames that are within the timestamps of these narrations. Formally, the semantic similarity distributions between the $\mathcal{S}$ and $\mathcal{N}$ of a video are computed as below:

$$\mathbf{A}_{SN} = \text{Softmax}\left(\frac{E_t(\mathcal{S}) \cdot E_t(\mathcal{N})^\top}{\tau}\right) \quad (4)$$

where $\mathbf{A}_{SN} \in \mathbb{R}^{L \times K}$, $E_t(\mathcal{S}) \in \mathbb{R}^{L \times D}$, and $E_t(\mathcal{N}) \in \mathbb{R}^{K \times D}$. Note that $E_t(\mathcal{S})$ and $E_t(\mathcal{N})$ are the L2 normalized embeddings of the $\mathcal{S}$ and $\mathcal{N}$ extracted using a pre-trained text encoder E_t, respectively. τ is the temperature. Then, the alignment score between LLM steps and video frames can be obtained as follows:

$$\mathbf{A}_{SNV} = \mathbf{A}_{SN} \cdot \mathbf{Y}^{NV} \in \mathbb{R}^{L \times T} \quad (5)$$

$\mathbf{Y}^{NV}$ is the pseudo matching matrix constructed based on the timestamps of narrations, as introduced in Eq. 2.

Since the temporal alignment derived from timestamps can be noisy, we incorporate additional information by directly aligning LLM steps and video segments. This is achieved by using a *long-term* video-text alignment model pre-trained using the instructional video data.

Specifically, we first use the text encoder E_t^L and video encoder E_v^L of the model to extract the embeddings of the $\mathcal{S}$ and $\mathcal{V}$, respectively. The cosine similarity between each LLM-step and video segment is calculated, resulting in the long-term step-to-video alignment score matrix $\mathbf{A}_{SV}^{long}$:

$$\mathbf{A}_{SV}^{long} = E_t^L(\mathcal{S}) \cdot E_v^L(\mathcal{V})^\top. \quad (6)$$

We also obtained additional direct alignment between LLM steps and videos, denoted as $\mathbf{A}_{SV}^{short}$, using the video foundation models pre-trained on general short-term video-text data. $\mathbf{A}_{SV}^{short}$ is obtained follow the same way as Eq. 6, but use the text encoders E_t^S and E_v^S of the video foundation model. This pathway complements the alignment by focusing on short-term, fine-grained information, such as human-object interactions. Furthermore, this pathway introduces knowledge learned from a broader domain.

The multiple pathways alignment score matrix $\mathbf{A}_{mult}$ is obtained by mean-pooling $\mathbf{A}_{SNV}$, $\mathbf{A}_{SV}^{long}$, and $\mathbf{A}_{SV}^{short}$.

Discussion. Previous works typically generate pseudo-labels by focusing on a single path, either $\mathbf{A}_{SNV}$ or $\mathbf{A}_{SV}^{long}$. It potentially results in labels that are affected by the noise present in each pathway. In contrast, our method leverages information from multiple paths, thereby enabling the generation of more reliable pseudo-alignments. Furthermore, our proposed $\mathbf{A}_{SV}^{short}$ introduces additional knowledge beyond the instructional video domain. Our method shares the same idea with TAN [18] that uses the mutual agreement between different models to filter noise in pseudo-labels. However, our approach leverages a broader range of complementary information through multiple alignment pathways and models pre-trained on a variety of datasets.

Learning LLM-Step Video Alignment. Given $\mathbf{A}_{mult}$, we construct the pseudo-matching matrix $\mathbf{Y}^{SV}$ as follows: For an LLM step s_i, we first identify the video segment v_k that has the highest alignment score with s_i. v_k is treated as the matched video segment for s_i, and thus we set $\mathbf{Y}^{SV}(i,k)$ to one. Additionally, for any video segment v_j, if its temporal distance from v_k (defined as $|j-k|$) falls within a predetermined window size W, it is also labeled as a matched segment to s_i, leading us to mark $\mathbf{Y}^{SV}(i,j)$ as ones accordingly. To ensure the reliability of pseudo-labels, we exclude LLM steps for which the highest score in $\mathbf{A}_{mult}$ falls below a specified threshold γ.

We use $\mathbf{Y}^{SV}$ as the supervision to train Φ to learn the alignment between $\mathcal{S}$ and $\mathcal{V}$. The model Φ takes $\mathcal{S}$ and $\mathcal{V}$ as inputs, and output a similarity matrix $\hat{\mathbf{A}}_{\mathbf{SV}}$ between $\mathcal{S}$ and $\mathcal{V}$:

$$\hat{\mathbf{A}}_{\mathbf{SV}} = \Phi(\mathcal{S}, \mathcal{V}), \tag{7}$$

where $\hat{\mathbf{A}}_{\mathbf{SV}} \in \mathbb{R}^{L \times T}$. The model Φ is trained to pull matched video segments and LLM-steps closely while pushing way unmatched ones, by minimizing the MIL-NCE loss $\mathcal{L}(\mathbf{Y}^{SV}, \hat{\mathbf{A}}_{\mathbf{SV}})$ shown in Eq. 2.

3.3 Architecture

Figure 2 illustrates the architecture of Φ. It consists of the video backbone and text encoder, the unimodal Transformers encoder [44] for video and text, and a joint-modal Transformer. Given the input video $\mathcal{V}$ and LLM-steps $\mathcal{S}$, we first extract their embeddings by using the video backbone f_b and text backbone g_b:

$$\mathbf{F}_V = f_b(\mathcal{V}), \quad \mathbf{F}_S = g_b(\mathcal{S}), \tag{8}$$

where $\mathbf{F}_V \in \mathbb{R}^{T \times D}$ and $\mathbf{F}_S \in \mathbb{R}^{L \times D}$. The extracted video segments and text features from the backbone are subsequently fed into the model-specific unimodal

Transformers, f_{trs} for video and g_{trs} for text, to refine the features by modeling the intra-modal relationships:

$$\overline{\mathbf{F}}_V = f_{trs}(\mathbf{F}_V + \mathbf{F}_{\mathrm{PE}}), \quad \overline{\mathbf{F}}_S = g_{trs}(\mathbf{F}_S), \tag{9}$$

where $\overline{\mathbf{F}}_V \in \mathbb{R}^{T \times D}$ and $\overline{\mathbf{F}}_S \in \mathbb{R}^{L \times D}$. Note that we add positional embedding (PE), denoted as $\mathbf{F}_{\mathrm{PE}}$, into the video segments to represent the time orders of video segments. We do not use PE with LLM-steps since the input procedure steps may not have sequential orders on downstream tasks [16,30].

Finally, the joint Transformer h_{trs} takes the sequential concatenation of $\hat{\mathbf{F}}_V$ and $\hat{\mathbf{F}}_S$ as inputs to further model the fine-grained cross-modal relationship between video segments and LLM-steps. It outputs the final feature representations of video segments $\hat{\mathbf{F}}_V$ and LLM-steps $\hat{\mathbf{F}}_S$.

$$[\hat{\mathbf{F}}_V; \hat{\mathbf{F}}_S] = h_{trs}([\overline{\mathbf{F}}_V; \overline{\mathbf{F}}_S]) \tag{10}$$

The similarity matrix $\hat{\mathbf{A}}_{SV}$ between LLM-steps and video segments are obtained by calculating the cosine similarity between $\hat{\mathbf{F}}_V$ and $\hat{\mathbf{F}}_S$:

$$\hat{\mathbf{A}}_{SV} = \hat{\mathbf{F}}_S \cdot \hat{\mathbf{F}}_V^{\top}. \tag{11}$$

4 Experiment

4.1 Dataset

HTM-370K (Training). Following the previous work [18,30] we train our model on the Food & Entertaining subset of Howto100M datasets [32]. It consists of about 370K videos collected from YouTube.

HT-Step (Evaluation). This dataset [30] consist of 600 videos selected from the Howto100M dataset [32]. Each video is manually matched to an instructional article of the wikiHow dataset [22]. The temporal boundaries of the procedure steps from the instructional articles are annotated if they are present in the videos. We use the dataset to evaluate the performance of models for *procedure step grounding*. Following the previous work [30], we report the R@1 metric. R@1 is defined as the ratio of successfully recalled steps to the total number of steps that occur, where a step is considered as successfully recalled if its most relevant video segment falls within the ground truth boundary.

CrossTask (Evaluation). This dataset [16] has about 4.7K instructional videos, which can be divided into 65 related tasks and 18 primary tasks. Following the previous study [30], we use the dataset to evaluate the model for zero-shot *action step localization* task. Action step localization is to localize the temporal positions for each occurring action step. Compared to the procedure steps in the HT-Step dataset, the action steps represent more atomic actions, and their text descriptions are more abstract, containing fewer details. The metric is Average Recall@1 (Avg. R@1) [30]. It's computed by first calculating the R@1 metric for each task and then averaging across tasks. We report the results averaging across 20 random sets, each containing 1850 videos from the primary tasks [30].

HTM-Align (Evaluation). This dataset [18] contains 80 videos with annotated temporal alignment between ASR transcriptions and the video contents. It is used to evaluate our model for narration grounding. We report the R@1 metric following the common practice [10,18].

4.2 Implementation Details

Following the common practice [18,30], we equally divide each input video into a sequence of 1-s non-overlapped segments and set the frame rate to 16 FPS. The pre-trained S3D video encoder, released by [31], is used as the video backbone f_b to extract a feature embedding from each segment. Following [18,30], f_b is kept fixed during training. The text backbone g_b is a Bag-of-Words (BoW) model based on Word2Vec embeddings [31]. Following [30], we use the TAN* models, a variance of the TAN model [18] that is pre-trained on long-term instructional videos, as E_e^L and E_v^L. We also use the text encoder of TAN* as E_t. In addition, we set E_t^S and E_v^S as the pre-trained text and video encoders of the InternVideo-MM-L14 model [45]. We use Llama2-7B [42] to extract LLM-steps from narrations. We set the temperature hyperparameters η and τ to 0.07. The model is trained on Nvidia A100 GPUs. We set the batch size to 32. The AdamW optimizer with an initial learning rate of 2×10^{-4} is employed for training. The learning rate follows a cosine decay over 12 epochs. For further details, please refer to the supplementary material.

Table 1. Results for the step grounding and action step localization on the HT-Step and CrossTask datasets when using different pathways to generate pseudo-labels. "ZS" indicates applying the correspondent pre-trained video-text alignment models to perform the downstream tasks in a zero-shot manner. The symbol $\checkmark$ indicates that the respective pathways are used.

Pathway Type			HT-Step	CrossTask
S-N-V	S-V-Long	S-V-Short	R@1 (%) ↑	Avg. R@1 (%) ↑
	ZS		30.3	32.3
		ZS	37.3	39.7
$\checkmark$			34.2	36.4
	$\checkmark$		35.0	39.1
		$\checkmark$	37.7	41.6
$\checkmark$	$\checkmark$		37.9	42.6
$\checkmark$		$\checkmark$	38.0	43.7
	$\checkmark$	$\checkmark$	38.9	42.5
$\checkmark$	$\checkmark$	$\checkmark$	**41.9**	**47.0**

4.3 Ablation Study

Effect of Multi-pathway Text-Video Alignment. We evaluate the effectiveness of our proposed MPTVA strategy by comparing the performances of models trained with labels generated through various pathways. The results are shown in Table 1. It is observed that models trained exclusively with S-V-Short or S-V-Long labels significantly outperform the pre-trained short-term and long-term video-text alignment models, respectively. The results suggest that the models learn knowledge surpassing the capabilities of the alignment models used for pseudo-label extraction. We can see that the model trained with S-N-V underperforms those trained with S-V-Short or S-V-Long. This may be attributed to the noise present in the timestamps. Moreover, we can observe that using any combination of two pathways achieves better performance improvement than using only one pathway. More importantly, the best improvement is achieved when considering all three pathways. The results demonstrate that each pathway provides complementary information. By fusing these pathways, we can leverage this combined information, resulting in more reliable pseudo-labels.

Table 2. Results for step grounding and action step localization on the HT-Step and CrossTask datasets when using different training data. "N", "W", and "S" denote narrations, wikiHow-steps, and LLM-steps, respectively. "TS" and "MPTVA" denote pseudo-labels generated from timestamps and our MPTVA strategy, respectively.

Training Data	HT-Step	CrossTask
	R@1 ↑	Avg. R@1 ↑
N-TS	30.3	32.3
N-MPTVA	32.1	38.0
W-MPTVA	31.2	37.8
N-MPTVA+W-MPTVA	35.9	40.6
S-MPTVA	41.9	47.0
N-MPTVA + S-MPTVA	**43.3**	**47.9**

Effectiveness of LLM-Steps. To evaluate the effectiveness of LLM-steps, we compare the model trained with different text inputs, including narrations, procedure steps extracted from the wikiHow database following [30], and our LLM-steps. The results are reported in Table 2. It can be observed that the N-TS model, which takes narrations as input and employs timestamps for supervision, exhibits the lowest performance. Using the pseudo-labels generated by our proposed MPTVA for narrations (N-MPTVA) leads to improved performance over N-TS, showing that our proposed MPTVA strategy is also effective for noisy narrations. Notably, this model (N-MPTVA) surpasses the one (W-MPTVA) utilizing wikiHow-steps as input. The results indicate that, when using the same supervision (MPTVA), wikiHow-steps are less informative than narrations. It may be attributed to mismatches between instruction articles and videos. Additionally, it could be due to the fact that even the matched instruction articles only cover a small portion of the procedure steps depicted in the videos, considering that a task can be performed in multiple ways. In addition, we observe that using narration and wikiHow articles together, as previous work [30], can enhance model performance. Nonetheless, training the models only with our LLM-steps outperforms this setting by 6% and 6.4% for step-grounding and action step localization tasks, respectively. These results underscore the effectiveness of LLM-steps in providing highly relevant and clean information for procedure step localization. In addition, we also note that incorporating the information of narration with LLM-steps can further boost the performance for 1.4% and 0.9%.

Table 3. Results of the models trained with the pseudo-labels that are generated with different filter thresholds γ and window sizes W.

γ	W	Valid Step Ratio	HT-Step	CrossTask
			R@1 (%)	Avg. R@1
0.60	5	37%	39.2	44.2
0.65	5	15%	40.6	45.2
0.70	5	3%	35.4	41.2
0.65	1	15%	40.3	45.7
0.65	2	15%	**41.9**	**47.0**
0.65	5	15%	40.6	45.2

Ablation of Filtering Threshold and Window Size for Pseudo-label Generation. We report the performance of models trained with pseudo-labels generated with different filtering thresholds (γ) and window sizes (W) in Table 3. We first fix the W as 5 to assess the influence of γ. We can see that the best performance is observed when γ is set to 0.65, and about 15% LLM steps are selected for training. Performance decreases as γ decreases from 0.65, likely due to the introduction of more noisy pseudo-labels. In addition, increasing γ from 0.65 significantly drops the performances, since a large percent of data (97%) is excluded. We then fix γ and evaluate the impact of varying W. We can the best performance is achieved when W is 2. Increasing W beyond this point reduces performance, likely due to the mislabeling of irrelevant frames as matched. Additionally, decreasing W from 2 to 1 results in a slight decrease in performance, as it causes some nearby relevant frames to be annotated as unmatched.

Table 4. Comparison with state-of-the-art methods for the narration grounding and step grounding on the HTM-Align and HT-Step datasets, respectively.

Method	HTM-Align R@1 ↑	HT-Step R@1 ↑
CLIP (ViT-B/32) [34]	23.4	–
MIL-NCE [31]	34.2	30.7
InternVideo-MM-L14 [45]	40.1	37.3
TAN [18]	49.4	–
TAN* (Joint, S1, PE+LC) [30]	63	31.2
VINA(w/o nar) [30]	–	35.6
VINA [30]	66.5	37.4
Ours	**69.3**	**43.3**

Table 5. Comparison with state-of-the-art methods for action step localization on CrossTask Dataset.

Method	↑Avg. R@1 (%)
HT100M [32]	33.6
VideoCLIP [46]	33.9
MCN [8]	35.1
DWSA [37]	35.3
MIL-NCE [31]	40.5
Zhukov [16]	40.5
VT-TWINS* [21]	40.7
UniVL [29]	42.0
VINA [30]	44.8
Ours	**47.9**

4.4 Comparison with State-of-the-Art Approaches

Results for Narration and Step Grounding. We compare our method with state-of-the-art approaches for narration and step grounding tasks on the HTM-Align and HT-Step datasets, and the results are presented in Table 4. The results for narration grounding on the HTM-Align dataset were obtained by training a model that adds positional embedding to the text input, following previous work [30]. We find that our method significantly outperforms previous approaches, achieving state-of-the-art performance for both narration and step-grounding tasks. Notably, our method improves the performance of VINA, which uses narrations and procedure steps from the wikiHow dataset for training, by 2.8% and 5.9% for narration and step grounding, respectively. Moreover, VINA [30] additionally uses narrations of videos during test time. In a fair comparison, where narrations are not available during testing, our method outperforms VINA (VINA w/o nar) by 7.7%. The results further demonstrate the effectiveness of our proposed methods.

Results for Action Step Localization. We further compare our model with other works on action step localization using the CrossTask dataset. The results are reported in Table 5. Our method improves the state-of-the-art performance by 3.1%. The results show the strong transfer capabilities of our learned models to different downstream datasets. In addition, the results indicate that our model can handle text with varying levels of information granularity, from abstract action steps to detailed, enriched procedure steps and various narrations.

5 Conclusions

In this work, we introduce a novel learning framework designed to train models for localizing procedure steps in noisy, narrated instructional videos. Initially, we utilize LLMs to filter out text information irrelevant to the tasks depicted in the videos and to summarize task-related procedural steps from the narrations. Subsequently, we employ our proposed Multi-Pathway Text-Video Alignment strategy to generate pseudo-alignments between LLM-steps and video segments for training purposes. Extensive experiments conducted across three datasets for three downstream tasks show that our method establishes new state-of-the-art benchmarks. Furthermore, ablation studies confirm that the procedural steps extracted by our framework serve as superior text inputs compared to both narration and procedural steps derived from human-constructed knowledge bases.

Acknowledgements. This research has been partially funded by research grants to Dimitris N. Metaxas through NSF: 2310966, 2235405, 2212301, 2003874, and FA9550-23-1-0417 and NIH 2R01HL127661.

References

1. Ahn, M., et al.: Do as i can, not as i say: grounding language in robotic affordances. arXiv preprint arXiv:2204.01691 (2022)
2. Bain, M., Nagrani, A., Varol, G., Zisserman, A.: Frozen in time: a joint video and image encoder for end-to-end retrieval. In: Proceedings of the IEEE/CVF International Conference on Computer Vision, pp. 1728–1738 (2021)
3. Brown, T., et al.: Language models are few-shot learners. In: Advances in Neural Information Processing Systems, vol. 33, pp. 1877–1901 (2020)
4. Caba Heilbron, F., Escorcia, V., Ghanem, B., Carlos Niebles, J.: ActivityNet: a large-scale video benchmark for human activity understanding. In: Proceedings of the IEEE Conference on Computer Vision and Pattern Recognition, pp. 961–970 (2015)
5. Cao, M., Yang, T., Weng, J., Zhang, C., Wang, J., Zou, Y.: LocVTP: video-text pretraining for temporal localization. In: Avidan, S., Brostow, G., Cissé, M., Farinella, G.M., Hassner, T. (eds.) ECCV 2022. LNCS, vol. 13686, pp. 38–56. Springer, Cham (2022). https://doi.org/10.1007/978-3-031-19809-0_3

6. Carreira, J., Zisserman, A.: Quo vadis, action recognition? A new model and the kinetics dataset. In: Proceedings of the IEEE Conference on Computer Vision and Pattern Recognition, pp. 6299–6308 (2017)
7. Chang, C.Y., Huang, D.A., Sui, Y., Fei-Fei, L., Niebles, J.C.: D3TW: discriminative differentiable dynamic time warping for weakly supervised action alignment and segmentation. In: Proceedings of the IEEE/CVF Conference on Computer Vision and Pattern Recognition, pp. 3546–3555 (2019)
8. Chen, B., et al.: Multimodal clustering networks for self-supervised learning from unlabeled videos. In: Proceedings of the IEEE/CVF International Conference on Computer Vision, pp. 8012–8021 (2021)
9. Chen, D., Liu, J., Dai, W., Wang, B.: Visual instruction tuning with polite flamingo. arXiv preprint arXiv:2307.01003 (2023)
10. Chen, L., et al.: Weakly-supervised temporal article grounding. arXiv preprint arXiv:2210.12444 (2022)
11. Dai, W., et al.: InstructBLIP: towards general-purpose vision-language models with instruction tuning. In: Oh, A., Naumann, T., Globerson, A., Saenko, K., Hardt, M., Levine, S. (eds.) Advances in Neural Information Processing Systems 36: Annual Conference on Neural Information Processing Systems 2023, NeurIPS 2023, New Orleans, LA, USA, 10–16 December 2023 (2023). http://papers.nips.cc/paper_files/paper/2023/hash/9a6a435e75419a836fe47ab6793623e6-Abstract-Conference.html
12. Driess, D., et al.: PaLM-E: an embodied multimodal language model. arXiv preprint arXiv:2303.03378 (2023)
13. Dvornik, M., Hadji, I., Derpanis, K.G., Garg, A., Jepson, A.: Drop-DTW: aligning common signal between sequences while dropping outliers. In: Advances in Neural Information Processing Systems, vol. 34, pp. 13782–13793 (2021)
14. Dvornik, N., Hadji, I., Zhang, R., Derpanis, K.G., Wildes, R.P., Jepson, A.D.: StepFormer: self-supervised step discovery and localization in instructional videos. In: Proceedings of the IEEE/CVF Conference on Computer Vision and Pattern Recognition, pp. 18952–18961 (2023)
15. Elhamifar, E., Naing, Z.: Unsupervised procedure learning via joint dynamic summarization. In: Proceedings of the IEEE/CVF International Conference on Computer Vision, pp. 6341–6350 (2019)
16. Gan, Z., et al.: Vision-language pre-training: basics, recent advances, and future trends. Found. Trends® Comput. Graph. Vis. **14**(3–4), 163–352 (2022)
17. Gilardi, F., Alizadeh, M., Kubli, M.: ChatGPT outperforms crowd-workers for text-annotation tasks. arXiv preprint arXiv:2303.15056 (2023)
18. Han, T., Xie, W., Zisserman, A.: Temporal alignment networks for long-term video. In: Proceedings of the IEEE/CVF Conference on Computer Vision and Pattern Recognition, pp. 2906–2916 (2022)
19. Jia, C., et al.: Scaling up visual and vision-language representation learning with noisy text supervision. In: International Conference on Machine Learning, pp. 4904–4916. PMLR (2021)
20. Kay, W., et al.: The kinetics human action video dataset. arXiv preprint arXiv:1705.06950 (2017)
21. Ko, D., et al.: Video-text representation learning via differentiable weak temporal alignment. In: Proceedings of the IEEE/CVF Conference on Computer Vision and Pattern Recognition, pp. 5016–5025 (2022)
22. Koupaee, M., Wang, W.Y.: WikiHow: a large scale text summarization dataset. arXiv preprint arXiv:1810.09305 (2018)

23. Lea, C., Vidal, R., Reiter, A., Hager, G.D.: Temporal convolutional networks: a unified approach to action segmentation. In: Hua, G., Jégou, H. (eds.) ECCV 2016, Part III. LNCS, vol. 9915, pp. 47–54. Springer, Cham (2016). https://doi.org/10.1007/978-3-319-49409-8_7
24. Lin, W., et al.: Match, expand and improve: unsupervised finetuning for zero-shot action recognition with language knowledge. arXiv preprint arXiv:2303.08914 (2023)
25. Lin, X., Petroni, F., Bertasius, G., Rohrbach, M., Chang, S.F., Torresani, L.: Learning to recognize procedural activities with distant supervision. In: Proceedings of the IEEE/CVF Conference on Computer Vision and Pattern Recognition, pp. 13853–13863 (2022)
26. Liu, H., Li, C., Li, Y., Lee, Y.J.: Improved baselines with visual instruction tuning. arXiv preprint arXiv:2310.03744 (2023)
27. Liu, H., Li, C., Wu, Q., Lee, Y.J.: Visual instruction tuning. In: Advances in Neural Information Processing Systems, vol. 36 (2024)
28. Lu, Z., Elhamifar, E.: Set-supervised action learning in procedural task videos via pairwise order consistency. In: Proceedings of the IEEE/CVF Conference on Computer Vision and Pattern Recognition, pp. 19903–19913 (2022)
29. Luo, H., et al.: UniVL: a unified video and language pre-training model for multimodal understanding and generation. arXiv preprint arXiv:2002.06353 (2020)
30. Mavroudi, E., Afouras, T., Torresani, L.: Learning to ground instructional articles in videos through narrations. arXiv preprint arXiv:2306.03802 (2023)
31. Miech, A., Alayrac, J.B., Smaira, L., Laptev, I., Sivic, J., Zisserman, A.: End-to-end learning of visual representations from uncurated instructional videos. In: Proceedings of the IEEE/CVF Conference on Computer Vision and Pattern Recognition, pp. 9879–9889 (2020)
32. Miech, A., Zhukov, D., Alayrac, J.B., Tapaswi, M., Laptev, I., Sivic, J.: HowTo100M: learning a text-video embedding by watching hundred million narrated video clips. In: Proceedings of the IEEE/CVF International Conference on Computer Vision, pp. 2630–2640 (2019)
33. Pu, X., Gao, M., Wan, X.: Summarization is (almost) dead. arXiv preprint arXiv:2309.09558 (2023)
34. Radford, A., et al.: Learning transferable visual models from natural language supervision. In: International Conference on Machine Learning, pp. 8748–8763. PMLR (2021)
35. Raffel, C., et al.: Exploring the limits of transfer learning with a unified text-to-text transformer. J. Mach. Learn. Res. **21**(1), 5485–5551 (2020)
36. Sakoe, H., Chiba, S.: Dynamic programming algorithm optimization for spoken word recognition. IEEE Trans. Acoust. Speech Sig. Process. **26**(1), 43–49 (1978)
37. Shen, Y., Wang, L., Elhamifar, E.: Learning to segment actions from visual and language instructions via differentiable weak sequence alignment. In: Proceedings of the IEEE/CVF Conference on Computer Vision and Pattern Recognition, pp. 10156–10165 (2021)
38. Shvetsova, N., Kukleva, A., Hong, X., Rupprecht, C., Schiele, B., Kuehne, H.: HowToCaption: prompting LLMS to transform video annotations at scale. arXiv preprint arXiv:2310.04900 (2023)
39. Tan, Z., et al.: Large language models for data annotation: a survey. arXiv preprint arXiv:2402.13446 (2024)
40. Tang, Y., et al.: COIN: a large-scale dataset for comprehensive instructional video analysis. In: Proceedings of the IEEE/CVF Conference on Computer Vision and Pattern Recognition, pp. 1207–1216 (2019)

41. Touvron, H., et al.: LLaMA: open and efficient foundation language models. arXiv preprint arXiv:2302.13971 (2023)
42. Touvron, H., et al.: LLaMA 2: open foundation and fine-tuned chat models. arXiv preprint arXiv:2307.09288 (2023)
43. Van Veen, D., et al.: Clinical text summarization: adapting large language models can outperform human experts. Res. Square (2023)
44. Vaswani, A., et al.: Attention is all you need. In: Advances in Neural Information Processing Systems, vol. 30 (2017)
45. Wang, Y., et al.: InternVideo: general video foundation models via generative and discriminative learning. arXiv preprint arXiv:2212.03191 (2022)
46. Xu, H., et al.: VideoCLIP: contrastive pre-training for zero-shot video-text understanding. arXiv preprint arXiv:2109.14084 (2021)
47. Xue, H., et al.: CLIP-ViP: adapting pre-trained image-text model to video-language representation alignment. arXiv preprint arXiv:2209.06430 (2022)
48. Ye, J., et al.: ZeroGen: efficient zero-shot learning via dataset generation. arXiv preprint arXiv:2202.07922 (2022)
49. Zhao, Q., et al.: AntGPT: can large language models help long-term action anticipation from videos? arXiv preprint arXiv:2307.16368 (2023)
50. Zhou, H., Martín-Martín, R., Kapadia, M., Savarese, S., Niebles, J.C.: Procedure-aware pretraining for instructional video understanding. In: Proceedings of the IEEE/CVF Conference on Computer Vision and Pattern Recognition, pp. 10727–10738 (2023)

Improving Hyperbolic Representations via Gromov-Wasserstein Regularization

Yifei Yang[1], Wonjun Lee[2], Dongmian Zou[3](✉), and Gilad Lerman[2]

[1] Wuhan University, Wuhan 430072, Hubei, China
yfyang@whu.edu.cn
[2] University of Minnesota, Minneapolis, MN 55455, USA
{lee01273,lerman}@umn.edu
[3] Duke Kunshan University, Kunshan 215316, Jiangsu, China
dongmian.zou@duke.edu

Abstract. Hyperbolic representations have shown remarkable efficacy in modeling inherent hierarchies and complexities within data structures. Hyperbolic neural networks have been commonly applied for learning such representations from data, but they often fall short in preserving the geometric structures of the original feature spaces. In response to this challenge, our work applies the Gromov-Wasserstein (GW) distance as a novel regularization mechanism within hyperbolic neural networks. The GW distance quantifies how well the original data structure is maintained after embedding the data in a hyperbolic space. Specifically, we explicitly treat the layers of the hyperbolic neural networks as a transport map and calculate the GW distance accordingly. We validate that the GW distance computed based on a training set well approximates the GW distance of the underlying data distribution. Our approach demonstrates consistent enhancements over current state-of-the-art methods across various tasks, including few-shot image classification, as well as semi-supervised graph link prediction and node classification.

Keywords: Gromov-Wasserstein distance · Hyperbolic neural networks · Few-shot learning · Semi-supervised learning

1 Introduction

The success of many machine learning applications relies on effectively capturing the geometric intricacies of complex data structures. This challenge necessitates a careful selection of the underlying space that aligns with the inherent properties of data. In this context, hyperbolic spaces have emerged as a popular choice for modeling data with hierarchical structures such as lexical databases [23] and phylogenetic trees [28]. This is attributed to their distinctive capacity to

---

Supplementary Information The online version contains supplementary material available at https://doi.org/10.1007/978-3-031-73007-8_13.

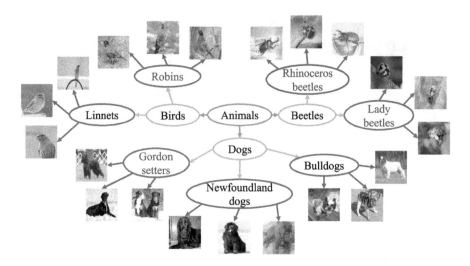

Fig. 1. Illustration of hierarchical structures in image data. The images are taken from the MiniImagenet dataset.

embody exponential growth and branching patterns, a feature rooted in their constant negative curvature. Such a geometric framework can reflect the tree-like architectures that are ubiquitous across numerous real-world networks.

Interestingly, hyperbolic spaces also play a significant role in computer vision [2,14], particularly in tasks like image classification and object recognition where hierarchical relationships among instances and concepts are common. For example, within the broader category of animals, there are dogs, which further subdivide into breeds such as bulldogs, Gordon setters, Newfoundland dogs, *etc.*. This is illustrated in Fig. 1. Hyperbolic embeddings can capture these hierarchies, thereby potentially improving classification performance and providing more interpretable models.

In crafting hyperbolic representations from non-hyperbolic features, two main strategies emerge: principled design and data-driven learning. The principled approach, exemplified by methods like Sarkar's algorithm [29], focuses on constructing embeddings that rigorously preserve the inherent geometric structure from data. This method particularly excels at maintaining hierarchical relationships within the data, ensuring that the embedded representation reflects structure of the original dataset with low distortion. On the other hand, the data-driven approach leverages the capabilities of end-to-end neural networks to learn the hyperbolic embedding directly from data. Such neural networks are well-known as hyperbolic neural networks (HNNs), which are pioneered in [10]. HNNs prioritize the development of hyperbolically coherent neural operations that are expressive and mirror the Euclidean counterparts. When dealing with input that is primarily Euclidean, HNNs employ straightforward operations, such as exponential maps, to transition features into the hyperbolic space, thereafter learning to extract deep features in a manner driven by the data. This method adapts the embedding to specific tasks, potentially unveiling intricate patterns and relationships inherent within data.

In our work, we aim to establish a methodology that integrates the benefits of low-distortion embedding into data-driven learning. In the context of learning where generalization is the goal, we can view hyperbolic embedding as a method of embedding the data distribution into hyperbolic spaces. Therefore, we need to compare the distributions before and after the embedding. This necessitates a comparison of data distributions before and after embedding, which is complicated by the fundamental differences in geometric principles, such as curvature and dimensionality, that govern Euclidean and hyperbolic spaces. Direct comparisons between such disparate spaces are inherently complex and challenging.

To address this challenge, we rely on the Gromov-Wasserstein (GW) distance [19], a tool proven effective in comparing distributions across diverse metric spaces. Specifically, we employ the Gromov-Monge (GM) formulation, an alternative form of the GW distance, to compute the explicit transport map between heterogeneous metric spaces. The HNN layers induce the embedding distribution, thus serving as the transport map. Consequently, the GM formulation is a more suitable approach for the corresponding HNN layers. However, implementing the GM distance poses computational challenges due to its complex constraints. Therefore, in light of this, we extend the GM-based embedding technique introduced in [16] to compute the geometry-preserving transport map from data distribution in Euclidean space to hyperbolic spaces.

The contribution of the current work is summarized as follows:

1. We propose a novel methodology that incorporates the GM distance into an end-to-end learning framework for hyperbolic representation learning. This approach ensures a more faithful representation of the intrinsic hierarchical structures. Moreover, we verify that our approach maintains computational efficiency by analyzing its time complexity.
2. We validate that the embedding faithfully represents an embedding of distributions from the original domain by providing an upper bound for the sampling error. Therefore, regularization on the training sample generalizes to the underlying data distribution.
3. We provide empirical evidence showing that our regularization strategy leads to consistent improvements over existing baseline methods when applied to datasets with hierarchical structures, including both image and graph data.

2 Related Works

Hyperbolic Embedding. The adoption of hyperbolic geometries for representing hierarchical connections between data points has attracted considerable attention. The exploration began when researchers considered low-distortion embedding of tree or tree-like graphs into the hyperbolic domain. In their seminal work, Sarkar [29] showed that a tree can be embedded into the Poincaré disk with arbitrarily low distortion. Sala et al. [28] generalized the embedding to high dimensional Poincaré balls and discussed embedding general graphs into trees. Sonthalia and Gilbert [32] bypassed the intermediary step of general graph structures, opting instead to directly infer tree architectures from the data itself. In addition to combinatorial methods, gradient-based methods are also used in hyperbolic embedding [23,24].

Hyperbolic Neural Networks. More recently, the advancement in hyperbolic representation learning has evolved to be more closely aligned with specific end tasks and specific datasets, such as text [34], images [14], biology [41], molecular [18], and knowledge graph [6]. This has led to the emergence of hyperbolic neural networks (HNNs) as a preferred framework for learning their representations. Numerous HNN architectures have emerged [3,5,10,11,18,24,31]. Comprehensive surveys on HNN include [26,40]. These HNNs define operations that conform with a choice of coordinate system in the hyperbolic space. However, when communicating between the Euclidean features and their hyperbolic embeddings, these HNNs take very simple approaches. For instance, Ganea *et al.* [10] directly used logarithmic and exponential maps at the origin, leading to unavoidable distortions. There are also very recent works addressing representations learned by HNNs. For instance, CO-SNE [12] reduces the dimensions of hyperbolic features for visualization, which is an analogy to t-SNE in the Euclidean domain. Parallel to this, Fan *et al.* [9] developed a systematic method for embedding data into nested hyperbolic spaces, which not only results in dimensionality reduction but also extends to design of HNNs. Nikolentzos *et al.* [25] applied the Weisfeiler-Leman algorithm to capture hierarchy in data and built HNNs accordingly. However, none of the above methods have considered comparing distributions before and after embedding into the hyperbolic space.

GW Distance. GW distance, a variant of optimal transport distance, has been widely utilized to quantify structural differences between different distributions. Memoli provided a detailed introduction and study of GW distance in [19,20]. Subsequently, Sturm [33] presented a more in-depth mathematical exposition of GW distance, focusing on its geodesic structure and gradient flow. Following these foundational works, GW distance has found applications in various fields, including computer vision [27,30], recommendation systems [17], natural language processing [1], generative models [4,16,22,35]. In particular, Lee *et al.* [16] employed the GM formulation to achieve geometry-preserving dimension reduction within a generative modeling framework. The discussion on the GM formulation can also be traced to [8,21]. Moreover, GW distance has been applied to specific data types such as graphs [38,39] and specific neural network architectures such as transformers [13]. To the best of our knowledge, GW distance has not yet been explored in the context of hyperbolic neural networks.

3 Methodology

3.1 Preliminaries of Hyperbolic Geometry

Hyperbolic geometry, characterized by its constant negative curvature, represents a non-Euclidean geometry essential for capturing complex hierarchical representations. We review operations that are defined in two different isometric coordinate systems, namely the Poincaré ball (Poincaré disk) model and the Lorentz (hyperboloid) model.

Poincaré Ball Model [10,31]. The Poincaré ball model for an n-dimensional hyperbolic space with curvature $-c$ is defined by

$$\mathbb{B}_c^n = \{\mathbf{x} \in \mathbb{R}^n : c \|\mathbf{x}\| < 1, c > 0\}, \tag{1}$$

with the corresponding Riemannian metric given by $\mathfrak{g}^\mathbb{B} = (\lambda_\mathbf{x}^c)^2 \mathbf{I}$, where $\lambda_\mathbf{x}^c = 2(1 - c \|\mathbf{x}\|^2)^{-1}$ is the conformal factor.

The analogy of a linear layer in HNN depends on two important operations, namely Möbius addition and multiplication. The Möbius addition $\oplus$ is defined as:

$$\mathbf{x} \oplus_c \mathbf{y} = \frac{\left(1 + 2c \langle \mathbf{x}, \mathbf{y} \rangle + c \|\mathbf{y}\|^2\right) \mathbf{x} + \left(1 - c \|\mathbf{x}\|^2\right) \mathbf{y}}{1 + 2c \langle \mathbf{x}, \mathbf{y} \rangle + c \|\mathbf{x}\|^2 \|\mathbf{y}\|^2}. \tag{2}$$

The Möbius multiplication $\mathbf{M} \otimes_c \mathbf{x}$ is defined as:

$$\mathbf{M} \otimes_c \mathbf{x} = \frac{1}{\sqrt{c}} \tanh\left(\frac{\|\mathbf{M}\mathbf{x}\|}{\mathbf{M}\mathbf{x}} \tanh^{-1}(\sqrt{c} \|\mathbf{x}\|)\right) \frac{\mathbf{M}\mathbf{x}}{\|\mathbf{M}\mathbf{x}\|}. \tag{3}$$

Finally, the geodesic distance between two points $\mathbf{x}$ and $\mathbf{y}$ in the Poincaré ball can be expressed as:

$$d_\mathbb{B}^c(\mathbf{x}, \mathbf{y}) = \frac{2}{\sqrt{c}} \tanh^{-1}\left(\sqrt{c} \|-\mathbf{x} \oplus_c \mathbf{y}\|\right). \tag{4}$$

Unless otherwise specified, we follow the common choice to take $c = 1$.

Lorentz Model [5,6]. The Lorentz model of an n-dimensional hyperbolic space $\mathbb{L}^n$ is a manifold embedded in a $(n + 1)$-dimensional Minkowski space. More specifically, this Minkowski space contains the same points as $\mathbb{R}^{n+1}$, but with an inner product $\langle \cdot, \cdot \rangle_\mathbb{L}$ defined by

$$\langle \mathbf{x}, \mathbf{y} \rangle_\mathbb{L} = -x_0 y_0 + \sum_{i=1}^n x_i y_i, \mathbf{x} = (x_0, \cdots, x_n), \mathbf{y} = (y_0, \cdots, y_n) \in \mathbb{R}^{n+1}. \tag{5}$$

The Lorentz model $\mathbb{L}^n$ contains points with $\langle \mathbf{x}, \mathbf{x} \rangle_\mathbb{L} = -c$, where $-c$ is the curvature. That is,

$$\mathbb{L}^n = \{\mathbf{x} = (x_0, \cdots, x_n) \in \mathbb{R}^{n+1} : \langle \mathbf{x}, \mathbf{x} \rangle_\mathbb{L} = -c, x_0 > 0\}, \tag{6}$$

The geodesic distance between two points $\mathbf{x}$ and $\mathbf{y}$ in the Lorentz model $\mathbb{L}^n$ is expressed as

$$d_\mathbb{L}^c(\mathbf{x}, \mathbf{y}) = \frac{1}{\sqrt{c}} \cosh^{-1}\left(-\langle \mathbf{x}, \mathbf{y} \rangle_\mathbb{L}\right). \tag{7}$$

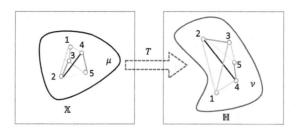

Fig. 2. Illustration of the transport map T. The feature distribution μ defined on $\mathbb{X}$ is pushed forward to $\nu = T_{\#}\mu$ on $\mathbb{H}$.

3.2 GW Regularization

To effectively utilize HNNs for learning from data, extending the geometry-preserving embedding technique introduced in [16], we employ a regularized learning approach focused on preserving the intrinsic geometric structures of the input features onto hyperbolic spaces. This involves employing a cost function capable of comparing two measures across distinct geometric spaces.

To bridge this gap, we consider the Gromov-Wasserstein (GW) distance [19] as it is adept at quantifying the similarity between probability distributions defined on distinct metric spaces. For two probability distributions μ and ν in a Euclidean feature space $\mathbb{X}$ and a hyperbolic space $\mathbb{H}$, respectively, the GW distance is defined as:

$$\text{GW}(\mu, \nu) := \min_{\pi \in \Pi(\mu,\nu)} \mathbb{E}_{((\mathbf{x},\mathbf{y}),(\mathbf{x}',\mathbf{y}')) \sim \pi^2} \left[|c_{\mathbb{X}}(\mathbf{x}, \mathbf{x}') - c_{\mathbb{H}}(\mathbf{y}, \mathbf{y}')|^2 \right] \quad (8)$$

where $c_{\mathbb{X}}$ and $c_{\mathbb{H}}$ represent the cost functions in the two spaces, and $\Pi(\mu, \nu)$ denotes the set of transport plans between μ and ν, which contains all joint distributions whose first and second marginals are given by μ and ν, respectively. By comparing the pairwise distances between two probability distributions to evaluate the similarity of geometric structures across different metric spaces.

When working with HNNs, an explicit map T from $\mathbb{X}$ to $\mathbb{H}$ is induced by the HNN layers, as illustrated in Fig. 2. The minimization problem presented in (8) can be reformulated as a Gromov-Monge (GM) distance minimization problem concerning a map T, defined as

$$\text{GM}(\mu, \nu) := \min_{T_{\#}\mu = \nu} \mathbb{E}_{(\mathbf{x},\mathbf{x}') \sim \mu^2} \left[|c_{\mathbb{X}}(\mathbf{x}, \mathbf{x}') - c_{\mathbb{H}}(T(\mathbf{x}), T(\mathbf{x}'))|^2 \right]. \quad (9)$$

In an HNN, when embedding the probability distribution μ from a Euclidean space to a hyperbolic space, we can consider minimizing the following cost over $T \in \mathcal{H}$, where $\mathcal{H}$ is a hypothesis class determined by the HNN:

$$\text{GM}(T; \mu) := \mathbb{E}_{(\mathbf{x},\mathbf{x}') \sim \mu^2} \left[|c_{\mathbb{X}}(\mathbf{x}, \mathbf{x}') - c_{\mathbb{H}}(T(\mathbf{x}), T(\mathbf{x}'))|^2 \right]. \quad (10)$$

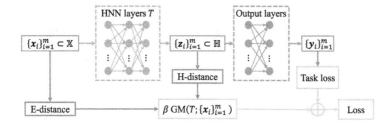

Fig. 3. The framework of our GW regularization.

Thus, we have $\min_{T \in \mathcal{H}} \mathrm{GM}(\mu, T_\#\mu) = \min_{T \in \mathcal{H}} \mathrm{GM}(T; \mu)$. While the GM problem described in (9) is highly challenging due to its pushforward constraint, minimizing the cost function in (10) is comparatively easier, as it does not require the pushforward constraint.

In practical implementation, we can employ the simple cost function $c_\mathbb{X}(\mathbf{x}, \mathbf{x}') = \|\mathbf{x} - \mathbf{x}'\|^2$ or other cost functions such as $c_\mathbb{X}(\mathbf{x}, \mathbf{x}') = \log\left(1 + \|\mathbf{x} - \mathbf{x}'\|^2\right)$. Additionally, the explicit form of $c_\mathbb{H}$ can be derived from Eqs. (4) and (7): we can use $c_\mathbb{H} = d_\mathbb{H}^c$ or $c_\mathbb{H} = \log\left(1 + d_\mathbb{H}^c\right)$, where $\mathbb{H} = \mathbb{B}$ or $\mathbb{H} = \mathbb{L}$.

When we have data points $\{\mathbf{x}_i\}_{i=1}^m$ sampled independently and identically distributed (i.i.d.) from μ, we consider an empirical version of the GM distance, namely,

$$\mathrm{GM}(T; \{\mathbf{x}_i\}_{i=1}^m) = \frac{1}{m(m-1)} \sum_{i,j=1}^m |c_\mathbb{X}(\mathbf{x}_i, \mathbf{x}_j) - c_\mathbb{H}(T(\mathbf{x}_i), T(\mathbf{x}_j))|^2. \quad (11)$$

The overall framework of our GW regularization is shown in Fig. 3. Here, $\{\mathbf{x}_i\}_{i=1}^m$ are the Euclidean features that are fed into the HNN layers denoted by T, $\{\mathbf{z}_i\}_{i=1}^m$ are the hyperbolic representations so that $\mathbf{z}_i = T(\mathbf{x}_i)$ for each i. The GM distance, calculated according to (11), is used as a penalty term, multiplied with a hyperparameter β and then added to the loss function associated with the specific task.

Generalization Analysis. In the tasks we consider, the ability to generalize beyond the given data points is essential. Although (11) only provides a formulation based on the training data, we anticipate that it will serve as an effective approximation of (10), provided that the training set $\{\mathbf{x}_i\}_{i=1}^m$ adequately captures the characteristics of the underlying distribution μ. This expectation is encapsulated in the theorem we present below, which formalizes the relationship between the pointwise formulation and its distributional approximation under the condition of a representative sample.

Theorem 1. *Given cost functions $c_\mathbb{X}$ and $c_\mathbb{H}$, suppose that there exists a constant $0 < \alpha \leq 1$ for which T satisfies the following bi-Lipschitz condition:*

$$\alpha c_\mathbb{X}(\mathbf{x}, \mathbf{x}') \leq c_\mathbb{H}(T(\mathbf{x}), T(\mathbf{x}')) \leq \frac{1}{\alpha} c_\mathbb{X}(\mathbf{x}, \mathbf{x}'), \tag{12}$$

for any $\mathbf{x}, \mathbf{x}' \in \mathbb{X}$. Let μ be a distribution defined on $\mathbb{X}$ and $\{\mathbf{x}_i\}_{i=1}^m$ be i.i.d. sampled from μ. Then,

$$|\mathrm{GM}(T; \{\mathbf{x}_i\}_{i=1}^m) - \mathrm{GM}(T; \mu)| \leq C \frac{(1/\alpha - 1)^2}{R^2} \tag{13}$$

holds with probability at least $1 - 2\exp\left(-\frac{mC}{8R^4}\right)$ where C is a constant depending on μ and $R := \max_{\mathbf{x}, \mathbf{x}' \in \mathrm{supp}(\mu)} c_\mathbb{X}(\mathbf{x}, \mathbf{x}')$.

The proof of Theorem 1 is presented in the appendix.

Complexity Analysis. Since the time complexity for calculating each Euclidean distance is $O(n)$ and the time complexity for calculating each hyperbolic distance (in either the Poincaré ball model or the Lorentz model) is $O(n)$, the time complexity for calculating the GW regularization term is $O(nm^2)$.

Note that, if we only consider linear layers in an L-layer Vanilla HNN, the time complexity is $O(mn^2 L)$. In the few-shot learning task and the semi-supervised graph node-level tasks that we consider in this paper, the number of data points m is usually smaller than or comparable with n. Moreover, in tasks such as link prediction, one also needs to compare pairs of data points, leading to a time complexity of $O(mn^2 L)$ in computing the task loss. Therefore, adding the regularization term will not lead to change of the order of time complexity.

4 Experiments

To validate the effectiveness of the GW regularization method, we conduct experiments on both image datasets and graph datasets. We present results on few-shot image classification in Sect. 4.1 and results on graph node classification and link prediction in Sect. 4.2. All the presented experiments are implemented on a server with RTX 4090 (24 GB) GPUs, where each implementation runs on a single GPU. The implementation of our proposed methods can be accessed at https://github.com/yyf1217/GW-Regularization.

4.1 Few-Shot Image Classification

We consider the important task of few-shot image classification where the goal is to recognize new categories with very few labeled examples per class. Unlike supervised image classification, this task aims to generalize from a small number of examples, typically one to five images per class. Learning structures of images is crucial since there is very limited labeled data.

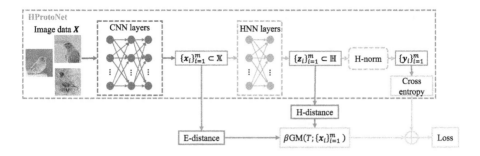

Fig. 4. The framework of the GW regularized hyperbolic ProtoNet model.

Datasets. We consider two widely used image datasets, namely *MiniImageNet* [36] and *Caltech-UCSD Birds-200-2011 (CUB)* [37]. These datasets have been used in [14], where the hyperbolic ProtoNet model excels Euclidean baselines. The MiniImageNet dataset is a subset of the ImageNet dataset [7]. It comprises images from 100 classes, with 600 images per class. The classes have a 64/16/20 split for training/validation/test. The CUB dataset consists of 200 categories of bird images, totaling 11,788 pictures. In this dataset, the training/validation/test split is 100/50/50.

Setting. We apply our GW regularization on the hyperbolic ProtoNet model [14], which primarily consists of convolutional neural network (CNN) layers used for feature extraction from Euclidean image data and subsequent HNN layers for classification. In [14], the HNN layers are taken to be a simple exponential map from the Euclidean space to the Poincaré ball. In our implementation, we utilize the publicly available code for the hyperbolic ProtoNet model from https://github.com/leymir/hyperbolic-image-embeddings.

For clarity, we illustrate the framework for this task in Fig. 4, which is based on Fig. 3. We take the features $\{\mathbf{x}_i\}_{i=1}^{m}$ from the output of the CNN layers, and then the HNN layers process them into $\{\mathbf{z}_i\}_{i=1}^{m}$. Following this, the hyperbolic distance between each hyperbolic embedding $\mathbf{z}_i$ and the origin of the Poincaré ball is computed according to (4), which is then used to calculate the maximum class probability. Finally, the cross-entropy loss is computed using the maximum class probability and the labels. We add $\beta\,\mathrm{GM}(T;\{\mathbf{x}_i\}_{i=1}^{m})$ to the cross-entropy loss to formulate our regularized loss function.

We consider various CNN architectures including Conv4, ResNet10, ResNet12, and ResNet18. The HNN layers within the model incorporate linear layers from [10]. Additionally, ReLU is used as the activation function across the model. For Conv4, the embedding dimension is set to 1,600, while for ResNet10, ResNet12, and ResNet18, the embedding dimension is set to 512. The initial learning rate is 0.001. We consider both the 1-Shot 5-Way and 5-Shot 5-Way settings, which means there are five classes involved for the task and only one/five examples per class for training. In all experiments for the 1-Shot 5-Way task, a consistent curvature parameter of -0.08 is employed. Similarly, for the 5-Shot

5-Way task, we uniformly set the curvature to -0.01 throughout all experiments. Both curvatures agree with the recommened settings in the original hyperbolic ProtoNet. The hyperparameter β is selected according to the validation set. We report the initial learning rate as well as the best β in Table 1.

Table 1. Hyperparameters used in few-shot image classification. "ResNet" abbreviated as "Res" due to space constraints.

		Initial learning rate (lr)				β			
		Conv4	Res10	Res12	Res18	Conv4	Res10	Res12	Res18
MiniImageNet	1-Shot 5-Way	0.0005	0.001	0.001	0.001	0.6	0.5	0.1	0.7
	5-Shot 5-Way	0.01	0.001	0.001	0.001	14	0.1	0.5	0.5
CUB	1-Shot 5-Way	0.001	0.001	0.001	0.001	0.02	0.015	0.15	0.1
	5-Shot 5-Way	0.01	0.001	0.001	0.001	0.07	0.01	0.25	0.1

Results. We present our numerical results for both the 1-Shot 5-Way and 5-Shot 5-Way settings in Table 2. We randomly sample 10,000 data points from the test set to evaluate the performance of the model and report 95% confidence intervals for all results. Each entry reports the average classification accuracy, including means and standard deviations.

Table 2. Accuracy (%) results for few-shot image classification.

	MiniImageNet		CUB	
	1-Shot 5-Way	5-Shot 5-Way	1-Shot 5-Way	5-Shot 5-Way
Conv4	53.77 ± 0.20	71.33 ± 0.16	64.66 ± 0.23	80.29 ± 0.16
Conv4+GW	55.57 ± 0.22	73.04 ± 0.16	68.67 ± 0.23	80.54 ± 0.16
ResNet10	56.45 ± 0.21	65.39 ± 0.17	72.01 ± 0.22	80.75 ± 0.15
ResNet10+GW	57.50 ± 0.22	67.62 ± 0.17	72.67 ± 0.22	83.73 ± 0.14
ResNet12	55.96 ± 0.22	70.32 ± 0.17	75.77 ± 0.22	82.90 ± 0.15
ResNet12+GW	56.95 ± 0.22	73.04 ± 0.16	77.24 ± 0.21	85.66 ± 0.14
ResNet18	57.04 ± 0.22	68.43 ± 0.16	71.86 ± 0.22	85.31 ± 0.13
ResNet18+GW	58.51 ± 0.22	71.64 ± 0.16	72.64 ± 0.22	85.83 ± 0.13

From the results, we observe that our GW regularization consistently boosts the accuracy of the model. This improvement is evident in both datasets and both scenarios, regardless of the underlying CNN architecture. For most CNN architectures, the improvement is more evident in the 5-Shot 5-Way scenario than in the 1-Shot 5-Way scenario. We believe that this can be attributed to the increased number of training instances available in the 5-Shot scenario, which provides a richer structure that GW regularization can leverage.

Runtime. We report the average run time for training one epoch in Table 3. As anticipated, incorporating GW regularization results in extended epoch training times. However, the increase in runtime is not significant, aligning with the time complexity analysis in Sect. 3.2. This observation underscores the efficiency of our GW regularization approach, ensuring that the added computational demand does not impose a significant burden on the overall training process.

Table 3. Average runtime (in seconds) for training one epoch in few-shot image classification.

	MiniImageNet		CUB	
	1-Shot 5-Way	5-Shot 5-Way	1-Shot 5-Way	5-Shot 5-Way
Conv4	31.2	25.2	32.4	32.9
Conv4+GW	33.6	26.4	35.1	32.9
ResNet10	37.2	28.8	38.4	34.9
ResNet10+GW	40.8	34.8	43.2	38.4
ResNet12	42.0	36.0	46.8	42.0
ResNet12+GW	45.6	40.8	50.4	45.6
ResNet18	55.2	43.2	60.0	52.8
ResNet18+GW	57.6	49.2	62.4	57.6

4.2 Semi-supervised Link Prediction and Node Classification

We consider graph link prediction and node classification, which are widely benchmarked semi-supervised tasks in graph deep learning literature.

Datasets. We consider nine datasets for node classification, namely *Cora, Disease, Airport, Cornell, Texas, Wisconsin, Chameleon, Squirrel,* and *Actor*. For the first three datasets, namely Disease, Airport, and Cora, we also perform link prediction. We briefly introduce the datasets in the appendix and summarize the statistics of the graphs in Table 4.

Table 4. Statistics of graph datasets.

	Disease-LP	Disease-NC	Airport	Cora	Cornell	Texas	Wisconsin	Chameleon	Squirrel	Actor
# nodes	2,665	1,044	3,188	2,708	183	183	251	2,277	5,201	7,600
# edges	2,265	2,265	15,837	4,488	280	295	466	31,421	198,493	26,752
# features	11	1,000	11	1,433	1,703	1,703	1,703	2,325	2,089	931
# classes	N/A	2	4	7	5	5	5	5	5	5

In link prediction tasks, we randomly split edges into 85%/5%/10% for training, validation, and test sets. For node classification tasks, we use a

70%/15%/15% splits for Airport, Cornell, Texas, Wisconsin, Chameleon, Squirrel, and Actor, we use 30%/10%/60% splits for Disease, and we use standard splits with 20 train examples per class for Cora. The above split settings agree with standard settings in the baseline papers.

Setting. For the graph datasets, we consider the following three HNNs: the vanilla HNN [10], HGCN [5], and HyboNet [6]. We utilize the publicly available code for HNN and HGCN from https://github.com/HazyResearch/hgcn and HyboNet from https://github.com/chenweize1998/fully-hyperbolic-nn.

For each model, we apply GW distance to the input node features and their hyperbolic representations after applying the Euclidean-to-hyperbolic operation. The regularization framework is the same as Fig. 3.

In these experiments, the HNN models share a common architecture consisting of HNN layers and output layers. These models employ different HNN layers to extract features from graph data and generate hyperbolic embeddings $\{\mathbf{z}_i\}_{i=1}^m$. Specifically, the Vanilla HNN and HGCN employ the Poincaré model to build linear layers and activation. HGCN also has aggregation operation, facilitating message passing. Differently, HyboNet employs the linear layers and aggregation in the Lorentz model. The output layers are used to accomplish various tasks. For node classification, the obtained hyperbolic embeddings are directly projected to the Euclidean space. A linear layer is then used to predict the class probabilities, with which the negative log-likelihood loss is computed. For link prediction, the hyperbolic distances between the hyperbolic embeddings are computed. Subsequently, the Fermi-Dirac algorithm [15] is employed to transform the hyperbolic distances into the edge probabilities. Finally, cross-entropy losses are computed using these edge probabilities and the true edge labels.

For simplicity, the curvature of the hyperbolic space is set as -1 for all experiments, which is a common setting for graph datasets. Tables 5 and 6 list hyperparameters such as initial learning rate, the number of the HNN layers, weight decay, dropout, and embedding dimensions. The hyperparameter β from validation is also listed for different datasets.

Table 5. Hyperparameters used in the Vanilla HNN and HGCN.

	Disease		Airport		Cora		Cornell	Texas	Wisconsin	Chameleon	Squirrel	Actor
	LP	NC	LP	NC	LP	NC	NC	NC	NC	NC	NC	NC
Initial lr	0.01	0.01	0.01	0.01	0.01	0.01	0.01	0.01	0.01	0.01	0.01	0.01
# layers	2	2	2	2	2	2	2	2	2	2	2	2
Weight decay	0.0	0.001	0.0005	0.0	0.005	0.001	0.001	0.001	0.001	0.001	0.001	0.001
Dropout	0.0	0.1	0.0	0.0	0.5	0.5	0.5	0.5	0.5	0.5	0.5	0.5
# embeddings	3	16	16	16	16	16	16	16	16	16	5	5
HNN β	0.9	0.08	2	1.25	0.5	0.35	0.25	0.1	0.15	0.02	0.001	0.03
HGCN β	0.3	0.09	2	1.5	0.1	0.6	0.25	0.15	0.2	0.08	0.01	0.03

Table 6. Hyperparameters used in HyboNet.

	Disease		Airport		Cora		Cornell	Texas	Wisconsin	Chameleon	Squirrel	Actor
	LP	NC	LP	NC	LP	NC	NC	NC	NC	NC	NC	NC
Initial lr	0.005	0.005	0.01	0.02	0.02	0.02	0.005	0.005	0.005	0.005	0.01	0.01
# layers	2	4	2	6	2	3	2	2	2	2	2	2
Weight decay	0.0	0.0	0.0	0.0001	0.001	0.01	0.0	0.0	0.0	0.0	0.001	0.001
Dropout	0.0	0.1	0.0	0.0	0.7	0.9	0.2	0.2	0.2	0.2	0.2	0.2
# embeddings	16	16	16	16	16	16	16	16	16	16	16	16
β	0.3	0.07	1.5	1.1	0.25	0.25	0.13	0.1	0.1	1.3	2.5	0.01

Results. For the link prediction task, we report the AUC scores in Table 7. For the node classification task, we report the F1 scores, in Table 8. Each score reports the average from three random runs as well as the standard deviation. For clarity, we just use "HNN" in place of the Vanilla HNN in the tables.

Table 7. AUC (%) results of the link prediction task.

	Disease	Airport	Cora
HNN	92.83 ± 1.83	93.56 ± 0.30	88.80 ± 1.29
HNN+GW	94.45 ± 0.58	94.23 ± 0.33	89.76 ± 2.07
HGCN	92.41 ± 1.78	93.42 ± 0.15	93.37 ± 0.15
HGCN+GW	93.79 ± 1.15	95.81 ± 0.02	93.48 ± 0.22
HYBONET	95.65 ± 0.57	96.18 ± 0.05	92.04 ± 0.37
HYBONET+GW	96.58 ± 0.57	96.44 ± 0.02	93.33 ± 0.22

From Table 7, it is evident that our GW regularization consistently enhances the AUC scores across different network architectures and across all datasets. Similarly, Table 8 showcases that GW regularization again demonstrates a positive impact on model performance across various datasets and architectures. In many scenarios of node classification, GW regularization actually achieves substantial improvement, such as HNN+GW for Disease, HGCN+GW for Cornell and HyboNet+GW for Wisconsin.

Runtime. We report the average run time for training one epoch for link prediction in Table 9. Similarly to the previous experiment, the runtime for GW regularized training is longer, but not significantly. This again aligns with the complexity analysis in Sect. 3.2.

Table 8. F1 (%) result of the node classification task.

	Disease	Airport	Cora	Chameleon	Squirrel
HNN	58.40 ± 2.53	87.40 ± 1.34	53.73 ± 0.91	71.43 ± 1.63	47.89 ± 2.21
HNN+GW	65.62 ± 1.98	89.38 ± 0.67	55.33 ± 0.35	72.53 ± 0.19	49.44 ± 1.05
HGCN	93.04 ± 1.64	86.96 ± 0.22	79.47 ± 1.01	77.72 ± 0.90	51.79 ± 0.58
HGCN+GW	95.28 ± 0.40	87.91 ± 0.72	80.00 ± 0.62	79.18 ± 1.64	52.70 ± 1.24
HYBONET	85.70 ± 0.60	93.13 ± 0.19	73.53 ± 0.75	82.05 ± 0.55	74.04 ± 2.42
HYBONET+GW	87.66 ± 1.21	94.46 ± 0.20	79.40 ± 0.95	83.58 ± 0.28	77.27 ± 0.78

	Cornell	Texas	Wisconsin	Actor
HNN	93.94 ± 1.31	94.69 ± 1.31	93.21 ± 3.86	42.63 ± 1.73
HNN+GW	95.45 ± 2.28	96.97 ± 1.32	98.15 ± 1.85	45.03 ± 1.73
HGCN	83.33 ± 3.47	78.79 ± 4.73	75.31 ± 2.83	36.39 ± 1.00
HGCN+GW	88.64 ± 2.28	81.82 ± 3.94	83.33 ± 1.86	39.19 ± 0.67
HYBONET	78.79 ± 4.73	67.42 ± 3.47	74.07 ± 1.86	45.74 ± 2.13
HYBONET+GW	81.82 ± 3.93	74.24 ± 1.31	80.87 ± 3.86	49.64 ± 0.38

Table 9. Average runtime (in seconds) for training one epoch in link prediction.

	Disease	Airport	Cora
HNN	0.0301	0.0881	0.0549
HNN+GW	0.0313	0.0884	0.0563
HGCN	0.0489	0.1075	0.0743
HGCN+GW	0.0505	0.1078	0.0747
HYBONET	0.0301	0.0631	0.0374
HYBONET+GW	0.0323	0.0639	0.0379

5 Conclusion

In this paper, we have delved into the integration of the GW distance as a novel regularization term within the realms of hyperbolic neural networks, with a particular emphasis on leveraging the GM formulation. This approach has allowed for a sophisticated comparison of distributions across the Euclidean space and the hyperbolic space, capturing the essence of the underlying structures while not increasing the order of the time complexity. We have demonstrated that the hyperbolic embeddings of the data points resonate closely with distributions where they are sampled from.

Our regularization method has been rigorously tested and validated across diverse datasets including image data for the few-shot classification task and graph data for the semi-supervised link prediction and node classification tasks, showcasing a uniform elevation in performance metrics.

Our future work involves more efficient methods in large scale settings, where the number of instances is much larger than feature dimensions. Additionally, we are intrigued by the potential of integrating our GW regularization framework with emerging geometric deep learning architectures for complex data structures, where the flexibility with the choice of underlying spaces is crucial.

Acknowledgement. YY and DZ acknowledge funding from National Natural Science Foundation of China (NSFC) under award number 12301117, WL acknowledges funding from the National Institute of Standards and Technology (NIST) under award number 70NANB22H021, and GL acknowledges funding from NSF award DMS 2124913.

References

1. Alvarez-Melis, D., Jaakkola, T.S.: Gromov-Wasserstein alignment of word embedding spaces. arXiv preprint arXiv:1809.00013 (2018)
2. Atigh, M.G., Schoep, J., Acar, E., van Noord, N., Mettes, P.: Hyperbolic image segmentation. In: IEEE Conference on Computer Vision and Pattern Recognition (2022)
3. Bachmann, G., Bécigneul, G., Ganea, O.: Constant curvature graph convolutional networks. In: International Conference on Machine Learning, pp. 486–496. PMLR (2020)
4. Bunne, C., Alvarez-Melis, D., Krause, A., Jegelka, S.: Learning generative models across incomparable spaces. In: International Conference on Machine Learning, pp. 851–861. PMLR (2019)
5. Chami, I., Ying, Z., Ré, C., Leskovec, J.: Hyperbolic graph convolutional neural networks. In: Advances in Neural Information Processing Systems, vol. 32, pp. 4868–4879 (2019)
6. Chen, W., et al.: Fully hyperbolic neural networks. In: Annual Meeting of the Association for Computational Linguistics, pp. 5672–5686 (2022)
7. Deng, J., Dong, W., Socher, R., Li, L.J., Li, K., Fei-Fei, L.: ImageNet: a large-scale hierarchical image database. In: IEEE Conference on Computer Vision and Pattern Recognition, pp. 248–255. IEEE (2009)
8. Dumont, T., Lacombe, T., Vialard, F.X.: On the existence of Monge maps for the Gromov-Wasserstein problem (2022)
9. Fan, X., Yang, C.H., Vemuri, B.C.: Nested hyperbolic spaces for dimensionality reduction and hyperbolic NN design. In: IEEE Conference on Computer Vision and Pattern Recognition (2022)
10. Ganea, O., Bécigneul, G., Hofmann, T.: Hyperbolic neural networks. In: Advances in Neural Information Processing Systems, vol. 31, pp. 5345–5355 (2018)
11. Gulcehre, C., et al.: Hyperbolic attention networks. In: International Conference on Learning Representations (2019). https://openreview.net/forum?id=rJxHsjRqFQ
12. Guo, Y., Guo, H., Yu, S.X.: Co-SNE: dimensionality reduction and visualization for hyperbolic data. In: IEEE Conference on Computer Vision and Pattern Recognition (2022)
13. Huang, Y., Zhang, T., Zhu, H.: Improving word alignment by adding Gromov-Wasserstein into attention neural network. In: Journal of Physics: Conference Series, vol. 2171, p. 012043. IOP Publishing (2022)

14. Khrulkov, V., Mirvakhabova, L., Ustinova, E., Oseledets, I., Lempitsky, V.: Hyperbolic image embeddings. In: IEEE Conference on Computer Vision and Pattern Recognition (2020)
15. Krioukov, D., Papadopoulos, F., Kitsak, M., Vahdat, A., Boguná, M.: Hyperbolic geometry of complex networks. Phys. Rev. E **82**(3), 036106 (2010)
16. Lee, W., Yang, Y., Zou, D., Lerman, G.: Monotone generative modeling via a Gromov-Monge embedding. arXiv e-prints (2023)
17. Li, X., et al.: Gromov-Wasserstein guided representation learning for cross-domain recommendation. In: ACM International Conference on Information & Knowledge Management, pp. 1199–1208 (2022)
18. Liu, Q., Nickel, M., Kiela, D.: Hyperbolic graph neural networks. In: Advances in Neural Information Processing Systems, vol. 32, pp. 8230–8241 (2019)
19. Mémoli, F.: On the use of Gromov-Hausdorff distances for shape comparison. In: PBG@Eurographics (2007)
20. Mémoli, F.: Gromov-Wasserstein distances and the metric approach to object matching. Found. Comput. Math. **11**, 417–487 (2011)
21. Mémoli, F., Needham, T.: Distance distributions and inverse problems for metric measure spaces. Stud. Appl. Math. **149**(4), 943–1001 (2022)
22. Nakagawa, N., Togo, R., Ogawa, T., Haseyama, M.: Gromov-Wasserstein autoencoders. In: International Conference on Learning Representations (2023). https://openreview.net/forum?id=sbS10BCtc7
23. Nickel, M., Kiela, D.: Poincaré embeddings for learning hierarchical representations. In: Advances in Neural Information Processing Systems, vol. 30 (2017)
24. Nickel, M., Kiela, D.: Learning continuous hierarchies in the lorentz model of hyperbolic geometry. In: International Conference on Machine Learning (2018)
25. Nikolentzos, G., Chatzianastasis, M., Vazirgiannis, M.: Weisfeiler and Leman go hyperbolic: learning distance preserving node representations. In: International Conference on Artificial Intelligence and Statistics, pp. 1037–1054 (2023)
26. Peng, W., Varanka, T., Mostafa, A., Shi, H., Zhao, G.: Hyperbolic deep neural networks: a survey. IEEE Trans. Pattern Anal. Mach. Intell. **44**(12), 10023–10044 (2021)
27. Peyré, G., Cuturi, M., Solomon, J.: Gromov-Wasserstein averaging of kernel and distance matrices. In: International Conference on Machine Learning, pp. 2664–2672. PMLR (2016)
28. Sala, F., De Sa, C., Gu, A., Ré, C.: Representation tradeoffs for hyperbolic embeddings. In: International Conference on Machine Learning (2018)
29. Sarkar, R.: Low distortion delaunay embedding of trees in hyperbolic plane. In: van Kreveld, M., Speckmann, B. (eds.) GD 2011. LNCS, vol. 7034, pp. 355–366. Springer, Heidelberg (2012). https://doi.org/10.1007/978-3-642-25878-7_34
30. Schmitzer, B., Schnörr, C.: Modelling convex shape priors and matching based on the Gromov-Wasserstein distance. J. Math. Imaging Vis. **46**, 143–159 (2013)
31. Shimizu, R., Mukuta, Y., Harada, T.: Hyperbolic neural networks++. In: International Conference on Learning Representations (2021)
32. Sonthalia, R., Gilbert, A.: Tree! I am no tree! I am a low dimensional hyperbolic embedding. In: Advances in Neural Information Processing Systems, vol. 33, pp. 845–856 (2020)
33. Sturm, K.T.: The Space of Spaces: Curvature Bounds and Gradient Flows on the Space of Metric Measure Spaces, vol. 290. American Mathematical Society (2023)
34. Tifrea, A., Bécigneul, G., Ganea, O.E.: Poincaré glove: hyperbolic word embeddings. In: International Conference on Learning Representations (2018)

35. Titouan, V., Flamary, R., Courty, N., Tavenard, R., Chapel, L.: Sliced Gromov-Wasserstein. In: Advances in Neural Information Processing Systems, vol. 32 (2019)
36. Vinyals, O., Blundell, C., Lillicrap, T., Wierstra, D., et al.: Matching networks for one shot learning. In: Advances in Neural Information Processing Systems, vol. 29 (2016)
37. Wah, C., Branson, S., Welinder, P., Perona, P., Belongie, S.: The Caltech-UCSD birds-200-2011 dataset (2011)
38. Xu, H., Luo, D., Carin, L.: Scalable Gromov-Wasserstein learning for graph partitioning and matching. In: Advances in Neural Information Processing Systems, vol. 32 (2019)
39. Xu, H., Luo, D., Zha, H., Carin, L.: Gromov-Wasserstein learning for graph matching and node embedding. In: International Conference on Machine Learning, pp. 6932–6941. PMLR (2019)
40. Yang, M., et al.: Hyperbolic graph neural networks: a review of methods and applications (2022)
41. Zhou, Y., Sharpee, T.O.: Hyperbolic geometry of gene expression. Iscience **24**(3), 102225 (2021)

VSViG: Real-Time Video-Based Seizure Detection via Skeleton-Based Spatiotemporal ViG

Yankun Xu[1], Junzhe Wang[1], Yun-Hsuan Chen[1], Jie Yang[1], Wenjie Ming[2], Shuang Wang[2], and Mohamad Sawan[1(✉)]

[1] CenBRAIN Neurotech, Westlake University, Hangzhou, China
{xuyankun,wangjunzhe,chenyunxuan,yangjie,sawan}@westlake.edu.cn
[2] Epilepsy Center, SAHZU, Zhejiang University, Hangzhou, China
{hflmwj,wangs77}@zju.edu.cn

Abstract. An accurate and efficient epileptic seizure onset detection can significantly benefit patients. Traditional diagnostic methods, primarily relying on electroencephalograms (EEGs), often result in cumbersome and non-portable solutions, making continuous patient monitoring challenging. The video-based seizure detection system is expected to free patients from the constraints of scalp or implanted EEG devices and enable remote monitoring in residential settings. Previous video-based methods neither enable all-day monitoring nor provide short detection latency due to insufficient resources and ineffective patient action recognition techniques. Additionally, skeleton-based action recognition approaches remain limitations in identifying subtle seizure-related actions. To address these challenges, we propose a novel Video-based Seizure detection model via a skeleton-based spatiotemporal Vision Graph neural network (VSViG) for its efficient, accurate and timely purpose in real-time scenarios. Our experimental results indicate VSViG outperforms previous state-of-the-art action recognition models on our collected patients' video data with higher accuracy (5.9% error), lower FLOPs (0.4G), and smaller model size (1.4M). Furthermore, by integrating a decision-making rule that combines output probabilities and an accumulative function, we achieve a 5.1 s detection latency after EEG onset, a 13.1 s detection advance before clinical onset, and a zero false detection rate. The project homepage is available at: https://github.com/xuyankun/VSViG/.

Keywords: Video-based seizure detection · Action recognition · Spatiotemporal vision GNN

Supplementary Information The online version contains supplementary material available at https://doi.org/10.1007/978-3-031-73007-8_14.

1 Introduction

As a common neurodegenerative disorder, epilepsy affects approximately 1% population worldwide [31,40]. A reliable epileptic seizure onset detection method can benefit patients significantly, because such a system equipped with accurate algorithms can promptly alert a seizure onset. According to previous studies [19,37,39], they predominantly focused on designing seizure detection algorithms based on electroencephalograms (EEGs). Although EEG can sense subtle brain changes just after a seizure begins, the use of scalp or implantable EEG devices often causes discomfort to patients and restricts them to hospital epilepsy monitoring units (EMUs). Consequently, there is a growing interest in developing an accurate video-based seizure detection system that could alleviate the discomfort associated with EEG helmets and facilitate remote monitoring of epileptic patients in residential settings [25]. However, the development of video-based seizure detection is hindered by several challenges as follows:

(1) Lack of datasets. Seizure-related video data collection is substantially time-consuming and requires doctoral expertise to annotate the different seizure-related periods. And, surveillance video recordings from the hospital EMUs contain highly sensitive information. Based on these reasons, there is no public video data intended for seizure study yet. Although [34] released a related dataset, they only implement action anticipation on ictal videos which cannot provide any false detection rate (FDR) information.

(2) Lack of effective analytic tools. Fundamentally, video-based seizure detection is an action recognition task, requiring the analysis of complex

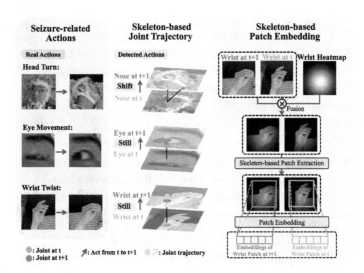

Fig. 1. Motivation of proposed skeleton-based patch embedding. The left shows real seizure-related actions; The middle shows challenges of traditional skeleton-based approaches; The right shows our strategy to address challenges.

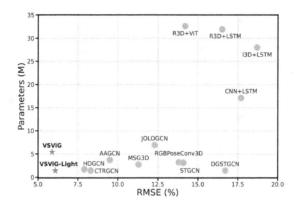

Fig. 2. Model comparison of achieved errors and the number of parameters.

seizure-related actions in epileptic patients. Prior research indicates that existing methods struggle with achieving short detection latency and are often limited to nocturnal seizures when using raw RGB frames [46,51]. While skeleton-based approaches powered by graph convolutional networks (GCNs) offer several advantages over RGB-based methods [11,23,52], there also encounter significant limitations in the context of seizure detection. Firstly, pre-trained pose estimation models [2,4,12,28,42] cannot be applied to epileptic patients' action recognition directly due to the complexity of epileptic patient scenarios, as shown in Fig. 6. Secondly, as shown in Fig. 1, there are inherent difficulties in recognizing certain seizure-related actions naively using coordinates as traditional skeleton-based approaches did.

(3) Detection latency and false detection rate. Detection latency is a critical metric in a seizure detection system, representing the time gap between the real onset of a seizure and the detected onset. A shorter detection latency, known as early seizure detection is highly desirable since it enables timely interventions prior to a serious seizure attack. Also, FDR is important [41,45], however, in video-based detection many normal patient behaviors closely resemble seizure-related actions. Previous studies are hard to distinguish them or even directly overlook this metric.

In this study, we propose a novel **V**ideo-based **S**eizure detection model via skeleton-based spatiotemporal **Vi**sion **G**raph neural network (VSViG) intended for efficient, accurate, and timely real-time scenarios. We address the aforementioned limitations by offering the following contributions: (1) We acquire a dataset of epileptic patient data from hospital EMUs and fine-tune the pose estimation model for epileptic patients; (2) As expressed in the Fig. 1, instead of using skeleton-based coordinates, our VSViG model utilizes skeleton-based patch embeddings as inputs to address mentioned challenges. Our proposed model including base and light versions outperforms the state-of-the-art models from the field of action recognition and video-based seizure detection in both accuracy and model size, as shown in Fig. 2; (3) We define the application as a regression task to generate the probability/likelihood outputs instead of naive binary classification on interictal (healthy) and ictal (unhealthy) status, thereby the model can accurately detect subtle or trending seizure-related actions in low likelihood for early detection propose.

2 Related Work

Video-Based Seizure Detection. There have been several efforts paid to video-based seizure detection or jerk detection over the past decade [8,15,21,46], but they neither achieved accurate performance nor supervised patients 24/7 by video monitoring. Recently, several works [22,30,34,51] made use of advanced deep learning (DL) models to improve the video-based seizure detection performance, however, there are many limitations among these works: (1) [22,51] cannot achieve short detection latency or require more video modality data than RGB; (2) [30,34] defined the early seizure detection task as a video-based action anticipation task, but they cannot achieve higher accuracy until utilizing at least 1/2 clip of ictal videos which cannot bring short latency, further action anticipation task cannot provide any FDR information.

Action Recognition. There are two mainstream strategies for action recognition tasks, one is 3D-CNN for RGB-based action recognition, and the other one is skeleton-based action recognition. Many 3D-CNN or ViT architectures [1,6,13,14,20,43,44,48] have been proven to be effective tools for learning spatiotemporal representations of RGB-based video streams, they are widely applied to video understanding and action recognition tasks. Compared to RGB-based approaches with a large number of trainable parameters, skeleton-based approaches perform better in both accuracy and efficiency because they only focus on the skeleton information which is more relevant to the actions, and can alleviate contextual nuisances from raw RGB frames. Several GCN-based approaches [7,27,35,38,50] were proposed to achieve great performance in skeleton-based action recognition. [11] proposed PoseConv3D to combine 3D-CNN architecture and skeleton-based heatmaps, and proposed its variant RGB-PoseConv3D to fuse RGB frames to obtain better performance. However, all aforementioned action recognition models show weaknesses in analyzing seizure-related actions, which include subtle behavioral changes when seizures begin. In this work, we propose VSViG model to address this challenge.

Vision Graph Neural Network. [16] first proposed vision graph neural networks (ViG) as an efficient and effective alternative backbone for image recognition. Since then many ViG variants [17,32,47,54] have been proposed to handle various applications. Inspired by ViG, we first propose VSViG to extend ViG to accomplish a skeleton-based action recognition task from seizure-related videos.

3 VSViG Framework

Figure 3 presents an overview of skeleton-based VSViG framework. The process begins with a video clip composed of consecutive RGB frames, we fine-tune the

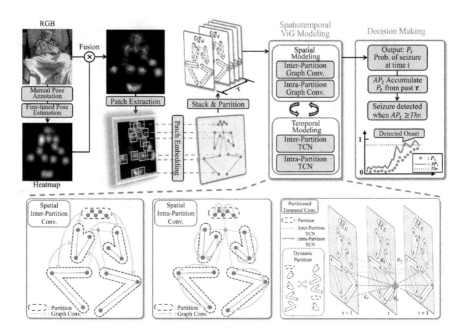

Fig. 3. Proposed skeleton-based VSViG framework. Starting from raw RGB frames, we extract skeleton-based patches around each joint by fusing RGB frames and pose heatmaps. Then features from patches are generated by a patch embedding. In spatiotemporal ViG modeling, a partition strategy with proposed inter-, intra-, and dynamic partition operations is used, as shown in the bottom three subfigures.

pose estimation model to generate joint heatmaps, then fuse these heatmaps and raw frames to construct skeleton-based patches. Before being processed by the VSViG model, these patches are transformed into feature vectors through a trainable patch embedding layer. During the spatiotemporal ViG processing phase, we conduct spatial and temporal modeling respectively, and utilize partition structure to learn inter-partition and intra-partition representations. VSViG model outputs probabilities instead of binary classes for video clips, then a decision-making rule integrating probabilities and accumulative function can achieve shorter detection latency and lower FDR. The subsequent subsections will provide detailed explanations of each step in this process.

3.1 Pose Estimation Model Fine-Tuning

The VSViG framework relies on a pose estimation algorithm to extract joint heatmaps at first. However, when we tested several mainstream pre-trained pose estimation models on our patient video data, the results were unsatisfactory, which are shown as Fig. 6. These models failed to accurately track the locations of patients and their joints. Therefore, we have to fine-tune the pose estimation model for epileptic patients.

We manually annotated 580 frames containing various behaviors during both ictal periods and healthy status across all patients, then utilized lightweight-openpose [33] as a base model to fine-tune due to its efficiency. Eventually, we achieved a fine-tuned pose estimation model can accurately track patients' joints across all patients, as shown in Fig. 6.

3.2 Patch Extraction and Patch Embedding

In custom pose estimation algorithm, each joint is associated with a heatmap generated by a 2D Gaussian map [5]. We make use of this heatmap as a filter to fuse the raw RGB frames, then extract a small patch around each joint from the fused image. However, determining the optimal size of the Gaussian maps to capture the regions of interest is a challenge during the training phase of the pose estimation model. To address this, we manually extract the patch $\mathbf{p}_{it}$ of i-th ($i = 1, 2..., N$) joint at frame t by generating a Gaussian map with adjustable σ for each joint to filter the raw RGB frame $\boldsymbol{I}_t$:

$$\mathbf{p}_{it} = \exp(-\frac{(m - x_{it})^2 + (n - y_{it})^2}{2\sigma^2}) \otimes \boldsymbol{I}_t(m, n) \tag{1}$$

where $|m - x_i| \leq H/2$, $|n - y_i| \leq W/2$. (H, W), (m, n) and (x_{it}, y_{it}) respectively stand for the size of patches, the coordinate of image pixels, and the location of i-th joint at frame t. The σ controls the shape of 2D Gaussian kernels. As a result, we are able to extract a sequence of skeleton-based patches $\mathbf{P} \in \mathbb{R}^{N \times T \times H \times W \times 3}$ from each video clip, then a patch embedding layer transforms patches into 1D feature vectors $\mathbf{X} \in \mathbb{R}^{N \times T \times C}$.

3.3 Graph Construction with Partition Strategy

Given a sequence of skeleton-based patch embeddings, we construct an undirected skeleton-based spatiotemporal vision graph as $\mathcal{G} = (\mathcal{V}, \mathcal{E})$ on this skeleton sequence across N joints and T stacked frames. Each node in the node set $\mathcal{V} = \{v_{it} | i = 1, 2, ..., N, t = 1, 2..., T\}$ is associated with feature vectors $\mathbf{x}_{it} \in \mathbb{R}^C$.

In this study, we proposed a partitioning strategy to construct graph edges between nodes. According to Openpose joint template, which contains 18 joints, we focus on 15 joints by excluding three (l_ear, r_ear, neck) as redundant. These 15 joints are divided into 5 partitions, each comprising three joints (**head**: nose, left/right eye; **right arm**: right wrist/elbow/shoulder; **right leg**: right hip/knee/ankle; **left arm**: left wrist/elbow/shoulder; **left leg**: left hip/knee/ankle). In our VSViG, we conduct both spatial and temporal modeling for the graph. Spatial modeling involves constructing two subsets of edges based on different partition strategies: inter-partition edge set and intra-partition edge set, which are denoted as:

$$\begin{aligned} \mathcal{E}_{inter} &= \{v_{it}v_{jt} | i \in \mathbf{Part}_{p_1}, j \in \mathbf{Part}_{p_2}, p_1 \neq p_2\} \\ \mathcal{E}_{intra} &= \{v_{it}v_{jt} | (i, j) \in \mathbf{Part}_p\} \end{aligned} \tag{2}$$

where $p/p_1/p_2 = 1, ..., 5$. As shown in the bottom left two schematic figures in Fig. 3, each node is connected to the nodes from all other partitions in the inter-partition step, and each node is only connected to the other nodes within the same partition in the intra-partition step. The bottom-left two figures of Fig. 3 visualize the graph construction with partitioning strategy.

For temporal modeling, we consider neighbors of each node within a $K_S \times K_T$ in both temporal and spatial dimensions. The temporal edge set is denoted as $\mathcal{E}_T = \{v_{it}v_{j(t+1)}|d(v_{it},v_{jt}) \leq K_S/2, d(v_{it},v_{i(t+1)}) \leq K_T/2\}$, where d is the distance between two nodes. To facilitate the partitioning strategy, we arrange the joints with their partitions in a 1D sequence. Thus, if $K_S = 3$, as shown in the bottom-right of Fig. 3, the middle joint of a partition aggregates only from intra-partition nodes over K_T frames, while the border joint of a partition aggregates from inter-partition nodes over K_T frames.

3.4 Partitioning Spatiotemporal Graph Modeling

Spatial Modeling. The proposed skeleton-based VSViG aims to process the input features $\mathbf{X} \in \mathbb{R}^{N \times T \times C}$ into the probabilities of seizure onset. Starting from input feature vectors, we first conduct spatial modeling to learn the spatial representations between nodes at each frame by inter-partition and intra-partition graph convolution operation as $\mathcal{G}' = F_{intra}(F_{inter}(\mathcal{G}, \mathcal{W}_{inter}), \mathcal{W}_{intra})$, where F is graph convolution operation, specifically we adopt max-relative graph convolution [16,26] to aggregate and update the nodes with fully-connected (FC) layer and activation function:

$$\mathbf{x}'_{it} = \sigma(Linear([\mathbf{x}_{it}, max(\mathbf{x}_{jt} - \mathbf{x}_{it}|j \in \mathcal{N}^t(\mathbf{x}_{it}))])) \qquad (3)$$

where $\mathcal{N}^t(\mathbf{x}_{it})$ is neighbor nodes of $\mathbf{x}_{it}$ at frame t, $Linear$ is a FC layer with learnable weights and σ is activation function, e.g. ReLU. The only difference between F_{inter} and F_{intra} depends on $\mathcal{N}^t(\mathbf{x}_{it})$ is from $\mathcal{E}_{inter}$ or $\mathcal{E}_{intra}$. Given input feature vectors at frame t, denoted as $\mathbf{X}_t \in \mathbb{R}^{N \times C}$, the graph convolution processing as Eq. (3) can be denoted as $\mathbf{X}'_t = MRGC(\mathbf{X}_t)$, then the partitioning spatial graph processing module in VSViG would be:

$$\begin{aligned} \mathbf{X}''_t &= \sigma(MRGC(\mathbf{X}_t W_{in\_inter})W_{out\_inter} + \mathbf{X}) \\ \mathbf{X}'_t &= \sigma(MRGC(\mathbf{X}''_t W_{in\_intra})W_{out\_intra} + \mathbf{X}) \end{aligned} \qquad (4)$$

We add original input $\mathbf{X}_t$ as a residual component to both two partitioning spatial modeling, which is intended to avoid smoothing and keep diversity of features during the model training [9,18]. We keep output dimension of every learnable layer is same as input dimension, so that we achieve output feature $\mathbf{X}' \in \mathbb{R}^{N \times T \times C}$ when we stack T frames $\mathbf{X}'_t \in \mathbb{R}^{N \times C}$ after spatial modeling.

Temporal Modeling. Given the dimension of output features after spatial modeling is $N \times T \times C$, and we order the partitions (containing joints) in a 1D sequence, so that we can naturally consider the input for the temporal modeling

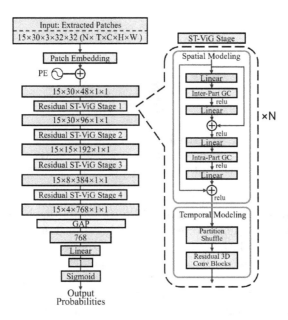

Fig. 4. VSViG architecture. VSViG consists of four residual spatiotemporal (ST) ViG stages and each stage contains several proposed spatial and temporal modeling layers. N, T, C denotes the number of joints, frames and channels, and H, W stand for the height and width of extracted patches.

as a 3D volume, denoted as $\mathbf{X} \in \mathbb{R}^{1 \times N \times T \times C}$. According to the construction of graph in temporal modeling $\mathcal{G} = (\mathcal{V}, \mathcal{E}_T)$, we can adopt convolution operation to aggregate and update nodes:

$$\mathbf{x}'_i = \sum_j^1 \sum_j^{K_S} \sum_j^{K_T} w_{ij} \mathbf{x}_i \tag{5}$$

Inspired by [11,18,50], we utilize residual 3D-CNN architecture, denoted as $Conv$, to conduct temporal graph modeling as Eq. (5) for its simplicity:

$$\mathbf{X}' = Conv(1)(Conv(1, K_S, K_T)(Conv(1)(\mathbf{X}))) + \mathbf{X} \tag{6}$$

where (1) and $(1, K_S, K_T)$ are kernel size for the 3D convolution operation.

3.5 VSViG Network Architecture

Figure 4 shows the detailed structure of proposed VSViG network including the shape of features at each layer. We keep feature size in spatial modeling, and implement downsampling at the last layer of each residual temporal block. After four residual spatiotemporal ViG stages, we make use of global average pooling to generate a 1D feature vector, then one FC layer and sigmoid function are

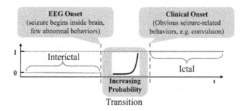

Fig. 5. Data labeling for a regression-based task. For each seizure, a video recording is categorized into 3 different periods: interictal (label: 0), ictal (label: 1), and transition (label: 0 to 1 in exponential function according to clinical phenomenon).

connected to generate output probabilities. In spatial modeling, $E = 12$ and $E = 2$ represent the number of edges (neighbors) in inter-partition and intra-partition graph convolution, respectively. The spatial modeling phase always keeps the output channels unchanged. In temporal modeling, we set $K_S = 3$ for simultaneous implementation of inter-partition and intra-partition temporal convolution network (TCN) operation, and set $K_T = 3$ for considering only one frame before and after the current frame. The residual 3D-CNN block is used to expand feature dimension and downsample temporal frames.

We propose two versions of VSViG models: VSViG and VSViG-Light respectively based on different model depths and feature expansions, they are designed for accurate or efficient purposes. Their details are illustrated in Appendix.

Positional Embedding. In previous studies, either skeleton-based STGCNs which input coordinate triplets or RGB-/Heatmap-based methods made use of specific joint coordinates or whole body context with global positional information. In this work, however, input patches in 1D sequence do not provide any positional information. Thus, we add a positional embedding to each node feature vector. Here we propose 3 different ways to inject positional information: (1) Add learnable weight: $\mathbf{x}_{it} = \mathbf{x}_{it} + \mathbf{e}_{it}$; (2) Concatenation with joint coordinates: $\mathbf{x}_{it} = [\mathbf{x}_{it}, (x_{it}, y_{it})]$; (3) Add embeddings of joint coordinates: $\mathbf{x}_{it} = \mathbf{x}_{it} + Stem(x_{it}, y_{it})$;

Dynamic Partitions. Since the partitions are arranged as a 1D sequence, each partition is only connected to 1 or 2 adjacent partitions, it cannot consider all other partitions as spatial modeling does. Thus, we obtain the dynamic partitions by conducting partition shuffle before temporal graph updating. It is noted that we only shuffle the partitions, and the order of joints within a partition is unchanged.

3.6 Seizure Onset Decision-Making

Inspired by [49], after generating output probabilities P_t by proposed skeleton-based VSViG model for video clips at time t, we can accumulate a period τ of previously detected probabilities as accumulative probabilities $AP_t = \sum_{i=t-\tau}^{t} P_i$ with detection rate r, then make a seizure onset decision at time t_{onset} when accumulative probabilities reach a decision threshold DT: $t_{onset} = AP_t > DT$. We measure the distance between t_{onset} and EEG onset as latency of EEG onset L_{EO}, and the distance between t_{onset} and clinical onset as latency of clinical onset L_{CO}.

4 Experiments

4.1 Dataset

We acquire surveillance video data of epileptic patients from the local hospital, this dataset includes 14 epileptic patients with 33 focal or generalized tonic-clonic seizures. As for each seizure, we utilize all ictal periods, and 5–30 min interictal periods before the EEG onset. We segment the raw video recording into video clips with 5 s for model training and manually label the video clips as likelihood for regression task according to Fig. 5. Naive binary classification of healthy and unhealthy status cannot accurately recognize the early occurrence of seizures, the introduced increasing probability way is consistent with the likelihood of real seizure attack risk. And an exponential way is more similar to the clinical phenomenon that the closer time to the clinical onset, the higher risk is obtained. We also test other functions to build increasing probability labels for this regression task, which is shown in Appendix.

Eventually, 2.78 h interictal period, 0.16 h transition period and 0.35 h ictal period video are obtained for the model training, and overlapping extraction operation is utilized to overcome the issue of data imbalance. Then we split 70% of each period for training, and the rest 30% for validation. More details information about the dataset is illustrated in Appendix.

Performance Metric. In this paper, we define a regression task intended for early seizure detection, RMSE metric measured between true video clip and predictive video clip is used to compare the model performance.

Data Usage. The data collection from hospital is carefully conducted and the use of data was approved by hospital Institutional Review Board. We will release this dataset when all sensitive and private information is removed. Code and fine-tuned pose estimation model will be also released.

4.2 Experimental Setting

We manually set a size of 32×32 ($H \times W$) for extracted patches based on the 1920×1080 raw frame resolution, and σ of gaussian kernel is 0.3. All generated outputs are connected with a Sigmoid function to map the outputs as probabilities from 0 to 1, the activation function used anywhere else is chosen as ReLU, and mean square error loss function is used to train the model. During the training phase, epoch and batch size are respectively set to 200 and 32, we choose Adam optimizer with 1e−4 learning rate with 0.1 gamma decay every 40 epochs and 1e−6 weight decay.

4.3 Seizure-Related Action Recognition Performance

According to Table 1, VSViG and VSViG-Light respectively obtain 5.9% and 6.1% errors across all patients, which outperform all previous state-of-the-art approaches. Furthermore, VSViG-Light only obtains 0.44G FLOPs and 1.4M model size which is desirable for real-time deployment.

We also reproduce several state-of-the-art action recognition and video-based seizure detection models to make comparisons with proposed VSViG model. There are four selective state-of-the-art video-based seizure detection approaches, they are all RGB-based strategies utilizing a CNN-based model to extract features that are followed by a sequence-based model. In terms of skeleton-based action recognition models, several state-of-the-art GCN-based models are chosen, and one RGB+skeleton fusion method is also considered.

Table 1. Model comparisons. We reproduce state-of-the-art approaches to make comparisons with proposed VSViG. (OF: optical flow, ♣: Video-based seizure detection models, ♠: Skeleton-based action recognition models, ★: This work)

Method	Input	RMSE (%)	GFLOPs	Param. (M)
♣ I3D+LSTM [22]	RGB	18.7	57.3	28.0
♣ CNN+LSTM [51]	RGB	17.7	5.0	17.1
♣ R3D+LSTM [34]	RGB	16.5	163.1	31.9
♣ R3D+ViT [30]	RGB	14.2	163.1	32.6
♠ DGSTGCN [10]	Skeleton	16.7	0.50	1.4
♠ STGCN [50]	Skeleton	14.1	1.22	3.1
♠ RGBPoseConv3D [11]	RGB+Skeleton	13.8	12.63	3.2
♠ JOLOGCN [3]	OF in Patches	12.3	37.30	6.9
♠ MSG3D [29]	Skeleton	11.3	1.82	2.7
♠ AAGCN [36]	Skeleton	9.5	1.38	3.7
♠ CTRGCN [7]	Skeleton	8.3	0.62	1.4
♠ HDGCN [24]	Skeleton	7.9	1.60	1.7
★ VSViG-Light	Patches	6.1	**0.44**	**1.4**
★ VSViG	Patches	**5.9**	1.76	5.4

We can see that RGB-based approaches perform worse in both error and FLOPs than skeleton-based approaches, it is consistent with RGB-based approach's drawbacks: large model size and can easily be disturbed by nuisances. Furthermore, several previous video-based seizure detection model rely on the pre-trained RGB-based 3D CNN model to extract features, which extremely increases the computational resources. As for skeleton-based approaches, most perform around 10% error with lower FLOPs, and CTRGCN performs the best among all skeleton-based approaches. These results underscore the superiority of our skeleton-based patch embedding VSViG models over both RGB-based and skeleton-based GCN approaches

Fig. 6. Pose estimation performance comparison between public pre-trained model and fine-tuned model. Black circles are used to mask sensitive information.

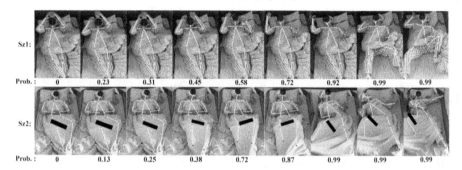

Fig. 7. Predictive probabilities/likelihoods of video clips from two different seizures are obtained by VSViG model. Black boxes are used to keep privacy.

4.4 Visualization

We visualize the pose estimation performance comparison between our fine-tuned model and the public pre-trained model, as shown in Fig. 6, thereby demonstrating the limitation of public pre-trained model for tracking patient skeletons. Secondly, we present the visualization of predictive probability of each video clip obtained by VSViG model in a successive seizure from two different seizures, which is shown as Fig. 7.

4.5 Ablation Study

We conduct ablation studies with VSViG to evaluate the effects of several modules and settings on the model performance, all shown values in Table 2 are RMSE errors on the validation set.

Effect of Spatial Modeling Module. There are several spatial modeling modules in proposed VSViG, we conduct an ablation study here to prove the effectiveness of inter-partition and intra-partition modules. We can see from Table 2a that proposed inter-intra module outperforms other conditions. Inter-partition and intra-partition fundamentally extract coarse and fine features respectively, and Inter-Intra strategy is consistent with coarse-to-fine feature extraction that can learn the representations better.

Effect of Temporal Modeling Kernel Size. As for the temporal modeling, a kernel size of (3, 3) stands for K_S, K_T in TCN operation, $K_S = 3$ means spatial neighbor joints are also considered, meanwhile $K_S = 1$ is not. And we also compare 3 and 5 convolution size in temporal dimension K_T. The result in Table 2b shows that considering neighbor nodes and $K_T = 3$ performs best.

Effect of Dynamic Partition. In Table 2c, we can see that Stem positional embedding way with dynamic partition achieves the best performance. Dynamic partition strategy shows better performance, which indicates it can effectively enhance the model to learn relationships between different partitions, thereby understanding the seizure-related actions.

Effect of Positional Embedding. Also, positional information can bring better performance for VSViG model, and Stem way outperforms all conditions. Further, no positional embedding with dynamic partition can also achieve a satisfactory result compared with other GCN-based approaches that utilize coordinates as input, this indicates that proposed VSViG model has the capability to complete an action recognition task without any specific or global positional information. In our point of view, the reason is that inter-partition and intra-partition representations contain enough information for action recognition no matter how these partitions are ordered or where they are.

Effect of Gaussian Kernel Size. During the phase of fine-tuning lightweight-openpose, σ size should be set in advance. Smaller size brings less RGB information in extracted patches, meanwhile larger size brings more contextual nuisances. In this ablation study Table 2d, 0.3 relative to the patch size is the best. It is noted that the chosen patch size depends on the video resolution and size of patient in the video, 32×32 is suitable for this scenario.

Table 2. Ablation studies on VSViG. All experiments are conducted with VSViG version, and values are RMSE.

(a) Spatial modeling module.

Inter	Intra	RMSE
✗	✗	13.0
✓	✗	11.7
✗	✓	9.2
✓	✓	**5.9**

(b) Temporal modeling kernel size.

(K_S, K_T)	RMSE
(1, 3)	9.9
(1, 5)	10.8
(3, 3)	**5.9**
(3, 5)	6.5

(c) Positional embedding and dynamic partition.

Positional embedding	Dynamic partition w/o	Dynamic partition w.
None	16.9	9.1
Cat.	14.8	8.9
Learnable	13.4	8.6
Stem (x, y)	8.4	**5.9**

(d) Gaussian kernel (σ) size.

σ	RMSE
0.1	10.1
0.2	8.5
0.3	**5.9**
0.4	8.6
0.5	7.6

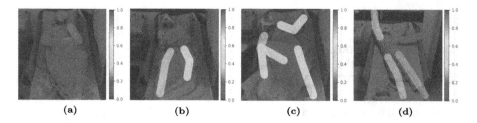

Fig. 8. Occlusion maps for model interpretability. Each image represents the end frame of the video clip. Redder partitions mean more salient to the seizure-related actions. (a)–(d) present a patient suffering a seizure from healthy status to convulsion.

4.6 Model Interpretability

We provide interpretability with occlusion maps [53] to demonstrate the capabilities of our VSViG model to learn representations of seizure-related actions within different partitions. We successively take 4 video clips of a patient from healthy status to convulsion, as shown in Fig. 8, each image represents the end frame of corresponding clips, higher value (redder) means the partition is more salient to seizure-related action, and lower value (bluer) means less relevant. We can see that, initially, when the patient is lying on the bed, all partitions are shown as less salient in Fig. 8a. As the seizure begins, the patient starts to exhibit abnormal movements, such as turning the head and moving the right arm. Correspondingly, these partitions appear redder in Fig. 8b and Fig. 8c, indicating their increased relevance to the seizure activity. Several seconds later, as the patient begins jerking, most partitions are shown as more salient in Fig. 8d.

4.7 Make Seizure Onset Decision

After obtaining probabilities from consecutive video clips by VSViG model, we set up $\tau = 3$ s, r $= 0.5$ s as mentioned in Sect. 3.6 to calculate L_{EO}, L_{CO} and FDR across all seizures. We also evaluate the effect of accumulative strategy on the detected latency, as shown in Table 3. We can see that performance achieved by accumulative strategy with $DT = 0.3$ can obtain much better performance where 5.1 s L_{EO}, -13.1 s L_{CO}, which means seizure can be detected only 5.1 s after seizure begins, and 13.1 s before patients tend to exhibit serious convulsions. And FDR of 0 is also obtained. The specific results of every patient are shown in Appendix.

Table 3. Seizure detection performance.

DT	FDR	Latency	Accumulative	
			w/o	w.
0.3	0	L_{EO}	11.4 s	**5.1 s**
		L_{CO}	-6.8 s	$\mathbf{-13.1\,s}$

5 Conclusion

In this work, we proposed a VSViG model utilizing skeleton-based patch embeddings to recognize the epileptic patients' actions effectively, experiments indicate proposed VSViG model outperforms state-of-the-art action recognition approaches. VSViG provides an accurate, efficient, and timely solution for video-based early seizure detection to fundamentally help epileptic patients.

In the future, we will collect more patient video data with various types of seizures to train a more generalized model for a large number of epileptic patients. And action recognition-based VSViG model can also be extended to detect other movement-based diseases, e.g. Parkinson's disease, fall detection.

Acknowledgement. This work was supported by STI2030-Major Projects (2022ZD0208805), "Pioneer" and "Leading Goose" R&D Program of Zhejiang (2024C03002), and the Key Project of Westlake Institute for Optoelectronics (Grant No. 2023GD004).

References

1. Arnab, A., Dehghani, M., Heigold, G., Sun, C., Lučić, M., Schmid, C.: ViViT: a video vision transformer. In: ICCV, pp. 6836–6846 (2021)
2. Bazarevsky, V., Grishchenko, I., Raveendran, K., Zhu, T., Zhang, F., Grundmann, M.: BlazePose: on-device real-time body pose tracking. arXiv preprint arXiv:2006.10204 (2020)
3. Cai, J., Jiang, N., Han, X., Jia, K., Lu, J.: JOLO-GCN: mining joint-centered lightweight information for skeleton-based action recognition. In: WACV, pp. 2735–2744 (2021)
4. Cao, Z., Hidalgo Martinez, G., Simon, T., Wei, S., Sheikh, Y.A.: OpenPose: realtime multi-person 2D pose estimation using part affinity fields. IEEE TPAMI **43**, 172–186 (2019)
5. Cao, Z., Simon, T., Wei, S.E., Sheikh, Y.: Realtime multi-person 2D pose estimation using part affinity fields. In: CVPR (2017)
6. Chen, J., Ho, C.M.: MM-ViT: multi-modal video transformer for compressed video action recognition. In: WACV, pp. 1910–1921 (2022)
7. Chen, Y., Zhang, Z., Yuan, C., Li, B., Deng, Y., Hu, W.: Channel-wise topology refinement graph convolution for skeleton-based action recognition. In: ICCV, pp. 13359–13368 (2021)
8. Cuppens, K., et al.: Using spatio-temporal interest points (STIP) for myoclonic jerk detection in nocturnal video. In: IEEE EMBC, pp. 4454–4457 (2012)
9. Dang, L., Nie, Y., Long, C., Zhang, Q., Li, G.: MSR-GCN: multi-scale residual graph convolution networks for human motion prediction. In: ICCV, pp. 11467–11476 (2021)
10. Duan, H., Wang, J., Chen, K., Lin, D.: DG-STGCN: dynamic spatial-temporal modeling for skeleton-based action recognition. arXiv preprint arXiv:2210.05895 (2022)
11. Duan, H., Zhao, Y., Chen, K., Lin, D., Dai, B.: Revisiting skeleton-based action recognition. In: CVPR, pp. 2969–2978 (2022)

12. Fang, H.S., et al.: AlphaPose: whole-body regional multi-person pose estimation and tracking in real-time. IEEE TPAMI **45**(6), 7157–7173 (2023)
13. Feichtenhofer, C.: X3D: expanding architectures for efficient video recognition. In: CVPR, pp. 203–213 (2020)
14. Feichtenhofer, C., Fan, H., Malik, J., He, K.: SlowFast networks for video recognition. In: CVPR, pp. 6202–6211 (2019)
15. Geertsema, E.E., et al.: Automated video-based detection of nocturnal convulsive seizures in a residential care setting. Epilepsia **59**, 53–60 (2018)
16. Han, K., Wang, Y., Guo, J., Tang, Y., Wu, E.: Vision GNN: an image is worth graph of nodes. In: NeurIPS, vol. 35, pp. 8291–8303 (2022)
17. Han, Y., Wang, P., Kundu, S., Ding, Y., Wang, Z.: Vision HGNN: an image is more than a graph of nodes. In: ICCV, pp. 19878–19888 (2023)
18. He, K., Zhang, X., Ren, S., Sun, J.: Deep residual learning for image recognition. In: CVPR, pp. 770–778 (2016)
19. Hussein, R., Palangi, H., Ward, R., Wang, Z.J.: Epileptic seizure detection: a deep learning approach. arXiv preprint arXiv:1803.09848 (2018)
20. Ji, S., Xu, W., Yang, M., Yu, K.: 3D convolutional neural networks for human action recognition. IEEE TPAMI **35**(1), 221–231 (2012)
21. Kalitzin, S., Petkov, G., Velis, D., Vledder, B., da Silva, F.L.: Automatic segmentation of episodes containing epileptic clonic seizures in video sequences. IEEE TBME **59**(12), 3379–3385 (2012)
22. Karácsony, T., Loesch-Biffar, A.M., Vollmar, C., Rémi, J., Noachtar, S., Cunha, J.P.S.: Novel 3D video action recognition deep learning approach for near real time epileptic seizure classification. Sci. Rep. **12**(1), 19571 (2022)
23. Kong, Y., Fu, Y.: Human action recognition and prediction: a survey. Int. J. Comput. Vis. **130**(5), 1366–1401 (2022)
24. Lee, J., Lee, M., Lee, D., Lee, S.: Hierarchically decomposed graph convolutional networks for skeleton-based action recognition. In: ICCV, pp. 10444–10453 (2023)
25. van der Lende, M., Cox, F.M.E., Visser, G.H., Sander, J.W., Thijs, R.D.: Value of video monitoring for nocturnal seizure detection in a residential setting. Epilepsia **57**(11), 1748–1753 (2016)
26. Li, G., Muller, M., Thabet, A., Ghanem, B.: DeepGCNs: can GCNs go as deep as CNNs? In: CVPR, pp. 9267–9276 (2019)
27. Li, M., Chen, S., Chen, X., Zhang, Y., Wang, Y., Tian, Q.: Actional-structural graph convolutional networks for skeleton-based action recognition. In: CVPR, pp. 3595–3603 (2019)
28. Liu, Z., et al.: Deep dual consecutive network for human pose estimation. In: CVPR, pp. 525–534 (2021)
29. Liu, Z., Zhang, H., Chen, Z., Wang, Z., Ouyang, W.: Disentangling and unifying graph convolutions for skeleton-based action recognition. In: CVPR, pp. 143–152 (2020)
30. Mehta, D., Sivathamboo, S., Simpson, H., Kwan, P., O'Brien, T., Ge, Z.: Privacy-preserving early detection of epileptic seizures in videos. In: Greenspan, H., et al. (eds.) MICCAI 2023. LNCS, vol. 14224, pp. 210–219. Springer, Cham (2023). https://doi.org/10.1007/978-3-031-43904-9_21
31. Moshé, S.L., Perucca, E., Ryvlin, P., Tomson, T.: Epilepsy: new advances. The Lancet **385**(9971), 884–898 (2015)
32. Munir, M., Avery, W., Marculescu, R.: MobileViG: graph-based sparse attention for mobile vision applications. In: CVPR, pp. 2210–2218 (2023)
33. Osokin, D.: Real-time 2D multi-person pose estimation on CPU: lightweight OpenPose. arXiv preprint arXiv:1811.12004 (2018)

34. Pérez-García, F., Scott, C., Sparks, R., Diehl, B., Ourselin, S.: Transfer learning of deep spatiotemporal networks to model arbitrarily long videos of seizures. In: de Bruijne, M., et al. (eds.) MICCAI 2021. LNCS, vol. 12905, pp. 334–344. Springer, Cham (2021). https://doi.org/10.1007/978-3-030-87240-3_32
35. Shi, L., Zhang, Y., Cheng, J., Lu, H.: Two-stream adaptive graph convolutional networks for skeleton-based action recognition. In: CVPR, pp. 12026–12035 (2019)
36. Shi, L., Zhang, Y., Cheng, J., Lu, H.: Skeleton-based action recognition with multi-stream adaptive graph convolutional networks. IEEE TIP **29**, 9532–9545 (2020)
37. Shoeb, A.H., Guttag, J.V.: Application of machine learning to epileptic seizure detection. In: ICML, pp. 975–982 (2010)
38. Song, Y.F., Zhang, Z., Shan, C., Wang, L.: Stronger, faster and more explainable: a graph convolutional baseline for skeleton-based action recognition. In: ACM MM, pp. 1625–1633 (2020)
39. Tang, S., et al.: Self-supervised graph neural networks for improved electroencephalographic seizure analysis. arXiv preprint arXiv:2104.08336 (2021)
40. Thijs, R.D., Surges, R., O'Brien, T.J., Sander, J.W.: Epilepsy in adults. The Lancet **393**(10172), 689–701 (2019)
41. Thodoroff, P., Pineau, J., Lim, A.: Learning robust features using deep learning for automatic seizure detection. In: Proceedings of the 1st Machine Learning for Healthcare Conference, vol. 56, pp. 178–190. PMLR (2016)
42. Toshev, A., Szegedy, C.: DeepPose: human pose estimation via deep neural networks. In: Proceedings of the IEEE Conference on Computer Vision and Pattern Recognition, pp. 1653–1660 (2014)
43. Tran, D., Bourdev, L., Fergus, R., Torresani, L., Paluri, M.: Learning spatiotemporal features with 3D convolutional networks. In: ICCV, pp. 4489–4497 (2015)
44. Tran, D., Wang, H., Torresani, L., Ray, J., LeCun, Y., Paluri, M.: A closer look at spatiotemporal convolutions for action recognition. In: CVPR, pp. 6450–6459 (2018)
45. Tzallas, A.T., et al.: Automated epileptic seizure detection methods: a review study. In: Epilepsy-Histological, Electroencephalographic and Psychological Aspects, pp. 2027–2036 (2012)
46. van Westrhenen, A., Petkov, G., Kalitzin, S.N., Lazeron, R.H., Thijs, R.D.: Automated video-based detection of nocturnal motor seizures in children. Epilepsia **61**, S36–S40 (2020)
47. Wu, J., et al.: PVG: progressive vision graph for vision recognition. In: ACM MM, pp. 2477–2486 (2023)
48. Xiang, W., Li, C., Wang, B., Wei, X., Hua, X.S., Zhang, L.: Spatiotemporal self-attention modeling with temporal patch shift for action recognition. In: Avidan, S., Brostow, G., Cissé, M., Farinella, G.M., Hassner, T. (eds.) ECCV 2022. LNCS, vol. 13663, pp. 627–644. Springer, Cham (2022). https://doi.org/10.1007/978-3-031-20062-5_36
49. Xu, Y., Yang, J., Ming, W., Wang, S., Sawan, M.: Shorter latency of real-time epileptic seizure detection via probabilistic prediction. Expert Syst. Appl. **236**, 121359 (2024)
50. Yan, S., Xiong, Y., Lin, D.: Spatial temporal graph convolutional networks for skeleton-based action recognition. In: AAAI, vol. 32 (2018)
51. Yang, Y., Sarkis, R.A., Atrache, R.E., Loddenkemper, T., Meisel, C.: Video-based detection of generalized tonic-clonic seizures using deep learning. IEEE JBHI **25**(8), 2997–3008 (2021)
52. Yue, R., Tian, Z., Du, S.: Action recognition based on RGB and skeleton data sets: a survey. Neurocomputing **512**, 287–306 (2022)

53. Zeiler, M.D., Fergus, R.: Visualizing and understanding convolutional networks. In: Fleet, D., Pajdla, T., Schiele, B., Tuytelaars, T. (eds.) ECCV 2014. LNCS, vol. 8689, pp. 818–833. Springer, Cham (2014). https://doi.org/10.1007/978-3-319-10590-1_53
54. Zhang, B., et al.: Factorized omnidirectional representation based vision GNN for anisotropic 3D multimodal MR image segmentation. In: ACM MM, pp. 1607–1615 (2023)

DiffSurf: A Transformer-Based Diffusion Model for Generating and Reconstructing 3D Surfaces in Pose

Yusuke Yoshiyasu[✉] and Leyuan Sun

National Institute of Advanced Industrial Science and Technology (AIST),
1-1-1 Umezono, Tsukuba, Japan
{yusuke-yoshiyasu,son.leyuansun}@aist.go.jp

Abstract. This paper presents DiffSurf, a transformer-based denoising diffusion model for generating and reconstructing 3D surfaces. Specifically, we design a diffusion transformer architecture that predicts noise from noisy 3D surface vertices and normals. With this architecture, DiffSurf is able to generate 3D surfaces in various poses and shapes, such as human bodies, hands, animals and man-made objects. Further, DiffSurf is versatile in that it can address various 3D downstream tasks including morphing, body shape variation and 3D human mesh fitting to 2D keypoints. Experimental results on 3D human model benchmarks demonstrate that DiffSurf can generate shapes with greater diversity and higher quality than previous generative models. Furthermore, when applied to the task of single-image 3D human mesh recovery, DiffSurf achieves accuracy comparable to prior techniques at a near real-time rate.

Keywords: Diffusion model · 3D surface · Human mesh recovery

1 Introduction

Creating and reconstructing 3D shape models in various shapes and poses is a significant challenge in computer vision and computer graphics, with extensive applications in gaming, augmented reality (AR) and virtual reality (VR). For such 3D content-based applications, a surface mesh is the most commonly used shape representation because of its efficiency during graphics rendering and user-friendliness for artists.

Over the past few years, diffusion models [28,65] have revolutionized the content creation paradigm in the image domain, particularly in the task of image generation from text prompts. Diffusion models can generate high-quality and diverse data by learning to reverse the diffusion process. This process gradually constructs desired data samples from noise whose dimensionality is higher than that of previous generative models such as generative adversarial networks (GANs). Additionally, diffusion models have been applied to the generation of 3D data, such as object point clouds [49,51,91], textured 3D models [41,63,81], scene radiance field [6] and 3D human pose [25]. Recent 3D shape diffusion models are able to generate 3D surfaces with complex geometry and topology by incorporating implicit functions, such as signed distance fields (SDF), and extracting their zero level-set surface using the marching cubes algorithm [13,71].

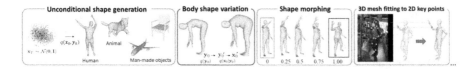

Fig. 1. DiffSurf addresses the unconditional generation of 3D surfaces in diverse poses. It can generate 3D surfaces of various objects types such as humans, mammals and man-made objects. Downstream tasks, including unconditional generation, morphing and fitting to 2D key points can be addressed with pre-trained DiffSurf models.

Yet, there remain several challenges in generating 3D surfaces based on diffusion models. **Firstly**, the current approaches do not consider pose. In the case of 3D human and animal generation in different poses, point-to-point correspondences between shapes are important but they are lost when employing the current 3D shape diffusion models. There exist recent works of diffusion models for generating 3D human poses and shapes conditioned on images [14,25,40,42] but generation of diverse body shapes and poses are not considered. **Secondly**, we want our generative model to handle a wide range of objects, such as human bodies, mammals and man-made objects. **Thirdly**, a framework that can deal with a wide variety tasks, e.g. interpolating two shapes, altering pose and manipulating shape, is highly sought after.

In this paper, we propose a transformer-based diffusion model for generating and reconstructing 3D surfaces (dubbed DiffSurf). To address the aforementioned challenges, we design a diffusion transformer architecture that predicts noise from noisy 3D surface vertex coordinates. By representing a surface with points and normals, processing them in diffusion transformer and then employing up-samplers dedicated for topologically fixed and varied cases, DiffSurf is able to handle diverse body poses, various object types and multiple different tasks as illustrated in Figs. 1 and 2. To our knowledge, DiffSurf is the first diffusion model that addresses generation of 3D surfaces in diverse body poses and shapes. The contributions of this paper are summarized as follows:

1. DiffSurf, a denoising diffusion transformer model that can generate 3D surfaces in various body shapes and poses.
2. It can generate 3D shapes of diverse object types, such as human bodies, mammals and man-made objects, using a diffusion transformer model that leverages point-normal representation.
3. It provides methodologies for addressing various 3D processing and image-to-3D downstream tasks by effectively utilizing pre-trained DiffSurf models based on score distillation sampling (SDS) and classifier-free guidance (CFG).

2 Related Work

Generative Models for 3D Shape and Pose. Previous generative models for 3D shape and pose generation predominantly utilize variational autoencoders (VAEs) [3,18,24,30,76,77,89,93], generative adversarial networks (GANs) [20,34] or normalizing flows [7,82,90]. SMPL-X [58] employs a VAE to learn a pose prior and enforces constraints on body joint angles. COMA [64] and CAPE [52] designed their VAEs based on graph neural networks to model facial expressions and clothing geometric deformations, respectively. Kanazawa et al. [34] introduced the human mesh recovery (HMR) technique that estimates a posed human body model from a single image by employing GANs to to provide body pose and shape priors. Other approaches focus on designing and obtaining latent encoders or representation using GANs [11,12,20]. Recent techniques for 3D human pose and shape estimation employ normalizing flows [7,39,82,90] in attempting to learn 3D priors from motion capture datasets. Pose-NDF instead represents and learns a pose space using neural distance fields [79].

Diffusion Models and 3D Generation. In 3D generation, Luo et al. [49] proposed the first 3D point cloud generation method based on diffusion models by extending Denoising Diffusion Probabilistic Models (DDPM) [28]. LION [91] and SLIDE [51] adopted Latent Diffusion Models (LDM) [65] in point cloud generation, aiming to reduce point cloud resolution for more efficient training and sampling. A learning-based surface reconstruction technique called ShapeAsPoint (SAP) [61] is then employed to convert the generated point clouds into volume functions and subsequently into meshes. To generate 3D surfaces, recent approaches use implicit functions in diffusion process such as SDF, and extract a mesh from 3D volume using marching cubes [13,71,88]. MeshDiffusion [47] uses DMTet [70] that combines a tetra mesh and SDF to represent the object shape to generate topologically and geometrically complex objects. Mo et al. proposed DiT-3D [56] which extends diffusion transformer [4,5,59,60] to voxelized 3D point clouds, accomplishing the generation of 3D objects. PolyDiff introduced a 3D diffusion model that can work with polygonal meshes [1]. Point-e [57] and Shape-e [33] extend [49] to colored point clouds using transformers.

3D Shape and Pose from Image. Human mesh recovery approaches [78] predict a 3D human body mesh from a single image or video frames, which can be roughly divided into parametric [8,34,92] and vertex-based approaches [16,38,43]. Parametric approaches regress the body shape and pose parameters of human body models like SMPL or SMPL-X. On the other hand, vertex-based approaches directly regress from an image to 3D vertex coordinates. Transformer architectures, which are known for their ability to capture long-range dependencies, have been employed in vertex-based human mesh recovery and shown strong performances [15,43,44,86,87]. The reconstruction of mammals from images has also been addressed in the field [67,83,94]. Research on 3D human body pose and shape generation via diffusion models has recently commenced but most being task-specific and conditional on 2D data. They include 3D human pose estimation methods from 2D keypoints [25,69], parametric human mesh recovery techniques [14,46] and vertex-based human mesh recovery methods [40,42].

3 Background: Diffusion Models

Diffusion models establish a Markov chain of diffusion steps by gradually adding random noise to data (i.e., a forward diffusion process) and learn to reverse this process to construct desired data samples from the noise (i.e., a reverse diffusion process). Considering a data point sampled from a data distribution $\mathbf{x}_0 \sim q(\mathbf{x})$, the forward process produces a sequence of noisy samples $\mathbf{x}_1 \ldots \mathbf{x}_T$ by adding a small amount of Gaussian noise to the sample in T steps:

$$q(\mathbf{x}_{1:T}|\mathbf{x}_0) = \prod_{t=1}^{T} q(\mathbf{x}_t|\mathbf{x}_{t-1}) \tag{1}$$

$$q(\mathbf{x}_t|\mathbf{x}_{t-1}) = \mathcal{N}(\mathbf{x}_t; \sqrt{1-\beta_t}\mathbf{x}_{t-1}, \beta_t \mathbf{I}) \tag{2}$$

where $\{\beta_t\}_1^T \in (0,1)$ is the noise variance schedule.

Data generation can be initiated from a Gaussian noise input $\mathbf{x}_T \sim \mathcal{N}(\mathbf{0}, \mathbf{I})$, provided that the aforementioned forward process is reversed to sample from $q(\mathbf{x}_{t-1}|\mathbf{x}_t)$. However, the calculation of $q(\mathbf{x}_{t-1}|\mathbf{x}_t)$ depends on the entire dataset and is not straightforward. To address this, a neural network model p_θ is used to approximate these conditional probabilities with a Gaussian model:

$$p_\theta(\mathbf{x}_{t-1}|\mathbf{x}_t) = \mathcal{N}(\mathbf{x}_{t-1}; \mu_\theta(\mathbf{x}_t, t), \sigma^2 \mathbf{I}) \tag{3}$$

where t is a timestep uniformly sampled from $1, 2, \ldots, T$. Then, $\mu_\theta(\mathbf{x}_t, t)$ can be rewritten with noise prediction $\epsilon_\theta(\mathbf{x}_t, t)$ as:

$$\mu_\theta(\mathbf{x}_t, t) = \frac{1}{\sqrt{\alpha_t}} \left(\mathbf{x}_t - \frac{1-\alpha_t}{\sqrt{1-\bar{\alpha}_t}} \epsilon_\theta(\mathbf{x}_t, t) \right) \tag{4}$$

where $\alpha_t = 1 - \beta_t$ and $\mathbf{x}_t = \sqrt{\bar{\alpha}_t}\mathbf{x}_0 - \sqrt{1-\bar{\alpha}_t}\epsilon$ with $\bar{\alpha}_t = \prod_{t=1}^{t} \alpha$. To learn to estimate $\epsilon_\theta(\mathbf{x}_t, t)$ between consecutive samples $\mathbf{x}_{t-1}$ and $\mathbf{x}_t$, the training loss is defined as follows:

$$L = \mathbb{E}_{t, \mathbf{x}_0, \epsilon} ||\epsilon - \epsilon_\theta(\mathbf{x}_t, t)||_2^2 \tag{5}$$

UniDiffuser. UniDiffuser [5] introduced an approach to handle multi-modal data distributions in the diffusion process in a unified manner. It defines the conditional expectations in a general form, $\mathbb{E}[\epsilon_x, \epsilon_y | \mathbf{x}_{t^x}, \mathbf{y}_{t^y}]$, for all $0 \leq t^x, t^y \leq T$, where t^x and t^y represent two potentially different timesteps. $\mathbf{x}_{t^x}$ and $\mathbf{y}_{t^y}$ are the corresponding perturbed data. With this formulation, marginal diffusion, conditional diffusion and joint diffusion can be achieved by setting $t^y = T$, $t^y = 0$ and $t^x = t^y = t$, respectively. To this end, a joint noise prediction network $\epsilon_\theta(\mathbf{x}_{t^x}, \mathbf{y}_{t^y}, t^x, t^y)$ is employed to predict noise $\epsilon_\theta = [\epsilon_\theta^x, \epsilon_\theta^y]$ injected into $\mathbf{x}_{t^x}$ and $\mathbf{y}_{t^y}$, which is trained by minimizing the following loss:

$$\mathcal{L}_{uni} = \mathbb{E}_{\mathbf{x}_0, \mathbf{y}_0, \epsilon_x, \epsilon_y, \mathbf{x}_{t^x}, \mathbf{y}_{t^y}} ||[\epsilon^x, \epsilon^y] - \epsilon_\theta(\mathbf{x}_{t^x}, \mathbf{y}_{t^y}, t^x, t^y)||_2^2$$

where $\mathbf{x}_0$ and $\mathbf{y}_0$ are the data points, ϵ_x and ϵ_y are sampled from Gaussian distributions, and t^x and t^y are independently and uniformly sampled from the range $1, 2, \ldots, T$. Furthermore, UniDiffuser is directly applicable to classifier-free guidance (CFG), which is a method introduced to enhance the sample quality of conditional diffusion models [27], without modifying the training loss. For instance, $\mathbf{x}_0$ conditioned on $\mathbf{y}_0$ can be generated as follows:

$$\hat{\epsilon}_\theta^x(\mathbf{x}_t, \mathbf{y}_0, t) = (1 + s_g)\epsilon_\theta^x(\mathbf{x}_t, \mathbf{y}_0, t, 0) - s_g \epsilon_\theta^x(\mathbf{x}_t, t, T) \tag{6}$$

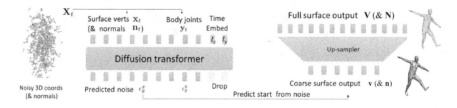

Fig. 2. Overview. DiffSurf consists of a diffusion transformer and an up-sampler. The diffusion transformer takes in the noisy 3D coordinates of surface vertices $\mathbf{x}_t \in \mathbb{R}^{N \times 3}$ and body joints $\mathbf{y}_t \in \mathbb{R}^{J \times 3}$. It processes these two modalities of data along with their corresponding timestep tokens t_x and t_y. The transformer then outputs noise predictions for vertex and joint tokens, ϵ_θ^x and ϵ_θ^y, respectively. For the 3D surface generation of man-made objects, we also input the noisy surface normals $\mathbf{n}_t \in \mathbb{R}^{N \times 3}$ corresponding to vertex tokens into the diffusion transformer. Once the 3D coordinates of surface vertices $\mathbf{v}$ (and normals $\mathbf{n}$) are generated, up-sampling is optionally performed to obtain the full dense surface output $\mathbf{V}$ (and $\mathbf{N}$).

where s_g is a guidance scale and $\epsilon_\theta^x(\mathbf{x}_t, \mathbf{y}_0, t, 0)$ and $\epsilon_\theta^x(\mathbf{x}_t, t, T)$ are the conditional and unconditional models, respectively.

Score Distillation Sampling (SDS). A loss calculation framework called Score Distillation Sampling (SDS) is proposed by DreamFusion [63] for utilizing pre-trained diffusion models in optimizing and regularizing a neural 3D scene model. The scene model is defined by a parametric function of the form $x = g(\phi)$ capable of generating an image x from the desired camera pose. In this context, g is a volumetric renderer such as NeRFs and ϕ is a Multi-Layer Perceptron (MLP) that models a 3D volume. Given a text condition $\mathbf{y}_0$, SDS derives the gradient to update ϕ in such a way that:

$$\nabla_\phi \mathcal{L}_{\text{SDS}}(\phi, g(\theta)) = \mathbb{E}_{t,\epsilon}\left[\omega(t)(\epsilon_\theta(\mathbf{x}_t, \mathbf{y}_0, t) - \epsilon)\frac{\partial \phi}{\partial x}\right]$$

where $\omega(t)$ is a weighting function. In practice, the conditional noise prediction model $\epsilon_\theta(\mathbf{x}_t, \mathbf{y}_0, t)$ is replaced with the classifier free guidance one, $\hat{\epsilon}_\theta(\mathbf{x}_t, \mathbf{y}_0, t)$.

4 Method

We introduce DiffSurf, a general network architecture for generating, editing and reconstructing 3D surfaces based on a plain diffusion transformer model, which is extendable to a wide range of object types. In addition, we introduce downstream methodologies to leverage pre-trained DiffSurf models for solving various 3D processing tasks.

For the generation of posed 3D surfaces, we propose to incorporate the vertex-based mesh recovery paradigm [15,43,44] from human mesh recovery into 3D shape generation. Then, point-to-point correspondences between shapes are ensured by learning from training meshes with the same connectivity. This is a straightforward yet effective strategy, which is overlooked by the recent 3D shape generation literature, when unconditionally generating human and animal 3D surfaces in different poses where correspondences across shapes are vital. To the best of our knowledge, DiffSurf is the first 3D diffusion model capable of unconditionally generating 3D surfaces of articulated objects in diverse poses.

As for generating man-made objects, we propose a strategy for capturing better geometry by incorporating surface normals into the diffusion process. This approach not only leads to better generation results but also provides a more informative input for the subsequent surface reconstruction post-processing [51,61,91] than using point sets alone, as shown in Fig. 3. This is par-

Fig. 3. w/o and with surface normals.

ticularly useful for generating 3D surfaces with complex geometry and topological differences, although point-to-point correspondences may be lost in this case. Unlike the recent diffusion models based on implicit functions [13,56], in the diffusion process, we avoid using a volumetric representation whose complexity grows cubically w.r.t voxel resolution. Instead, we directly work on an explicit point representation through the diffusion transformer to exploit long-range dependencies between points. As a result, DiffSurf only needs a single diffusion model as opposed to previous hierarchical latent diffusion models which rely on two diffusion models [91], making our model computationally more efficient.

4.1 Network Architecture

We describe our diffusion transformer architecture for generating 3D surfaces. It draws inspiration from vertex-based human mesh recovery approaches [15,43,44] and point cloud latent diffusion models [49,51,91], which process explicit 3D point-based representations in neural network models. Specifically, we have designed a UniDiffuser-like multi-modal diffusion transformer architecture [84] that predicts noise from noisy 3D coordinates of surface vertices and body joints, treating them as two distinct modalities. The incorporation of body joints not only facilitates more effective training [43] but also provides landmark controls [51] for manipulating shape and pose. Consequently, we introduce a simple yet versatile transformer architecture for generating and reconstructing 3D surfaces of articulated objects in various shapes and poses, thereby enabling various downstream 3D processing tasks as illustrated in Fig. 2. DiffSurf consists of 1) a diffusion transformer and 2) a mesh up-sampler.

Diffusion Transformer Blocks. The inputs to the diffusion transformer consist of noisy 3D coordinates for a set of joint query tokens $Q_J = \{Q_J^1 \ldots Q_J^J\}$ and coarse vertex query tokens $Q_V = \{Q_V^1 \ldots Q_V^N\}$, corresponding to an articulated body mesh comprising J joints and N vertices. The input noisy 3D coordinates of surface vertices, body joints and their concatenations are respectively denoted as $\mathbf{x}_t \in \mathbb{R}^{N \times 3}$, $\mathbf{y}_t \in \mathbb{R}^{J \times 3}$ and $\mathbf{X}_t \in \mathbb{R}^{(J+N) \times 3}$. The diffusion transformer processes these two modalities of data and their corresponding timesteps t_x and t_y as tokens. It outputs noise predictions for vertices and joints, ϵ_θ^x and ϵ_θ^y. For the generation of man-made objects, we concatenate the noisy 3D coordinates of vertices $\mathbf{x}_t$ with the corresponding surface normals $\mathbf{n}_t \in \mathbb{R}^{N \times 3}$ to construct an $N \times 6$ matrix, which is then input to the diffusion transformer. Our diffusion transformer consists of 7 layers of transformer blocks and input/output MLP layers. Each transformer block has hidden layers with the dimension of 256 channels. The input MLP layer converts $\mathbf{x}_t$ and $\mathbf{y}_t$ into 256-dimensional embedding features and the output MLP layer converts the features processed by transformer into ϵ_θ^x and ϵ_θ^y.

Up-Sampling. After a coarse surface comprising N vertices, $\mathbf{v} \in \mathbb{R}^{N \times 3}$ and corresponding surface normals $\mathbf{n} \in \mathbb{R}^{N \times 3}$ are produced from the noise prediction ϵ_θ^x by the diffusion transformer, we apply an upsampling operation. For human and animal generation, where point-to-point correspondences i.e. mesh connectivity is fixed, an upsampling technique based on MLPs similar to [43,44,86] is adopted to obtain a dense mesh (see Fig. 2) with M vertices, $\mathbf{V} \in \mathbb{R}^{M \times 3}$. For the surface generation of man-made objects, refinement and upsampling based on the improved PointNet++ model [50] are applied to increase the number of points and normals by a factor of $\times 5$ [51]. Subsequently, a learning-based surface reconstruction technique called Shape-As-Points (SAP) [61] is employed to convert the upsampled points $\mathbf{V}$ and normals $\mathbf{N}$ into a mesh.

4.2 Downstream Methodologies

Here, we demonstrate that DiffSurf is capable of performing a series of downstream tasks in 3D surface editing and reconstruction.

Pose Conditioned Mesh Generation. DiffSurf can generate a mesh that is conditioned on 3D skeleton landmark locations, such as those obtained using motion capture and image-based 3D pose regressors. Essentially, this process of conditional mesh generation involves feeding 3D body joint locations as queries into the diffusion transformer and setting the timestep to $t_y = 0$. Naively feeding 3D joint locations into DiffSurf results in slight discrepancies between the mesh and the joints. To address this, we leverage CFG (Eq. (6)) to push the mesh toward joint locations and improve alignments between them. We found that setting the CFG weight to around $s_g = 1$ effectively improves alignment while preserving the mesh structure. Excessively increased CFG weights, e.g., $s_g > 3$, can result in distortion of the mesh (as shown in the Appendix).

Body Shape Variation. By feeding a skeleton to DiffSurf as a condition and performing sampling with varying random noise, we can generate meshes in different body shapes, as shown in Fig. 1 (right). However, this approach does not allow for changes in body heights and segment lengths. To address this, we adopt a two-step strategy. The first step involves unimodal generation to create a batch of skeletons with different body poses and styles. The second step then adjusts the segment lengths of one of generated skeletons based on others in the batch. By inputting these modified skeletons into DiffSurf, we can generate meshes in diverse body styles while maintaining the pose.

Shape Morphing. DiffSurf is capable of morphing between two meshes with different poses and shapes. This is achieved by blending two meshes represented as the Gaussian noise, $\mathbf{x}_T^1$ and $\mathbf{x}_T^2$, through spherical linear interpolation (SLERP) [72], $\hat{\mathbf{x}}_T = \text{SLERP}(\mathbf{x}_T^1, \mathbf{x}_T^2, w)$, where w is an interpolation weight within the range $[0, 1]$. Setting w outside this range $[0, 1]$ e.g., $[-0.25, 1.25]$, results in extrapolation. Given the new noise $\hat{\mathbf{x}}_T$, a mesh is then sampled from it using DiffSurf. It is noteworthy that DiffSurf has the capability to simultaneously handle both shape and pose variations.

Shape Refinement. Instead of starting from Gaussian noise to generate a mesh, DiffSurf can refine a mesh exhibiting noise and distortions by leveraging the SDS gradients as calculated in Eq. (3). Analogous to mesh fairing [21], a sequence of refined meshes can be constructed by explicitly applying the SDS gradients to a mesh:

$$\mathbf{X}^{l+1} = \mathbf{X}^l - \nabla_\phi \mathcal{L}_{\text{SDS}}(\phi, \mathbf{X}^l) \tag{7}$$

Fig. 4. Control point deformation. Left: comparison of the loss terms. Right: deformation examples obtained using five control points (head, wrists and ankles, marked in red). The process starts with the rest pose and a human mesh is deformed to align with the control points. DiffSurf can deform a mesh into extremely different poses, such as a forward-bending posture. (Color figure online)

It should be noted that, in contrast to DreamFusion [63], $\mathbf{X}^l$ is not an output from a neural network model but rather the 3D coordinates of surface vertices and body joints.

Control Point Deformation. DiffSurf facilitates data-driven mesh deformation, similar to prior shape editing techniques [22,23,75], by creating a mesh in a plausible shape and pose through the specification of a set of control points and fitting the mesh toward them. Building upon [63], we devise a loss function derived from differential mesh properties of the SDS target mesh to preserve local geometries of the mesh. We minimize the following total loss L_{def} to optimize $\mathbf{X}$:

$$L_{\text{def}}(\mathbf{X}) = L_{\text{SDS}} + L_{\text{SDS}}^{\text{edge}} + L_{\text{SDS}}^{\text{lap}} + L_{\text{consist}} + L_{\text{CP}} \quad (8)$$

where L_{SDS}, $L_{\text{SDS}}^{\text{edge}}$ and $L_{\text{SDS}}^{\text{lap}}$ represent losses defined by the distances between the SDS targets and predictions for the 3D vertex coordinates, edges and Laplacian coordinates of the coarse mesh, respectively. L_{consist} maintains the consistency between the joint and mesh predictions, defined by the distances between the optimized joints and the regressed joints, which are calculated from the coarse mesh vertices using the joint regressor. The regressed joints $\mathbf{j}_{\text{reg}} \in \mathbb{R}^{J \times 3}$ are calculated from the mesh vertices using the joint regressor, $\mathbf{j}_{\text{reg}} = \mathcal{J}\mathbf{V}$, where $\mathcal{J} \in \mathbb{R}^{J \times M}$ is a joint regressor matrix [48,66,94]. L_{CP} quantifies the distances between the optimized joint locations and the control points (see the Appendix for more details).

Figure 4 (left) shows the comparisons of the loss terms. Using L_{SDS} and L_{CP}, DiffSurf is able to fit a human mesh towards control points, but there remain some distances between them. Incorporating L_{consist} improves the fit but introduces distortions around the control points. Adding $L_{\text{SDS}}^{\text{edge}}$ and $L_{\text{SDS}}^{\text{lap}}$ remedies this issue by considering the differential properties of the coarse mesh to preserve its local geometry.

2D Keypoint Fitting. While previous research [8,39] has addressed the challenge of fitting a parametric model to 2D landmarks, we propose an alternative vertex-based fitting approach for this task. Consequently, our method is applicable to both parametric and vertex-based mesh recovery approaches and improves their mesh recovery results.

Starting from the initial solution derived from a mesh recovery approach, DiffSurf optimizes mesh vertices and body joints to improve their alignment with 2D keypoint locations. The loss function for 2D keypoint fitting is defined by modifying Eq. (8) slightly as follows:

$$L_{\text{fit}}(\mathbf{X}) = L_{\text{SDS}} + L_{\text{SDS}}^{\text{edge}} + L_{\text{SDS}}^{\text{lap}} + L_{\text{consist}} + L_{\text{2D}} \quad (9)$$

where L_{2D} measures the discrepancies between the ground truth and predicted 2D keypoints. These 2D keypoint predictions are obtained by projecting 3D joint positions using the camera parameter predictions from the mesh recovery approach. In addition, as DiffSurf is trained on the dataset with global position and rotation aligned at the root, we rigidly align the mesh prediction with the canonical orientation before subtracting the SDS gradients, such that: $T(\mathbf{X} - \nabla_\phi \mathcal{L}_{\text{SDS}}(\phi, T^{-1}(\mathbf{X})))$, where T is the global transformation of the mesh recovery result w.r.t the canonical orientation. The global rotation can be the root orientation predicted by the mesh recovery approach when the global pose is available. Otherwise, a coordinate frame defined from the body joint predictions can be used to obtain T e.g., the x-axis and y-axis are defined from the unit vectors emanating from the pelvis to the neck and from the right hip to the left hip.

Mesh Generation from 3D Keypoints. DiffSurf is capable of reconstructing a 3D mesh from 3D joint locations by employing the pre-trained DiffSurf model for pose-conditioned surface generation. To achieve this, we first predict 3D body joint locations from an image using a 3D pose regressor, which produces the 3D positions of 14 body joints. Subsequently, these 3D joint positions are inputted into DiffSurf to perform conditional mesh generation with Eq. (6). Similar to the SDS-based 2D keypoint fitting, this approach requires aligning the mesh with the canonical orientation prior to mesh sampling. It is noteworthy that this modular human mesh recovery design, based on DiffSurf, decouples an image-based 3D pose regressor from a mesh generator, enabling its training even when image-mesh paired data are not available. The architecture of our 3D pose regressor used here is transformer-based (see the Appendix).

5 Experiments

5.1 Dataset and Metrics

3D Generation. We trained our DiffSurf models separately on publicly available 3D datasets: SURREAL [80], AMASS [53], FreiHAND [17], BARC [67], Animal3D [83] and ShapeNet [10]. We follow 3D-CODED [26] for the SURREAL train/test split definition. The global positions of meshes used in the training are aligned at the root position and oriented to face forward. The body joints are obtained from meshes using joint regressors [48,94] for humans and animals. For ShapeNet objects, we feed sparse latent points generated by SLIDE [51] as body joint tokens to transformer.

Evaluation of 3D human generation was conducted on the SURREAL testset (200 meshes) and DFAUST [9] (800 meshes). The standard metric used for evaluating 3D generation is the 1-NNA metric [85], which quantifies the distributional similarity between generated shapes and the validation set. This metric assesses both the quality and diversity of the generated results. For human generation, given the differences in global orientations between validation and training meshes, we first perform a rigid alignment of the predicted mesh with the validation meshes before calculating the 1-NNA metric. We refer to this modified metric as Rigid Aligned 1-NNA (RA-1-NNA).

Human Mesh Recovery. We trained our 3D pose regressor using publicly available datasets, adopting the mixed dataset training strategies as outlined in [36,43]. The datasets used include Human3.6M [29], MPI-INF-3DHP [55], COCO [45], MPII [2]

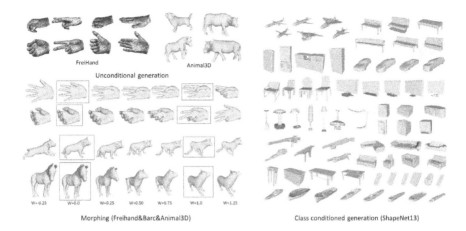

Fig. 5. Left: Unconditional generation and morphing of hands, dogs and animals. Right: class conditioned generation of man-made objects.

and LSPET [31]. For training, we utilized the 3D joint labels from Human 3.6M and 2D keypoint labels from all the dataset. Additionally, we conduct another training experiment using 3D body joint labels obtained from pseudo 3D meshes produced by EFT [32] on the in-the-wild image datasets. Unlike recent mesh transformer approaches [15,43], we did not use 3DPW [54] as training data for fine-tuning on 3DPW, but instead only performed evaluations on its test set.

We used the following three standard metrics for evaluation: MPJPE, PA-MPJPE and MPVE. Mean-Per-Joint-Position-Error (MPJPE) measures the Euclidean distances between the ground truth and the predicted joints. The PA-MPJPE metric, where PA stands for Procrustes Analysis, measures the error of the reconstruction after removing the effects of scale and rotation. Mean-Per-Vertex-Error (MPVE) measures the Euclidean distances between the ground truth and the predicted vertices.

5.2 Training and Sampling

The training of DiffSurf involves two steps: training of the diffusion model and the up-samplers are done separately. We use pre-trained up-sampler models for fixed and varied topology cases (see the Appendix for the details on the network architectures and their training). Our diffusion transformer model is trained with a batch size of 256 for 400 epochs on 4 NVIDIA V100 GPUs for the SURREAL dataset, and for 200 epochs on 8 NVIDIA A100 GPUs for the AMASS dataset. It takes about 1 day for both cases. For the BARC, Animal3D and ShapeNet objects, we extend the training of diffusion transformer to 4000–8000 epochs because they contain fewer meshes than SURREAL and AMASS. The dataset statistics are provided in Appendix. We use the Adam optimizer for training our models, while reducing the learning rate by a factor of 10 after $1/2$ of the total training epochs beginning from 1×10^{-4}. For the training objective of DiffSurf, we adopt the v-prediction parameterization [62,68] and employ the DDIM [73] sampler along with a sigmoid variance scheduler. We set the diffusion time step to $T = 1000$ and tested sampling steps in the range [1–250].

Fig. 6. Example results of human mesh recovery by DiffSurf on the 3DPW dataset. Compared to METRO, DiffSurf produces less distorted results, especially in occluded situations.

5.3 Downstream Applications

Unconditional Shape Generation. Figures 1 (left) and 5 (left) depict unconditional 3D mesh generation results of humans, hands, dogs, mammals and man-made objects based on DiffSurf. DiffSurf trained on FreiHAND can generate 3D hand meshes in a variety of poses, including thumbs up, victory (peace), open and close. The results obtained using the BARC and Animal3D dataset indicate that DiffSurf can handle a range of dog breeds, from small to large, as well as different species. Our approach can achieve class-conditioned generation of ShapeNet 13 objects by inputting class labels to the transformer as in U-ViT, which includes generation of topologically different shapes such as the lamp examples. These results demonstrate the ability of DiffSurf to generate 3D meshes in diverse shapes and poses.

Pose Conditioned Generation and Body Shape Variations. Figure 1 and the Appendix demonstrate the outcomes of pose-conditional mesh generation and body shape variation using DiffSurf. When provided with different 3D joint locations while sampling from the same mesh noise input, DiffSurf can generate various poses of the same body mesh. Variations in body shape are realized by altering the mesh noise input.

Shape Morphing. In Figs. 1, 5 (left), 7 (right) and the Appendix, we present morphing results produced by DiffSurf, where two meshes are interpolated in the Gaussian noise space. In contrast to linear interpolation of 3D vertex coordinates in the 3D Euclidean space, which results in a straight-line interpolation trajectory leading to the artifacts such as arm shrinkage and hand expansions, DiffSurf yields visually plausible interpolation outcomes (see Fig. 7 right). Since LIMP is an approach that learns from a small amount of meshes (in static poses), its pose representation capability is limited. As shown in Fig. 5 (left), this approach can morph between two objects with different shapes, such as horse and dog, which showcases the ability of DiffSurf for handling body shapes and poses together and its potential for becoming a viable alternative to prior nonlinear and data-driven morphing techniques [22,23].

Control Point Deformation. Figure 4 (right) illustrates the results of control point deformation. In these examples, five control points are specified at the head, wrists and ankles and the deformation begins with the rest pose. Through gradient-based optimization and progressively decreasing diffusion time steps, the mesh is refined to a pose that conforms to the control points without distortions.

Human Mesh Recovery from Image. Figure 6 visualizes the mesh recovery results on the 3DPW dataset obtained by DiffSurf. Even though DiffSurf is trained without image-mesh paired data, it produces visually pleasing results without noticeable artifacts.

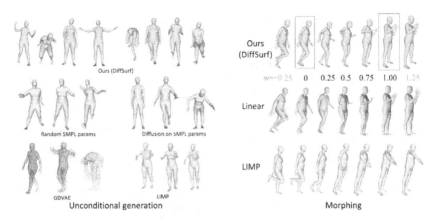

Fig. 7. Qualitative comparisons with previous techniques are presented. Left: comparison of unconditional generation results. Right: comparison of morphing results against linear interpolation in 3D Euclidean space and LIMP.

Table 1. Comparisons with baseline models on unconditional human generation. The RA-1-NNA metric [%] assesses the diversity and quality of generated results. A lower value on this metric signifies superior performance. SD indicates the standard deviation.

	SURREAL	DFAUST
Random SMPL (SD×0.2)	81.1	92.8
Random SMPL (SD×1.0)	67.2	71.4
Random SMPL (SD×3.0)	71.6	95.4
VPoser [58]	60.7	70.2
Parametric Diffusion	59.6	76.2
*Trained on AMASS		

Train \ Test		SURREAL	DFAUST
GDVAE [3]	SURREAL	93.8	98.1
LIMP [18]	FAUST	81.3	93.3
DiffSurf	SURREAL	54.4	69.6
DiffSurf	AMASS	**54.0**	**69.5**

5.4 Comparisons

Unconditional Human Mesh Generation. Here, DiffSurf is compared against five baselines. We employ parametric baseline approaches: Random SMPL which draws SMPL body shape/pose parameters randomly from ×[0.2, 1.0, 3.0] the standard deviations of AMASS parameter collections to generate human body meshes; VPoser [58] which learns pose priors with VAEs; parametric diffusion transformer that generates SMPL parameters. We also compared DiffSurf with the previous generative models for surfaces based on VAEs: GDVAE [3] and LIMP [18].

Figure 7 presents a qualitative comparison, while Table 1 provides quantitative comparisons using the RA-1-NNA metric. As shown in Table 1, naively producing body meshes from random SMPL parameters proved to be far inferior to our approach. DiffSurf also outperforms VPoser and parametric diffusion, which use rotational parametrization of pose that is usually difficult to learn with neural networks. Since GDVAE relies on a point cloud representation, the results exhibit outliers especially around hands and feet (see Fig. 7). LIMP is based on mesh representation and preserves mesh structure by maintaining both extrinsic and intrinsic surface properties. However, LIMP's diversity in body poses and shapes appears to be constrained, likely due to its training strategy relying on a limited dataset. As both methods are based on MLP-based VAEs, their generated sample quality is not as high as that produced by DiffSurf.

Human Mesh Recovery from Image. Table 2 presents a comparison of our method with previous human mesh recovery approaches, which are divided into parametric and vertex-based categories, on the 3DPW and Human3.6M datasets. Note that none of methods used the 3DPW dataset in training. DiffSurf achieves top-level performances among vertex-based approaches. Our method is also comparable to recent diffusion based methods [14,42], even though ours is not trained end-to-end on image-mesh paired dataset. Since DiffSurf does not explicitly relate its generation to an image, performance is affected by random input mesh noise ϵ_θ^x that possibly alters body styles and twisting joint angles of generations. We show in the Appendix how multiple hypotheses on the input mesh noise can further possibly improve DiffSurf's performance.

3D Human Mesh Fitting to 2D Keypoints. Table 3 shows a comparison of optimization approaches that fits a mesh with ground truth (GT) 2D keypoints. With our SDS-based approach, the MPVPE and PA-MPJPE errors for both parametric and vertex-based mesh recovery techniques decrease, outperforming the previous fitting approaches that utilized GT keypoints [8,32,39,74]. In fact, PA-MPJPE dropped by approximately 2pts and 9pts for HMR-EFT [32] and METRO [43], respectively. These results suggest that DiffSurf can potentially aid in the creation of an image-mesh paired dataset with improved alignment between images and meshes.

Table 2. Comparisons with other 3D human mesh recovery approaches on 3DPW. No fine-tuning on 3DPW performed.

	Method	3DPW		Human 3.6M	
		MPVE ↓	PA-MPJPE ↓	MPJPE ↓	PA-MPJPE ↓
Parametric	SPIN [37]	116.4	59.2	62.5	41.1
	ProHMR [39]	–	59.8	–	41.2
	OCHHuman [35]	107.1	58.3	–	–
	DiffHMR [14]	110.9	56.5	–	–
	HMR-EFT [32]	–	54.3	–	–
	PARE [36]	**97.9**	**50.9**	76.8	50.6
Vertex-based	METRO [43]	119.1	63.0	54.0	36.7
	Pose2Mesh [16]	106.3	58.3	64.9	46.3
	GATOR [87]	104.5	56.8	64.0	44.7
	HMDiff [42]	–	–	49.3	**32.4**
	DiffSurf	108.0	53.7	**48.9**	36.1
	DiffSurf-EFT	102.6	52.6	50.1	36.9

Table 3. Comparisons with previous fitting approaches on 3DPW. GT 2D keypoints are used.

Method	MPVE ↓	PA-MPJPE ↓
SMPLify [8]	–	106.1
LearnedGD [74]	–	55.9
ProHMR + fitting [39]	–	55.1
HMR-EFT + EFT fitting [32]	–	53.7
HMR-EFT [32] + SDS fitting	98.6	52.1
METRO [43] + SDS fitting	103.0	54.2
DiffSurf + SDS fitting	**93.7**	**48.7**

5.5 Ablation Studies

Network Architectures. To conduct the ablation study on our diffusion transformer model's components, we modified the elements within it and compared their performances. The basic network architecture from which we started is an adaptation of U-ViT [4], tailored to handle 3D mesh and body joint tokens (see Table 4a, top row). We investigated the impacts of long skip connections, different methods of incorporating

Table 4. Ablation studies: (a) Network components. Errors are measured for unconditional human generation using the RA-1-NNA metric on the SURREAL dataset; (b) Sampling time steps. Errors are measured for 3D human mesh recovery on the 3DPW dataset with a CFG weight $s_g = 1.0$; (c) CFG scale factor s_g. The error is measured by PA-MPJPE↓ on the 3DPW and H3.6M datasets with 10 DDIM sampling time steps.

(a) Ablation study on network components.

Layer	Pos emb	Time emb	Long skip	1NNA ↓
Transformer	Learned	Token	Yes	55.6
Transformer	Learned	Token	No	54.4
Transformer	Learned	Add	Yes	55.4
Transformer	3D Template	Token	Yes	54.8
MLP	Learned	Token	Yes	92.6

(b) Ablation study on sampling time steps.

	1	3	5	10	20	30	100
MPVE ↓	230.4	115.8	107.3	**105.4**	105.8	106.1	106.9
PA-MPJPE ↓	143.9	60.2	55.6	54.3	**54.1**	54.2	54.7
fps	35.9	32.05	25.25	21.2	13.6	8.81	3.38

(c) Ablation study on CFG scale factor.

$s_g =$	0.0	0.1	0.3	0.5	1.0
3DPW	**52.6**	52.6	52.9	53.3	54.3
H3.6M	37.9	37.0	36.2	**36.1**	36.7

time embedding, position embedding constructions and the effect of varying the network layer types (transformer or MLPs). As shown in Table 4a, the switch in the layer type from transformer to MLPs yields the most significant impact, indicating that the most critical component of DiffSurf is the transformer layer. As opposed to U-ViT [4] for image generation, the use of long skip connection does not contribute to improving 3D human mesh generation. This may be attributed to the fact that the current form of the diffusion transformer in DiffSurf primarily processing coarse-level mesh vertices. Expressive human body generation that incorporates fine-grained details like finger poses and facial expressions could potentially benefit from long-skip connections.

Sampling Timesteps. Table 4b presents an ablation study on the sampling timesteps for 3D human mesh recovery from an image. It is observed that an increase in the sampling timesteps leads to a decrease in MPVPE and PA-MPJPE, reaching optimal performance at approximately 10–20 steps. We also measured the inference time with respect to sampling timesteps. In this configuration, DiffSurf operates at nearly real-time speed, approximately 20 FPS, when performing DDIM sampling with 10 steps. Furthermore, the sampling speed of unconditional human generation with 431 vertices were 37, 9 and 1.5 fps for 10, 50 and 250 DDIM sampling steps, respectively. The sampling speed of man-made objects generation (2048 points) were 17.8, 3.5 and 1.8 fps for 10, 50 and 100 DDIM sampling steps, respectively, which is roughly 7× faster than LION [91] with DDIM sampling. Our experiments were conducted on an NVIDIA A100 GPU with the Flash Attention layer enabled [19].

CFG Scale Factor. Table 4c presents the ablation study on the CFG scaling factor s_g for 3D human mesh recovery from an image. For Human 3.6M, a value up to around 1 leads to improvements. On the other hand, we observed that increasing the CFG factor negatively impacts mesh reconstruction on the 3DPW dataset. This discrepancy is likely due to the difference in the accuracy of 3D pose regressors and the frequency of occlusions associated with each dataset. In general, the 3D joint estimation results on Human3.6M are more accurate and reliable than those on 3DPW. This is because the Human3.6M dataset is captured in a controlled experimental environment and is included in the training, whereas 3DPW is an in-the-wild dataset and not used as the training data. DiffSurf provides a method to consider the accuracy and reliability of the 3D joint prediction by balancing between diffusion model priors and 3D body joint conditions through the CFG scaling factor.

6 Conclusion

We presented DiffSurf, a denoising diffusion transformer model for generating 3D surfaces in diverse body shapes and poses. DiffSurf can generate 3D surfaces for a wide range of object types and solve various downstream 3D processing tasks. In future work, we aim to extend DiffSurf towards the generation of expressive human body meshes with fine-grained details, such as facial expressions and finger poses. It would also be intriguing to design a foundational 3D generative model by increasing the capacity of DiffSurf and training it on a larger-scale 3D data.

Acknowledgements. This work was supported by JSPS KAKENHI Grant Number 20K19836, 22H00545, 23H03426 and 23K28116 in Japan. This work was supported by NEDO JPNP20006 (New Energy and Industrial Technology Development Organization) in Japan.

References

1. Alliegro, A., Siddiqui, Y., Tommasi, T., Nießner, M.: PolyDiff: generating 3D polygonal meshes with diffusion models (2023)
2. Andriluka, M., Pishchulin, L., Gehler, P., Bernt, S.: 2D human pose estimation: new benchmark and state of the art analysis. In: CVPR (2014)
3. Aumentado-Armstrong, T., Tsogkas, S., Jepson, A., Dickinson, S.: Geometric disentanglement for generative latent shape models. In: ICCV, pp. 8180–8189 (2019)
4. Bao, F., et al.: All are worth words: a ViT backbone for diffusion models. In: CVPR (2023)
5. Bao, F., et al.: One transformer fits all distributions in multi-modal diffusion at scale (2023)
6. Bautista, M.A., et al.: GAUDI: a neural architect for immersive 3D scene generation. arXiv (2022)
7. Biggs, B., Ehrhart, S., Joo, H., Graham, B., Vedaldi, A., Novotny, D.: 3D multibodies: fitting sets of plausible 3D models to ambiguous image data. In: NeurIPS (2020)
8. Bogo, F., Kanazawa, A., Lassner, C., Gehler, P., Romero, J., Black, M.J.: Keep it SMPL: automatic estimation of 3D human pose and shape from a single image. In: Leibe, B., Matas, J., Sebe, N., Welling, M. (eds.) ECCV 2016. LNCS, vol. 9909, pp. 561–578. Springer, Cham (2016). https://doi.org/10.1007/978-3-319-46454-1_34
9. Bogo, F., Romero, J., Pons-Moll, G., Black, M.J.: Dynamic FAUST: registering human bodies in motion. In: CVPR (2017)
10. Chang, A.X., et al.: ShapeNet: an information-rich 3D model repository (2015)
11. Chen, H., Tang, H., Shi, H., Peng, W., Sebe, N., Zhao, G.: Intrinsic-extrinsic preserved GANs for unsupervised 3D pose transfer. In: ICCV, pp. 8610–8619 (2021)
12. Cheng, S., Bronstein, M.M., Zhou, Y., Kotsia, I., Pantic, M., Zafeiriou, S.: MeshGAN: nonlinear 3D morphable models of faces. CoRR **abs/1903.10384** (2019)
13. Cheng, Y.C., Lee, H.Y., Tulyakov, S., Schwing, A.G., Gui, L.Y.: SDFusion: multimodal 3D shape completion, reconstruction, and generation. In: CVPR, pp. 4456–4465 (2023)
14. Cho, H., Kim, J.: Generative approach for probabilistic human mesh recovery using diffusion models (2023)

15. Cho, J., Youwang, K., Oh, T.H.: Cross-attention of disentangled modalities for 3D human mesh recovery with transformers. In: Avidan, S., Brostow, G., Cissé, M., Farinella, G.M., Hassner, T. (eds.) ECCV 2022. LNCS, vol. 13661, pp. 342–359. Springer, Cham (2022). https://doi.org/10.1007/978-3-031-19769-7_20
16. Choi, H., Moon, G., Lee, K.M.: Pose2Mesh: graph convolutional network for 3D human pose and mesh recovery from a 2D human pose. In: Vedaldi, A., Bischof, H., Brox, T., Frahm, J.-M. (eds.) ECCV 2020. LNCS, vol. 12352, pp. 769–787. Springer, Cham (2020). https://doi.org/10.1007/978-3-030-58571-6_45
17. Christian, Z., Duygu, C., Jimei, Y., Russel, B., Argus, M., Brox, T.: FreiHAND: a dataset for markerless capture of hand pose and shape from single RGB images. In: ICCV (2019)
18. Cosmo, L., Norelli, A., Halimi, O., Kimmel, R., Rodolà, E.: LIMP: learning latent shape representations with metric preservation priors. In: Vedaldi, A., Bischof, H., Brox, T., Frahm, J.-M. (eds.) ECCV 2020. LNCS, vol. 12348, pp. 19–35. Springer, Cham (2020). https://doi.org/10.1007/978-3-030-58580-8_2
19. Dao, T., Fu, D.Y., Ermon, S., Rudra, A., Ré, C.: FlashAttention: fast and memory-efficient exact attention with IO-awareness. In: NeurIPS (2022)
20. Davydov, A., Remizova, A., Constantin, V., Honari, S., Salzmann, M., Fua, P.: Adversarial parametric pose prior. In: CVPR, pp. 10987–10995 (2022)
21. Desbrun, M., Meyer, M., Schröder, P., Barr, A.H.: Implicit fairing of irregular meshes using diffusion and curvature flow. In: Proceedings of the 26th Annual Conference on Computer Graphics and Interactive Techniques, SIGGRAPH 1999, pp. 317–324 (1999)
22. Fröhlich, S., Botsch, M.: Example-driven deformations based on discrete shells. Comput. Graph. Forum **30**(8), 2246–2257 (2011)
23. Gao, L., Lai, Y.K., Yang, J., Zhang, L.X., Xia, S., Kobbelt, L.: Sparse data driven mesh deformation. IEEE Trans. Vis. Comput. Graph. **27**(3), 2085–2100 (2021)
24. Georgakis, G., Li, R., Karanam, S., Chen, T., Košecká, J., Wu, Z.: Hierarchical kinematic human mesh recovery. In: Vedaldi, A., Bischof, H., Brox, T., Frahm, J.-M. (eds.) ECCV 2020. LNCS, vol. 12362, pp. 768–784. Springer, Cham (2020). https://doi.org/10.1007/978-3-030-58520-4_45
25. Gong, J., Foo, L.G., Fan, Z., Ke, Q., Rahmani, H., Liu, J.: DiffPose: toward more reliable 3D pose estimation. In: CVPR (2023)
26. Groueix, T., Fisher, M., Kim, V.G., Russell, B.C., Aubry, M.: 3D-CODED: 3D correspondences by deep deformation. In: Ferrari, V., Hebert, M., Sminchisescu, C., Weiss, Y. (eds.) ECCV 2018. LNCS, vol. 11206, pp. 235–251. Springer, Cham (2018). https://doi.org/10.1007/978-3-030-01216-8_15
27. Ho, J.: Classifier-free diffusion guidance. arXiv abs/2207.12598 (2022)
28. Ho, J., Jain, A., Abbeel, P.: Denoising diffusion probabilistic models. arXiv preprint arxiv:2006.11239 (2020)
29. Ionescu, C., Papava, D., Olaru, V., Sminchisescu, C.: Human3.6M: large scale datasets and predictive methods for 3D human sensing in natural environments. IEEE TPAMI **36**(7), 1325–1339 (2014)
30. Jiang, B., Zhang, J., Cai, J., Zheng, J.: Disentangled human body embedding based on deep hierarchical neural network. IEEE Trans. Vis. Comput. Graph. **26**(8), 2560–2575 (2020)
31. Johnson, S., Everingham, M.: Learning effective human pose estimation from inaccurate annotation. In: CVPR, pp. 1465–1472 (2011)
32. Joo, H., Neverova, N., Vedaldi, A.: Exemplar fine-tuning for 3D human pose fitting towards in-the-wild 3D human pose estimation. In: 3DV (2020)
33. Jun, H., Nichol, A.: Shap-E: generating conditional 3D implicit functions (2023)
34. Kanazawa, A., Black, M.J., Jacobs, D.W., Malik, J.: End-to-end recovery of human shape and pose. In: CVPR (2018)

35. Khirodkar, R., Tripathi, S., Kitani, K.: Occluded human mesh recovery. In: CVPR, pp. 1715–1725 (2022)
36. Kocabas, M., Huang, C.H.P., Hilliges, O., Black, M.J.: PARE: part attention regressor for 3D human body estimation. In: ICCV, pp. 11127–11137 (2021)
37. Kolotouros, N., Pavlakos, G., Black, M.J., Daniilidis, K.: Learning to reconstruct 3D human pose and shape via model-fitting in the loop. In: ICCV (2019)
38. Kolotouros, N., Pavlakos, G., Daniilidis, K.: Convolutional mesh regression for single-image human shape reconstruction. In: CVPR (2019)
39. Kolotouros, N., Pavlakos, G., Jayaraman, D., Daniilidis, K.: Probabilistic modeling for human mesh recovery. In: ICCV (2021)
40. Li, L., Zhuo, L., Zhang, B., Bo, L., Chen, C.: DiffHand: end-to-end hand mesh reconstruction via diffusion models (2023)
41. Lin, C.H., et al.: Magic3D: high-resolution text-to-3D content creation. In: CVPR (2023)
42. Lin, G.F., Jia, G., Hossein, R., Jun, L.: Distribution-aligned diffusion for human mesh recovery. In: ICCV (2023)
43. Lin, K., Wang, L., Liu, Z.: End-to-end human pose and mesh reconstruction with transformers. In: CVPR (2021)
44. Lin, K., Wang, L., Liu, Z.: Mesh graphormer. In: ICCV (2021)
45. Lin, T., et al.: Microsoft COCO: common objects in context. CoRR abs/1405.0312 (2014)
46. Liu, Y., Yang, J., Gu, X., Guo, Y., Yang, G.Z.: EgoHMR: egocentric human mesh recovery via hierarchical latent diffusion model. In: 2023 IEEE International Conference on Robotics and Automation (ICRA), pp. 9807–9813 (2023)
47. Liu, Z., Feng, Y., Black, M.J., Nowrouzezahrai, D., Paull, L., Liu, W.: MeshDiffusion: score-based generative 3D mesh modeling. In: International Conference on Learning Representations (2023)
48. Loper, M., Mahmood, N., Romero, J., Pons-Moll, G., Black, M.J.: SMPL: a skinned multi-person linear model. ACM TOG **34**(6), 248:1–248:16 (2015)
49. Luo, S., Hu, W.: Diffusion probabilistic models for 3D point cloud generation. In: CVPR (2021)
50. Lyu, Z., Kong, Z., Xu, X., Pan, L., Lin, D.: A conditional point diffusion-refinement paradigm for 3D point cloud completion (2022)
51. Lyu, Z., Wang, J., An, Y., Zhang, Y., Lin, D., Dai, B.: Controllable mesh generation through sparse latent point diffusion models. In: CVPR (2023)
52. Ma, Q., et al.: Learning to dress 3D people in generative clothing. In: CVPR (2020)
53. Mahmood, N., Ghorbani, N., Troje, N.F., Pons-Moll, G., Black, M.J.: AMASS: archive of motion capture as surface shapes. In: ICCV (2019)
54. von Marcard, T., Henschel, R., Black, M.J., Rosenhahn, B., Pons-Moll, G.: Recovering accurate 3D human pose in the wild using IMUs and a moving camera. In: Ferrari, V., Hebert, M., Sminchisescu, C., Weiss, Y. (eds.) ECCV 2018. LNCS, vol. 11214, pp. 614–631. Springer, Cham (2018). https://doi.org/10.1007/978-3-030-01249-6_37
55. Mehta, D., et al.: Monocular 3D human pose estimation in the wild using improved CNN supervision. In: 3DV. IEEE (2017)
56. Mo, S., Xie, E., Chu, R., Hong, L., Nießner, M., Li, Z.: DiT-3D: exploring plain diffusion transformers for 3D shape generation. arXiv preprint arXiv: 2307.01831 (2023)
57. Nichol, A., Jun, H., Dhariwal, P., Mishkin, P., Chen, M.: Point-E: a system for generating 3D point clouds from complex prompts (2022)
58. Pavlakos, G., et al.: Expressive body capture: 3D hands, face, and body from a single image. In: CVPR (2019)
59. Peebles, W., Radosavovic, I., Brooks, T., Efros, A., Malik, J.: Learning to learn with generative models of neural network checkpoints. arXiv preprint arXiv:2209.12892 (2022)

60. Peebles, W., Xie, S.: Scalable diffusion models with transformers. arXiv preprint arXiv:2212.09748 (2022)
61. Peng, S., Jiang, C.M., Liao, Y., Niemeyer, M., Pollefeys, M., Geiger, A.: Shape as points: a differentiable Poisson solver. In: NeurIPS (2021)
62. Phil, W.: denoising-diffusion-pytorch (2023). https://github.com/lucidrains/denoising-diffusion-pytorch
63. Poole, B., Jain, A., Barron, J.T., Mildenhall, B.: DreamFusion: text-to-3D using 2D diffusion. arXiv (2022)
64. Ranjan, A., Bolkart, T., Sanyal, S., Black, M.J.: Generating 3D faces using convolutional mesh autoencoders. In: Ferrari, V., Hebert, M., Sminchisescu, C., Weiss, Y. (eds.) ECCV 2018. LNCS, vol. 11207, pp. 725–741. Springer, Cham (2018). https://doi.org/10.1007/978-3-030-01219-9_43
65. Rombach, R., Blattmann, A., Lorenz, D., Esser, P., Ommer, B.: High-resolution image synthesis with latent diffusion models. In: CVPR, pp. 10684–10695 (2022)
66. Romero, J., Tzionas, D., Black, M.J.: Embodied hands: modeling and capturing hands and bodies together. ACM TOG **36**(6), 1–17 (2017)
67. Rueegg, N., Zuffi, S., Schindler, K., Black, M.J.: BARC: learning to regress 3D dog shape from images by exploiting breed information. In: CVPR (2022)
68. Salimans, T., Ho, J.: Progressive distillation for fast sampling of diffusion models (2022)
69. Shan, W., et al.: Diffusion-based 3D human pose estimation with multi-hypothesis aggregation. arXiv preprint arXiv:2303.11579 (2023)
70. Shen, T., Gao, J., Yin, K., Liu, M.Y., Fidler, S.: Deep marching tetrahedra: a hybrid representation for high-resolution 3D shape synthesis. In: Advances in Neural Information Processing Systems (NeurIPS) (2021)
71. Shim, J., Kang, C., Joo, K.: Diffusion-based signed distance fields for 3D shape generation. In: CVPR, pp. 20887–20897 (2023)
72. Shoemake, K.: Animating rotation with quaternion curves. In: Proceedings of the 12th Annual Conference on Computer Graphics and Interactive Techniques, SIGGRAPH 1985, pp. 245–254. Association for Computing Machinery, New York (1985)
73. Song, J., Meng, C., Ermon, S.: Denoising diffusion implicit models. arXiv:2010.02502 (2020)
74. Song, J., Chen, X., Hilliges, O.: Human body model fitting by learned gradient descent. In: Vedaldi, A., Bischof, H., Brox, T., Frahm, J.-M. (eds.) ECCV 2020. LNCS, vol. 12365, pp. 744–760. Springer, Cham (2020). https://doi.org/10.1007/978-3-030-58565-5_44
75. Sumner, R.W., Zwicker, M., Gotsman, C., Popovic, J.: Mesh-based inverse kinematics. ACM TOG **24**(3), 488–495 (2005)
76. Sun, X., et al.: Learning semantic-aware disentangled representation for flexible 3D human body editing. In: CVPR (2023)
77. Tan, Q., Gao, L., Lai, Y.K., Xia, S.: Variational autoencoders for deforming 3D mesh models. In: CVPR, pp. 5841–5850 (2018)
78. Tian, Y., Zhang, H., Liu, Y., Wang, L.: Recovering 3D human mesh from monocular images: a survey. arXiv preprint arXiv:2203.01923 (2022)
79. Tiwari, G., Antić, D., Lenssen, J.E., Sarafianos, N., Tung, T., Pons-Moll, G.: Pose-NDF: modeling human pose manifolds with neural distance fields. In: Avidan, S., Brostow, G., Cissé, M., Farinella, G.M., Hassner, T. (eds.) ECCV 2022. LNCS, vol. 13665, pp. 572–589. Springer, Cham (2022). https://doi.org/10.1007/978-3-031-20065-6_33
80. Varol, G., et al.: Learning from synthetic humans. In: CVPR (2017)
81. Wang, Z., et al.: ProlificDreamer: high-fidelity and diverse text-to-3D generation with variational score distillation. arXiv preprint arXiv:2305.16213 (2023)

82. Xu, H., Bazavan, E.G., Zanfir, A., Freeman, W.T., Sukthankar, R., Sminchisescu, C.: GHUM & GHUML: generative 3D human shape and articulated pose models. In: CVPR, pp. 6184–6193 (2020)
83. Xu, J., et al.: Animal3D: a comprehensive dataset of 3D animal pose and shape. arXiv preprint arXiv:2308.11737 (2023)
84. Xu, X., Wang, Z., Zhang, G., Wang, K., Shi, H.: Versatile diffusion: text, images and variations all in one diffusion model. In: ICCV, pp. 7754–7765 (2023)
85. Yang, G., Huang, X., Hao, Z., Liu, M.Y., Belongie, S., Hariharan, B.: PointFlow: 3D point cloud generation with continuous normalizing flows. arXiv (2019)
86. Yoshiyasu, Y.: Deformable mesh transformer for 3D human mesh recovery. In: CVPR, pp. 17006–17015 (2023)
87. You, Y., Liu, H., Li, X., Li, W., Wang, T., Ding, R.: GATOR: graph-aware transformer with motion-disentangled regression for human mesh recovery from a 2D pose. In: ICASSP 2023 - 2023 IEEE International Conference on Acoustics, Speech and Signal Processing (ICASSP), pp. 1–5 (2023)
88. Yu, Z., et al.: Surf-D: high-quality surface generation for arbitrary topologies using diffusion models. arXiv preprint arXiv:2311.17050 (2023)
89. Yuan, Y.J., Lai, Y.K., Yang, J., Duan, Q., Fu, H., Gao, L.: Mesh variational autoencoders with edge contraction pooling. In: CVPRW, pp. 274–275 (2020)
90. Zanfir, A., Bazavan, E.G., Xu, H., Freeman, W.T., Sukthankar, R., Sminchisescu, C.: Weakly supervised 3D human pose and shape reconstruction with normalizing flows. In: Vedaldi, A., Bischof, H., Brox, T., Frahm, J.-M. (eds.) ECCV 2020. LNCS, vol. 12351, pp. 465–481. Springer, Cham (2020). https://doi.org/10.1007/978-3-030-58539-6_28
91. Zeng, X., et al.: LION: latent point diffusion models for 3D shape generation. In: Advances in Neural Information Processing Systems (NeurIPS) (2022)
92. Zhang, H., et al.: PyMAF: 3D human pose and shape regression with pyramidal mesh alignment feedback loop. In: ICCV (2021)
93. Zhou, K., Bhatnagar, B.L., Pons-Moll, G.: Unsupervised shape and pose disentanglement for 3D meshes. In: Vedaldi, A., Bischof, H., Brox, T., Frahm, J.-M. (eds.) ECCV 2020. LNCS, vol. 12367, pp. 341–357. Springer, Cham (2020). https://doi.org/10.1007/978-3-030-58542-6_21
94. Zuffi, S., Kanazawa, A., Jacobs, D., Black, M.J.: 3D menagerie: modeling the 3D shape and pose of animals. In: CVPR (2017)

Exploiting Supervised Poison Vulnerability to Strengthen Self-supervised Defense

Jeremy Styborski[1], Mingzhi Lyu[2]([✉]), Yi Huang[1], and Adams Kong[1]

[1] College of Computing and Data Science, Nanyang Technological University, Singapore, Singapore
{styb0001,AdamsKong}@ntu.edu.sg
[2] Rapid-Rich Object Search (ROSE) Lab, Interdisciplinary Graduate Programme, Nanyang Technological University, Singapore, Singapore
lyum0002@ntu.edu.sg

Abstract. Availability poisons exploit supervised learning (SL) algorithms by introducing class-related shortcut features in images such that models trained on poisoned data are useless for real-world datasets. Self-supervised learning (SSL), which utilizes augmentations to learn instance discrimination, is regarded as a strong defense against poisoned data. However, by extending the study of SSL across multiple poisons on the CIFAR-10 and ImageNet-100 datasets, we demonstrate that it often performs poorly, far below that of training on clean data. Leveraging the vulnerability of SL to poison attacks, we introduce adversarial training (AT) on SL to obfuscate poison features and guide robust feature learning for SSL. Our proposed defense, designated VESPR (Vulnerability Exploitation of Supervised Poisoning for Robust SSL), surpasses the performance of six previous defenses across seven popular availability poisons. VESPR displays superior performance over all previous defenses, boosting the minimum and average ImageNet-100 test accuracies of poisoned models by 16% and 9%, respectively. Through analysis and ablation studies, we elucidate the mechanisms by which VESPR learns robust class features.
Code: https://github.com/JStyborski/VESPR.

Keywords: Adversarial Training · Availability Poisoning · Poison Defense · Self-Supervised Learning

1 Introduction

The proliferation of open-source labeled data on the Internet has fueled the rapid development and application of deep neural networks across various domains.

J. Styborski, M. Lyu, Y. Huang—Equal contribution.

However, reliance on supervised learning in tasks like image classification exposes neural networks to shortcut learning, whereby models learn 'easy' features that are spuriously correlated with task labels in place of useful image features for accurate representation. Shortcut learning is an inherent behavior for neural networks [29,56,69,72]. Availability poisons exploit this vulnerability by injecting malicious data into real-world data sources, thereby introducing shortcuts and reducing the effectiveness of trained models. Crucially, availability poisons are imperceptible; they only perturb images slightly and do not modify image labels, such that poisoned images appear clean to human observers.

Models trained on poisoned data exhibit poor generalization on real-world clean data, posing significant security risks in critical applications like self-driving vehicles [61], anomaly detection [66] and medical AI [2,3]. For instance, a self-driving system trained with poisoned images of stop signs may fail to recognize real images of stop signs in practical scenarios. A model trained on poisoned data will classify images based on the poisoned patterns rather than genuine features. Consequently, the model becomes incapable of correctly classifying benign images, jeopardizing its reliability in real-world applications.

Currently, there are various proposed defense methods against availability poisons [18,21,49,53,57,63,85,87]. However, these methods are limited in their ability to defend against many popular availability poisons, leaving significant vulnerabilities to certain poisons while also suffering from reduced average performance. According to the surveys of poison defenses in Tables 7 and 8 (see Supplemental Material Sect. A), not all of these defense methods have been fully assessed on state-of-the-art poisons. Though most of these baseline defenses achieve robust performance on small-image datasets with few classes (e.g., CIFAR-10 [48] and SVHN [59]), our experiments reveal that they often fail to achieve comparable performance on high-resolution datasets with many classes (e.g., ImageNet-100 [20]), highlighting a critical research gap.

Poisoned perturbations establish shortcuts between labels and poisonous patterns in the images, prompting us to explore alternatives to traditional SL. SSL methods, such as contrastive learning [12,34,60], focus on learning image representations that are invariant to image transformations [58], rather than explicitly modeling the correlation between features and labels. In SSL, the combined effect of augmentations and the instance-discrimination objective filter out nuisance features introduced by poisons. Consequently, encoders trained with SSL tend to extract more semantic information from images and avoid class collapse [80]. In extending the study of SSL-based defense to multiple ImageNet-100 poisons, our experiments reveal that SSL unfortunately learns poison information alongside genuine features, challenging observations made largely on CIFAR-10 [32].

Motivated by the discrepancy in poison robustness between SL and SSL, we propose leveraging the poison weakness of SL to generate adversarial noise in order to improve augmentations for SSL, thereby preferentially capturing genuine image features and dismissing poisoned features. Thus, we introduce Vulnerability Exploitation of Supervised Poisoning for Robust SSL (VESPR). As illustrated in Fig. 2, VESPR trains an encoder using a multi-task loss function, while adversarial inputs are generated using gradients computed with the supervised cross-entropy loss. We evaluate the performance of VESPR against seven state-of-the-art poisoning methods [26,27,32,40,67,79,83] and compare it with six baseline defense methods [21,32,49,53,57,85,87] on the ImageNet-100 [20] and CIFAR-10 [48] datasets. Additionally, we conduct comprehensive ablation studies to verify and understand the effectiveness of VESPR.

Our work advances the state-of-the-art for poison defense methods through the following contributions:

- We extend the study of multiple defense methods, including SSL, to seven poisons on the CIFAR-10 and ImageNet-100 datasets. For all defenses, we find that the discrepancy between clean and poison training is exacerbated for large-image datasets.
- We exploit the vulnerability of SL to availability poisons in order to enhance the robustness of SSL training through our proposed method, VESPR. Integrating AT for SL within the SSL framework, VESPR generates optimized adversarial augmentations that preferentially emphasize robust image features and avoid poison shortcuts.
- Through extensive experiments, we demonstrate that VESPR achieves state-of-the-art performance against multiple prevalent poisons, improving both minimum and average poison defense beyond all other baselines.

The remainder of this paper is organized as follows: Sect. 2 reviews other works relevant to our research, Sect. 3 analyzes trends in shortcut learning and introduces the VESPR framework, and Sect. 4 covers experimental results, subsequent ablations, and corresponding analyses.

2 Related Works

2.1 Data Poisoning

The concept of data poisoning for computer vision was borne out of adversarial examples [30,49,57,68], wherein images could be imperceptibly altered yet decimate supervised classification performance. Importantly, adversarial examples can transfer across models [19], such that generating poisons against unknown training methods becomes feasible. Adversarial Poisons (AP) is an early poisoning method that simply uses adversarial examples [26].

Synthetic poisons [67,79,83] are designed specifically for poisoning supervised classification methods. Synthetic poisons are crafted heuristically and commonly applied class-wise. They aim to introduce consistent class-wise shortcut features. The simplest synthetic poisoning method is the One-Pixel Shortcut [79], wherein a single pixel is altered to a consistent color for all images of a class. Min-max poisons [8,25,84] directly apply the data poisoning objective: find perturbations that minimize training loss on poison data while maximizing test loss on clean data. Min-min poisons [27,32,40] create 'unlearnable examples', such that a network trained on the poisoned samples will quickly learn subtle shortcuts without learning useful image features.

Unlike availability poisons, dirty-label attacks [6,52] do not typically modify images but instead inject mislabelled images into datasets. Dirty-label are effective but easily detectable; they can only poison datasets with little or no human supervision. We do not consider dirty-label attacks in this paper and instead focus on poisons that perturb image features.

2.2 Poison Defenses

Poison defenses for SL are heretofore largely based on augmentations and adversarial training. Most poisons demonstrate reduced effectiveness against models trained with data augmentations like Cutout [21], Mixup [87], CutMix [85], and RandAugment [18]. Grayscale and JPEG-compression augmentations may be applied to counteract low-frequency and high-frequency poisons, respectively [53]. UEraser samples multiple augmentations and selects the one that maximizes classification loss [63]. Augmentation-based defenses filter out some non-robust poisonous patterns, but falter against poisons such as AP and CUDA [26,67]. In fact, multiple poisoning papers [26,32,79] demonstrate their triumph over augmentation-based methods as a proof of their effectiveness. Gaussian noise augmentation applied to training data reduces the effectiveness of adversarial examples [16]. Unlike random noise, AT [57,74,82] generates optimized adversarial noise during training to improve model robustness. AT is largely considered a state-of-the-art poison defense, improving performance against multiple poisons. However, multiple poison methods still enforce a significant accuracy drop for models defended by AT [32,67,79], and some poisons even anticipate the use of AT during training [27]. Indeed, we demonstrate in Sect. 4.3 that these SL-based defenses underperform on multiple poisons.

SSL has also been uncovered as a method for learning robust image features from poisoned data sets [32]. Unfortunately, prior studies are limited in the extent to which they test various poisons, exploring SSL as poison defense against only one ImageNet-100 poison and three CIFAR-10 poisons (see Tables 7 and 8 in Supplemental Material Sect. A). We extend the study of SSL across seven poisons for CIFAR-10 and ImageNet-100. Furthermore, we are the first to investigate adversarial SSL methods [1,11,24,42,45,46,55] applied to poison defense.

2.3 Self-supervised Learning

Autoencoders [23,33,36,37,47,54,75,81] are considered the one of the first SSL methods, utilizing reconstruction-based loss with an encoder-decoder architecture to learn compressed image embeddings. Recently, invariance-based SSL methods have gained popularity. Contrastive methods [12,15,34,58,64,70,73,77] typically rely on the InfoNCE loss function [60] to encourage all projections of views from the same source image to be similar, yet different from projections of views from other images. Non-contrastive methods [14,31,71,78,88–90] eliminate the requirement for negative samples, focusing their loss functions entirely on reducing the distance between same-source projections.

Like SL, SSL is vulnerable to feature suppression [43,50,65,69,78,86,90] and shortcut learning [13,28]. Multiple works have proposed modifications to architecture or training objectives in order to mitigate feature suppression in SSL [39,44,51,65,76]. To ensure that SSL does not favor poison shortcuts, we employ AT on SL in order to generate adversarially-perturbed augmentations that shift focus towards robust image features.

2.4 Multi-task Learning

More recently, multiple groups have proposed combining architectures and loss objectives in order to improve learned representation quality and downstream performance. Multiple works investigate the combination of instance discrimination and reconstruction objectives [4,9,22]. Islam et al. [41] empirically investigate the performance of supervised losses combined with contrastive SSL. Chen et al. [10] and Xue et al. [80] both demonstrate that combining supervised and self-supervised contrastive learning objectives avoids class-collapse and mitigates feature suppression. To our knowledge, no one has yet studied multi-task architectures or combined objectives for data poisoning.

3 Method

3.1 Formulation of Availability Poison

We describe the availability poison objective in Eq. 1. We assume a clean dataset $D_c = \{(x_{c,1}, y_1), (x_{c,2}, y_2), ..., (x_{c,N}, y_N)\}$ with distribution $\mathcal{P}_c$, where $x_{c,i}$ is a benign image and y_i is the corresponding class label. The corresponding test dataset is $D'_t = \{(x'_{c,1}, y'_1), (x'_{c,2}, y'_2), ..., (x'_{c,M}, y'_M)\}$, where each pair $(x'_{c,i}, y'_i)$ is drawn from distribution $\mathcal{P}_c$ but not included in D_c. The goal of data poisoning is to generate poison dataset $D_p = \{(x_{p,1}, y_1), (x_{p,2}, y_2), ..., (x_{p,N}, y_N)\}$ such that a classification model $CLS_{\theta_1^*}(\cdot)$ trained upon poison data D_p will perform poorly on clean test data D'_t. Poison sample $x_{p,i}$ is generated from corresponding clean sample $x_{c,i}$ via poisoning process $PSN_{\theta_2^*}(x_{c,i})$.

$$\theta_2^* = \arg\max_{\theta_2} \sum_{i=0}^{M} L(CLS_{\theta_1^*}(x'_{c,i}), y'_i)$$

$$\text{s.t.} \quad \theta_1^* = \arg\min_{\theta_1} \sum_{i=0}^{N} L(CLS_{\theta_1}(x_{p,i}), y_i), \qquad (1)$$

$$x_{p,i} = PSN_{\theta_2}(x_{c,i}), \quad d(x_{p,i}, x_{c,i}) \leq \epsilon$$

L is the supervised learning loss function. In the context of classification, L is often set as the cross entropy (CE) loss, $L_{CE}(CLS_{\theta_1}(x), y) = y \log CLS_{\theta_1}(x)$, where $CLS_{\theta_1}(x)$ is the logit of input x with label y. Distance metric $d(\cdot)$ between $x_{p,i}$ and $x_{c,i}$ is bounded by a (small) threshold ϵ such that poisoned images retain high image quality and appear similar to their corresponding clean images. Equation 1 says that poisoned images contain imperceptible perturbations yet maximally mislead poison-trained models to perform poorly on clean data.

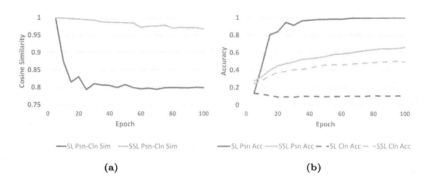

Fig. 1. (a) Mean of cosine similarity between representations of clean images and poison images (Psn-Cln Sim, $cossim[f(x_{p,i}), f(x_{c,i})]$). (b) Poison (Psn) and clean (Cln) classification accuracies tested with $x_{p,i}$ and $x_{c,i}$, respectively. We display curves for SL and SSL models, both trained on adversarial poison [26] data. (Color figure online)

3.2 Analysis of SL and SSL Against Availability Poisons

Intuitively, if there exist features with high task correlation and low variance across training instances, deep learning models will grant greater weight on these features. 'Easy' features that are spuriously correlated with the training task become shortcuts, which are quickly learned by SL models. Simplicity bias due to stochastic gradient descent may strengthen the preference for learning shortcuts [56,80]. Availability poisons exploit the shortcut-learning weakness in order to mislead model training.

In order to demonstrate the effects of shortcut learning, we train a ResNet18 [35] encoder with an adversarial poison [26] via SL and compare representations of clean and poison images. As shown in Fig. 1a, representation similarity between a benign image $x_{c,i}$ and its poisoned version $x_{p,i}$ reduces rapidly (solid blue line in Fig. 1a), suggesting that the encoder quickly learns distinct representations based on the shortcut signals in the poisoned images. Though poison accuracy (solid blue line in Fig. 1b) quickly maximizes near 100%, clean accuracy never improves (dashed blue line in Fig. 1b).

Since availability poisons are often generated to confuse image classifiers, we explore SSL methods, which bypass the requirement for class labels. Invariance-based SSL techniques are designed to learn representations of images that are robust to input transformations (e.g., cropping, color shifts), yet distinct from representations of other images.

Contrastive learning is a classic invariance-based SSL framework, backed by extensive research [12,15,34,58,70] and proven improvements against some availability poisons [32]. Thus, we primarily utilize contrastive learning for SSL. As noted by Wang and Isola [77], the contrastive learning objective is optimized by two measures: representation alignment between positive samples and uniformity across all representations. Because SSL is **not** motivated by class labels, poison features shared by same-class images are less relevant. Additionally, the alignment objective of contrastive learning encourages augmentation invariance; poison features are obfuscated and representation learning favors clean features.

Table 1. Models are trained on poisoned datasets with SL, SL with SSL-style augmentations, and SSL. Subsequent columns display average cosine similarity between representations of poison images of the same class (Psn In-Cls Sim, $cossim[f(x_{p,i}), f(x_{p,j})]$, where $y_i = y_j$), average cosine similarity between of poison image representations and their corresponding clean image representations (Psn-Cln Sim, $cossim[f(x_{p,i}), f(x_{c,i})]$), and poison representation feature diversity (Psn E-Rank, see [88,90]). We average results across seven different poisons (see Supplemental Material Sect. B.1).

Method	Psn In-Cls Sim	Psn-Cln Sim	Psn E-Rank
SL	0.78	0.70	16.67
SL (SSL Aug)	0.79	0.79	16.43
SSL	0.47	0.86	59.61

In Table 1, we explore feature suppression in SL and SSL by analyzing representation characteristics for their encoders across seven different poisons. We instantiate the SSL objective L_{CL} with the commonly-used InfoNCE loss [60]. We begin by ablating the effect of augmentation: compared to SL, using SSL-style augmentations during SL barely affects Psn In-Cls Sim and Psn E-Rank, though there is a significant improvement in Psn-Cln Sim. SSL delivers significantly lower Psn In-Cls Sim than SL trained with or without SSL-style augmentation, demonstrating that the SSL training objective (and not SSL-style augmentation) is crucial for encoding instance-specific information rather than learning class-specific shortcuts. Likewise, higher Psn E-Rank and Psn-Cln Sim shows that SSL representations contain more features and are less influenced by poison signals, likely resulting in better generalization on clean test data.

Nevertheless, SSL models cannot inherently differentiate semantic information and poison information, and therefore cannot avoid poison features entirely. This is verified by Fig. 1b, where SSL still permits a significant gap between clean and poison accuracies. Therefore, the SSL model has learned some poisoned features along with true image features, leading the downstream classifiers to misclassify data lacking poisonous patterns.

3.3 VESPR

Motivated by these observations, we seek methods to improve augmentations for SSL that can better obscure poison features, thereby allowing the learning algorithm to learn true features. Noting the discrepancies between SL and SSL in representation similarities (Table 1) and performance (Fig. 1b), we conclude that the classification network largely learns poison features, distilling them from other image features. This betrays a weakness of availability poisons: once processed by a poisoned encoder, they are no longer subtle nor imperceptible! Drawing from theory for adversarial examples [26,30], we hypothesize that poisoned

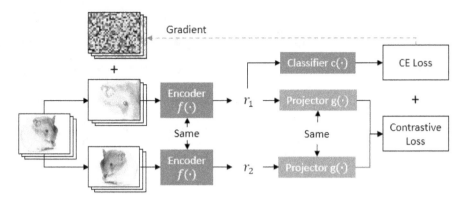

Fig. 2. Overview of VESPR architecture. For each input image, VESPR generates two views through augmentation and encodes them with encoder $f(\cdot)$ as representations r_1 and r_2, respectively. Cross-entropy (CE) loss is calculated from the output of the classifier head, $c(r_1)$. The CE loss gradient is used to generate adversarial perturbations to the first view using projected gradient descent [49,57]. The adversarial images are encoded to replace the original representations. Contrastive loss is calculated from projected representations, $g(r_1)$ and $g(r_2)$, using projections of other images as negative samples. The VESPR network is trained from the combined CE and contrastive losses.

classifiers can be exploited to generate adversarial examples that maximally muddle poison signals. We further hypothesize that incorporating these adversarial examples as augmentations will improve representations learned by SSL, thereby improving poison defense. Our final architecture is a combined SSL+SL framework, utilizing AT on the supervised loss. We designate our method as VESPR, Vulnerability Exploitation of Supervised Poisons for Robust SSL.

As shown in Fig. 2, VESPR contains feature encoder $f(\cdot)$, representation projector $g(\cdot)$, and classification head $c(\cdot)$. During training, each image of a batch is augmented into two different views. One view is taken as the source image to generate adversarial examples using the following objective function:

$$\delta^* = \arg\max_{\delta} L_{CE}(c(f(x+\delta)), y) \qquad (2)$$
$$\text{s.t.} \quad l_p(\delta) \leq \epsilon_{AT}$$

where δ is an adversarial perturbation, x is an input view, $l_p(\cdot)$ denotes the p-norm, and ϵ_{AT} is the AT threshold, with $p = \infty$ in our case.

The encoder generates representations of the input views, $f(x+\delta^*)$, $f(x^+)$, and $f(x^-)$. The combined loss function L_{VESPR} is calculated from the outputs of the projector $g(\cdot)$ and classifier $c(\cdot)$ networks:

$$\begin{aligned}L_{VESPR}(x+\delta^*, x^+, x^-, y) = &\alpha L_{CL}[g(f(x+\delta^*)), g(f(x^+)), g(f(x^-))] \\ &+ \beta L_{CE}[c(f(x+\delta^*)), y]\end{aligned} \qquad (3)$$

with inputs as view x with label y, corresponding positive view x^+ (same source image), and negative views x^- (different source image). α and β weight the balance between L_{CL} and L_{CE}.

As CE loss is integrated into L_{VESPR}, VESPR leverages class-specific features for training. Simultaneously, the contrastive loss component of L_{VESPR} not only focuses on features that distinguish different images, but also mitigates class collapse and poison shortcut learning induced by CE loss [10,80]. Finally, adversarial samples generated by the classification loss create perturbed augmentations that maximally obscure poison features.

4 Experiments

4.1 Experiment Setup

We evaluate the performance of VESPR and other baseline defenses on seven popular poisoning methods: Adversarial Poisons (AP8) [26], Contrastive Poisoning (CP8) [32], Convolution-based Unlearnable Datasets (CUDA) [67], Linearly Separable Poisons (LSP16) [83], One-Pixel Shortcut (OPS) [79], Robust Unlearnable Examples (RUE8) [27], and Unlearnable Examples (UE8) [40]. Further poison descriptions can be found in Supplemental Material Sect. B.1.

We primarily train models on poisoned versions of a 100-class subset of the ImageNet dataset [20], each with a total of 130,000 training images and 5,000 test images. For benchmarking purposes, we also train models on poisoned CIFAR-10 datasets [48], each with a total of 50,000 training images and 10,000 test images. Due to the lengthy optimization process required by some poisons on large images, AP8, CP8, and UE8 poisons for ImageNet-100 are generated with 20% of the dataset, containing 100 classes with a total of 26,000 training images.

To benchmark the effectiveness of VESPR, we compare against six state-of-the-art poison defense methods: adversarial training (SL+AT), Cutout [21], Mixup [87], CutMix [85], ISS [53], and SSL (SimCLR) [12].

For all models, we employ ResNet18 encoders [35] with an encoding dimension of 512. When running experiments on CIFAR-10, we follow the architecture of CIFAR ResNet models, as in [17]. For all datasets, all SSL methods utilize a 3-layer MLP projection network following the MoCo projector structure [34] with hidden and output dimensions of 2048. For classification, we attach a single linear layer to the encoder output. Adversarial examples are generated using projected gradient descent [49,57] with 10 steps of size 0.6/255, where perturbations are L_∞-bounded at $\epsilon_{AT} = 4/255$. Additional training details are deferred to Supplemental Material Sect. B.

We evaluate performance of trained models by measuring classification accuracy on clean test data. For each defense method, we measure the minimum (Psn Min) and average (Psn Avg) accuracies across all poisons. An adversary will choose the most effective poison given state-of-the-art defenses, so increasing Psn Min is crucial to improving model robustness in real-world scenarios.

Table 2. ImageNet-100 clean test set accuracies for VESPR and six defense benchmarks. Bold numbers indicate the best defense method for each poison, while underlined numbers indicate second-best methods.

Poison	SL	SL+AT	Cutout [21]	Mixup [87]	CutMix [85]	ISS [53]	SSL [12]	VESPR
Clean	77.66	69.76	78.04	80.38	**81.08**	71.58	70.96	74.84
AP8 [26]	8.18	42.88	7.68	9.68	8.38	24.22	41.80	**48.68**
CP8 [32]	57.46	48.76	58.20	61.76	**62.28**	50.18	14.80	54.40
CUDA [67]	6.10	36.34	8.66	5.16	5.70	3.58	26.12	**45.20**
LSP16 [83]	6.62	28.92	5.06	5.50	7.84	33.70	**62.06**	56.34
OPS [79]	51.00	52.56	53.86	48.82	**65.12**	57.36	62.22	63.26
RUE8 [27]	14.18	57.34	15.60	33.08	16.10	67.52	65.30	**70.36**
UE8 [40]	6.32	43.26	6.42	12.38	5.80	41.18	**50.02**	49.34
Psn Min	6.10	28.92	5.06	5.16	5.70	3.58	14.80	**45.20**
Psn Avg	21.41	44.29	22.21	25.20	24.46	39.68	46.05	**55.37**

Table 3. CIFAR-10 clean test set accuracies for VESPR and six defense benchmarks.

Poison	SL	SL+AT	Cutout [21]	Mixup [87]	CutMix [85]	ISS [53]	SSL [12]	VESPR
Clean	93.86	89.57	95.62	95.46	**95.78**	82.40	90.55	89.35
AP8 [26]	12.45	85.67	9.70	33.27	8.38	81.10	75.77	**86.78**
CP8 [32]	93.59	88.43	**94.51**	93.36	93.77	82.21	73.08	89.20
CUDA [67]	21.89	48.58	23.46	21.75	24.04	21.94	66.58	**80.31**
LSP16 [83]	21.47	85.67	18.47	22.76	21.68	79.33	88.87	**88.96**
OPS [79]	27.24	20.10	65.12	37.56	85.27	72.14	**87.79**	84.74
RUE8 [27]	18.91	34.45	19.67	23.27	27.46	80.04	86.70	**87.82**
UE8 [40]	32.53	89.31	36.18	54.19	38.40	82.12	88.45	**89.39**
Psn Min	18.91	20.10	9.70	21.75	8.38	21.94	66.58	**80.31**
Psn Avg	32.58	64.60	38.16	40.88	42.71	71.27	81.03	**86.74**

4.2 VESPR Performance

Table 2 presents clean test accuracy on ImageNet-100 for VESPR as well as multiple baseline defenses. VESPR debuts as a leading defense method, achieving the highest Psn Min and Psn Avg over all previous defenses by 16% and 9%, respectively, as well as maintaining strong performance when trained on clean data. VESPR frequently attains the highest clean test accuracy (AP8, CUDA, and RUE) or the second-highest clean test accuracy (LSP16, OPS, and UE8).

Though augmentation-based defenses occasionally achieve top performance (e.g., CutMix on CP8 and OPS), they are often overshadowed by other methods. Indeed the Psn Min values for augmentation-based methods are all less than 10%. SL+AT and SSL are consistently high performers, achieving average test accuracies of 44.29% and 46.05% across all poisons. However, their performance when trained on clean data alone is underwhelming, compared to other methods. This is unfortunate, as it suggests a smaller upper bound for clean test accuracy after poison training; intuitively, a model trained on poison data will not outperform the same model trained on clean data. As expected, SSL is particularly weak to CP8, a poison designed specifically to counter SSL.

Table 3 shows that the trends observed on ImageNet-100 performance are reflected in CIFAR-10. On CIFAR-10, VESPR outperforms other defense methods on multiple poisons (AP8, CUDA, LSP16, RUE8, and UE8). VESPR maintains the highest minimum and average performance across poisons by 13% and 5% respectively. For all defenses, the performance gap between clean and poison training is reduced relative to ImageNet-100, likely due to fewer classes and the ease of classifying smaller images.

4.3 Ablation Studies

Architecture Ablation In Table 4, we present the poison defense performance of all combinations for SL, SSL, and AT on ImageNet-100. Prior to this work, SSL+AT and SSL+SL methods have not been evaluated for poison defense. As measured by Psn Min and Psn Avg performance, VESPR outperforms all ablations by 13% and 4%, respectively.

Table 4. ImageNet-100 clean test set accuracies for VESPR and multiple ablations to architecture and training procedure.

Poison	SL	SL (SSL Aug)	SL+AT	SSL [12]	SSL+AT [42]	SSL+SL	VESPR
Clean	77.66	75.48	69.76	70.96	60.86	**79.66**	74.84
AP8 [26]	8.18	16.54	42.88	41.80	40.46	28.82	**48.68**
CP8 [32]	57.46	**60.40**	48.76	14.80	44.00	58.38	54.40
CUDA [67]	6.10	17.30	36.34	26.12	31.78	21.82	**45.20**
LSP16 [83]	6.62	49.30	28.92	**62.06**	49.00	52.36	56.34
OPS [79]	51.00	63.68	52.56	62.22	53.16	**70.14**	63.26
RUE8 [27]	14.18	49.08	57.34	65.30	54.30	70.02	**70.36**
UE8 [40]	6.32	34.46	43.26	50.02	40.94	**56.68**	49.34
Psn Min	6.10	16.54	28.92	14.80	31.78	21.82	**45.20**
Psn Avg	21.41	41.54	44.29	46.05	44.81	51.17	**55.37**

In order to assess the impact of augmentations, we additionally evaluate SL with SSL-style augmentations. For ease of comparison, we compile all augmentation-based defense method results for ImageNet-100 in Table 10 of the Supplemental Material Sect. C.1. As shown in Table 10, SL with SSL-style augmentations achieves the highest Psn Min and Psn Avg over the other augmentation-based methods by 10% and 2%, respectively. However, when compared to VESPR, SSL+SL, SSL, and SL+AT in Table 4, SL with SSL-style augmentations consistently underperforms. We hypothesize that existing augmentation-based defenses are weak to data poisoning due to their reliance on image cropping; most poisons are spread throughout images such that all crops are poisoned. This hypothesis is further supported by the fact that all augmentation-based defenses are particularly resistant to OPS, the only poison which does not spread throughout images. This ablation underscores that augmentations cannot fully explain the advances observed with VESPR.

As illustrated in Table 4, integrating AT with SL enhances performance over SL alone for all poisons except CP8. Though SL+AT, which trains using bounded additive noise, is particularly effective against bounded additive poisons (e.g., AP8, UE8), improvement is relatively limited on non-additive poisons (e.g. OPS) or poisons with large perturbations (e.g., LSP16). Additionally, applying AT to SSL degrades performance relative to SSL for most poisons. This highlights the nuanced nature of AT application: as poisons aim to undermine supervised classification accuracy, the adversarial signals from SSL alone have limited capability to ignore poisonous patterns.

Aligning with expectations from prior literature, [10,80], SSL+SL achieves top performance when trained on clean data, as the combined objectives of SSL and SL prevent the failure modes of either individual objective. Nevertheless, the intuition from prior literature does not necessarily apply for poisoned training, where SSL+SL often performs worse than other defenses, including SSL (e.g., AP8, CUDA, and LSP16). This signifies that the SL component of SSL+SL is still weak to poison patterns, dragging down the performance of SSL+SL. Furthermore, it emphasizes that the additional AT component of VESPR is crucial. For **all** poisons where SSL+SL underperforms SSL, VESPR utilizes SL-guided adversarial samples to outperform SSL+SL. This validates the claim that poisoned-SL can provide useful guidance for adversarial example generation by obfuscating poison features.

We defer the corresponding architecture ablation for CIFAR-10 to Table 11 of the Supplemental Material Sect. C.2. The trends seen for ImageNet-100 in Table 4 are largely reflected in CIFAR-10, and VESPR attains top performance for Psn Min and Psn Avg across all architecture ablations.

Architecture Ablation Analysis We expand our previous analysis on learned representations (Table 1) to all architecture ablations in Table 5. We additionally evaluate representation roughness (i.e., variability) using a local Lipschitz metric (Psn Local Lip) [82], where lower values correspond to smoother, more robust representation landscapes.

As illustrated in Table 5, VESPR maintains Psn In-Cls Sim values that are similar to those of SSL+SL and substantially higher than SSL or SSL+AT. Furthermore, utilizing AT consistently increases Psn-Cln Sim and reduces Psn Local Lip across all methods, including VESPR, compared to those without AT. In particular, the Psn Local Lip of VESPR decreases by an order of magnitude, from 8546 to 712, over SSL+SL. This improvement in smoothness is greater than

Table 5. Columns display Psn In-Cls Sim, Psn-Cln Sim, and Psn E-Rank, as in Table 1. We also present poison image representation roughness (Psn Local Lip, see [82]). We average results across seven different poisons (see Supplemental Material Sect. B.1)

Method	Psn In-Cls Sim	Psn-Cln Sim	Psn E-Rank	Psn Local Lip
SL	0.78	0.70	16.67	6180
SL (SSL Aug)	0.79	0.79	16.43	5868
SL+AT	0.76	0.85	31.64	3312
SSL	0.47	0.86	59.61	317
SSL+AT	0.60	0.97	23.47	40
SSL+SL	0.77	0.83	34.07	8546
VESPR	0.72	0.91	30.29	712

that of any other architecture that incorporates AT, underscoring the superior effect of AT on VESPR. Taken together, the trends in Table 5 suggest that SSL+SL aids VESPR in establishing distinct class-wise clusters of representations while VESPR's SL-guided AT ensures that these representations are reliant on robust, clean features. These findings justify the superior clean-classification performance of VESPR across multiple poisons.

Trends in representation similarity from Table 5 can also be identified visually. Figure 3 displays 2D t-SNE plots [38] of the 512D representations for three classes of poison and clean images and their CUDA-poisoned counterparts for various training methods. Slight mixing between encodings is expected as all three classes are types of fish, yet successful classification requires that clusters are largely distinct. Both SL and SSL tend to tightly cluster poison representations, yet corresponding clean images are encoded to distant representations. Furthermore, all clean image representations from SL and SSL are clustered together, regardless of class, suggesting that clean images are treated as an out-of-distribution class. In contrast, VESPR achieves superior poison-clean alignment, with representations of poison and clean samples overlapping for most samples. This verifies the high Psn-Cln Sim value for VESPR found in Table 5.

VESPR Adversarial Ablation In Table 6, we ablate the method of adversarial sample generation. Specifically, we compare the effect of random Gaussian noise (SSL+SL+GN) to methods of adversarial perturbation (VESPR and variants). VESPR remains the top-performing implementation.

For SSL+SL+GN, Gaussian noise with 4/255 standard deviation is applied to images with 50% probability. As the permissible Gaussian perturbation is higher than the AT perturbation bound, the domain of possible samples generated by Gaussian noise subsumes the domain of samples generated by AT. SSL+SL+GN slightly outperforms SSL+SL on most poisons, but overall performs similarly. Notably, the performance of SSL+SL+GN is conspicuously distinct from that of VESPR, suggesting that gradient-based perturbations from AT are crucial for

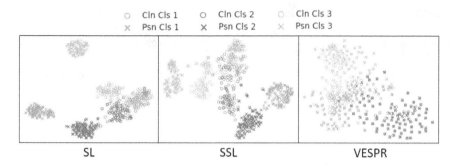

Fig. 3. 2-dimensional t-SNE plots [38] of clean and CUDA-poisoned image representations for SL, SSL, and VESPR models trained on CUDA data.

learning robust features. We note that VESPR consistently outperforms VESPR-SSL and VESPR-Both, with notable improvements on CUDA and RUE8 in particular. VESPR also attains the highest minimum and average clean test accuracies. This suggests that VESPR better exploits the adversarial signal from SL, allowing the encoder to learn robust features with SSL.

Table 6. ImageNet-100 clean test set accuracy for different methods of sample perturbation. SSL+SL+GN is similar to SSL+SL, but with additional Gaussian noise augmentation. VESPR-SSL uses AT with SSL loss, VESPR-Both uses AT with the combined VESPR loss in Eq. 3, and VESPR uses AT with SL loss.

Poison	SSL+SL	SSL+SL+GN	VESPR-SSL	VESPR-Both	VESPR
Clean	**79.66**	78.86	73.74	73.96	74.84
AP8 [26]	28.82	29.92	47.56	48.02	**48.68**
CP8 [32]	58.38	**59.14**	54.20	53.64	54.40
CUDA [67]	21.82	24.02	36.64	41.26	**45.20**
LSP16 [83]	52.36	53.00	54.36	55.88	**56.34**
OPS [79]	70.14	**71.12**	64.30	63.14	63.26
RUE8 [27]	70.02	69.90	66.38	68.58	**70.36**
UE8 [40]	56.68	**58.80**	49.28	48.20	49.34
Psn Min	21.82	24.02	36.64	41.26	**45.20**
Psn Avg	51.17	52.27	53.25	54.10	**55.37**

5 Conclusion

We extend the investigation of poison defense to seven poisons on CIFAR-10 and ImageNet-100 datasets, revealing limited performance of current defense methods, including SSL. Drawing from insights on SL and SSL poison defense, we introduce VESPR, which leverages supervised poison vulnerability to enhance self-supervised defense. Through extensive experiments, we demonstrate that VESPR achieves state-of-the-art defense against multiple prevalent poisons, surpassing all other baselines in both minimum and average poison performance.

Acknowledgements. This research is supported by the National Research Foundation, Singapore and Infocomm Media Development Authority under its Trust Tech Funding Initiative and Strategic Capability Research Centres Funding Initiative. Any opinions, findings and conclusions or recommendations expressed in this material are those of the author(s) and do not reflect the views of National Research Foundation, Singapore and Infocomm Media Development Authority.

References

1. Addepalli, S., Jain, S., Babu, R.V.: Efficient and effective augmentation strategy for adversarial training (2022)
2. Adnan, M., Ioannou, Y., Tsai, C.Y., Galloway, A., Tizhoosh, H.R., Taylor, G.W.: Monitoring shortcut learning using mutual information (2022)
3. Alkhunaizi, N., Kamzolov, D., Takáč, M., Nandakumar, K.: Suppressing poisoning attacks on federated learning for medical imaging. In: Wang, L., Dou, Q., Fletcher, P.T., Speidel, S., Li, S. (eds.) MICCAI 2022. LNCS, vol. 13438, pp. 673–683. Springer, Cham (2022). https://doi.org/10.1007/978-3-031-16452-1_64
4. Baier, F., Mair, S., Fadel, S.G.: Self-supervised siamese autoencoders. In: Miliou, I., Piatkowski, N., Papapetrou, P. (eds.) IDA 2024. LNCS, vol. 14641, pp. 117–128. Springer, Cham (2023). https://doi.org/10.1007/978-3-031-58547-0_10
5. Bardes, A., Ponce, J., LeCun, Y.: VICReg: variance-invariance-covariance regularization for self-supervised learning (2022)
6. Carlini, N., Terzis, A.: Poisoning and backdooring contrastive learning. In: International Conference on Learning Representations (2022). https://openreview.net/forum?id=iC4UHbQ01Mp
7. Caron, M., et al.: Emerging properties in self-supervised vision transformers (2021)
8. Chan-Hon-Tong, A.: An algorithm for generating invisible data poisoning using adversarial noise that breaks image classification deep learning. Mach. Learn. Knowl. Extract. **1**(1), 192–204 (2019). https://www.mdpi.com/2504-4990/1/1/11
9. Chen, K., Liu, Z., Hong, L., Xu, H., Li, Z., Yeung, D.Y.: Mixed autoencoder for self-supervised visual representation learning (2024)
10. Chen, M.F., et al.: Perfectly balanced: improving transfer and robustness of supervised contrastive learning (2022)
11. Chen, T., Liu, S., Chang, S., Cheng, Y., Amini, L., Wang, Z.: Adversarial robustness: from self-supervised pre-training to fine-tuning (2020)
12. Chen, T., Kornblith, S., Norouzi, M., Hinton, G.: A simple framework for contrastive learning of visual representations (2020)
13. Chen, T., Luo, C., Li, L.: Intriguing properties of contrastive losses (2021)
14. Chen, X., He, K.: Exploring simple siamese representation learning (2020)
15. Chen, X., Xie, S., He, K.: An empirical study of training self-supervised vision transformers (2021)
16. Cohen, J.M., Rosenfeld, E., Kolter, J.Z.: Certified adversarial robustness via randomized smoothing (2019)

17. da Costa, V.G.T., Fini, E., Nabi, M., Sebe, N., Ricci, E.: Solo-learn: a library of self-supervised methods for visual representation learning. J. Mach. Learn. Res. **23**(56), 1–6 (2022)
18. Cubuk, E.D., Zoph, B., Shlens, J., Le, Q.V.: RandAugment: practical automated data augmentation with a reduced search space (2019)
19. Demontis, A., et al.: Why do adversarial attacks transfer? Explaining transferability of evasion and poisoning attacks (2019)
20. Deng, J., Dong, W., Socher, R., Li, L.J., Li, K., Fei-Fei, L.: ImageNet: a large-scale hierarchical image database. In: CVPR 2009 (2009)
21. DeVries, T., Taylor, G.W.: Improved regularization of convolutional neural networks with cutout (2017)
22. Dippel, J., Vogler, S., Höhne, J.: Towards fine-grained visual representations by combining contrastive learning with image reconstruction and attention-weighted pooling (2022)
23. Eastwood, C., Williams, C.K.I.: A framework for the quantitative evaluation of disentangled representations. In: International Conference on Learning Representations (2018). https://openreview.net/forum?id=By-7dz-AZ
24. Fan, L., Liu, S., Chen, P.Y., Zhang, G., Gan, C.: When does contrastive learning preserve adversarial robustness from pretraining to finetuning? (2021)
25. Feng, J., Cai, Q.Z., Zhou, Z.H.: Learning to confuse: generating training time adversarial data with auto-encoder (2019)
26. Fowl, L., Goldblum, M., Chiang, P., Geiping, J., Czaja, W., Goldstein, T.: Adversarial examples make strong poisons (2021)
27. Fu, S., He, F., Liu, Y., Shen, L., Tao, D.: Robust unlearnable examples: protecting data privacy against adversarial learning. In: International Conference on Learning Representations (2022). https://openreview.net/forum?id=baUQQPwQiAg
28. Geirhos, R., Narayanappa, K., Mitzkus, B., Bethge, M., Wichmann, F.A., Brendel, W.: On the surprising similarities between supervised and self-supervised models (2020)
29. Geirhos, R., Rubisch, P., Michaelis, C., Bethge, M., Wichmann, F.A., Brendel, W.: ImageNet-trained CNNs are biased towards texture; increasing shape bias improves accuracy and robustness. In: International Conference on Learning Representations (2019). https://openreview.net/forum?id=Bygh9j09KX
30. Goodfellow, I.J., Shlens, J., Szegedy, C.: Explaining and harnessing adversarial examples (2015)
31. Grill, J.B., et al.: Bootstrap your own latent: a new approach to self-supervised learning (2020)
32. He, H., Zha, K., Katabi, D.: Indiscriminate poisoning attacks on unsupervised contrastive learning. In: The Eleventh International Conference on Learning Representations (2023). https://openreview.net/forum?id=f0a_dWEYg-Td
33. He, K., Chen, X., Xie, S., Li, Y., Dollár, P., Girshick, R.: Masked autoencoders are scalable vision learners (2021)
34. He, K., Fan, H., Wu, Y., Xie, S., Girshick, R.: Momentum contrast for unsupervised visual representation learning (2020)
35. He, K., Zhang, X., Ren, S., Sun, J.: Deep residual learning for image recognition (2015)
36. Higgins, I., et al.: beta-VAE: learning basic visual concepts with a constrained variational framework. In: International Conference on Learning Representations (2017). https://openreview.net/forum?id=Sy2fzU9gl

37. Hinton, G.E., Salakhutdinov, R.R.: Reducing the dimensionality of data with neural networks. Science **313**(5786), 504–507 (2006). https://doi.org/10.1126/science.1127647. https://www.science.org/doi/abs/10.1126/science.1127647
38. Hinton, G.E., Roweis, S.: Stochastic neighbor embedding. In: Becker, S., Thrun, S., Obermayer, K. (eds.) Advances in Neural Information Processing Systems, vol. 15. MIT Press (2002). https://proceedings.neurips.cc/paper_files/paper/2002/file/6150ccc6069bea6b5716254057a194ef-Paper.pdf
39. Hua, T., Wang, W., Xue, Z., Ren, S., Wang, Y., Zhao, H.: On feature decorrelation in self-supervised learning (2021)
40. Huang, H., Ma, X., Erfani, S.M., Bailey, J., Wang, Y.: Unlearnable examples: making personal data unexploitable (2021)
41. Islam, A., Chen, C.F., Panda, R., Karlinsky, L., Radke, R., Feris, R.: A broad study on the transferability of visual representations with contrastive learning (2021)
42. Jiang, Z., Chen, T., Chen, T., Wang, Z.: Robust pre-training by adversarial contrastive learning (2020)
43. Jing, L., Vincent, P., LeCun, Y., Tian, Y.: Understanding dimensional collapse in contrastive self-supervised learning (2022)
44. Kahana, J., Hoshen, Y.: A contrastive objective for learning disentangled representations. In: Avidan, S., Brostow, G., Cissé, M., Farinella, G.M., Hassner, T. (eds.) ECCV 2022. LNCS, vol. 13686, pp. 579–595. Springer, Cham (2022). https://doi.org/10.1007/978-3-031-19809-0_33
45. Kim, M., Ha, H., Son, S., Hwang, S.J.: Effective targeted attacks for adversarial self-supervised learning (2023)
46. Kim, M., Tack, J., Hwang, S.J.: Adversarial self-supervised contrastive learning (2020)
47. Kingma, D.P., Welling, M.: Auto-encoding variational bayes (2022)
48. Krizhevsky, A.: Learning multiple layers of features from tiny images (2009). https://api.semanticscholar.org/CorpusID:18268744
49. Kurakin, A., Goodfellow, I., Bengio, S.: Adversarial examples in the physical world (2017)
50. Li, A.C., Efros, A.A., Pathak, D.: Understanding collapse in non-contrastive siamese representation learning. In: Avidan, S., Brostow, G., Cissé, M., Farinella, G.M., Hassner, T. (eds.) ECCV 2022. LNCS, vol. 13691, pp. 490–505. Springer, Cham (2022). https://doi.org/10.1007/978-3-031-19821-2_28
51. Li, T., et al.: Addressing feature suppression in unsupervised visual representations (2021)
52. Liu, H., Jia, J., Gong, N.Z.: PoisonedEncoder: poisoning the unlabeled pre-training data in contrastive learning (2023). https://arxiv.org/abs/2205.06401
53. Liu, Z., Zhao, Z., Larson, M.: Image shortcut squeezing: countering perturbative availability poisons with compression (2023)
54. Locatello, F., et al.: Challenging common assumptions in the unsupervised learning of disentangled representations (2019)
55. Luo, R., Wang, Y., Wang, Y.: Rethinking the effect of data augmentation in adversarial contrastive learning (2023)
56. Lyu, K., Li, Z., Wang, R., Arora, S.: Gradient descent on two-layer nets: margin maximization and simplicity bias (2021)
57. Madry, A., Makelov, A., Schmidt, L., Tsipras, D., Vladu, A.: Towards deep learning models resistant to adversarial attacks (2019)
58. Misra, I., van der Maaten, L.: Self-supervised learning of pretext-invariant representations (2019)

59. Netzer, Y., Wang, T., Coates, A., Bissacco, A., Wu, B., Ng, A.Y.: Reading digits in natural images with unsupervised feature learning. In: NIPS Workshop on Deep Learning and Unsupervised Feature Learning 2011 (2011). http://ufldl.stanford.edu/housenumbers/nips2011_housenumbers.pdf
60. van den Oord, A., Li, Y., Vinyals, O.: Representation learning with contrastive predictive coding (2019)
61. Patel, N., Krishnamurthy, P., Garg, S., Khorrami, F.: Bait and switch: online training data poisoning of autonomous driving systems. arXiv preprint arXiv:2011.04065 (2020)
62. Qin, T., Gao, X., Zhao, J., Ye, K., Xu, C.Z.: APBench: a unified benchmark for availability poisoning attacks and defenses (2023)
63. Qin, T., Gao, X., Zhao, J., Ye, K., Xu, C.Z.: Learning the unlearnable: adversarial augmentations suppress unlearnable example attacks (2023)
64. Radford, A., et al.: Learning transferable visual models from natural language supervision (2021)
65. Robinson, J., Sun, L., Yu, K., Batmanghelich, K., Jegelka, S., Sra, S.: Can contrastive learning avoid shortcut solutions? (2021)
66. Rubinstein, B.I., et al.: ANTIDOTE: understanding and defending against poisoning of anomaly detectors. In: Proceedings of the 9th ACM SIGCOMM Conference on Internet Measurement, pp. 1–14 (2009)
67. Sadasivan, V.S., Soltanolkotabi, M., Feizi, S.: CUDA: convolution-based unlearnable datasets (2023)
68. Szegedy, C., et al.: Intriguing properties of neural networks (2014)
69. Tian, Y., Sun, C., Poole, B., Krishnan, D., Schmid, C., Isola, P.: What makes for good views for contrastive learning? (2020)
70. Tian, Y.: Understanding deep contrastive learning via coordinate-wise optimization (2022)
71. Tian, Y., Chen, X., Ganguli, S.: Understanding self-supervised learning dynamics without contrastive pairs (2021)
72. Tishby, N., Pereira, F.C., Bialek, W.: The information bottleneck method (2000)
73. Tsai, Y.H.H., Ma, M.Q., Yang, M., Zhao, H., Morency, L.P., Salakhutdinov, R.: Self-supervised representation learning with relative predictive coding (2021)
74. Uesato, J., Alayrac, J.B., Huang, P.S., Stanforth, R., Fawzi, A., Kohli, P.: Are labels required for improving adversarial robustness? (2019)
75. Vincent, P., Larochelle, H., Bengio, Y., Manzagol, P.A.: Extracting and composing robust features with denoising autoencoders. In: Proceedings of the 25th International Conference on Machine Learning, pp. 1096–1103 (2008). https://doi.org/10.1145/1390156.1390294
76. Wang, T., Yue, Z., Huang, J., Sun, Q., Zhang, H.: Self-supervised learning disentangled group representation as feature (2021)
77. Wang, T., Isola, P.: Understanding contrastive representation learning through alignment and uniformity on the hypersphere (2022)
78. Wang, X., Chen, X., Du, S.S., Tian, Y.: Towards demystifying representation learning with non-contrastive self-supervision (2022)
79. Wu, S., Chen, S., Xie, C., Huang, X.: One-pixel shortcut: on the learning preference of deep neural networks (2023)
80. Xue, Y., Joshi, S., Gan, E., Chen, P.Y., Mirzasoleiman, B.: Which features are learnt by contrastive learning? On the role of simplicity bias in class collapse and feature suppression (2023)
81. Yang, W., Kirichenko, P., Goldblum, M., Wilson, A.G.: Chroma-VAE: mitigating shortcut learning with generative classifiers (2022)

82. Yang, Y.Y., Rashtchian, C., Zhang, H., Salakhutdinov, R., Chaudhuri, K.: A closer look at accuracy vs. robustness (2020)
83. Yu, D., Zhang, H., Chen, W., Yin, J., Liu, T.Y.: Availability attacks create shortcuts. In: Proceedings of the 28th ACM SIGKDD Conference on Knowledge Discovery and Data Mining. ACM (2022). https://doi.org/10.1145/3534678.3539241
84. Yuan, C.H., Wu, S.H.: Neural tangent generalization attacks. In: Meila, M., Zhang, T. (eds.) Proceedings of the 38th International Conference on Machine Learning. Proceedings of Machine Learning Research, vol. 139, pp. 12230–12240. PMLR (2021). https://proceedings.mlr.press/v139/yuan21b.html
85. Yun, S., Han, D., Oh, S.J., Chun, S., Choe, J., Yoo, Y.: CutMix: regularization strategy to train strong classifiers with localizable features (2019)
86. Zhang, C., Zhang, K., Zhang, C., Pham, T.X., Yoo, C.D., Kweon, I.S.: How does simsiam avoid collapse without negative samples? A unified understanding with self-supervised contrastive learning (2022)
87. Zhang, H., Cisse, M., Dauphin, Y.N., Lopez-Paz, D.: mixup: beyond empirical risk minimization (2018)
88. Zhang, Y., Tan, Z., Yang, J., Huang, W., Yuan, Y.: Matrix information theory for self-supervised learning (2023)
89. Zhou, J., Dong, L., Gan, Z., Wang, L., Wei, F.: Non-contrastive learning meets language-image pre-training (2022)
90. Zhuo, Z., Wang, Y., Ma, J., Wang, Y.: Towards a unified theoretical understanding of non-contrastive learning via rank differential mechanism (2023)

Dense Hand-Object (HO) GraspNet with Full Grasping Taxonomy and Dynamics

Woojin Cho[1]([✉]), Jihyun Lee[1], Minjae Yi[1], Minje Kim[1], Taeyun Woo[1], Donghwan Kim[1], Taewook Ha[1], Hyokeun Lee[3], Je-Hwan Ryu[4], Woontack Woo[1], and Tae-Kyun Kim[1,2]

[1] KAIST, Daejeon, South Korea
woojin.cho@kaist.ac.kr
[2] Imperial College London, London, UK
[3] Kwangwoon University, Seoul, South Korea
[4] Surromind, Seoul, South Korea

Abstract. Existing datasets for 3D hand-object interaction are limited either in the data cardinality, data variations in interaction scenarios, or the quality of annotations. In this work, we present a comprehensive new training dataset for hand-object interaction called HOGraspNet. It is the only real dataset that captures full grasp taxonomies, providing grasp annotation and wide intraclass variations. Using grasp taxonomies as atomic actions, their space and time combinatorial can represent complex hand activities around objects. We select 22 rigid objects from the YCB dataset and 8 other compound objects using shape and size taxonomies, ensuring coverage of all hand grasp configurations. The dataset includes diverse hand shapes from 99 participants aged 10 to 74, continuous video frames, and a 1.5M RGB-Depth of sparse frames with annotations. It offers labels for 3D hand and object meshes, 3D keypoints, contact maps, and *grasp labels*. Accurate hand and object 3D meshes are obtained by fitting the hand parametric model (MANO) and the hand implicit function (HALO) to multi-view RGBD frames, with the MoCap system only for objects. Note that HALO fitting does not require any parameter tuning, enabling scalability to the dataset's size with comparable accuracy to MANO. We evaluate HOGraspNet on relevant tasks: grasp classification and 3D hand pose estimation. The result shows performance variations based on grasp type and object class, indicating the potential importance of the interaction space captured by our dataset. The provided data aims at learning universal shape priors or foundation models for 3D hand-object interaction. Our dataset and code are available at https://hograspnet2024.github.io/.

Keywords: hand-object interaction · grasp taxonomy · 3D shape and pose estimation · new benchmark

---

Supplementary Information The online version contains supplementary material available at https://doi.org/10.1007/978-3-031-73007-8_17.

© The Author(s), under exclusive license to Springer Nature Switzerland AG 2025
A. Leonardis et al. (Eds.): ECCV 2024, LNCS 15140, pp. 284–303, 2025.
https://doi.org/10.1007/978-3-031-73007-8_17

1 Introduction

The importance of modeling and inferring 3D hand-object interactions is growing. While earlier works focused on single object instances [25,64,68,69,74,76, 81,86,87,89,91,94–96], recent efforts have been made on multiple 3D objects and their complex interactions [4,8,16,19,24,28,33,38,39,54,57,58,71,73,80,85,93]. The human hand is the most dexterous and important testbed, and its research is extendable to human bodies or faces in similar articulated and deformable categories. We observe a few new benchmarks on hand-object interaction each latest year. However, existing datasets are limited either in the cardinality of data, the amount of data variations in hands/objects, or the quality of annotations. See Table 1, where HO3D [28] and DexYCB [8] include only 15 and 14 (out of 33) grasping taxonomies respectively. YCB Affordance [16] is only an existing benchmark that represents all grasp taxonomies and provides grasp labels, but synthetic; ARCTIC places visible markers on hands in RGB images, and OakInk (the closest to ours) does not provide grasp labels with fewer subjects but more objects. More comparison with OakInk is shown in the t-SNE plot Fig. 5.

Fig. 1. (left) **Diverse samples in HOGraspNet (best viewed with zoom-in).** HOGraspNet captures all hand-object grasp taxonomies with high-quality 3D annotations. (right) **Grasp Taxonomy t-SNE.** It covers well the grasp taxonomy space with intra-class variations.

We introduce HOGraspNet, an extensive multi-view RGBD training dataset for hand, object, and their interaction with grasp annotations. Based on the existing hand grasping taxonomy [20], our design redefines 28 of the 33 grasps by merging geometrically similar or uncommon poses. Our dataset is the only real dataset covering all grasp taxonomies, including grasp labels and a wide range of intraclass variations. We exploit 22 rigid objects from the YCB dataset [5] with 8 other compound/articulated objects. As an example shown in Fig. 4, 3 distinct hand grasps are performed for each object, totaling 90 interaction scenarios. Note that 30 objects are chosen enough to cover all grasp taxonomies, while textures and shapes beyond grasp areas can be synthetically augmented with 3D models. The dataset comprises a diverse range of hand identities from 99 participants aged 10 to 74. Overall, the dataset contains 1.5M RGB-Depth

Table 1. Comparison of hand-object interaction datasets. Interaction info is the key criteria utilized in terms of hand-object interaction.

Dataset	Type	#image	#views	#obj	#subj	#Grasps in [20]	Real Video	Marker-less hand	Dynamic interaction	Hand-obj contactmap	Grasp variation	Grasp annotation	Interaction info.
Obman(CVPR19) [33]	RGBD	154k	1	3k	20		✗	✗	✓	✗	✗	✗	✗	
YCB-Affordance(CVPR20) [16]	RGB	133k	1	58	-	100%	✗	✓	✓	✗	✗	✓	✓	Grasp
FreiHAND(ICCV18) [96]	RGB	37k	8	2	32		✓	✗	✓	✗	✗	✗	✗	
MOW(ICCV21) [6]	RGB	500	1	~500	-	82%	✓	✗	✓	✗	✗	✗	✗	
DexYCB(CVPR21) [8]	RGBD	582k	8	20	10	42%	✓	✓	✓	✗	✗	✗	✗	
FPHA(CVPR18) [24]	RGBD	105k	1	4	6		✓	✓	✗	✓	✗	✗	✗	Action
HO3D(CVPR20) [28]	RGBD	78k	1	10	10	45%	✓	✓	✓	✓	✗	✗	✗	
SHOWMe(ICCVW23) [72]	RGBD	87k	1	42	15	61%	✓	✓	✓	✗	✗	✗	✗	
ContactPose(ECCV20) [4]	RGBD	2.9M	3	25	50		✓	✓	✓	✓	✓	✗	✗	Intent
H2O(ICCV21) [41]	RGBD	571k	5	8	4		✓	✓	✓	✓	✗	✓	✗	
ARCTIC(CVPR23) [19]	RGBD	2.1M	9	10	9		✓	✓	✗	✓	✓	✓	✗	Intent
OakInk(CVPR22) [83]	RGBD	230k	4	100	12		✓	✓	✓	✓	✓	✓	✗	
Ours	RGBD	1.5M	4	30	99	85%	✓	✓	✓	✓	✓	✓	✓	Grasp

frames from 4 viewpoints, with annotations for 3D meshes, 3D keypoints, contact maps, and *grasp labels*. We adopt the hand parametric model (MANO [67]) and the hand implicit function (HALO [40]) individually to annotate the hand mesh. Presenting the novel annotation pipeline using the hand implicit function that requires simpler settings (i.e., less hyper-parameters) than MANO with accuracy and continuous shape representation. Considering the small objects in HOGrasp-Net, we utilize optical markers for MoCap only to obtain object 6D pose. We report experimental results of grasp classification, and SOTA hand-object 3D pose estimation methods. The new dataset demonstrates its comprehensiveness and potential (Fig. 1).

Further possibilities from the presented dataset are to 1) synthetically augment the data by changing backgrounds or object textures and shapes beyond grasp areas, 2) provide environments for learning a grasping agent in physical simulators, 3) extend to non-grasping actions, e.g., pushing, throwing, or deformed processes of non-rigid objects by hand. We hope the new dataset serves as a basis for understanding and modeling diverse inference models of hand-object interactions and that learned knowledge applies to human-object or human-human interactions.

2 Survey on Interaction Datasets

This section provides a comprehensive overview of existing datasets on the interaction of 3D shapes, i.e., single hand, hand-object, hand-hand, human-object, and human-human. We also briefly discuss the existing literature on hand-object reconstruction, which is used for benchmarking our dataset (in Sect. 4). Further survey results are available in the supplementary materials.

Single-Hand Datasets. Earlier research efforts to build a hand dataset have focused on capturing single hands from RGB [25,68,91,94,96], depth [64,74, 76,81,87], or RGBD [69,86,89,95], stimulating various learning-based methods for hand reconstruction [7,23,46,59,62]. These datasets can be categorized via three characteristics: (1) whether the captured hand frames are synthetic [68, 95] or real [25,64,68,69,74,76,81,86,87,89,91,94,96], (2) whether the hand is annotated as sparse keypoints [25,68,74,76,81,87,91,95,96] or mesh [86,94], and (3) whether the annotation is obtained via marker-based [25,64,74,87] or markerless system [68,69,76,81,86,89,91,94–96]. More recently, various hand datasets aim to capture hands in interaction with an object [3,4,8,16,17,19,24,28,33,58, 72,73] or another hand [47,56,57,78,97]. Since the goal of our work is to collect a dataset that comprehensively captures hand-object interactions, we focus on discussing the existing hand-object datasets in the following.

Hand-Object Datasets. Recently, various hand-object datasets [3,4,8,16, 17,19,24,28,33,58,72,73] have been proposed. Regarding **(1) annotation method**, most of the earlier datasets collect synthetic RGB and/or depth images rendered from a parametric hand model (MANO [67]) and template object models [16,33,58], or collect real images with markers [4,19,73] or magnetic sensors [24] to obtain hand annotations. However, these samples lack realism due to the rendering of synthetic models or the presence of visible sensors. Thus, many recent datasets use a markerless system to fit the MANO model to RGB-D images captured in a multi-view setup while using a minimal number of markers to obtain object poses [8,28,72]. Our work also follows such marker-less capture system to provide MANO-based hand annotations while additionally fitting an implicit function-based hand model (HALO [40]) to provide supplementary hand shape information. Regarding the **(2) characteristics of captured data**, existing datasets are limited either in data cardinality, the number of object categories or hand identities, or interaction taxonomies (please refer to Table 1). For example, HO3D [28] and DexYCB [8] (which are the most widely used hand-object datasets) only consider 10 object categories and capture 10 and 20 hand identities, respectively. While ObMan [33] and SHOWMe [72] capture more diverse object categories, they are limited in the number of hand identities (20 and 15, respectively) and the data cardinality (154K and 87K, respectively). ARCTIC [19] and OakInk [83] are recently proposed datasets that capture dexterous interactions between hands and objects, containing a range of motion variations. However, they do not cover the diverse grasp poses for each object, as they instruct participants to assume poses based on their intent to interact with the object. Our work aims to collect a training dataset that is more comprehensive in terms of interaction scenarios based on grasps, object categories, hand identities, and data cardinality. We also note that most of the existing datasets do not provide a grasping type of each sample, which can further provide a useful prior for the captured hand-object interaction [16,26,49]. Our work also carefully identifies a taxonomy of 33 grasping types and provides grasping type annotation for each sample. For comparison with other datasets, we conducted a thorough survey to ascertain the number of grasp classes present among the 33 grasp taxonomies in Feix et al. [20] and reported in Table 1. The detailed results are provided in the supplementary material.

Table 2. Comparison of Human-object interaction datasets.

Dataset	Type	#image	#views	#obj	#subj	#Kinects	Label	Contact annotation	Whole body interact.	Marker-less hand	Natural scene	Scene interact.	Dynamic hand	Articulated object
EgoBody(ECCV22) [90]	RGBD	220K	3~5	15	36	3~5	SMPL-X	✗	✓	✓	✓	✓	✗	✗
GRAB(ECCV20) [73]	Mesh	1.6M	-	51	10	-	SMPL-X	✓	✓	✗	✗	✗	✓	✗
ARCTIC(CVPR23) [19]	RGB	2.1M	9	10	9	-	SMPL-X	✓	✓	✗	✗	✗	✓	✓
CHAIRS(ICCV23) [37]	RGBD	1.7M	4	81	46	-	SMPL-X	✓	✓	✗	✗	✗	✗	✓
BEHAVE(CVPR22) [2]	RGBD	15K	5	20	8	-	SMPL	✓	✓	✗	✓	✗	✗	✗
InterCap(IJCV24) [36]	RGBD	67K	6	10	10	6	SMPL-X	✓	✓	✓	✓	✗	✗	✗
PROX(ICCV19) [30]	RGBD	100K	3	12	20	1	SMPL-X	✓	✓	✓	✓	✓	✗	✗
RICH(CVPR22) [35]	RGBD	540K	6~8	5	22	1	SMPL-X	✓	✓	✓	✓	✓	✗	✗

Two-Hand Datasets. Similar to hand-object datasets, various interacting two-hand datasets have been proposed. HIC [78] and RGB2Hands [79] are some of the earliest two-hand interaction datasets, but their data cardinality and interaction diversity are small compared to more recent datasets. InterHand2.6M [57] is the most widely-used large-scale dataset, which captures interacting hands in a multi-view markerless motion capture setup. More recently, Re:InterHand [56] is proposed to capture more diverse two-hand interactions in terms of image appearances and interaction poses via the use of environment maps and hand relighting. TwoHand500K [97] is another recently proposed dataset consisting of (1) real data captured using a marker-based system and (2) synthetic data obtained via combining poses randomly sampled from single-hand datasets. These datasets have inspired various methods for interacting two-hand reconstruction [43,45,88] and generation [42,47].

Interacting Human Datasets. In addition, research attention to interacting hands and several datasets contain interacting humans. These can be categorized as human-object interaction, human-human interaction, and human-scene interaction. **(1) Human-object interaction**: As demonstrated in Table 2, several datasets have been released to depict various aspects of human-object interactions. GRAB [73] and ARCTIC [19] propose datasets with 3d human mesh which focus on hand-object interaction. These three datasets [2,36,37] extend the interacting region to the whole body. Leveraging SMPL-X as mesh templates, [2,19,36] contain contact supervision, and [19,37] focus on articulated objects. **(2) Human-human interaction**: PanopticStudio [38], MuCo-3DHP [54] and MuPoTS-3D [54] propose datasets with 3d human sparse keypoints. More recently, there has been increased interest on reconstructing 3d human mesh [21,39,44,60,63,71,80,85,93]. The majority of them employ parametric models like SMPL [51] or SMPL-X [61], except some datasets [60,63,80,85,93] that provides textured scans, which is beneficial to represent geometric details. **(3) Human-scene interaction**: Certain datasets [30,35,90] broaden the scope of the interaction to include scenes, utilizing SMPL-X as a mesh template. PROX [30] and RICH [35] provide contact supervision between humans and scenes. RICH [35], in particular, extends its scope to outdoor scenes, and EgoBody [90] provides motion text labels.

Hand-Object Reconstruction. Reconstructing hand and object in interaction has been actively explored. Most of the recent works can be categorized into optimization or learning-based methods. Optimization-based methods [6,27,32,77,84] typically fit MANO [67] hand and template object models based on contact or other physical constraints (e.g., attraction and repulsion [84], friction [34]). Learning-based methods [10–14, 18, 29, 31, 33, 50, 75] directly regress hand and object poses via a neural network, while focusing on exploring an effective architecture for feature learning [10, 18, 29, 31, 33, 50, 75] and/or shape representation [11, 12]. In our work, we benchmark the RGB-based reconstruction task on our HOGraspNet dataset using HFL-Net [48], which is the most recent state-of-the-art method.

3 HOGraspNet

3.1 Dataset Overview

The dataset includes continuous video images and 1,489,112 annotated RGB-D frames, covering 28 hand grasp classes. We redefined the grasp classes by merging visually similar configurations (see supplementary) using 30 objects. The frames were captured at 4 distinct viewpoints and performed by 99 participants aged 10 to 74. Along with diverse hand shapes, a good scope of intra-class variations within each grasp class has been collected, which is important as a training dataset. Each RGB-D frame is annotated with the 3D hand pose for 21 joints and mesh, corresponding grasp class, 6D object pose, and contact map between the hand and object. Hand mesh models are obtained fitting MANO [67] and/or HALO [40], while all object mesh models (3d shapes and textures) are pre-scanned and provided. In Fig. 2, we present examples of the data types included in the dataset.

3.2 Object Categories and Grasp Taxonomy

While interacting with various objects, we often take specific grasp poses based on the object's shape and intention. To capture a wide range of hand pose space, especially to cover all grasping taxonomies, we identified 22 types of objects from the YCB dataset [5] and 8 other daily objects. These objects are selected considering factors, such as the primitive shapes of objects (cylinder, sphere, disk, cube), especially grasp areas, object sizes, and additional articulated/compound objects (see Fig. 3 (right)). Referring to previous studies on hand-object grasp taxonomy [1,15,20,49,70], we captured the three most common grasp classes for each object. Example per-object taxonomies are shown in Fig. 4. Also, some grasp classes out of 33 are seemingly redundant, visually hard to distinguish, and geometrically close; we redefined them to 28 grasp classes, with their indices following those presented in [20]. Compared to the existing benchmarks (MOW [6] and OakInk [83]), we have a relatively smaller set of objects. However, we span more grasp space with large intra-class variations, thanks to diverse hand

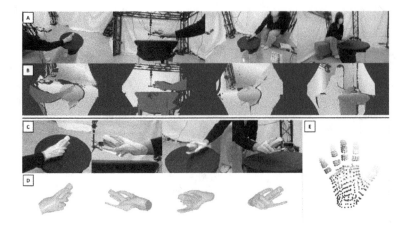

Fig. 2. Structure of HOGraspNet. It captures diverse hand-object grasping at 4 different viewpoints. Example RGB images (A) and depth images (B) are shown, while the fitted hand and object meshes are visualized in (C) and (D). (E) shows the contact map.

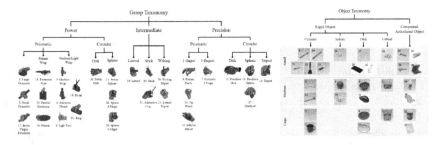

Fig. 3. (left) 33 hand grasping taxonomies, (right) 30 objects used in the dataset. The object types are cylinder, sphere, disk, cuboid, or compound/articulated. They are further dividend to small/medium/large sizes, purporting to cover all grasp taxonomies.

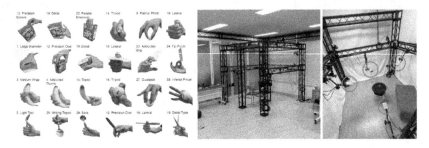

Fig. 4. (left) Per-object taxonomy examples (right) System setup. The full list is shown in the supplementary.

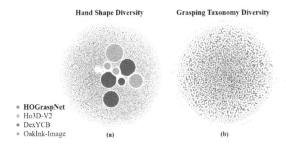

Fig. 5. t-SNE [53] visualization of (left) MANO [67] shape parameter distributions and (right) grasp feature distributions.

shapes and the number of frames. We consider synthetic object augmentation in textures and shapes beyond grasp areas and background augmentation as future work.

3.3 Hardware Setup and Data Collection

Sensors. Figure 4 shows the recording studio setup, where 4 temporally synchronized RGB-D cameras (Azure Kinect) are positioned around the designated space. The one in the backside is roughly at users' eye locations, imitating a fixed egocentric view. The cameras capture RGB and depth at 1920 × 1080 resolution and 30 FPS. For object poses, 8 IR cameras with a frame rate of 120 FPS were set, and 3 to 5 optical markers (3 mm) were attached to each object. Notably, no markers were used on hands to maintain their realistic appearance, thereby minimizing the potential degradation of image features in networks trained using our dataset due to RGB image contamination (cf. ARCTIC [19]). However, markers on objects can still limit hand poses, so we minimized this impact by placing markers on regions least likely to be grasped (e.g., the blade of scissors). Note that all objects were symmetrical enough to place markers while avoiding contact areas. Temporal synchronization between the RGB-D and IR cameras was obtained by manually aligning the starting frames during each recording session through a start blink of the LED.

Data Acquisition. We conducted data capture involving 99 participants with diverse hand sizes, shapes, and textures. Detailed instructions regarding grasp classes for each object were provided, and participants were requested to grasp each object with their right hand according to the specified grasp while freely performing pose variations such as translation and rotation. Each participant completed the procedure 2 to 4 times, with each trial recorded for 20 s to adequately capture actions ranging from reaching for the object to freely manipulating it in the air and eventually placing it back down. This way, diverse intra-class variations were captured.

3.4 Data Distributions

To further demonstrate that our dataset captures more comprehensive hand grasps, we visualize our data distribution in comparison to HO3D [28], DexYCB [8], and OakInk [83]. In Fig. 5(a), we show t-SNE [53] visualizations of the four datasets in the MANO [67] shape parameter space. Our dataset captures more hand shape diversity than the others, as a larger number of hand identities were included (as shown in Table 1). Figure 5(b) shows the t-SNE visualizations in the grasp feature space. For grasp feature extraction, we train a hand auto-encoder with mesh reconstruction loss and the auxiliary contact reconstruction and grasp classification losses (see Sect. 4.2 for more details) to obtain features that capture hand pose and grasp configurations. Ours is shown to be significantly more diverse than the other datasets in this feature space as well, thanks to our data acquisition process associated with carefully determined grasping types. We hope that the comprehensiveness of our dataset can serve as an effective prior for the downstream tasks related to hand-object interaction.

3.5 MANO and Object Annotation

For MANO [67] hand and object annotation, we use an automatic annotation and verification pipeline inspired by prior studies [8,28]. In the following subsections, we discuss each step of our pipeline, while more details can be found in the supplementary.

Data Preprocessing. We downsample the captured RGB-D frames from 30FPS to 10FPS to filter temporally redundant samples. We also prepare the hand and object segmentation masks using the DeepLabv3 [9] model, which is fine-tuned using our data with a few manually annotated segmentation masks for each object (Fig. 6).

Initial Hand Keypoint Estimation. To prepare the initial hand keypoints used for MANO fitting, we use MediaPipe [52] hand pose estimator, which is known to have high generalization ability. We estimate 2.5D hand keypoints from each multi-view frame and lift them to 3D keypoints via triangulation. However, such keypoint estimates may be noisy for viewpoints with high hand-object occlusion. To overcome this, we introduce a novel bootstrapping procedure to achieve a better 3D keypoint lifting quality. Given the 2.5D keypoint estimates from our four viewpoints $\{vp_i\}_{i=0,1,2,3}$, we obtain the lifted 3D keypoints $\{\hat{J}_i\}_{i=0,1,2,3}$, where $\hat{J}_i \in \mathbb{R}^{21\times 3}$ denotes 3D keypoints lifted using three viewpoints while *excluding* vp_i. We assume that if the MPJPE between (1) $\hat{J}_i$ projected onto vp_i and (2) the original 2.5D keypoint estimates from vp_i is above a threshold τ, then the original estimates from vp_i is an outlier. In this way, we filter out noisy 2.5D keypoints during 3D lifting procedure to obtain more robust 3D keypoints per frame. The valid hand poses for each viewpoint and the visibility v_i for each joint i (computed using depth maps) are also stored to serve as pseudo-ground truth (GT) data in the following steps.

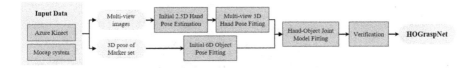

Fig. 6. MANO [67] and object annotation pipeline (Sect. 3.5).

Initial Object Pose Estimation. To obtain the initial 6D poses of an object, we attach optical sensors to the predefined surface locations of each object. Using multiple IR cameras, the 3D positions of each optical sensor are collected through in-built software. Then, the object's 3D rotation and translation are computed via Least-Squares Fitting to the marker positions.

Multi-view Multi-frame Gradual Hand-Object Model Fitting. In this stage, our goal is to fit MANO [67] hand and object template models to multi-view RGB-D frames and the initial hand and object poses. To this end, we formulate an optimization-based fitting scheme similar to previous works [8,96]. To avoid local minima, we further propose to gradually fit the MANO parameters, such that our optimization consists of three stages: (1) fitting global hand transformation, (2) fitting partial hand poses extended from the wrist, and (3) fitting the full hand and object pose (see the supplementary for details).

Our overall loss function for the MANO pose $\theta \in \mathbb{R}^{48}$ and shape $\beta \in \mathbb{R}^{10}$ parameters and the object 6D pose $\phi \in \mathbb{R}^6$ can be written as:

$$\mathcal{L} = \lambda_h^{2D} \mathcal{L}_h^{2D} + \lambda_o^{3D} \mathcal{L}_o^{3D} + \lambda_{seg} \mathcal{L}_{seg} + \lambda_{depth} \mathcal{L}_{depth} + \lambda_{reg} \mathcal{L}_{reg} + \lambda_{phy} \mathcal{L}_{phy}. \quad (1)$$

$\mathcal{L}_h^{2D}$ measures the L2 distance between the pseudo GT 2D joints and the 2D projection of the MANO 3D joints weighted by the visibility v_i. $\mathcal{L}_o^{3D}$ computes the L2 distance between the 3D marker positions and the corresponding vertex position of an object model. $\mathcal{L}_{seg}$ and $\mathcal{L}_{depth}$ measures the L1 distance between the GT and the rendered segmentation masks and depth maps, respectively.

Following [96], we also incorporate a regularization term $\mathcal{L}_{reg} = ||\tilde{\theta}||_2 + ||\tilde{\beta}||_2 + ||\theta_t - \theta_{t-1}||_2 + ||\beta_t - \beta_{t-1}||_2$, which (1) penalizes MANO pose and shape parameters that deviate too much from the mean zero vectors and (2) encourages the previous and current hand parameters to be close for temporal consistency. To additionally regularize the fitted hand and object meshes to be physically plausible, we incorporate another regularization term $\mathcal{L}_{phy}$, which is designed as a weighted sum of penetration loss and contact loss: $\mathcal{L}_{phy} = \lambda_{pen} \mathcal{L}_{pen} + \lambda_{contact} \mathcal{L}_{contact}$. For penetration loss λ_{pen}, we use a vertex normal projection-based technique used in [28]. For contact loss $\mathcal{L}_{contact}$, we minimize the distances between hand and object vertices below a distance threshold τ to encourage physical contact. In Eq. 1, $\{\lambda_i\}_{i=h, o, seg, depth, reg, phy}$ is a set of scalar values to control the weighting between the loss terms. Please see the supplementary for more details about the annotation procedure.

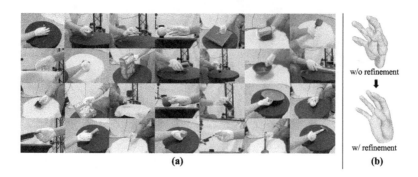

Fig. 7. HALO [40] fitting results. (a) Annotated HALO hand examples. (b) Comparisons between the HALO shapes with and without applying inverse kinematics-based keypoint refinement [11].

Post-verification. We conduct both automatic and manual verification steps to further filter out the noisy annotations. We compute the Intersection over Union (IoU) between the pseudo-GT and the rendered segmentation masks, filtering out annotations with an IoU below 0.6 in any view. Subsequently, we perform manual verification through crowdsourcing using LabelOn (https://www.labelon.kr/). Each crowdsourcer identifies misprocessed data that significantly deviates from the hand and object meshes or results from operational errors.

3.6 HALO Annotation

HALO [40] Fitting. For hand shape annotation, we additionally provide the hand implicit surface based on HALO [40], which is a neural implicit representation that parameterizes an articulated occupancy field [55] with 3D hand keypoints. Thus, a straightforward approach to fit HALO to our collected data would be to use the 3D hand keypoints lifted from the multi-view 2D keypoint estimates (as described in Sect. 3.5) as an input to the HALO model. However, we observe that it leads to a less plausible implicit hand surface since the keypoints are not guaranteed to form a valid kinematic structure of the hand. Thus, we postprocess the lifted keypoints to the nearest keypoints on the hand space learned by MANO via the inverse kinematics algorithm in [11]. Our HALO fitting results are shown in Fig. 7.

Comparisons with MANO [67] Fitting. As HALO [40] is an occupancy function that takes hand keypoints as input, it does not require many hyperparameters (except for an occupancy threshold [55]) for model fitting, while MANO fitting typically requires numerous loss weighting terms (i.e., λ_* in Eq. 1) for defining the optimization objective. Thus, HALO can be more convenient and scalable for annotating a large-scale dataset. Also, HALO can produce hand shapes in a resolution higher than MANO due to its resolution-independent nature (see Fig. 8(a)). However, HALO is shown to capture less hand shape variation than MANO (see the red circle in Fig. 8(b)) due to its keypoint-based

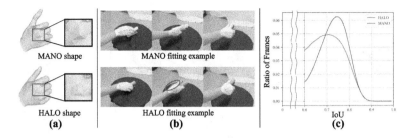

Fig. 8. Comparisons between **HALO** [40] and **MANO** [67] fitting results. (a) Hand shapes. (b) Fitting examples. (c) IoU distributions after post-verification stage.

parameterization for hand shape. As shown in Fig. 8(c)), the IoU distribution of HALO is marginally worse than that of MANO, but it still achieves comparable mean IoU results (MANO: 0.739, HALO: 0.719) despite its simple annotation procedure.

4 Experimental Results

In this section, we first report the split protocols of HOGraspNet (Sect. 4.1). We then present our experimental results on grasp classification (Sect. 4.2) and hand-object pose estimation (Sect. 4.3) using our dataset.

4.1 Split Protocols

For the evaluation setup, we generated five distinct train/test splits based on key components within our dataset:

- **S0 (default).** This split encompasses all subjects, views, objects, and grasp classes. The dataset is split by sequences, with the first sequence of each subject selected as the test set and the remaining sequences used for training.
- **S1 (unseen subjects).** The dataset is split by subjects, following a 7:3 train/test ratio.
- **S2 (unseen views).** The dataset is split by camera views, following a 3:1 train/test ratio.
- **S3 (unseen objects).** The dataset is split by objects. 7 objects that collectively represent all 28 grasps are selected as the test set, while the other 23 objects are used for training.
- **S4 (unseen taxonomy).** The dataset is divided by the grasping taxonomy. All the *intermediate* grasp types in Fig. 3 are selected as the test set, while others are used for training.

Note that we will release the exact split configurations through code. Also, refer to the supplementary for the benchmarking hand-object reconstruction results for each split.

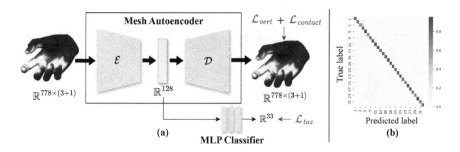

Fig. 9. (left) The network architecture for grasp classification, (right) confusion matrix.

4.2 Grasp Classification

We evaluate grasp classification performance on HOGraspNet using our S0 (Sect. 4.1) benchmark setup. For the classification network, we modify the existing convolutional mesh autoencoder (CoMA [66]) to take as input a hand mesh with per-vertex contact value as an additional vertex feature. The bottleneck feature of the autoencoder is fed to an MLP-based classifier to predict a grasp type. To train our model, we use L1 loss ($\mathcal{L}_{vert}$) that learns vertex reconstruction, and two cross-entropy losses that learn grasp taxonomy classification ($\mathcal{L}_{tax}$) and contact classification [27] ($\mathcal{L}_{contact}$), where the range of contact value [0,1] is split into 10 bins. Our overall network architecture is shown in Fig. 9. Note that we utilize auxiliary reconstruction losses for the classification task to obtain richer grasp features, which are also utilized for t-SNE visualization in Sect. 3.4. Our model achieves 0.95 in f1 score for contact map reconstruction and 0.88 in accuracy for taxonomy classification. This experimental validation demonstrates that our grasping taxonomy can be delineated using hand meshes with contact maps without considering an object as input, indicating that our grasp annotation is generalized well across the samples.

4.3 Hand-Object Pose Estimation

In this section, we present the benchmarking results on hand-object pose estimation on HOGraspNet using S0 split. We use HFL-Net [48] as a baseline, as it is the current state-of-the-art network on hand-object reconstruction. HFL-Net jointly estimates a hand mesh and 6D pose of an object from the input image via attention modules (please refer to [48] and Sect. 2 for more details).

In Fig. 10, we visualize the hand pose estimation results in PA-MPJPE for each grasp classes (left). The baseline achieves high accuracy across all grasp types with mean PA-MPJPE value of 5.67 mm, which is comparable to the state-of-the-art hand pose estimation results on other widely-used hand-object datasets [22,65,82,92]. This further verifies that the quality of our hand annotation is decent, which allows for effective learning of the downstream hand pose-related task. We found that three of the top four grasp classes with the highest

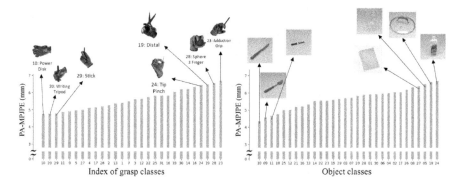

Fig. 10. Hand pose estimation results in PA-MPJPE (mm) (left) per grasp class and (right) object class.

Table 3. 6D object pose estimation results in ADD-0.1D per object class using HOGraspNet.

	ADD-0.1D(↑)		ADD-0.1D(↑)
1: cracker_box	88.79	16: golf_ball	46.27
2: potted_meat_can	59.47	17: credit_card	34.06
3: banana	58.21	18: dice	2.44
4: apple	75.74	19: disk_lid	99.29
5: wine_glass	97.13	20: smartphone	52.22
6: bowl	94.64	21: mouse	41.40
7: mug	72.04	22: tape	62.34
8: plate	99.42	23: master_chef_can	88.75
9: spoon	50.60	24: scrub_cleanser_bottle	89.80
10: knife	38.17	25: large_marker	34.59
11: small_marker	28.29	26: stapler	61.40
12: spatula	67.16	27: note	88.70
13: flat_screwdriver	63.52	28: scissors	54.34
14: hammer	82.99	29: foldable_phone	25.02
15: baseball	73.72	30: cardboard_box	77.80
Avg	63.61		

errors were not included in the DexYCB [8] and HO3D [28] datasets, respectively. This implies that our dataset's newly introduced real grasp poses might be challenging for the hand pose estimation model trained on existing datasets. The supplementary materials provide details of the missing grasp classes per dataset. In Fig. 10 (right), we also shows the hand pose metric per object classes. As expected, larger objects with more occlusion showed higher errors. Table 3 shows the object pose estimation results in ADD-0.1D per object class. We again achieve reasonable results that are comparable to the state-of-the-art object pose

Table 4. Cross-benchmark results on hand pose estimation using HFL-Net [48].

Train Set	Test Set	MPJPE (mm)	PA-MPJPE (mm)
HO3D [28]	DexYCB [8]	57.31	10.31
HOGraspNet	DexYCB [8]	**42.65**	**9.36**

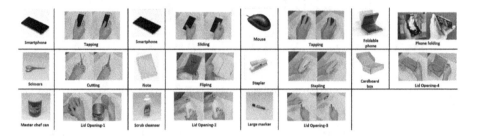

Fig. 11. Non-grasping action sequences in our dataset.

estimation performance on the other datasets [8,48], except for *Dice* class. As the *Dice* has small visible regions due to hand-object occlusions, increasing the ill-posedness of the pose estimation task. Additional results using other split protocols (S1-S4) can be found in the supplementary material.

4.4 Cross-Benchmark Results on Hand Pose Estimation

We additionally report the cross-validation results on hand pose estimation, following the experimental setup used in [83]. Since HFL-Net [48] is an object-aware network, we conducted experiments on samples with object classes that mutually exist in all the datasets to perform fair comparisons. In Table 4, the network trained on HOGraspNet achieves better estimation accuracy than the network trained on HO3D [28], indicating the comprehensiveness of HOGraspNet.

5 Conclusions

We have proposed a real RGB-D dataset, HOGraspNet, featuring comprehensive grasp labels. We have also presented the experimental results on grasp classification, and hand-object pose estimation. Our dataset captures diverse hand-object interactions involving 30 objects, 99 participants, and 90 interaction scenarios. It includes MANO [67] and HALO [40] 3D hand meshes, 3D keypoints, object meshes, contact maps, and grasp annotations for every sequence. The benchmark notably improves accuracy across datasets by a broader range of interaction scenarios compared to the existing datasets.

Limitations and Future Work. We aimed to incorporate various compound and articulated objects to capture dynamic actions. However, we currently treat them as rigid objects. Nevertheless, interaction actions like tapping, folding, and opening have already been recorded, as illustrated in Fig. 11. We plan to update the dataset with the object articulation annotations in the future. Furthermore, the dataset can be improved by including non-grasping actions such as pushing, throwing, squeezing, or deforming non-rigid objects like plastic bottles and sponges, which will be addressed as our future work.

Acknowledgement. This work was in part sponsored by NST grant (CRC 21011, MSIT), IITP grant (No. 2019-0-01270 and RS-2023-00228996, MSIT).

References

1. Arapi, V., Della Santina, C., Averta, G., Bicchi, A., Bianchi, M.: Understanding human manipulation with the environment: a novel taxonomy for video labelling. IEEE Robot. Autom. Lett. **6**(4), 6537–6544 (2021)
2. Bhatnagar, B.L., Xie, X., Petrov, I., Sminchisescu, C., Theobalt, C., Pons-Moll, G.: Behave: dataset and method for tracking human object interactions. In: CVPR (2022)
3. Brahmbhatt, S., Ham, C., Kemp, C.C., Hays, J.: ContactDB: analyzing and predicting grasp contact via thermal imaging. In: CVPR (2019)
4. Brahmbhatt, S., Tang, C., Twigg, C.D., Kemp, C.C., Hays, J.: ContactPose: a dataset of grasps with object contact and hand pose. In: ECCV (2020)
5. Calli, B., Singh, A., Walsman, A., Srinivasa, S., Abbeel, P., Dollar, A.M.: The YCB object and model set: towards common benchmarks for manipulation research. In: ICAR (2015)
6. Cao, Z., Radosavovic, I., Kanazawa, A., Malik, J.: Reconstructing hand-object interactions in the wild. In: ICCV (2021)
7. Caramalau, R., Bhattarai, B., Kim, T.K.: Active learning for Bayesian 3D hand pose estimation. In: Proceedings of the IEEE/CVF Winter Conference on Applications of Computer Vision, pp. 3419–3428 (2021)
8. Chao, Y.W., et al.: DexYCB: a benchmark for capturing hand grasping of objects. In: CVPR (2021)
9. Chen, L.C., Papandreou, G., Schroff, F., Adam, H.: Rethinking atrous convolution for semantic image segmentation. arXiv preprint arXiv:1706.05587 (2017)
10. Chen, Y., et al.: Joint hand-object 3D reconstruction from a single image with cross-branch feature fusion. TIP (2021)
11. Chen, Z., Chen, S., Schmid, C., Laptev, I.: gSDF: geometry-driven signed distance functions for 3D hand-object reconstruction. In: CVPR (2023)
12. Chen, Z., Hasson, Y., Schmid, C., Laptev, I.: AlignSDF: pose-aligned signed distance fields for hand-object reconstruction. In: ECCV (2022)
13. Cho, W., Park, G., Woo, W.: Tracking an object-grabbing hand using occluded depth reconstruction. In: ISMAR-Adjunct (2018)
14. Cho, W., Park, G., Woo, W.: Bare-hand depth inpainting for 3D tracking of hand interacting with object. In: ISMAR (2020)
15. Cini, F., Ortenzi, V., Corke, P., Controzzi, M.: On the choice of grasp type and location when handing over an object. Sci. Robot. **4**(27), eaau9757 (2019)
16. Corona, E., Pumarola, A., Alenya, G., Moreno-Noguer, F., Rogez, G.: Ganhand: predicting human grasp affordances in multi-object scenes. In: CVPR (2020)

17. Damen, D., et al.: Rescaling egocentric vision: collection, pipeline and challenges for epic-kitchens-100. IJCV (2022)
18. Doosti, B., Naha, S., Mirbagheri, M., Crandall, D.J.: Hope-net: a graph-based model for hand-object pose estimation. In: CVPR (2020)
19. Fan, Z., et al.: ARCTIC: a dataset for dexterous bimanual hand-object manipulation. In: CVPR (2023)
20. Feix, T., Romero, J., Schmiedmayer, H.B., Dollar, A.M., Kragic, D.: The grasp taxonomy of human grasp types. IEEE Trans. Hum.-Mach. Syst. **46**(1), 66–77 (2015)
21. Fieraru, M., Zanfir, M., Oneata, E., Popa, A.I., Olaru, V., Sminchisescu, C.: Three-dimensional reconstruction of human interactions. In: CVPR (2020)
22. Fu, Q., Liu, X., Xu, R., Niebles, J.C., Kitani, K.M.: Deformer: dynamic fusion transformer for robust hand pose estimation. arXiv preprint arXiv:2303.04991 (2023)
23. Garcia-Hernando, G., Johns, E., Kim, T.K.: Physics-based dexterous manipulations with estimated hand poses and residual reinforcement learning. In: IROS (2020)
24. Garcia-Hernando, G., Yuan, S., Baek, S., Kim, T.K.: First-person hand action benchmark with RGB-D videos and 3d hand pose annotations. In: CVPR (2018)
25. Gomez-Donoso, F., Orts-Escolano, S., Cazorla, M.: Large-scale multiview 3D hand pose dataset. IVC (2019)
26. Goyal, M., Modi, S., Goyal, R., Gupta, S.: Human hands as probes for interactive object understanding. In: CVPR (2022)
27. Grady, P., Tang, C., Twigg, C.D., Vo, M., Brahmbhatt, S., Kemp, C.C.: ContactOpt: optimizing contact to improve grasps. In: CVPR (2021)
28. Hampali, S., Rad, M., Oberweger, M., Lepetit, V.: Honnotate: a method for 3D annotation of hand and object poses. In: CVPR (2020)
29. Hampali, S., Sarkar, S.D., Rad, M., Lepetit, V.: Keypoint transformer: solving joint identification in challenging hands and object interactions for accurate 3d pose estimation. In: CVPR (2022)
30. Hassan, M., Choutas, V., Tzionas, D., Black, M.J.: Resolving 3D human pose ambiguities with 3D scene constraints. In: ICCV (2019)
31. Hasson, Y., Tekin, B., Bogo, F., Laptev, I., Pollefeys, M., Schmid, C.: Leveraging photometric consistency over time for sparsely supervised hand-object reconstruction. In: CVPR (2020)
32. Hasson, Y., Varol, G., Schmid, C., Laptev, I.: Towards unconstrained joint hand-object reconstruction from RGB videos. In: 3DV (2021)
33. Hasson, Y., et al.: Learning joint reconstruction of hands and manipulated objects. In: Proceedings of the IEEE/CVF Conference on Computer Vision and Pattern Recognition, pp. 11807–11816 (2019)
34. Hu, H., Yi, X., Zhang, H., Yong, J.H., Xu, F.: Physical interaction: reconstructing hand-object interactions with physics. In: SIGGRAPH Asia (2022)
35. Huang, C.H.P., et al.: Capturing and inferring dense full-body human-scene contact. In: CVPR (2022)
36. Huang, Y., Taheri, O., Black, M.J., Tzionas, D.: InterCap: joint markerless 3D tracking of humans and objects in interaction from multi-view RGB-D images. IJCV (2024)
37. Jiang, N., et al.: Full-body articulated human-object interaction. In: ICCV (2023)
38. Joo, H., et al.: Panoptic studio: a massively multiview system for social motion capture. In: ICCV (2015)

39. Joo, H., Neverova, N., Vedaldi, A.: Exemplar fine-tuning for 3D human pose fitting towards in-the-wild 3D human pose estimation. In: 3DV (2020)
40. Karunratanakul, K., Spurr, A., Fan, Z., Hilliges, O., Tang, S.: A skeleton-driven neural occupancy representation for articulated hands. In: 3DV (2021)
41. Kwon, T., Tekin, B., Stühmer, J., Bogo, F., Pollefeys, M.: H2O: two hands manipulating objects for first person interaction recognition. In: ICCV (2021)
42. Lee, J., Saito, S., Nam, G., Sung, M., Kim, T.K.: InterHandGen: two-hand interaction generation via cascaded reverse diffusion. In: Proceedings of the IEEE/CVF Conference on Computer Vision and Pattern Recognition, pp. 527–537 (2024)
43. Lee, J., Sung, M., Choi, H., Kim, T.K.: Im2hands: learning attentive implicit representation of interacting two-hand shapes. In: CVPR (2023)
44. Leroy, V., Weinzaepfel, P., Brégier, R., Combaluzier, H., Rogez, G.: SMPLy benchmarking 3D human pose estimation in the wild. In: 3DV (2020)
45. Li, M., et al.: Interacting attention graph for single image two-hand reconstruction. In: CVPR (2022)
46. Lin, K., Wang, L., Liu, Z.: Mesh graphormer. In: ICCV (2021)
47. Lin, P., et al.: HandDiffuse: generative controllers for two-hand interactions via diffusion models. In: CoRR, vol. abs/2312.04867 (2023)
48. Lin, Z., Ding, C., Yao, H., Kuang, Z., Huang, S.: Harmonious feature learning for interactive hand-object pose estimation. In: CVPR (2023)
49. Liu, J., Feng, F., Nakamura, Y.C., Pollard, N.S.: A taxonomy of everyday grasps in action. In: 2014 IEEE-RAS International Conference on Humanoid Robots, pp. 573–580. IEEE (2014)
50. Liu, S., Jiang, H., Xu, J., Liu, S., Wang, X.: Semi-supervised 3D hand-object poses estimation with interactions in time. In: CVPR (2021)
51. Loper, M., Mahmood, N., Romero, J., Pons-Moll, G., Black, M.J.: SMPL: a skinned multi-person linear model. ACM TOG (2015)
52. Lugaresi, C., et al.: MediaPipe: a framework for building perception pipelines. arXiv preprint arXiv:1906.08172 (2019)
53. Van der Maaten, L., Hinton, G.: Visualizing data using t-SNE. J. Mach. Learn. Res. (2008)
54. Mehta, D., et al.: Single-shot multi-person 3D pose estimation from monocular RGB. In: 3DV (2018)
55. Mescheder, L., Oechsle, M., Niemeyer, M., Nowozin, S., Geiger, A.: Occupancy networks: learning 3D reconstruction in function space. In: CVPR (2019)
56. Moon, G., et al.: A dataset of relighted 3d interacting hands. In: NeurIPS (2024)
57. Moon, G., Yu, S.I., Wen, H., Shiratori, T., Lee, K.M.: InterHand2.6M: a dataset and baseline for 3D interacting hand pose estimation from a single RGB image. In: ECCV (2020)
58. Mueller, F., Mehta, D., Sotnychenko, O., Sridhar, S., Casas, D., Theobalt, C.: Real-time hand tracking under occlusion from an egocentric RGB-D sensor. In: ICCV (2017)
59. Park, G., Kim, T.K., Woo, W.: 3D hand pose estimation with a single infrared camera via domain transfer learning. In: ISMAR (2020)
60. Patel, P., Huang, C.H.P., Tesch, J., Hoffmann, D.T., Tripathi, S., Black, M.J.: AGORA: avatars in geography optimized for regression analysis. In: CVPR (2021)
61. Pavlakos, G., et al.: Expressive body capture: 3D hands, face, and body from a single image. In: CVPR (2019)
62. Pavlakos, G., Shan, D., Radosavovic, I., Kanazawa, A., Fouhey, D., Malik, J.: Reconstructing hands in 3D with transformers. In: CVPR (2024)

63. Pumarola, A., Sanchez, J., Choi, G., Sanfeliu, A., Moreno-Noguer, F.: 3DPeople: modeling the geometry of dressed humans. In: ICCV (2019)
64. Qian, C., Sun, X., Wei, Y., Tang, X., Sun, J.: Realtime and robust hand tracking from depth. In: CVPR (2014)
65. Qu, W., et al.: Novel-view synthesis and pose estimation for hand-object interaction from sparse views. In: ICCV (2023)
66. Ranjan, A., Bolkart, T., Sanyal, S., Black, M.J.: Generating 3D faces using convolutional mesh autoencoders. In: Proceedings of the European Conference on Computer Vision (ECCV) (2018)
67. Romero, J., Tzionas, D., Black, M.J.: Embodied hands: modeling and capturing hands and bodies together. ACM TOG (2017)
68. Simon, T., Joo, H., Matthews, I., Sheikh, Y.: Hand keypoint detection in single images using multiview bootstrapping. In: CVPR (2017)
69. Sridhar, S., Oulasvirta, A., Theobalt, C.: Interactive markerless articulated hand motion tracking using RGB and depth data. In: ICCV (2013)
70. Stival, F., Michieletto, S., Cognolato, M., Pagello, E., Müller, H., Atzori, M.: A quantitative taxonomy of human hand grasps. J. Neuroeng. Rehabil. **16**, 1–17 (2019)
71. Sun, Y., Liu, W., Bao, Q., Fu, Y., Mei, T., Black, M.J.: Putting people in their place: monocular regression of 3D people in depth. In: CVPR (2022)
72. Swamy, A., et al.: SHOWMe: benchmarking object-agnostic hand-object 3D reconstruction. In: ICCV (2023)
73. Taheri, O., Ghorbani, N., Black, M.J., Tzionas, D.: Grab: a dataset of whole-body human grasping of objects. In: ECCV 2020 (2020)
74. Tang, D., Jin Chang, H., Tejani, A., Kim, T.K.: Latent regression forest: structured estimation of 3D articulated hand posture. In: CVPR (2014)
75. Tekin, B., Bogo, F., Pollefeys, M.: H+O: unified egocentric recognition of 3D hand-object poses and interactions. In: CVPR (2019)
76. Tompson, J., Stein, M., Lecun, Y., Perlin, K.: Real-time continuous pose recovery of human hands using convolutional networks. ACM TOG (2014)
77. Tse, T.H.E., Zhang, Z., Kim, K.I., Leonardis, A., Zheng, F., Chang, H.J.: S2 contact: graph-based network for 3D hand-object contact estimation with semi-supervised learning. In: ECCV (2022)
78. Tzionas, D., Ballan, L., Srikantha, A., Aponte, P., Pollefeys, M., Gall, J.: Capturing hands in action using discriminative salient points and physics simulation. IJCV (2016)
79. Wang, J., et al.: RGB2Hands: real-time tracking of 3D hand interactions from monocular RGB video. ACM TOG (2020)
80. Wen, G., Xiaoyu, B., Xavier, A.P., Francesc, M.N.: Multi-person extreme motion prediction. In: CVPR (2022)
81. Xu, C., Cheng, L.: Efficient hand pose estimation from a single depth image. In: ICCV (2013)
82. Xu, H., Wang, T., Tang, X., Fu, C.W.: H2ONet: hand-occlusion-and-orientation-aware network for real-time 3D hand mesh reconstruction. In: CVPR (2023)
83. Yang, L., et al.: OakInk: a large-scale knowledge repository for understanding hand-object interaction. In: CVPR (2022)
84. Yang, L., Zhan, X., Li, K., Xu, W., Li, J., Lu, C.: CPF: learning a contact potential field to model the hand-object interaction. In: ICCV (2021)
85. Yin, Y., Guo, C., Kaufmann, M., Zarate, J., Song, J., Hilliges, O.: Hi4D: 4D instance segmentation of close human interaction. In: CVPR (2023)

86. Yu, Z., Yang, L., Chen, S., Yao, A.: Local and global point cloud reconstruction for 3D hand pose estimation. In: BMVC (2021)
87. Yuan, S., Ye, Q., Stenger, B., Jain, S., Kim, T.K.: BigHand2.2M benchmark: hand pose dataset and state of the art analysis. In: CVPR (2017)
88. Zhang, B., et al.: Interacting two-hand 3D pose and shape reconstruction from single color image. In: ICCV (2021)
89. Zhang, J., Jiao, J., Chen, M., Qu, L., Xu, X., Yang, Q.: 3D hand pose tracking and estimation using stereo matching. In: ICIP (2017)
90. Zhang, S., et al.: EgoBody: human body shape and motion of interacting people from head-mounted devices. In: ECCV (2022)
91. Zhang, X., et al.: Hand image understanding via deep multi-task learning. In: ICCV (2021)
92. Zheng, X., Wen, C., Xue, Z., Ren, P., Wang, J.: HaMuCo: hand pose estimation via multiview collaborative self-supervised learning. In: ICCV (2023)
93. Zheng, Y., et al.: Deepmulticap: performance capture of multiple characters using sparse multiview cameras. In: ICCV (2021)
94. Zimmermann, C., Argus, M., Brox, T.: Contrastive representation learning for hand shape estimation. In: GCPR (2021)
95. Zimmermann, C., Brox, T.: Learning to estimate 3D hand pose from single RGB images. In: ICCV (2017)
96. Zimmermann, C., Ceylan, D., Yang, J., Russell, B., Argus, M., Brox, T.: FreiHAND: a dataset for markerless capture of hand pose and shape from single RGB images. In: ICCV (2019)
97. Zuo, B., Zhao, Z., Sun, W., Xie, W., Xue, Z., Wang, Y.: Reconstructing interacting hands with interaction prior from monocular images. In: ICCV (2023)

Human Pose Recognition via Occlusion-Preserving Abstract Images

Saad Manzur[✉] and Wayne Hayes

University of California Irvine, Irvine, CA 92697, USA
{smanzur,whayes}@uci.edu

Abstract. Existing 2D-to-3D pose lifting networks suffer from poor performance in cross-dataset benchmarks. Although 2D keypoints joined by "stick-figure" limbs is the dominant trend, stick-figures do not preserve occlusion information that is inherent in an image, resulting in significant ambiguities that are ruled out when occlusion information is present. In addition, datasets with ground truth 3D poses are much harder to obtain in contrast to similar human annotated 2D datasets. To address these issues, we propose to replace stick figures with *abstract images*—figures with opaque limbs that preserve occlusion information while implicitly encoding joint locations. We then break down the pose estimation task into two stages: (1) Generating an abstract image from a real image, and (2) garnering the pose from the abstract image. Crucially, given the GT 3D keypoints for a particular pose, we can *synthesize* an arbitrary number of abstract images of the same pose as seen from arbitrary cameras, even without a part map. Given a *set* of 3D GT keypoints, this allows training of Stage (2) on an unlimited dataset without over-training, which in turn allows us to correctly interpret poses from arbitrary viewpoints not included in the original dataset. Additionally, our unlimited training of Stage 2 allows good generalizations across datasets, demonstrated through a significant improvement in cross-dataset benchmarks, while still showing competitive performance in same-dataset benchmark.

Keywords: 3D Human Pose Estimation · Cross-Dataset Generalization · Viewpoint Encoding

1 Introduction

Recent work on 3D human pose estimation from still images can be classified into two main groups: direct-from-image methods [17,18,26,27,32,33,47], and methods that use intermediate 2D/2.5D keypoints [4,5,7,10,11,16,20,42,45,50]. Keypointing has resulted in a number of methods that "lift" the 3D pose from the intermediate 2D keypoints and their associated "stick-figure" limbs (cf. Fig. 1).

Supplementary Information The online version contains supplementary material available at https://doi.org/10.1007/978-3-031-73007-8_18.

© The Author(s), under exclusive license to Springer Nature Switzerland AG 2025
A. Leonardis et al. (Eds.): ECCV 2024, LNCS 15140, pp. 304–321, 2025.
https://doi.org/10.1007/978-3-031-73007-8_18

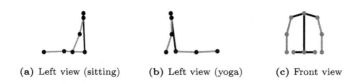

(a) Left view (sitting)   (b) Left view (yoga)   (c) Front view

Fig. 1. Poses that are indistinguishable without occlusion. Crucially, this implies that the "stick-figure"-to-pose mapping is, mathematically, not a 1-to-1 function.

Unfortunately, stick figures are "see-through"—they do not preserve occlusion information that is clearly visible in real images [26,46]. This introduces myriad problems, for example failing to correctly distinguish left-from-right viewpoints when limbs from opposite sides overlap (Figs. 1a and 1b). Current methods also perform poorly across datasets [10,39] because they depend on both camera viewing angle, and z-score normalization [16,23,30,32,36,37,45,50], which likely changes with the prior between datasets. Solving these problems requires us to (1) estimate the viewpoint accurately, (2) avoid discarding occlusion information, and (3) avoid camera-dependent metrics derived from the training set.

To simultaneously solve all of these problems, we propose training on synthetic "abstract" images of real human poses viewed from a virtually unlimited number of viewpoints. We address the viewpoint bias (Fig. 1) with a novel encoding that creates a mathematical 1-to-1 mapping between the camera viewpoint and the input image; a similar 1-to-1 encoding defines the pose. Both encodings support fully-convolutional training. As depicted in Fig. 2, our method uses the abstract image as input to two networks: one for viewpoint, and another for pose. At inference time, we (1) generate the abstract image from a real image, and (2) take the predicted viewpoint and pose from the abstract image to reconstruct the 3D pose. Since reconstruction does not ensure the correct forward facing direction of the subject, the ground-truth target pose will be related to the reconstructed pose by a simple rotation to compare with other methods.

A key observation is that the camera viewpoint as seen from the subject, and the subject's observed pose as seen from the camera, are *independent*: although they are intimately tied together in the sense that both are needed to fully reconstruct an abstract image—and thus a pose—they answer completely separate questions. More explicitly: (1) the location of the camera as viewed from the subject is completely independent of the subject's pose; and (2) the pose of the subject is completely independent of where the camera is located.

The "abstract-to-pose" part of our method (Stage 2) decomposes 3D human pose recognition into the above two orthogonal components: (1) the camera's location in subject-centered coordinates, and (2) the observed pose of the subject in camera coordinates. Note that identical three-dimensional poses from different viewpoints will change *both* answers, but combining the answers should always allow us to reconstruct the same subject-centered pose.

This, then, is our "secret sauce": by incorporating occlusion information, we can independently train two fully convolutional systems: one that learns

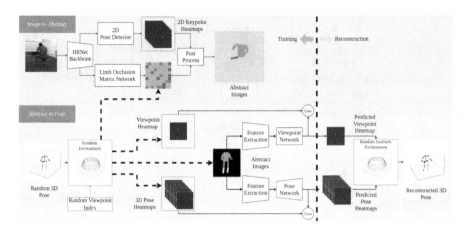

Fig. 2. Overview Dashed arrows indicate our custom supervision targets: limb occlusion matrix, viewpoint heatmap, 3D pose heatmap, and abstract images. **Stage 1** has two phases: first, the "image-to-abstract" network optimizes a binary cross-entropy loss for the limb occlusion matrix and MSE loss for the 2D keypoint heatmaps; the second draws the abstract image with limb occlusion matrix as a z-buffer and 2D keypoints as the core structure. **Stage 2**: generate abstract images (both flat and cube variants) from a random viewpoint and 3D pose pairs using the synthetic environment. The viewpoint and pose heatmaps—generated by the synthetic environment—are used as supervision targets for the abstract to viewpoint and pose networks. The "abstract to pose and viewpoint" network optimize the L2 loss on the output of viewpoint and pose network. The **Reconstruction** stage takes viewpoint and pose heatmaps from Stage 2, using a random synthetic environment to reconstruct the 3D pose.

a 1-to-1 mapping between images and the subject-centered camera viewpoint, and another that learns a 1-to-1 mapping between images and camera-centered pose. The final ingredient is to train these two CNN's using a virtually unlimited set of abstract images, with occlusion, generated from randomly chosen camera viewpoints observing the ground-truth 3D joint locations of real humans in real poses. Given a sufficiently large (synthetic) dataset of abstract images, we are able to independently train two CNNs that reliably encode the two 1-to-1 mappings. After the above CNNs are trained, the last ingredient we need for a true image-to-pose is a method to create an abstract image from a real input image. We outline our "image-to-abstract" method in Sects. 3.2 and 4 which shows competitive performance indicating adaptability in real-world scenario.

The key contributions of our paper are:

1. We replace "stick figures" with *abstract images* as the intermediate representation between images and poses. We represent limbs as opaque, solid, rectangular blocks that preserve occlusion and part-mapping. Using 2D/3D GT keypoints, we can generate *synthetic* abstract images from an unlimited number of camera viewpoints.

2. Novel viewpoint and pose encoding schemes, which facilitate learning a 1-to-1 mapping with input while preserving a spherical prior; and
3. We significantly improve state-of-the-art performance in cross-dataset benchmark without relying on dataset dependent normalization, and only marginal effects on same-dataset performance.

2 Related Work

3D Pose Estimation generally involves regression [4,7,10,16,20,23,42,45,50] ending with a fully-connected layer, or a voxel-based approach [24,26,27,46] with fully-convolutional supervision, the latter with a target space of size $w \times h \times d \times N$, where w is the width, h height, d depth, and N is the number of joints. However, position regression requires a training-set-dependent normalization (*e.g.* z-score) [16,23,30,32,36,37,45,50]. Both the graph convolution based approach [7,20,42,45,50] and hypothesis generation approach [16,30,37] rely on z-score normalization to improve same-, and most crucially, cross- dataset performance. To address missing depth information, Pavlakos *et al.* [26] include ordinal relations in training; Zhou *et al.* [46] use a heatmap triplet-based intermediate representation per part.

Part Based Approach. Kundu *et al.* [15] applies an unsupervised part-guided approach to 3D pose estimation. From an image, they generate part-segmentation with the help of intermediate 3D pose and a 2D part dictionary.

Viewpoint. Viewpoint estimation generally boils down to regressing some form of (θ, ϕ) [8], rotation matrix [35,49], or quaternions [39]. Regardless of the approach, everyone agrees on viewpoint estimation relative to the subject. However, relative subject rotation makes it harder to estimate viewpoint accurately.

Relation to Previous Work. We train on synthetically-generated abstract images of ground-truth 3D human poses. The abstract image contains opaque limbs that are uniquely color-coded (implicitly defining a part-map). We believe the abstract image contains the minimum information required to completely describe a human pose. Opaque 3D humanoid limbs have been used previously; for example their 3D bodies have been ray-traced as a means of determining occlusion [6]; but so far as we know such opaque limbs have never been used to train a CNN directly to estimate 3D pose, nor to augment image datasets by transforming 3D GT keypoints into an unlimited number of synthetic viewpoints (both used to aid training our Stage 2). Our own early tests showed that regression on 3D joint positions performs extremely badly across datasets when the same z-score parameters are used for both training and test sets, and improves only marginally if the normalization parameters are independently computed for both training and test sets (which is infeasible in the field, but is reported in Table 3 below). Conversely, voxel regression [27] presents a trade-off in performance vs. memory footprint as voxel resolution is increased. Our pose encoding (1) does not require training set dependent normalization, (2) takes much less memory than a voxel-based representation, and (3) being heatmap-based, it

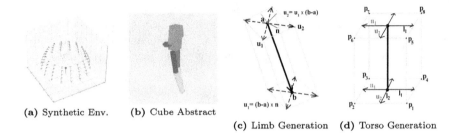

(a) Synthetic Env.   (b) Cube Abstract   (c) Limb Generation   (d) Torso Generation

Fig. 3. (a) Synthetic environment setup: Cameras are arranged spherically, all pointing to f (magenta dot near the center). (b) Blue-coloring the left forearm and femur allows easy identification of the subject's front—in contrast to "stick figures" that omit occlusion information, introducing ambiguity in determining the subject's "front". (c) Limb generation from a vector; (d) Torso generation from right and forward vectors. (Color figure online)

integrates well in a fully-convolutional setup. Finally, most methods encode the viewpoint using a rotation matrix, sine and cosines, or quaternions; all of these methods suffer from a discontinuous mapping at 2π. In contrast, our method avoids discontinuities by mapping both pose and viewpoint heatmaps to a cylinder that wraps around at 2π.

3 Method

3.1 Synthetic Environment

The environment we use to generate an unlimited supply of synthetic abstract images from any given pose consists of a room full of cameras all pointing to the same fixed point at the center of the room. We define $\mathbf{T} \in \mathbb{R}^{X \times Y \times 3}$, the translation/position of the cameras in X columns and Y rows. As shown in Fig. 3a, the fixed point is defined as, $f = \frac{c}{XY} \sum \mathbf{T}$, where the constant $c < 0.5$ defines the desired height of the "center"; each camera is related to the room via a rotation matrix, $\mathbf{R} \in \mathbb{R}^{X \times Y \times 3 \times 3}$. We compute the look vector as $l_{ij} = f - \mathbf{T}_{ij}$ for camera (i, j) and take a cross-product with $-\hat{z}$ as the up vector to compute the right vector r, all of which are fine-tuned to satisfy orthonormality by a series of cross-products. Refer to Sect. 4, for predefined values.

3.2 Abstract Shape Representation

Our abstract shapes come in two variants: Cube and Flat. Using a mixture of both helps the network learn the underlying pose structure without overfitting. We show later that the flat variant is easy to obtain from images. To ensure that occlusion information is clear in both variants, our robot's 8 limbs and torso use 9 easily-distinguishable, high-contrast colors (Fig. 3b).

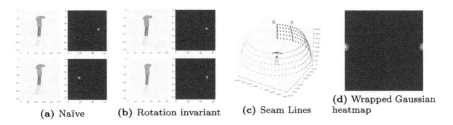

(a) Naïve (b) Rotation invariant (c) Seam Lines (d) Wrapped Gaussian heatmap

Fig. 4. (a) Naïve approach of encoding viewpoint. As can be seen, for a rotated subject, we have same image but different viewpoint encoding. (b) Rotation invariant approach makes sure we have the same encoding if the image is same even if the subject is rotated. (c) Seam lines after computing cosine distances. Camera indices (black=0, white=63), rotate with subject. Seam line A (red) is the original starting point of the indices. Seam line B (purple) is the new starting point consistent with subject's rotation. (d) A Gaussian heatmap warped horizontally. (Color figure online)

The Cube Variant has 3D limbs and torso formed by cuboids with orthogonal edges formed via appropriate cross-products; limbs (Fig. 3c) have a long axis (a to b) along the bone with a square cross-section, while the torso (Fig. 3d) is longest along the spine and has a rectangular cross-section. While the limb cuboid is generated from a single vector (a to b), the torso is generated by a body-centered coordinate system [39] (see Supplementary for details). All endpoints are compiled into a matrix $\mathbf{X}_{3D} \in \mathbb{R}^{3 \times N}$, where N is the number of vertices. We project these points to $\mathbf{X}_{2D} \in \mathbb{R}^{2 \times N}$ using the focal length f_{cam} and camera center c_{cam} (predefined for a synthetic room). Using the QHull algorithm [2], we compute the convex hull of the projected 2D points for each limb. We compute the Euclidean distance between each part's midpoint and the camera. Next, we iterate over the parts in order of longest distance, extract the polygon from hull points, and assign limb colors. While obtaining the 3D variant this way, we obtain a binary limb occlusion matrix for l limbs, $\mathbf{L} \in \mathbb{Z}^{l \times l}$, where each entry (u, v) determines whether limb u is occluding limb v if there is polygonal overlap above certain threshold.

The Flat Variant utilizes the limb occlusion matrix $\mathbf{L}$ and 2D keypoints $\mathbf{X}_{2D}$ to render the abstract image. $\mathbf{L}$ is used to topologically sort the order to render the limbs farthest to nearest. The limbs in this variant can be easily obtained by rendering a rectangle with the 2D endpoints forming a principal axis. If the rectangle area is small—for example if the torso is sideways or a limb points directly at the camera—we inflate to make the limbs more visible. We follow a similar approach while rendering the torso with four endpoints (two hips and two shoulders). The detailed algorithm is presented in the supplementary.

3.3 Viewpoint Encoding

Figure 4a shows a naive encoding of azimuth (θ) and elevation (ϕ), which can result in two viewpoints generating the same image—a serious problem because not even a neural network can circumvent the fact that the viewpoint and pose problems must be 1-to-1 mappings with the input image-*i.e.*, mathematically *injective*. Our solution preserves the 1-to-1 nature of poses to images (cf. Fig. 5b).

The matrix representation of our cylindrical coordinates wrap around at the *seam line*; crucially, we arrange for the viewpoint seam to always lie *behind* the subject (Fig. 4c), which ensures the coordinates on the matrix always stay in a fixed point related to the subject's orientation.

Formally, we compute the cosine distance between the subject's forward vector $\mathbf{F}_s$ projected onto xy-plane of the room $\mathbf{F}_{sp}$, and camera's forward vector $\mathbf{F}_c$ and place the seam line (index 0 and 63 of the matrix) directly behind the subject. Figure 4b reflects the improvement from Fig. 4a. Note for the same input, we have same viewpoint encoding now.

To learn a spherical mapping, we have to make the network understand the spherical positioning of the cameras. In general, a normal heatmap-based regression will clip the Gaussian at the border of the matrix. On the contrary, we allow the Gaussian heatmaps in the matrix to wrap around at the boundaries—corresponding to the seam line. Our Gaussian is

$$\mathcal{G}(x, y, \mu_x, \mu_y) = \exp^{-\frac{(x-\mu_x)^2+(y-\mu_y)^2}{2\sigma^2}}. \quad (1)$$

The associated heatmap is

$$\mathcal{H}^v[i,j] = \begin{cases} \mathcal{G}(j, i, \mu_x, \mu_y), & \text{if } |\mu_x - j| < W_k, \\ \mathcal{G}(j - I_w, i, \mu_x, \mu_y), & \text{if } |j - I_w - \mu_x| < W_k, \\ \mathcal{G}(j + I_w, i, \mu_x, \mu_y), & \text{if } |\mu_x - I_w - j| < W_k, \end{cases} \quad (2)$$

where (μ_x, μ_y) is the index of the viewpoint in our rotated synthetic room, I_w is the image size, and $W_k = 2\sigma + 1$ is the kernel width, with $\sigma = 1$. Algorithm 1 is used to rotate the camera indices in the synthetic room to ensure the camera position is consistent with the subject. This encodes the camera position in subject space and addition of Gaussian heatmap relaxes the area for network to optimize on (i.e. picking an almost approximate neighboring camera).

Algorithm 1: Rotate Camera Array

Data: S_e (Synthetic Environment), $\mathbf{F}_s$ (Subject Forward Vector)
Result: T', R'
1 $\mathbf{F}_c \leftarrow S_e$.camera_forwards;
2 $\mathbf{F}_{sp} \leftarrow \mathbf{F}_s - (\mathbf{F}_s \cdot \hat{z})\hat{z}$;
3 $D \leftarrow \mathbf{F}_c \cdot \mathbf{F}_{sp}$;
4 $S \leftarrow \arg\max D$;
5 $\mathcal{I} \leftarrow S_e$.original_index_array;
6 $\mathcal{I}_r \leftarrow$ rotate_index_array($\mathcal{I}, S$);
7 $T \leftarrow S_e$.camera_position;
8 $R \leftarrow S_e$.camera_rotation;
9 $T' \leftarrow T[\mathcal{I}_r]$;
10 $R' \leftarrow R[\mathcal{I}_r]$;

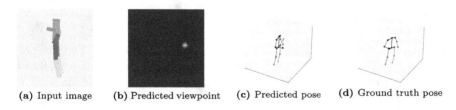

(a) Input image (b) Predicted viewpoint (c) Predicted pose (d) Ground truth pose

Fig. 5. A sample output from our network. (a) is the abstract image (cube variant) fed into the network. (b) the predicted viewpoint heatmap. (c) is the reconstructed pose from the pose and viewpoint heatmaps. The arrow shooting out the pose's left indicates the camera was left of the subject. (d) is the ground-truth 3D pose which is the reconstructed pose related with a rotation and the visible difference is due to camera viewpoint, not actual pose, which has error 37 mm.

3.4 Pose Encoding

We decompose the pose into bone vectors $\mathcal{B}_r$, and bone lengths B_r, both relative to parent joint. Let the synthetic environment's selected camera rotation matrix be R_{ij}, and $\mathcal{B}_{r_{ij}} = R_{ij}'\mathcal{B}_r$ be the bone vectors in R_{ij}'s coordinate space. Then, we normalize the spherical angles (θ, ϕ) of $\mathcal{B}_{r_{ij}}$ from range $[-2\pi, 2\pi]$ to range $[0, 127]$. Note that this encoding is not dependent on any normalization of the training- and by implication, is also independent of any normalization of the test set. We now have (θ, ϕ) normalized on a 128 × 128 grid. We take a similar approach to viewpoint encoding and allow the Gaussian heatmap generated around the matrix locations to wrap around at the boundaries. Note that in the viewpoint encoding, we only need to account for horizontal wrapping, whereas in pose we wrap both coordinates. For joint i and $k_1, k_2 \in [-\frac{W_k}{2}, \frac{W_k}{2})$,

$$\mathcal{H}_i^p[h, g] = \mathcal{G}(k_1, k_2, 0, 0) \tag{3}$$

where $h = \mu_y + k_2(\mathrm{mod} I_w)$ and $g = \mu_y + k_1(\mathrm{mod} I_w)$. Thus, we have another heatmap-based encoding for the pose. This encoding $\mathcal{H}^p \in \mathbb{R}^{128 \times 128 \times N}$, where N is the number of joints. Figure 4d shows a wrapped version of the heatmap.

3.5 Pose Reconstruction in Stage 2

Since the camera viewpoint is encoded in a subject-based coordinate system, the first step of pose reconstruction is to transform the camera's position from subject-centered coordinates to world coordinates. The mathematical details are in the Supplementary, but the upshot is shown in Fig. 5, which shows the unseen test output from our actual network. Specifically, in Fig. 5c, note how the reconstructed pose is rotated from the ground-truth pose in Fig. 5d. The arrow shooting out from the subject's left in Fig. 5c, indicates the relative position of the camera when the picture was taken.

4 Implementation

We calculated average bone lengths for the H36M training set [13]. The viewpoint was discretized into 24×64 indices, encoded to occupy rows $[21, 45]$ of a 64×64 heatmap matrix, giving angular resolution $5.625°$. Our synthetic environment uses fixed point scalar (cf. Sect. 3.1) of 0.4, with radius 5569 mm. (In principle, we could easily include cameras covering the entire sphere, for example to account for images of astronauts floating in the ISS viewed from *any* angle.) The pose was normalized to fall into the range $[0, 128]$ to occupy a $13 \times 128 \times 128$ matrix, with 13 being the number of limbs since we have 14 joints. For simplicity, we followed similar network architecture for all of the networks. We always use HRNet [31] pretrained on MPII [1], and COCO [19] for feature extraction.

The "image-to-abstract" network outputs 2D keypoint heatmaps and a binary limb occlusion matrix of dimension 9×9. We keep the original architecture for the 2D pose prediction and add a separate branch with a series of three interleaved convolution and batch normalization block with kernel size 3×3 to reduce the resolution. The reduced output is flattened and passed through two fully connected blocks to output the limb occlusion matrix as an 81D vector. We optimized the mean squared error loss on the 2D heatmaps and binary cross-entropy loss for the limb occlusion matrix.

The pose network consists two Convolution and Batch Normalization block pairs, followed by a transposed Convolution to match the output size of 128×128. All the convolution blocks use a 3×3 kernel with padding and stride set to 1. The final transposed convolution uses stride 2 and outputs a $13 \times 128 \times 128$ size tensor. For viewpoint estimation, we apply only one Convolution and Batch Normalization pair on the output of HRNet. The final stage is a regular convolution block that shrinks the output channel to 1 and outputs a $1 \times 64 \times 64$ size tensor. Since our target is a heatmap, we applied the standard L2 loss.

All training used batch size 64, with Adam [14] as our optimizer using Cosine Annealing with warm restart [21] and a learning rate warming from 1×10^{-9} to 1×10^{-3}. The viewpoint network ran for 200 epochs (2 days), and the pose network ran for 300 epochs (4 days), both on an RTX 3090, though it was stopped early due to convergence. When training the "abstract to pose and viewpoint" network, on every epoch we pick a random set of camera indices and render an abstract image from one of the two variants with equal probability. Thus, no two epochs are the same. This randomness minimizes overfitting—in turn helping with generalization—though it results in longer time-to-convergence.

5 Experiments

After we explain the datasets, metrics and models used in Sect. 5.1, we first show the results from our method under same dataset benchmark in Sect. 5.2. Then, we shift our focus to generalization capability across datasets in Sect. 5.3. We also report qualitative results in Sect. 5.4 and reflect on other experiments in Sect. 5.5.

5.1 Datasets, Metrics and Models

Datasets. We train our HRNet backbone framework on both MPII [1] and COCO [19] first on 2D keypoint prediction task. For the rest of the task, we only train on Human3.6M dataset [13]. We report cross-dataset results on the test sets of Geometric Pose Affordance Dataset (GPA) [38], 3D Poses in the Wild Dataset [22], and SURREAL Dataset [34] dataset. For the details of the dataset please refer to [39].

Evaluation Metrics. We report Mean Per Joint Position Error (MPJPE) in millimeters, which we call Protocol #1 and MPJPE after Procrustes Alignment (PA-MPJPE) as Protocol #2, following convention. Since the reconstructed pose is related with the ground truth by rotation only, we report it under Protocol #1. Further, PA-MPJPE reduces the error since the reconstruction uses preset bone-lengths. This becomes important in cross-dataset benchmarks.

Models. First, we train our "abstract-to-pose" network with a uniform mixture of Cube and Flat variant of the abstract images, which we will refer to as **"Model (Hybrid)"**. We also trained the same network for fewer epochs on the Cube variant separately which we will refer to as **"Model (Cube)"**. The former is tuned to handle both the flat and cube variant of the abstract image. Since we get the "Flat Abstract Image" from the real world image through the Stage 1 of our approach, the hybrid model can reconstruct the pose from this input. However, the Flat variant is a simplified version of the Cube variant, which is more robust and preserve occlusion. Pitting these two models in contrast to each other helps us further assess the capabilities of both variants and models.

5.2 Conventional Comparisons - Same Dataset

During training Stage 1, "image to abstract", we fine-tuned the trained HRNet network with H36M's training set along with limb ordering annotations. In Stage 2, "abstract to pose and viewpoint", we take each GT pose and pair it with the corresponding GT synthetically generated abstract image from a randomly chosen synthetic viewpoint taken from our 64×24 synthetic cameras; for testing, we use only poses and cameras from the real dataset.

Results are in Table 1 and Table 2. Our method is competitive with both image (Stage 1 & 2) and abstract image (Stage 2 only) as input. Since the latter is comparable to methods using GT 2D keypoints as input, we compare against such methods when available. We believe this is a fair comparison with our work, since the GT 2D keypoints are projected from 3D joints. During reconstruction, we always use a preset bone-length. PA-MPJPE score on Table 2, which includes rigid transformation, accounts for bone-length variation and reduces the error even more. Note that, in both these table we do not claim the lead even though we lead in quite a few cases. Our focus is to prove viability of using abstract image as a replacement for 2D keypoints in lifting networks. Our method is comparable, and sometimes superior, across the board.

Table 1. Quantitative comparisions of MPJPE (Protocol #1) between the ground truth 3D pose and reconstructed 3D pose. Our method shows competitive performance when image is used as an input. Hybrid abstract-to-pose model is trained on mixed Flat and Cube abstract images. Note: the results of others, obtained from GT 2D keypoints, are marked with an asterisk. Multi-frame methods [12,29,41,48] not included.

Method	Dir.	Disc.	Eat	Greet	Phone	Photo	Pose	Purch.	Sit	SitD.	Smoke	Wait	WalkD.	Walk	WalkT.	Avg.
Moreno [25]	69.5	80.1	78.2	87	100.7	102.7	76	69.6	104.7	113.9	89.7	98.5	79.2	82.4	77.2	87.3
Chen [4]	71.6	66.6	74.7	79.1	70.1	93.3	67.6	89.3	90.7	195.6	83.5	71.3	85.6	55.7	62.5	82.7
Martinez [23]	51.8	56.2	58.1	59	69.5	78.4	55.2	58.1	74	94.6	62.3	59.1	65.1	49.5	52.4	62.9
Yang [40]	51.5	58.9	50.4	57	62.1	65.4	49.8	52.7	69.2	85.2	57.4	58.4	43.6	60.1	47.7	58.6
Sharma [30]	48.6	54.5	54.2	55.7	62.6	72	50.5	54.3	70	78.3	58.1	55.4	61.4	45.2	49.7	58
Zhao [45]	47.3	60.7	51.4	60.5	61.1	49.9	47.3	68.1	86.2	55	67.8	61	42.1	60.6	45.3	57.6
Pavlakos [26]	48.5	54.4	54.4	52	59.4	65.3	49.9	52.9	65.8	71.1	56.6	52.9	60.9	44.7	47.8	56.2
Ci [7]	46.8	52.3	44.7	50.4	52.9	68.9	49.6	46.4	60.2	78.9	51.2	50	54.8	40.4	43.3	52.7
Li [16]	43.8	48.6	49.1	49.8	57.6	61.5	45.9	48.3	62	73.4	54.8	50.6	56	43.4	45.5	52.7
Martinez [23]*	37.7	44.4	40.3	42.1	48.2	54.9	44.4	42.1	54.6	58	45.1	46.4	47.6	36.4	40.4	45.5
Zhao [45]*	37.8	49.4	37.6	40.9	45.1	41.4	40.1	48.3	50.1	42.2	53.5	44.3	40.5	47.3	39	43.8
Zhou [46]	34.4	42.4	36.6	42.1	38.2	39.8	34.7	40.2	45.6	60.8	39	42.6	42	29.8	31.7	39.9
Gong [10]*	-	-	-	-	-	-	-	-	-	-	-	-	-	-	-	38.2
Zhai [43]*	31.3	34.0	28.0	32.0	33.1	42.1	34.1	28.1	33.6	39.8	31.7	32.9	33.8	26.7	28.9	32.7
Ours (Hybrid)	40.7	52.1	43.8	48.2	46.9	54.4	49.1	51.4	55.6	65.2	47	49.5	44.1	42.1	42.3	48.8
Ours (Cube)	29.0	32.5	29.1	32.0	29.9	38.8	32.1	32.5	38.7	49.5	33.2	33.7	32.1	29.9	29.9	33.5

Table 2. Quantitative comparison of PA-MPJPE (Protocol #2) between the ground truth 3D pose and reconstructed 3D pose. We follow the same notation as Table 1. Lower is better.

Method	Dir.	Disc.	Eat	Greet	Phone	Photo	Pose	Purch.	Sit	SitD.	Smoke	Wait	WalkD.	Walk	WalkT.	Avg.
Moreno [25]	66.1	61.7	84.5	73.7	65.2	67.2	60.9	67.3	103.5	74.6	92.6	69.6	71.5	78	73.2	74
Martinez [23]	39.5	43.2	46.4	47	51	56	41.4	40.6	56.5	69.4	49.2	45	49.5	38	43.1	47.7
Li [16]	35.5	39.8	41.3	42.3	46	48.9	36.9	37.3	51	60.6	44.9	40.2	44.1	33.1	36.9	42.6
Ci [7]	36.9	41.6	38	41	41.9	51.1	38.2	37.6	49.1	62.1	43.1	39.9	43.5	32.2	37	42.2
Pavlakos [26]	34.7	39.8	41.8	38.6	42.5	47.5	38	36.6	50.7	56.8	42.6	39.6	43.9	32.1	36.5	41.8
Sharma [30]	35.3	35.9	45.8	42	40.9	52.6	36.9	35.8	43.5	51.9	44.3	38.8	45.5	29.4	34.3	40.9
Zhou [46]	21.6	27	29.7	28.3	27.3	32.1	23.5	30.3	30	37.7	30.1	25.3	34.2	19.2	23.2	27.9
Ours (Hybrid)	32.8	38.0	35.3	38.2	38.1	44.8	37.9	37.9	47.1	59.4	40.4	39.7	38.1	33.3	34.4	39.7
Ours (Cube)	23.8	27.6	25.5	27.1	26.1	34.5	27.0	28.7	34.4	46.8	28.9	29.0	28.4	24.4	23.9	29.1

5.3 Cross-Dataset Generalization

Our primary focus is improving cross-dataset performance through extensive training on synthetically generated images. For this experiment, we only consider the case where we train on H36M dataset without any *domain adaptation training* (*e.g.* [44]). To test generalization capabilities, we render the images from GPA, 3DPW, and SURREAL dataset. We report results obtained from all three variants of our models.

To our knowledge, Wang *et al.* [39], is the only work with an extensive cross-dataset analysis on 4 datasets. To add more contenders, we re-implemented methods from Martinez et al. [23] and Zhao et al. [45], since their code was available and easily adapted. Both rely on z-score normalization of the testing set separately from the training set—though we note that such a normalization is impossible "in the wild". Even with this advantage, our method still leads in most cases by 30%–40% in cross-dataset performance (cf. Table 3). Goel et al. [9] outperforms us in 3DPW dataset. We believe this can be attributed to training a transformer based network on two 3D pose datasets. Gong et al. [10] reported cross-dataset performance for few networks on 3DPW dataset in PA-MPJPE. As shown in Table 3, we outperform the other methods by about a factor of 2.

Table 3. Cross-Dataset results on GPA, 3DPW, SURREAL in MPJPE and PA-MPJPE. We show significant improvement across all dataset except 3DPW. Asterisk marks our own experiment. Note: Goel et al. [9] trained their network on two 3D pose datasets. All the other networks were trained on H36M.

Method	MPJPE				PA-MPJPE		
	H36M	GPA	3DPW	SURR.	GPA	3DPW	SURR.
Martinez [23]*	55.52	117.37	135.53	108.63			
Zhao [45]*	53.59	115.01	154.3	103.75			
Wang [39]	52	98.3	124.2	114			
Goel [9] (H36M+3DHP Training)	44.8	-	**70.0**	-			
Zhao [45]					-	152.3	-
Martinez [23]					-	145.2	-
ST-GCN [3] (1-Frame)					-	154.3	-
VPose [28] (1-Frame)					-	146.3	-
Zhao + Gong [10]					-	140	-
Martinez + Gong [10]					-	130.3	-
ST-GCN (1-Frame) + Gong [10]					-	129.7	-
VPose (1-Frame) + Gong [10]					-	129.7	-
Goel [9]					-	**44.5**	-
Ours (Hybrid)	40.01	99.43	106.27	80.13	70.1	71.39	59.06
Ours (Cube)	**33.52**	**92.31**	95.83	**65.62**	**69.48**	64.28	**51.53**

Fairness of Comparison. We choose to report all the results both obtained through GT 2D keypoints and images, because cross-dataset results are hard to obtain. Both Martinez et al. [23] and Zhao et al. [45] pass only 2D keypoints to the later 3D pose estimation phase. Note the PA-MPJPE reported by Gong et al. [10] is higher than the MPJPE score we obtained for Martinez et al. [23] and Zhao et al. [45], which shows that we have given them all possible advantages.

(a) Input Image (b) Abstract Image (c) Ground Truth (d) Prediction

Fig. 6. Qualitative results on H36M dataset. (Color figure online)

5.4 Qualitative Results

Figure 6 shows the qualitative performance of our network on H36M. We see viewpoint estimation indicated by the blue arrow on the second column of each test sample. This, indeed shows the accuracy and efficacy of our method on separating viewpoint from pose.

5.5 Ablative and Other Experiments

Number of Vertical Bins. Our synthetic environment has hundreds of cameras placed systematically in "levels" on the sphere, pointing inwards (cf. Sect. 3.1). For this study, we reduced the number of levels. Intuitively, decreasing the number of levels should reduce performance, which is what we see in Fig. 7a. Increase in number of vertical bins shows a global improvement in cross-dataset performance. In Figs. 7b and 7c, we have plotted the azimuth and elevation distribution across all datasets, demonstrating the generalization and augmentation ability of our approach. Notice how our approach that relies on random viewpoints, have an almost uniform distribution. In contrast, notice how majority of the datasets have more data points distributed at the level of the subject.

Angular Error vs Bin Size. With an $N_g \times N_g$ heatmap, we expect performance to decrease with decreasing N_g. From our experiments, the angular error goes

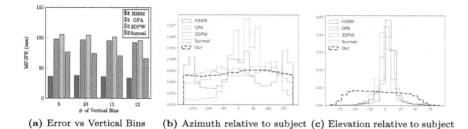

(a) Error vs Vertical Bins (b) Azimuth relative to subject (c) Elevation relative to subject

Fig. 7. (a) shows how the global cross-dataset error changes as the number of vertical bins is increased. (b) and (c) demonstrate how our synthetic training set using (unlimited in principle) synthetic viewpoints levels the distribution of trained viewpoints in (b) azimuth and (c) elevation. Notice that majority of the datasets have a relative distribution that is at the level of the subject, whereas ours follow a uniform distribution.

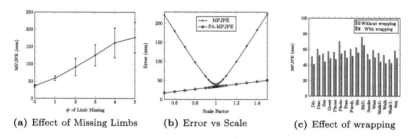

(a) Effect of Missing Limbs (b) Error vs Scale (c) Effect of wrapping

Fig. 8. (a) error (MPJPE) vs number of missing parts. As more parts are missing from the input image, error and uncertainty increases. (b) error as a function of scaled bone length in the synthetic environment. As expected, error increases as the scale deviates 1.0. (Note that PA-MPJPE applies affine transformation which compensates for scale difference.) (c) error with (blue) and without (red) wrapping. The errors are plotted across different actions on H36M dataset. Wrapping is clearly better. (Color figure online)

up as we decrease the grid resolution. We see the biggest jump from 16×16 ($13.35°$) to 32×32 ($9.24°$), with a smaller change when going to 64×64 ($8.44°$), implying diminishing returns at higher resolutions.

Limb Ablation: we randomly skip rendering a subset of the limbs, causing an expected increase in error and uncertainty (cf. Fig. 8a).

Variation in Reconstruction Scale. For this experiment, we change the bone lengths that are used to reconstruct the pose, deviating 50% scale up and down from the average bone lengths of an adult person. When comparing we keep the ground-truth at default scale (cf. Fig. 8b. Obviously, the reconstructed pose will not match in scale and the error will increase when deviating from the average bone lengths. However, at the bottom of the V-shaped curve the error deviation is small – which means, if the subject scale varies by a small amount, the error is negligible. As Procrustes alignment is performed first before measuring the error in PA-MPJPE, the curve forms a straight line and stays below the V curve.

Effect of Wrapping on Error. We observe the impact of wrapping on the error by computing pose estimation error on all the actions of H36M dataset with and without wrapping enabled. As can be seen in Fig. 8c, we achieve an average of 9.4 mm of improvement - from 58.2 mm (red) to 48.8 mm (blue). In addition, when we toggle wrapping in predicting viewpoint (*i.e.* naïve vs our viewpoint encoding), we see an angular errors are respectively 55.44° *vs.* 8.44°— an improvement factor of 6.6.

6 Conclusion

We have shown that using abstract images in the task of lifting the pose into 3D gives better cross-dataset results than using 2D keypoints, at negligible cost to same-dataset results. Moreover, abstract images also enable augmenting the "lifting" part of the framework to train from a virtually unlimited number of viewpoints. We have also presented an approach to obtain abstract image from real-world images, which shows the applicability of our approach.

References

1. Andriluka, M., Pishchulin, L., Gehler, P., Schiele, B.: 2D human pose estimation: new benchmark and state of the art analysis. In: IEEE Conference on Computer Vision and Pattern Recognition (CVPR) (2014)
2. Barber, C.B., Dobkin, D.P., Huhdanpaa, H.: The quickhull algorithm for convex hulls. ACM Trans. Math. Softw. **22**(4), 469–483 (1996). https://doi.org/10.1145/235815.235821
3. Cai, Y., et al.: Exploiting spatial-temporal relationships for 3D pose estimation via graph convolutional networks. In: 2019 IEEE/CVF International Conference on Computer Vision (ICCV), pp. 2272–2281 (2019). https://doi.org/10.1109/ICCV.2019.00236
4. Chen, C.H., Ramanan, D.: 3D human pose estimation= 2D pose estimation+ matching. In: Proceedings of the IEEE Conference on Computer Vision and Pattern Recognition, pp. 7035–7043 (2017)
5. Chen, C.H., et al.: Unsupervised 3D pose estimation with geometric self-supervision. In: Proceedings of the IEEE/CVF Conference on Computer Vision and Pattern Recognition, pp. 5714–5724 (2019)
6. Cheng, Y., Yang, B., Wang, B., Yan, W., Tan, R.T.: Occlusion-aware networks for 3D human pose estimation in video. In: Proceedings of the IEEE/CVF International Conference on Computer Vision (ICCV) (2019)
7. Ci, H., Wang, C., Ma, X., Wang, Y.: Optimizing network structure for 3D human pose estimation. In: Proceedings of the IEEE/CVF International Conference on Computer Vision, pp. 2262–2271 (2019)
8. Ghezelghieh, M.F., Kasturi, R., Sarkar, S.: Learning camera viewpoint using CNN to improve 3D body pose estimation. In: 2016 Fourth International Conference on 3D Vision (3DV), pp. 685–693 (2016). https://doi.org/10.1109/3DV.2016.75

9. Goel, S., Pavlakos, G., Rajasegaran, J., Kanazawa, A., Malik, J.: Humans in 4D: reconstructing and tracking humans with transformers. In: Proceedings of the IEEE/CVF International Conference on Computer Vision (ICCV), pp. 14783–14794 (2023)
10. Gong, K., Zhang, J., Feng, J.: PoseAug: a differentiable pose augmentation framework for 3D human pose estimation. In: Proceedings of the IEEE/CVF Conference on Computer Vision and Pattern Recognition, pp. 8575–8584 (2021)
11. Habibie, I., Xu, W., Mehta, D., Pons-Moll, G., Theobalt, C.: In the wild human pose estimation using explicit 2D features and intermediate 3D representations. In: Proceedings of the IEEE/CVF Conference on Computer Vision and Pattern Recognition, pp. 10905–10914 (2019)
12. Hu, W., Zhang, C., Zhan, F., Zhang, L., Wong, T.T.: Conditional directed graph convolution for 3D human pose estimation. In: Proceedings of the 29th ACM International Conference on Multimedia, pp. 602–611 (2021)
13. Ionescu, C., Papava, D., Olaru, V., Sminchisescu, C.: Human3.6M: large scale datasets and predictive methods for 3D human sensing in natural environments. IEEE Trans. Pattern Anal. Mach. Intell. **36**(7), 1325–1339 (2014). https://doi.org/10.1109/TPAMI.2013.248
14. Kingma, D.P., Ba, J.: Adam: a method for stochastic optimization. arXiv preprint arXiv:1412.6980 (2014)
15. Kundu, J.N., Seth, S., Jampani, V., Rakesh, M., Babu, R.V., Chakraborty, A.: Self-supervised 3D human pose estimation via part guided novel image synthesis. In: Proceedings of the IEEE/CVF Conference on Computer Vision and Pattern Recognition, pp. 6152–6162 (2020)
16. Li, C., Lee, G.H.: Generating multiple hypotheses for 3D human pose estimation with mixture density network. In: Proceedings of the IEEE/CVF Conference on Computer Vision and Pattern Recognition, pp. 9887–9895 (2019)
17. Li, S., Chan, A.B.: 3D human pose estimation from monocular images with deep convolutional neural network. In: Cremers, D., Reid, I., Saito, H., Yang, M.-H. (eds.) ACCV 2014. LNCS, vol. 9004, pp. 332–347. Springer, Cham (2015). https://doi.org/10.1007/978-3-319-16808-1_23
18. Li, S., Zhang, W., Chan, A.B.: Maximum-margin structured learning with deep networks for 3D human pose estimation. In: Proceedings of the IEEE International Conference on Computer Vision, pp. 2848–2856 (2015)
19. Lin, T.-Y., et al.: Microsoft COCO: common objects in context. In: Fleet, D., Pajdla, T., Schiele, B., Tuytelaars, T. (eds.) ECCV 2014. LNCS, vol. 8693, pp. 740–755. Springer, Cham (2014). https://doi.org/10.1007/978-3-319-10602-1_48
20. Liu, K., Ding, R., Zou, Z., Wang, L., Tang, W.: A comprehensive study of weight sharing in graph networks for 3D human pose estimation. In: Vedaldi, A., Bischof, H., Brox, T., Frahm, J.-M. (eds.) ECCV 2020. LNCS, vol. 12355, pp. 318–334. Springer, Cham (2020). https://doi.org/10.1007/978-3-030-58607-2_19
21. Loshchilov, I., Hutter, F.: SGDR: stochastic gradient descent with warm restarts. arXiv preprint arXiv:1608.03983 (2016)
22. von Marcard, T., Henschel, R., Black, M., Rosenhahn, B., Pons-Moll, G.: Recovering accurate 3D human pose in the wild using imus and a moving camera. In: European Conference on Computer Vision (ECCV) (2018)
23. Martinez, J., Hossain, R., Romero, J., Little, J.J.: A simple yet effective baseline for 3D human pose estimation. In: Proceedings of the IEEE International Conference on Computer Vision, pp. 2640–2649 (2017)
24. Mehta, D., et al.: VNect: real-time 3D human pose estimation with a single RGB camera. ACM Trans. Graph. (ToG) **36**(4), 1–14 (2017)

25. Moreno-Noguer, F.: 3D human pose estimation from a single image via distance matrix regression. In: Proceedings of the IEEE Conference on Computer Vision and Pattern Recognition, pp. 2823–2832 (2017)
26. Pavlakos, G., Zhou, X., Daniilidis, K.: Ordinal depth supervision for 3D human pose estimation. In: Proceedings of the IEEE Conference on Computer Vision and Pattern Recognition, pp. 7307–7316 (2018)
27. Pavlakos, G., Zhou, X., Derpanis, K.G., Daniilidis, K.: Coarse-to-fine volumetric prediction for single-image 3D human pose. In: Proceedings of the IEEE Conference on Computer Vision and Pattern Recognition, pp. 7025–7034 (2017)
28. Pavllo, D., Feichtenhofer, C., Grangier, D., Auli, M.: 3D human pose estimation in video with temporal convolutions and semi-supervised training. In: Proceedings of the IEEE/CVF Conference on Computer Vision and Pattern Recognition (CVPR) (2019)
29. Shan, W., et al.: Diffusion-based 3D human pose estimation with multi-hypothesis aggregation. In: Proceedings of the IEEE/CVF International Conference on Computer Vision, pp. 14761–14771 (2023)
30. Sharma, S., Varigonda, P.T., Bindal, P., Sharma, A., Jain, A.: Monocular 3D human pose estimation by generation and ordinal ranking. In: Proceedings of the IEEE/CVF International Conference on Computer Vision, pp. 2325–2334 (2019)
31. Sun, K., Xiao, B., Liu, D., Wang, J.: Deep high-resolution representation learning for human pose estimation. In: CVPR (2019)
32. Sun, X., Shang, J., Liang, S., Wei, Y.: Compositional human pose regression. In: Proceedings of the IEEE International Conference on Computer Vision, pp. 2602–2611 (2017)
33. Tekin, B., Katircioglu, I., Salzmann, M., Lepetit, V., Fua, P.: Structured prediction of 3D human pose with deep neural networks. arXiv preprint arXiv:1605.05180 (2016)
34. Varol, G., et al.: Learning from synthetic humans. In: CVPR (2017)
35. Wandt, B., Little, J.J., Rhodin, H.: ElePose: unsupervised 3d human pose estimation by predicting camera elevation and learning normalizing flows on 2D poses. In: Proceedings of the IEEE/CVF Conference on Computer Vision and Pattern Recognition (CVPR), pp. 6635–6645 (2022)
36. Wandt, B., Rosenhahn, B.: RepNet: weakly supervised training of an adversarial reprojection network for 3D human pose estimation. In: Proceedings of the IEEE/CVF Conference on Computer Vision and Pattern Recognition, pp. 7782–7791 (2019)
37. Wang, M., Chen, X., Liu, W., Qian, C., Lin, L., Ma, L.: DRPose3D: depth ranking in 3D human pose estimation. arXiv preprint arXiv:1805.08973 (2018)
38. Wang, Z., Chen, L., Rathore, S., Shin, D., Fowlkes, C.: Geometric pose affordance: 3D human pose with scene constraints. arXiv (2019)
39. Wang, Z., Shin, D., Fowlkes, C.C.: Predicting camera viewpoint improves cross-dataset generalization for 3D human pose estimation. In: Bartoli, A., Fusiello, A. (eds.) ECCV 2020. LNCS, vol. 12536, pp. 523–540. Springer, Cham (2020). https://doi.org/10.1007/978-3-030-66096-3_36
40. Yang, W., Ouyang, W., Wang, X., Ren, J., Li, H., Wang, X.: 3D human pose estimation in the wild by adversarial learning. In: Proceedings of the IEEE Conference on Computer Vision and Pattern Recognition, pp. 5255–5264 (2018)
41. Yu, B.X., Zhang, Z., Liu, Y., Zhong, S., Liu, Y., Chen, C.W.: GLA-GCN: global-local adaptive graph convolutional network for 3D human pose estimation from monocular video. In: Proceedings of the IEEE/CVF International Conference on Computer Vision, pp. 8818–8829 (2023)

42. Zeng, A., Sun, X., Huang, F., Liu, M., Xu, Q., Lin, S.: SRNet: improving generalization in 3D human pose estimation with a split-and-recombine approach. In: Vedaldi, A., Bischof, H., Brox, T., Frahm, J.-M. (eds.) ECCV 2020. LNCS, vol. 12359, pp. 507–523. Springer, Cham (2020). https://doi.org/10.1007/978-3-030-58568-6_30
43. Zhai, K., Nie, Q., Ouyang, B., Li, X., Yang, S.: HopFIR: hop-wise graphformer with intragroup joint refinement for 3D human pose estimation. In: Proceedings of the IEEE/CVF International Conference on Computer Vision (ICCV), pp. 14985–14995 (2023)
44. Zhang, X., Wong, Y., Kankanhalli, M.S., Geng, W.: Unsupervised domain adaptation for 3D human pose estimation. In: Proceedings of the 27th ACM International Conference on Multimedia, MM 2019, pp. 926–934. Association for Computing Machinery, New York (2019). https://doi.org/10.1145/3343031.3351052
45. Zhao, L., Peng, X., Tian, Y., Kapadia, M., Metaxas, D.N.: Semantic graph convolutional networks for 3D human pose regression. In: Proceedings of the IEEE/CVF Conference on Computer Vision and Pattern Recognition, pp. 3425–3435 (2019)
46. Zhou, K., Han, X., Jiang, N., Jia, K., Lu, J.: HEMlets pose: learning part-centric heatmap triplets for accurate 3D human pose estimation. In: Proceedings of the IEEE/CVF International Conference on Computer Vision, pp. 2344–2353 (2019)
47. Zhou, X., Sun, X., Zhang, W., Liang, S., Wei, Y.: Deep kinematic pose regression. In: Hua, G., Jégou, H. (eds.) ECCV 2016. LNCS, vol. 9915, pp. 186–201. Springer, Cham (2016). https://doi.org/10.1007/978-3-319-49409-8_17
48. Zhu, W., Ma, X., Liu, Z., Liu, L., Wu, W., Wang, Y.: MotionBERT: a unified perspective on learning human motion representations. In: Proceedings of the IEEE/CVF International Conference on Computer Vision, pp. 15085–15099 (2023)
49. Zimmermann, C., Brox, T.: Learning to estimate 3D hand pose from single RGB images. In: Proceedings of the IEEE International Conference on Computer Vision (ICCV) (2017)
50. Zou, Z., Tang, W.: Modulated graph convolutional network for 3D human pose estimation. In: Proceedings of the IEEE/CVF International Conference on Computer Vision, pp. 11477–11487 (2021)

DA-BEV: Unsupervised Domain Adaptation for Bird's Eye View Perception

Kai Jiang[1], Jiaxing Huang[2], Weiying Xie[1], Jie Lei[4], Yunsong Li[1], Ling Shao[3], and Shijian Lu[2(✉)]

[1] State Key Laboratory of Integrated Services Networks, Xidian University, Xi'an 710071, China
kjiang_19@stu.xidian.edu.cn
[2] S-Lab, School of Computer Science and Engineering, Nanyang Technological University, Singapore, Singapore
{Jiaxing.Huang,Shijian.Lu}@ntu.edu.sg
[3] UCAS-Terminus AI Lab, University of Chinese Academy of Sciences, Beijing, China
[4] School of Electrical and Data Engineering at the University of Technology Sydney, Ultimo, Australia

Abstract. Camera-only Bird's Eye View (BEV) has demonstrated great potential in environment perception in a 3D space. However, most existing studies were conducted under a supervised setup which cannot scale well while handling various new data. Unsupervised domain adaptive BEV, which effective learning from various unlabelled target data, is far under-explored. In this work, we design DA-BEV, the first domain adaptive camera-only BEV framework that addresses domain adaptive BEV challenges by exploiting the complementary nature of image-view features and BEV features. DA-BEV introduces the idea of query into the domain adaptation framework to derive useful information from image-view and BEV features. It consists of two query-based designs, namely, query-based adversarial learning (QAL) and query-based self-training (QST), which exploits image-view features or BEV features to regularize the adaptation of the other. Extensive experiments show that DA-BEV achieves superior domain adaptive BEV perception performance consistently across multiple datasets and tasks such as 3D object detection and 3D scene segmentation.

Keywords: Bird's eye view perception · Unsupervised domain adaptation · Self-training · Adversarial learning

---

K. Jiang and J. Huang—These authors contributed equally to this work.

Supplementary Information The online version contains supplementary material available at https://doi.org/10.1007/978-3-031-73007-8_19.

1 Introduction

3D visual perception [32,45] aims at sensing and understanding surrounding environments in 3D space, which plays an important role in various applications such as mobile robotics [3,20], autonomous driving [61,66], virtual reality [54,82], etc. Despite the remarkable progress of monocular and LiDAR-based 3D perception [4,11,12,19,29,34,58,60,71,75,76,85], camera-only 3D perception in Bird's Eye View (BEV) [28,37,39,40,70,73] has drawn increasing attention in recent years thanks to its advantages in comprehensive 3D understanding, rich semantic information, high computational efficiency, and low deployment cost. On the other hand, camera-only BEV models [63] trained over a source domain usually experience clear performance degradation when applied on a target domain due to clear cross-domain discrepancy as illustrated in Fig. 1.

Unsupervised Domain adaptation (UDA) [68] aims to address the cross-domain discrepancy by transferring a deep network model trained over a labelled source domain towards an unlabelled target domain. It has been extensively explored for various 2D computer vision tasks [16,46,80,84] that perceive environments in a 2D space. However, how to mitigate the cross-domain discrepancy for camera-only BEV perception is much more challenging but largely under explored.

We explore unsupervised domain adaptation for camera-only BEV perception, and we start from a simple observation: Camera-only BEV perception generally works in two stages by first encoding multi-camera images into image-view features separately and then fusing the image-view features with the corresponding camera configurations into BEV features. As shown in Fig. 2, BEV features can capture global 3D information over the BEV space but they usually suffer from more severe inter-domain discrepancy due to explicit changes of camera configurations beyond the change of image styles and appearance within the image-view features. As a comparison, the image-view features are encoded from images only and would not be explicitly affected by camera configurations, which suffer from less inter-domain discrepancy but capture largely local information within each single image and lack global 3D information. Hence,

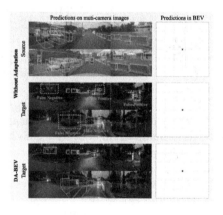

Fig. 1. Domain adaptive bird's eye view perception (DA-BEV). The first two rows show the detection by a source-only model (trained with source data with no adaptation) over a source scene and a target scene, respectively. The yellow 3D boxes indicate correct detection while the red dotted boxes highlight false-positive and false-negative detection. The third row shows the detection by our DA-BEV on the same target scene. The two columns visualize the 3D predictions in multi-camera view and bird's eye view, respectively.

one effective way of bridging the inter-domain discrepancy of camera-only BEV perception is by exploiting the complementary nature of image-view features and BEV features.

To this end, we design DA-BEV, the first domain adaptive camera-only BEV perception framework that addresses BEV domain adaptation challenges by exploiting the complementary nature of image-view features and BEV features. We introduce learnable queries to facilitate the interplay between image-view and BEV features while adapting them across domains. Specifically, the global 3D information from BEV features helps adapt image-view features while the 2D information in image-view features with less domain variation helps adapt BEV features. Following this philosophy, we design two query-based domain adaptation techniques including query-based adversarial learning (QAL) and query-based self-training (QST): QAL/QST exploits the useful information queried from image-view features or BEV features to regularize the "adversarial learning"/"self-training" of the another.

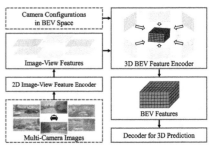

Fig. 2. The architecture of camera-only BEV models. Boxes in dash lines denote model inputs including multi-camera images and camera configurations. Boxes in solid lines stand for encoding/decoding processes and intermediate representations.

The proposed DA-BEV has three desirable features: 1) It introduces a novel query-based domain adaptation strategy that exploits the complementary nature of image-view features and BEV features for unsupervised BEV perception adaptation. The query-based domain adaptation strategy is well aligned with the architecture of camera-only BEV perception models; 2) It designs query-based adversarial learning and query-based self-training that achieve UDA by aligning inter-domain features and maximizing the target-domain likelihood, respectively. The two designs complement each other which jointly achieve robust unsupervised BEV perception adaptation effectively; 3) It is generic and can work for different camera-only BEV perception tasks such as 3D object detection and 3D scene segmentation.

The contributions of this work can be summarized in three major aspects. *First*, we explore unsupervised camera-only BEV perception adaptation and propose a query-based technique that mitigates the inter-domain discrepancy by exploiting the complementary nature of image-view features and BEV features. To the best of our knowledge, this is the first work that exploits the specific BEV network architecture for unsupervised BEV perception adaptation. *Second*, we design DA-BEV that introduces query-based adversarial learning and query-based self-training that jointly tackle domain adaptive BEV perception effectively. *Third*, extensive experiments show that DA-BEV achieves superior BEV perception adaptation consistently across different datasets and tasks such as 3D object detection and 3D scene segmentation.

2 Related Work

2.1 Camera-Only BEV Perception

Camera-only Bird's Eye View (BEV) perception [32,45] has recently attracted increasing attention thanks to its rich semantic information, high computational efficiency, and low deployment cost. It captures multi-view images of a scene and formulates unified BEV feature in ego-car coordinates for various 3D vision tasks such as 3D detection, map segmentation, etc. According to the way of view transformation, most existing camera-only BEV methods can be grouped into four categories [45], namely, homograph-based, depth-based, multilayer perception (MLP)-based, and transformer-based. Homograph-based methods [17,22,43,49,56] directly transform the multi-view image features into BEV features with explicitly constructed homography matrices. The recent are computationally efficient but exhibit limited performance in complex real-world scenarios. Depth-based methods [2,10,24–26,35,57,64,69,81] estimate depth distribution and apply it to generate BEV features from multi-view 2D features. MLP methods [21,33,47,50,52,74,86] learn a transformation between multi-view features and BEV features, aiming to relieve the inherent inductive biases under a calibrated camera setup. The recent transformer-based methods [6,8,9,13,28,37,39,40,48,51,55,62,67,70,72,73] design a set of BEV queries to retrieve the corresponding image features via cross-attention, ultimately constructing BEV features in a top-down manner. Differently, we explore unsupervised domain adaptation for camera-only BEV perception.

2.2 Domain Adaptation

Domain Adaptation aims to adapt source-trained models towards various target domains. Most existing work focuses on unsupervised domain adaptation (UDA) for 2D vision tasks, which minimizes the domain discrepancy by discrepancy minimization [41,59], adversarial training [18,44,59], or self-training [30,78,79]. However, most existing 2D UDA methods do not work well in 3D vision tasks that aim to understand objects and scenes in three-dimensional space. 3D UDA has been explored recently. For example, STMono3D [36] introduces self-training for cross-domain monocular 3D object detection. [5,53] and [1] exploit lidar and multimodal data in 3D UDA but they cannot work for camera-only BEV perception with extra data modality. Recently, M^2ATS [38] studies domain adaptation for multi-view 3D detection but still relies on depth supervision from Lidar. Unsupervised domain adaptation has been largely neglected for the task of camera-only BEV perception

3 Domain Adaptive BEV Perception

3.1 Preliminary

Task Definition. This work focuses on the task of unsupervised domain adaptation of camera-only BEV perception. It involves a labelled source domain

$D_S = \{x_S^n, k_S^n, y_S^n\}_{n=1}^{N_S}$ and an unlabelled target domain $D^T = \{x_T^n, k_T^n\}_{n=1}^{N_T}$, where x denotes a set of multi-camera images, k stands for the camera configurations, and y denotes the 3D object annotations in the BEV space as represented by x and k. The goal is to learn a BEV perception model that performs well on the unlabelled target domain.

A Revisit of BEV Models. Camera-only BEV model typically works in two stages as illustrated in Fig. 2. It first encodes multi-camera images into image-view features and then fuses these image-view features with the corresponding camera configurations to produce BEV features for 3D perception. Specifically, camera-only BEV model consists of a 2D image-view feature encoder $\mathcal{E}^{iv}$, a 3D BEV feature encoder $\mathcal{E}^{bev}$, a BEV feature decoder $\mathcal{D}^{bev}$, and a 3D detection head $\mathcal{H}_{det}$. Given a set of multi-camera images x and their corresponding camera configurations k, camera-only BEV model first encodes x into image-view features with the $\mathcal{E}^{iv}$ as follows:

$$f^{iv} = \mathcal{E}^{iv}(x), \tag{1}$$

The image features are then encoded with the corresponding camera configurations k into BEV features with the $\mathcal{E}^{bev}$:

$$f^{bev} = \mathcal{E}^{bev}(f^{iv}, k). \tag{2}$$

The BEV feature decoder $\mathcal{D}^{bev}$ then decodes the BEV features f^{bev} with a set of learnable BEV queries q^{bev}:

$$\tilde{q}^{bev} = \mathcal{D}^{bev}(q^{bev}, f^{bev}). \tag{3}$$

where the BEV queries q^{bev} interact with the BEV features f^{bev} and aggregate the 3D location information of objects in BEV, leading to the decoded query features $\tilde{q}^{bev}$. Given the decoded query features $\tilde{q}^{bev}$, 3D detection head $\mathcal{H}_{det}$ predicts 3D object bounding boxes p^{det} by:

$$p^{det} = \mathcal{H}_{det}(\tilde{q}^{bev}). \tag{4}$$

Under the setting of supervised learning, the camera-only BEV model can be formulated as follows:

$$L^{bev} = \mathcal{L}_{det}(p^{det}, y). \tag{5}$$

where $\mathcal{L}_{det}$ denotes a 3D detection loss function.

3.2 Framework of DA-BEV

Our method tackles BEV domain adaptation challenges by exploiting the complementary nature of image-view features and BEV features as illustrated in Fig. 3. Specifically, we design novel query-based domain adaptation that introduces learnable queries to enable interaction between the image-view and BEV features and as well as their synergetic adaptation. Intuitively, the global 3D

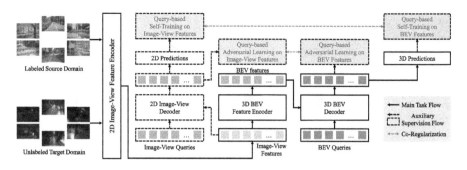

Fig. 3. The overall framework of DA-BEV, which exploits the complementary nature of image-view and BEV features for unsupervised BEV perception adaptation. To this end, DA-BEV first introduces an additional 2D Image-View Decoder into the BEV perception model to capture image-view features with local 2D information, which complement BEV features that capture rich global 3D information. The training of DA-BEV comprises two designs including query-based adversarial learning (QAL) and query-based self-training (QST). The former exploits the complementary information from image-view or BEV features to regularize the adversarial learning of the another, while the latter exploits the complementary information from both image-view and BEV features to regularize their self-training. Note, all the auxiliary supervision flows are utilized during network training but discarded after adaptation, which introduces slight computation overhead during model training but has little effect on testing.

information from BEV features helps adapt image-view features while the local 2D information in image-view features with less domain variation helps adapt BEV features. Based on this philosophy, we design two query-based domain adaptation techniques, namely, query-based adversarial learning (QAL) and query-based self-training (QST). QAL/QST exploits the useful information queried from image-view features or BEV features to regularize the "adversarial learning"/"self-training" of the another as illustrated in Fig. 3.

To capture 2D information in the image-view features with less inter-domain, we introduce an image-view feature decoder $\mathcal{D}^{iv}$ with a set of learnable image-view queries q^{iv}. The interaction between q^{iv} f^{iv} thus produces the image-view query features $\tilde{q}^{iv}$ as follows:

$$\tilde{q}^{iv} = \mathcal{D}^{iv}(q^{iv}, f^{iv}). \qquad (6)$$

The query feature $\tilde{q}^{iv}$ is then fed into a multi-label classification head $\mathcal{H}_{cls}$ to predict the probability of each object category, where $\mathcal{H}_{cls}$ is trained by a multi-label classification loss function L_{cls} as follows:

$$p^{cls} = \mathcal{H}_{cls}(\tilde{q}^{iv}), \quad L^{iv} = \mathcal{L}_{cls}(p^{cls}, y^{iv}), \qquad (7)$$

where y^{iv} denotes the image-view multi-label classification annotations. Note $\mathcal{L}_{cls}$ is computed for both source and target data. For source data x_S, k_S, y_S, we generate the image-view multi-label classification annotation y_S^{iv} by merging the object category labels in y_S. For target data x_T, k_T, we generate y_T^{iv} based on the model predictions as defined in Eq. 14.

Hence, the decoded image-view query features $\tilde{q}^{iv}$ captures rich 2D semantic and location information from the image-view features. They contain much less inter-domain discrepancy than the BEV features and can therefore benefit BEV feature adaptation greatly. In addition, the multi-label classification training objective of image-view query q^{iv} encourages capturing more balanced information among different categories, which facilitates the adaptation of BEV feature which are trained with one-hot object category annotation and thus often suffer from clear class imbalance issues [14].

To capture the global 3D information in BEV features, we directly employ the off-the-shelf BEV queries q^{bev} in Eq. 3, which interact with BEV features to generate the decoded BEV query features $\tilde{q}^{bev}$. As BEV features are encoded with camera configurations and q^{bev} is trained with 3D object annotations, the decoded BEV query feature $\tilde{q}^{bev}$ captures rich global 3D information including the object position in 3D BEV space. They can help the adaptation of image-view features that capture little global 3D information in BEV space.

3.3 Query-Based Adversarial Learning

Our proposed QAL exploits the useful information queried from image-view or BEV features to regularize the adversarial learning of the another. As illustrated in Fig. 3, QAL employs two domain classifiers to measure the inter-domain distances of the image-view query features and the BEV query features respectively, and utilizes the measured inter-domain distance for mutual regularization.

Hence, QAL simultaneously mitigates the inter-domain discrepancies of the local 2D information in image-view features and the global 3D information in BEV features, both being critical to locate and recognize objects and backgrounds in a 3D space. Besides, the adversarial learning of 2D image-view query features involves little 3D information, where the BEV query features can regularize it effectively by providing rich global 3D information, i.e., enhancing 2D image-view adversarial learning for the samples with significant 3D BEV inter-domain discrepancies while diminishing 2D image-view adversarial learning for the samples with minor 3D BEV inter-domain discrepancies. Similarly, the adversarial learning of 3D BEV query features can be regularized by the image-view query features capturing rich 2D semantic and location information.

Specifically, we employ domain classifiers $\mathcal{C}^{iv}$ and $\mathcal{C}^{bev}$ to measure the inter-domain distance of 2D image-view and 3D BEV features as follows:

$$\begin{aligned}\lambda^{iv} &= \exp\left(-\log\mathcal{C}^{iv}(\tilde{q}_s^{iv}) - \log\left(1 - \mathcal{C}^{iv}(\tilde{q}_t^{iv})\right)\right), \\ \lambda^{bev} &= \exp\left(-\log\mathcal{C}^{bev}(\tilde{q}_s^{bev}) - \log\left(1 - \mathcal{C}^{bev}(\tilde{q}_t^{bev})\right)\right),\end{aligned} \quad (8)$$

where $\tilde{q}_s^{iv}$ and $\tilde{q}_t^{iv}$ denote the decoded image-view query features in Eq. (6) that interacted with the source and target image-view features, respectively. $\tilde{q}_s^{bev}$ and $\tilde{q}_t^{bev}$ denote the decoded BEV query features in Eq. (3) that interacted with source and target domain BEV features, respectively.

The mutual regularization of QAL between image-view and BEV features can be formulated as follows,

$$L_{qal} = \lambda^{iv} \cdot \mathcal{L}_{al}(\tilde{q}_s^{iv}, \tilde{q}_t^{iv})) + \lambda^{bev} \cdot \mathcal{L}_{al}(\tilde{q}_s^{bev}, \tilde{q}_t^{bev}), \tag{9}$$

where $\mathcal{L}_{al}(\cdot)$ is a widely adopted adversarial learning loss function for cross-domain alignment [15].

3.4 Query-Based Self-training

Our proposed QST exploits the useful information queried from both image-view and BEV features to regularize their self-training, as shown in Fig. 3. Intuitively, the decoded image-view query features capture rich 2D semantic and location information that has less inter-domain discrepancy, while the decoded BEV query features capture rich global 3D information in the BEV space. Hence, the two types of features complement each other and together regularize self-training effectively with comprehensive 2D and 3D information.

Specifically, QST first leverages the predictions from image-view or BEV features to denoise the predictions from the other:

$$\hat{p}^{cls} = \max\left(p^{det}\right) \cdot p^{cls}, \quad \hat{p}^{det} = p^{det} \cdot p^{cls}, \tag{10}$$

where $p^{cls}/\hat{p}^{cls}$ and $p^{det}/\hat{p}^{det}$ denotes image-view and BEV predictions before/after cross-view denoising. The operation $\max(\cdot)$ stands for selecting the maximum probability in BEV predictions for each category. Please note we conduct the above cross-view denoising operations for each camera perspective, respectively.

Then, QST acquires global class distribution by accumulating the denoised predictions that capture comprehensive 2D and 3D information, and further exploit it to facilitate pseudo label generation. Specifically, we model the global class probability distribution as Gaussian distributions $\mathcal{N}(\mu_c, \sigma_c^2)$, where c denotes the category index. We estimate $\hat{\mu}_c$ and $\hat{\sigma}_c^2$ by maintaining a first-in-first-out queue of historical predictions:

$$\hat{\mu}_c = \frac{1}{B}\sum_{j=1}^{B} \hat{p}_c^{det}(j), \quad \hat{\sigma}_c^2 = \frac{1}{B-1}\sum_{j=1}^{B}\left(\hat{p}_c^{det}(j) - \hat{\mu}_c\right)^2, \tag{11}$$

where B denotes the queue length and $\hat{p}_c^{det}(j)$ stands for the j-th historical prediction in the queue. Note we use $\hat{p}^{det}$ only as it already includes the 2D information in $\hat{p}^{cls}$ after cross-denoising.

We accumulate $\hat{\mu}_c$ and $\hat{\sigma}_c^2$ along the training process for more stable estimation of $\mathcal{N}(\mu_c, \sigma_c^2)$:

$$\mu_c = \gamma \cdot \mu_c + (1-\gamma) \cdot \hat{\mu}_c, \quad \sigma_c^2 = \gamma \cdot \sigma_c^2 + (1-\gamma) \cdot \hat{\sigma}_c^2, \tag{12}$$

where γ denotes the Exponential Moving Average parameter for smooth update.

According to the estimated global class probability distribution, we select the top v percent predictions as the pseudo labels for each category. The pseudo label threshold τ_c can be determined by solving the following equation:

$$\frac{\Phi(1|\mu_c, \sigma_c) - \Phi(\tau_c|\mu_c, \sigma_c)}{\Phi(1|\mu_c, \sigma_c) - \Phi(0|\mu_c, \sigma_c)} = v, \tag{13}$$

where $\Phi(\cdot)$ refers to the cumulative density function of $\mathcal{N}(\mu_c, \sigma_c^2)$. Note we apply thresholds $\{\tau_c|c=1,...,C\}$ to both image-view and BEV predictions $\hat{p}^{cls}$ and $\hat{p}^{det}$ to acquire image-view and BEV pseudo labels $\hat{y}^{cls}$ and $\hat{y}^{det}$.

Our pseudo label generation method thus has three desirable features: 1) the thresholds are update-to-date as they are dynamically determined along the training process according to the complementary 2D and 3D information captured by image-view and BEV features; 2) it mitigates class imbalance issue by selecting the same percentage of pseudo labels for each class; 3) it works online and does not need an additional inference round as in [27,87].

The training loss of QST can thus be formulated by:

$$L_{qst} = \mathcal{L}_{cls}(p^{cls}, \hat{y}^{iv}) + \mathcal{L}_{det}(p^{det}, \hat{y}^{bev}). \tag{14}$$

3.5 Overall Objective

In summary, the overall training objective of the proposed DA-BEV can be formulated as follows:

$$\min_{\mathcal{E},\mathcal{D},\mathcal{H}} \max_{\mathcal{C}} L^{iv} + L^{bev} + L_{qst} + L_{qal}, \tag{15}$$

where $\mathcal{E} = \{\mathcal{E}^{2d}, \mathcal{E}^{3d}\}$, $\mathcal{D} = \{\mathcal{D}^{iv}, \mathcal{D}^{bev}\}$, $\mathcal{H} = \{\mathcal{H}_{cls}, \mathcal{H}_{det}\}$, $\mathcal{C} = \{\mathcal{C}^{iv}, \mathcal{C}^{bev}\}$.

4 Experiments

4.1 Datasets

We benchmark our DA-BEV extensively over various BEV adaptation scenarios including adaptation across different illumination conditions, different weather conditions, different cities, and different camera systems. The four BEV adaptation scenarios involve two 3D perception datasets including nuScenes [7] and Lyft [23].

nuScences [7] is a large-scale autonomous driving 3D dataset, which comprises 6-camera images captured from 1000 scenes, including the scenes with daytime/nighttime and "clear weather"/"rainy weather" conditions, and from Singapore/Boston cities. For 3D object detection, it includes 1.4M annotated 3D bounding boxes with 10 categories. For 3D scene segmentation [40,83], it contains 200K frames with fine-grained segmentation annotations in 3D space.

Lyft [23] is a self-driving dataset for 3D perception, which has a diverse set of cameras with distinct intrinsic and extrinsic parameters. It comprises 366 scenes with 1.3M 3D object detection annotations, which are captured from California, and under a different camera system as compared with nuScences.

With nuScences [7] and Lyft [23], the six BEV adaptation scenarios are formulated. We formulate "nuScences Daytime → nuScences Nighttime" for cross-illumination adaptation, "nuScences Clear Weather → nuScences Rainy Weather" for cross-weather adaptation, and "nuScences Singapore → nuScences Boston" for cross-city adaptation. Moreover, "Lyft → nuScences", "Lyft → nuScences Nighttime", and "Lyft → nuScences Rainy Weather" are formulated for adaptation across different camera configurations.

Table 1. Unsupervised domain adaptation of camera-only BEV perception across different illumination conditions, weather conditions, cities.

Daytime → Nighttime												
Methods	Average Precision↑										mAP↑	NDS↑
	Car	Truck	Cons. Vehicle	Bus	Trailer	Barrier	Motorcycle	Bicycle	Pedestrian	Cone		
PETR [39] (Baseline)	40.79	28.88	0.41	29.71	0.00	9.15	22.00	0.71	24.66	2.84	15.92	22.44
SFA [65]	40.70	27.05	0.00	30.48	0.00	11.04	21.52	0.42	24.21	4.69	16.01	23.95
MTTrans [77]	41.48	34.39	0.01	30.49	0.00	12.48	21.40	0.69	24.94	4.16	17.00	24.48
STM3D [36]	41.77	33.19	0.27	29.22	0.00	13.28	21.94	0.28	24.21	4.36	16.85	24.41
DA-BEV (Ours)	45.07	41.96	1.91	38.12	0.00	16.66	25.74	0.54	28.05	4.65	20.27	26.98
Clear Weather → Rainy Weather												
Methods	Average Precision↑										mAP↑	NDS↑
	Car	Truck	Cons. Vehicle	Bus	Trailer	Barrier	Motorcycle	Bicycle	Pedestrian	Cone		
PETR [39] (Baseline)	41.68	18.19	2.24	41.58	7.07	38.09	13.16	21.57	33.31	30.66	24.75	32.84
SFA [65]	42.23	18.87	2.12	49.21	6.26	37.61	11.11	21.59	33.09	31.78	25.39	33.18
MTTrans [77]	41.04	20.08	3.51	50.42	7.07	37.78	11.56	20.91	33.57	34.08	26.00	33.68
STM3D [36]	42.45	19.22	2.68	47.62	7.00	38.52	13.39	21.44	33.85	31.97	25.81	33.56
DA-BEV (Ours)	46.72	25.06	3.38	51.62	8.70	46.65	17.46	26.85	39.37	37.80	30.36	36.81
Singapore → Boston												
Method	Average Precision↑										mAP↑	NDS↑
	Car	Truck	Cons. Vehicle	Bus	Trailer	Barrier	Motorcycle	Bicycle	Pedestrian	Cone		
PETR [39] (Baseline)	41.29	8.51	0.00	29.84	0.00	11.23	4.06	5.80	28.59	35.80	16.51	21.74
SFA [65]	42.75	10.51	0.00	29.25	0.00	12.78	3.68	8.43	31.70	37.04	17.61	23.43
MTTrans [77]	42.68	11.24	0.00	29.97	0.00	13.91	3.42	8.60	32.56	38.01	18.04	24.69
STM3D [36]	43.44	10.69	0.00	28.29	0.00	13.03	3.83	8.75	32.60	39.50	18.01	24.68
DA-BEV (Ours)	45.35	14.69	0.00	31.68	0.00	19.58	6.63	9.78	34.20	42.12	20.40	28.35

4.2 Implementation Details

We implement our DA-BEV upon PETR [39] with backbone VoVNetV2 [31]. Following PETR, we adopt AdamW [42] optimizer with an initial learning rate of 1×10^{-4} and a weight decay of 0.01. All models are trained for 50k iterations on one 2080Ti GPU with a batch size of 2. For 3D object detection, the transformer decoder consist of 6 layers with 900 object queries. For 3D scene segmentation, we follow the settings in [40,83], where 625 segmentation queries along with a 6 layer transformer decoder are adopted for semantic segmentation result.

4.3 Main Results

Since there is little study on domain adaptive camera-only BEV, we compare DA-BEV with domain adaptation methods that achieved state-of-the-art performance in 2D object detection [65,77] and monocular 3D object detection [36].

These compared methods can be easily applied on domain adaptive camera-only BEV perception by directly introducing their adversarial learning and/or

Table 2. Unsupervised domain adaptation of camera-only BEV perception across different camera systems.

Lyft → nuScenes Nighttime								
Methods	Average Precision↑						mAP↑	NDS*↑
	Car	Truck	Bus	Motorcycle	Bicycle	Pedestrian		
PETR [39] (Baseline)	15.55	1.00	1.83	8.53	0.43	4.26	5.27	10.54
SFA [65]	16.50	0.78	1.45	8.56	0.44	4.41	5.35	10.41
MTTrans [77]	16.86	1.62	2.01	8.11	0.76	2.87	5.37	10.20
STM3D [36]	16.10	1.06	2.50	10.48	0.50	4.06	5.78	11.06
DA-BEV (Ours)	21.69	1.57	2.24	13.79	0.56	7.75	7.93	13.62
Lyft → nuScenes Rainy Weather								
Methods	Average Precision↑						mAP↑	NDS*↑
	Car	Truck	Bus	Motorcycle	Bicycle	Pedestrian		
PETR [39] (Baseline)	22.91	1.95	2.56	9.15	0.88	7.01	7.41	12.25
SFA [65]	25.52	1.92	2.00	9.95	0.55	9.88	7.97	13.19
MTTrans [77]	24.84	2.57	2.08	10.18	1.63	9.63	8.48	13.84
STM3D [36]	26.28	1.66	1.92	12.21	1.51	10.59	9.03	14.41
DA-BEV (Ours)	30.46	1.57	2.74	15.27	1.53	14.55	11.02	18.10
Lyft → nuScenes								
Methods	Average Precision↑						mAP↑	NDS*↑
	Car	Truck	Bus	Motorcycle	Bicycle	Pedestrian		
PETR [39] (Baseline)	28.43	3.70	4.30	12.88	0.65	10.15	10.02	16.44
SFA [65]	33.17	3.55	4.18	12.44	0.62	11.96	10.98	17.87
MTTrans [77]	33.19	3.98	4.00	13.03	0.75	11.06	11.00	17.98
STM3D [36]	36.51	4.34	4.63	15.96	0.51	14.93	12.81	18.58
DA-BEV (Ours)	38.45	3.93	4.23	23.53	0.75	22.85	15.62	24.78

Table 3. Ablation study of the proposed DA-BEV. The experiments are conducted on Daytime → Nighttime adaptation.

Methods	QAL			QST			mAP↑	NDS↑
	Image-View	BEV	Co-Regularization	Image-View	BEV	Co-Regularization		
PETR [39] (Baseline)	-	-	-	-	-	-	15.93	22.41
	✓						16.45	22.93
	✓	✓					16.96	23.21
	✓	✓	✓				17.80	24.07
	✓	✓	✓	✓			18.53	25.06
	✓	✓	✓	✓	✓		19.39	26.30
DA-BEV (Ours)	✓	✓	✓	✓	✓	✓	20.27	26.98

self-training techniques into camera-only BEV perception model. The comparisons are conducted over four challenging BEV adaptation scenarios as detailed in the following text.

Tables 1 reports unsupervised domain adaptation of camera-only BEV perception across different illuminations, weathers and cities, respectively. We can observe that DA-BEV achieves substantial improvements upon the baseline on these three adaptation scenarios, including 4.35%, 5.61% and 3.89% improve-

Table 4. Camera-only BEV perception of 3D scene segmentation on Daytime → Nighttime domain adaptation.

Method	IoU			mIoU
	Drive	Lane	Vehicle	
PETR [39] (Baseline)	39.33	25.91	18.78	28.01
DA-BEV (Ours)	51.37	30.43	24.98	35.59

ments in mAP, and 4.54%, 3.97% and 6.61% improvement in NDS. In addition, DA-BEV outperforms the state-of-the-art UDA methods clearly and consistently across these three adaptation scenarios. The superior BEV adaptation performance of DA-BEV is largely attributed to two factors: 1) DA-BEV introduces query-based domain adaptation that ingeniously exploits the complementary nature of image-view features and BEV features in BEV model, which helps tackle the domain gaps in BEV perception effectively. 2) DA-BEV introduces two query-based domain adaptation methods, i.e., QAL and QST, which complement each other and jointly achieve robust BEV perception adaptation effectively.

Table 2 reports unsupervised domain adaptation of camera-only BEV perception across different camera systems. Lyft and nuScences data are collected with very different camera systems that have clear domain gaps in terms of camera intrinsic parameters including camera focal lengths and field of views (FOVs) as well as camera extrinsic parameters such as camera poses and camera positions. It can be observed that DA-BEV outperforms both the baseline on three adaptation scenarios substantially and consistently in mAP and NDC metrics, including over 2.66%, 3.61% and 5.60% improvements in mAP and 3.08%, 5.85% and 8.34% improvements in NDS. These experiments demonstrate that DA-BEV can well handle the BEV adaptation across different camera systems that suffers from severe domain gaps from camera configuration inputs beyond those from image inputs. On the other hand, it shows that our DA-BEV outperforms both baselines and other methods clearly on "Lyft → nuScenes Nighttime" and "Lyft → nuScenes Rainy Weather", demonstrating the effectiveness and robustness of DA-BEV under the cross-domain changes of camera configurations beyond image styles and appearance.

4.4 Ablation Study

We conduct extensive ablation studies to investigate how different components in DA-BEV contribute. The experiments are conducted on Daytime → Nighttime adaptation. As Table 3 shows, *the baseline* does not perform well due to the domain gaps. As a comparison, including query-based adversarial learning on image-view features and/or BEV features improves the baseline clearly, indicating that QAL can simultaneously mitigate the inter-domain discrepancies in both 2D image-view features and 3D BEV features. In addition, introducing co-regularization between image-view and BEV query-based adversarial learning further improves the performance clearly, showing that image-view and BEV features encode complementary cross-domain discrepancy information that effectively regularize the adversarial learning of each other.

In addition, applying query-based self-training on image-view and/or BEV features brings clear improvements, indicating that QST helps learn effective 2D and 3D information of unlabeled target data from image-view and BEV, respectively. Moreover, the co-regularization of query-based self-training of image-view and BEV features brings further improvement, demonstrating that image-view and BEV features provide complementary and comprehensive target information that effectively regularizes their self-training. Further, the combination of QAL and QST performs the best clearly, showing that QAL and QST complement each other and jointly achieve robust unsupervised BEV perception adaptation effectively.

4.5 Discussion

Generalzation Across 3D Perception Tasks. We study how DA-BEV generalizes across 3D perception tasks by evaluating it over two representative tasks including 3D object detection and 3D scene segmentation. Experiments in Tables 1 and 4 show that DA-BEV achieves superior performance consistently over 3D detection and segmentation tasks, demonstrating the generalization ability of DA-BEV across different 3D perception tasks.

Parameter Studies. As defined in Eq. 13 parameter v controls how many predictions we select as the pseudo labels. We study v by changing it from 10% to 30% with a step of 5%. The experiments in the top part of Table 5 show that v does not affect domain adaptation much while it changes from 15% to 30%. The performance drops more when v is set as 10%, largely because the adaptation is biased toward the source when very limited pseudo labels of target samples are selected. As mentioned in Eq. 12, parameter γ controls the update speed of the estimated global class probability distribution. We study γ by changing it from 0.9 to 0.99999. The experiments in the bottom part of Table 5 show that DA-BEV is tolerant to parameter γ when it varies from 0.99 to 0.9999. The performance drops more when γ is set as 0.9 or 0.99999, largely because these two values lead to too fast or too slow update, resulting in unstable or nearly fixed estimation of global class probability distribution and degraded performance.

Generalzation Across Network Backbones. We study how DA-BEV generalizes across network backbones by evaluating it over three backbones of different sizes, including ResNet-50-C5 (46.9M), ResNet-50-P4 (46.9M) and VoV-P4 (96.9M). Results in Table 5 show that DA-BEV improves consistently over both small and large backbones, demonstrating the generalization ability of DA-BEV across different network backbones.

Generalization Across Different Baselines. We study how our DA-BEV generalizes across different baselines by evaluating it over three types of mainstream BEV perception solutions, including PETR [39], BEVFormer [73] and

Table 5. Left: Parameter analysis on Daytime → Nighttime adaptation. Right: Generalization across different backbones.

v	10%	15%	20%	25%	30%
mAP	18.24	19.83	20.27	20.15	19.57
γ	0.9	0.99	0.999	0.9999	0.99999
mAP	17.20	20.03	20.27	19.48	18.86

Backbone		R50-C5	R50-P4	VoV-P4
PETR [39] (Baseline)	mAP	10.63	11.61	15.93
DA-BEV (Ours)	mAP	13.03	14.22	20.27

Table 6. Generalization across different baselines.

Daytime → Nighttime											
Methods	Average Precision↑									mAP↑	NDS↑
	Car	Truck	Cons. Vehicle	Bus	Trailer	Barrier	Motorcycle	Bicycle	Pedestrian	Cone		
PETR [39] (Baseline)	40.79	28.88	0.41	29.71	0.00	9.15	22.00	0.71	24.66	2.84	15.92	22.44
DA-BEV (Ours)	45.07	41.96	1.91	38.12	0.00	16.66	25.74	0.54	28.05	4.65	20.27	26.98
BEVFormer [73] (Baseline)	39.31	27.98	0.18	26.95	0.00	11.04	21.84	0.61	26.50	2.96	15.73	21.97
DA-BEV (Ours)	41.74	36.41	0.08	36.38	0.00	19.47	24.27	0.40	28.93	3.79	19.15	24.14
CAPE [70] (Baseline)	41.68	27.95	0.18	29.57	0.00	11.49	23.01	0.98	24.85	3.20	16.29	23.54
DA-BEV (Ours)	44.81	39.91	0.19	38.68	0.00	17.27	25.98	0.65	28.09	4.93	20.05	26.06

CAPE [70]. For a fair comparison, all of them use VoVNetV2 [31] as the backbone, and only single frame information is used for BEV perception. Experimental results in Table 6 show that DA-BEV brings clear improvements consistently over various baselines, demonstrating the generalization ability of DA-BEV across different BEV perception solutions.

Computation Efficiency Analysis. DA-BEV introduces an additional image-view decoder into the backbone of detector during training, and this image-view decoder is discarded in inference during testing. We study how this additional image-view decoder affects training and testing efficiency by benchmarking DA-BEV with several other methods in training time, testing time, training memory and testing memory. Table 7 reports the experiments on "nuScences Daytime → nuScenes Nighttime" adaptation over PETR [39]. It can be observed that incorporating an additional image-view decoder into PETR introduces slight computation overhead during model training but has little effect on model testing, largely because the image-view encoder is utilized during network training but discarded after adaptation.

Table 7. Training and inference time analysis of all the compared methods. The experiments are conducted on one RTX 2080Ti.

Method	Training Time (hours)	Training Memory (MB)	Testing Speed (images per second)	Testing Memory (MB)
PETR [39] (Baseline)	-	-	5.5551	3075
SFA [65]	53.4966	4967	5.5551	3077
MMTrans [77]	56.3645	10120	5.5551	3068
STM3D [36]	53.6178	4854	5.5551	3055
DA-BEV (Ours)	67.2670	5540	5.5551	3089

Fig. 4. Qualitative illustration of DA-BEV on 3D object detection for cross-weather domain adaptation (i.e., Clear Weather → Rainy Weather).

Qualitative Results. Figure 4 provides qualitative illustrations which show that DA-BEV can produce accurate 3D object detection under weather-induced domain gaps, e.g., the leftmost car in Front Left camera becomes fuzzy due to rains and our DA-BEV can still detect it accurately.

5 Conclusion

In this paper, we present DA-BEV, the first domain adaptive camera-only BEV framework that addresses domain adaptive BEV challenges by exploiting the complementary nature of image-view features and BEV features. DA-BEV introduces query-based adversarial learning (QAL) and query-based self-training (QST), where QAL/QST exploits the useful information queried from image-view features or BEV features to regularize the "adversarial learning"/"self-training" of the another. Extensive experiments showcase the superior domain adaptive BEV perception performance of DA-BEV across various datasets and tasks. Moving forward, we will further explore the complementary nature of image-view and BEV features by introducing the temporal information of them.

Acknowledgements. This work was supported in part by the National Natural Science Foundation of China under Grant 62121001, Grant 62322117 and Grant 62371365; in part by Young Elite Scientist Sponsorship Program by the China Association for Science and Technology under Grant 2020QNRC001.

References

1. Acuna, D., Philion, J., Fidler, S.: Towards optimal strategies for training self-driving perception models in simulation. In: Advances in Neural Information Processing Systems, vol. 34, pp. 1686–1699 (2021)
2. Akan, A.K., Güney, F.: StretchBEV: stretching future instance prediction spatially and temporally. In: Avidan, S., Brostow, G., Cissé, M., Farinella, G.M., Hassner, T. (eds.) ECCV 2022. LNCS, vol. 13698, pp. 444–460. Springer, Cham (2022). https://doi.org/10.1007/978-3-031-19839-7_26
3. Antonello, M., Carraro, M., Pierobon, M., Menegatti, E.: Fast and robust detection of fallen people from a mobile robot. In: 2017 IEEE/RSJ International Conference on Intelligent Robots and Systems (IROS), pp. 4159–4166. IEEE (2017)

4. Bai, X., et al.: Transfusion: robust lidar-camera fusion for 3D object detection with transformers. In: Proceedings of the IEEE/CVF Conference on Computer Vision and Pattern Recognition, pp. 1090–1099 (2022)
5. Barrera, A., Beltrán, J., Guindel, C., Iglesias, J.A., García, F.: Cycle and semantic consistent adversarial domain adaptation for reducing simulation-to-real domain shift in lidar bird's eye view. In: 2021 IEEE International Intelligent Transportation Systems Conference (ITSC), pp. 3081–3086. IEEE (2021)
6. Bartoccioni, F., Zablocki, É., Bursuc, A., Pérez, P., Cord, M., Alahari, K.: LaRa: latents and rays for multi-camera bird's-eye-view semantic segmentation. In: Conference on Robot Learning, pp. 1663–1672. PMLR (2023)
7. Caesar, H., et al.: nuScenes: a multimodal dataset for autonomous driving. In: Proceedings of the IEEE/CVF Conference on Computer Vision and Pattern Recognition, pp. 11621–11631 (2020)
8. Chen, L., et al.: PersFormer: 3D lane detection via perspective transformer and the OpenLane benchmark. In: Avidan, S., Brostow, G., Cissé, M., Farinella, G.M., Hassner, T. (eds.) ECCV 2022. LNCS, vol. 13698, pp. 550–567. Springer, Cham (2022). https://doi.org/10.1007/978-3-031-19839-7_32
9. Chen, S., Wang, X., Cheng, T., Zhang, Q., Huang, C., Liu, W.: Polar parametrization for vision-based surround-view 3D detection. arXiv preprint arXiv:2206.10965 (2022)
10. Chen, Y., Liu, S., Shen, X., Jia, J.: DSGN: deep stereo geometry network for 3D object detection. In: Proceedings of the IEEE/CVF Conference on Computer Vision and Pattern Recognition, pp. 12536–12545 (2020)
11. Chen, Z., et al.: AutoAlign: pixel-instance feature aggregation for multi-modal 3D object detection. arXiv preprint arXiv:2201.06493 (2022)
12. Chen, Z., Li, Z., Zhang, S., Fang, L., Jiang, Q., Zhao, F.: AutoAlignV2: deformable feature aggregation for dynamic multi-modal 3D object detection. arXiv preprint arXiv:2207.10316 (2022)
13. Chen, Z., Li, Z., Zhang, S., Fang, L., Jiang, Q., Zhao, F.: Graph-DETR3D: rethinking overlapping regions for multi-view 3D object detection. In: Proceedings of the 30th ACM International Conference on Multimedia, pp. 5999–6008 (2022)
14. Chen, Z., Luo, Y., Wang, Z., Baktashmotlagh, M., Huang, Z.: Revisiting domain-adaptive 3D object detection by reliable, diverse and class-balanced pseudo-labeling. In: Proceedings of the IEEE/CVF International Conference on Computer Vision, pp. 3714–3726 (2023)
15. Ganin, Y., et al.: Domain-adversarial training of neural networks. J. Mach. Learn. Res. **17**(1), 2096–2030 (2016)
16. Gao, Z., Huang, K., Zhang, R., Liu, D., Ma, J.: Towards better robustness against common corruptions for unsupervised domain adaptation. In: Proceedings of the IEEE/CVF International Conference on Computer Vision (ICCV), pp. 18882–18893 (2023)
17. Garnett, N., Cohen, R., Pe'er, T., Lahav, R., Levi, D.: 3D-LaneNet: end-to-end 3D multiple lane detection. In: Proceedings of the IEEE/CVF International Conference on Computer Vision, pp. 2921–2930 (2019)
18. Gong, R., Li, W., Chen, Y., Gool, L.V.: DLOW: domain flow for adaptation and generalization. In: Proceedings of the IEEE/CVF Conference on Computer Vision and Pattern Recognition, pp. 2477–2486 (2019)
19. Guo, X., Shi, S., Wang, X., Li, H.: Liga-stereo: learning lidar geometry aware representations for stereo-based 3D detector. In: Proceedings of the IEEE/CVF International Conference on Computer Vision, pp. 3153–3163 (2021)

20. Gutmann, J.S., Fukuchi, M., Fujita, M.: 3D perception and environment map generation for humanoid robot navigation. Int. J. Robot. Res. **27**(10), 1117–1134 (2008)
21. Hendy, N., et al.: Fishing net: future inference of semantic heatmaps in grids. arXiv preprint arXiv:2006.09917 (2020)
22. Hou, Y., Zheng, L., Gould, S.: Multiview detection with feature perspective transformation. In: Vedaldi, A., Bischof, H., Brox, T., Frahm, J.-M. (eds.) ECCV 2020. LNCS, vol. 12352, pp. 1–18. Springer, Cham (2020). https://doi.org/10.1007/978-3-030-58571-6_1
23. Houston, J., et al.: One thousand and one hours: self-driving motion prediction dataset. In: Conference on Robot Learning, pp. 409–418. PMLR (2021)
24. Hu, A., et al.: Fiery: future instance prediction in bird's-eye view from surround monocular cameras. In: Proceedings of the IEEE/CVF International Conference on Computer Vision, pp. 15273–15282 (2021)
25. Huang, J., Huang, G.: BEVDet4D: exploit temporal cues in multi-camera 3D object detection. arXiv preprint arXiv:2203.17054 (2022)
26. Huang, J., Huang, G., Zhu, Z., Ye, Y., Du, D.: BEVDet: high-performance multi-camera 3D object detection in bird-eye-view. arXiv preprint arXiv:2112.11790 (2021)
27. Jamal, M.A., Brown, M., Yang, M.H., Wang, L., Gong, B.: Rethinking class-balanced methods for long-tailed visual recognition from a domain adaptation perspective. In: Proceedings of the IEEE/CVF Conference on Computer Vision and Pattern Recognition, pp. 7610–7619 (2020)
28. Jiang, Y., et al.: PolarFormer: multi-camera 3D object detection with polar transformer. In: Proceedings of the AAAI Conference on Artificial Intelligence, vol. 37, pp. 1042–1050 (2023)
29. Jiao, Y., Jie, Z., Chen, S., Chen, J., Ma, L., Jiang, Y.G.: MSMDFusion: fusing lidar and camera at multiple scales with multi-depth seeds for 3D object detection. In: Proceedings of the IEEE/CVF Conference on Computer Vision and Pattern Recognition, pp. 21643–21652 (2023)
30. Lee, D.H.: Pseudo-label: the simple and efficient semi-supervised learning method for deep neural networks. In: Workshop on Challenges in Representation Learning, ICML, vol. 3, p. 2 (2013)
31. Lee, Y., Park, J.: CenterMask: real-time anchor-free instance segmentation. In: Proceedings of the IEEE/CVF Conference on Computer Vision and Pattern Recognition, pp. 13906–13915 (2020)
32. Li, H., et al.: Delving into the devils of bird's-eye-view perception: a review, evaluation and recipe. arXiv preprint arXiv:2209.05324 (2022)
33. Li, Q., Wang, Y., Wang, Y., Zhao, H.: HDMapNet: an online HD map construction and evaluation framework. In: 2022 International Conference on Robotics and Automation (ICRA), pp. 4628–4634. IEEE (2022)
34. Li, Y., et al.: DeepFusion: lidar-camera deep fusion for multi-modal 3D object detection. In: Proceedings of the IEEE/CVF Conference on Computer Vision and Pattern Recognition, pp. 17182–17191 (2022)
35. Li, Y., et al.: BEVDepth: acquisition of reliable depth for multi-view 3D object detection. In: Proceedings of the AAAI Conference on Artificial Intelligence, vol. 37, pp. 1477–1485 (2023)
36. Li, Z., et al.: Unsupervised domain adaptation for monocular 3D object detection via self-training. In: Avidan, S., Brostow, G., Cissé, M., Farinella, G.M., Hassner, T. (eds.) ECCV 2022. LNCS, vol. 13669, pp. 245–262. Springer, Cham (2022). https://doi.org/10.1007/978-3-031-20077-9_15

37. Li, Z., et al.: BEVFormer: learning bird's-eye-view representation from multi-camera images via spatiotemporal transformers. In: Avidan, S., Brostow, G., Cissé, M., Farinella, G.M., Hassner, T. (eds.) ECCV 2022. LNCS, vol. 13669, pp. 1–18. Springer, Cham (2022). https://doi.org/10.1007/978-3-031-20077-9_1
38. Liu, J., et al.: Multi-latent space alignments for unsupervised domain adaptation in multi-view 3D object detection. arXiv preprint arXiv:2211.17126 (2022)
39. Liu, Y., Wang, T., Zhang, X., Sun, J.: PETR: position embedding transformation for multi-view 3D object detection. In: Avidan, S., Brostow, G., Cissé, M., Farinella, G.M., Hassner, T. (eds.) ECCV 2022. LNCS, vol. 13687, pp. 531–548. Springer, Cham (2022). https://doi.org/10.1007/978-3-031-19812-0_31
40. Liu, Y., et al.: PETRv2: a unified framework for 3D perception from multi-camera images. In: Proceedings of the IEEE/CVF International Conference on Computer Vision, pp. 3262–3272 (2023)
41. Long, M., Cao, Y., Wang, J., Jordan, M.: Learning transferable features with deep adaptation networks. In: International Conference on Machine Learning, pp. 97–105 (2015)
42. Loshchilov, I., Hutter, F.: Decoupled weight decay regularization. arXiv preprint arXiv:1711.05101 (2017)
43. Loukkal, A., Grandvalet, Y., Drummond, T., Li, Y.: Driving among flatmobiles: bird-eye-view occupancy grids from a monocular camera for holistic trajectory planning. In: Proceedings of the IEEE/CVF Winter Conference on Applications of Computer Vision, pp. 51–60 (2021)
44. Luo, Y., Liu, P., Zheng, L., Guan, T., Yu, J., Yang, Y.: Category-level adversarial adaptation for semantic segmentation using purified features. IEEE Trans. Pattern Anal. Mach. Intell. **44**(8), 3940–3956 (2021)
45. Ma, Y., et al.: Vision-centric BEV perception: a survey. arXiv preprint arXiv:2208.02797 (2022)
46. Oza, P., Sindagi, V.A., Sharmini, V.V., Patel, V.M.: Unsupervised domain adaptation of object detectors: a survey. IEEE Trans. Pattern Anal. Mach. Intell. (2023)
47. Pan, B., Sun, J., Leung, H.Y.T., Andonian, A., Zhou, B.: Cross-view semantic segmentation for sensing surroundings. IEEE Robot. Autom. Lett. **5**(3), 4867–4873 (2020)
48. Peng, L., Chen, Z., Fu, Z., Liang, P., Cheng, E.: BEVSegFormer: bird's eye view semantic segmentation from arbitrary camera rigs. In: Proceedings of the IEEE/CVF Winter Conference on Applications of Computer Vision, pp. 5935–5943 (2023)
49. Reiher, L., Lampe, B., Eckstein, L.: A sim2real deep learning approach for the transformation of images from multiple vehicle-mounted cameras to a semantically segmented image in bird's eye view. In: 2020 IEEE 23rd International Conference on Intelligent Transportation Systems (ITSC), pp. 1–7. IEEE (2020)
50. Roddick, T., Cipolla, R.: Predicting semantic map representations from images using pyramid occupancy networks. In: Proceedings of the IEEE/CVF Conference on Computer Vision and Pattern Recognition, pp. 11138–11147 (2020)
51. Roh, W., et al.: ORA3D: overlap region aware multi-view 3D object detection. arXiv preprint arXiv:2207.00865 (2022)
52. Saha, A., Mendez, O., Russell, C., Bowden, R.: Enabling spatio-temporal aggregation in birds-eye-view vehicle estimation. In: 2021 IEEE International Conference on Robotics and Automation (ICRA), pp. 5133–5139. IEEE (2021)
53. Saleh, K., et al.: Domain adaptation for vehicle detection from bird's eye view lidar point cloud data. In: Proceedings of the IEEE/CVF International Conference on Computer Vision Workshops (2019)

54. Schuemie, M.J., Van Der Straaten, P., Krijn, M., Van Der Mast, C.A.: Research on presence in virtual reality: a survey. Cyberpsychol. Behav. **4**(2), 183–201 (2001)
55. Shi, Y., et al.: SRCN3D: sparse R-CNN 3D surround-view camera object detection and tracking for autonomous driving. arXiv preprint arXiv:2206.14451 (2022)
56. Song, L., Wu, J., Yang, M., Zhang, Q., Li, Y., Yuan, J.: Stacked homography transformations for multi-view pedestrian detection. In: Proceedings of the IEEE/CVF International Conference on Computer Vision, pp. 6049–6057 (2021)
57. Sun, Y., Liu, W., Bao, Q., Fu, Y., Mei, T., Black, M.J.: Putting people in their place: monocular regression of 3D people in depth. In: Proceedings of the IEEE/CVF Conference on Computer Vision and Pattern Recognition, pp. 13243–13252 (2022)
58. Vora, S., Lang, A.H., Helou, B., Beijbom, O.: Pointpainting: sequential fusion for 3D object detection. In: Proceedings of the IEEE/CVF Conference on Computer Vision and Pattern Recognition, pp. 4604–4612 (2020)
59. Vu, T.H., Jain, H., Bucher, M., Cord, M., Pérez, P.: Advent: adversarial entropy minimization for domain adaptation in semantic segmentation. In: Proceedings of the IEEE Conference on Computer Vision and Pattern Recognition, pp. 2517–2526 (2019)
60. Wang, C., Ma, C., Zhu, M., Yang, X.: Pointaugmenting: cross-modal augmentation for 3D object detection. In: Proceedings of the IEEE/CVF Conference on Computer Vision and Pattern Recognition, pp. 11794–11803 (2021)
61. Wang, L., et al.: Multi-modal 3D object detection in autonomous driving: a survey and taxonomy. IEEE Trans. Intell. Veh. (2023)
62. Wang, S., Jiang, X., Li, Y.: Focal-PETR: embracing foreground for efficient multi-camera 3D object detection. arXiv preprint arXiv:2212.05505 (2022)
63. Wang, S., et al.: Towards domain generalization for multi-view 3D object detection in bird-eye-view. In: Proceedings of the IEEE/CVF Conference on Computer Vision and Pattern Recognition, pp. 13333–13342 (2023)
64. Wang, T., Lian, Q., Zhu, C., Zhu, X., Zhang, W.: MV-FCOS3D++: multi-view camera-only 4D object detection with pretrained monocular backbones. arXiv preprint arXiv:2207.12716 (2022)
65. Wang, W., et al.: Exploring sequence feature alignment for domain adaptive detection transformers. In: Proceedings of the 29th ACM International Conference on Multimedia, pp. 1730–1738 (2021)
66. Wang, Y., et al.: Multi-modal 3D object detection in autonomous driving: a survey. Int. J. Comput. Vis. 1–31 (2023)
67. Wang, Y., Guizilini, V.C., Zhang, T., Wang, Y., Zhao, H., Solomon, J.: DETR3D: 3D object detection from multi-view images via 3D-to-2D queries. In: Conference on Robot Learning, pp. 180–191. PMLR (2022)
68. Wilson, G., Cook, D.J.: A survey of unsupervised deep domain adaptation. ACM Trans. Intell. Syst. Technol. (TIST) **11**(5), 1–46 (2020)
69. Xie, E., et al.: M 2 BEV: multi-camera joint 3D detection and segmentation with unified birds-eye view representation. arXiv preprint arXiv:2204.05088 (2022)
70. Xiong, K., et al.: Cape: camera view position embedding for multi-view 3D object detection (2023)
71. Xu, J., Zhang, R., Dou, J., Zhu, Y., Sun, J., Pu, S.: RPVNet: a deep and efficient range-point-voxel fusion network for lidar point cloud segmentation. In: Proceedings of the IEEE/CVF International Conference on Computer Vision, pp. 16024–16033 (2021)

72. Xu, R., Tu, Z., Xiang, H., Shao, W., Zhou, B., Ma, J.: CoBEVT: cooperative bird's eye view semantic segmentation with sparse transformers. arXiv preprint arXiv:2207.02202 (2022)
73. Yang, C., et al.: BEVFormer v2: adapting modern image backbones to bird's-eye-view recognition via perspective supervision. In: Proceedings of the IEEE/CVF Conference on Computer Vision and Pattern Recognition, pp. 17830–17839 (2023)
74. Yang, W., et al.: Projecting your view attentively: monocular road scene layout estimation via cross-view transformation. In: Proceedings of the IEEE/CVF Conference on Computer Vision and Pattern Recognition, pp. 15536–15545 (2021)
75. Yin, T., Zhou, X., Krahenbuhl, P.: Center-based 3D object detection and tracking. In: Proceedings of the IEEE/CVF Conference on Computer Vision and Pattern Recognition, pp. 11784–11793 (2021)
76. Yoo, J.H., Kim, Y., Kim, J., Choi, J.W.: 3D-CVF: generating joint camera and LiDAR features using cross-view spatial feature fusion for 3D object detection. In: Vedaldi, A., Bischof, H., Brox, T., Frahm, J.-M. (eds.) ECCV 2020. LNCS, vol. 12372, pp. 720–736. Springer, Cham (2020). https://doi.org/10.1007/978-3-030-58583-9_43
77. Yu, J., et al.: MTTrans: cross-domain object detection with mean teacher transformer. In: Avidan, S., Brostow, G., Cissé, M., Farinella, G.M., Hassner, T. (eds.) ECCV 2022. LNCS, vol. 13669, pp. 629–645. Springer, Cham (2022). https://doi.org/10.1007/978-3-031-20077-9_37
78. Zhang, P., Zhang, B., Zhang, T., Chen, D., Wang, Y., Wen, F.: Prototypical pseudo label denoising and target structure learning for domain adaptive semantic segmentation. In: Proceedings of the IEEE/CVF Conference on Computer Vision and Pattern Recognition, pp. 12414–12424 (2021)
79. Zhang, Q., Zhang, J., Liu, W., Tao, D.: Category anchor-guided unsupervised domain adaptation for semantic segmentation. In: Advances in Neural Information Processing Systems, pp. 433–443 (2019)
80. Zhang, Y.: A survey of unsupervised domain adaptation for visual recognition. arXiv preprint arXiv:2112.06745 (2021)
81. Zhang, Y., et al.: BEVerse: unified perception and prediction in birds-eye-view for vision-centric autonomous driving. arXiv preprint arXiv:2205.09743 (2022)
82. Zhao, Q.: A survey on virtual reality. Sci. China Ser. F: Inf. Sci. **52**(3), 348–400 (2009)
83. Zhou, B., Krähenbühl, P.: Cross-view transformers for real-time map-view semantic segmentation. In: Proceedings of the IEEE/CVF Conference on Computer Vision and Pattern Recognition, pp. 13760–13769 (2022)
84. Zhu, J., Bai, H., Wang, L.: Patch-mix transformer for unsupervised domain adaptation: a game perspective. In: Proceedings of the IEEE/CVF Conference on Computer Vision and Pattern Recognition (CVPR), pp. 3561–3571 (2023)
85. Zhu, X., et al.: Cylindrical and asymmetrical 3D convolution networks for lidar segmentation. In: Proceedings of the IEEE/CVF Conference on Computer Vision and Pattern Recognition, pp. 9939–9948 (2021)
86. Zou, J., et al.: HFT: lifting perspective representations via hybrid feature transformation. arXiv preprint arXiv:2204.05068 (2022)
87. Zou, Y., Yu, Z., Kumar, B., Wang, J.: Unsupervised domain adaptation for semantic segmentation via class-balanced self-training. In: Proceedings of the European Conference on Computer Vision (ECCV), pp. 289–305 (2018)

SlimFlow: Training Smaller One-Step Diffusion Models with Rectified Flow

Yuanzhi Zhu[1], Xingchao Liu[2], and Qiang Liu[2(✉)]

[1] ETH Zurich, 8092 Zurich, Switzerland
yuazhu@student.ethz.ch
[2] UT Austin, Austin, TX 78712, USA
xcliu@utexas.edu, lqiang@cs.utexas.edu

Abstract. Diffusion models excel in high-quality generation but suffer from slow inference due to iterative sampling. While recent methods have successfully transformed diffusion models into one-step generators, they neglect model size reduction, limiting their applicability in compute-constrained scenarios. This paper aims to develop small, efficient one-step diffusion models based on the powerful rectified flow framework, by exploring joint compression of inference steps and model size. The rectified flow framework trains one-step generative models using two operations, reflow and distillation. Compared with the original framework, squeezing the model size brings two new challenges: (1) the initialization mismatch between large teachers and small students during reflow; (2) the underperformance of naive distillation on small student models. To overcome these issues, we propose Annealing Reflow and Flow-Guided Distillation, which together comprise our SlimFlow framework. With our novel framework, we train a one-step diffusion model with an FID of 5.02 and 15.7M parameters, outperforming the previous state-of-the-art one-step diffusion model (FID = 6.47, 19.4M parameters) on CIFAR10. On ImageNet 64 × 64 and FFHQ 64 × 64, our method yields small one-step diffusion models that are comparable to larger models, showcasing the effectiveness of our method in creating compact, efficient one-step diffusion models.

Keywords: Diffusion Models · Flow-based Models · One-Step Generative Models · Efficient Models

1 Introduction

In recent years, diffusion models [14,56] have revolutionized the field of image generation [4,32,41,46], surpassing the quality achieved by traditional approaches such as Generative Adversarial Networks (GANs) [9,19], Normalizing Flows (NFs) [5,42] and Variational Autoencoders (VAEs) [22,47]. Its success

---

Supplementary Information The online version contains supplementary material available at https://doi.org/10.1007/978-3-031-73007-8_20.

© The Author(s), under exclusive license to Springer Nature Switzerland AG 2025
A. Leonardis et al. (Eds.): ECCV 2024, LNCS 15140, pp. 342–359, 2025.
https://doi.org/10.1007/978-3-031-73007-8_20

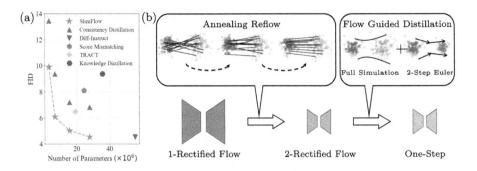

Fig. 1. (a) Comparison of different one-step diffusion models on the CIFAR10 dataset. (b) To get powerful one-step diffusion model, our SlimFlow framework designs two stages: **Annealing Reflow** provides a warm-start for the small 2-Rectified Flow model by gradually shifting from training with random pairs to teacher pairs; **Flow Guided Distillation** enhances the one-step small model by distillation from 2-Rectified Flow with both off-line generated data using precise ODE solver and online generated data using 2-step Euler solver.

even extends to other modalities, including audio [17,23], video [8,13,40] and 3D content generation [30,45,58,61].

Despite these advancements, diffusion models face significant challenges in terms of generation efficiency. Their iterative sampling process and substantial model size pose considerable obstacles to widespread adoption, particularly in resource-constrained environments, such as edge devices. To enhance the inference speed of diffusion models, recent research has focused on two primary strategies. The first strategy aims at reducing the number of inference steps, which can be achieved through either adoption of fast solvers [6,34,52,66,70] or distillation [10,29,35,49,53]. The second strategy focuses on lowering the computational cost within each inference step, by model structure modification [2,7,44,63], quantization [11,15,26,27,50,57], and caching [38,60]. Recently, these techniques have been applied to various large-scale diffusion models, e.g., Stable Diffusion [48], to significantly improve their inference speed and reduce their cost [33,36,39,62,65].

This paper focuses on advancing the rectified flow framework [29,31], which has demonstrated promising results in training few/one-step diffusion models [33]. While the framework has shown success in reducing inference steps, it has not addressed the challenge of model size reduction. We aim to enhance the rectified flow framework by simultaneously reducing both the number of inference steps and the network size of diffusion models. The rectified flow framework straightens the trajectories of pre-trained generative probability flows through a process called reflow, thereby decreasing the required number of inference steps. This procedure also refines the coupling between noise and data distributions. High-quality one-step generative models are then obtained by distilling from the straightened flow. Unlike typical rectified flow applications that maintain an invariant model structure during the entire process, we aim to reduce the network size in reflow and distillation, targeting efficient one-step diffusion models. It introduces two key challenges: (1) The reflow operation typically initializes the student flow with the pre-trained teacher's weights to inherit knowledge and accelerate convergence. However, this strategy is inapplicable when the student network has a different structure. (2) The smaller student network suffers from reduced capacity, causing naive distillation to underperform. To address these challenges, we propose SlimFlow, comprising two stages: Annealing Reflow and Flow-Guided Distillation (Fig. 1). Annealing Reflow provides a warm-start initialization for the small student model by smoothly interpolating between training from scratch and reflow. Flow-Guided Distillation introduces a novel regularization that leverages guidance from the learned straighter student flow, resulting in better distilled one-step generators. SlimFlow achieves state-of-the-art Frechet Inception Distance (FID) among other one-step diffusion models with a limited number of parameters. Furthermore, when applied to ImageNet 64×64 and FFHQ 64×64, SlimFlow yields small one-step diffusion models that are comparable to larger models.

2 Background

2.1 Diffusion Models

Diffusion models define a forward diffusion process that maps data to noise by gradually perturbing the input data with Gaussian noise. Then, in the reverse process, they generate images by gradually removing Gaussian noise, with the intuition from non-equilibrium thermodynamics [51]. We denote the data $\mathbf{x}$ at t as $\mathbf{x}_t$. The forward process can be described by an Itô SDE [56]:

$$\mathrm{d}\mathbf{x}_t = \mathbf{f}(\mathbf{x}_t, t)\mathrm{d}t + g(t)\mathrm{d}\mathbf{w}, \tag{1}$$

where $\mathbf{w}$ is the standard Wiener process, $\mathbf{f}(\cdot, t)$ is a vector-valued function called the drift coefficient, and $g(\cdot)$ is a scalar function called the diffusion coefficient.

For every diffusion process in Eq. (1), there exists a corresponding deterministic Probability Flow Ordinary Differential Equation (PF-ODE) which induces the same marginal density as Eq. (1):

$$\frac{\mathrm{d}\mathbf{x}_t}{\mathrm{d}t} = \mathbf{f}(\mathbf{x}_t, t) - \frac{1}{2}g^2(t)\nabla_{\mathbf{x}_t} \log p_t(\mathbf{x}_t), \tag{2}$$

where $p_t(\cdot)$ is the marginal probability density at time t. $\nabla_{\mathbf{x}_t} \log p_t(\mathbf{x}_t)$ is called the score function, and can be modelled as $\mathbf{s}_\theta(\mathbf{x}, t)$ using a neural network θ. Usually, the network is trained by score matching [16,54,55]. Starting with samples from an initial distribution π_T such as a standard Gaussian distribution, we can generate data samples by simulating Eq. (2) from $t = T$ to $t = 0$. We call Eq. (2) with the approximated score function $\mathbf{s}_\theta(\mathbf{x}_t, t)$ the empirical PF-ODE, written as $\frac{\mathrm{d}\mathbf{x}_t}{\mathrm{d}t} = \mathbf{f}(\mathbf{x}_t, t) - \frac{1}{2}g^2(t)\mathbf{s}_\theta(\mathbf{x}_t, t)$. It is worth noting that the deterministic PF-ODE gives a deterministic correspondence between the initial noise distribution and the generated data distribution.

2.2 Rectified Flows

Rectified flow [28,29,31] is an ODE-based generative modeling framework. Given the initial distribution π_T and the target data distribution π_0, rectified flow trains a velocity field parameterized by a neural network with the following loss function,

$$\mathcal{L}_{\mathrm{rf}}(\theta) := \mathbb{E}_{\mathbf{x}_T \sim \pi_T, \mathbf{x}_0 \sim \pi_0} \left[\int_0^T \|\mathbf{v}_\theta(\mathbf{x}_t, t) - (\mathbf{x}_T - \mathbf{x}_0)\|_2^2 dt \right], \quad (3)$$

where $\mathbf{x}_t = (1 - t/T)\mathbf{x}_0 + t\mathbf{x}_T/T$.

Without loss of generality, T is usually set to 1. Based on the trained rectified flow, we can generate samples by simulating the following ODE from $t = 1$ to $t = 0$,

$$\frac{d\mathbf{x}_t}{dt} = \mathbf{v}_\theta(\mathbf{x}_t, t). \quad (4)$$

PF-ODEs transformed from pre-trained diffusion models can be seen as special forms of rectified flows. For a detailed discussion of the mathematical relationship between them, we refer readers to [29,31]. In computer, Eq. (4) is approximated by off-the-shelf ODE solvers, e.g., the forward Euler solver,

$$\mathbf{x}_{t-\frac{1}{N}} = \mathbf{x}_t - \frac{1}{N}\mathbf{v}_\theta(\mathbf{x}_t, t), \quad \forall t \in \{1, 2, \ldots, N\}/N. \quad (5)$$

Here, the ODE is solved in N steps with a step size of $1/N$. Large N leads to accurate but slow simulation, while small N gives fast but inaccurate simulation. Fortunately, straight probability flows with uniform speed enjoy one-step simulation with no numerical error because $\mathbf{x}_t = \mathbf{x}_1 - (1-t)\mathbf{v}_\theta(\mathbf{x}_1, 1)$.

Reflow. In the rectified flow framework, a special operation called reflow is designed to train straight probability flows,

$$\mathcal{L}_{\mathrm{reflow}}(\phi) := \mathbb{E}_{\mathbf{x}_1 \sim \pi_1} \left[\int_0^T \|\mathbf{v}_\phi(\mathbf{x}_t, t) - (\mathbf{x}_1 - \hat{\mathbf{x}}_0)\|_2^2 dt \right], \quad (6)$$

where $\mathbf{x}_t = (1-t)\hat{\mathbf{x}}_0 + t\mathbf{x}_1$ and $\hat{\mathbf{x}}_0 = \mathrm{ODE}[\mathbf{v}_\theta](\mathbf{x}_1)$.

Compared with Eq. (3), $\hat{\mathbf{x}}_0$ is not a random sample from distribution π_0 that is independent with $\mathbf{x}_1$ anymore. Instead, $\hat{\mathbf{x}}_0 = \mathrm{ODE}[\mathbf{v}_\phi](\mathbf{x}_1) := \mathbf{x}_1 + \int_1^0 \mathbf{v}_\theta(\mathbf{x}_t, t)dt$ is induced from the pre-trained flow $\mathbf{v}_\theta$. After training, the new flow $\mathbf{v}_\phi$ has straighter trajectories and requires fewer inference steps for generation. A common practice in reflow is initializing $\mathbf{v}_\phi$ with the pre-trained $\mathbf{v}_\theta$ for faster convergence and higher performance [31]. Following previous works, we name $\mathbf{v}_\theta$ as 1-rectified flow and $\mathbf{v}_\phi$ as 2-rectified flow. The straightness of the learned flow $\mathbf{v}$ can be defined as:

$$S(\mathbf{v}) = \int_0^1 \mathbb{E}\left[\|\mathbf{v}(\mathbf{x}_t, t) - (\mathbf{x}_1 - \mathbf{x}_0)\|^2\right] dt, \quad (7)$$

Distillation. Given a pre-trained probability flow, for example, $\mathbf{v}_\phi$, we can further enhance their one-step generation via distillation,

$$\mathcal{L}_{\text{distill}}(\phi') := \mathbb{E}_{\mathbf{x}_1 \sim \pi_1} \left[\mathbb{D}(\text{ODE}[\mathbf{v}_\phi](\mathbf{x}_1), \mathbf{x}_1 - \mathbf{v}_{\phi'}(\mathbf{x}_1, 1)) \right]. \tag{8}$$

In this equation, $\mathbb{D}$ is a discrepancy loss that measures the difference between two images, e.g., ℓ_2 loss or the LPIPS loss [67]. Through distillation, the distilled model $\mathbf{v}_{\phi'}$ can use one-step Euler discretization to approximate the result of the entire flow trajectory. Note that reflow can refine the deterministic mapping between the noise and the generated samples defined by the probability flow [31, 33]. Consequently, distillation from $\mathbf{v}_\phi$ gives better one-step generators than distillation from $\mathbf{v}_\theta$.

3 SlimFlow

In this section, we introduce the SlimFlow framework for learning small, efficient one-step generative models. SlimFlow enhances the rectified flow framework by improving both the reflow and distillation stages. For the reflow stage, we propose Annealing Reflow. This novel technique provides a warm-start initialization for the small student model. It smoothly transitions from training a 1-rectified flow to training a 2-rectified flow, accelerating the training convergence of the smaller network. For the distillation stage, we propose Flow-Guided Distillation, which leverages the pre-trained 2-rectified flow as an additional regularization during distillation to improve the performance of the resulting one-step model. By combining these two techniques, SlimFlow creates a robust framework for training efficient one-step diffusion models that outperform existing methods in both model size and generation quality.

3.1 Annealing Reflow

The reflow procedure in the rectified flow framework trains a new, straighter flow $\mathbf{v}_\phi$ from the pre-trained 1-rectified flow $\mathbf{v}_\theta$, as described in Eq. (6). Typically, when $\mathbf{v}_\phi$ and $\mathbf{v}_\theta$ have the same model structure, $\mathbf{v}_\phi$ is initialized with $\mathbf{v}_\theta$'s weights to accelerate training. However, our goal of creating smaller models cannot adopt this approach, as $\mathbf{v}_\phi$ and $\mathbf{v}_\theta$ have different structures. A naive solution would be to train a new 1-rectified flow using the smaller model with Eq. (3) and use this as initialization. However, this strategy is time-consuming and delays the process of obtaining an efficient 2-rectified flow with the small model.

To address this challenge, we propose Annealing Reflow, which provides an effective initialization for the small student model without significantly increasing training time. This method starts the training process with random pairs, as in Eq. (3), then gradually shifts to pairs generated from the teacher flow, as in Eq. (6).

Formally, given the teacher velocity field $\mathbf{v}_\theta$ defined by a large neural network θ, the objective of Annealing Reflow is defined as,

$$\mathcal{L}^k_{\text{a-reflow}}(\phi) := \mathbb{E}_{\mathbf{x}_1,\mathbf{x}_1'\sim\pi_1}\left[\int_0^T \left\|\mathbf{v}_\phi(\mathbf{x}_t^{\beta(k)},t) - \left(\mathbf{x}_1^{\beta(k)} - \hat{\mathbf{x}}_0\right)\right\|_2^2 dt\right],$$

$$\text{where}\quad \mathbf{x}_t^{\beta(k)} = (1-t)\hat{\mathbf{x}}_0 + t\mathbf{x}_1^{\beta(k)},$$
$$\mathbf{x}_1^{\beta(k)} = \left(\sqrt{1-\beta^2(k)}\mathbf{x}_1 + \beta(k)\mathbf{x}_1'\right), \quad (9)$$
$$\hat{\mathbf{x}}_0 = \text{ODE}[\mathbf{v}_\theta](\mathbf{x}_1) = \mathbf{x}_1 + \int_1^0 \mathbf{v}_\theta(\mathbf{x}_t,t) dt.$$

In Eq. (9), $\mathbf{x}_1$ and $\mathbf{x}_1'$ are independent random samples from the initial distribution π_1. k refers to the number of training iterations, and $\beta(k) : \mathbb{N} \to [0,1]$ is a function that maps k to a scalar between 0 and 1. The function $\beta(k)$ must satisfy two key conditions:

1. $\beta(0) = 1$: This initial condition reduces $\mathcal{L}^0_{\text{a-reflow}}$ to the rectified flow objective (Eq. (3)), with $\hat{\mathbf{x}}_0$ generated from the well-trained teacher model instead of being randomly sampled from π_0. Note that when $\mathbf{v}_\theta$ is well-trained, $\hat{\mathbf{x}}_0$ also follows the target data distribution π_0.
2. $\beta(+\infty) = 0$: This asymptotic condition ensures that $\mathcal{L}^{+\infty}_{\text{a-reflow}}$ converges to the reflow objective (Eq. (6)), guaranteeing that the small model will eventually become a valid 2-rectified flow.

By carefully choosing the schedule β, we create a smooth transition from training the small student model using random pairs to using pairs generated from the large teacher flow. This approach provides an appropriate initialization for the student model and directly outputs a small 2-rectified flow model, balancing efficiency and effectiveness in the training process.

Leveraging the Intrinsic Symmetry in Reflow Pairs. To perform reflow, a dataset of pair $(\mathbf{x}_1, \hat{\mathbf{x}}_0)$ is usually generated by the teacher flow before training the student model. We find that the hidden symmetry in the pairs can be exploited to create new pairs without simulating the teacher flow. Specifically, if the initial noise $\mathbf{x}_1$ is horizontally flipped (referred to as H-Flip), the corresponding image $\hat{\mathbf{x}}_0$ is also horizontally flipped (Fig. 2). As a result, for each generated pair $(\mathbf{x}_1, \hat{\mathbf{x}}_0)$, we can also add (H-Flip($\mathbf{x}_1$), H-Flip($\hat{\mathbf{x}}_0$)) as a valid pair to double the sample size of our reflow dataset.

3.2 Flow-Guided Distillation

Distillation is the other vital step in training one-step diffusion models with the rectified flow framework. To distill the student flow, naive distillation (Eq. (8)) requires simulating the entire student 2-rectified flow with off-the-shelf solvers, e.g., Runge-Kutta 45 (RK45) solver, to get the distillation target ODE[$\mathbf{v}$]($\mathbf{x}_1$). In most times, practitioners generate a dataset of M teacher samples in advance

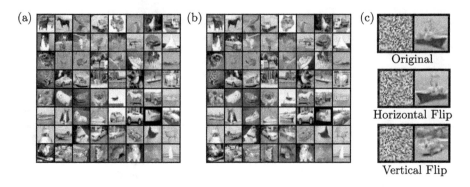

Fig. 2. (a) Generation from 1-Rectified Flow trained without data augmentation. (b) Generation from the 1-Rectified Flow model in (a) after applying horizontal flip to the same set of random noises in (a). (c) Horizontally flipping the noise results in horizontally flipped generated image, but vertical flip does not result in vertically flipped generated image.

to training the student model, denoted as $\mathcal{D}_{distill} = \left\{ \left(\mathbf{x}_1^{(i)}, \text{ODE}[\mathbf{v}](\mathbf{x}_1^{(i)}) \right) \right\}_{i=1}^{M}$. Such simulation takes tens of steps, leading to excessive time and storage for the distilled one-step model. On the one hand, for small models in our scenario, we want to use large M to reach satisfying one-step models due to their limited capacity. On the other hand, we also expect to avoid costly data generation in the distillation stage.

To balance this trade-off, we propose flow-guided distillation, where we complement the expensive direct distillation with an additional regularization. This regularization is based on few-step generation from the 2-rectified flow $\mathbf{v}_\phi$. When estimating the integration result of $\text{ODE}[\mathbf{v}_\phi](\mathbf{x}_1)$, we can adopt the forward Euler solver,

$$\mathbf{x}_{t_i} = \mathbf{x}_{t_{i+1}} + (t_i - t_{i+1})\mathbf{v}_\phi(\mathbf{x}_{t_{i+1}}, t_{i+1}), \quad \forall i \in \{0, 1, \ldots, N-1\}, \tag{10}$$

where the N time steps $\{t_i\}_{i=0}^N$ are defined by the user and $t_0 = 0, t_N = 1$. To find an appropriate supervision for one-step simulation without cumbersome precise simulation with advanced solvers, a feasible choice is two-step simulation, which is,

$$\hat{\mathbf{x}}_0 = \mathbf{x}_1 - (1-t)\mathbf{v}_\phi(\mathbf{x}_1, 1) - t\mathbf{v}_\phi(\mathbf{x}_t, t). \tag{11}$$

Here, t can be an intermediate time point between $(0, 1)$. Using this two-step approximation, we get another distillation loss,

$$\mathcal{L}_{\text{2-step}}(\phi') := \mathbb{E}_{\mathbf{x}_1 \sim \pi_1} \left[\int_0^1 \mathbb{D}(\mathbf{x}_1 - (1-t)\mathbf{v}_\phi(\mathbf{x}_1, 1) - t\mathbf{v}_\phi(\mathbf{x}_t, t), \mathbf{x}_1 - \mathbf{v}_{\phi'}(\mathbf{x}_1, 1))dt \right]. \tag{12}$$

While this two-step generation is not as accurate as the precise generation $\text{ODE}[\mathbf{v}_\phi](\mathbf{x}_1)$ with Runge-Kutta solvers, it can serve as a powerful additional regularization in the distillation process to boost the performance of the student

Algorithm 1. Flow-Guided Distillation

Require: Pre-trained 2-rectified flow v_ϕ, dataset $\mathcal{D}_{distill}$ generated with v_ϕ.
1: Initialize the one-step student model $\mathbf{v}_{\phi'}$ with the weights of $\mathbf{v}_\phi$.
2: **repeat**
3:     Randomly sample $(\mathbf{x}_1, \text{ODE}[\mathbf{v}_\phi](\mathbf{x}_1)) \sim \mathcal{D}_{distill}$.
4:     Compute $\mathcal{L}_{\text{distill}}(\phi')$ with Eq. (8).
5:     Randomly sample $\mathbf{x}_1 \sim \pi_1$.
6:     Compute $\mathcal{L}_{\text{2-step}}(\phi')$ with Eq. (12).
7:     Compute $\mathcal{L}_{\text{combined}}(\phi') = \mathcal{L}_{\text{distill}}(\phi') + \mathcal{L}_{\text{2-step}}(\phi')$.
8:     Optimize ϕ' with an gradient-based optimizer using $\nabla_{\phi'}\mathcal{L}_{\text{combined}}$.
9: **until** $\mathcal{L}_{\text{combined}}$ converges.
10: **Return** one-step model $v_{\phi'}$.

one-step model. Because two-step generation is fast, we can compute this term online as extra supervision to the student one-step model without the need to increase the size of $\mathcal{D}_{distill}$. Our final distillation loss is,

$$\mathcal{L}_{\text{combined}}(\phi') = \mathcal{L}_{\text{distill}}(\phi') + \mathcal{L}_{\text{2-step}}(\phi'). \tag{13}$$

It is worth mentioning that we can use more simulation steps to improve the accuracy of the few-step regularization within the computational budget. In our experiments, using two-step generation as our regularization already gives satisfying improvement. We leave the discovery of other efficient regularization as our future work. The whole procedure of our distillation stage is captured in Algorithm 1.

4 Experiments

In this section, we provide empirical evaluation of SlimFlow and compare it with prior arts. The source code is available at https://github.com/yuanzhi-zhu/SlimFlow.

4.1 Experimental Setup

Datasets and Pre-trained Teacher Models. Our experiments are performed on the CIFAR10 [24] 32×32 and the FFHQ [19] 64×64 datasets to demonstrate the effectiveness of SlimFlow. Additionally, we evaluated our method's performance on the ImageNet [3] 64×64 dataset, focusing on conditional generation tasks. The pre-trained large teacher models are adopted from the official checkpoints from previous works, Rectified Flow [29] and EDM [18].

Implementation Details. For experiments on CIFAR10 32×32 and the FFHQ 64×64, we apply the U-Net architecture of NCSN++ proposed in [56]. For experiments on ImageNet 64×64, we adopt a different U-Net architecture proposed in [4]. In the CIFAR10 experiments, we executed two sets of experiments with

Table 1. Comparison on CIFAR10. '*' refers to reproduced results.

Category	Method	#Params	NFE (↓)	FID (↓)	MACs (↓)
Teacher Model	EDM [18]	55.7M	35	1.96	20.6G
	1-Rectified Flow [31]	61.8M	127	2.58	10.3G
Diffusion + Pruning	Proxy Pruning [25]	38.7M	100	4.21	
	Diff-Pruning [7]	27.5M	100	4.62	5.1G
	Diff-Pruning [7]	19.8M	100	5.29	3.4G
	Diff-Pruning [7]	14.3M	100	6.36	2.7G
Diffusion + Fast Samplers	AMED-Solver [70]	55.7M	5	7.14	20.6G
	GENIE [6]	61.8M	10	5.28	10.3G
	3-DEIS [66]	61.8M	10	4.17	10.3G
	DDIM [52]	35.7M	10	13.36	6.1G
	DPM-Solver-2 [34]	35.7M	10	5.94	6.1G
	DPM-Solver-Fast [34]	35.7M	10	4.70	6.1G
Diffusion + Distillation	DSNO [69]	65.8M	1	3.78	
	Progressive Distillation [49]	60.0M	1	9.12	
	Score Mismatching [64]†	24.7M	1	8.10	
	TRACT [1]	19.4M	1	6.47	
	Diff-Instruct [37]†	55.7M	1	4.53	20.6G
	TRACT [1]	55.7M	1	3.78	20.6G
	DMD [65]†	55.7M	1	3.77	20.6G
	Consistency Distillation (EDM teacher) [53]	55.7M	1	3.55	20.6G
	1-Rectified Flow (+distill) [29]	61.8M	1	6.18	10.3G
	2-Rectified Flow (+distill) [29]	61.8M	1	4.85	10.3G
	3-Rectified Flow (+distill) [29]	61.8M	1	5.21	10.3G
	Consistency Distillation (EDM teacher) [53]*	27.9M	1	6.83	6.6G
	SlimFlow (EDM teacher)	27.9M	1	4.53	6.6G
	Knowledge Distillation [35]	35.7M	1	9.36	6.1G
	Consistency Distillation (EDM teacher) [53]*	15.7M	1	7.21	3.7G
	SlimFlow (EDM teacher)	15.7M	1	5.02	3.7G
	SlimFlow (1-Rectified Flow teacher)	15.7M	1	5.81	3.7G

two different teacher models: one is the pre-trained 1-Rectitied Flow [31] and the other is the pre-trained EDM model [18]. For all the other experiments, we adopt only the pre-trained EDM model as the large teacher model, unless specifically noted otherwise. All the experiments are conducted on 4 NVIDIA 3090 GPUs. In Distillation, we found replacing $v_\phi(\mathbf{x}_1, 1)$ with $v_{\phi'}(\mathbf{x}_1, 1)$ in Eq. (12) leads to high empirical performance as $v_{\phi'}$ is a better one-step generator and speed up the training by saving one forward of the 2-rectified flow, so we keep that in our practice. More details can be found in the Appendix.

Evaluation Metrics. We use the Fréchet inception distance (FID) [12] to evaluate the quality of generated images. In our experiments, we calculate the FID by comparing 50,000 generated images with the training dataset using Clean-FID [43]. We also report the number of parameters (#Params), Multiply-Add Accumulation (MACs), and FLoating-point OPerations per second (FLOPs) as metrics to compare the computational efficiency of different models. It is important to note that in this paper, both MACs and FLOPs refer specifically to the computation required for a single forward inference pass through the deep neural network. We use Number of Function Evaluations (NFEs) to denote the number of inference steps.

Fig. 3. Random generation from our best one-step small models on three different datasets. Left: CIFAR10 32 × 32 (#Params = 27.9M). Mid: FFHQ 64 × 64 (#Params = 27.9M). Right: ImageNet 64 × 64 (#Params = 80.7M).

Table 2. Comparison on FFHQ 64 × 64 and ImageNet 64 × 64.

Dataset	Method	#Params	NFE (↓)	FID (↓)	MACs (↓)	FLOPs (↓)
FFHQ 64 × 64	EDM [18]	55.7M	79	2.47	82.7G	167.9G
	DDIM [52]	55.7M	10	18.30	82.7G	167.9G
	AMED-Solver [70]	55.7M	5	12.54	82.7G	167.9G
	BOOT [10]	66.9M	1	9.00	25.3G	52.1G
	SlimFlow (EDM teacher)	27.9M	1	7.21	26.3G	53.8G
	SlimFlow (EDM teacher)	15.7M	1	7.70	14.8G	30.4G
ImageNet 64 × 64	EDM [18]	295.9M	79	2.37	103.4G	219.4G
	DDIM [52]	295.9M	10	16.72	103.4G	219.4G
	AMED-Solver [70]	295.9M	5	13.75	103.4G	219.4G
	DSNO [69]	329.2M	1	7.83		
	Progressive Distillation [49]	295.9M	1	15.39	103.4G	219.4G
	Diff-Instruct [37]	295.9M	1	5.57	103.4G	219.4G
	TRACT [1]	295.9M	1	7.43	103.4G	219.4G
	DMD [65]	295.9M	1	2.62	103.4G	219.4G
	Consistency Distillation [53]	295.9M	1	6.20	103.4G	219.4G
	Consistency Training [53]	295.9M	1	13.00	103.4G	219.4G
	BOOT [10]	226.5M	1	16.30	78.2G	157.4G
	SlimFlow (EDM teacher)	80.7M	1	12.34	31.0G	67.8G

4.2 Empirical Results

SlimFlow on CIFAR10. We report the results on CIFAR10 in Table 1 and Fig. 3. For all the SlimFlow models, we train them using Annealing Reflow with the data pairs generated from the large teacher model for 800,000 iterations, then distilling them to one-step generator with flow-guided distillation for 400,000 iterations. For the consistency distillation baselines, we train them for 1,200,000 iterations with the same EDM teacher to ensure a fair comparison. With only 27.9M parameters, one-step SlimFlow outperforms the 61.8M 2-Rectified Flow+Distill model (FID: 4.53 ↔ 4.85). With the less than 20M parameters, SlimFlow gives better FID (5.02, #Params = 15.7M) than TRACT (6.48, #Params = 19.4M) and Consistency Distillation (7.21,

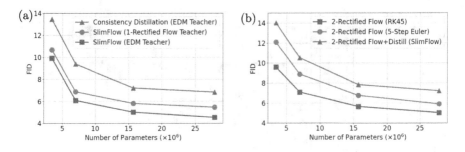

Fig. 4. (a) Comparison of models trained with different methods on CIFAR10. (b) Comparison between 2-rectified flow and the distilled one-step generator on CIFAR10.

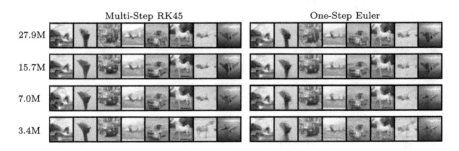

Fig. 5. CIFAR10 samples from 2-rectified flow models trained with Annealing Reflow. All images are generated with the same set of random noises.

#Params = 15.7M). Additional SlimFlow results with different parameters can be found in Fig. 4a for comparison with Consistency Distillation [53] and in Fig. 4b for comparison with 2-rectified flow.

SlimFlow on FFHQ and ImageNet. We report the results of SlimFlow on FFHQ 64 × 64 and ImageNet 64 × 64 in Table 2 and Fig. 3. For ImageNet 64 × 64, the models are trained in the conditional generation scenarios where class labels are provided. In FFHQ, our SlimFlow models surpasses BOOT with only 15.7M parameters. In ImageNet, SlimFlow obtains comparable performance as the 295.9M models trained with Consistency Training, BOOT and Progressive Distillation, using only 80.7M parameters. These results demonstrates the effectiveness of SlimFlow in training efficient one-step generative models.

Analysis of Annealing Reflow. We examine the straightening effect of Annealing Reflow. We measured the straightness in the Annealing Reflow stage of models with different sizes on both the CIFAR10 32×32 and the FFHQ 64×64 dataset. Here, straightness is defined as in Eq. (7) following [31,33]. In Fig. 6a, straightness decreases as $\beta(k)$ gradually approaches 0. In Fig. 6b, we observe that the straightness of the resulting 2-rectified flows decreases as their number of parameters increase. In Fig. 5, samples generated with four 2-rectified flows

Table 3. Ablation study on the reflow strategy of SlimFlow.

Annealing	H-Flip	FID ($\downarrow$)
-	-	5.84
-	✓	5.06
✓	-	5.46
✓	✓	4.51

Table 4. Ablation study on the annealing strategy of SlimFlow. Recall that k is the number of training iterations. We set $K_{step} = 50,000$.

$\beta(k)$	FID ($\downarrow$)
0	5.06
$\exp(-k/K_{step})$	4.78
$\cos(\pi \min(1, k/2K_{step})/2)$	4.79
$(1 + \cos(\pi \min(1, k/2K_{step})))/2$	4.70
$1 - \min(1, k/6K_{step})$	4.51

Table 5. Ablation study on the distillation stage of SlimFlow on CIFAR10.

Initialize from v_ϕ	Source of $\mathcal{D}_{distill}$	Loss $\mathbb{D}$	with $\mathcal{L}_{\text{2-step}}$	FID
-	1-rectified flow v_θ	ℓ_2	-	12.09
✓	1-rectified flow v_θ	ℓ_2	-	8.98
-	2-rectified flow v_ϕ	ℓ_2	-	9.19
✓	2-rectified flow v_ϕ	ℓ_2	-	7.90
✓	2-rectified flow v_ϕ	ℓ_2	✓	7.30
✓	2-rectified flow v_ϕ	LPIPS	-	6.43
✓	2-rectified flow v_ϕ	LPIPS	✓	5.81

with different samplers are presented. These results demonstrate the effectiveness of Annealing Reflow in learning straight generative flows.

4.3 Ablation Study

Annealing Reflow. We examine the design choices in Annealing Reflow. We train 2-rectified flow for 800,000 iterations and measure its FID with RK45 solver. In Table 3, we report the influence of the Annealing Reflow strategy and the effectiveness of exploiting the intrinsic symmetry of reflow. It can be observed that both the annealing strategy and the intrinsic symmetry improve the performance of 2-rectified flow, and their combination gives the best result. In Table 4, we analyze the schedule of $\beta(k)$. When $\beta(k) = 0$, it is equivalent to training 2-rectified flow with random initialization. All other schedules output lower FID than $\beta(k) = 0$, showing the usefulness of our annealing strategy. We adopt $\beta(k) = 1 - \min(1, k/6K_{step})$ as our default schedule in our experiments.

Flow-Guided Distillation. In Table 5, we analyze the influence of different choices in our distillation stage. We found that using our 2-step regularization boosts the FID the distilled one-step model. Using ℓ_2 loss, it improves the FID from 7.90 to 7.30. Using LPIPS loss, it improves the FID from 6.43 to 5.81. We use the model architecture with 15.7M parameters for all ablation studies.

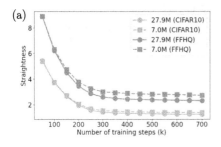 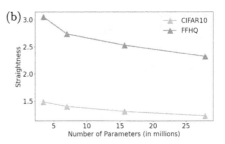

Fig. 6. (a) Straightness of 2-rectified flows with different sizes during Annealing Reflow. (b) Final straightness of 2-rectified flows with different numbers of parameters.

5 Related Work

Reducing the Inference Steps of Diffusion Models. Diffusion models generate new samples through an iterative process, where a noisy image is repeatedly denoised by a neural network. To accelerate this process, researchers have proposed various methods, which can be categorized into two main approaches.

The first category is training-free methods, which aim to reduce the number of inference steps for existing diffusion models by optimizing the sampling process. By reformulating diffusion models into PF-ODEs [56], numerous techniques are introduced to accelerate sampling while minimizing quality loss. Some of the notable examples in this area include DDIM [52], EDM [18], DEIS [66] and DPM-solver [34]. A recent advancement in this category is the AMED-Solver [70], which leverages the mean value theorem to minimize discretization error, achieving high-quality generation with even fewer function evaluations.

Beyond these fast ODE salvers, distillation-based methods aim to achieve few-step sampling, or even one-step sampling, by training a new student model with the pre-trained multi-step diffusion models as teachers. Progressive Distillation [49] proposed to repeatedly train a student network whose step size is twice as the step size of the teacher and set the student as a new teacher for the next round. BOOT [10] suggested distilling the knowledge of the teacher models in a data-free manner with the help of the signal-ODE. Distribution Matching Distillation [65] extended the idea of Variational Score Distillation [59] to train a one-step generator by alternatively updating the one-step generator and a fake data score function. Consistency models [36,53] are a new family of generative models that trains few-step diffusion models by applying consistency loss. Additionally, several methods have incorporated adversarial training to enhance the performance of one-step diffusion models [21,62,64].

Reducing the Size of Diffusion Models. The increasing demand for low-budget and on-device applications necessitates the development of compact diffusion models, as current state-of-the-art models like Stable Diffusion are typically very large. To address this challenge, researchers have explored various compression techniques, primarily focusing on network pruning and quantization methods.

Network pruning involves selectively removing weights from the network and subsequently fine-tuning the pruned model to maintain performance comparable to the pre-trained version. Diff-Pruning [7] utilizes Taylor expansion over pruned timesteps to identify non-contributory diffusion steps and important weights through informative gradients. BK-SDM [20] discovers that block pruning combined with feature distillation is an efficient and sufficient strategy for obtaining lightweight models.

Another line of work is model quantization, which aims to reduce the storage and computational requirements of diffusion models during deployment [11,15,26,27,50,57]. Q-diffusion [26] introduces timestep-aware calibration and split short-cut quantization, tailoring post-training quantization (PTQ) methods specifically for diffusion models. EfficientDM [11] proposes a quantization-aware variant of the low-rank adapter (QALoRA), achieving Quantization-Aware Training (QAT)-level performance with PTQ-like efficiency.

In summary, SlimFlow advances the state of the art by addressing the dual challenge of minimizing both inference steps and neural network size, with the ultimate goal of developing the most efficient diffusion models. While our approach shares some similarities with MobileDiffusion [68], which also explores efficient structure design and acceleration of diffusion models, SlimFlow distinguishes itself in several key aspects: (1) SlimFlow is built upon the rectified flow framework, which provides more stable training dynamics than the GAN-based training used in MobileDiffusion; (2) Unlike MobileDiffusion's focus on fine-grained specific network structure optimization, SlimFlow offers a general framework applicable to a wide range of efficient network architectures. This flexibility allows our approach to leverage advancements in efficient network design across the field, including MobileDiffusion's efficient text-to-image network.

6 Limitations and Future Works

While our approach demonstrates advancements in efficient one-step diffusion models, we acknowledge several limitations and areas for future research. The quality of our one-step model is inherently bounded by the capabilities of the teacher models, specifically the quality of synthetic data pairs they generate. This limitation suggests a direct relationship between teacher model performance and the potential of our approach. In the future, we will leverage more advanced teacher models and real datasets. Besides, we plan to extend our method to more network architectures, e.g., transformers and pruned networks. Finally, given sufficient computational resources, we aim to apply our approach to more complex diffusion models, such as Stable Diffusion.

7 Conclusions

In this paper, we introduced SlimFlow, an innovative approach to developing efficient one-step diffusion models. Our method can significantly reduce model complexity while preserving the quality of one-step image generation, as evidenced by our results on CIFAR-10, FFHQ, and ImageNet datasets. This work paves the way for faster and more resource-efficient generative modeling, broadening the potential for real-world applications of diffusion models.

Acknowledgements. We would like to thank Zewen Shen for his insightful discussion during this work.

References

1. Berthelot, D., et al.: Tract: denoising diffusion models with transitive closure time-distillation. arXiv preprint arXiv:2303.04248 (2023)
2. Crowson, K., Baumann, S.A., Birch, A., Abraham, T.M., Kaplan, D.Z., Shippole, E.: Scalable high-resolution pixel-space image synthesis with hourglass diffusion transformers. arXiv preprint arXiv:2401.11605 (2024)
3. Deng, J., Dong, W., Socher, R., Li, L.J., Li, K., Fei-Fei, L.: ImageNet: a large-scale hierarchical image database. In: 2009 IEEE Conference on Computer Vision and Pattern Recognition, pp. 248–255. IEEE (2009)
4. Dhariwal, P., Nichol, A.: Diffusion models beat GANs on image synthesis. In: Advances in Neural Information Processing Systems, vol. 34, pp. 8780–8794 (2021)
5. Dinh, L., Sohl-Dickstein, J., Bengio, S.: Density estimation using real NVP. arXiv preprint arXiv:1605.08803 (2016)
6. Dockhorn, T., Vahdat, A., Kreis, K.: Genie: higher-order denoising diffusion solvers. In: Advances in Neural Information Processing Systems, vol. 35, pp. 30150–30166 (2022)
7. Fang, G., Ma, X., Wang, X.: Structural pruning for diffusion models. In: Advances in Neural Information Processing Systems, vol. 36 (2024)
8. Geyer, M., Bar-Tal, O., Bagon, S., Dekel, T.: TokenFlow: consistent diffusion features for consistent video editing. arXiv preprint arXiv:2307.10373 (2023)
9. Goodfellow, I., et al.: Generative adversarial networks. Commun. ACM **63**(11), 139–144 (2020)
10. Gu, J., Zhai, S., Zhang, Y., Liu, L., Susskind, J.: Boot: data-free distillation of denoising diffusion models with bootstrapping. arXiv preprint arXiv:2306.05544 (2023)
11. He, Y., Liu, J., Wu, W., Zhou, H., Zhuang, B.: EfficientDM: efficient quantization-aware fine-tuning of low-bit diffusion models. arXiv preprint arXiv:2310.03270 (2023)
12. Heusel, M., Ramsauer, H., Unterthiner, T., Nessler, B., Hochreiter, S.: GANs trained by a two time-scale update rule converge to a local Nash equilibrium. In: Advances in Neural Information Processing Systems, vol. 30 (2017)
13. Ho, J., et al.: Imagen video: high definition video generation with diffusion models. arXiv preprint arXiv:2210.02303 (2022)
14. Ho, J., Jain, A., Abbeel, P.: Denoising diffusion probabilistic models. In: Advances in Neural Information Processing Systems, vol. 33, pp. 6840–6851 (2020)
15. Huang, Y., Gong, R., Liu, J., Chen, T., Liu, X.: TFMQ-DM: temporal feature maintenance quantization for diffusion models. arXiv preprint arXiv:2311.16503 (2023)
16. Hyvärinen, A., Dayan, P.: Estimation of non-normalized statistical models by score matching. J. Mach. Learn. Res. **6**(4) (2005)

17. Jeong, M., Kim, H., Cheon, S.J., Choi, B.J., Kim, N.S.: Diff-TTS: a denoising diffusion model for text-to-speech. arXiv preprint arXiv:2104.01409 (2021)
18. Karras, T., Aittala, M., Aila, T., Laine, S.: Elucidating the design space of diffusion-based generative models. arXiv preprint arXiv:2206.00364 (2022)
19. Karras, T., Laine, S., Aila, T.: A style-based generator architecture for generative adversarial networks. In: Proceedings of the IEEE/CVF Conference on Computer Vision and Pattern Recognition, pp. 4401–4410 (2019)
20. Kim, B.K., Song, H.K., Castells, T., Choi, S.: BK-SDM: a lightweight, fast, and cheap version of stable diffusion. arXiv preprint arXiv:2305.15798 (2023)
21. Kim, D., et al.: Consistency trajectory models: learning probability flow ode trajectory of diffusion. arXiv preprint arXiv:2310.02279 (2023)
22. Kingma, D.P., Welling, M.: Auto-encoding variational bayes. arXiv preprint arXiv:1312.6114 (2013)
23. Kong, Z., Ping, W., Huang, J., Zhao, K., Catanzaro, B.: DiffWave: a versatile diffusion model for audio synthesis. arXiv preprint arXiv:2009.09761 (2020)
24. Krizhevsky, A., Hinton, G., et al.: Learning multiple layers of features from tiny images (2009)
25. Li, W., et al.: Not all steps are equal: efficient generation with progressive diffusion models. arXiv preprint arXiv:2312.13307 (2023)
26. Li, X., et al.: Q-diffusion: quantizing diffusion models. In: Proceedings of the IEEE/CVF International Conference on Computer Vision, pp. 17535–17545 (2023)
27. Li, Y., Xu, S., Cao, X., Sun, X., Zhang, B.: Q-DM: an efficient low-bit quantized diffusion model. In: Advances in Neural Information Processing Systems, vol. 36 (2024)
28. Lipman, Y., Chen, R.T., Ben-Hamu, H., Nickel, M., Le, M.: Flow matching for generative modeling. arXiv preprint arXiv:2210.02747 (2022)
29. Liu, Q.: Rectified flow: a marginal preserving approach to optimal transport. arXiv preprint arXiv:2209.14577 (2022)
30. Liu, R., Wu, R., Van Hoorick, B., Tokmakov, P., Zakharov, S., Vondrick, C.: Zero-1-to-3: Zero-shot one image to 3D object. In: Proceedings of the IEEE/CVF International Conference on Computer Vision, pp. 9298–9309 (2023)
31. Liu, X., Gong, C., Liu, Q.: Flow straight and fast: learning to generate and transfer data with rectified flow. arXiv preprint arXiv:2209.03003 (2022)
32. Liu, X., Wu, L., Ye, M., Liu, Q.: Let us build bridges: understanding and extending diffusion generative models. arXiv preprint arXiv:2208.14699 (2022)
33. Liu, X., Zhang, X., Ma, J., Peng, J., Liu, Q.: Instaflow: one step is enough for high-quality diffusion-based text-to-image generation. arXiv preprint arXiv:2309.06380 (2023)
34. Lu, C., Zhou, Y., Bao, F., Chen, J., Li, C., Zhu, J.: DPM-solver: a fast ode solver for diffusion probabilistic model sampling in around 10 steps. In: Advances in Neural Information Processing Systems, vol. 35, pp. 5775–5787 (2022)
35. Luhman, E., Luhman, T.: Knowledge distillation in iterative generative models for improved sampling speed. arXiv preprint arXiv:2101.02388 (2021)
36. Luo, S., Tan, Y., Huang, L., Li, J., Zhao, H.: Latent consistency models: synthesizing high-resolution images with few-step inference. arXiv preprint arXiv:2310.04378 (2023)
37. Luo, W., Hu, T., Zhang, S., Sun, J., Li, Z., Zhang, Z.: Diff-instruct: a universal approach for transferring knowledge from pre-trained diffusion models. arXiv preprint arXiv:2305.18455 (2023)
38. Ma, X., Fang, G., Wang, X.: DeepCache: accelerating diffusion models for free. arXiv preprint arXiv:2312.00858 (2023)

39. Meng, C., et al.: On distillation of guided diffusion models. In: Proceedings of the IEEE/CVF Conference on Computer Vision and Pattern Recognition, pp. 14297–14306 (2023)
40. Molad, E., et al.: Dreamix: video diffusion models are general video editors. arXiv preprint arXiv:2302.01329 (2023)
41. Nichol, A.Q., Dhariwal, P.: Improved denoising diffusion probabilistic models. In: International Conference on Machine Learning, pp. 8162–8171. PMLR (2021)
42. Papamakarios, G., Nalisnick, E., Rezende, D.J., Mohamed, S., Lakshminarayanan, B.: Normalizing flows for probabilistic modeling and inference. J. Mach. Learn. Res. **22**(57), 1–64 (2021)
43. Parmar, G., Zhang, R., Zhu, J.Y.: On aliased resizing and surprising subtleties in GAN evaluation. In: Proceedings of the IEEE/CVF Conference on Computer Vision and Pattern Recognition, pp. 11410–11420 (2022)
44. Pernias, P., Rampas, D., Richter, M.L., Pal, C., Aubreville, M.: Würstchen: an efficient architecture for large-scale text-to-image diffusion models. In: The Twelfth International Conference on Learning Representations (2023)
45. Poole, B., Jain, A., Barron, J.T., Mildenhall, B.: DreamFusion: text-to-3D using 2D diffusion. arXiv preprint arXiv:2209.14988 (2022)
46. Ramesh, A., Dhariwal, P., Nichol, A., Chu, C., Chen, M.: Hierarchical text-conditional image generation with clip latents. arXiv preprint arXiv:2204.06125 (2022)
47. Razavi, A., Van den Oord, A., Vinyals, O.: Generating diverse high-fidelity images with VQ-VAE-2. In: Advances in neural information processing systems, vol. 32 (2019)
48. Rombach, R., Blattmann, A., Lorenz, D., Esser, P., Ommer, B.: High-resolution image synthesis with latent diffusion models. In: Proceedings of the IEEE/CVF Conference on Computer Vision and Pattern Recognition, pp. 10684–10695 (2022)
49. Salimans, T., Ho, J.: Progressive distillation for fast sampling of diffusion models. arXiv preprint arXiv:2202.00512 (2022)
50. Shang, Y., Yuan, Z., Xie, B., Wu, B., Yan, Y.: Post-training quantization on diffusion models. In: Proceedings of the IEEE/CVF Conference on Computer Vision and Pattern Recognition, pp. 1972–1981 (2023)
51. Sohl-Dickstein, J., Weiss, E., Maheswaranathan, N., Ganguli, S.: Deep unsupervised learning using nonequilibrium thermodynamics. In: International Conference on Machine Learning, pp. 2256–2265. PMLR (2015)
52. Song, J., Meng, C., Ermon, S.: Denoising diffusion implicit models. arXiv preprint arXiv:2010.02502 (2020)
53. Song, Y., Dhariwal, P., Chen, M., Sutskever, I.: Consistency models. arXiv preprint arXiv:2303.01469 (2023)
54. Song, Y., Ermon, S.: Generative modeling by estimating gradients of the data distribution. In: Advances in Neural Information Processing Systems, vol. 32 (2019)
55. Song, Y., Garg, S., Shi, J., Ermon, S.: Sliced score matching: a scalable approach to density and score estimation. In: Uncertainty in Artificial Intelligence, pp. 574–584. PMLR (2020)
56. Song, Y., Sohl-Dickstein, J., Kingma, D.P., Kumar, A., Ermon, S., Poole, B.: Score-based generative modeling through stochastic differential equations. arXiv preprint arXiv:2011.13456 (2020)
57. Wang, C., Wang, Z., Xu, X., Tang, Y., Zhou, J., Lu, J.: Towards accurate data-free quantization for diffusion models. arXiv preprint arXiv:2305.18723 (2023)

58. Wang, H., Du, X., Li, J., Yeh, R.A., Shakhnarovich, G.: Score Jacobian chaining: lifting pretrained 2D diffusion models for 3D generation. In: Proceedings of the IEEE/CVF Conference on Computer Vision and Pattern Recognition, pp. 12619–12629 (2023)
59. Wang, Z., et al.: ProlificDreamer: high-fidelity and diverse text-to-3d generation with variational score distillation. arXiv preprint arXiv:2305.16213 (2023)
60. Wimbauer, F., et al.: Cache me if you can: accelerating diffusion models through block caching. arXiv preprint arXiv:2312.03209 (2023)
61. Wu, L., Gong, C., Liu, X., Ye, M., Liu, Q.: Diffusion-based molecule generation with informative prior bridges. In: Advances in Neural Information Processing Systems, vol. 35, pp. 36533–36545 (2022)
62. Xu, Y., Zhao, Y., Xiao, Z., Hou, T.: UFOGen: you forward once large scale text-to-image generation via diffusion GANs. arXiv preprint arXiv:2311.09257 (2023)
63. Yang, X., Zhou, D., Feng, J., Wang, X.: Diffusion probabilistic model made slim. In: Proceedings of the IEEE/CVF Conference on Computer Vision and Pattern Recognition, pp. 22552–22562 (2023)
64. Ye, S., Liu, F.: Score mismatching for generative modeling. arXiv preprint arXiv:2309.11043 (2023)
65. Yin, T., et al.: One-step diffusion with distribution matching distillation. In: Proceedings of the IEEE/CVF Conference on Computer Vision and Pattern Recognition, pp. 6613–6623 (2024)
66. Zhang, Q., Chen, Y.: Fast sampling of diffusion models with exponential integrator. arXiv preprint arXiv:2204.13902 (2022)
67. Zhang, R., Isola, P., Efros, A.A., Shechtman, E., Wang, O.: The unreasonable effectiveness of deep features as a perceptual metric. In: CVPR (2018)
68. Zhao, Y., Xu, Y., Xiao, Z., Hou, T.: MobileDiffusion: subsecond text-to-image generation on mobile devices. arXiv preprint arXiv:2311.16567 (2023)
69. Zheng, H., Nie, W., Vahdat, A., Azizzadenesheli, K., Anandkumar, A.: Fast sampling of diffusion models via operator learning. In: International Conference on Machine Learning, pp. 42390–42402. PMLR (2023)
70. Zhou, Z., Chen, D., Wang, C., Chen, C.: Fast ode-based sampling for diffusion models in around 5 steps. arXiv preprint arXiv:2312.00094 (2023)

PhysGen: Rigid-Body Physics-Grounded Image-to-Video Generation

Shaowei Liu, Zhongzheng Ren, Saurabh Gupta, and Shenlong Wang(✉)

University of Illinois Urbana-Champaign, Champaign, USA
shenlong@illinois.edu
https://stevenlsw.github.io/physgen/

Abstract. We present PhysGen, a novel image-to-video generation method that converts a single image and an input condition (*e.g.*, force and torque applied to an object in the image) to produce a realistic, physically plausible, and temporally consistent video. Our key insight is to integrate model-based physical simulation with a data-driven video generation process, enabling plausible image-space dynamics. At the heart of our system are three core components: (i) an image understanding module that effectively captures the geometry, materials, and physical parameters of the image; (ii) an image-space dynamics simulation model that utilizes rigid-body physics and inferred parameters to simulate realistic behaviors; and (iii) an image-based rendering and refinement module that leverages generative video diffusion to produce realistic video footage featuring the simulated motion. The resulting videos are realistic in both physics and appearance and are even precisely controllable, showcasing superior results over existing data-driven image-to-video generation works through quantitative comparison and comprehensive user study. PhysGen's resulting videos can be used for various downstream applications, such as turning an image into a realistic animation or allowing users to interact with the image and create various dynamics.

1 Introduction

Looking at the images in Fig. 1, we can effortlessly predict or visualize the potential outcomes of various physical effects applied to the car, the curling stone, and the stack of dominos. This understanding of physics empowers our imagination of the counterfactual: we can mentally imagine various consequences of an action from an image without experiencing them. We aim to provide computers with similar capabilities – understanding and simulating physics from a single image and creating realistic animations. We expect the resulting video to produce realistic animations with physically plausible dynamics and interactions.

S. Gupta and S. Wang—Equal advising.

Supplementary Information The online version contains supplementary material available at https://doi.org/10.1007/978-3-031-73007-8_21.

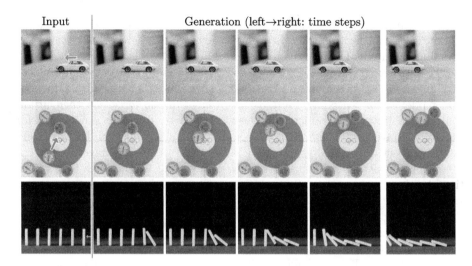

Fig. 1. Given a single image, training-free PhysGen generates future frames controlled by physics and initial state. Semantics, geometry and dynamics are reasoned during generation and the result videos are both physics-grounded and photo-realistic.

Tremendous progress has been made in generating realistic and physically plausible video footage. Conventional graphics leverage model-based dynamics to simulate plausible motions and utilize a graphics renderer to simulate images. Such methods provide realistic and controllable dynamics. However, both the dynamics of physics and lighting physics are predefined, making it hard to achieve our goal of animating an image captured in the real world. On the other hand, data-driven image-to-video (I2V) generation techniques [6,8,56] learn to create realistic videos from a single image through training diffusion models over internet-scale data. However, despite advancements in video generative models [6,8,56], the incorporation of real-world physics principles into the video generation process remains largely unexplored and unsolved. Consequently, the synthesized videos often lack temporal coherence and fail to replicate authentic object motions observed in reality. Furthermore, such text-driven methods also lack fine-grained controllability. For instance, they cannot simulate the consequences of different forces and torques applied to an object.

In light of this gap, we propose a paradigm shift in video generation through the introduction of *model-based video generative models*. In contrast to existing purely generative methods, which are trained in a data-driven manner and rely on diffusion models to learn image space dynamics via scaling laws, we propose grounding object dynamics explicitly using rigid body physics, thereby integrating fundamental physical principles into the generation process. As classical mechanics theory reveals, the motion of an object is determined by its physical properties (*e.g.*, mass, elasticity, surface roughness) and external factors (*e.g.*, external forces, environmental conditions, boundary conditions).

We tackle our image-to-video generation problem via a computational framework, PhysGen. PhysGen consists of three stages: 1) image-based physics understanding; 2) model-based dynamics simulation; and 3) generative video rendering with dynamics guidance. Our method first infers object compositions and physical parameters from the input image through large visual foundation model-based reasoning (Sect. 3.1). Given an input force, we then utilize the inferred physical parameters to perform realistic dynamic simulations for each object, accounting for their rigid body dynamics, collisions, frictions, and elasticity (Sect. 3.2). Finally, guided by motion and the input image, we present a novel generative video diffusion-based renderer that outputs the final realistic video footage (Sect. 3.3). Several illustrative examples of our framework are shown in Fig. 1. As shown in this figure, our approach combines the best of the worlds of data-driven generative modeling's appearance realism and model-based dynamics's verified physical plausibility. Notably, our generation process is highly controllable, allowing users to specify underlying physics parameters and initial conditions. This capability facilitates a range of interactive applications. Moreover, our proposed generation pipeline operates solely during inference time, eliminating the need for any training.

We evaluate PhysGen generation ability on multiple data source including the web and self-captured images. We compare against both state-of-the-art image-to-video models [17,85,93] as well as image editing method [27] quantitatively and qualitatively. We also perform user-study to evaluate the physical-realism and photo-realism of the generated video. Our method demonstrates physics-informed results for controllable image-to-video generation, producing realistic video sequences with grounded rigid-body dynamics and interaction. PhysGen combines learning-based generative approaches with traditional model-based physics to deliver visually appealing results without any training. Our approach harnesses a plethora of ingredients from previous works, particularly recent progress in large foundational visual models for segmentation [39,47,60,91], physical understanding [2,55], normal estimation [22,25], relighting, and generative rendering [8,17]. While these individual ingredients exist, our system combines them in a novel manner to enable an exciting new capacity, featuring physical plausibility for video generation at an unprecedented level.

Our contributions can be summarized as follows: 1) We propose a novel image-space dynamics model that can understand physical parameters from a single image and simulate realistic, controllable, and interactive dynamics. 2) We present a novel image-to-video system by integrating our presented dynamics model with generative diffusion-based video priors. The model is training-free, and the resulting videos are highly realistic, physically plausible, and controllable, consistently surpassing existing state-of-the-art video generative models.

2 Related Work

Image-Based Physics Reasoning and Simulation. Our image-based dynamics model is built upon physical simulation. Symbolic physics simulators or physics engines have been extensively studied for various physical processes, including rigid body dynamics, fluid simulation, and deformable objects [35–37,40,48,54,71]. Such methods have been widely used across graphics, scientific discovery, robotics, and video effects. However, these rule-based or solver-based simulators face limitations in terms of expressiveness, efficiency, generalizability, and parameter tuning. To overcome these challenges, neural physics, which combines traditional physics simulators/engines with deep neural networks, has been developed to model dynamic processes such as planar pushing and bouncing [4], as well as object interactions [3]. Domain-specific network architectures [7,44,53,62] have been developed to address various physical problems. Additionally, the integration of neural scene representations or large language models with physics simulators [42,49,84] has also been investigated recently. Nevertheless, most forward physics models, whether neural or not, have not resolved the dependency on pre-defined physical parameters, which are often specified by users.

Reasoning physical parameters from visual data allows us to simulate without manually defined physics, further enhancing convenience and realism for image-based physics simulation. To enable this, subsequent research efforts [20,42,43,79–81,83,86] have introduced pipelines that first extract scene representations of the physical world from images or videos using neural networks, and then utilize these representations for either physics reasoning or simulation. Despite the promise of these approaches, they primarily rely on synthetic data for training, which might suffer from generalization issues, or on inverse physics, which requires inference from observed physical phenomena, often requiring video observations instead of a single image. In our work, we aim to generate physics-grounded, high-fidelity predictive videos from static input images of real-world objects with complex backgrounds. Contrary to existing work, we investigate leveraging the power of large pretrained visual foundational models [2,39,55] for physics reasoning in a zero-shot manner.

Video Generative Model. Video generative models have experienced remarkable advancements in recent years [6,8,11,17,26,28,30,32,41,56,64,73,75,85,90,93], with the state-of-the-art framework [56] achieving the capability to generate photo-realistic and coherent imaginative videos from text instructions using diffusion models [33,57,61,65,67]. Despite progress, challenges remain, such as the need for extensive training data, object permanence, realistic lighting, and reducing unrealistic motion and physics violations. Overcoming these is key for advancing video generative modeling and its applications, especially in physics-accurate areas like scientific discovery and robotics. Our work addresses some issues by proposing a novel image-to-video framework that merges a model-based approach with a pretrained video prior, grounding the generated content with real-world physics.

Image Animation. Our method is heavily inspired by image-based animation [18,21,34,63,69,76,87] and cinemagraphs. One common way to improve quality is to incorporate class-specific heuristics [18,38], which can be physically simulated and integrated into the generation process. Recent advancements in

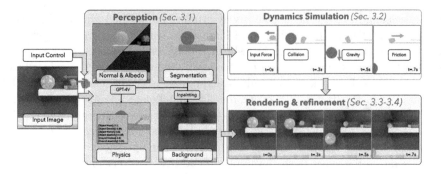

Fig. 2. Method overview. Our framework consists of three interleaved components: the perception module, the dynamics simulation module, and the rendering module. The perception module interprets the semantics, geometry, and physical parameters in the given image. The dynamics simulation module simulates the rigid-body motion and interactions of each instance in the scene, governed by Newton's Laws and physical constraints. The rendering module renders the final outcome, leveraging an off-the-shelf relighting model and diffusion-model-based video priors.

deep learning have led to data-driven solutions in this domain, where temporal neural networks are trained on video datasets to directly predict consecutive video frames from an input image [10,16,23,29,34,74]. Moreover, there is growing interest in interactive [9,10,45] and controllable [1,21] image-to-video synthesis. To further improve temporal coherence, more priors such as motion fields [23,34,50,51,68,87] or flows [13,27,94], 3D geometry information [59,78], and user annotations [45] have been subsequently introduced. In this work, we propose a model-based solution and tackle the image animation task from a new perspective, where our method takes a single image and user-specific initial force or torque as input, and models object dynamics explicitly via rigid physics simulation. Our key insight is that physical parameters can be inferred from a single image automatically through large foundation models such as GPT-4V [2,55].

3 Approach

Given a single input image, our goal is to generate a realistic T-frame video featuring rigid body dynamics and interactions. Our generation can be conditioned on user-specific inputs in the form of an initial force $\mathbf{F}_0$ and/or torque τ_0. Our key insight is to incorporate physics-informed simulation into the generation process, ensuring the dynamics are realistic and interactive, thus producing large, plausible object motion. To achieve this, we present a novel framework consisting of three stages: physical understanding, dynamics generation, and generative rendering. The perception module aims to reason about the semantics, collision geometry, and physics of the objects presented in the image (Sect. 3.1). The dynamics module then leverages the input force/torque, as well as the inferred shape and geometry, to simulate the rigid-body motions of the objects, as well as object-object and object-scene interactions, such as friction and collisions (Sect. 3.2). Finally, we convert the simulated dynamics into pixel-level motion and combine image-based warping and generative video to render the final video (Sect. 3.3). Figure 2 depicts the overall framework.

3.1 Perception

Simulating dynamics on an image requires a holistic understanding of object compositions, materials, geometry, and physics. Toward this goal, we designed our perception module to infer such properties from a single image. Our key insight is to harness readily available, large pretrained models [2,14,25,55,60] to achieve this goal, as shown in Fig. 2.

Segmentation. Identifying and segmenting each individual physical entity lays the foundation for image-based dynamics. Inspired by the recent success of large pretrained segmentation models, we incorporate GPT-4V [2,55] to recognize all image categories and send to Grounded-SAM [60] to detect and segment each individual instance-level objects $\{\mathbf{o}^i \in \mathbb{R}^{W \times H}\}_i^N$, where $\mathbf{o}^i$ is the binary mask for i-th object. We also query the GPT-4V to classify the objects into foreground and background based on their movability. The foreground objects are send to the physical simulation once user inputs are applied. We extract collision boundaries and supporting edges (e.g., ground, walls, etc.) from the background objects. The details are presented in the supplementary.

Physical Properties Reasoning. Realistic physics dependents on accurate physical parameters, such as surface friction, mass, and elasticity. Unlike prior physics-based image dynamics work [18], our work leverages visual foundational models to reason physical properties directly. Our solution is simple yet effective. Inspired by the success of mask-based prompting [2,55,88], we directly ask GPT-4V [55] for certain physical properties, providing an object mask overlaid on the input image. Following [83,92], we send the crafted prompt to GPT-4V, which contains GPT-4V instructions to return a quantitative measure of each queried property in a metric unit. For each object, we query its mass, elasticity, and friction coefficient (M_i, E_i, μ_i). We use the Coulomb friction model [58] for friction modeling. The elasticity coefficient is a scalar, where 0.0 results in no bounce, and 1.0 results in a perfect bounce. Given each object mass M_i in grams and its shape primitive, we compute the rotational inertia $\mathbf{I}_i$ accordingly.

Geometry Primitives. Rasterized instance masks are not suitable for physical simulation; hence, we need to convert each object into vectorized shape primitives. We fit two types of primitive shapes: a circle and generic polygons. For each object, we choose the primitive type with the maximum coverage of its segmentation. For circles, we fit the center and radius to cover the segmentation mask. Circles allow us to realistically simulate rolling motions. For non-circle objects, we instead fit the generic polygons. Specifically, we perform contour extraction on the segmentation masks to obtain the polygon vertices.

Intrinsic Decomposition. The dynamic movement of the objects will also result in comprehensive changes in shading, as their position wrt the lighting environment changes. To compensate for this effect during rendering, we must perform image decomposition to infer albedo, normal as well as lighting from the single image for each object ($\mathbf{A}_0^i, \mathbf{N}_0^i, \mathbf{L}_0$). To achieve this we leverage off-the-shelf intrinsic decomposition model [14] to compute $\mathbf{A}_0^i$ and $\mathbf{L}_0$ and exploits pretrained surface normal estimator for the computation of object normal $\mathbf{N}_0^i$.

Background Inpainting. Once the foreground objects move, they will leave holes in the background. To produce realistic and complete video footage, we leverage an off-the-shelf generative image inpainting model [91] to recover the complete background scene without foreground objects. The inpainted image $\mathbf{B} \in \mathbb{R}^{W \times H \times 3}$ will be used for composition in the rendering stage.

3.2 Image Space Dynamics Simulation

Given the foreground objects with physical properties, we use rigid-body physics for dynamics simulation in image spaces. We choose to perform simulations in image spaces for three reasons: 1) image-space dynamics are better coupled with our output video; 2) a full 3D simulation requires complete 3D scene reconstruction and understanding, which remains an unsolved problem from a single image; 3) image-space dynamics can already cover various object dynamics and have been widely used in prior work [4,18,34,45,79]. Specifically, at a given time t, each rigid object i is characterized by its 2D pose and velocity at its center of mass. The position includes a translation $\mathbf{t}^i(t) \in \mathbb{R}^2$ and a rotation $\mathbf{R}^i(t) \in \mathbb{SO}(2)$ specified in a world coordinate. The velocity includes linear velocity $\boldsymbol{\nu}^i(t) \in \mathbb{R}^2$ and angular velocity $\boldsymbol{\omega}^i(t) \in \mathbb{R}^2$. Hence, the state of the each object i at time t can be represented as $\mathbf{q}^i(t) = [\mathbf{t}^i(t), \mathbf{R}^i(t), \boldsymbol{\nu}^i(t), \boldsymbol{\omega}^i(t)]$.

Following [24], the rigid body motion dynamics is given by Eq. (1) for each object. For simplicity, we omit the object index i in the following:

$$\frac{d}{dt}\mathbf{q}(t) = \frac{d}{dt}\begin{bmatrix}\mathbf{t}(t)\\\mathbf{R}(t)\\\boldsymbol{\nu}(t)\\\boldsymbol{\omega}(t)\end{bmatrix} = \begin{bmatrix}\mathbf{v}(t)\\\boldsymbol{\omega}(t) \times \mathbf{R}(t)\\\frac{\mathbf{F}(t)}{M}\\\mathbf{I}(t)^{-1}(\boldsymbol{\tau} - \boldsymbol{\omega}(t) \times \mathbf{I}(t)\boldsymbol{\omega}(t))\end{bmatrix} \quad (1)$$

where $\mathbf{F}$ is the force, $\boldsymbol{\tau}$ is the torque, M is the mass of the object, $\mathbf{I}(t) = \mathbf{R}(t)\mathbf{IR}(t)^T$ is the rotation inertia in world coordinates, $\mathbf{I} \in \mathbb{R}^{2 \times 2}$ is the inertia matrix in its body coordinates. This inertia indicates its resistance against rotational motion. The state $\mathbf{q}(t)$ is given by integrating Eq. (1):

$$\mathbf{q}(t) = \mathbf{q}(0) + \int_0^t \frac{d}{dt}\mathbf{q}(t)dt = \mathbf{q}(0) + \sum_{i=1}^{T}\frac{d}{dt}\mathbf{q}(t)|_{t=t_i}\Delta_t \quad (2)$$

where $\mathbf{q}(0)$ is the initial condition specified in the input image. We compute the integral using numerical ODE integrations, e.g. Euler method [19]. Given the initial state of the objects, the physical motion simulation module synthesis each foreground object future motion. The location of each object is updated by the affine transformation $\mathbf{T}(t) = [\mathbf{R}(t), \mathbf{t}(t)]$ from the initial location.

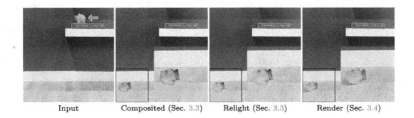

Input   Composited (Sec. 3.3)   Relight (Sec. 3.3)   Render (Sec. 3.4)

Fig. 3. Rendered video comparison. We show a toy example of piggy bank. The left shows the input frame, and the rest 3 are future frame generations. The composited piggy bank hasn't been aware of the light change. The relighted output synthesized no shadows beneath. The rendered output from diffusion model is most photo realistic.

External Force and Torque. For each object, the external forces $\mathbf{F}(t)$ and torques τ include gravity, friction, and elasticity between the object's surface and the environment. Specifically, we consider the initial external force and torque from user input, as well as gravity, rolling friction, and sliding friction.

Collision. When objects move, they might collide with each other or with the scene boundary (e.g., hitting a wall). Therefore, collision checking is necessary at every time step. A collision will result in offset forces being applied from its center of mass, producing τ and causing the object to rotate. In collision reactions, changes in energy and momentum are minimized to adhere to Newton's conservation laws. In the case of an object falling, it could bounce back or start rolling on the ground due to the elasticity of both the object and the ground.

3.3 Rendering

Given the simulated motion sequence $\{\mathbf{q}^i(0), \ldots, \mathbf{q}^i(t), \ldots\}_i^N$ and the input image $\mathbf{C}$, the rendering module outputs the rendered video $\{\ldots, \mathbf{C}(t), \ldots\}$. There are a few desiderata for generating realistic video: it needs to be temporally consistent, complete, reflect the lighting changes as the object moves, and accurately represent the simulated image space dynamics. To achieve this goal, we design a novel motion-guided rendering pipeline consisting of an image-based transformation and composition module to ensure motion awareness, a relighting module to ensure the plausibility of lighting physics and a generative video editing module that retouches the composed and relit video to ensure realism.

Composited Video. Given the object state from the physical motion module, we compose an initial video by alpha-blending the foreground scene with the static inpainted background from Sect. 3.1. The foreground scene is rendered by performing forward-warping from the input image to future frames using the affine transformation $\mathbf{T}(t)$. The alpha channel is computed using the same procedure, with the input being the segmentation mask in Sect. 3.1:

$$\hat{\mathbf{X}}(t) = \texttt{composite}(\mathbf{B}, \texttt{warp}(\mathbf{X}^0, \mathbf{T}^0(t))\ldots\texttt{warp}(\mathbf{X}^i, \mathbf{T}^i(t))\ldots..)$$

where $\mathbf{X}^i$ are segmented input image of i-th object. Apart from the RGB sequence $\hat{\mathbf{V}} = \{\ldots\hat{\mathbf{X}}(t)\ldots\}$, we also use the same affine warping and composition to compute the blended albedo map $\hat{\mathbf{A}}(t)$, which is later used for relighting.

Relighting. Our relighting module simulates the changes in shading due to object movement. Specifically, it takes as input the transformed albedo $\hat{\mathbf{A}}(t)$ and RGB image $\hat{\mathbf{X}}(t)$ at each time t, as well as the estimated lighting $\mathbf{L}$. We first compute the surface normal $\hat{\mathbf{N}}(t)$ from the composited RGB image $\hat{\mathbf{X}}(t)$ using an off-the-shelf normal estimator [22]. We then perform reshading using the parametric illumination model [14]: $\tilde{\mathbf{X}}(t) = f(\hat{\mathbf{X}}(t), \hat{\mathbf{A}}(t), \hat{\mathbf{N}}(t), \mathbf{L})$, which returns the relit image $\tilde{\mathbf{X}}(t)$ for each frame t.

3.4 Generative Refinement with Latent Diffusion Models

One limitation of the single-image based relighting model is that it neglects the background and cannot handle complicated lighting effects due to complex object-object interactions, such as ambient occlusion and cast shadows. Moreover, the composition results in unnatural boundaries. To address these issues and enhance realism, we incorporated a diffusion-based video to refine the relit video, denoted as $\tilde{\mathbf{V}} = \{\ldots, \tilde{\mathbf{X}}(t), \ldots\}$, and obtain the final video output.

Specifically, we incorporate a pretrained video diffusion model to refine our video $\tilde{\mathbf{V}}$. We first use the pretrained latent diffusion encoder [61] to encode the guidance video $\tilde{\mathbf{V}}$ into a latent code $\mathbf{z}$. Inspired by SDEdit [52], we add noise to the guided latent code and gradually denoise it using the pretrained video diffusion. Given that the content in the guided latent code is already satisfactory, we do not perform the denoising process from scratch. Instead, we define an initial noise strength s where $s \in [0,1]$ controls the amount of noise added to $\mathbf{z}$. In practice, we find that a certain approach finds a good trade-off between fidelity and realism. At each denoising step, we ensure the foreground objects are as consistent as possible with the guidance (copied from the guided latent code) while synthesizing new content (e.g., shadows) in the background (using the denoised latent code). To this end, we use a fusion weight w to control how much of the generated latent to inject into the foreground; the fusion weight is gradually increased during the denoising as the perturbation of noise decreases. We also define a fusion timestamp δ as the signal to stop the fusion. The final denoised latent code $\mathbf{z}_*$ is then decoded to our final video output $\mathbf{V}$. Please see the supplementary material for a detailed algorithm.

Figure 3 depicts the composed video $\hat{\mathbf{V}}$, the relit video $\tilde{\mathbf{V}}$, and our final output $\mathbf{V}$, illustrating the effectiveness of each rendering component.

4 Experiments

Implementation Details. For physical body simulation, we use a 2D rigid body physics simulator Pymunk [12]. For each experiment, we run 120 simulation steps and uniformly sample 16 frames to form the videos. We set the generation resolution at 512×512 so that frames can easily fit into the diffusion models. For video diffusion prior model, we use SEINE [17]. For inference, DDIM sampling [66] is used and the noise strength i set to $s = 0.5$, *i.e.*, executing 25/50 steps for denoising. The latents fusion stops at timestamp $\delta = 5$.

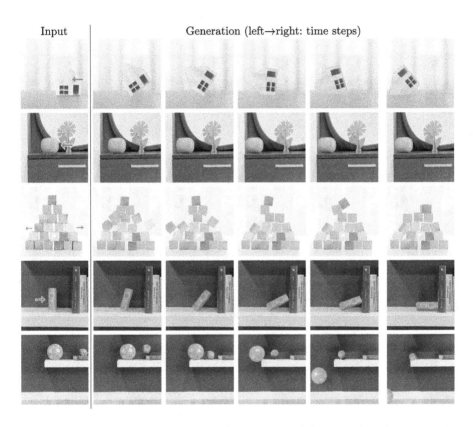

Fig. 4. Video generation results. Left: Input initial frame with red arrows representing the initial force or speed on the object. Right: Generated future frames.

Speed. PhysGen's run-time speed is 3 min for one image-to-video gen on a Nvidia A40 GPU. Image understanding and relighting takes the most time.

Data. To show the generalizability and robustness of our method, we use both internet data and self-captured indoor images from a cell phone. The dataset is diverse, with variations in lighting, object count, geometries, physical attributes, and environmental boundaries. We compare different approaches on 15 photos that require physical reasoning. Additionally, we collect 50 recorded videos of a given scene as ground truth by varying input conditions for quantitative comparison. Randomly selected sequences are shown in the supplementary material.

Baselines. We compare against two streams of approaches. The first stream is current state-of-the-art I2V models: SEINE [17], I2VGen-XL [93] and DynamiCrafter [85]. These methods take both an image and a text prompt as input for generation. The text prompt is generated by GPT-4V [55] to describe the future events. Another stream we compare is the image-based manipulation approach Motion Guidance [27]. This method uses an image and optical flow as inputs to predict future movement and generate corresponding images. We employ computed motion flow from our physical motion module as the target input for Motion Guidance to optimize each video frame.

Table 1. Human evaluation. We evaluate both physical-realism and photo-realism of the three competing image-to-video models. The user is asked to evaluate the generated video using a 5-point scale from strongly disagree (1) to strongly agree (5) that the video is physical-realistic and photo-realistic. Our method ranks the first in both terms.

Methods	Physical-realism↑	Photo-realism↑
SEINE [17]	1.39	1.86
DynamiCrafter [85]	1.68	1.81
I2VGen-XL [93]	2.11	2.25
Ours	**4.14**	**3.86**

Table 2. Quantitative evaluation. We measure the Image-FID and Motion-FID of videos generated by four methods against the collected GT videos for the given scene. Our method achieves both low Image-FID and Motion-FID.

Methods	Image-FID↓	Motion-FID↓
SEINE [17]	138.89	38.76
DynamiCrafter [85]	**99.64**	109.61
I2VGen-XL [93]	104.62	82.04
Ours	105.70	**30.20**

4.1 Results

We show our generated results on 8 different physical procedures from multiple sources in Fig. 4. Our method could simulate reasonable complex physical procedures which involves object-object interaction, object-scene interaction with different physical properties and initial conditions.

Visual Comparisons. The comparison results are shown in Fig. 5. As can be seen, image-video generation methods could not produce any physically plausible

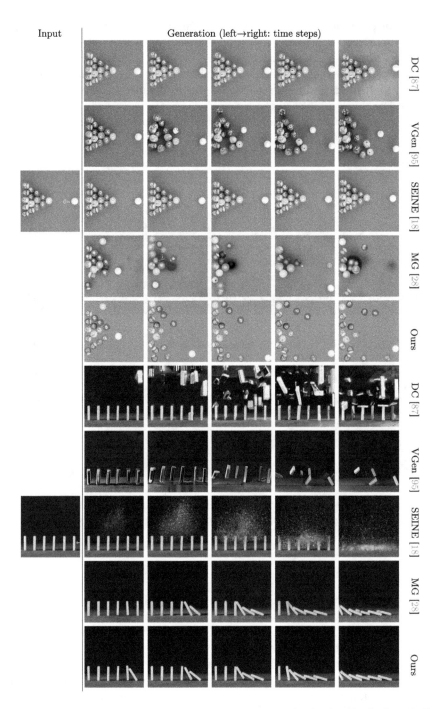

Fig. 5. Qualitative comparison. against DynamiCrafter(**DC**) [85], I2VGen-XL(**VGen**) [93], **SEINE** [17], Motion Guidance(**MG**) [27].

futures. For each method, even we carefully tune the input prompt, the models couldn't understand the direction, physical meanings and interactions. All 3 models fail to synthesize realistic videos. For motion guidance with accurate motion flow provided, the results is better but brings a lot inconsistency. When there are multiple objects interaction, the method suffer from introducing new contents. For the domino case, the generation is not smooth and contain artifacts as it could not accurately model the physical procedure.

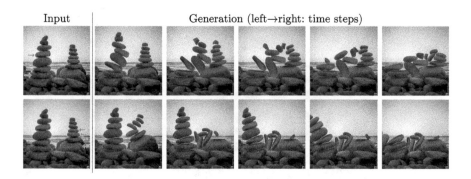

Fig. 6. Controllability. PhysGen generates diverse results reflecting initial forces.

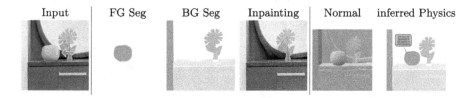

Fig. 7. Qualitative results from the perception module. The results include foreground and background mask, inpainting output, normal map and inferred physics.

4.2 Human Evaluation

To comprehensively assess both the **physical-realism** and **photo-realism** of the generated videos, we conducted human evaluations following a similar methodology to prior work [15,77,82,89]. Here by physical-realism we mean whether the content of the generated video is realistic in its dynamics and object interactions and accurately reflects the instructed movements.

Specifically, we include 15 videos with diverse objects and conditions for evaluation, compared against other three I2V model generations (SEINE [17], DynamiCrafter [85], I2VGen-XL [93]). For each compared baseline method, we meticulously adjusted the input prompts and conducted multiple runs to select the most plausible output. In contrast, our method did not undergo specific tuning for each video. We randomized the presentation order and asked 14 participants to rate the videos on a five-point scale, from strongly disagree (1) to strongly agree (5) for physical-realism and photo-realism. The results in Table 1 show that our method achieves the highest ratings. Our average scores for physical-realism (4.14) and photo-realism (3.86) fall within the Agree level, while other methods perform poorly.

4.3 Quantitative Evaluation

We quantitatively evaluate the collected GT videos of a given scene. The GT data includes 833 images. For each method, we generate 50 videos with random samples, collecting 800 images per method. Our method varies initial conditions to ensure all generated videos are different. Following [11], we adopt the Fréchet Inception Distance (FID) [31] as evaluation metric. We didn't use Fréchet Video Distance (FVD) [72] due to the small sample size and the goal to disentangle evaluation of appearance and motion. We evaluate appearance using **Image-FID** and motion using **Motion-FID**. For Motion-FID, we use RAFT [70] to extract and colorize optical flow [5], then compute FID from these rendered images. The results are shown in Table 2. We notice other methods which either generate minimal motion, leading to low Image-FID and high Motion-FID (e.g., DynamiCrafter [85]), or produce realistic motion but fail to maintain image consistency, resulting in low Motion-FID and high Image-FID (e.g., SEINE [17]). Our approach achieves both low Image-FID and Motion-FID compared to others.

4.4 Analysis

Perception Evaluation. We evaluate the perception module using 10 complex open-world images with 118 annotated movable instances (COCO [46] format). Our system achieves **0.93** precision, **0.82** recall at 0.5 IoU. More details are shown in the supplementary. Figure 7 shows intermediate outputs from our perception module and simulation module.

Physical Reasoning. We compare the physical parameters estimated by GPT-4V to random values sampled from human-estimated ranges. Without GPT-4V, Image-FID increases from **105.70** to **111.01**, Motion-FID from **30.20** to **36.60**. It shows the physical reasoning are crucial for appearance and motion realism.

Controllability. We showcase the controllability and diversity of our method in Fig. 6 by varying the initial conditions. For different rows, we adjust the direction and magnitude of initial speed and forces. The results show diverse physical-plausible trajectories. This demonstrates the potential of applying our method for diverse physical world generations.

Limitations. Our method is an initial step towards physics-based video generation but has two main limitations: it focuses mainly on rigid objects, limiting its application to non-rigid ones, and it lacks a comprehensive 3D understanding, making it unable to handle out-of-plane motions. We leave leveraging deformable physics and full 3D understanding as future work.

5 Conclusion

We present PhysGen, a novel image-to-video generation method that turns a static image into a high-fidelity, physics-grounded, and temporally coherent videos. The main module is a model-based physical simulator with a data-driven video generation process. PhysGen enables a series of controllable and interactive downstream applications. PhysGen provide a new image-to-video generation paradigm and hopefully can inspire more follow-up works.

Acknowledgements. This project is supported by NSF Awards #2331878, #2340254, and #2312102, the IBM IIDAI Grant, and an Intel Research Gift. We greatly appreciate the NCSA for providing computing resources. We thank Tianhang cheng for helpful discussions. We thank Emily Chen and Gloria Wang for proofreading.

References

1. Aberman, K., Weng, Y., Lischinski, D., Cohen-Or, D., Chen, B.: Unpaired motion style transfer from video to animation. ACM TOG (2020)
2. Achiam, J., et al.: GPT-4 technical report. arXiv preprint arXiv:2303.08774 (2023)
3. Ajay, A., et al.: Combining physical simulators and object-based networks for control. In: ICRA (2019)
4. Ajay, A., et al.: Augmenting physical simulators with stochastic neural networks: case study of planar pushing and bouncing. In: IROS (2018)
5. Baker, S., Scharstein, D., Lewis, J.P., Roth, S., Black, M.J., Szeliski, R.: A database and evaluation methodology for optical flow. IJCV (2011)
6. Bar-Tal, O., et al.: Lumiere: a space-time diffusion model for video generation. arXiv preprint arXiv:2401.12945 (2024)
7. Battaglia, P., Pascanu, R., Lai, M., Jimenez Rezende, D., et al.: Interaction networks for learning about objects, relations and physics. In: NeurIPS (2016)
8. Blattmann, A., et al.: Stable video diffusion: scaling latent video diffusion models to large datasets. arXiv preprint arXiv:2311.15127 (2023)
9. Blattmann, A., Milbich, T., Dorkenwald, M., Ommer, B.: iPOKE: poking a still image for controlled stochastic video synthesis. In: ICCV (2021)
10. Blattmann, A., Milbich, T., Dorkenwald, M., Ommer, B.: Understanding object dynamics for interactive image-to-video synthesis. In: CVPR (2021)
11. Blattmann, A., et al.: Align your latents: high-resolution video synthesis with latent diffusion models. In: CVPR (2023)
12. Blomqvist, V.: Pymunk (2023). https://pymunk.org
13. Bowen, R.S., Tucker, R., Zabih, R., Snavely, N.: Dimensions of motion: monocular prediction through flow subspaces. In: 3DV (2022)
14. Careaga, C., Miangoleh, S.M.H., Aksoy, Y.: Intrinsic harmonization for illumination-aware compositing. arXiv preprint arXiv:2312.03698 (2023)
15. Chen, H., et al.: Videocrafter1: open diffusion models for high-quality video generation. arXiv preprint arXiv:2310.19512 (2023)
16. Chen, X., et al.: Livephoto: real image animation with text-guided motion control. arXiv preprint arXiv:2312.02928 (2023)
17. Chen, X., et al.: Seine: short-to-long video diffusion model for generative transition and prediction. arXiv preprint arXiv:2310.20700 (2023)

18. Chuang, Y.Y., Goldman, D.B., Zheng, K.C., Curless, B., Salesin, D.H., Szeliski, R.: Animating pictures with stochastic motion textures. ACM TOG (2005)
19. Ciarlet, P.G., Lions, J.L.: Handbook of Numerical Analysis. Gulf Professional Publishing (1990)
20. Davis, A., Bouman, K.L., Chen, J.G., Rubinstein, M., Durand, F., Freeman, W.T.: Visual vibrometry: estimating material properties from small motion in video. In: CVPR (2015)
21. Davis, A., Chen, J.G., Durand, F.: Image-space modal bases for plausible manipulation of objects in video. ACM TOG (2015)
22. Eftekhar, A., Sax, A., Malik, J., Zamir, A.: Omnidata: a scalable pipeline for making multi-task mid-level vision datasets from 3D scans. In: ICCV (2021)
23. Endo, Y., Kanamori, Y., Kuriyama, S.: Animating landscape: self-supervised learning of decoupled motion and appearance for single-image video synthesis. arXiv preprint arXiv:1910.07192 (2019)
24. Friedland, B.: Control system design: an introduction to state-space methods. Courier Corporation (2012)
25. Fu, X., et al.: Geowizard: unleashing the diffusion priors for 3D geometry estimation from a single image. arXiv preprint arXiv:2403.12013 (2024)
26. Ge, S., et al.: Preserve your own correlation: a noise prior for video diffusion models. In: ICCV (2023)
27. Geng, D., Owens, A.: Motion guidance: diffusion-based image editing with differentiable motion estimators. In: ICLR (2024)
28. Girdhar, R., et al.: Emu video: factorizing text-to-video generation by explicit image conditioning. arXiv preprint arXiv:2311.10709 (2023)
29. Guo, Y., et al.: Animatediff: animate your personalized text-to-image diffusion models without specific tuning (2023)
30. Gupta, A., et al.: Photorealistic video generation with diffusion models. arXiv preprint arXiv:2312.06662 (2023)
31. Heusel, M., Ramsauer, H., Unterthiner, T., Nessler, B., Hochreiter, S.: GANs trained by a two time-scale update rule converge to a local Nash equilibrium. In: NeurIPS (2017)
32. Ho, J., et al.: Imagen video: high definition video generation with diffusion models. arXiv preprint arXiv:2210.02303 (2022)
33. Ho, J., Jain, A., Abbeel, P.: Denoising diffusion probabilistic models. In: NeurIPS (2020)
34. Holynski, A., Curless, B.L., Seitz, S.M., Szeliski, R.: Animating pictures with Eulerian motion fields. In: CVPR (2021)
35. Hu, Y., et al.: Difftaichi: differentiable programming for physical simulation. In: ICLR (2020)
36. Hu, Y., et al.: A moving least squares material point method with displacement discontinuity and two-way rigid body coupling. ACM TOG (2018)
37. Hu, Y., Li, T.M., Anderson, L., Ragan-Kelley, J., Durand, F.: Taichi: a language for high-performance computation on spatially sparse data structures. ACM TOG (2019)
38. Jhou, W.C., Cheng, W.H.: Animating still landscape photographs through cloud motion creation. IEEE Trans. Multimed. (2015)
39. Kirillov, A., et al.: Segment anything. arXiv:2304.02643 (2023)
40. Koenig, N., Howard, A.: Design and use paradigms for gazebo, an open-source multi-robot simulator. In: IROS (2004)
41. Kondratyuk, D., et al.: VideoPoet: a large language model for zero-shot video generation. arXiv preprint arXiv:2312.14125 (2023)

42. Li, X., et al.: PAC-NeRF: physics augmented continuum neural radiance fields for geometry-agnostic system identification. In: ICLR (2023)
43. Li, Y., et al.: Visual grounding of learned physical models. In: ICML (2020)
44. Li, Y., Wu, J., Tedrake, R., Tenenbaum, J.B., Torralba, A.: Learning particle dynamics for manipulating rigid bodies, deformable objects, and fluids. In: ICLR (2019)
45. Li, Z., Tucker, R., Snavely, N., Holynski, A.: Generative image dynamics. arXiv preprint arXiv:2309.07906 (2023)
46. Lin, T.-Y., et al.: Microsoft COCO: common objects in context. In: Fleet, D., Pajdla, T., Schiele, B., Tuytelaars, T. (eds.) ECCV 2014. LNCS, vol. 8693, pp. 740–755. Springer, Cham (2014). https://doi.org/10.1007/978-3-319-10602-1_48
47. Liu, S., et al.: Grounding DINO: marrying DINO with grounded pre-training for open-set object detection. arXiv preprint arXiv:2303.05499 (2023)
48. Liu, T., Bargteil, A.W., O'Brien, J.F., Kavan, L.: Fast simulation of mass-spring systems. ACM TOG (2013)
49. Lv, J., et al.: GPT4motion: scripting physical motions in text-to-video generation via blender-oriented GPT planning. In: CVPRW, pp. 1430–1440 (2024)
50. Mahapatra, A., Kulkarni, K.: Controllable animation of fluid elements in still images. In: CVPR (2022)
51. Mallya, A., Wang, T.C., Liu, M.Y.: Implicit warping for animation with image sets. In: NeurIPS (2022)
52. Meng, C., et al.: SDEdit: guided image synthesis and editing with stochastic differential equations. In: International Conference on Learning Representations (2022)
53. Mrowca, D., et al.: Flexible neural representation for physics prediction. In: NeurIPS (2018)
54. NVIDIA: Nvidia Physx (2019). https://developer.nvidia.com/physx-sdk
55. OpenAI: GPT-4v(ISION) system card (2023)
56. OpenAI: Creating video from text (2024). https://openai.com/sora
57. Peebles, W., Xie, S.: Scalable diffusion models with transformers. arXiv preprint arXiv:2212.09748 (2022)
58. Popova, E., Popov, V.L.: The research works of coulomb and amontons and generalized laws of friction. Friction (2015)
59. Qiu, H., et al.: Relitalk: relightable talking portrait generation from a single video (2023)
60. Ren, T., et al.: Grounded SAM: assembling open-world models for diverse visual tasks (2024)
61. Rombach, R., Blattmann, A., Lorenz, D., Esser, P., Ommer, B.: High-resolution image synthesis with latent diffusion models. In: CVPR (2022)
62. Sanchez-Gonzalez, A., et al.: Graph networks as learnable physics engines for inference and control. In: ICML (2018)
63. Schödl, A., Szeliski, R., Salesin, D.H., Essa, I.: Video textures. In: PACMCGIT (2000)
64. Singer, U., et al.: Make-a-video: text-to-video generation without text-video data. arXiv preprint arXiv:2209.14792 (2022)
65. Sohl-Dickstein, J., Weiss, E., Maheswaranathan, N., Ganguli, S.: Deep unsupervised learning using nonequilibrium thermodynamics. In: ICML (2015)
66. Song, J., Meng, C., Ermon, S.: Denoising diffusion implicit models. arXiv preprint arXiv:2010.02502 (2020)
67. Song, Y., Ermon, S.: Generative modeling by estimating gradients of the data distribution. In: NeurIPS (2019)

68. Sugimoto, R., He, M., Liao, J., Sander, P.V.: Water simulation and rendering from a still photograph. In: SIGGRAPH Asia (2022)
69. Szummer, M., Picard, R.W.: Temporal texture modeling. In: ICIP (1996)
70. Teed, Z., Deng, J.: Raft: recurrent all-pairs field transforms for optical flow. In: ECCV, pp. 402–419 (2020)
71. Todorov, E., Erez, T., Tassa, Y.: Mujoco: a physics engine for model-based control. In: IROS (2012)
72. Unterthiner, T., Van Steenkiste, S., Kurach, K., Marinier, R., Michalski, M., Gelly, S.: Towards accurate generative models of video: a new metric & challenges. arXiv preprint arXiv:1812.01717 (2018)
73. Villegas, R., et al.: Phenaki: variable length video generation from open domain textual description. arXiv preprint arXiv:2210.02399 (2022)
74. Wang, X., et al.: Videocomposer: compositional video synthesis with motion controllability. arXiv preprint arXiv:2306.02018 (2023)
75. Wang, Y., et al.: Lavie: high-quality video generation with cascaded latent diffusion models. arXiv preprint arXiv:2309.15103 (2023)
76. Wei, L.Y., Levoy, M.: Fast texture synthesis using tree-structured vector quantization. In: PACMCGIT (2000)
77. Wei, Y., et al.: Dreamvideo: composing your dream videos with customized subject and motion. arXiv preprint arXiv:2312.04433 (2023)
78. Weng, C.Y., Curless, B., Kemelmacher-Shlizerman, I.: Photo wake-up: 3D character animation from a single photo. In: CVPR (2019)
79. Wu, J., Lim, J.J., Zhang, H., Tenenbaum, J.B., Freeman, W.T.: Physics 101: learning physical object properties from unlabeled videos. In: BMVC (2016)
80. Wu, J., Lu, E., Kohli, P., Freeman, W.T., Tenenbaum, J.B.: Learning to see physics via visual de-animation. In: NeurIPS (2017)
81. Wu, J., Yildirim, I., Lim, J.J., Freeman, W.T., Tenenbaum, J.B.: Galileo: perceiving physical object properties by integrating a physics engine with deep learning. In: NeurIPS (2015)
82. Wu, R., Chen, L., Yang, T., Guo, C., Li, C., Zhang, X.: Lamp: learn a motion pattern for few-shot-based video generation. arXiv preprint arXiv:2310.10769 (2023)
83. Xia, H., Lin, Z.H., Ma, W.C., Wang, S.: Video2game: real-time, interactive, realistic and browser-compatible environment from a single video. In: CVPR (2024)
84. Xie, T., et al.: PhysGaussian: physics-integrated 3d gaussians for generative dynamics. In: CVPR (2024)
85. Xing, J., et al.: DynamiCrafter: animating open-domain images with video diffusion priors. arXiv preprint arXiv:2310.12190 (2023)
86. Xu, Z., Wu, J., Zeng, A., Tenenbaum, J.B., Song, S.: DensePhysNet: learning dense physical object representations via multi-step dynamic interactions. In: RSS (2019)
87. Xue, T., Wu, J., Bouman, K.L., Freeman, W.T.: Visual dynamics: stochastic future generation via layered cross convolutional networks. T-PAMI (2018)
88. Yang, J., Zhang, H., Li, F., Zou, X., Li, C., Gao, J.: Set-of-mark prompting unleashes extraordinary visual grounding in GPT-4v. arXiv preprint arXiv:2310.11441 (2023)
89. Yu, J., et al.: Animatezero: video diffusion models are zero-shot image animators. arXiv preprint arXiv:2312.03793 (2023)
90. Yu, L., et al.: Language model beats diffusion–tokenizer is key to visual generation. arXiv preprint arXiv:2310.05737 (2023)
91. Yu, T., et al.: Inpaint anything: segment anything meets image inpainting. arXiv preprint arXiv:2304.06790 (2023)

92. Zhai, A.J., et al.: Physical property understanding from language-embedded feature fields. In: CVPR (2024)
93. Zhang, S., et al.: I2VGen-XL: high-quality image-to-video synthesis via cascaded diffusion models. arXiv preprint arXiv:2311.04145 (2023)
94. Zhao, J., Zhang, H.: Thin-plate spline motion model for image animation. In: CVPR (2022)

Depth-Aware Blind Image Decomposition for Real-World Adverse Weather Recovery

Chao Wang[1](✉), Zhedong Zheng[2], Ruijie Quan[3], and Yi Yang[3]

[1] ReLER Lab, AAII, University of Technology Sydney, Ultimo, Australia
chao.wang-11@student.uts.edu.au
[2] FST and ICI, University of Macau, Zhuhai, China
zhedongzheng@um.edu.mo
[3] ReLER Lab, CCAI, Zhejiang University, Hangzhou, China
{quanruijie,yangyics}@zju.edu.cn

Abstract. In this paper, we delve into Blind Image Decomposition (BID) tailored for real-world scenarios, aiming to uniformly recover images from diverse, unknown weather combinations and intensities. Our investigation uncovers one inherent gap between the controlled lab settings and the complex real-world environments. In particular, existing BID methods and datasets usually overlook the physical property that adverse weather varies with scene depth rather than a uniform depth, thus constraining their efficiency on real-world photos. To address this limitation, we design an end-to-end Depth-aware Blind Network, namely **DeBNet**, to explicitly learn the depth-aware transmissivity maps, and further predict the depth-guided noise residual to jointly produce the restored output. Moreover, we employ neural architecture search to adaptively find optimal architectures within our specified search space, considering significant shape and structure differences between multiple degradations. To verify the effectiveness, we further introduce two new BID datasets, namely BID-CityScapes and BID-GTAV, which simulate depth-aware degradations on real-world and synthetic outdoor images, respectively. Extensive experiments on both existing and proposed benchmarks show the superiority of our method over state-of-the-art approaches.

Keywords: Image decomposition · Scene depth · Weather recovery

1 Introduction

Image restoration against adverse weather remains a classical yet challenging task for many real-world applications, *e.g.*, auto-driving car, with increasing demands on safety and robustness. Traditional methods usually focus on

Supplementary Information The online version contains supplementary material available at https://doi.org/10.1007/978-3-031-73007-8_22.

Fig. 1. Example images in two proposed depth-aware datasets (**left:** BID-CityScapes *(real-world)*, **right:** BID-GTAV *(synthesized)*). Different from existing datasets, we explicitly involve depth into the simulation process. For instance, rain exists in different formats such as rain streak and raindrop. Rain streak usually appears in the remote area, while raindrops are closer to the camera. Similarly, we also consider snow, haze and dark illumination, which often co-occur during rain. When testing, BID setting demands the model to handle images with an arbitrary combination of multiple adverse weathers, which is challenging yet more overarching.

specific pre-defined weather conditions, such as image deraining [10,22,46,53], dehazing [17,24,43,63], desnowing [3,4,34], low-light enhancement [14,19,28,35], adherent raindrop removal [39,42,59], *etc.* In an attempt to increase model scalability, researchers further propose all-in-one networks [2,31,55,60,61] to universally handle different degradations with multiple model ensembles. However, this line of methods requires extra individual training for each specific weather task. Considering efficiency, some researchers resort to a unified architecture [5,26,30,48,54]. Such an approach can handle a single type of corruption at one time, but still suffers from multiple adverse weather combinations. In nature, adverse weather conditions tend to occur simultaneously in a random combination, *e.g.*, rain usually comes with fog and dim lighting. Therefore, in this work, we study Blind Image Decomposition (BID) task [15,49], which takes a step closer to real-world practice. BID considers the corrupted images as an arbitrary combination of degradation layers and separates the superimposed image into constituent underlying images in a **blind** setting, *i.e.*, both the source components involved in mixing as well as the mixing mechanism are unknown.

Existing methods and datasets for BID [15,49] have achieved competitive performance in the public academic datasets, but usually ignored an inherent property that adverse weather varies with scene depth instead of a uniform depth, limiting the scalability to real-world cases. For instance, objects closer to the camera are mainly affected by the rain streaks with more light reflected into the camera, while objects far away are affected more heavily by the fog and the low-light condition. There remain two problems: (1) How to leverage the depth? Despite a few works [18,45] have discussed degradation modeling combined with depth information, they are still limited to specific weather conditions and fail to work under the BID setting. (2) The scarcity of adverse weather data with depth. The lack of depth-aware BID datasets comprising different weather combinations, intensities and their corresponding ground truth also impedes decomposition algorithm development, which is closer to real-world applications.

To address these two limitations, we propose a new depth-aware network, dubbed **DeBNet** and two large-scale datasets, *i.e.*, BID-CityScapes and BID-GTAV. (1) In particular, DeBNet simultaneously restores arbitrary hybrid adverse weather conditions in a generic framework. DeBNet is an end-to-end network encapsulating the underlying physics principles behind the formation of depth-aware BID. Specifically, DeBNet takes the corrupted image as inputs, predicts a depth map, and then produces a clean image as the output. Considering the features of different weathers usually do not share the same characteristics, we further resort to a Neural Architecture Search (NAS) method to achieve the optimal accuracy-efficiency trade-offs. Specifically, we design a specific BID search space that consists of several effective fundamental restoration operations, such as multi-scale convolutions [47] and self-attention modules [62]. (2) Meanwhile, in order to enable depth-aware BID training, we first construct a BID dataset in real-world autonomous driving scenes based on the CityScape [6] dataset, dubbed **BID-CityScapes**, including five types of adverse weather mixed under BID setting, as well as their corresponding depth maps and clean images. Moreover, to enrich the diversity and enable DeBNet to generalize well on real-world images with random viewpoints, we further generate a high-resolution synthetic dataset using a commercial video game Grand Theft Auto (GTAV) [44], called **BID-GTAV**. Examples of the two datasets are illustrated in Fig. 1. Unlike existing methods that are limited to type-specific or depth-independent degradation modeling, we analyze the physical properties of weather degradations and formulate a depth-aware BID setting including five different weather conditions: rain streak, fog, raindrop, snow and low-light degradation. Our contributions are as follows:

- **What is the remaining gap between lab and real-world images against adverse weather?** We identify one overlooked practical problem between the depth property and the corrupted observation, and introduce a new Depth-aware Blind Network, named **DeBNet**, to explore the inherent depth prior of multiple adverse weathers. The key idea underpinning the network design follows the reverse process to disentangle the noise with the aid of depth prediction. In particular, we leverage the neural architecture search to adaptively find the optimal architecture for processing depth and visual features in an end-to-end manner.
- To verify the effectiveness of the proposed method, we introduce two new BID datasets, called BID-CityScapes and BID-GTAV. To our knowledge, the datasets are the first two depth-aware BID datasets including different types of degradations captured across various scenes and viewpoints. Extensive experiments on two proposed BID benchmarks substantiate that our learned model achieves competitive PSNR and SSIM scores, and is scalable to other existing image restoration benchmarks, including SPAdata [53], DeRaindrop [39], SOTS [25], Snow100K [34], SICE [1] and RainDS [41].

2 Related Work

Blind Image Decomposition. Aiming at the adverse weather removal task, several restoration works [9,32,39,57] have achieved satisfactory results on various degradations. However, these methods still require individual training for each type of degradation, impeding the real-world all-in-one implementation. Based on blind source separation problem [13,23], Han et al. first propose the "Blind Image Decomposition" (BID) [15], regarding rain and other real-world weather corruptions as superimposing and separable to a clean image. Wang et al.further introduce a two-stage BID learning paradigm [49] to utilize a pretrained masked autoencoder (MAE) [16] for efficient fine-tuning on the restoration network. However, these BID methods remain heavily dependent on tedious multi-scale multi-head reconstruction or time-consuming pertaining, and neglect the inherent depth-related physical properties of weather degradations. In comparison, our end-to-end method explicitly leverages the depth prior for a better understanding of the corrupted image against adverse weathers.

Neural Architecture Search (NAS). Neural Architecture Search [21,33,38, 64] automates the designing of neural network architectures for optimal performance while minimizing human hours and efforts. For instance, Li et al.first applies NAS to all-in-one weather removal task [30], while Gou et al.further propose an efficient search space [12] with three task-flexible modules. Quan et al.introduce a two-stage searching and training strategy [41] for the joint removal of raindrops and rain streaks. Yet these methods all tend to search the combinations of each individual operation for specific degradations, which are not suitable for the BID problem. Our work is closely related to [7,30,33] to further formulate the searching task into an end-to-end optimization problem. Different from previous methods, we adopt a unified multi-branch search space to explore the weightings of different operations, considering the inherent patterns of varying feature interference among multiple weathers.

3 Method

3.1 Problem Formulation

Existing methods are usually type-specified and ignore the scene depth during weather formulation. In this paper, we first propose a uniform BID imaging model inherited with depth annotations. Without loss of generality, the degraded image I_{noisy} can be formulated as a composition of a clean background I_{clean} and multiple weather degradation layers, including raindrop (RD), snowflakes (SN), lighting (L), rain streak (RS), and fog (A). The formulation can be written as:

$$I_{noisy} = \underbrace{\overbrace{\underbrace{L}_{Lighting} \odot [\underbrace{I_{clean}(1 - RS - A)}_{Transparency} + \underbrace{RS + A_0 A}_{Scene\ Occlusion}]}^{Depth-variant} \odot \overbrace{\underbrace{(1 - M_{rd} \odot M_{sn})}_{Non-occluded\ Area} + \underbrace{RD + SN}_{Lens\ Occlusion}}^{Depth-invariant}}. \quad (1)$$

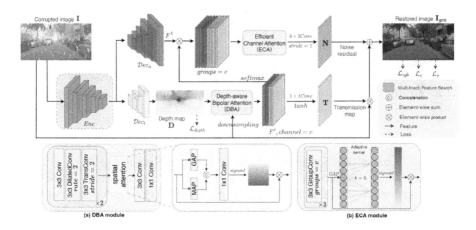

Fig. 2. The architecture of our **D**epth-aware **B**lind image decomposition **Net**work (**DeBNet**). Given the corrupted image I_{noisy} with adverse weather, we extract the visual features via a searchable multi-branch encoder Enc (See Sect. 3.3 and Fig. 3). Then two decoder branches (Sect. 3.2) are deployed to progressively upsample the feature map with skip-connected features from the encoder. In particular, the transmittance decoder branch Dec_t is to predict the scene depth D. The noise branch Dec_n combines the learned attention weights with reconstructed high-resolution features. As shown in (a), we apply Depth-aware Bipolar Attention (DBA) module to compare the depth and the input image as the transmissivity map T. Besides, we also adopt an Efficient Channel Attention (ECA) module to fuse the depth and visual features to produce the depth-guided noise residual N as shown in (b). Finally, we fuse T, N, and I_{noisy} for joint restoration following Eq. (3).

where $\odot$ denotes the element-wise multiplication. From left to right, $L \in [0, 1]$ represents the illumination intensity. $RS \in [0, 1]$, $A \in [0, 1]$ are single-channel soft masks for rain streak and fog with a larger value indicating a higher weather intensity (lower transparency). A_0 is the atmospheric light intensity [45]. $M_{rd} \in \{0, 1\}$ and $M_{sn} \in \{0, 1\}$ denote the masks of raindrop and snow area, which are usually transparent to some extent. RD is the visual occlusion brought by adherent raindrops on lens, representing the blurred imagery reflected by the environment. Similarly, SN represents snowflakes on lens, which totally occlude the scene. It is worth noting that the scene area is usually depth-variant, while the occlusion on lens is depth-invariant. Therefore, for the first term, we should involve the depth prior, and not consider depth for the occlusion on lens. In particular, for **rain streak & fog:** we follow the formulation [18] to simulate the observation changes according to the depth; for **low-light**, we follow existing low-light image enhancement works [8,27] to regard inverted low-lighted inputs as haze images. We also introduce depth information into low-light modeling, which aligns with human intuition that the farther the distance, the heavier the fog, and the weaker the illumination. for **raindrop & snow:** we follow the degradation in [34,39], to split occlusion as two different types, i.e., scene and lens. More details on the degradation modeling are given in the **suppl**.

Considering the multiplication and addition operation in Eq. (1), we can simplify it as follows:
$$I_{noisy} = WI_{clean} + b, \tag{2}$$
where W and b denote the multiplicative and additive factors, respectively. Both factors are partially related to the scene depth. Our depth-aware BID task is a reverse prediction task, and thus can be formulated as:
$$I_{gen} = TI_{noisy} + N, \tag{3}$$
where the **Transmittance** map $T = W^{-1}$ and the **Noise** map $N = -W^{-1}b$. In general, T denotes the positional and intensity information (e.g., the density of rain and fog), while N tends to characterize the intrinsic information of the degradation itself (e.g., blurry or obstruction effects of the raindrops). As shown in Fig. 2, this paper leverages the deep neural network to predict the value of T and N. Different from previous methods that directly learn an additive and non-convex degradation residual [18,41], our method explicitly separates the learning of transmittance and noise factors, which is more conducive for feature learning under multiple overlapping weather conditions.

3.2 Depth-Aware BID Network

Depth-Aware Transmittance Map. As shown in the bottom branch of Fig. 2, the transmittance decoder Dec_t predicts the scene depth with direct supervision. Following [18], we calculate the depth reconstruction loss, which is the $\mathcal{L}_2$ distance between the predicted depth values D_{gen} and ground truth depth D_{gt}:
$$\mathcal{L}_{depth} = \|D_{gen} - D_{gt}\|_2, \tag{4}$$
where D_{gen} and D_{gt} are normalized into $[0, 1]$ with size of a quarter of the input image. To further utilize the learned depth, we propose a Depth-aware Bipolar Attention (DBA) module to produce the transmittance map, i.e., T in Eq. (3). Considering that different weather conditions lead to particular pixel enhancement (e.g., fog, rain) or attenuation (e.g., low light), we adopt a continuous bipolar mask between $[-1, 1]$ for the transmittance map T, indicating both contextual information and coarse weather types underlying the images. As shown in Fig. 2(a), the proposed DBA module takes both downsampled input image and depth map as inputs, containing two cascaded convolution blocks followed by an attention block [56] to emphasize the spatial information. Finally, we apply tanh function to normalize features and generate the transmittance map T.

Depth-Guided Noise Residual. According to Eq. (3), the additive noise map N is also related to the scene depth D, thus we further introduce the noise decoder branch Dec_n to generate the final high-resolution feature map F^h, and then combine it with the learned transmittance features F^t from transmittance branch to produce the noise map N.

Similar to [18], we consider the output of the last convolutional block in the DBA module as a set of un-normalized attention weights, representing the complete transmittance features. In general, each weight $F_i^t (i = 1, 2, ..., c)$ corresponds to a certain type of weather degradation after softmax. We divide the features F^h into c groups as $F_i^h (i = 1, 2, ..., c)$, where c is the channel number of our transmittance features F^t ($c = 32$ in this work). Then, we re-weight each submap F_i^h through an element-wise multiplication with F_i^t to produce the final noise map. After the multiplication, we further perform group convolutions [58] in c groups to individually refine the features of different types of degradation. Since the channel number of a weather condition varies under different combinations, we further adopt an efficient channel attention mechanism (ECA) [52] with the adaptive kernel size k as $|\frac{\log_2(C)+1}{2}|_{odd}$, where C is the channel number of the input features. $|t|_{odd}$ indicates the nearest odd number of t. As shown in Fig. 2(b), ECA can adaptively learn a local cross-channel interaction, thus improving the robustness against different hybrid weather conditions. Finally, we merge all the features from different groups using a 1×1 convolution to produce the noise residual N, with which we combine the transmittance map T to produce the restored clean image I_{gen} following Eq. (3). The reconstruction loss on the restored image can be written as:

$$\mathcal{L}_{rgb} = \|I_{gen} - I_{gt}\|_2, \tag{5}$$

where I_{gen} is the restored image and I_{gt} denotes the ground truth.

Conditional Adversarial Training. To mimic the distributions of the clean image I_{gt}, we first introduce a discriminator Dis to distinguish whether the input image is real or generated. Our source adversarial loss $\mathcal{L}_s$ is written as:

$$\mathcal{L}_s = \mathbb{E}\left[\log\left(1 - Dis\left(I_{gen}\right)\right)\right] + \mathbb{E}\left[\log Dis\left(I_{gt}\right)\right], \tag{6}$$

where $Dis(I)$ predicts the possibility that the image I is uncorrupted. For this discriminator, we hope $Dis(I_{gt}) = 1$ and $Dis(I_{gen}) = 0$, so we maximize the $\mathcal{L}_s$. Meanwhile, we also deploy a degradation classifier Cls to classify the weather combinations from the restored image. The multi-class conditional loss $\mathcal{L}_c$ can be defined as:

$$\mathcal{L}_c = -\sum_{i=0}^{N-1} \log\left(1 - Cls(I_{gen})_i\right), \tag{7}$$

where N denotes the number of degenerate types, $Cls(I)_i$ denote the possibility that the image I suffer from degeneration type i. In this paper, we select five common weather types, e.g., rainstreak, raindrop, snow, fog and low-light condition. It is worth noting that we have multiple combined weather during training. Therefore, we apply 5-dimensional multi-class vectors to indicate the combined weather type. The weather type is not fixed but randomly generated during training. The overall loss function $\mathcal{L}_{gen}$ for DeBNet is formulated as:

$$\mathcal{L}_{gen} = \mathcal{L}_{depth} + \mathcal{L}_{rgb} + \lambda_s \mathcal{L}_s + \mathcal{L}_c, \tag{8}$$

here we empirically set $\lambda_s = 0.1$ to balance the relative weight of $\mathcal{L}_s$.

3.3 Multi-branch Architecture Search

Multi-branch feature connection is a potential solution for the BID task [51], since image restoration is a classical position-sensitive vision problem requiring precise spatial information and rich semantics. To adaptively restore images with various weather combinations, we adopt a multi-branch search method [7] to explore the optimal architecture for various weather combinations. Different from simply utilizing Neural Architecture Search (NAS) to select features from different encoders [30,41], we deeply involve NAS to extract depth and visual features from different convolution kernels and the transformer branch in an efficient way, considering different types of weather correspond to varying sizes of receptive field. In this way, we introduce a channel/token-wise fine-grained search strategy embedded in a multi-branch high-resolution feature space.

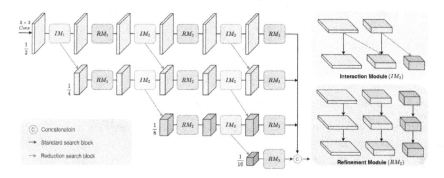

Fig. 3. Illustration of the multi-branch feature search, which contains three-stage of alternated Interaction Module IM and Refinement Module RM. In this work, we deploy the multi-branch learning after a standard 3×3 convolution layer to decrease the resolution by half. Each black line represents a standard search block, and the red line denotes a reduction search block for downsampling. Taking the second stage IM_2, RM_2 as an example, IM_2 generates an extra branch from the previous lowest-resolution branch by a reduction searching block (red line), while RM_2 further refine the features using cascaded standard search blocks (black line) within the same resolution. (Color figure online)

Search Space. Our multi-branch search space is embedded in the encoder Enc. Specifically, the network consists of two modules: the refinement module RM and the interaction module IM, as shown in Fig. 3. Each module is composed of several search blocks operating at different resolutions. We alternate between using two modules to construct a multi-branch search space. The interaction module $IM_i(i = 1, 2, 3, ..., n)$ achieves high-to-low resolution feature transformation to maintain more semantic details, while the refinement module $RM_i(i = 1, 2, 3, ..., n)$ obtains larger receptive fields and multi-scale features by stacking searching blocks in each branch. Finally, multi-branch features are resized and concatenated together, connected to the double-branch decoder.

Search Block. Unlike previous NAS-based restoration methods [30,41] that are designed for specific tasks, we aim to customize the network for various weather combinations. Our searching block contains two paths: a MixConv [47] path for multi-scale feature extraction, and a lightweight Transformer [7] to provide more global contexts in a residual manner. Instead of selecting different sizes or stacking order of the single-scale convolution kernels, Mix-Conv divides all channels into groups and applies multi-scale convolution kernels to each group in a depth-wise complementary way, which can extract features with different receptive field sizes. The number of convolutional channels and the number of tokens in the Transformer are searchable parameters as channels in depth-wise convolutions are independent in the searching block [7,36], any convolution channels or transformer queries can be easily removed without affecting the other search blocks. More details on the search block are given in the **suppl**.

Search Algorithm. Following Darts [33], we use the importance factors which are learned jointly with the network weights of each search block. We also adopt a resource-aware $\mathcal{L}_1$ penalty [7] to push the importance factors of high computational costs to zero. More details on the resource-aware penalty can be found in the **suppl**. Combing with this resource-aware penalty term $\mathcal{L}_{l_1}$ with an attenuation coefficient λ set as 0.01, the overall training loss is:

$$\mathcal{L}_{total} = \mathcal{L}_{gen} + \lambda \mathcal{L}_{\ell 1}. \tag{9}$$

4 Datasets

BID-CityScapes. There are several large-scale synthetic datasets [15,18,29] available for training restoring networks. However, none of them considers depth effects under the BID setting, thus impeding the performance for real-world images. To enable the depth supervision for DeBNet, we introduce a new depth-aware BID dataset named BID-CityScapes, using images from CityScapes [6] dataset as background. We first generate synthetic degradation layers based on the provided camera parameters and scene depth. We apply the rendering strategy in [18,45] to smooth the degradation layers, and then generate the final corrupted images according to Eq. (1). Altogether, our BID-CityScapes dataset has 6,000 training image pairs and 800 pairs for testing. Each pair contains a clean background image, a depth map from the original dataset, and 14 types of different weather combinations (see Fig. 1 left), including different types and intensities. Although BID-CityScapes dataset has effectively expanded the practicality for BID tasks, its imaging perspective is restricted to autonomous driving scenes, which limits the generalization on real-world photos with various imaging scenes and viewpoints. More details and examples are given in the **suppl**.

BID-GTAV. We further propose the **BID-GTAV** dataset with more diverse scenes as a supplement (see Fig. 1 right). The dataset is rendered from Grand Theft Auto V (GTAV) [44], an open-world game with large-scale city models. We capture the images and corresponding ground-truth depth maps from the game with two plugins, Script Hook V and Script Hook V.NET [20]. We endeavor to leverage the rich virtual worlds created for major video games to simulate real-world scenarios with high-level fidelity and various viewpoints. We extract 8,000 pairs of images for training and 1,000 pairs for testing. The extracted pairs contain various weather conditions using graphics debugging mods, as well as their clear counterparts and precise depth maps.

5 Experiment

5.1 Implementation Details

Our model is trained in an end-to-end manner following [30,33]. The training set is split into a search training part Set_1 (70%) and a search validation part Set_2 (30%). We simultaneously optimize the architecture parameters on Set_1 and the network parameters on Set_2. Each search block contains two operating paths. For the MixConv path, we set the expansion rate as 4. In the reduction block, we set the stride of the depthwise convolution as 2. For the Transformer path, we set $s = 8$. The number of attention heads is 1, and the hidden dimension of the attended subspaces is set as 64. We resize the features into half size of

Table 1. Quantitative results on our synthetic BID-CityScapes and BID-GTAV datasets. We evaluate the performance in Peak Signal-to-Noise Ratio (PSNR) and Structural Similarity (SSIM) under 5 BID cases, which are (1) heavy rain + heavy fog, (2) light rain + light fog + snow, (3) medium rain + medium fog + raindrop, (4) light rain + light fog + light dark, (5) light rain + light fog + light dark + raindrop. The best performance under each case is marked in **bold**.

case	BID-CityScapes								BID-GTAV							
	TransWeather		BIDeN		CPNet		DeBNet		TransWeather		BIDeN		CPNet		DeBNet	
	PSNR	SSIM	PSNR	SSIM	PSNR	SSIM	PSNR	SSIM	PSNR	SSIM	PSNR	SSIM	PSNR	SSIM	PSNR	SSIM
(1)	23.07	0.845	25.70	0.869	26.97	0.902	**29.51**	**0.916**	18.45	0.623	19.55	0.699	22.30	0.725	**23.88**	**0.751**
(2)	24.33	0.877	26.05	0.880	27.18	0.904	**29.72**	**0.920**	18.88	0.631	19.80	0.720	22.57	0.731	**24.31**	**0.763**
(3)	24.20	0.865	25.61	0.867	27.05	0.901	**29.54**	**0.917**	18.42	0.622	19.61	0.700	22.31	0.727	**24.05**	**0.757**
(4)	22.39	0.801	25.11	0.861	25.87	0.869	**28.35**	**0.909**	18.33	0.619	18.78	0.643	20.90	0.713	**23.02**	**0.744**
(5)	22.10	0.792	24.07	0.849	25.44	0.857	**28.18**	**0.907**	18.01	0.609	18.51	0.631	20.67	0.701	**22.71**	**0.737**

Fig. 4. Qualitative comparisons on BID-CityScapes and BID-GTAV under several mixed cases. DeBNet produces clean and precise restored images without visible artifacts or color shifts, especially for depth-related scenarios. Please zoom in to see details.

the former resolution during inverse projection in Transformer to achieve downsampling. We progressively remove the search units with less importance [33] and re-calibrate the running statistics of BN layers after every 5 epochs. After the architecture search, DeBNet can fully explore global and local information across different weather degradations and restore clean images. Besides, since the network training and architecture search are conducted in a unified end-to-end manner, the resulting network can be used directly without fine-tuning. More implementation details are given in the **suppl**. Code will be released in https://github.com/Oli-iver/Depth-BID.

5.2 Synthetic BID Analysis

Instead of directly combining existing single-degradation datasets [30], we further propose two depth-aware BID datasets, BID-CityScapes and BID-GTAV, across various scenes and viewpoints by incorporating depth information during simulation. We conduct extensive experiments on the proposed two datasets to evaluate the performance of the proposed DeBNet and the existing state-of-the-art BID methods BIDeN [15] and CPNet [49], as well as an all-in-one method TransWeather [48]. We select five common weather combinations existing in the real world for evaluation as: (1) heavy rain + heavy fog, (2) light rain + light fog + snow, (3) medium rain + medium fog + raindrop, (4) light rain + light

Table 2. Quantitative results for benchmarking the proposed DeBNet and the state-of-the-art methods on real-world deraining SPAdata [53] and dehazing SOTS-outdoor [25] test sets. The original pre-trained weights of all these models are directly used for evaluation. We have also trained these methods on our BID-CityScapes (red) and BID-GTAV (blue) datasets. Results of the two-stage refined DeBNet$^+$ are marked in green. The best performance is marked in **bold**.

Method	SPAdata [53]			SOTS-outdoor [25]		
	RCDNet [50]	AirNet [26]	DeBNet	FFA-Net [40]	AirNet [26]	DeBNet
PSNR	34.08/36.72/34.11	34.05/35.32/34.40	36.69/34.78/**36.77**	33.07/34.60/31.25	33.09/34.51/31.10	33.50/31.57/**33.82**
SSIM	0.953/0.969/0.958	0.948/0.963/0.960	0.965/0.954/**0.973**	0.980/0.982/0.975	0.979/0.983/0.968	0.986/0.982/**0.989**

| Corrupted Image I_{in} | Transmission map T | Noise map N | Restored Image I_{gen} | Corrupted Image I_{in} | Transmission map T | Noise map N | Restored Image I_{gen} |

Fig. 5. Visualization of the learned Transmittance map T and Noise map N on different weather scenarios. T and N are re-normalized into $[0,1]$ for better visualization.

fog + light dark, (5) light rain + light fog + light dark + raindrop. All the methods are trained and evaluated under the same settings for fair comparisons. The quantitative results on the two proposed synthetic datasets are reported in Table 1. It can be found that our DeBNet delivers state-of-the-art performance under the BID setting on both datasets. Notably, our method performs favorably against other counterparts under the scenes where, e.g., DeBNet exceeds CPNet by 2.54dB on PSNR under heavy rain and fog combinations. This phenomenon can be attributed to the explicit utilization of scene depth in DeBNet, which encourages the model to leverage spatial information while extracting features. We also show some qualitative comparisons in Fig. 4. Due to the page limitation, more analyses and comparisons on synthetic datasets are given in the **suppl**.

5.3 Real-World BID Analysis

To validate the effectiveness of our approach, we further conducted experiments on real-world images. As existing real-world datasets are limited with single-type degradation and lack the ground-truth scene depth, we directly evaluated the generalization ability of models trained respectively on BID-CityScapes and BID-GTAV datasets with real-world test sets [25,53]. To further validate the effectiveness of our BID-GTAV dataset constructed for multi-view natural image restoration, we further prepared a two-stage learned model (DeBNet$^+$) which is first trained on the BID-CityScapes dataset, and then fine-tune the searched model on the BID-GTAV dataset. Table 2 shows the quantitative results of different restoration tasks. It can be found that though the performance of DeBNet is constrained by the BID training setting, our method remains competitive (DeBNet$_a$) or even better performance (DeBNet$^+$) against other task-specific counterparts. This experiment also indicates the quality of our proposed datasets, as the depth-aware modeling method effectively improves the generalization ability on real-world images. More results on other real-world restoration tasks can be found in the **suppl**.

5.4 Discussion and Ablation Study

Map Visualization. To validate the effectiveness of our proposed depth-aware BID modeling approach, we visualized the learned transmittance map T and noise map N under real-world scenarios. As shown in Fig. 5, both T and N are highly spatially related to the scene depth. **(1).** In highly depth-related scenes like low-light and fog, the transmittance map T tends to capture low-frequency patterns while the noise map N focuses on high-frequency details, e.g., small points. **(2).** For raindrops and snow scenes, it is challenging to distinguish transmittance and noise maps from a frequency perspective, as the chaotic distribution of such noise constitutes low-frequency interference.

Depth Branch. The design of the depth branch is motivated by two points: (1) The off-the-shelf depth models are not robust to complicated weather, due to various occlusions and illumination. Therefore, we decide to re-train the depth estimation task from scratch. (2) Combining with the depth estimation is complementary to our main task, i.e., image decomposition, since the low-level feature of depth tasks focuses on the distance and object edges. As shown in Fig. 6, we further select real images from KITTI dataset [11]

Fig. 6. Predicted depths on real images.

under autonomous driving scenarios, and randomly generate outputs using the modeling method as Eq. (3) for the scene depth D visualization. It can be found that our method can not only achieve ideal BID restoration but also predict reasonable scene depths.

Comparisons with Diffusion-Based Method. We also provide comparisons with another diffusion-based method WeatherDiff [37] in Table 3a. We retrained the official model from scratch on our proposed dataset and tested it on two real-world datasets. It can be observed that, benefiting from the depth guidance and NAS targeted at various degradation combinations, our method DeBNet clearly outperforms the compared method, particularly in real-world datasets and complex scenarios.

Datasets. To further validate the effectiveness, we conduct experiments on three real-world datasets [1,34,39]. As shown from Table 3b, we could observe two points. **(1).** Since existing BID dataset [15] does not consider the impact of depth on weather imaging, there is a notable performance loss with $DeBNet_1$. In contrast, $DeBNet_2$ trained on our dataset generalizes well to real-world images, indicating our dataset's efficacy in prompting depth-related feature learning which is vital for real-world applications. **(2).** Pre-training on our dataset and fine-tuning on downstream datasets also yields better performance and shows scalability to different real-world scenarios. For instance, ft $DeBNet_2$ pretrained on our dataset surpassed the performance of models pretrained on existing datasets or from scratch, which is also evidenced in Table 2.

DBA and ECA Modules. In our modeling with Eq. (1) and Eq. (3), the transmittance map primarily characterizes the location and density information of adverse weather conditions, which are intimately associated with depth. Thus we design a bipolar transmittance mechanism in DBA to fully exploit the learned depth features. Moreover, DBA further embeds diverse weather characteristics into the reconstructed image features through the ECA module. To understand the contributions of these two attention modules, we respectively replace the DBA and ECA modules into 1×1 convolution to build two variants. The quan-

Table 3. More experiment results on different settings. The best performance is marked in **bold**. case-1: heavy rain + heavy fog, case-2: light rain + light fog + snow, case-5: light rain + light fog + light dark + raindrop. (a) Comparison with the diffusion-based method in terms of PSNR and SSIM. (b) More experiment results on different training datasets in terms of PSNR and SSIM. *ft* DeBNet means the model is pre-trained on target dataset/previous BIDeN/BID-CityScapes and fine-tuned on the training set of each evaluated dataset. (c) Ablation on DBA and ECA modules with BID-CityScapes dataset in terms of PSNR and SSIM. (d) Ablation on hyperparameters with BID-CityScapes dataset in terms of PSNR and number of parameters.

(a)

Methods	BID-CityScapes case-1	BID-CityScapes case-5	SPAdata [53] real-world rain	SOTS-outdoor [25] real-world haze
TransWeather [48]	23.07/0.845	22.10/0.792	33.75/0.942	29.89/0.959
WeatherDiff [37]	28.11/0.910	25.07/0.887	35.32/0.955	33.17/0.979
DeBNet	**29.51/0.916**	**28.18/0.907**	**36.69/0.965**	**33.50/0.986**

(c)

Ablation	w/o DBA	w/o ECA	DeBNet (full)
case-1	28.87 / 0.906	29.23 / 0.911	**29.51 / 0.916**
case-5	27.50 / 0.868	27.95 / 0.882	**28.18 / 0.907**

(b)

Variants	Pre-training Dataset	Deraindrop [39]	Snow100k-M [34]	SICE [1]
DeBNet₁	BIDeN dataset [15]	29.75/0.931	31.15/0.933	20.03/0.719
DeBNet₂	BID-CityScapes	32.79/0.942	33.80/0.948	21.79/0.730
ft DeBNet	from scratch	33.05/0.942	**33.99/0.953**	22.10/0.738
ft DeBNet₁	BIDeN dataset	32.89/0.941	33.75/0.942	22.03/0.735
ft DeBNet₂	BID-CityScapes	**33.11/0.945**	33.96/0.953	**22.12/0.738**

(d)

Variants		case-2	case-5
	Uformer [55]	25.64 (50.90M)	22.05 (50.90M)
NAS $\mathcal{L}_{l_1}$	$\lambda = 0.05$	26.82 (**30.55M**)	24.91 (**35.23M**)
	$\lambda = 0.005$	29.73 (44.73M)	**28.19** (47.53M)
	w/o NAS	**29.79** (53.15M)	27.88 (53.15M)
GAN $\mathcal{L}_s$	$\lambda_s = 0.05$	29.70 (39.08M)	**28.20** (42.50M)
	$\lambda_s = 0.5$	29.68 (38.95M)	28.05 (43.11M)
DeBNet ($\lambda = 0.01, \lambda_s = 0.1$)		29.72 (39.15M)	28.18 (42.37M)

titative results in Table 3c show that the absence of these two attention modules in linking depth information with image restoration leads to a significant decline in performance, particularly in complex composite scenes. In conclusion, high-quality BID requires spatial attention (DBA) to capture common features (*e.g.*, scene depth), as well as channel attention (ECA) to enhance distinctive features of different degradations.

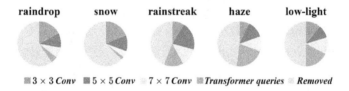

Fig. 7. Visualization of the searched architectures under different scenarios. For simplicity, we only show the averaged searching proportions in refinement modules when the feature size is scaled to 1/8 (see Fig. 3). More visualizations are given in the **suppl**.

Hyperparameters. 1. NAS $\mathcal{L}_{l_1}$. In practice, we can obtain different searched models using different λ values. As shown in Table 3d, we observe two points: (1) We usually could yield a small-size NAS model. Our model ($\lambda = 0.05$) easily achieves a better PSNR than a raw end-to-end Uformer [55] with much fewer parameters. (2) For the simple scenes case-2, the proposed model ($\lambda = 0.005$) could achieve a competitive performance compared with a bigger baseline model w/o NAS, while achieving a significant PSNR improvement in a challenging weather combination (case-5). Our DeBNet achieves a favorable trade-off between performance and efficiency. **2. GAN Loss** $\mathcal{L}_s$. We also conduct experiments on the hyperparameter λ_s of the source adversarial loss, it can be found that our trained model is not sensitive to the GAN loss weight λ_s.

Architecture Search. Our multi-branch feature search aims to find the best structure to comprehensively handle BID in various weather combinations. To verify the role of different operations in our search space, we further use a fixed combination dataset for the search. As shown in Fig. 7, our method exhibits adaptability, finding dynamic architectures for different weather degradations. As for local weather degradations (*e.g.*, raindrop and snow), the model tends to favor smaller convolutional kernels. In comparison, for globally corrupted weather types (*e.g.*, haze and low-light), the model prefers utilizing larger-sized convolutions and attention mechanisms to achieve a broader receptive field.

6 Conclusion

In this paper, we explore the visual effects of various adverse weather conditions subject to scene depth and formulate a depth-aware BID imaging model. Considering the depth information within images, we propose a novel BID model, namely DeBNet, to restore arbitrary hybrid adverse weather conditions in a unified framework. Taking advantage of neural architecture search and the specifically designed restoring search space, we achieved an effective deraining network to remove various types of rain. To bridge the domain gap between real and synthetic images, we present two depth-aware BID datasets BID-CityScapes and BID-GTAV under real-world and synthetic scenes, respectively. Extensive experiments on our benchmarks and other real-world datasets demonstrate the effectiveness and superiority of our unified network. We hope that our dataset would take a step closer for transferring the research of blind image decomposition to real-world applications.

Acknowledgement. The paper is supported by Start-up Research Grant at the University of Macau (SRG2024-00002-FST).

References

1. Cai, J., Gu, S., Zhang, L.: Learning a deep single image contrast enhancer from multi-exposure images. IEEE Trans. Image Process. **27**(4), 2049–2062 (2018)
2. Chen, L., Chu, X., Zhang, X., Sun, J.: Simple baselines for image restoration. In: Avidan, S., Brostow, G., Cissé, M., Farinella, G.M., Hassner, T. (eds.) ECCV 2022 Part VII. LNCS, vol. 13667, pp. 17–33. Springer, Cham (2022). https://doi.org/10.1007/978-3-031-20071-7_2
3. Chen, W.-T., Fang, H.-Y., Ding, J.-J., Tsai, C.-C., Kuo, S.-Y.: JSTASR: joint size and transparency-aware snow removal algorithm based on modified partial convolution and veiling effect removal. In: Vedaldi, A., Bischof, H., Brox, T., Frahm, J.-M. (eds.) ECCV 2020 XXI. LNCS, vol. 12366, pp. 754–770. Springer, Cham (2020). https://doi.org/10.1007/978-3-030-58589-1_45

4. Chen, W.T., et al.: All snow removed: single image desnowing algorithm using hierarchical dual-tree complex wavelet representation and contradict channel loss. In: Proceedings of the IEEE/CVF International Conference on Computer Vision, pp. 4196–4205 (2021)
5. Chen, W.T., Huang, Z.K., Tsai, C.C., Yang, H.H., Ding, J.J., Kuo, S.Y.: Learning multiple adverse weather removal via two-stage knowledge learning and multi-contrastive regularization: toward a unified model. In: Proceedings of the IEEE/CVF Conference on Computer Vision and Pattern Recognition, pp. 17653–17662 (2022)
6. Cordts, M., et al.: The cityscapes dataset for semantic urban scene understanding. In: Proceedings of the IEEE Conference on Computer Vision and Pattern Recognition, pp. 3213–3223 (2016)
7. Ding, M., et al.: Hr-nas: searching efficient high-resolution neural architectures with lightweight transformers. In: Proceedings of the IEEE/CVF Conference on Computer Vision and Pattern Recognition, pp. 2982–2992 (2021)
8. Dong, X., Pang, Y., Wen, J.: Fast efficient algorithm for enhancement of low lighting video. In: ACM SIGGRApH 2010 posters, pp. 1–1 (2010)
9. Fan, Y., et al.: Neural sparse representation for image restoration. Adv. Neural. Inf. Process. Syst. **33**, 15394–15404 (2020)
10. Fu, X., Huang, J., Zeng, D., Huang, Y., Ding, X., Paisley, J.: Removing rain from single images via a deep detail network. In: Proceedings of the IEEE Conference on Computer Vision and Pattern Recognition, pp. 3855–3863 (2017)
11. Geiger, A., Lenz, P., Urtasun, R.: Are we ready for autonomous driving? the kitti vision benchmark suite. In: Conference on Computer Vision and Pattern Recognition (CVPR) (2012)
12. Gou, Y., Li, B., Liu, Z., Yang, S., Peng, X.: Clearer: multi-scale neural architecture search for image restoration. Adv. Neural. Inf. Process. Syst. **33**, 17129–17140 (2020)
13. Gu, S., Meng, D., Zuo, W., Zhang, L.: Joint convolutional analysis and synthesis sparse representation for single image layer separation. In: ICCV (2017)
14. Guo, X., Li, Y., Ling, H.: Lime: low-light image enhancement via illumination map estimation. IEEE Trans. Image Process. **26**(2), 982–993 (2016)
15. Han, J., et al.: Blind image decomposition. In: Avidan, S., Brostow, G., Cissé, M., Farinella, G.M., Hassner, T. (eds.) ECCV 2022. LNCS, vol. 13678, pp. 218–237. Springer, Cham (2022). https://doi.org/10.1007/978-3-031-19797-0_13
16. He, K., Chen, X., Xie, S., Li, Y., Dollár, P., Girshick, R.: Masked autoencoders are scalable vision learners. In: Proceedings of the IEEE/CVF Conference on Computer Vision and Pattern Recognition, pp. 16000–16009 (2022)
17. He, K., Sun, J., Tang, X.: Single image haze removal using dark channel prior. IEEE Trans. Pattern Anal. Mach. Intell. **33**(12), 2341–2353 (2010)
18. Hu, X., Fu, C.W., Zhu, L., Heng, P.A.: Depth-attentional features for single-image rain removal. In: IEEE Conference on Computer Vision Pattern Recognition (2019)
19. Janner, M., Wu, J., Kulkarni, T.D., Yildirim, I., Tenenbaum, J.: Self-supervised intrinsic image decomposition. In: Advances in Neural Information Processing Systems, vol. 30 (2017)
20. Johnson-Roberson, M., Barto, C., Mehta, R., Sridhar, S.N., Rosaen, K., Vasudevan, R.: Driving in the matrix: Can virtual worlds replace human-generated annotations for real world tasks? arXiv preprint arXiv:1610.01983 (2016)
21. Jozefowicz, R., Zaremba, W., Sutskever, I.: An empirical exploration of recurrent network architectures. In: International Conference on Machine Learning, pp. 2342–2350. PMLR (2015)

22. Kang, L.W., Lin, C.W., Fu, Y.H.: Automatic single-image-based rain streaks removal via image decomposition. IEEE Trans. Image Process. **21**(4), 1742–1755 (2011)
23. Levin, A., Weiss, Y.: User assisted separation of reflections from a single image using a sparsity prior. TPAMI **29**(9), 1647–1654 (2007)
24. Li, B., Peng, X., Wang, Z., Xu, J., Feng, D.: Aod-net: All-in-one dehazing network. In: Proceedings of the IEEE International Conference on Computer Vision, pp. 4770–4778 (2017)
25. Li, B., et al.: Benchmarking single-image dehazing and beyond. IEEE Trans. Image Process. **28**(1), 492–505 (2018)
26. Li, B., Liu, X., Hu, P., Wu, Z., Lv, J., Peng, X.: All-in-one image restoration for unknown corruption. In: Proceedings of the IEEE/CVF Conference on Computer Vision and Pattern Recognition, pp. 17452–17462 (2022)
27. Li, L., Wang, R., Wang, W., Gao, W.: A low-light image enhancement method for both denoising and contrast enlarging. In: 2015 IEEE International Conference on Image Processing (ICIP), pp. 3730–3734. IEEE (2015)
28. Li, M., Liu, J., Yang, W., Sun, X., Guo, Z.: Structure-revealing low-light image enhancement via robust retinex model. IEEE Trans. Image Process. **27**(6), 2828–2841 (2018)
29. Li, R., Cheong, L.F., Tan, R.T.: Heavy rain image restoration: Integrating physics model and conditional adversarial learning. In: Proceedings of the IEEE/CVF Conference on Computer Vision and Pattern Recognition, pp. 1633–1642 (2019)
30. Li, R., Tan, R.T., Cheong, L.F.: All in one bad weather removal using architectural search. In: Proceedings of the IEEE/CVF Conference on Computer Vision and Pattern Recognition, pp. 3175–3185 (2020)
31. Liang, J., Cao, J., Sun, G., Zhang, K., Van Gool, L., Timofte, R.: Swinir: image restoration using swin transformer. In: Proceedings of the IEEE/CVF International Conference on Computer Vision, pp. 1833–1844 (2021)
32. Lin, D., et al.: Generative status estimation and information decoupling for image rain removal. In: Advances in Neural Information Processing Systems, vol. 35, pp. 4612–4625 (2022)
33. Liu, H., Simonyan, K., Yang, Y.: Darts: differentiable architecture search. arXiv preprint arXiv:1806.09055 (2018)
34. Liu, Y.F., Jaw, D.W., Huang, S.C., Hwang, J.N.: Desnownet: context-aware deep network for snow removal. IEEE Trans. Image Process. **27**(6), 3064–3073 (2018)
35. Lore, K.G., Akintayo, A., Sarkar, S.: LLNet: a deep autoencoder approach to natural low-light image enhancement. Pattern Recogn. **61**, 650–662 (2017)
36. Mei, J., et al.: Atomnas: fine-grained end-to-end neural architecture search. arXiv preprint arXiv:1912.09640 (2019)
37. Özdenizci, O., Legenstein, R.: Restoring vision in adverse weather conditions with patch-based denoising diffusion models. IEEE Trans. Pattern Anal. Mach. Intell. (2023)
38. Pham, H., Guan, M., Zoph, B., Le, Q., Dean, J.: Efficient neural architecture search via parameters sharing. In: International Conference on Machine Learning, pp. 4095–4104. PMLR (2018)
39. Qian, R., Tan, R.T., Yang, W., Su, J., Liu, J.: Attentive generative adversarial network for raindrop removal from a single image. In: Proceedings of the IEEE Conference on Computer Vision and Pattern Recognition, pp. 2482–2491 (2018)
40. Qin, X., Wang, Z., Bai, Y., Xie, X., Jia, H.: FFA-Net: feature fusion attention network for single image dehazing. In: Proceedings of the AAAI Conference on Artificial Intelligence, vol. 34, pp. 11908–11915 (2020)

41. Quan, R., Yu, X., Liang, Y., Yang, Y.: Removing raindrops and rain streaks in one go. In: Proceedings of the IEEE/CVF Conference on Computer Vision and Pattern Recognition, pp. 9147–9156 (2021)
42. Quan, Y., Deng, S., Chen, Y., Ji, H.: Deep learning for seeing through window with raindrops. In: Proceedings of the IEEE/CVF International Conference on Computer Vision, pp. 2463–2471 (2019)
43. Ren, W., et al.: Gated fusion network for single image dehazing. In: Proceedings of the IEEE Conference on Computer Vision and Pattern Recognition, pp. 3253–3261 (2018)
44. Richter, S.R., Vineet, V., Roth, S., Koltun, V.: Playing for data: ground truth from computer games. In: Leibe, B., Matas, J., Sebe, N., Welling, M. (eds.) ECCV 2016 Part II. LNCS, vol. 9906, pp. 102–118. Springer, Cham (2016). https://doi.org/10.1007/978-3-319-46475-6_7
45. Sakaridis, C., Dai, D., Van Gool, L.: Semantic foggy scene understanding with synthetic data. Int. J. Comput. Vis. **126**, 973–992 (2018)
46. Shao, M.W., Li, L., Meng, D.Y., Zuo, W.M.: Uncertainty guided multi-scale attention network for raindrop removal from a single image. IEEE Trans. Image Process. **30**, 4828–4839 (2021)
47. Tan, M., Le, Q.V.: Mixconv: mixed depthwise convolutional kernels. arXiv preprint arXiv:1907.09595 (2019)
48. Valanarasu, J.M.J., Yasarla, R., Patel, V.M.: Transweather: transformer-based restoration of images degraded by adverse weather conditions. In: Proceedings of the IEEE/CVF Conference on Computer Vision and Pattern Recognition, pp. 2353–2363 (2022)
49. Wang, C., Zheng, Z., Quan, R., Sun, Y., Yang, Y.: Context-aware pretraining for efficient blind image decomposition. In: IEEE Conference on Computer Vision and Pattern Recognition (2023)
50. Wang, H., Xie, Q., Zhao, Q., Meng, D.: A model-driven deep neural network for single image rain removal. In: Proceedings of the IEEE/CVF Conference on Computer Vision and Pattern Recognition, pp. 3103–3112 (2020)
51. Wang, J., et al.: Deep high-resolution representation learning for visual recognition. IEEE Trans. Pattern Anal. Mach. Intell. **43**(10), 3349–3364 (2020)
52. Wang, Q., Wu, B., Zhu, P., Li, P., Zuo, W., Hu, Q.: ECA-Net: efficient channel attention for deep convolutional neural networks. In: Proceedings of the IEEE/CVF Conference on Computer Vision and Pattern Recognition, pp. 11534–11542 (2020)
53. Wang, T., Yang, X., Xu, K., Chen, S., Zhang, Q., Lau, R.W.: Spatial attentive single-image deraining with a high quality real rain dataset. In: Proceedings of the IEEE/CVF Conference on Computer Vision and Pattern Recognition, pp. 12270–12279 (2019)
54. Wang, T., Zheng, Z., Sun, Y., Yan, C., Yang, Y., Chua, T.S.: Multiple-environment self-adaptive network for aerial-view geo-localization. Pattern Recogn. **152**, 110363 (2024)
55. Wang, Z., Cun, X., Bao, J., Zhou, W., Liu, J., Li, H.: Uformer: a general u-shaped transformer for image restoration. In: Proceedings of the IEEE/CVF Conference on Computer Vision and Pattern Recognition, pp. 17683–17693 (2022)
56. Woo, S., Park, J., Lee, J.Y., Kweon, I.S.: Cbam: convolutional block attention module. In: Proceedings of the European Conference on Computer Vision (ECCV), pp. 3–19 (2018)
57. Xiao, J., Fu, X., Wu, F., Zha, Z.J.: Stochastic window transformer for image restoration. Adv. Neural. Inf. Process. Syst. **35**, 9315–9329 (2022)

58. Xie, S., Girshick, R., Dollár, P., Tu, Z., He, K.: Aggregated residual transformations for deep neural networks. In: Proceedings of the IEEE Conference on Computer Vision and Pattern Recognition, pp. 1492–1500 (2017)
59. You, S., Tan, R.T., Kawakami, R., Mukaigawa, Y., Ikeuchi, K.: Adherent raindrop modeling, detection and removal in video. IEEE Trans. Pattern Anal. Mach. Intell. **38**(9), 1721–1733 (2015)
60. Zamir, S.W., Arora, A., Khan, S., Hayat, M., Khan, F.S., Yang, M.H.: Restormer: efficient transformer for high-resolution image restoration. In: Proceedings of the IEEE/CVF Conference on Computer Vision and Pattern Recognition, pp. 5728–5739 (2022)
61. Zamir, S.W., et al.: Multi-stage progressive image restoration. In: Proceedings of the IEEE/CVF Conference on Computer Vision and Pattern Recognition, pp. 14821–14831 (2021)
62. Zhang, H., Goodfellow, I., Metaxas, D., Odena, A.: Self-attention generative adversarial networks. In: International Conference on Machine Learning, pp. 7354–7363. PMLR (2019)
63. Zhang, H., Patel, V.M.: Densely connected pyramid dehazing network. In: Proceedings of the IEEE Conference on Computer Vision and Pattern Recognition, pp. 3194–3203 (2018)
64. Zoph, B., Vasudevan, V., Shlens, J., Le, Q.V.: Learning transferable architectures for scalable image recognition. In: Proceedings of the IEEE Conference on Computer Vision and Pattern Recognition, pp. 8697–8710 (2018)

DreamSampler: Unifying Diffusion Sampling and Score Distillation for Image Manipulation

Jeongsol Kim[1]([✉]), Geon Yeong Park[1], and Jong Chul Ye[2]

[1] Department of Bio and Brain Engineering, KAIST, Daejeon, South Korea
{jeongsol,pky3436}@kaist.ac.kr
[2] Kim Jae Chul AI Graduate School, KAIST, Daejeon, South Korea
jong.ye@kaist.ac.kr

Abstract. Reverse sampling and score-distillation have emerged as main workhorses in recent years for image manipulation using latent diffusion models (LDMs). While reverse diffusion sampling often requires adjustments of LDM architecture or feature engineering, score distillation offers a simple yet powerful model-agnostic approach, but it is often prone to mode-collapsing. To address these limitations and leverage the strengths of both approaches, here we introduce a novel framework called *DreamSampler*, which seamlessly integrates these two distinct approaches through the lens of regularized latent optimization. Similar to score-distillation, DreamSampler is a model-agnostic approach applicable to any LDM architecture, but it allows both distillation and reverse sampling with additional guidance for image editing and reconstruction. Through experiments involving image editing, SVG reconstruction and etc., we demonstrate the competitive performance of DreamSampler compared to existing approaches, while providing new applications.

Keywords: Latent diffusion model · Generation · Score distillation

1 Introduction

Diffusion models [6,26,27] have been extensively studied as powerful generative models in recent years. These models operate by generating clean images from Gaussian noise through a process termed ancestral sampling. This involves progressively reducing noise by utilizing an estimated score function to guide the generation process from a random starting point towards the distribution of natural images. Reverse diffusion sampling introduces stochasticity through the reverse Wiener process within the framework of SDE [27], contributing to the prevention of mode collapse in generated samples while enhancing the fidelity [3].

J. Kim and G. Y. Park—Equal contribution.

Supplementary Information The online version contains supplementary material available at https://doi.org/10.1007/978-3-031-73007-8_23.

Fig. 1. DreamSampler can be used for vectorized image restoration, editing, text-guided inpainting, etc. Code: https://github.com/DreamSampler/dream-sampler

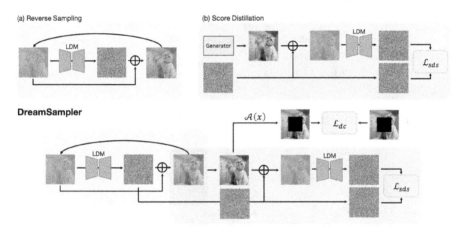

Fig. 2. DreamSampler vs (a) reverse diffusion and (b) score distillation.

Furthermore, the reverse diffusion can be flexibly regularized by various guidance gradients. For instance, classifier-guidance [3] applies classifier gradients to intermediate noisy samples, facilitating conditional image generation. Additionally, DPS [1] utilizes approximated likelihood gradients to constrain the sampling process, ensuring data consistency between the current estimated solution and given observation, thereby solving noisy inverse problems in a zero-shot manner (Figs. 1 and 2).

On the other hand, another type of approach, called *score distillation*, utilizes diffusion model as prior knowledge for image generation and editing. For example, DreamFusion [21] leveraged 2D text-conditioned diffusion models for text-guided 3D representation learning via NeRF. Here, the diffusion model serves as the teacher model, generating gradients by comparing its predictions with the label noises and guiding the generator as the student model. The strength of the score distillation method lies in its ability to leverage the pre-trained diffusion model in a black-box manner without requiring any feature engineering, such as adjustments to the model architecture. Unfortunately, the score distillation is more often prone to mode collapsing compared to the reverse diffusion.

Although both algorithms are grounded in the same principle of diffusion models, the approaches appear to be different, making it unclearly how to synergistically combine the two approaches. To address this, we introduce a unified framework called *DreamSampler*, that seamlessly integrates two distinct approaches and take advantage of the both worlds through the lens of regularized latent optimization. Specifically, DreamSampler is model-agnostic and does not require any feature engineering, such as adjustments to the model architecture. Moreover, in contrast to the score distillation, DreamSampler is free of mode collapsing thanks to the stochastic nature of the sampling.

The pioneering aspect of DreamSampler is rooted in two pivotal insights. First, we demonstrate that the process of latent optimization during reverse diffusion can be viewed as a proximal update from the posterior mean by Tweedie's formula. This interpretation allows us to integrate additional regularization terms, such as measurement consistency in inverse problems, to steer the sampling procedure. Moreover, we illustrate that the loss associated with the proximal update can be conceptualized as the score distillation loss. This insight bridges a natural connection between the score-distillation methodology and reverse sampling strategies, culminating in their harmonious unification. In subsequent sections, we will explore various applications emerging from this integrated framework and demonstrate the efficacy of DreamSampler through empirical evidence.

2 Motivations

2.1 Preliminaries

In LDMs [23], the encoder $\mathcal{E}_\phi$ and the decoder $\mathcal{D}_\varphi$ are trained as auto-encoder, satisfying $x = \mathcal{D}_\varphi(\mathcal{E}_\phi(x)) = \mathcal{D}_\varphi(z_0)$ where x denotes clean image and z_0 denotes encoded latent vector. Then, the diffusion process is defined on the latent space, which is a range space of $\mathcal{E}_\phi$. Specifically, the forward diffusion process is

$$z_t = \sqrt{\bar{\alpha}_t} z_0 + \sqrt{1 - \bar{\alpha}_t} \epsilon \tag{1}$$

where $\bar{\alpha}_t$ denotes pre-defined coefficient that manages noise scheduling, and $\epsilon \sim \mathcal{N}(0, I)$ denotes a noise sampled from normal distribution. The reverse diffusion process requires a score function via a neural network (i.e. diffusion model, ϵ_θ) trained by denoising score matching [6,27]:

$$\min_\theta \mathbb{E}_{t, \epsilon \sim \mathcal{N}(0, I)} \| \epsilon - \epsilon_\theta(z_t, t) \|_2^2. \tag{2}$$

According to formulation of DDIM [2,26], the reverse sampling from the posterior distribution $p(z_{t-1} | z_t, z_0)$ could be described as

$$z_{t-1} = \sqrt{\bar{\alpha}_{t-1}} \hat{z}_{0|t} + \sqrt{1 - \bar{\alpha}_{t-1}} \tilde{\epsilon} \tag{3}$$

where

$$\hat{z}_{0|t} = (z_t - \sqrt{1-\bar{\alpha}_t}\epsilon_\theta(z_t,t))/\sqrt{\bar{\alpha}_t} \tag{4}$$

$$\tilde{\epsilon} = \frac{\sqrt{1-\bar{\alpha}_{t-1}-\eta^2\beta_t^2}\epsilon_\theta(z_t,t)}{\sqrt{1-\bar{\alpha}_{t-1}}} + \frac{\eta\beta_t\epsilon}{\sqrt{1-\bar{\alpha}_{t-1}}}. \tag{5}$$

Here, $\hat{z}_{0|t}$ refers to the denoised latent through Tweedie's formula, and $\tilde{\epsilon}$ is the noise term composed of both deterministic $\epsilon_\theta(z_t,t)$ and stochastic term $\epsilon \sim \mathcal{N}(0,\boldsymbol{I})$. Note that η and β_t denote variables that controls the stochastic property of sampling. When $\eta\beta_t = 0$, the sampling is deterministic.

For text conditioning, classifier-free-guidance (CFG) [7] is widely leveraged. The estimated noise is computed by

$$\epsilon_\theta^\omega(z_t,t,c_{ref}) = \epsilon_\theta(z_t,t,c_\varnothing) + \omega[\epsilon_\theta(z_t,t,c_{ref}) - \epsilon_\theta(z_t,t,c_\varnothing)] \tag{6}$$

where ω denotes the guidance scale, $c_\varnothing$ refers to the null-text embedding, and c_{ref} is the conditioning text embedding, which are encoded by pre-trained text encoder such as CLIP [22]. For simplicity, we will interchangeably use the terms $\epsilon_\theta(z_t)$, $\epsilon_\theta(z_t,t)$ and $\epsilon_\theta^\omega(z_t,t,c_{ref})$ unless stated otherwise.

2.2 Key Observations

Score distillation sampling (SDS) [21] and reverse diffusion [6,26,27] represent two distinct methodologies, each with its own pros and cons. SDS is an optimization method that focuses on minimizing score distillation loss, while reverse diffusion utilizes ancestral sampling, which stems from SDE or DDPM formulations. Although SDS is straightforward and model-agnostic, it often suffers from mode collapse due to its non-stochastic nature. Conversely, ancestral sampling typically avoids mode collapsing and generates more diverse outputs, but it is non-trivial to generalize the ancestral sampling with generic parameter space, e.g. NeRF MLP. The primary contribution of our study is the integration of these two approaches through an optimization perspective based on following key observations. The first key insight of DreamSampler is that the DDIM sampling can be interpreted as the solution of following optimization problem:

$$z_{t-1} = \sqrt{\bar{\alpha}_{t-1}}\bar{z} + \sqrt{1-\bar{\alpha}_{t-1}}\tilde{\epsilon}, \quad \text{where} \quad \bar{z} = \arg\min_z \|z - \hat{z}_{0|t}\|^2 \tag{7}$$

Although looks trivial, one of the important implications of (7) is that we can now extend the solution of (7) to include additional regularization term,

$$\min_z \|z - \hat{z}_{0|t}\|_2^2 + \lambda_{reg}\mathcal{R}(z) \tag{8}$$

where λ_{reg} is scalar weight for the regularization function $\mathcal{R}(z)$. For example, we can use data consistency loss to ensure that the updated variable agrees with the given observation during inverse problem solving [9].

Second key insight arises from the important connection between (7) and the score-distillation loss. Specifically, using the forward diffusion to generate z_t from the clean latent z:

$$z_t = \sqrt{\bar{\alpha}_t}z + \sqrt{1-\bar{\alpha}_t}\epsilon, \tag{9}$$

the objective function in (7) can be converted to:

$$\|z - \hat{z}_{0|t}\|^2 = \left\| \frac{z_t - \sqrt{1-\bar{\alpha}_t}\epsilon}{\sqrt{\bar{\alpha}_t}} - \frac{z_t - \sqrt{1-\bar{\alpha}_t}\epsilon_\theta(z_t,t)}{\sqrt{\bar{\alpha}_t}} \right\|^2 \tag{10}$$

$$= \frac{1-\bar{\alpha}_t}{\bar{\alpha}_t}\|\epsilon - \epsilon_\theta(z_t,t)\|^2, \tag{11}$$

which is equivalent to the score-distillation loss up to a constant scaling factor.

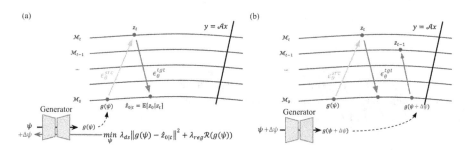

Fig. 3. Unified framework of DreamSampler. (a) Distillation step where the gradient is computed from regularized latent optimization problem. (b) Reverse sampling step where estimated noise by diffusion model is added to the updated generation.

3 DreamSampler

From Sect. 2.2, it is evident that DDIM sampling inherently includes 'score-distillation' optimization, although it is conducted with z instead of generic parameters ψ. Inspired by this observation, we aim to generalize this optimization-based sampling with arbitrary generic parameter ψ, as in conventional score distillation sampling protocols.

3.1 General Formulation

Suppose that $g(\psi)$ denotes a generated data by an arbitrary generator g and parameter ψ. Inspired by the two key insights described in the previous section, the sampling process of DreamSampler at timestep t is given by[1]

$$z_t = \sqrt{\bar{\alpha}_t}g(\psi_t) + \sqrt{1-\bar{\alpha}_t}\tilde{\epsilon}, \tag{12}$$

where the noise $\tilde{\epsilon}$ is defined as in (5), and

$$\psi_t = \arg\min_\psi \|g(\psi) - \hat{z}_{0|t}\|^2 + \lambda_{reg}\mathcal{R}(g(\psi)), \tag{13}$$

$$\hat{z}_{0|t} = \left(z_t - \sqrt{1-\bar{\alpha}_t}\epsilon_\theta(z_t,t)\right)/\sqrt{\bar{\alpha}_t}. \tag{14}$$

---
[1] Here, we omit the encoder $\mathcal{E}_\phi$ in $\mathcal{E}_\phi(g(\psi))$ for notational simplicity. The encoder maps the generated image $g(\psi)$ to the latent space.

It implies that the score distillation sampling and reverse sampling can be integrated based on this generalized latent optimization framework with proper generator and regularization functions. In the following sections, we further delineate the special cases of DreamSampler.

3.2 DreamSampler with External Generators

Similar to DreamFusion [21], for any differentiable generator g, DreamSampler can feasibly update parameters ψ by leveraging the diffusion model and sharing the same sampling process. To emphasize the distinctions between the original score distillation algorithms and DreamSampler, we conduct a line-by-line comparison of the pseudocode in Algorithm 1 and Algorithm 2.

Algorithm 1. Score Distillation	**Algorithm 2** DreamSampler
Require: $T, \zeta, g, \psi, \mathcal{E}_\phi, \{\bar{\alpha}_t\}_{t=1}^T$	**Require:** $T, \zeta, g, \psi, \mathcal{E}_\phi, \{\bar{\alpha}_t\}_{t=1}^T$
1: $z_0 \leftarrow \mathcal{E}_\phi(x_0)$	1: $z_0 \leftarrow \mathcal{E}_\phi(x_0), \epsilon_\theta(z_{T+1}) := \epsilon \sim \mathcal{N}(0, I)$
2: **for** $i = T$ to 1 **do**	2: **for** $i = T$ to 1 **do**
3: $\quad t \sim U[0, T]$	3: $\quad t \leftarrow i, \epsilon \sim \mathcal{N}(0, I)$
4: $\quad \tilde{\epsilon} \sim \mathcal{N}(0, I)$	4: $\quad \tilde{\epsilon} \leftarrow \frac{\sqrt{1-\bar{\alpha}_{t-1}-\eta^2\beta_t^2}\hat{\epsilon}_\theta + \eta\beta_t\epsilon}{\sqrt{1-\bar{\alpha}_t}}$
5: $\quad z_t \leftarrow \sqrt{\bar{\alpha}_t}z_0 + \sqrt{1-\bar{\alpha}_t}\tilde{\epsilon}$	5: $\quad z_t \leftarrow \sqrt{\bar{\alpha}_t}z_0 + \sqrt{1-\bar{\alpha}_t}\tilde{\epsilon}$
6: $\quad \hat{\epsilon}_\theta \leftarrow \epsilon_\theta^\omega(z_t, t, c)$	6: $\quad \hat{\epsilon}_\theta \leftarrow \epsilon_\theta^\omega(z_t, t, c)$
7: $\quad \nabla_\psi \mathcal{L}_{ds} \leftarrow \tilde{\epsilon} - \hat{\epsilon}_\theta$	7: $\quad \nabla_\psi \mathcal{L}_{ds} \leftarrow \tilde{\epsilon} - \hat{\epsilon}_\theta$
8: $\quad \psi \leftarrow \psi - \zeta \nabla_\psi \mathcal{L}_{ds}$	8: $\quad \psi \leftarrow \psi - \zeta[\nabla_\psi \mathcal{L}_{ds} + \lambda_{reg}\nabla_\psi \mathcal{R}(z)]$
9: $\quad z_0 \leftarrow \mathcal{E}_\phi(g(\psi))$	9: $\quad z_0 \leftarrow \mathcal{E}_\phi(g(\psi))$
10: **end for**	10: **end for**
11: **return** ψ	11: **return** ψ

First, DreamSampler follows the timestep schedule of the reverse sampling process, while distillation sampling algorithms use uniformly random timestep for optimization. This provides us with a novel potential for further refinement in utilizing various time schedulers to improve reconstruction quality or accelerate sampling. Second, as DreamSampler is built upon the general proximal optimization framework, it is compatible with additional regularization functions. From this design, one can explore various applications of the proposed distillation sampling. For example, by defining the regularization function as a data consistency term, one can constrain the generator to reconstruct the true image that aligns with the given measurement for inverse imaging.

Figure 3 illustrates the sampling process of DreamSampler. At each timestep, the generated image $g(\psi)$ is mapped to a noisy manifold by incorporating the estimated noise from the previous timestep, and new noise is subsequently estimated by the diffusion model. The distillation gradient is then computed between these two estimated noises and utilized to update the generator parameters.

3.3 DreamSampler for Image Editing

As DreamSampler is a general framework, we can reproduce other existing algorithms by properly defining $g(\psi)$, $\hat{z}_{0|t}$, and $\mathcal{R}$. As a representative example, here we derive Delta Denoising Score (DDS) [4] for image editing task and discuss its potential extension from the perspective of DreamSampler.

The main assumption of DDS is to decompose the SDS gradient [21] into text component and bias component, where only the text component contains information to be edited according to given text prompt while the bias component includes preserved information. To remove the bias component from distillation gradient, DDS leverages the difference of two conditional predicted noises, $\epsilon_\theta(z_t, t, c_{tgt}) - \epsilon_\theta(z_t, t, c_{src})$, where c_{tgt} denotes description of editing direction and c_{src} denotes description of the original image. Specifically, DDS update to clean latent reads[2]

$$\bar{z} = z - \gamma[\epsilon_\theta(z_t, t, c_{tgt}) - \epsilon_\theta(z_t, t, c_{src})]. \tag{15}$$

In the context of Dreamsampler, let the generator $g(\psi) := z$ be a clean latent. Then, the following Theorem 1 shows that the one-step DDS update can be reproduced with DreamSampler, by defining the regularization function $\mathcal{R}(z)$ as Euclidean distance from the posterior mean conditioned on text.

Theorem 1. *Supposed c_{src} in (15) be defined as the null-text, i.e. $c_{src} = c_\varnothing$ and consider text-conditioned posterior mean:*

$$\hat{z}_{0|t}(c) = \mathbb{E}[z_0|z_t, c] = (z_t - \sqrt{1 - \bar{\alpha}_t}\epsilon_\theta(z_t, t, c))/\sqrt{\bar{\alpha}_t}. \tag{16}$$

Then, DDS update in (15) can be obtained from the following latent optimization:

$$\min_z \|z - \hat{z}_{0|t}(c_\varnothing)\|^2 + \gamma R(z), \quad \text{where} \quad R(z) := \frac{\|z - \hat{z}_{0|t}(c_{tgt})\|^2}{(1 - \gamma)} \tag{17}$$

Furthermore, it is equivalent to Tweedie's formula with CFG, i.e.:

$$\hat{z}_{0|t}^\gamma(c_{tgt}) := \frac{z_t - \sqrt{1 - \bar{\alpha}_t}\epsilon_\theta^\gamma(z_t, t, c_{tgt})}{\sqrt{\bar{\alpha}_t}} \tag{18}$$

where $\epsilon_\theta^\gamma(z_t, t, c_{tgt}) = \epsilon_\theta(z_t, t, c_\varnothing) + \gamma[\epsilon_\theta(z_t, t, c_{tgt}) - \epsilon_\theta(z_t, t, c_\varnothing)]$.

Theorem 1 reveals that the one step latent optimization (17) of DreamSampler reproduces the DDS update (15). That being said, the main advantages of DreamSampler stems from the added noise to updated source image z_0. Specifically, in contrast to the original DDS method that adds newly sampled Gaussian noise to z_0, DreamSampler adds the estimated noise by ϵ_θ in the previous timestep of reverse sampling. Initiated from the inverted noise, reverse sampling do not deviate significantly from the reconstruction trajectory even

---
[2] The update term is equivalent to (3) in [4] when $\theta = z$, $\hat{y} = c_{src}$, and $y = c_{tgt}$.

though source prompt is not given, because the latent optimization (17) represents a proximal problem that regulates the sampling process. Consequently, DreamSampler only requires text prompt that describe the editing direction for the real imaging editing.

Finally, it is noteworthy that the equivalent interpretation (18) could be readily extended to spatially localized distillation by computing the solution as

$$\hat{z}_{0|t}^{\gamma} = \left(z_t - \sqrt{1-\bar{\alpha}_t} \sum_i \mathcal{M}_i \odot \epsilon_\theta^{\gamma}(z_t, t, c_{tgt}^{(i)})\right)/\sqrt{\bar{\alpha}_t} \tag{19}$$

where $\mathcal{M}_i$ denotes the pixel-wise mask, $\odot$ is element-wise multiplication, and $c_{tgt}^{(i)}$ denote ith mask and corresponding text prompt pair. The entire process for the real image editing via DreamSampler is described in Algorithm 3. Remark that the Algorithm 3 is equivalent to the Algorithm 2 through Theorem 1, by setting $\eta\beta_t = 0$ to achieve the deterministic sampling, and by setting $\omega = 0$ to ensure that the latent optimization problem is solved during unconditional sampling. Line 9 of the Algorithm 2 is disappeared since we are assuming that $\psi = z_0$ and g as identity mapping.

3.4 DreamSampler for Inverse Problems

DreamSampler can leverage multiple regularization terms to precisely constrain the sampling process to solve inverse problems. For example, we can solve the text-guided image inpainting task by defining the regularization function as

$$\mathcal{R}(z) = (1-\gamma)\|\mathcal{M} \odot (z - \hat{z}_{0|t}(c_{tgt}))\|^2 + \gamma\|y - \mathcal{AD}_\varphi(z)\|^2, \tag{20}$$

where $\mathcal{M}$ denotes the operation to create the measurement by masking out the target region. This regularization term implies that the sample image satisfies data consistency for regions where the true signal is preserved, while guiding the masked region to reflect the target text prompt. When we solve the entire latent optimization problem, we follows two-step approach of TReg [9]. For the details, refer to the appendix. The main difference with TReg is that we separate the text-guidance from the data consistency term. In other words, we initialize the z for the data consistency term as $\hat{z}_{0|t}(c_\varnothing)$ while TReg uses $(1-\omega)\hat{z}_{0|t}(c_\varnothing) + \omega\hat{z}_{0|t}(c_{tgt})$ where the ω denotes the CFG scale. This difference allows DreamSampler to solve the inpainting problem with different text-guidance for each masked region, by combining localized distillation approach introduced in Sect. 3.3.

4 Experimental Results

4.1 Image Restoration Through Vectorization

As a novel application that other methods have not explored, here we present an image vectorization from blurry measurement using DreamSampler. In this scenario, the generator $g(\cdot)$ corresponds to a differentiable rasterizer (DiffVG [11]),

Algorithm 3 DreamSampler for Image Editing

Require: source image x, image encoder $\mathcal{E}_\phi$, latent diffusion model ϵ_θ, null-text embedding $c_\varnothing$, conditioning text embedding c_{tgt}.
$z_0 \leftarrow \mathcal{E}_\phi(x)$
$z_T \leftarrow \text{Inversion}(z_0)$
for $t \in [T, 0]$ **do**
　　$\hat{\epsilon}_\theta \leftarrow \epsilon_\theta(z_t, t, c_\varnothing) + \gamma[\epsilon_\theta(z_t, t, c_{tgt}) - \epsilon_\theta(z_t, t, c_\varnothing)]$
　　$\bar{z} \leftarrow (z_t - \sqrt{1 - \bar{\alpha}_t}\hat{\epsilon}_\theta)/\sqrt{\bar{\alpha}_t}$
　　$z_t \leftarrow \sqrt{\bar{\alpha}_{t-1}}\bar{z} + \sqrt{1 - \bar{\alpha}_{t-1}}\epsilon_\theta(z_t, t, c_\varnothing)$
end for

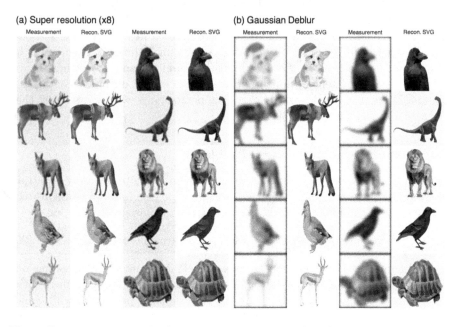

Fig. 4. Representative results for image vectorization task with image reconstruction.

and the parameters ψ of the generator consist of path parameters comprising Scalable Vector Graphics (SVG).

Specifically, we address a text-guided SVG inverse problem, acknowledging that low-quality, noisy measurements can detract from the detail and aesthetic quality of vector designs. Here, our goal is to accurately reconstruct SVG paths for the parameter ψ that, when rasterized, align closely with the provided measurements y and text conditions c_y, related through the forward measurement operator $\mathcal{A}$. In this context, the regularizer $\mathcal{R}$ in (14) corresponds to the data consistency term. Forward measurement operators are specified as follows: (**a**) For super-resolution, bicubic downsampling is performed with scale ×8. (**b**) For Gaussian blur, the kernel has size 61 × 61 with a standard deviation of 5. Then, the latent optimization framework of the SVG inverse problem is defined as:

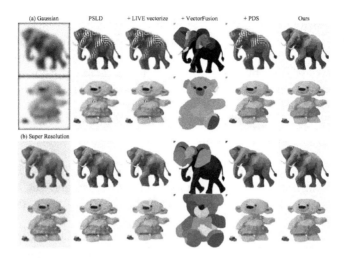

Fig. 5. Qualitative comparison of SVG reconstruction. For baselines, we first obtain an initial reconstruction using PSLD [25], vectorize it with LIVE [15], and refine the output vector with VectorFusion [8] or PDS [10]. DreamSampler outperforms this multi-step approach by simultaneously solving the inverse problem and updating SVG parameters via score distillation.

$$\min_{\psi}(1-\gamma)\lambda_{SDS}\|\mathcal{E}(g(\psi)) - \hat{z}_{0|t}(c_y)\|^2 + \gamma\lambda_{DC}\|y - \mathcal{A}g(\psi)\|^2, \qquad (21)$$

where we found out that $\gamma = \bar{\alpha}_t$ works well in practice. SVG primitives ψ are initialized with radius 20, random fill color, and opacity uniformly sampled between 0.7 and 1, following [8]. In this paper, we use closed Bézier curves for iconography artistic style. For the optimization of (21), we use Adam optimizer with (β_1, β_2)=(0.9, 0.9). For the text condition c_y, we append a suffix to the object[3]. Additional experimental details are provided in the appendix.

Comparatively, the proposed framework is evaluated against a multi-stage baseline approach involving initial rasterized image reconstruction using the state-of-the-art solver (PSLD [24]) that is based on the latent diffusion model. Then, we vectorize the reconstruction using the off-the-shelf Layer-wise Image Vectorization program (LIVE [15]). An optional step includes refining the SVG output via latent score distillation sampling algorithms, exemplified by VectorFusion [8] and Posterior Distillation Sampling (PDS, [10]).

Figure 4 illustrates that the proposed framework achieves high-quality SVG reconstructions with semantic alignment closely matching the specified text condition c_y. In contrast, Fig. 5 highlights the deficiencies of the multi-stage baseline methods, particularly its inability to retain detailed fidelity. Errors accumulate during the initial restoration process, resulting in blurriness and undesirable path overlap in the vector outputs. While the solver may achieve adequate reconstruc-

---

[3] e.g. "{*a cute fox*}, *minimal flat 2d vector icon. lineal color. on a white background. trending on artstation.*".

tions, the subsequent vectorization step disregards text caption c_y, leading to the potential loss of details and semantic coherence. Additional VectorFusion fine-tuning loses consistency with the measurement. Conversely, DreamSampler effectively restores SVGs by directly utilizing latent-space diffusion and text conditions for both vectorization and reconstruction, ensuring the preservation of detail and contextual relevance.

4.2 Real Image Editing

For real image editing via DreamSampler, we leverage the Stable-Diffusion v1.5 provided by HuggingFace. We use linear time schedule and set NFE to 200 for both the DDIM inversion and reverse sampling. For more details on hyperparameter setting, please refer to the appendix.

To demonstrate the ability of DreamSampler in real image editing, we leverage source images from various domains, including photographs of animals, human faces and drawings. All results in Fig. 6 show that DreamSampler accurately reflects the provided text prompts for editing. Specifically, DreamSampler does not change bias components, which are intended not to be edited

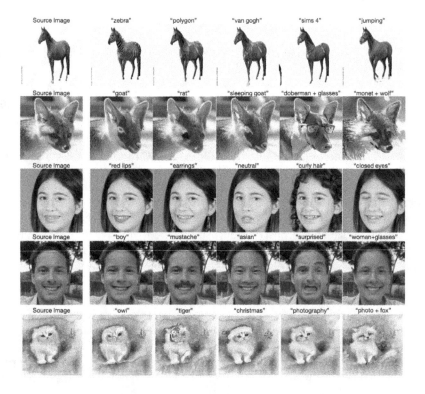

Fig. 6. Representative results for real image editing via distillation through reverse sampling. We leverage source images from various domains and the caption above each image denotes text prompt reflecting the editing direction. The result demonstrates that distillation could be effectively conducted during the reverse sampling.

by text prompt. For instance, in 3rd and 4th rows of Fig. 6, features such as braces and background including striped shirt, are well-maintained while the text prompts are accurately reflected. Moreover, DreamSampler effectively reflects multiple editing directions simultaneously such as "doberman" + "wearing glasses", "woman" + "wearing glasses" or "photography" + "fox". We next conduct a qualitative comparison with the original DDS algorithm to show the improvement from integration of distillation into reverse sampling. As illustrated in Fig. 7, DreamSampler is capable of editing according to text prompt robustly across various domains of source images. Furthermore, DreamSampler achieves better fidelity and bias component preservation compared to the original DDS.

For the quantitative comparison, we compare DreamSampler against diffusion-based editing algorithms [5,16,18], following the experimental setups of [19,20]. We use the prompt "a photograph of {}" with the source and tar-

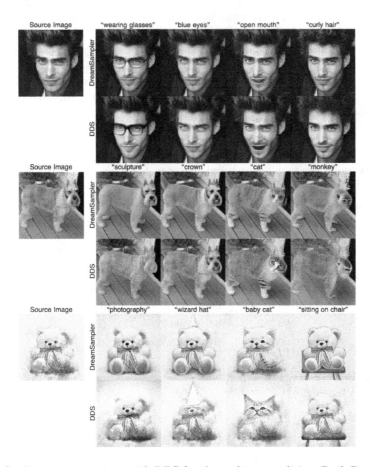

Fig. 7. Qualitative comparison with DDS for the real image editing. Both DreamSampler and DDS edit images following target text, but DreamSampler achieves higher fidelity with preserving bias component.

get objects inserted. For the CFG scale, we use $0.15\bar{\alpha}_t$ across all cases. Table 1 demonstrates that DreamSampler outperforms most baselines in image editing tasks. Specifically, DreamSampler generates glasses more naturally in the "cat → cat w/ glasses" task, improves fidelity in source image edits compared to baseline algorithms as shown in Figure S10. Some baselines achieve high CLIP accuracy by focusing on the generation of glasses, regardless of its natural appearance.

4.3 Text-Guided Image Inpainting

For the text-guided image inpainting task, we also use Stable-Diffusion v1.5, linear time schedule, and 200 NFE. In addition to solving (20), we apply DPS [1] steps during sampling by following TReg [9] to enhance the consistency of masked region and other regions. We generate measurements by masking out two rectangular regions on the eyes and mouth, where the masked region is determined based on the averaged face of the 1k FFHQ validation set. The size of measurement is 512 × 512. We solve the inpainting problem by giving text prompts "a photography of face wearing glasses" and "a photography of face with smile". Figure 8 shows that DreamSampler fills masked region by reflecting given text prompt accurately. While the text guidance is applied inside the mask, data consistency gradients combined with the reconstruction by inversion is applied outside the mask, which results in superior fidelity of the output image. Note that we display the image output as is without any post-processing such as

Table 1. Comparison to diffusion-based editing methods. Dist for DINO-ViT Structure Distance. Baseline results are from [19].

Method	Cat→Dog		Horse→Zebra		Cat→ Cat w/ glasses	
	CLIP-Acc ↑	Dist ↓	CLIP-Acc ↑	Dist ↓	CLIP-Acc ↑	Dist ↓
SDEdit + word swap	71.2%	0.081	92.2%	0.105	34.0%	0.082
DDIM + word swap	72.0%	0.087	94.0%	0.123	37.6%	0.085
prompt to prompt	66.0%	0.080	18.4%	0.095	69.6%	0.081
p2p-zero	92.4%	0.044	75.2%	0.066	71.2%	0.028
EBCA	93.7%	0.040	90.4%	0.061	81.1%	0.052
DreamSampler (Ours)	90.3%	**0.029**	**95.2%**	**0.038**	48.3%	**0.025**

Table 2. Comparison to text-conditioned diffusion-based inpainting solvers. **Bold**: the best score, Underline: the second best.

Method	Glasses			Smile		
	PSNR ↑	FID ↓	CLIP-sim ↑	PSNR ↑	FID ↓	CLIP-sim ↑
Stable-Inpaint	19.82	<u>54.26</u>	<u>0.281</u>	25.33	**19.22**	<u>0.249</u>
TReg	<u>21.97</u>	61.04	**0.288**	<u>26.71</u>	24.48	<u>0.249</u>
DreamSampler (Ours)	**24.61**	**27.10**	0.263	**27.90**	<u>24.33</u>	0.242

projection[4]. The bottom row of Fig. 8 depict the solution for inpainting problem with two different masks and distinct text guidance. Through localized distillation gradient, DreamSampler generates solutions according to the provided guidance. DreamSampler generates masked regions with better robustness than TReg. Additionally, in the case of multiple masks, DreamSampler successfully reflects text in the correct regions via localized distillation gradients, whereas TReg fails to meet one of the conditions. For more results, refer to appendix.

We also evaluate both the quality of reconstructions (PSNR, FID) and the accuracy of the text guidance (CLIP similarity) on 1k FFHQ validation set. For baselines, we select diffusion-based image inpainting models [14,23] and a text-guided inverse problem solver [9]. Specifically, Stable-Inpaint [23] is a fine-tuned StableDiffusion model specialized for inpainting task. Table 2 shows that DreamSampler outperforms the baselines in terms of PSNR and FID score while achieving comparable CLIP similarity. Stable-Inpaint achieves a lower FID score, while DreamSampler, serving as a zero-shot inpainting solver, achieves a higher PSNR. This suggests superior reconstruction quality of Dreamsampler through data consistency update.

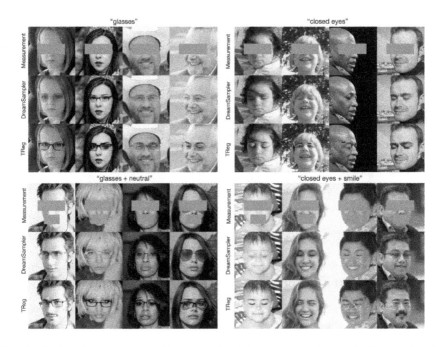

Fig. 8. Qualitative comparison for text guided image inpainting task. DreamSampler can generate more realistic images according to given text prompt.

---

[4] For linear operator $\mathcal{A}$ and measurement y, the projection means $\mathcal{A}^\top \mathcal{A} y + (I - \mathcal{A}^\top \mathcal{A}) x$.

5 Conclusion

We presented DreamSampler, a unified framework of reverse sampling and score distillation by taking the advantages of each algorithms. Specifically, we connected two distinct algorithms under the perspective of latent optimization problem. Consequently, we introduced a generalized optimization framework, which offers new design space to solve various applications. Especially, DreamSampler enables to combine various regularization functions to constrain the sampling process. Additionally, we provided three applications including image vectorization with reconstruction, real image editing, and image inpainting. DreamSampler could be extended to other algorithms by defining appropriate regularization functions. The codebase is available to public at https://github.com/DreamSampler/dream-sampler.

Potential Negative Social Impact. The performance of algorithms established on DreamSampler heavily depends on the prior of diffusion model. Hence, the proposed method basically influenced by the potential negative impacts of LDM itself. Thus, proper political regulation is required to mitigate these risks.

Acknowledgments. This work was supported by the National Research Foundation of Korea under Grant RS-2024-00336454, by Institute of Information & communications Technology Planning & Evaluation (IITP) grant funded by the Korea government (MSIT)) (No. RS-2022-II220984, Development of Artificial Intelligence Technology for Personalized Plug-and-Play Explanation and Verification of Explanation, and No. 2019-0-00075, Artificial Intelligence Graduate School Program (KAIST)), by Culture, Sports and Tourism R&D Program through the Korea Creative Content Agency grant funded by the Ministry of Culture, Sports and Tourism in 2023, by Field-oriented Technology Development Project for Customs Administration funded by the Korea government (the Ministry of Science & ICT and the Korea Customs Service) through the National Research Foundation (NRF) of Korea under Grant NRF2021M3I1A1097910.

References

1. Chung, H., Kim, J., Mccann, M.T., Klasky, M.L., Ye, J.C.: Diffusion posterior sampling for general noisy inverse problems. arXiv preprint arXiv:2209.14687 (2022)
2. Chung, H., Lee, S., Ye, J.C.: Fast diffusion sampler for inverse problems by geometric decomposition. arXiv preprint arXiv:2303.05754 (2023)
3. Dhariwal, P., Nichol, A.: Diffusion models beat gans on image synthesis. In: Advance Neural Information Processing System, vol. 34, pp. 8780–8794 (2021)
4. Hertz, A., Aberman, K., Cohen-Or, D.: Delta denoising score. In: Proceedings of the IEEE/CVF International Conference on Computer Vision, pp. 2328–2337 (2023)
5. Hertz, A., Mokady, R., Tenenbaum, J., Aberman, K., Pritch, Y., Cohen-Or, D.: Prompt-to-prompt image editing with cross attention control. arXiv preprint arXiv:2208.01626 (2022)
6. Ho, J., Jain, A., Abbeel, P.: Denoising diffusion probabilistic models. In: Advance Neural Information Processing System, vol. 33, pp. 6840–6851 (2020)

7. Ho, J., Salimans, T.: Classifier-free diffusion guidance. arXiv preprint arXiv:2207.12598 (2022)
8. Jain, A., Xie, A., Abbeel, P.: Vectorfusion: text-to-SVG by abstracting pixel-based diffusion models. In: Proceedings of the IEEE/CVF Conference on Computer Vision and Pattern Recognition, pp. 1911–1920 (2023)
9. Kim, J., Park, G.Y., Chung, H., Ye, J.C.: Regularization by texts for latent diffusion inverse solvers. arXiv preprint arXiv:2311.15658 (2023)
10. Koo, J., Park, C., Sung, M.: Posterior distillation sampling. In: Proceedings of the IEEE/CVF Conference on Computer Vision and Pattern Recognition, pp. 13352–13361 (2024)
11. Li, T.M., Lukáč, M., Gharbi, M., Ragan-Kelley, J.: Differentiable vector graphics rasterization for editing and learning. ACM Trans. Graph. (TOG) **39**(6), 1–15 (2020)
12. Liu, G.H., Vahdat, A., Huang, D.A., Theodorou, E.A., Nie, W., Anandkumar, A.: I^{2}sb: image-to-image Schrodinger bridge. arXiv preprint arXiv:2302.05872 (2023)
13. Liu, Y., et al.: Syncdreamer: generating multiview-consistent images from a single-view image. arXiv preprint arXiv:2309.03453 (2023)
14. Lugmayr, A., Danelljan, M., Romero, A., Yu, F., Timofte, R., Van Gool, L.: Repaint: inpainting using denoising diffusion probabilistic models. In: Proceedings of the IEEE/CVF Conference on Computer Vision and Pattern Recognition, pp. 11461–11471 (2022)
15. Ma, X., et al.: Towards layer-wise image vectorization. In: Proceedings of the IEEE/CVF Conference on Computer Vision and Pattern Recognition, pp. 16314–16323 (2022)
16. Meng, C., Song, Y., Song, J., Wu, J., Zhu, J.Y., Ermon, S.: Sdedit: image synthesis and editing with stochastic differential equations. arXiv preprint arXiv:2108.01073 (2021)
17. Mildenhall, B., Srinivasan, P.P., Tancik, M., Barron, J.T., Ramamoorthi, R., Ng, R.: Nerf: representing scenes as neural radiance fields for view synthesis. Commun. ACM **65**(1), 99–106 (2021)
18. Mokady, R., Hertz, A., Aberman, K., Pritch, Y., Cohen-Or, D.: Null-text inversion for editing real images using guided diffusion models. In: Proceedings of the IEEE/CVF Conference on Computer Vision and Pattern Recognition, pp. 6038–6047 (2023)
19. Park, G.Y., Kim, J., Kim, B., Lee, S.W., Ye, J.C.: Energy-based cross attention for bayesian context update in text-to-image diffusion models. In: Advances in Neural Information Processing Systems , vol. 36 (2024)
20. Parmar, G., Kumar Singh, K., Zhang, R., Li, Y., Lu, J., Zhu, J.Y.: Zero-shot image-to-image translation. In: ACM SIGGRAPH 2023 Conference Proceedings, pp. 1–11 (2023)
21. Poole, B., Jain, A., Barron, J.T., Mildenhall, B.: Dreamfusion: text-to-3d using 2d diffusion. arXiv preprint arXiv:2209.14988 (2022)
22. Radford, A., et al.: Learning transferable visual models from natural language supervision. In: International Conference on Machine Learning, pp. 8748–8763. PMLR (2021)
23. Rombach, R., Blattmann, A., Lorenz, D., Esser, P., Ommer, B.: High-resolution image synthesis with latent diffusion models. In: Proceedings of the IEEE/CVF Conference on Computer Vision and Pattern Recognition, pp. 10684–10695 (2022)
24. Rout, L., Raoof, N., Daras, G., Caramanis, C., Dimakis, A., Shakkottai, S.: Solving linear inverse problems provably via posterior sampling with latent diffusion models. In: Advances in Neural Information Processing Systems, vol. 36 (2024)

25. Rout, L., Raoof, N., Daras, G., Caramanis, C., Dimakis, A.G., Shakkottai, S.: Solving linear inverse problems provably via posterior sampling with latent diffusion models. arXiv preprint arXiv:2307.00619 (2023)
26. Song, J., Meng, C., Ermon, S.: Denoising diffusion implicit models. arXiv preprint arXiv:2010.02502 (2020)
27. Song, Y., Sohl-Dickstein, J., Kingma, D.P., Kumar, A., Ermon, S., Poole, B.: Score-based generative modeling through stochastic differential equations. arXiv preprint arXiv:2011.13456 (2020)
28. Zhang, R., Isola, P., Efros, A.A., Shechtman, E., Wang, O.: The unreasonable effectiveness of deep features as a perceptual metric. In: Proceedings of the IEEE Conference on Computer Vision and Pattern Recognition, pp. 586–595 (2018)
29. Zhu, J., Zhuang, P.: HiFA: high-fidelity text-to-3d with advanced diffusion guidance. arXiv preprint arXiv:2305.18766 (2023)

Reshaping the Online Data Buffering and Organizing Mechanism for Continual Test-Time Adaptation

Zhilin Zhu[1,2(✉)], Xiaopeng Hong[1,2(✉)], Zhiheng Ma[3,4,5], Weijun Zhuang[1,2], Yaohui Ma[1,5], Yong Dai[2], and Yaowei Wang[1,2]

[1] Harbin Institute of Technology, Harbin, China
{zhuzhilin,weijunzhuang}@stu.hit.edu.cn, hongxiaopeng@ieee.org
[2] Pengcheng Laboratory, Shenzhen, China
wangyw@pcl.ac.cn
[3] Shenzhen University of Advanced Technology, Shenzhen, China
[4] Guangdong Provincial Key Laboratory of Computility Microelectronics, Shenzhen, China
[5] Shenzhen Institute of Advanced Technology, Chinese Academy of Science, Shenzhen, China
{zh.ma,yh.ma}@siat.ac.cn

Abstract. Continual Test-Time Adaptation (CTTA) involves adapting a pre-trained source model to continually changing unsupervised target domains. In this paper, we systematically analyze the challenges of this task: online environment, unsupervised nature, and the risks of error accumulation and catastrophic forgetting under continual domain shifts. To address these challenges, we reshape the online data buffering and organizing mechanism for CTTA. We propose an uncertainty-aware buffering approach to identify and aggregate significant samples with high certainty from the unsupervised, single-pass data stream. Based on this, we propose a graph-based class relation preservation constraint to overcome catastrophic forgetting. Furthermore, a pseudo-target replay objective is used to mitigate error accumulation. Extensive experiments demonstrate the superiority of our method in both segmentation and classification CTTA tasks. Code is available at https://github.com/z1358/OBAO.

Keywords: Continual test-time adaptation · Unsupervised learning · Continual learning · Catastrophic forgetting

1 Introduction

Deep learning models have exhibited remarkable advancements in scenarios characterized by a consistent data distribution between training and test environments. However, the deployment of machine perception systems in the

---

Supplementary Information The online version contains supplementary material available at https://doi.org/10.1007/978-3-031-73007-8_24.

real world frequently exposes them to non-stationary and constantly changing environments [14,21,47,53,60]. When such phenomena occur (*e.g.*, unexpected domain shifts over time), these models are prone to severe performance degradation [18,37,51]. Consequently, the development of techniques for continual adaptation during the unsupervised deployment phase is essential for enhancing the reliability of machine perception systems.

Table 1. Comparison of different adaptation settings. We differentiate along four dimensions: the available form of the source domain data and target domain data, the distribution and access mode of the target domain data stream during adaptation.

Setting	Data		Target domain data stream	
	Source	Target	Distribution	Access mode
Fine-tuning	-	$\mathcal{X}^T, \mathcal{Y}^T$	stationary	offline (multi-epoch)
Domain Generalization	$\mathcal{X}^S, \mathcal{Y}^S$	-	-	-
Domain Adaptation	$\mathcal{X}^S, \mathcal{Y}^S$	$\mathcal{X}^T$	stationary	offline (multi-epoch)
Test-Time Training	$\mathcal{X}^S, \mathcal{Y}^S$	$\mathcal{X}^T$	stationary	online (single-pass)
Test-Time Adaptation	-	$\mathcal{X}^T$	stationary	offline or online
Continual Test-Time Adaptation	-	$\mathcal{X}_1^T \to \cdots \to \mathcal{X}_n^T$	continually changing	online (single-pass)

With sufficient source labeled data accessible, the fields of domain generalization [22,23,63,66] and domain adaptation [20,24,25,28] have witnessed considerable success. However, a particularly challenging scenario arises when no source data is available, and unsupervised adaptation must happen at test time when unlabeled test inputs arrive. This task is known as test-time adaptation (TTA) [5,7,45]. Given the rarity of encountering only a singular domain shift in practical applications, the community has recently been increasingly focusing on continual test-time adaptation (CTTA) [2,12,30,51,58], *i.e.*, evaluating the ability of methods to adapt to continual domain shifts. We summarize the differences between existing adaptation settings in Table 1. Obviously, the considered CTTA setting is the most practical and challenging.

Within this setting, we expose the key challenges as follows: **(1) Online environment.** This denotes a scenario where data is processed in a single pass, necessitating immediate predictions from the model. Due to the characteristics of the online environment, it is difficult for the model to obtain a sufficient number of informative samples, thereby impeding the efficacy of adaptation; **(2) Unsupervised nature.** Throughout adaptation, ground-truth labels for input data are unavailable. Moreover, due to privacy concerns or legal restrictions, the source domain data may no longer be available and we only have access to the source pre-trained model; **(3) Error accumulation and catastrophic forgetting.** The dynamic distribution shift when adapting to continually changing target domains can render pseudo-labels unreliable and impede knowledge retention about the source domain, leading to error accumulation [51,54] and triggering catastrophic forgetting [6,30,58].

Addressing the first challenge, an intuitive solution is to employ a buffer to store high-value samples. However, existing methods that utilize such a buffer store source domain samples with ground-truth labels [12,43], which is difficult to satisfy in real-world scenarios and conflicts with the source-free constraint in the second challenge. Therefore, the issue of effectively *buffering* and *organizing* high-value samples in an *unsupervised single-pass* data stream has not been fully explored. Current CTTA methods primarily address the third challenge, focusing mainly on alignments in either the parameter space [2,35,40,51] or the feature space [4,12] of the source model to mitigate knowledge forgetting.

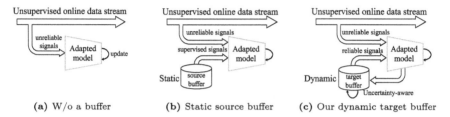

Fig. 1. Comparison of different buffering mechanisms. (a) Without a buffer, only unreliable signals from the current unsupervised input data are available. (b) Some methods [12,43] use a static source buffer that stores source samples with ground-truth labels to provide supervised signals. (c) We work on a dynamic target buffer, which can provide reliable signals for the adaptation process.

In this paper, we endeavor to address these three challenges simultaneously. To this end, we reshape the online data buffering and organizing mechanism for CTTA. As shown in Fig. 1, unlike previous methods which directly sample from the source domain [12,43], we are targeted at source-free, unsupervised, and online buffering. Through a specially designed uncertainty-aware buffering approach, we dynamically evaluate and aggregate reliable information in the unsupervised, single-pass data stream (Challenges **1** and **2**). Based on this, we efficiently organize these samples to assist the model's unsupervised adaptation process. Utilizing pseudo-labels generated from these low-uncertainty samples, we consider them as anchors of classes and organize them into a class relation graph (CRG). To combat catastrophic forgetting under continually changing domains, we introduce a novel graph-based class relation preservation constraint to maintain the intrinsic class relations obtained on the source domain (Challenge **3**). Specifically, we penalize topological changes within the CRG to mitigate knowledge forgetting, while allowing reasonable vertex movements to enhance adaptability. Additionally, these samples are incorporated into a pseudo-target replay objective to reduce the risk of error accumulation from misclassified samples (Challenge **3**). In summary, our main contributions are as follows:

- To address the three revealed challenges simultaneously, we reshape the online data buffering and organizing mechanism for CTTA.
- To frame the buffering mechanism, we design an uncertainty-aware approach to identify and aggregate reliable data from unsupervised online data stream.

- Through a well-designed graph-based class relation preservation constraint and a pseudo-target replay objective, we effectively mitigate catastrophic forgetting and error accumulation in CTTA.
- We extensively evaluate our proposed method and the results on four challenging datasets show that the method clearly improves the CTTA performance on both segmentation and classification benchmarks, indicating its effectiveness and universality.

2 Related Work

2.1 Domain Adaptation

Domain adaptation (DA) [38,65] aims to generalize a model from a well-annotated source domain to an unlabeled target domain following the transductive learning principle. The underlying assumption in DA is that both labeled source domain and unlabeled target domain data are concurrently accessible. Consequently, the prevalent methods in DA encompass the utilization of domain adversarial training [8,11,16,19,29,49] or the minimization of domain discrepancy [32,33,39]. These methods work towards harmonizing the feature distribution [16,32,49] or the input space [19,59] between the two domains.

2.2 Continual Test-Time Adaptation

Test-Time Adaptation (TTA). TTA focuses on a more challenging setting after model deployment where only a pre-trained source model and unlabeled target data are available. Some approaches [5,27] focus on the offline setting, where multiple epochs of access to the entire set of test data are provided for adaptation, while others [1,7,31,52] focus on the online setting with a single-pass data stream. To name a few, TENT [50] adapts the model by using entropy minimization loss to update a few trainable parameters in batch normalization layers. This approach has also inspired subsequent works such as EATA [36] and RoTTA [61] to investigate the robustness of normalized layers. Due to the domain gap, TSD [52] promotes feature alignment and uniformity from the perspective of feature revision, while TIPI [34] proposes invariance regularizers to find the transformation space that can simulate the domain shift.

Continual Test-Time Adaptation (CTTA). In recent years, CTTA has received increasing research attention [15,44,51,58] due to its practicality. CoTTA [51] stands out as the first method specifically tailored to address this task. It utilizes weight-averaged and augmentation-averaged predictions to reduce error accumulation, while stochastically restoring the weights of a small fraction of neurons to the source model during each iteration to mitigate catastrophic forgetting. Building upon this, PETAL [2] further introduces a data-driven parameter recovery technique to align with the source model in the parameter space, while RMT [12] and SANTA [4] seek alignment with the source model in the feature space by optimizing an additional contrastive learning objective.

Additionally, DSS [54] considers sample security to mitigate the risk of using incorrect information. Some approaches [15,30,44] also prevent forgetting by freezing a majority or all of the model parameters, but they require optimization along with the model during training using the source domain data, thereby relaxing the unsupervised challenge and not being truly source-free. Under our proposal, we simultaneously address three challenges in CTTA.

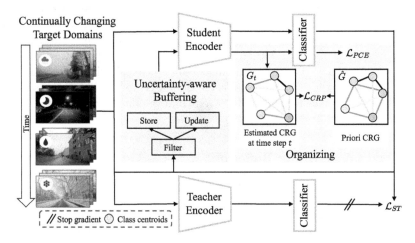

Fig. 2. The framework of our proposed method. At time step t, the incoming data and sampled buffer data are used for adaptation. The proposed class relation preservation loss $\mathcal{L}_{CRP}$ ensures topological consistency between the Class Relation Graph (CRG) G_t estimated at the current time step t and $\hat{G}$ from the source domain, thereby effectively preserving intrinsic semantic information. Concurrently, we integrate a pseudo-target replay loss $\mathcal{L}_{PCE}$ to mitigate error accumulation.

3 Methodology

3.1 Preliminaries and Overall Framework

In continual test-time adaptation (CTTA), we are given a source model f_{θ_0} pre-trained from the labeled source domain $(\mathcal{X}^S, \mathcal{Y}^S)$ and a series of sequentially changing unlabeled target domains such as $[\mathcal{X}_1^T, \mathcal{X}_2^T, \cdots, \mathcal{X}_n^T]$. Each target domain shares the same label space as the source domain (for close-set adaptation). Note that the model has no information about when the domain changes happen during the adaptation process. The goal of CTTA is to adapt the model to a continually changing unsupervised online data stream of target domains without source data. To be more specific, at inference time step t, the unlabeled target test data $x(t)$ is given to the model f_{θ_t}, and the model needs to make predictions and adapt itself accordingly ($f_{\theta_t} \to f_{\theta_{t+1}}$) for the upcoming test inputs $x(t+1)$.

We adopt a teacher-student framework following previous works [2,12,51] as illustrated in Fig. 2. Leveraging the proposed uncertainty-aware buffering mechanism, our method dynamically evaluates and buffers significant high-certainty samples in the unsupervised online data stream, providing reliable information for the model's adaptation process. Building upon this, we further introduce a novel class relation preservation constraint ($\mathcal{L}_{CRP}$) for preserving domain-invariant semantic information in dynamic environments, effectively mitigating the catastrophic forgetting problem. Additionally, we address error accumulation through a pseudo-target replay objective ($\mathcal{L}_{PCE}$) to minimize the impact of miscalibration. A self-training loss ($\mathcal{L}_{ST}$) is used to stabilize the adaptation process of the student model (adapted model). In the adopted framework, both the teacher model $f_{\theta'}$ and the student model f_θ are initialized with the source model f_{θ_0}. Subsequently, the teacher model is continuously updated by the exponential moving average of the student model. We elaborate our method in the following subsections.

3.2 Uncertainty-Aware Buffering

In the unsupervised online data stream within CTTA, the model necessitates adaptation utilizing the data at hand, typically characterized by a scarcity of information. While certain methods [12,43] endeavor to incorporate additional *supervised* information by formulating a static buffer from the source domain, fulfilling this requirement proves challenging within practical application scenarios. Conversely, our proposition involves sampling and buffing throughout the *unsupervised* adaptation process by a designed uncertainty-aware approach.

At the outset, the dynamic buffer is initialized as empty. As the adaptation progresses, the buffer undergoes continual updates to reflect the evolving nature of the data stream encountered. Specifically, at any given inference time step t, the model calculates the uncertainty associated with the current batch of samples. The quantification of sample uncertainty manifests as the entropy of the sample, denoted as:

$$H(x(t)) = -f_{\theta_t}(x(t))^\top (\log f_{\theta_t}(x(t))), \tag{1}$$

where $f_{\theta_t}(x(t))$ corresponds the softmax predictions of the student model for the current input data. Samples exhibiting an entropy value less than a predefined threshold H_0 are deemed to possess a high degree of certainty and are thus considered candidates for inclusion in the dynamic buffer. H_0 is defined as:

$$H_0 = \alpha \ln C, \tag{2}$$

where α serves as a threshold coefficient and C denotes the total number of classes in the dataset. The process of storing samples into the buffer, or updating existing entries, is contingent upon the buffer's capacity. High-certainty samples identified from the present batch, along with their derived pseudo-labels from the teacher model, are added to the buffer if space permits. Conversely, in scenarios where the buffer is at capacity, these new samples replace the ones exhibiting the highest uncertainty to maintain an optimal blend of timeliness and reliability. Upon completion of these updates, the resulting buffer $\mathcal{M}_t$ stands ready to facilitate the model's adaptation in subsequent iterations.

Notably, our approach to buffering is entirely devoid of reliance on source data, underscoring its *source-free* nature. This strategy enables the model to discern and prioritize data based on their uncertainty. By dynamically buffering high-certainty samples within the data stream, our approach not only circumvents the limitations posed by reliance on static source buffers but also enriches the adaptation process with valuable, reliable data. The intuitive idea is to utilize CE loss to replay samples in the buffer, thereby emphasizing reliable data and effectively suppressing the risk and impact of miscalibration due to error accumulation. In each test-time adaptation iteration, we randomly sample examples and labels $(\tilde{x}(t), \tilde{y}(t))$ from the buffer. The number of sampled examples is equal to the batch size and the corresponding pseudo-target replay loss is:

$$\mathcal{L}_{PCE} = -\sum_{c=1}^{C} \tilde{y}_c(t) \log \tilde{p}_c, \qquad (3)$$

where $\tilde{p}$ is the softmax prediction of the student model for $\tilde{x}(t)$. For simplicity, we omit the summation of all samples in the batch. Moreover, these sampled examples also serve as anchors and are organized in a class relation graph. On this basis, we introduce a novel class relation preservation constraint in Sect. 3.3 to preserve domain-invariant knowledge and overcome forgetting.

3.3 Class Relation Preservation Constraint

In online scenarios where domains are constantly changing, neural networks tend to "take shortcuts" and focus on simplistic features [17,55] from the single-pass data stream. This phenomenon results in the model's forgetting of intrinsic semantic information. In this work, we propose to overcome catastrophic forgetting by preserving the domain-invariant class relation, which is consistent across domains regardless of distribution discrepancy [64]. Therefore, it is reasonable to utilize the intrinsic class relation obtained on the source domain with large labeled samples to constrain the model adaptation process on target domains.

We describe here how to obtain the defined class relation graph (CRG). For computational stability, we normalize the feature space and adopt the cosine similarity metric. Let $\bar{\cdot}$ denotes the normalization operation, where $\bar{f} = f/\|f\|$. Given the normalized feature space $\bar{\mathcal{F}}$, the CRG G is defined as $G = \langle V, E \rangle$, where $V = \{\bar{v}_1, \cdots, \bar{v}_C | \bar{v}_i \in \bar{\mathcal{F}}\}$ is the set of C vertices representative of $\bar{\mathcal{F}}$, and E is the edge set describing the neighborhood relations of vertices in V. Each edge e_{ij} is assigned with a weight s_{ij}, which is the similarity between $\bar{v}_i$ and $\bar{v}_j$: $s_{ij} = \bar{v}_i^\top \bar{v}_j$. Each vertex $\bar{v}_i$ is the centroid vector of class i. The intrinsic CRG $\hat{G}$ is obtained by modeling the source-domain class relation. We demonstrate in experiments that the construction of our CRG $\hat{G}$ can be derived from different data conditions, including class prototypes and classifier weights.

At time step t, given CRG $G_t = \langle V_t, E_t \rangle$, when catastrophic forgetting occurs, G_t is distorted as the semantic relations among categories are confused. To mitigate forgetting, the model should be adapted to strive towards preserving the *topology* in G_t the same as that in $\hat{G}$. This is achieved by restricting the neighboring relations of vertices described by the weights of edges (*i.e.*, similarity between classes) in E_t. One way to do this is to keep the ranking of the edge weights constant during the adaptation process. However optimizing a non-smooth global ranking is difficult and inefficient [3,46]. An alternative and more feasible approach involves assessing the extent of change in neighboring relations [13,48] by evaluating the correlation between the initial edge weights and their current observation. A lower correlation indicates a higher probability that the rank of an edge has changed during adaptation, and this should be penalized. On this basis, we define the Class Relation Preservation constraint (CRP) as follows:

$$\mathcal{L}_{CRP} = -\frac{\sum_{i,j}^{C} s_{ij} \tilde{s}_{ij}}{\sqrt{\sum_{i,j}^{C} s_{ij}^2} \sqrt{\sum_{i,j} \tilde{s}_{ij}^2}}, \qquad (4)$$

where $S = \{s_{ij} | 1 \leq i, j \leq C\}$ and $\tilde{S} = \{\tilde{s}_{ij} | 1 \leq i, j \leq C\}$ are the sets of the initial and observation values of the edge weights in $\hat{E}$ and E_t, respectively.

To achieve end-to-end optimization, the ideal situation is using source-domain images with *ground-truth* labels to obtain an estimate of the current CRG G_t under time step t. That is, extracting features from these images by f_{θ_t}, which in turn yields the class prototypes and ultimately the class relation. However, it is often impractical to store a certain number of source-domain samples in the source-free CTTA setting. In this work, we leverage reliable data from our proposed uncertainty-aware buffering mechanism to accurately estimate CRG G_t and achieve end-to-end optimization in an unsupervised test data stream.

From the perspective of viewing S as a vector, the proposed CRP term constrains the consistency of the direction between two vectors during the adaptation process. Opting for the origin moment instead of the central moment allows the CRP term to prioritize direction consistency between vectors. This makes it more robust to scale variations or magnitude disparities in edge weights, which may arise from shifts in data distribution. Some previous methods [4,12] directly penalize the movement of target domain samples in the feature space relative to the source prototypes. This can be broadly interpreted as penalizing the movement of corresponding elements between $\hat{V}$ and V_t in our defined CRG. However, in addition to suffering from classification errors when assigning source prototypes, constraints on features can also impede the effectiveness of adaptation in target domains. Instead of penalizing the movement of vertices in G in the feature space, the proposed CRP term penalizes the change of topological relations between vertices, while allowing reasonable movement of vertices to enhance adaptation. To summarize, by preserving the domain-invariant class relation during the model adaptation process, the proposed CRP term effectively overcomes catastrophic forgetting caused by semantic confusion.

3.4 Optimization Objective

At time step t, the teacher model $f_{\theta'_t}$ first generates pseudo-labels to help the learning process of the student model f_{θ_t}. Specifically, we compute the symmetric cross-entropy loss [12] between the outputs of the student model and the pseudo-labels. The self-training loss can be formulated as follows:

$$\mathcal{L}_{ST} = -\sum_{c=1}^{C} q_c \log p_c - \sum_{c=1}^{C} p_c \log q_c, \quad (5)$$

where q is the softmax prediction of the teacher model, and p is the softmax prediction of the student model, with C being the total number of classes. Finally, the total loss function $\mathcal{L}_T$ can be formulated as follows:

$$\mathcal{L}_T = \mathcal{L}_{ST} + \mathcal{L}_{PCE} + \lambda_{CRP}\mathcal{L}_{CRP}, \quad (6)$$

where λ_{CRP} is a trade-off hyperparameter.

4 Experiments

We conduct comprehensive experiments under the CTTA setting as in [12,51], encompassing three mainstream datasets ImageNet-to-ImageNet-C, CIFAR10-to-CIFAR10C, and CIFAR100-to-CIFAR100C for image classification, as well as Cityscapes-to-ACDC for semantic segmentation.

4.1 Datasets and Experimental Settings

CIFAR10C, CIFAR100C, ImageNet-C. These datasets are originally developed to evaluate the robustness of classification networks [18]. Each dataset contains 15 distinct types of corruption at varying severity levels ranging from 1 to 5. These corruptions are applied to the test and validation images of CIFAR and ImageNet, respectively. We employ the clean training sets on CIFAR10, CIFAR100, and ImageNet as source domain datasets, while CIFAR10C, CIFAR100C, and ImageNet-C serve as the corresponding target domain datasets. Following the continuous benchmark setup in [51], we sequentially adapt the source model to 15 target domains under the largest severity level 5. The entire adaptation process is performed in an online manner, without assuming knowledge of when the domain changes. For each corruption, we utilize 10,000 images for datasets derived from CIFAR and 5,000 images for ImageNet-C.

Cityscapes-to-ACDC. This dataset is designed for the continual semantic segmentation task [51] in autonomous driving. The Cityscapes dataset [9] is used as the source domain to provide an off-the-shelf pre-trained segmentation model, while the Adverse Conditions Dataset (ACDC) [41], which contains

Table 2. Semantic segmentation results (mIoU in %) on the Cityscapes-to-ACDC CTTA task. The four test conditions are repeated ten times to evaluate the long-term adaptation performance. For brevity, we only show the continual adaptation results solely in the first, fourth, seventh, and final rounds. All results are evaluated based on the Segformer-B5 architecture. Bold text indicates the best performance.

Time	t																					All
Round	1					4					7					10						
Condition	Fog	Night	Rain	Snow	Mean	Fog	Night	Rain	Snow	Mean	Fog	Night	Rain	Snow	Mean	Fog	Night	Rain	Snow	Mean		Mean↑
Source	69.1	40.3	59.7	57.8	56.7	69.1	40.3	59.7	57.8	56.7	69.1	40.3	59.7	57.8	56.7	69.1	40.3	59.7	57.8	56.7		56.7
BN Stats Adapt	62.3	38.0	54.6	53.0	52.0	62.3	38.0	54.6	53.0	52.0	62.3	38.0	54.6	53.0	52.0	62.3	38.0	54.6	53.0	52.0		52.0
TENT-cont. [50]	69.0	40.2	60.1	57.3	56.7	66.5	36.3	58.7	54.0	53.9	64.2	32.8	55.3	50.9	50.8	61.8	29.8	51.9	47.8	47.8		52.3
CoTTA [51]	70.9	41.2	62.4	59.7	58.6	70.9	41.2	62.4	59.7	58.6	70.9	41.2	62.4	59.7	58.6	70.9	41.2	62.4	59.7	58.6		58.6
Ours	71.2	42.3	65.0	62.0	**60.1**	72.8	43.6	66.7	63.3	**61.6**	72.5	42.5	66.8	63.3	**61.3**	72.5	42.9	66.7	63.0	**61.3**		**61.3**

images collected under four different weather conditions: Fog, Night, Rain, and Snow, is used to represent the changing target domains. 400 unlabeled images from each condition are used for adaptation. Note that the ACDC dataset shares the same semantic classes as the evaluation classes of Cityscapes. To simulate continual distribution shifts encountered in real-world scenarios, we cyclically repeat the same sequence of target domains ten times (e.g., in total 40: Fog→Night→Rain→Snow→Fog→ ...). This also provides a long-term perspective on the performance evaluation of different adaptation methods.

Implementation Details. To ensure consistency and comparability, we follow the implementation details from previous works [12,51]. We adopt pre-trained WideResNet-28 [62] model for CIFAR10-to-CIFAR10C, pre-trained ResNeXt-29 [57] model for CIFAR100-to-CIFAR100C, and standard pre-trained ResNet-50 model for ImageNet-to-ImagenetC from Robustbench [10] for all methods. Our batch size is set to 64 for ImageNet-C and 200 for other classification experiments. We evaluate the model using online predictions obtained immediately as the data is encountered. For segmentation CTTA, we employ a transformer-based architecture Segformer-B5 [56] trained on Cityscapes as our segmentation model, and use down-sampled images from ACDC with a resolution of 960×540 as network inputs. We utilize the Adam optimizer with a learning rate of 1e-4 and set the batch size to 1. To generate robust pseudo-labels, we use multi-scale input with flipping as the augmentation method following [51] for the segmentation experiments. When obtaining the intrinsic class relation $\hat{G}$, we utilize the weights of the classifier as proxies for classes, whereas in classification scenarios where feature prototypes are easily accessible [4,12], we utilize the prototypes to do so. All experiments are conducted using an NVIDIA RTX4090 GPU.

4.2 Experiments on Segmentation CTTA

We report results based on the mean intersection over union (mIoU) metric for the complex continual test-time semantic segmentation Cityscapes-to-ACDC

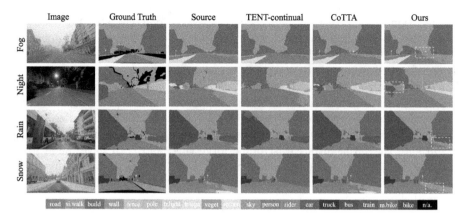

Fig. 3. Qualitative comparison of semantic segmentation on the Cityscapes-to-ACDC CTTA task. The results for all methods come from the final round.

task. As detailed in Table 2, a discernible decline in the performance of TENT across long-term sequences is observed, highlighting the pervasive issue of model degradation over time. Conversely, our proposed method effectively mitigates catastrophic forgetting and error accumulation, thereby consistently maintaining a high mean mIoU level across repeated sequences of target domains. Particularly noteworthy is our method's achievement of the highest mIoU score, showcasing an impressive relative improvement of 4.6% in mIoU compared to the previous sota method CoTTA [51], averaged across ten rounds. This significant improvement demonstrates the ability of our method to adapt to dynamic environments within dense prediction tasks over the long term. Additionally, there is a trend of incremental improvement in the average mIoU metric during the initial rounds, indicating the capacity of our method to extract and leverage the reliable information embedded within the unsupervised data stream. This capability facilitates further refinement of the adaptation performance. Comprehensive results are available in the supplementary material.

For a more intuitive validation of the effectiveness of our method, we conduct qualitative comparison experiments and show the results obtained through the CTTA process in Fig. 3. Compared to other state-of-the-art methods, our results exhibit superior clarity and accuracy, yielding the best segmentation maps in all four target domains. Our method provides consistent improvements for most of the categories (as shown by the white boxes), while the comparison methods mostly suffer from severe semantic confusion. These observations demonstrate the effectiveness of our method in complex real-world environments.

4.3 Experiments on Classification CTTA

ImageNet-to-ImageNet-C. Given the source model pre-trained on ImageNet, we sequentially execute CTTA on the 15 domains of ImageNet-C. As shown in Table 3, when directly testing the source model on target domains, the average classification error rate is as high as 82.0%, which demonstrates the need for adaptation. Leveraging the batch normalization statistics from the input data of the current iteration yielded the BN Stats Adapt method [26,42], a relatively straightforward approach that reduces the average classification error rate to 68.6%. While the TENT-based method [50] initially helps the model to continuously adapt to target domains, it is prone to severe catastrophic forgetting and error accumulation over long periods. CoTTA [51] and RMT [12] align to the source domain either in parameter space or feature space, respectively, intending to mitigate the problem of catastrophic forgetting, and both have made some progress. DSS [54] has noted the existence of high-quality and low-quality samples in the data stream, but it lacks further exploration on how to organize and utilize this information.

Table 3. Classification error rate (%, lower is better) for the standard ImageNet-to-ImageNet-C CTTA task. All results are evaluated with the largest corruption severity level 5 in an online manner. Bold text indicates the best performance.

Time	t $\longrightarrow$															
Method	Gaussian	shot	impulse	defocus	glass	motion	zoom	snow	frost	fog	brightness	contrast	elastic	pixelate	jpeg	Mean↓
Source	97.8	97.1	98.2	81.7	89.8	85.2	78.0	83.5	77.1	75.9	41.3	94.5	82.5	79.3	68.6	82.0
BN Stats Adapt	85.0	83.7	85.0	84.7	84.3	73.7	61.2	66.0	68.2	52.1	34.9	82.7	55.9	51.3	59.8	68.6
TENT-cont. [50]	81.6	74.6	72.7	77.6	73.8	65.5	55.3	61.6	63.0	51.7	38.2	72.1	50.8	47.4	53.3	62.6
CoTTA [51]	84.7	82.1	80.6	81.3	79.0	68.6	57.5	60.3	60.5	48.3	36.6	66.1	47.2	41.2	46.0	62.7
RMT [12]	80.2	76.4	74.5	77.1	74.4	66.2	57.6	57.0	59.1	48.0	39.1	60.6	47.3	42.5	43.4	60.2
DSS [54]	84.6	80.4	78.7	83.9	79.8	74.9	62.9	62.8	62.9	49.7	37.4	71.0	49.5	42.9	48.2	64.6
Ours	78.5	75.3	73.0	75.7	73.1	64.5	56.0	55.8	58.1	47.6	38.5	58.5	46.1	42.0	43.4	**59.0**

In our method, we reshape the online data buffering mechanism in CTTA, enabling dynamic aggregation of high-value samples from the unsupervised data stream. Based on this, we introduce a novel graph-based class relation preservation constraint alongside a pseudo-target replay objective. As a result, our method achieves a substantial reduction in the average classification error rate from 64.6% observed in the DSS method to 59.0%. Notably, our method outperforms all previous methods and reduces the error rate by a relative 2.0% compared to the previous sota method, averaged across 15 domains. Moreover, our method demonstrates the best performance on the vast majority of corruption types, highlighting its effectiveness in mitigating catastrophic forgetting and error accumulation.

CIFAR10-To-CIFAR10C, CIFAR100-to-CIFAR100C. To further demonstrate the effectiveness of the proposed method, we evaluate it on CIFAR10-to-CIFAR10C and CIFAR100-to-CIFAR100C CTTA tasks that have different numbers of categories. The experimental results are summarized in Table 4. Once again, our method outperforms all previous competing methods and reduces the average error rate to 15.8% and 29.0%, respectively. Compared with both the previous sota parameter space recovery methods (PETAL [2]) and feature space alignment methods (SANTA [4]), our method shows substantial improvements, demonstrating the superiority of the proposed class relation preservation mechanism in overcoming catastrophic forgetting. The consistency enhancement across datasets with varying levels of complexity proves the effectiveness and universality of our method.

Experiments on Gradually Changing Setup. Following [4,12,51], we also experiment with a gradually changing setup. Different from the previous standard setup where the corruption type changes abruptly at the largest severity level, the change process under the gradually changing setup can be represented as follows: $\underbrace{\ldots \to 2 \to 1}_{before} \xrightarrow{change\ type} \underbrace{1 \to 2 \to 3 \to 4 \to 5 \to 4 \to 3 \to 2 \to 1}_{current\ type,\ gradually\ changing\ severity} \xrightarrow{change\ type} \underbrace{1 \to 2 \to \ldots,}_{after}$ where each number denotes the severity of corruption. The exper-

Table 4. Classification error rate (%, lower is better) for the standard CTTA tasks on CIFAR10-to-CIFAR10C, CIFAR100-to-CIFAR100C. All results are evaluated with the largest corruption severity level 5 in an online manner. Bold text indicates the best.

	Time	t ⟶														
	Method	Gaussian	shot	impulse	defocus	glass	motion	zoom	snow	frost	fog	brightness	contrast	elastic	pixelate	jpeg	Mean↓
CIFAR10C	Source	72.3	65.7	72.9	46.9	54.3	34.8	42.0	25.1	41.3	26.0	9.3	46.7	26.6	58.5	30.3	43.5
	BN Stats Adapt	28.1	26.1	32.5	13.2	35.3	14.2	17.3	12.7	17.3	15.3	8.4	12.6	23.8	19.7	27.3	20.4
	TENT-cont. [50]	24.8	20.6	28.6	14.4	31.1	16.5	14.1	19.1	18.6	18.6	12.2	20.3	25.7	20.8	24.9	20.7
	CoTTA [51]	24.3	21.3	26.6	11.6	27.6	12.2	10.3	14.8	14.1	12.4	7.5	10.6	18.3	13.4	17.3	16.2
	RMT [12]	24.0	20.4	25.6	12.6	25.4	14.2	12.2	15.4	15.1	14.1	10.3	13.7	17.1	13.5	16.0	16.7
	PETAL [2]	23.4	21.1	25.7	11.7	27.2	12.2	10.3	14.8	13.7	12.7	7.4	10.5	18.1	13.4	16.8	16.0
	DSS [54]	24.1	21.3	25.4	11.7	26.9	12.2	10.5	14.5	14.1	12.5	7.8	10.8	18.0	13.1	17.3	16.0
	SANTA [4]	23.9	20.1	28.0	11.6	27.4	12.6	10.2	14.1	13.2	12.2	7.4	10.3	19.1	13.3	18.5	16.1
	Ours	23.6	19.9	26.0	11.8	25.3	13.2	10.9	14.3	13.5	12.7	9.0	11.9	17.1	12.7	15.9	**15.8**
CIFAR100C	Source	73.0	68.0	39.4	29.3	54.1	30.8	28.8	39.5	45.8	50.3	29.5	55.1	37.2	74.7	41.2	46.4
	BN Stats Adapt	42.1	40.7	42.7	27.6	41.9	29.7	27.9	34.9	35.0	41.5	26.5	30.3	35.7	32.9	41.2	35.4
	TENT-cont. [50]	37.2	35.8	41.7	37.9	51.2	48.3	48.5	58.4	63.7	71.1	70.4	82.3	88.0	88.5	90.4	60.9
	CoTTA [51]	40.1	37.7	39.7	26.9	38.0	27.9	26.4	32.8	31.8	40.3	24.7	26.9	32.5	28.3	33.5	32.5
	RMT [12]	40.5	36.1	36.3	27.7	33.9	28.5	26.4	29.0	29.0	32.5	25.1	27.4	28.2	26.3	29.3	30.4
	PETAL [2]	38.3	36.4	38.6	25.9	36.7	27.2	25.4	32.0	30.8	38.7	24.4	26.4	31.5	26.9	32.5	31.5
	DSS [54]	39.7	36.0	37.2	26.3	35.6	27.5	25.1	31.4	30.0	37.8	24.2	26.0	30.0	26.3	31.1	30.9
	SANTA [4]	36.5	33.1	35.1	25.9	34.9	27.7	25.4	29.5	29.9	33.1	23.6	26.7	31.9	27.5	35.2	30.4
	Ours	38.8	35.0	35.4	26.7	33.2	27.4	25.0	27.4	26.8	29.8	24.1	25.1	26.9	24.9	28.0	**29.0**

Table 5. Gradually changing CIFAR10-to-CIFAR10C results. The severity level changes gradually between the lowest and the highest.

Avg. Error (%)	Source	BN Stats Adapt	TENT-cont. [50]	CoTTA [51]	RMT [12]	SANTA [4]	Our
CIFAR10C	24.7	13.7	20.4	10.9	10.4	10.7	**9.9**

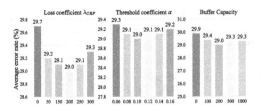

Fig. 4. Hyperparameter sensitivity analysis.

Table 6. Ablation experiments.

Buffering		$\mathcal{L}_{PCE}$	$\mathcal{L}_{CRP}$	Mean↓
Ex_1	-	-	-	30.7
Ex_2	Reservoir	✓	-	30.5
Ex_3	FIFO	✓	-	30.6
Ex_4	Uncertainty	✓	-	29.7
Ex_5	-	-	✓	30.0
Ex_6	Uncertainty	✓	✓	29.0

imental results are shown in Table 5, and our proposed method also achieves superior performance in this setup, reducing the average error rate to 9.9% on the gradual CIFAR10-to-CIFAR10C benchmark.

4.4 Ablation Study

Effectiveness of Each Component. We conduct the ablation study on the CIFAR100-to-CIFAR100C scenario and evaluate the effectiveness of each component in Table 6. $Ex1$ denotes the performance using only self-training loss in the teacher-student framework, yielding a relatively high average error rate. From $Ex2$ to $Ex4$, we demonstrate the importance of our proposed uncertainty-aware buffering mechanism. In contrast, the reservoir and FIFO buffering mechanisms do not have the ability to actively perform information aggregation, and thus they provide only a negligible improvement when combined with our pseudo-target replay loss. $Ex5$ demonstrates the effectiveness of our proposed class relation preservation constraint. $Ex6$ shows a significant overall improvement, suggesting that our proposed components are complementary.

Hyperparameter Sensitivity. Here we analyze the sensitivity of hyperparameters within the CIFAR100-to-CIFAR100C scenario. We consider three hyperparameters as illustrated in Fig. 4. **Loss coefficient λ_{CRP}.** The hyperparameter λ_{CRP} in Eq. (6) controls the strength of the proposed CRP constraint. We vary it within the range of $\{0, 50, 150, 200, 250, 300\}$. It can be observed that increasing λ_{CRP} over a reasonably broad range leads to a reduction in the average classification error rate of CTTA across 15 domains, indicating the effectiveness of CRP. While excessively large values of λ_{CRP} (e.g., 300) may compromise the model's adaptability. **Threshold coefficient α.** The hyperparameter α in Eq. (2) is utilized to adjust the threshold for screening low-uncertainty samples in the streaming data. It is evident that it is stable over specific ranges and achieves the lowest average error rate among all domains at a value of 0.1. **Buffer capacity.** Here we examine the capacity of the buffer, which is employed to dynamically aggregate high-value samples from the online unsupervised data stream. As demonstrated in Fig. 4, our method achieves optimal results with a buffer capacity of 200, imposing only a negligible storage overhead.

5 Conclusion

In this paper, we revisit the key challenges in CTTA and address all of them simultaneously. To this end, we reshape the online data buffering and organizing mechanism, endowing our dynamic target buffer with the ability to aggregate reliable information in the unsupervised data stream. Through our designed class relation preservation constraint and a pseudo-target replay objective, the problem of catastrophic forgetting and error accumulation is significantly alleviated. Extensive experiments demonstrate the effectiveness of the proposed method. Regarding the adoption of uncertainty, our method employs the mainstream practice of sample entropy. In our framework, there is potential for investigating more diverse forms of uncertainty modeling, such as those based on energy functions. We leave this exploration to future work.

Acknolwedgement. This work was funded in part by the National Natural Science Foundation of China (62076195, 62376070, 62206271, U20B2052), in part by the Fundamental Research Funds for the Central Universities (AUGA5710011522), and by the Pengcheng Laboratory Research Project No. PCL2023A08.

References

1. Boudiaf, M., Mueller, R., Ben Ayed, I., Bertinetto, L.: Parameter-free online test-time adaptation. In: Proceedings of the IEEE/CVF Conference on Computer Vision and Pattern Recognition, pp. 8344–8353 (2022)
2. Brahma, D., Rai, P.: A probabilistic framework for lifelong test-time adaptation. In: Proceedings of the IEEE/CVF Conference on Computer Vision and Pattern Recognition (CVPR), pp. 3582–3591 (2023)
3. Burges, C., Ragno, R., Le, Q.: Learning to rank with nonsmooth cost functions. In: Advances in Neural Information Processing Systems, vol. 19 (2006)
4. Chakrabarty, G., Sreenivas, M., Biswas, S.: SANTA: source anchoring network and target alignment for continual test time adaptation. Trans. Mach. Learn. Res. (2023). https://openreview.net/forum?id=V7guVYzvE4
5. Chen, D., Wang, D., Darrell, T., Ebrahimi, S.: Contrastive test-time adaptation. In: Proceedings of the IEEE/CVF Conference on Computer Vision and Pattern Recognition, pp. 295–305 (2022)
6. Chen, H., Wang, Y., Hu, Q.: Multi-granularity regularized re-balancing for class incremental learning. IEEE Trans. Knowl. Data Eng. **35**(7), 7263–7277 (2022)
7. Choi, S., Yang, S., Choi, S., Yun, S.: Improving test-time adaptation via shift-agnostic weight regularization and nearest source prototypes. In: Avidan, S., Brostow, G., Cissé, M., Farinella, G.M., Hassner, T. (eds.) ECCV 2022. LNCS, vol. 13693, pp. 440–458. Springer, Cham (2022). https://doi.org/10.1007/978-3-031-19827-4_26
8. Cicek, S., Soatto, S.: Unsupervised domain adaptation via regularized conditional alignment. In: Proceedings of the IEEE/CVF International Conference on Computer Vision, pp. 1416–1425 (2019)
9. Cordts, M., et al.: The cityscapes dataset for semantic urban scene understanding. In: Proceedings of the IEEE Conference on Computer Vision and Pattern Recognition (CVPR), pp. 3213–3223 (2016)

10. Croce, F., et al.: Robustbench: a standardized adversarial robustness benchmark. In: Thirty-fifth Conference on Neural Information Processing Systems Datasets and Benchmarks Track (Round 2) (2021)
11. Cui, S., Wang, S., Zhuo, J., Su, C., Huang, Q., Tian, Q.: Gradually vanishing bridge for adversarial domain adaptation. In: Proceedings of the IEEE/CVF Conference on Computer Vision and Pattern Recognition, pp. 12455–12464 (2020)
12. Döbler, M., Marsden, R.A., Yang, B.: Robust mean teacher for continual and gradual test-time adaptation. In: Proceedings of the IEEE/CVF Conference on Computer Vision and Pattern Recognition (CVPR), pp. 7704–7714 (2023)
13. Dong, S., Hong, X., Tao, X., Chang, X., Wei, X., Gong, Y.: Few-shot class-incremental learning via relation knowledge distillation. In: Proceedings of the AAAI Conference on Artificial Intelligence, pp. 1255–1263 (2021)
14. Fan, Y., Wang, Y., Zhu, P., Hu, Q.: Dynamic sub-graph distillation for robust semi-supervised continual learning. In: Proceedings of the AAAI Conference on Artificial Intelligence, pp. 11927–11935 (2024)
15. Gan, Y., et al.: Decorate the newcomers: visual domain prompt for continual test time adaptation. In: Proceedings of the AAAI Conference on Artificial Intelligence, pp. 7595–7603 (2023)
16. Ganin, Y., Lempitsky, V.: Unsupervised domain adaptation by backpropagation. In: International Conference on Machine Learning, pp. 1180–1189. PMLR (2015)
17. Geirhos, R., et al.: Shortcut learning in deep neural networks. Nat. Mach. Intell. **2**(11), 665–673 (2020)
18. Hendrycks, D., Dietterich, T.: Benchmarking neural network robustness to common corruptions and perturbations. In: International Conference on Learning Representations (2019). https://openreview.net/forum?id=HJz6tiCqYm
19. Hoffman, J., et al.: Cycada: cycle-consistent adversarial domain adaptation. In: International Conference on Machine Learning, pp. 1989–1998. PMLR (2018)
20. Hwang, S., Lee, S., Kim, S., Ok, J., Kwak, S.: Combating label distribution shift for active domain adaptation. In: Avidan, S., Brostow, G., Cissé, M., Farinella, G.M., Hassner, T. (eds.) ECCV 2022. LNCS, vol. 13693, pp. 549–566. Springer, Cham (2022). https://doi.org/10.1007/978-3-031-19827-4_32
21. Koh, P.W., et al.: Wilds: a benchmark of in-the-wild distribution shifts. In: International Conference on Machine Learning, pp. 5637–5664. PMLR (2021)
22. Lee, K., Kim, S., Kwak, S.: Cross-domain ensemble distillation for domain generalization. In: Avidan, S., Brostow, G., Cissé, M., Farinella, G.M., Hassner, T. (eds.) ECCV 2022. LNCS, vol. 13685, pp. 1–20. Springer, Cham (2022). https://doi.org/10.1007/978-3-031-19806-9_1
23. Li, H., Pan, S.J., Wang, S., Kot, A.C.: Domain generalization with adversarial feature learning. In: Proceedings of the IEEE Conference on Computer Vision and Pattern Recognition, pp. 5400–5409 (2018)
24. Li, S., Xie, B., Lin, Q., Liu, C.H., Huang, G., Wang, G.: Generalized domain conditioned adaptation network. IEEE Trans. Pattern Anal. Mach. Intell. **44**(8), 4093–4109 (2021)
25. Li, S., Xie, M., Gong, K., Liu, C.H., Wang, Y., Li, W.: Transferable semantic augmentation for domain adaptation. In: Proceedings of the IEEE/CVF Conference on Computer Vision and Pattern Recognition, pp. 11516–11525 (2021)
26. Li, Y., Wang, N., Shi, J., Liu, J., Hou, X.: Revisiting batch normalization for practical domain adaptation. arXiv preprint arXiv:1603.04779 (2016)
27. Liang, J., Hu, D., Feng, J.: Do we really need to access the source data? source hypothesis transfer for unsupervised domain adaptation. In: International Conference on Machine Learning, pp. 6028–6039. PMLR (2020)

28. Lin, H., et al.: Prototype-guided continual adaptation for class-incremental unsupervised domain adaptation. In: Avidan, S., Brostow, G., Cissé, M., Farinella, G.M., Hassner, T. (eds.) ECCV 2022. LNCS, vol. 13693, pp. 351–368. Springer, Cham (2022). https://doi.org/10.1007/978-3-031-19827-4_21
29. Liu, H., Long, M., Wang, J., Jordan, M.: Transferable adversarial training: a general approach to adapting deep classifiers. In: International Conference on Machine Learning, pp. 4013–4022. PMLR (2019)
30. Liu, J., et al.: ViDA: homeostatic visual domain adapter for continual test time adaptation. In: The Twelfth International Conference on Learning Representations (2024). https://openreview.net/forum?id=sJ88Wg5Bp5
31. Liu, Y., Kothari, P., Van Delft, B., Bellot-Gurlet, B., Mordan, T., Alahi, A.: Ttt++: When does self-supervised test-time training fail or thrive? In: Advance in Neural Information Processing System, vol. 34, pp. 21808–21820 (2021)
32. Long, M., Cao, Y., Wang, J., Jordan, M.: Learning transferable features with deep adaptation networks. In: International Conference on Machine Learning, pp. 97–105. PMLR (2015)
33. Long, M., Zhu, H., Wang, J., Jordan, M.I.: Deep transfer learning with joint adaptation networks. In: International Conference on Machine Learning, pp. 2208–2217. PMLR (2017)
34. Nguyen, A.T., Nguyen-Tang, T., Lim, S.N., Torr, P.H.: Tipi: test time adaptation with transformation invariance. In: Proceedings of the IEEE/CVF Conference on Computer Vision and Pattern Recognition, pp. 24162–24171 (2023)
35. Niloy, F.F., Ahmed, S.M., Raychaudhuri, D.S., Oymak, S., Roy-Chowdhury, A.K.: Effective restoration of source knowledge in continual test time adaptation. In: Proceedings of the IEEE/CVF Winter Conference on Applications of Computer Vision, pp. 2091–2100 (2024)
36. Niu, S., et al.: Efficient test-time model adaptation without forgetting. In: International Conference on Machine Learning, pp. 16888–16905. PMLR (2022)
37. Niu, S., et al.: Towards stable test-time adaptation in dynamic wild world. In: The Eleventh International Conference on Learning Representations (2023). https://openreview.net/forum?id=g2YraF75Tj
38. Patel, V.M., Gopalan, R., Li, R., Chellappa, R.: Visual domain adaptation: a survey of recent advances. IEEE Signal Process. Mag. **32**(3), 53–69 (2015)
39. Peng, X., Bai, Q., Xia, X., Huang, Z., Saenko, K., Wang, B.: Moment matching for multi-source domain adaptation. In: Proceedings of the IEEE/CVF International Conference on Computer Vision, pp. 1406–1415 (2019)
40. Press, O., Schneider, S., Kümmerer, M., Bethge, M.: Rdumb: a simple approach that questions our progress in continual test-time adaptation. In: Advances in Neural Information Processing Systems, vol. 36 (2023)
41. Sakaridis, C., Dai, D., Van Gool, L.: ACDC: the adverse conditions dataset with correspondences for semantic driving scene understanding. In: Proceedings of the IEEE/CVF International Conference on Computer Vision, pp. 10765–10775 (2021)
42. Schneider, S., Rusak, E., Eck, L., Bringmann, O., Brendel, W., Bethge, M.: Improving robustness against common corruptions by covariate shift adaptation. In: Advance in Neural Information Processing System, vol. 33, pp. 11539–11551 (2020)
43. Sójka, D., Cygert, S., Twardowski, B., Trzciński, T.: AR-TTA: a simple method for real-world continual test-time adaptation. In: Proceedings of the IEEE/CVF International Conference on Computer Vision (ICCV) Workshops, pp. 3491–3495 (2023)

44. Song, J., Lee, J., Kweon, I.S., Choi, S.: Ecotta: memory-efficient continual test-time adaptation via self-distilled regularization. In: Proceedings of the IEEE/CVF Conference on Computer Vision and Pattern Recognition, pp. 11920–11929 (2023)
45. Tan, M., et al.: Uncertainty-calibrated test-time model adaptation without forgetting. arXiv preprint arXiv:2403.11491 (2024)
46. Tao, X., Chang, X., Hong, X., Wei, X., Gong, Y.: Topology-preserving class-incremental learning. In: Vedaldi, A., Bischof, H., Brox, T., Frahm, J.-M. (eds.) ECCV 2020 XIX. LNCS, vol. 12364, pp. 254–270. Springer, Cham (2020). https://doi.org/10.1007/978-3-030-58529-7_16
47. Tao, X., Hong, X., Chang, X., Dong, S., Wei, X., Gong, Y.: Few-shot class-incremental learning. In: Proceedings of the IEEE/CVF Conference on Computer Vision and Pattern Recognition (CVPR) (2020)
48. Tao, X., Hong, X., Chang, X., Gong, Y.: Bi-objective continual learning: learning 'new' while consolidating 'known'. In: Proceedings of the AAAI Conference on Artificial Intelligence, pp. 5989–5996 (2020)
49. Tsai, Y.H., Hung, W.C., Schulter, S., Sohn, K., Yang, M.H., Chandraker, M.: Learning to adapt structured output space for semantic segmentation. In: Proceedings of the IEEE Conference on Computer Vision and Pattern Recognition, pp. 7472–7481 (2018)
50. Wang, D., Shelhamer, E., Liu, S., Olshausen, B., Darrell, T.: Tent: fully test-time adaptation by entropy minimization. In: International Conference on Learning Representations (2021). https://openreview.net/forum?id=uXl3bZLkr3c
51. Wang, Q., Fink, O., Van Gool, L., Dai, D.: Continual test-time domain adaptation. In: Proceedings of the IEEE/CVF Conference on Computer Vision and Pattern Recognition (CVPR), pp. 7201–7211 (2022)
52. Wang, S., Zhang, D., Yan, Z., Zhang, J., Li, R.: Feature alignment and uniformity for test time adaptation. In: Proceedings of the IEEE/CVF Conference on Computer Vision and Pattern Recognition, pp. 20050–20060 (2023)
53. Wang, Y., Ma, Z., Huang, Z., Wang, Y., Su, Z., Hong, X.: Isolation and impartial aggregation: a paradigm of incremental learning without interference. In: Proceedings of the AAAI Conference on Artificial Intelligence, pp. 10209–10217 (2023)
54. Wang, Y., et al.: Continual test-time domain adaptation via dynamic sample selection. In: Proceedings of the IEEE/CVF Winter Conference on Applications of Computer Vision (WACV), pp. 1701–1710 (2024)
55. Wei, Y., Ye, J., Huang, Z., Zhang, J., Shan, H.: Online prototype learning for online continual learning. In: Proceedings of the IEEE/CVF International Conference on Computer Vision (ICCV), pp. 18764–18774 (2023)
56. Xie, E., Wang, W., Yu, Z., Anandkumar, A., Alvarez, J.M., Luo, P.: Segformer: simple and efficient design for semantic segmentation with transformers. In: Beygelzimer, A., Dauphin, Y., Liang, P., Vaughan, J.W. (eds.) Advances in Neural Information Processing Systems (2021). https://openreview.net/forum?id=OG18MI5TRL
57. Xie, S., Girshick, R., Dollár, P., Tu, Z., He, K.: Aggregated residual transformations for deep neural networks. In: Proceedings of the IEEE Conference on Computer Vision and Pattern Recognition, pp. 1492–1500 (2017)
58. Yang, X., Gu, Y., Wei, K., Deng, C.: Exploring safety supervision for continual test-time domain adaptation. In: Proceedings of the Thirty-Second International Joint Conference on Artificial Intelligence, IJCAI-23, pp. 1649–1657. International Joint Conferences on Artificial Intelligence Organization (2023). https://doi.org/10.24963/ijcai.2023/183

59. Yang, Y., Soatto, S.: Fda: fourier domain adaptation for semantic segmentation. In: Proceedings of the IEEE/CVF Conference on Computer Vision and Pattern Recognition, pp. 4085–4095 (2020)
60. Yao, X., et al.: Socialized learning: making each other better through multi-agent collaboration. In: Forty-First International Conference on Machine Learning (2024)
61. Yuan, L., Xie, B., Li, S.: Robust test-time adaptation in dynamic scenarios. In: Proceedings of the IEEE/CVF Conference on Computer Vision and Pattern Recognition, pp. 15922–15932 (2023)
62. Zagoruyko, S., Komodakis, N.: Wide residual networks. In: British Machine Vision Conference 2016. British Machine Vision Association (2016)
63. Zhang, J., Qi, L., Shi, Y., Gao, Y.: MVDG: a unified multi-view framework for domain generalization. In: Avidan, S., Brostow, G., Cissé, M., Farinella, G.M., Hassner, T. (eds.) ECCV. LNCS, vol. 13687, pp. 161–177. Springer, Cham (2022). https://doi.org/10.1007/978-3-031-19812-0_10
64. Zhang, Y., Wang, Z., He, W.: Class relationship embedded learning for source-free unsupervised domain adaptation. In: Proceedings of the IEEE/CVF Conference on Computer Vision and Pattern Recognition, pp. 7619–7629 (2023)
65. Zhang, Y., Wang, Z., Li, J., Zhuang, J., Lin, Z.: Towards effective instance discrimination contrastive loss for unsupervised domain adaptation. In: Proceedings of the IEEE/CVF International Conference on Computer Vision, pp. 11388–11399 (2023)
66. Zhou, K., Liu, Z., Qiao, Y., Xiang, T., Loy, C.C.: Domain generalization: a survey. IEEE Trans. Pattern Anal. Mach. Intell. (2022)

Personalized Privacy Protection Mask Against Unauthorized Facial Recognition

Ka-Ho Chow[1], Sihao Hu[2], Tiansheng Huang[2], and Ling Liu[2]

[1] The University of Hong Kong, Pok Fu Lam, Hong Kong
kachow@cs.hku.hk
[2] Georgia Institute of Technology, Atlanta, USA
{sihaohu,thuang374}@gatech.edu, lingliu@cc.gatech.edu

Abstract. Face recognition (FR) can be abused for privacy intrusion. Governments, private companies, or even individual attackers can collect facial images by web scraping to build an FR system identifying human faces without their consent. This paper introduces Chameleon, which learns to generate a user-centric personalized privacy protection mask, coined as P3-Mask, to protect facial images against unauthorized FR with three salient features. First, we use a cross-image optimization to generate one P3-Mask for each user instead of tailoring facial perturbation for each facial image of a user. It enables efficient and instant protection even for users with limited computing resources. Second, we incorporate a perceptibility optimization to preserve the visual quality of the protected facial images. Third, we strengthen the robustness of P3-Mask against unknown FR models by integrating focal diversity-optimized ensemble learning into the mask generation process. Extensive experiments on two benchmark datasets show that Chameleon outperforms three state-of-the-art methods with instant protection and minimal degradation of image quality. Furthermore, Chameleon enables cost-effective FR authorization using the P3-Mask as a personalized deobfuscation key, and it demonstrates high resilience against adaptive adversaries.

1 Introduction

Face recognition (FR) has long been investigated due to its potential for enhancing security and convenience in various domains [22,28,29]. Many pretrained FR models are available online [6,7]. Once a face database (a.k.a. the gallery) with facial images for each person of interest is provided, these pretrained models can be used to recognize them [34].

While empowering many life-enriching applications, FR can be abused to cause serious privacy issues [2]. Privacy intruders can build a face database of victims of interest from publicly available facial images on the Internet by web

---

Supplementary Information The online version contains supplementary material available at https://doi.org/10.1007/978-3-031-73007-8_25.

scraping (Fig. 1 (top)). By utilizing this database, the adversary can perform unauthorized FR of individual users for stalking victims [3], intruding on victims' privacy by flooding targeted ads [1], or facilitating criminals to commit identity fraud [5]. This is a real threat. For example, companies like Clearview [4] and PimEyes [8] have collected billions of online images and can recognize millions of citizens without their consent.

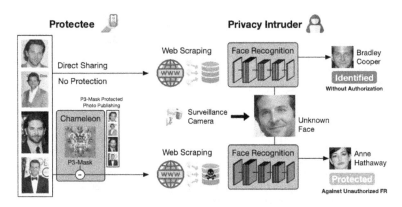

Fig. 1. (Top) Without protection, the privacy intruder can build a face database by web scraping and identify the unknown face. (Bottom) Chameleon learns the facial signature of the user (protectee) to generate a P3-Mask, which can be applied to protect any facial images before sharing them online against unauthorized FR.

Such a facial privacy threat has motivated the development of anti-FR technologies [36], which allow users (protectees) to preprocess their facial images before sharing them online. We argue that a competitive anti-FR solution should possess the following four properties. First, from the robustness perspective, the protected facial images should not be matched by an FR model to a query (probe) image with a correct identity (e.g., Bradley Cooper is recognized as Anne Hathaway in Fig. 1 (bottom)), including those FR models possibly unknown during the preprocessing but are used by the privacy intruder. Second, from the efficiency perspective, the protection should be done instantly on any device with or without AI accelerators to maintain user experience and support equitable access to privacy protection. Third, from the perception usability perspective, the protected image should (i) preserve the visual quality rather than using excessive noise and (ii) be visually recognizable by humans to be the same person as the one in the original unprotected photo. Finally, the service usability for authorized FR providers is another important property. Existing approaches treat all FR models as unauthorized. We argue that anti-FR solutions should also allow users to grant FR permission to authorized models *cost-effectively*. This user-initiated de-obfuscation capability is critical to those, e.g., who publish their photos on a social media platform with privacy protection and, at the same time, wish to authorize the platform to perform some FR services (e.g., face tagging [32]) without sending and storing multiple versions of the same photo.

Based on the above objectives, we introduce Chameleon, a user-centric facial privacy protection system with three contributions. First, we develop an optimization algorithm to construct a Personalized Privacy Protection mask, coined as P3-Mask, for each user. The P3-Mask of a user can be used to protect any facial images of the same user with minimal impact on image quality, including those facial images unseen during the P3-Mask generation process. Second, it boosts the robustness of P3-Mask against unknown FR models by a principled approach, leveraging ensemble learning with models of high focal diversity. The per-user P3-Mask provides higher robustness against unknown FR models while preserving perception usability and the service usability for authorized FR providers. Chameleon users can utilize their own P3-Mask to obfuscate their facial images prior to public release, and the same P3-Mask can be used as a personalized de-obfuscation key to grant authorized FR service providers the ability to restore the facial signature of photos for correct recognition. Third, we conduct extensive experiments on two FR benchmark datasets [20,26] to analyze the shortcomings of state-of-the-art anti-FR solutions [31,39,40] and show that Chameleon can better protect against unknown FR models with a high success rate. The protection is lightweight and in real-time. It remains effective under adaptive privacy intruders deploying various strategies to counter Chameleon.

2 Related Work

Existing anti-FR approaches can be broadly classified into two categories. Synthesis-based approaches [11,16,19,21,33,41] use generative adversarial networks (GANs) [25] to synthesize a face to replace the original one for protection. While the synthetic faces look realistic, they appear to be strangers, not recognizable even by the users themselves, failing to meet the perception usability requirement.

Chameleon falls into the second category, which applies small changes to facial images [12]. Fawkes [31] formulates an untargeted attack to push the facial image away from the original location in the embedding space; TIP-IM [39] improves the image quality with MMD [9]; LowKey [10] improves the robustness under the image processing pipeline in an end-to-end ML system by incorporating it into the optimization process. These techniques can preserve the visual identity of protected faces. However, they both require iterative optimization for each image. Even for a GPU server, it can take over 100 s to perturb one facial image (Sect. 6.3). In contrast, Chameleon provides instant protection on any device, with protection even stronger than those spending minutes to find the best perturbation for each image. OPOM [40] attempts to improve efficiency by finding the embedding subspace enclosing facial images of the same person and generates a privacy protection mask to push any images away from it. However, it fails to maintain protection effectiveness when image processing operations are applied to protected images (Sect. 6.1), and the mask degrades the image quality much more significantly than Chameleon (Sect. 6.3).

3 Overview

To build an FR system with a pretrained model, the owner first collects a gallery dataset (face database) $\mathcal{D}$, which includes the facial images of individuals of interest. Then, the pretrained FR model F is used to map each gallery image $\tilde{x} \in \mathcal{D}$ to the embedding $F(\tilde{x})$, and those embeddings should form clusters corresponding to different people. When a facial image x of an unknown identity, called the probe image, is given, the FR system can use the same FR model F to map it to an embedding $F(x)$ and use the identity of the nearest gallery image to be the identity of the unknown person. The identity, $\mathcal{FR}(x; F, \mathcal{D})$, of the probe image x given the FR model F and the gallery dataset $\mathcal{D}$ is defined as:

$$\mathcal{FR}(x; F, \mathcal{D}) = \mathcal{I}\bigl(\arg\min_{\tilde{x} \in \mathcal{D}} \text{Dist}(F(x), F(\tilde{x}))\bigr), \tag{1}$$

where $\mathcal{I}(\tilde{x})$ denotes the identity of the gallery image $\tilde{x}$, which is known to the FR system, and $\text{Dist}(\cdot, \cdot)$ is a distance function such as Euclidean distance.

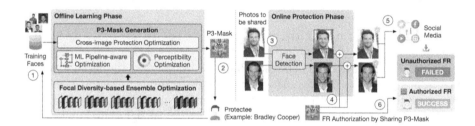

Fig. 2. An overview of Chameleon's two-phase workflow for offline learning to generate P3-Mask and online protection with P3-Mask for personalized facial signature masking. The P3-Mask can protect *any* facial images of the same person without further learning.

Threat Model. We consider a threat model where a privacy intruder conducts web scraping to collect images containing citizens' faces. Web scraping is large-scale and untargeted. Many citizens are included in the face database, and the privacy intruder may not know them. Given a probe image taken by, e.g., a stalker's camera, the privacy intruder uses an FR model on the face database to search for matched image(s). Then, the privacy intruder can analyze the associated metadata, such as the web page from where it was downloaded or even the exact identity. The goal of Chameleon is to allow the user to preprocess her facial images before posting them online. Even if scraped and included in the face database, they will not be matched to a probe image of her.

3.1 Chameleon Design

Chameleon performs preprocessing of the user's photos before sharing them online by applying the user-specific P3-Mask. This mask is generated offline for a Chameleon user. Figure 2 provides an overview of Chameleon's workflow.

Offline Learning Phase. Chameleon learns the unique facial signature of a user from a few facial images of her using the P3-Mask generator (Sect. 4), as shown in Fig. 2 ①. P3-Mask is designed to apply directly to any facial images of the same user, including those unseen during offline learning, and is robust against lossy image processing operations in an end-to-end ML system without compromising the image quality. We use a focal diversity-optimized ensemble learning method to find a team of FR models such that the P3-Mask generated against them has strong robustness and is effective in countering other FR models that are unknown during offline learning.

Online Protection Phase. After the offline learning, ② the P3-Mask can be sent to the user's local device (e.g., the mobile phone). For a photo of the user to be protected, ③ a lightweight face detection model, such as MediaPipe [23], can be used by the Chameleon user running on the user's device to locate the face region and ④ instantly protect it with the P3-Mask before the user shares it online ⑤. The user can also authorize some trusted entities to conduct FR on their shared photos protected with P3-Mask. The user-specific de-obfuscation key ⑥ allows the protected photos to be restored for authorized FR.

4 P3-Mask Generation

For a user $\mathcal{P}$, Chameleon generates a P3-Mask by learning the facial signature $\mathcal{M}_\mathcal{P}$ from a set of facial images the user provides. Several key factors need to be accomplished. (1) We need to promote cross-image protection as an optimization objective to generate the most representative facial signature applicable to protect any facial images of the user $\mathcal{P}$, including those unknown ones during the generation process (Sect. 4.1). (2) We need to control the amount of perturbation introduced by the P3-Mask. When applied to an unprotected image by removing the learned facial signature, it ensures minimal perception loss to preserve the visual quality (Sect. 4.2). (3) We need to keep the protected image far away from the original image in the embedding space. Multiple FR models should be used to generate alternative embedding representations to enhance generalizability against unknown models (Sect. 4.3).

4.1 Cross-Image Protection Optimization

The cross-image protection capability comes from iterative learning on a set of training images the user provides. The idea is to keep fine-tuning the P3-Mask so that it can offer protection simultaneously to training images while preserving image quality. At the t-th iteration, it samples a mini-batch $\mathcal{B}$ from the training dataset Ω containing photos of user $\mathcal{P}$. Then, we find the modification to the P3-Mask that can lead to better protection on each image in the mini-batch with better visual quality with the following operations:

$$\mathcal{M}_\mathcal{P}^{t+1} = \text{CLIP}_{[-\epsilon,\epsilon]}(\mathcal{M}_\mathcal{P}^t - \eta \text{SIGN}(\frac{1}{|\mathcal{B}|} \sum_{\mathcal{X} \in \mathcal{B}} \nabla_{\mathcal{M}_\mathcal{P}^t} \mathcal{L}(\mathcal{X}, \mathcal{M}_\mathcal{P}^t; \mathcal{T}, \omega))). \quad (2)$$

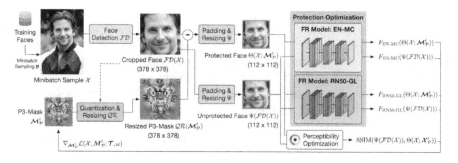

Fig. 3. The iterative generation of P3-Mask for a user. Chameleon goes through multiple facial images of the same user to optimize a P3-Mask with ML pipeline awareness.

Specifically, for each image $\mathcal{X} \in \mathcal{B}$, we use the P3-Mask learned up to the current iteration, $\mathcal{M}_\mathcal{P}^t$, to protect it and compute the loss designed with dual goals:

$$\mathcal{L}(\mathcal{X}, \mathcal{M}_\mathcal{P}^t; \mathcal{T}, \omega) = \mathcal{L}_{\text{Protect}}(\mathcal{X}, \mathcal{M}_\mathcal{P}^t; \mathcal{T}) + \mathcal{L}_{\text{Percept}}(\mathcal{X}, \mathcal{M}_\mathcal{P}^t; \omega). \quad (3)$$

It uses $\mathcal{L}_{\text{Protect}}$ to learn how to adjust the P3-Mask $\mathcal{M}_\mathcal{P}^t$ from the previous iteration to better protect against a team of pre-selected FR models $\mathcal{T}$ and $\mathcal{L}_{\text{Percept}}$ to preserve the image quality controlled by ω. We use the SIGN of the gradients to update the P3-Mask with a learning rate η. A clipping function $\text{CLIP}_{[-\epsilon,\epsilon]}$ is applied to ensure the changes made by the P3-Mask on the facial image are bounded by ϵ. Such cross-image protection is not limited to those in the training dataset Ω but also unseen images of the same user. We next discuss the design of $\mathcal{L}_{\text{Protect}}$, and $\mathcal{L}_{\text{Percept}}$ will be detailed in Sect. 4.2.

For privacy protection, P3-Mask maximizes the distance between the protected and unprotected images in the embedding spaces of a team $\mathcal{T}$ of pre-selected FR models. We incorporate image processing operations into the optimization to avoid those lossy operations degrading P3-Mask's protection. Consider the raw training image $\mathcal{X}$ in the mini-batch in Fig. 3. We first conduct face detection to produce the cropped face $\mathcal{FD}(\mathcal{X})$. Quantization and resizing are executed on the P3-Mask to match the resolution of the cropped face, denoted by $\mathcal{QR}(\mathcal{M}_\mathcal{P}^t)$. Then, we can apply the mask to the face for protection:

$$\Theta(\mathcal{X}; \mathcal{M}_\mathcal{P}^t) = \Psi(\text{CLIP}_{[0,255]}(\mathcal{FD}(\mathcal{X}) - \mathcal{QR}(\mathcal{M}_\mathcal{P}^t))), \quad (4)$$

where Ψ is a padding and resizing function to meet the input resolution requirement of an FR model (e.g., (112×112) in ArcFace [17]). Note that to simplify the notation, we removed certain arguments without causing confusion.

The protection optimization in P3-Mask maximizes the average arccosine distance [17], ARCCOS, between the embedding of the protected face $F(\Theta(\mathcal{X}; \mathcal{M}_\mathcal{P}^t))$ and the embedding of its unprotected counterpart $F(\Psi(\mathcal{FD}(\mathcal{X})))$ extracted by every FR model $F \in \mathcal{T}$ in the team:

$$\mathcal{L}_{\text{Protect}}(\mathcal{X}, \mathcal{M}_\mathcal{P}^t; \mathcal{T}) = \frac{-1}{|\mathcal{T}|} \sum_{F \in \mathcal{T}} \text{ARCCOS}(F(\Psi(\mathcal{FD}(\mathcal{X}))), F(\Theta(\mathcal{X}; \mathcal{M}_\mathcal{P}^t))). \quad (5)$$

4.2 Perceptibility Optimization

Preserving the visual quality of the facial image is necessary to maintain the usability of the protected image in practice. Hence, we incorporate a perceptibility optimization term in Eq. 3. The idea is to minimize the perceptual difference between unprotected and protected images while learning a user's facial signature, such that removing the facial signature from a given facial image will have minimal impact on the visual quality of the protected version of the original image. We use the structural similarity (SSIM) [35] to capture perceptual differences, as it has been shown to align with the human perception system:

$$\mathcal{L}_{\text{Percept}}(\mathcal{X}, \mathcal{M}_\mathcal{P}^t; \omega) = \lambda_{\text{SSIM}} \max\left[\frac{1 - \text{SSIM}(\Psi(\mathcal{FD}(\mathcal{X})), \Theta(\mathcal{X}; \mathcal{M}_\mathcal{P}^t))}{2} - \omega, 0\right]. \tag{6}$$

The parameters, ω and λ_{SSIM}, enable two features. ω controls the SSIM degradation. Any SSIM degradation greater than 2ω will cause the term to be non-zero, making the optimization adjust the P3-Mask reducing its impact on image quality. λ_{SSIM} balances the importance of privacy protection and perceptibility. We use dynamic scheduling [31] such that it will be adjusted automatically.

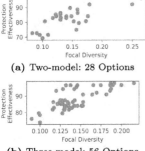

(a) Two-model: 28 Options

(b) Three-model: 56 Options

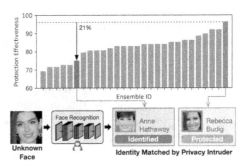

Fig. 4. Focal diversity provides a strong indicator of protection effectiveness for selecting an ensemble.

Fig. 5. Out of 28 options of two-model ensembles from a pool of eight models, the ensemble selected by our approach (green) leads to much better protection than a randomly selected one (red). (Color figure online)

4.3 Focal Diversity Ensemble Optimization

For each original image and its protected version by removing the facial signature learned up to the current iteration, we use two FR models in Fig. 3 to generate two embeddings, which serve as alternative channels to learn a high-quality facial signature. Choosing FR models that best complement each other can achieve better protection than randomly selected FR models. We use an ensemble optimization method to select the best k-model ensemble among a pool of N FR models. The key idea is to find an ensemble with members making decorrelated mistakes. To quantify this behavior, we leverage the focal diversity framework [13,14,37]. As shown in Fig. 4, the higher the focal diversity of the ensemble (a green dot), the stronger the P3-Mask will be in protecting against other FR models *not* used during the P3-Mask generation process.

Given a collection of N FR models $\{F_1, ..., F_N\}$, we first identify the negative samples for each FR model F by locating validation images that F fails to recognize their true identity. To rank ensembles of size S, we enumerate all $\binom{N}{S}$ combinations. For each combination (ensemble) $\mathcal{T}$, we consider each member to be the focal model F_{focal} and use its negative samples to statistically estimate the level of negative correlation $\lambda_{\text{focal}}(\mathcal{T}; F_{\text{focal}})$ between F_{focal} and the remaining models in $\mathcal{T}$, computed by measuring the degree of disagreements using the generalized non-pairwise measure [27]. The same procedure is repeated by considering each member in the ensemble as the focal model, and the focal diversity of the ensemble $\mathcal{T}$ is finalized as $d_{\text{focal}}(\mathcal{T}) = \frac{1}{S}\sum_{F_{\text{focal}} \in \mathcal{T}}[1 - \lambda_{\text{focal}}(\mathcal{T}; F_{\text{focal}})]$. Given a team size S, the ensemble with the highest focal diversity can be deployed. As shown in Fig. 5, the selected two-model team (green bar) is indeed the one leading to the strongest protection among all 28 options, which is over 20% stronger than a randomly selected ensemble (red bar). Note that we only need to analyze the failures of individual FR models, which is significantly more efficient than generating P3-Mask for each option and conducting evaluation.

5 Online Image Protection and Refinement

The user $\mathcal{P}$ obtains her P3-Mask $\mathcal{M}_\mathcal{P}$ from Chameleon. Given an image $\mathcal{X}$, she can obfuscate her facial identity on her device by applying the P3-Mask:

$$\text{MASK}(\mathcal{X}; \mathcal{M}_\mathcal{P}) = \text{CLIP}_{[0,255]}(\mathcal{FD}(\mathcal{X}) - \mathcal{QR}(\mathcal{M}_\mathcal{P})), \tag{7}$$

which is the protected facial region to be put in the original image $\mathcal{X}$ to produce the protected version $\mathcal{X}'$ for sharing.

Chameleon supports authorizing trusted third parties to conduct FR on a user's facial images in a space-time efficient manner without transmitting and storing multiple versions of the same photo (one protected for public sharing and one unprotected for internal FR). By sharing the P3-Mask $\mathcal{M}_\mathcal{P}$ as the key, the third parties can de-obfuscate the facial signature by unmasking:

$$\text{UNMASK}(\mathcal{X}'; \mathcal{M}_\mathcal{P}) = \text{CLIP}_{[0,255]}(\mathcal{FD}(\mathcal{X}') + \mathcal{QR}(\mathcal{M}_\mathcal{P})). \tag{8}$$

In Sect. 6.2, we will show two properties of this process: (i) the signature-restored photos can be used for FR with no accuracy degradation, and (ii) the protection can only be done by the P3-Mask of the same person in the facial image, and the restoration is successful only if the same key is used.

6 Experimental Evaluation

We conduct experiments on FaceScrub [26] and LFW [20]. To provide a detailed analysis, ten celebrities are randomly selected as users on FaceScrub (Table 1). For each user, we split their facial images into three parts: (i) 10% are used as probe images (2nd column), (ii) 70% are used as gallery images and are used by Chameleon to train the mask (3rd column), and (iii) 20% are also used as gallery images but unseen during the training process (4th column). The last part is crucial to evaluate the protection of unseen images of the user. In total, we have 123 facial images with "unknown identities." Following relevant works [31,40], all facial images of other people are included in the gallery dataset as noise. The same splitting method is also used for LFW. By default, the results reported in this paper focus on FaceScrub due to similar observations on LFW. The source code is available at https://github.com/git-disl/Chameleon.

Table 1. Ten example celebrities (users) on FaceScrub for analysis.

Name	Probe Images	Gallery Images		
		Seen	Unseen	Total
Morena Baccarin	11	82	23	105
Bradley Cooper	12	89	25	114
America Ferrera	16	113	32	145
Gerard Butler	11	82	24	106
Eva Longoria	12	91	26	117
Melissa Egan	12	87	24	111
Kim Cattrall	13	96	27	123
Allison Janney	12	89	26	115
Roma Downey	11	78	22	100
Steve Carell	13	96	28	124
Others (Noise)	/	/	/	48368
Total Number of Gallery Images:				49528

Table 2. The collection of publicly available pretrained FR models.

Model ID	Neural Arch.	Training Dataset	FR Acc.
EN-MC	EfficientNet	MS-Celeb-1M	94.94%
RN50-GL	ResNet50	Glint360K	96.68%
RN50-MC	ResNet50	MS-Celeb-1M	89.25%
RN50-VF	ResNet50	VGG-Face2	94.30%
RN50-WF	ResNet50	WebFace600K	96.72%
RN18-MC	ResNet18	MS-Celeb-1M	82.64%
RN34-MC	ResNet34	MS-Celeb-1M	84.49%
RN100-MC	ResNet100	MS-Celeb-1M	91.85%

Public clouds (e.g., Azure) now require manual approval to use their FR services. Hence, we believe that privacy intruders will opt for deploying pretrained FR models, as many high-quality ones are available on the Internet and can be used out of the box. Due to the large number of possible FR algorithms and architectures, we first focus on the eight FR models listed in Table 2, which are based on ArcFace [17], the state-of-the-art FR algorithm, with varying neural architectures and training datasets. Chameleon can be easily extended to incorporate any FR models by simply adding them to the collection. Nonetheless, in Sect. 6.4, we will show that thanks to the focal diversity-optimized teaming, the P3-Mask generated from a collection of ArcFace-only models can also be effective in protecting against privacy intruders using models of other FR algorithms, such as FaceNet [30] and MagFace [24]. By default, we use the two-model team (EN-MC, RN50-GL) with the highest focal diversity (Sect. 4.3). The P3-Mask is trained for 50 epochs on NVIDIA RTX 2080 SUPER GPU with $\eta = 0.001$, $|\mathcal{B}| = 4$, $\epsilon = 0.063$, and $\omega = 0.03$. We provide details for reproducibility and additional results in the appendix.

6.1 Protection Effectiveness

Table 3 shows the personalized mask for seven users on FaceScrub (other users are available in the appendix). Different users have distinct facial signatures (P3-Mask) learned by Chameleon. We apply these masks to their respective gallery images and test the FR accuracy using probe images. All probe images can be correctly identified when no protection mechanism is used. Under Chameleon, the probe images should be matched to an incorrect identity, resulting in a low FR accuracy. Hence, we define an evaluation metric, Protection Success Rate (PSR), to be (100 − FR ACCUACY) reported in percentages.

Table 3. The P3-Mask (rescaled) for seven users on FaceScrub. Chameleon learns the distinct facial signature for each of them.

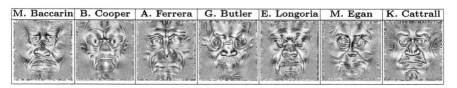

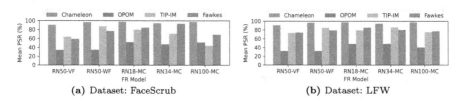

(a) Dataset: FaceScrub   (b) Dataset: LFW

Fig. 6. Protection effectiveness on two benchmarks using Chameleon, OPOM [40], TIP-IM [39], and Fawkes [31] against FR models unknown to them.

Chameleon can protect against unknown FR models. We compare the cross-model protection of Chameleon with OPOM [40], the only anti-FR approach generating one privacy protection mask for each person as Chameleon, and both TIP-IM [39] and Fawkes [31], which conduct per-image optimization. For a fair comparison, all methods are set with the same perturbation budget. Figure 6 summarizes the results on both datasets. All five FR models are *unknown* to both protection mechanisms. We make two observations. First, Chameleon consistently outperforms OPOM by a large margin. We found that OPOM masks do not transfer well, especially considering the end-to-end ML pipeline. Its PSR can drop by 33% because of those lossy operations. Second, Chameleon can be as competitive as those methods conducting per-image optimization. TIP-IM and Fawkes spend minutes to optimize the protection for *each* image. Still, Chameleon outperforms them and can be applied instantly.

Table 4 reports the detailed PSR for each user Chameleon achieves when the privacy intruder uses different FR models (2nd to 9th columns). We make two

Table 4. Using the two-model team with high focal diversity (2nd and 3rd columns), Chameleon can offer privacy protection even when the privacy intruder uses FR models unseen during the mask generation process (4th to 9th columns).

Name	Protection Success Rate - PSR (%)									
	EN-MC	RN50-GL	RN50-MC	RN50-VF	RN50-WF	RN18-MC	RN34-MC	RN100-MC	Mean	Std
M. Baccarin	100.00	100.00	100.00	72.73	90.91	100.00	90.91	100.00	94.32	9.64
B. Cooper	100.00	100.00	83.33	100.00	100.00	100.00	91.67	91.67	95.83	6.30
A. Ferrera	100.00	100.00	93.75	75.00	100.00	100.00	93.75	100.00	95.31	8.68
G. Butler	100.00	100.00	90.91	100.00	100.00	90.91	90.91	100.00	96.59	4.70
E. Longoria	100.00	100.00	100.00	100.00	100.00	100.00	100.00	100.00	100.00	0.00
M. Egan	100.00	100.00	100.00	100.00	100.00	100.00	100.00	91.67	98.96	2.95
K. Cattrall	100.00	100.00	100.00	84.62	100.00	100.00	100.00	100.00	98.08	5.44
A. Janney	100.00	100.00	91.67	100.00	91.67	100.00	100.00	100.00	97.92	3.86
R. Downey	100.00	100.00	100.00	72.73	90.91	100.00	90.91	100.00	94.32	9.64
S. Carell	100.00	100.00	84.62	92.31	92.31	92.31	76.92	100.00	92.31	8.22
Mean	100.00	100.00	93.52	90.65	96.58	97.41	94.42	97.42		
Std	0.00	0.00	6.40	11.16	4.43	4.18	7.41	4.15		

observations. First, when the privacy intruder uses an FR model known to the generation process, a PSR of 100% can be consistently achieved (2nd to 3rd columns). Second, Chameleon can protect against unknown FR models (4th to 9th columns). Even when the FR model used by the intruder is trained on a different dataset (i.e., RN50-VF and RN50-WF) or with a different neural architecture (i.e., RN18-MC, RN34-MC, and RN100-MC) than any FR models known by Chameleon, it still provides a PSR over 90.65%, meaning that facial images of a user are matched to gallery images of a different person. To demonstrate such a mismatch, in Table 5, we show the probe images from two users and their most similar gallery images found by four unknown FR models with two settings: (i) the "Unprotected" scenario where no one employs protection and (ii) the "Chameleon" scenario where the intruder scraped photos protected by our solution. Taking M. Baccarin as an example, the most similar gallery images found in the unprotected scenario belong to her. In contrast, when Chameleon is used, the same probe image is misidentified, e.g., as L. Hartley by RN50-MC.

6.2 FR Service Usability

Chameleon allows users to grant FR permission to trusted third parties by sharing their P3-Mask to de-obfuscate the protected image. There is no need to transmit and store an unprotected photo for internal use and a protected one for public visibility, doubling network and storage costs. Table 6 shows the results in FR accuracy. First, using the correct mask (i.e., the one used to protect the images) to reverse the protection can restore FR accuracy. Taking M. Baccarin as an example, the FR accuracy increases from 6.82% before unmasking (b) to 100% after unmasking (c). Second, using an incorrect mask for unmasking cannot restore the FR accuracy and can lead to even worse performance. It drops from 6.82% to 1.14% after unmasking (d).

Table 5. The most similar gallery images on FaceScrub found by different FR models unknown to Chameleon.

Probe Image	The Most Similar Gallery Image & Its Identity Found by Unseen FR Models							
	RN50-MC		RN50-VF		RN34-MC		RN100-MC	
	Unprotected	Chameleon	Unprotected	Chameleon	Unprotected	Chameleon	Unprotected	Chameleon
M. Baccarin	M. Baccarin	L. Hartley	M. Baccarin	L. Hartley	M. Baccarin	C. Electra	M. Baccarin	L. Hartley
B. Cooper	B. Cooper	P. Walker	B. Cooper	M. Vartan	B. Cooper	M. Vartan	B. Cooper	M. Vartan

6.3 Protection Cost Analysis

Speed and Resources. Figure 7 compares Chameleon with OPOM, TIP-IM, and Fawkes regarding the protection time per face, compute, and storage costs. Chameleon and OPOM produce masks that are applicable to any facial image of the same user. The protection can be completed in 0.0076 s. Note that Chameleon

Table 6. The protection by Chameleon can reduce FR accuracy against FR models (b), but the user can provide the P3-Mask used to protect her images as the key for the authorized third party to reverse the protection process (c), which leads to restored FR performance. However, an incorrect key cannot restore FR and may worsen it (d).

Scenario	Face Recognition Accuracy (%)									
	EN-MC	RN50-GL	RN50-MC	RN50-VF	RN50-WF	RN18-MC	RN34-MC	RN100-MC	Mean	Std
User: M. Baccarin										
(a) No Protection	100.00	100.00	100.00	100.00	100.00	100.00	100.00	100.00	100.00	0.00
(b) Protected	0.00	0.00	9.09	18.18	9.09	9.09	0.00	9.09	6.82	6.43
(c) Correctly Unmasked	100.00	100.00	100.00	100.00	100.00	100.00	100.00	100.00	100.00	0.00
(d) Incorrectly Unmasked	0.00	0.00	0.00	0.00	9.09	0.00	0.00	0.00	1.14	3.21
User: B. Cooper										
(a) No Protection	100.00	100.00	100.00	100.00	100.00	100.00	100.00	100.00	100.00	0.00
(b) Protected	0.00	0.00	16.67	0.00	0.00	0.00	8.33	8.33	4.17	6.30
(c) Correctly Unmasked	100.00	100.00	100.00	100.00	100.00	100.00	100.00	100.00	100.00	0.00
(d) Incorrectly Unmasked	0.00	0.00	0.00	0.00	0.00	0.00	0.00	0.00	0.00	0.00

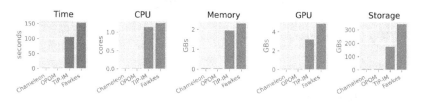

Fig. 7. Chameleon offers much better privacy protection than OPOM without sacrificing speed and resource usages. It can protect instantly and require negligible compute resources as OPOM, while TIP-IM and Fawkes are slow and costly.

is significantly more effective than OPOM, as described in Fig. 6. Instead, TIP-IM and Fawkes require per-image optimization and take 105.12 s to complete the protection with GPUs for acceleration and storage for FR models. Chameleon only needs to conduct simple arithmetic operations. Only the P3-Mask needs to be stored on the user device. Hence, it can be deployed with instant protection even on an edge device with weak computing power.

Image Quality. Chameleon can well preserve image quality. Table 7 provides two examples for four users, contrasting the facial images with no protection (3rd and 5th columns) with the ones with their facial signature removed by Chameleon (4th and 6th columns). They capture different variations of facial images: M. Baccarin's images have different lighting conditions, B. Cooper's have different face sizes, E. Longoria's have different postures, and G. Bulter's have different expressions. The protected images are visually similar to the unprotected counterparts, but they will not be matched to the clean images of the corresponding person. Indeed, Chameleon can generate higher-quality images than existing methods. As reported in Table 8, Chameleon not only offers better protection (Fig. 6) with a similar cost (Fig. 7) as OPOM but also achieves much better image quality measured in SSIM (0.9493 vs 0.8839).

Table 7. The P3-Mask can be applied to any facial images of the same person while preserving image quality.

Table 8. Chameleon offers much better protection while maintaining image quality.

6.4 Focal Diversity-Based Teaming

Chameleon automatically selects a high-quality team for deployment with a specified budget. Table 9 compares the detailed PSR using the most diverse and the least diverse teams with (a) two and (b) three models. In both cases, the most diverse team outperforms the least diverse one, which is only effective when the privacy intruder uses the FR model known in the team. The selection of high-quality teams is non-trivial because we observe that composing a team of FR models with the highest FR accuracy is not always a good option, and a team of models having different neural architectures is not always the top priority [13].

In Fig. 8, we further show that the P3-Mask generated from a carefully-chosen team can protect against FR models of unknown algorithms. Specifically, we consider the most and the least diverse three-model teams in Table 9(b). Even though both teams include only FR models based on ArcFace [17], the most diverse one can effectively protect against FR models based on FaceNet [30] or MagFace [24]. The protection effectiveness can be further strengthened by incorporating more FR models into the collection.

Table 9. The most diverse and the least diverse teams with (a) two or (b) three models identified by our focal diversity-based teaming method. The most diverse teams offer significantly better protection in terms of protection success rates.

Protection Team	Protection Success Rate - PSR (%)									
	EN-MC	RN50-GL	RN50-MC	RN50-VF	RN50-WF	RN18-MC	RN34-MC	RN100-MC	Mean	Std
(a) Two-model Teams										
Most Diverse: (EN-MC, RN50-GL)	100.00	100.00	93.52	90.65	96.58	97.41	94.42	97.42	**96.25**	**3.23**
Least Diverse: (RN18-MC, RN34-MC)	11.38	60.16	100.00	55.28	73.17	100.00	100.00	80.49	72.56	28.54
(b) Three-model Teams										
Most Diverse: (EN-MC, RN50-GL, RN50-WF)	100.00	100.00	100.00	100.00	92.68	93.50	100.00	100.00	**98.27**	**3.00**
Least Diverse: (RN18-MC, RN34-MC, RN50-MC)	2.44	78.86	100.00	52.03	84.55	100.00	100.00	93.50	76.42	31.82

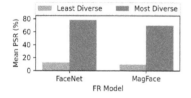

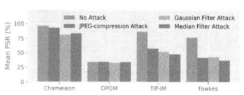

Fig. 8. The most diverse teams allow Chameleon to be effective against FR models of unknown algorithms.

Fig. 9. The PSR of Chameleon against adaptive intruders using different strategies to wash out the patterns in protected images.

6.5 Chameleon Against Adaptive Adversaries

An adaptive adversary may run Chameleon to produce the P3-Mask for each person and de-obfuscate their photos. When probe images need to be identified, they will be matched against a "clean" face database. However, it is infeasible, as we show in Sect. 6.2, that the FR accuracy can only be restored when the masks used for obfuscation and de-obfuscation are identical. It is impossible to reproduce the same P3-Mask used by the user because it requires the same set of clean images and random states.

Alternatively, an adaptive adversary may wash out patterns introduced by Chameleon before performing FR. We consider three popular lightweight and task-agnostic methods studied in the adversarial example domain [18] in Fig. 9, including (a) JPEG compression [15], (b) Gaussian Filter, and (d) Median Filter [38]. The results show that their effectiveness is limited on Chameleon. The consideration of the end-to-end ML pipeline increases the robustness of P3-Mask.

7 Conclusions

We have presented Chameleon against unauthorized FR. Chameleon generates a P3-Mask for each user with cross-image and perceptibility optimizations to offer (i) instant and lightweight protection on any facial images of the same user and (ii) preservation of image quality. Also, we have shown that P3-Mask enables cost-effective de-obfuscation for authorizing FR services. Such an authorization process can only be conducted by the one with the identical mask used for protection. Chameleon also features a focal diversity-optimized teaming method to select a high-quality FR team to generate P3-Mask with strong robustness against unknown FR models.

Acknowledgments. This research is partially sponsored by the NSF CISE grants 2302720, 2312758, 2038029, an IBM faculty award, and a grant from CISCO Edge AI program. It is part of the PhD dissertation of the first author, who graduated from Georgia Tech in Spring 2023. The first author acknowledges the support of the IBM PhD Fellowship in 2022–2023 and the support from the HKU-CS Start-up Fund.

References

1. Brazilian retailer quizzed over facial recognition tech (2019). https://www.zdnet.com/article/brazilian-retailer-quizzed-over-facial-recognition-tech/
2. Facial recognition in the united states: Privacy concerns and legal developments (2021). https://www.asisonline.org/security-management-magazine/monthly-issues/security-technology/archive/2021/december/facial-recognition-in-the-us-privacy-concerns-and-legal-developments/
3. The secretive company that might end privacy as we know it (2021). https://www.nytimes.com/2020/01/18/technology/clearview-privacy-facial-recognition.html
4. Clearview AI (2023). https://www.clearview.ai/
5. Facial recognition and identity risk (2023). https://www.equifax.co.uk/resources/identity-protection/facial-recognition-and-identity-risk.html
6. HuggingFace (2023). https://huggingface.co
7. Insightface: 2D and 3D face analysis project (2023). https://github.com/deepinsight/insightface
8. PimEyes (2023). https://pimeyes.com/
9. Borgwardt, K.M., Gretton, A., Rasch, M.J., Kriegel, H.P., Schölkopf, B., Smola, A.J.: Integrating structured biological data by kernel maximum mean discrepancy. Bioinformatics **22**(14), e49–e57 (2006)

10. Cherepanova, V., et al.: Lowkey: leveraging adversarial attacks to protect social media users from facial recognition. In: International Conference on Learning Representations (ICLR) (2021)
11. Choi, Y., Choi, M., Kim, M., Ha, J.W., Kim, S., Choo, J.: StarGAN: unified generative adversarial networks for multi-domain image-to-image translation. In: Proceedings of the IEEE Conference on Computer Vision and Pattern Recognition, pp. 8789–8797 (2018)
12. Chow, K.H., Hu, S., Huang, T., Ilhan, F., Wei, W., Liu, L.: Diversity-driven privacy protection masks against unauthorized face recognition. Proc. Priv. Enhancing Technol. 4, 381–392 (2024)
13. Chow, K.H., Liu, L.: Robust object detection fusion against deception. In: Proceedings of the 27th ACM SIGKDD Conference on Knowledge Discovery & Data Mining, pp. 2703–2713 (2021)
14. Chow, K.H., Liu, L.: Boosting object detection ensembles with error diversity. In: 2022 IEEE International Conference on Data Mining (ICDM), pp. 903–908. IEEE (2022)
15. Das, N., et al.: Shield: fast, practical defense and vaccination for deep learning using jpeg compression. In: Proceedings of the 24th ACM SIGKDD International Conference on Knowledge Discovery & Data Mining, pp. 196–204 (2018)
16. Deb, D., Zhang, J., Jain, A.K.: AdvFaces: adversarial face synthesis. In: 2020 IEEE International Joint Conference on Biometrics (IJCB), pp. 1–10. IEEE (2020)
17. Deng, J., Guo, J., Xue, N., Zafeiriou, S.: ArcFace: additive angular margin loss for deep face recognition. In: Proceedings of the IEEE/CVF Conference on Computer Vision and Pattern Recognition, pp. 4690–4699 (2019)
18. Guo, C., Rana, M., Cisse, M., Van Der Maaten, L.: Countering adversarial images using input transformations. arXiv preprint arXiv:1711.00117 (2017)
19. Hu, S., et al.: Protecting facial privacy: generating adversarial identity masks via style-robust makeup transfer. In: Proceedings of the IEEE/CVF Conference on Computer Vision and Pattern Recognition, pp. 15014–15023 (2022)
20. Huang, G.B., Ramesh, M., Berg, T., Learned-Miller, E.: Labeled faces in the wild: a database for studying face recognition in unconstrained environments. Technical report. 07-49, University of Massachusetts, Amherst (2007)
21. Li, T., Lin, L.: AnonymousNet: natural face de-identification with measurable privacy. In: Proceedings of the IEEE/CVF Conference on Computer Vision and Pattern Recognition Workshops (2019)
22. Liu, L., Zhou, B., Zou, Z., Yeh, S.C., Zheng, L.: A smart unstaffed retail shop based on artificial intelligence and IoT. In: 2018 IEEE 23rd International Workshop on Computer Aided Modeling and Design of Communication Links and Networks (CAMAD), pp. 1–4. IEEE (2018)
23. Lugaresi, C., et al.: MediaPipe: a framework for building perception pipelines. arXiv preprint arXiv:1906.08172 (2019)
24. Meng, Q., Zhao, S., Huang, Z., Zhou, F.: MagFace: a universal representation for face recognition and quality assessment. In: Proceedings of the IEEE/CVF Conference on Computer Vision and Pattern Recognition, pp. 14225–14234 (2021)
25. Mirza, M., Osindero, S.: Conditional generative adversarial nets. arXiv preprint arXiv:1411.1784 (2014)
26. Ng, H.W., Winkler, S.: A data-driven approach to cleaning large face datasets. In: 2014 IEEE International Conference on Image Processing (ICIP), pp. 343–347. IEEE (2014)
27. Partridge, D., Krzanowski, W.: Software diversity: practical statistics for its measurement and exploitation. Inf. Softw. Technol. **39**(10), 707–717 (1997)

28. Pinto, N., Stone, Z., Zickler, T., Cox, D.: Scaling up biologically-inspired computer vision: a case study in unconstrained face recognition on Facebook. In: CVPR 2011 Workshops, pp. 35–42. IEEE (2011)
29. Sahani, M., Nanda, C., Sahu, A.K., Pattnaik, B.: Web-based online embedded door access control and home security system based on face recognition. In: 2015 International Conference on Circuits, Power and Computing Technologies, ICCPCT 2015, pp. 1–6. IEEE (2015)
30. Schroff, F., Kalenichenko, D., Philbin, J.: FaceNet: a unified embedding for face recognition and clustering. In: Proceedings of the IEEE Conference on Computer Vision and Pattern Recognition, pp. 815–823 (2015)
31. Shan, S., Wenger, E., Zhang, J., Li, H., Zheng, H., Zhao, B.Y.: Fawkes: protecting privacy against unauthorized deep learning models. In: Proceedings of the 29th USENIX Security Symposium (2020)
32. Stone, Z., Zickler, T., Darrell, T.: AutoTagging Facebook: social network context improves photo annotation. In: 2008 IEEE Computer Society Conference on Computer Vision and Pattern Recognition Workshops, pp. 1–8. IEEE (2008)
33. Sun, Q., Tewari, A., Xu, W., Fritz, M., Theobalt, C., Schiele, B.: A hybrid model for identity obfuscation by face replacement. In: Proceedings of the European Conference on Computer Vision (ECCV), pp. 553–569 (2018)
34. Wang, M., Deng, W.: Deep face recognition: a survey. Neurocomputing **429**, 215–244 (2021)
35. Wang, Z., Bovik, A.C., Sheikh, H.R., Simoncelli, E.P.: Image quality assessment: from error visibility to structural similarity. IEEE Trans. Image Process. **13**(4), 600–612 (2004)
36. Wenger, E., Shan, S., Zheng, H., Zhao, B.Y.: SoK: anti-facial recognition technology. In: 2023 IEEE Symposium on Security and Privacy (SP), pp. 134–151 (2023)
37. Wu, Y., Liu, L., Xie, Z., Chow, K.H., Wei, W.: Boosting ensemble accuracy by revisiting ensemble diversity metrics. In: Proceedings of the IEEE/CVF Conference on Computer Vision and Pattern Recognition, pp. 16469–16477 (2021)
38. Xu, W., Evans, D., Qi, Y.: Feature squeezing: detecting adversarial examples in deep neural networks (2018)
39. Yang, X., et al.: Towards face encryption by generating adversarial identity masks. In: Proceedings of the IEEE/CVF International Conference on Computer Vision, pp. 3897–3907 (2021)
40. Zhong, Y., Deng, W.: OPOM: customized invisible cloak towards face privacy protection. IEEE Trans. Pattern Anal. Mach. Intell. **45**, 3590–3603 (2022)
41. Zhu, Z.A., Lu, Y.Z., Chiang, C.K.: Generating adversarial examples by makeup attacks on face recognition. In: 2019 IEEE International Conference on Image Processing (ICIP), pp. 2516–2520. IEEE (2019)

PosterLlama: Bridging Design Ability of Language Model to Content-Aware Layout Generation

Jaejung Seol, Seojun Kim, and Jaejun Yoo(✉)

Laboratory of Advanced Imaging Technology (LAIT), Ulsan National Institute of Science and Technology (UNIST), Ulsan, South Korea
{tjfwownd,seojun.kim,jaejun.yoo}@unist.ac.kr

Abstract. Visual layout plays a critical role in graphic design fields such as advertising, posters, and web UI design. The recent trend toward content-aware layout generation through generative models has shown promise, yet it often overlooks the semantic intricacies of layout design by treating it as a simple numerical optimization. To bridge this gap, we introduce PosterLlama, a network designed for generating visually and textually coherent layouts by reformatting layout elements into HTML code and leveraging the rich design knowledge within language models. Furthermore, we enhance the robustness of our model with a unique depth-based poster augmentation strategy. This ensures our generated layouts remain semantically rich but also visually appealing, even with limited data. Our extensive evaluations across several benchmarks demonstrate that PosterLlama outperforms existing methods in producing authentic and content-aware layouts. It supports an unparalleled range of conditions, including but not limited to content-aware layout generation, element conditional layout generation, and layout completion, among others, serving as a highly versatile user manipulation tool. Project webpage: PosterLlama.

Keywords: Content-Aware Layout Generation · Graphic Design · Language Model

1 Introduction

Layout is a fundamental element in graphic design, harmoniously arranging design elements such as logos and text to capture the reader's attention and effectively convey essential information. Its significance extends across various applications, ranging from graphic design, such as web UI, posters, and document typesetting, to downstream tasks like Human-Object Interaction(e.g., region-controlled image generation [28,45] and layout-guided video generation [29]).

Supplementary Information The online version contains supplementary material available at https://doi.org/10.1007/978-3-031-73007-8_26.

© The Author(s), under exclusive license to Springer Nature Switzerland AG 2025
A. Leonardis et al. (Eds.): ECCV 2024, LNCS 15140, pp. 451–468, 2025.
https://doi.org/10.1007/978-3-031-73007-8_26

Due to its versatile applications, layout design has garnered widespread attention by offering the potential to replace manual efforts for experts and achieve cost savings. Consequently, considerable research has been dedicated to creating controllable, high-quality layouts, aiming to enhance both aesthetic appeal and functional efficiency in design processes. [1,4,10,16–20,22,26,27,35,44,46]

As the significance of visual content awareness in layout generation has gained increasing recognition, predicting layout structure that ensures text readability and maintains visual balance on the canvas has become more important. This is referred to as the poster layout generation problem. Following the pioneering work of ContentGAN [49], the first model to blend visual and textual data for generating layout structure, subsequent models have introduced sophisticated techniques to refine this process. CGL-GAN [50] and DS-GAN [14] enhance spatial information encoding using multi-scale CNNs, employing transformers and CNN-LSTM architectures for decoding in layout generation, respectively. RADM [25] has expanded this paradigm by incorporating visual and textual content considerations into poster generation through diffusion models.

Despite advancements in layout generation, existing methods have limitations because they treat layout elements as simple numerical values for prediction, converting semantically rich categories into 0, 1, and so on. This approach restricts the network's ability to learn beyond the distribution of layouts and fails to capture the semantic relationships among elements. For example, indicating that "The text is placed over the underlay" offers a more interpretable representation compared to stating "Element0 is placed over Element1". To tackle this challenge, recent efforts, such as LayoutPromtper [30], Layout GPT [7], and LayoutNUWA [42], have been focused on leveraging the powerful capabilities of language models. While these approaches show promise in producing high-quality layouts, they still face challenges in considering fine visual content.

In this paper, we introduce PosterLlama, a novel model designed for generating poster layouts aware of both visual canvas and textual contents, integrating semantically rich layout elements. To achieve this, following the previous work [30,42], we reformat layout elements into HTML code to leverage the design knowledge embedded in language models. We additionally incorporate text descriptions into code format encouraging the model to learn text awareness. To ensure the language model takes visual content into account, we design a two-stage training process inspired by the efficient vision-language training method [51]. In the first stage, we train an adapter to connect the visual encoder with the LLM, and in the second stage, we train the model to generate HTML sequences. Given the challenges in assembling a large dataset due to copyright issues and other constraints with poster datasets, we propose a depth-based augmentation method that primarily focuses on the presence of salient objects within the poster. Lastly, to bridge the gap between layout generation tasks and real-world industrial applications, we propose a pipeline for generating advertisement posters that utilize a scene-text generation module.

We conduct a comprehensive evaluation of PosterLlama using various methods. Experimental results demonstrate that PosterLlama is efficient with dataset size, showing state-of-the-art performance in nearly all metrics. Notably, our model exhibits almost identical performance to real layouts regarding layout quality measurement, thanks to leveraging design knowledge from LLMs. To the best of our knowledge, PosterLlama is the first model that can handle all types of content-aware layout generation tasks, ranging from layout elements conditional generation to layout completion and refinement. This capability suggests the applicability of PosterLlama to practical content-aware layout generation tasks.

2 Related Work

Content-Agnostic Layout Generation. Content-agnostic layout generation aims to create layouts without being restricted by specific content, like a canvas. Early works in layout generation [2,23,34,38] rely on professionally designed templates or heuristic rules for generating layouts. These methods require task-specific professional knowledge and are often unable to generate a wide variety of layouts. To overcome these issues, a data-driven approach utilizing deep generative models is proposed. LayoutGAN [26] is the first approach introducing GAN to synthesize semantic and geometric layout elements. It adopted a differentiable wireframe rendering to incorporate visual attributes into the layout generation process. Subsequently, LayoutGAN has been enhanced to support attribute-conditioned design tasks [27]. Following the initial approaches, methods involving Variational Autoencoders [1,19,20,35,44], Diffusion models [4,16,17,46], and transformers [10,18,22] have been introduced for both constrained and unconstrained layout generation tasks. Meanwhile, recent developments in LLM-based layout generation methods [7,30,42] have shown results that match those of traditional models. These advancements highlight the limitations of current layout generation approaches, specifically their insufficient understanding of semantic information among layout elements, as they mainly rely on numerical optimization methods. Among these methods, LayoutNUWA [42] is the most closely related to ours, achieving state-of-the-art performance by fine-tuning LLMs to generate layouts in HTML format. However, unlike our approach, LayoutNUWA's scope is limited to content-agnostic layout generation.

Content-Aware Layout Generation. With the advancement of content-agnostic layout generation methods, research incorporating content such as images and text into layout generation has received increased attention, particularly in generative [3,14,25,49,50] manner. ContentGAN [49] first considered the relationship not only between layout elements but also between layout and images. Although they benefit from content information, as they neglect composition like spatial information, they encounter the occlusion problem in poster layout generation. CGL-GAN [50] and DS-GAN [14] employ an encoder-decoder architecture, utilizing a standard transformer and a CNN-LSTM for the decoder, respectively. By leveraging a ResNet-FPN network as the visual encoder and additionally injecting a saliency map, they demonstrate robust capabilities in generating a visual content-aware layout. RADM [25] stands out as the first

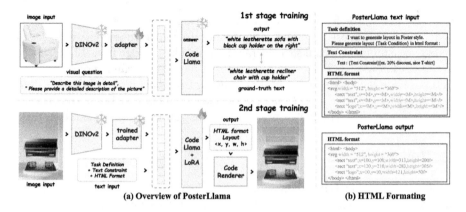

Fig. 1. (a) The overall training step of PosterLlama. The first stage of training (top) involves training the adapter to learn visual alignment. In the second stage of training (bottom), we fine-tune the model for HTML formatted layout generation, utilizing the frozen adapter. (b) Illustration of input-output sequence in PosterLlama

paper to integrate textual content, a highly significant attribute, into visual elements for layout generation, considering spatial composition. Through a refined design of the relation-aware module within the denoising diffusion model, RADM achieves realistic poster layout generation. Recently, leveraging the power of LLM, LayoutPrompter [30] has been introduced as a content-aware layout generation method. However, it has a limitation in achieving fine visual composition as it solely represents visual content information using the bounding box of salient objects. In contrast, our work takes into account visual encoding, enabling high-quality content-aware layout generation with a focus on fine visual composition.

3 Method

In this section, we introduce PosterLlama, an LLM-based multi-modal layout generation model. We redesign layout generation tasks as HTML sequence generation to leverage LLM's design knowledge effectively (Sect. 3.1). To incorporate visual understanding into LLM, we propose a two-stage training method for multi-modal layout generation (Sect. 3.2). We also develop a Layout Augmentation Module for enhanced layout robustness (Sect. 3.3) (Fig. 1).

3.1 Input Output Sequence Formatting

Layout Formation. The goal of content-aware layout generation is to generate layouts constrained on given content condition $\mathcal{C}$. In our poster layout generation, the condition $\mathcal{C}$ is defined as multi-modal content (e.g., poster canvas, textual explanation). The layout is represented by a set of N elements $\{e_i\}_{i=1}^N$, where $e_i = (t_i, s_i, c_i)$ consists of bounding box location $t_i = (x_i, y_i)$, size $s_i = (w_i, h_i)$, and category c_i. Following [18], in addition to conditions by content, the subset of layout elements can be also applied as a constraint. For example, Gen-IT [18] generates a layout constrained by element category types.

HTML Formatting. To leverage the extensive knowledge encapsulated in LLM for layout design, we represent layouts in the form of HTML sequence formations [30,42]. This method not only allows us to utilize design priors embedded in the LLM's training data, such as those from web-UI designs, but also offers a stronger expressive capability, compared to merely representing layout attributes as numerical values. In line with the approach [42], we develop a template for generating text-aware layouts by constructing the model input sequence with a task definition, HTML formatting, and adding Text Constraints. The task definition, identified by {Task Condition}, specifies the conditions of the input sequence (for instance, "according to the categories and image" in Gen-IT [18]). We employ HTML formatting to encapsulate the diverse tags characterizing the Web-UI layout, such as the <rect> tag, to wrap around the layout elements. To facilitate conditional layout generation, we introduce a mask token <M>, prompting the LLM to predict this masked token. As we can readily imagine, layout elements do not inherently possess a specific order. However, organizing the input and output of the layout into a predefined order of mask tokens during the learning process with limited data and diverse conditions can lead to overfitting. Hence, we introduce a random permutation to the layout order while maintaining synchronization between input and output elements. Additionally, for efficient training with a focus on reducing the overall token length, We discretize the attributes of each element similarly to previous work [4,17,46].

3.2 Training Method

We employ LLM for the poster layout generation. To leverage the vision-language capabilities of the model for layout generation, we adopt a two-stage training approach inspired by the efficient Visual Question Answering training method of Mini-GPT4 [51] with instruction tuning. In the first stage, we utilize a linear layer as an adapter to align the image encoder with the LLM, training only the adapter while keeping others frozen. This training is performed on an extensive collection of aligned image-text pairs. The encoded image feature of the encoder is encapsulated within <img> tokens and treated as a text token along with textual instructions: "<img><ImageFeature></img> Describe this image in detail." For the visual encoder, we adopt DINOv2 [32], drawing inspiration from recent advancements in visual embedding progress [41]. In the second stage, while keeping the visual adapter frozen, we fine-tune the LLM to generate layouts using an HTML-formatted dataset described in Sect. 3.1. To address the challenge of catastrophic forgetting and optimize the LLM fine-tuning process, our approach involves the utilization of LoRA as proposed by [15]. We train our model using cross-entropy loss as the objective function.

3.3 Depth-Guided Poster Augmentation

The performance of generative models fundamentally improves with diverse and rich data. However, poster datasets significantly lack the volume compared to the vast data pools, like LAION [39] that are typically utilized for training foundational generative models. This scarcity, coupled with copyright issues related to the processed nature of poster images, makes the collection of large datasets challenging. To address the limitations, we introduce a novel poster augmentation illustrated in Fig. 2. It leverages depth-based augmentation, closely connected with salient objects [48], and a top-k similarity selection in a two-step process, drawing inspiration from recent content-aware layout generation works that utilize salient object information as a visual condition [14,50]. In the depth-based augmentation phase, we employ ControlNet [47]-Depth, a diffusion-based generative model conditioned on text and depth maps. To construct the text condition, we generate captions from original images using the high-quality captioning network, Blip-2, and formulate scripts like "Please generate {Caption} in advertisement poster." Depth maps are estimated using readily available estimation networks. Despite ControlNet's ability to synthesize high-quality images, recent studies have shown that diffusion-generated images have undesirable artifacts [9]. Especially, artifacts that appear on the salient objects, can detrimentally affect the correlation between layout and image canvas, hindering the learning process of the network. To mitigate this problem, we introduce DreamSIM [8], a similarity measurement sensitive to layout and semantic content, to select the top-k samples from N generated samples. Here, we use $N = 10$ and $k = 3$; *i.e.*, per each image, three samples are selected for augmentation out of 10 generated samples. The resultant images illustrate the generation of high-

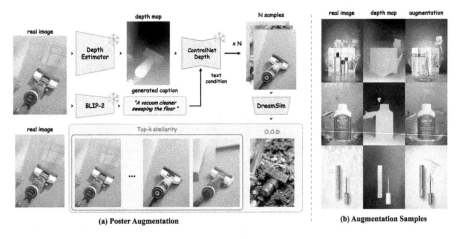

Fig. 2. (a) The illustration of depth-guided poster augmentation. The estimated depth map and caption undergo multiple augmentations with the ControlNet, and the top-k similarity samples with real images are selected as augmentation samples. (b) The illustration of augmented samples.

Table 1. Quantitative comparison with baselines on content-aware layout generation task. Text-aware Poster Llama is annotated as Poster Llama-T.

Method	Graphic						Content	
	val↑	ove↓	ali↓	und_l↑	und_s↑	FD↓	rea↓	occ↓
CGL-Dataset								
Real-data	0.9839	0.0002	0.0017	0.9937	0.9884	-	0.2059	0.1399
DS GAN	0.8545	0.0379	0.0033	0.9193	0.6717	41.45	0.2121	0.1491
LayoutPrompter	0.9950	0.0035	0.0025	0.4473	0.3201	4.048	0.2314	0.3011
RADM	0.9962	0.0702	0.0008	0.9854	0.9277	**0.729**	0.1961	**0.1391**
Poster Llama (Ours)	0.9986	0.0052	0.0004	**0.9982**	**0.9909**	2.518	0.2072	0.1480
Poster Llama-T (Ours)	**0.9988**	0.0024	**0.0003**	0.9918	0.9905	2.082	0.2042	0.1418
PKU-Dataset								
Real-data	0.9997	0.0013	0.0021	0.9974	0.9909	-	0.1729	0.1828
DS GAN	0.8740	0.0336	0.0037	0.8688	0.5746	39.96	0.2035	0.2256
LayoutPromtper	0.9995	0.0036	0.0032	0.5789	0.4139	**2.753**	0.2124	0.2992
Poster Llama (Ours)	**1.000**	0.0032	0.0009	**0.9998**	**0.9910**	12.44	**0.1875**	0.2087

quality synthetic data with minimal changes, preserving the overall composition and salient objects. Figure 2(b) shows augmented examples.

4 Experiments

4.1 Experimental Setting

Datasets. We utilize two publicly available datasets, CGL and PKU, which are poster data collected from e-commerce platforms. The PKU dataset consists of three elements - Logo, Text, and Underlay, while the CGL dataset includes Embellishment as an additional element. CGL provides a total of 60,548 annotated pairs of poster-layout, along with 1,000 unannotated posters. On the other hand, the PKU dataset offers 9,974 annotated pairs and 905 unannotated posters. As CGL does not supply inpainted posters separately, similar to prior work [25,50], we inpainted the dataset using an inpainting network. Additionally, because both CGL and PKU datasets do not provide text annotations, we opt for the CGL-v2 dataset to facilitate layout generation that incorporates both textual and visual content. Lastly, since PKU and CGL datasets do not provide annotated poster splits for validation and testing, we approximately divide the data into train/val/test sets in an 8:1:1.

Baselines. We compare our method with four Layout Generation studies. (1) DS-GAN [14] utilizes a CNN encoder and a CNN-LSTM decoder. Due to its specific element sorting algorithm, this model is unsuitable for constrained generation tasks. (2) LayoutPrompter [30] is a method for generating layouts through in-context learning without the need for training. Similar to our approach, it uses LLM for layout generation. While the original paper employed GPT-3 text-davinci-003, our study utilized the GPT-3.5 turbo instruct model, as the former is currently unavailable. (3) RADM [25] is a diffusion-based model capable of handling visual-textual content for layout generation. (4) RALF [13] is an autoregressive model that leverages a Transformer decoder and sophisticated retrieval augmentation. Since RALF uses a different dataset processing method from ours, we use the publicly released weights without retraining and evaluate on an augmented dataset at analysis which is a completely separate setting.

Implementation Details. Our PosterLlama model is built upon a language model and utilizes the ViT [6] as its visual encoder. For the visual encoding process, we select the DINOv2-base [32] model, which is capable of processing images with a resolution of 224 × 224. The CodeLlaMA [37] model serves as the LLM component. The rationale behind choosing the visual encoder and LLM is thoroughly examined in the ablation study detailed in Sect. 4.4. For the first stage training, we use a combined 2M image-captioning dataset that includes images from LAION [39], Conceptual Captions [40], and SBU [33]. To train our model efficiently, we employ the DeepSpeed [36] Library, conducting the training for five days on two A100 GPUs, with a batch size of 32 and leveraging gradient accumulation techniques [24]. During inference, we utilize top-p [12] sampling with a p value of 0.9 and a sampling temperature of 0.7.

4.2 Evaluation Metrics

we follow the evaluation metrics in previous work [14,21,50], including two aspects: graphic measures, and composition-relevant measures.

Graphic Measures. This measure assesses graphic quality by considering the relationships between layout elements, without taking the canvas into account. We evaluated six commonly used metrics for layout graphic measure: Validity(val), Alignment (ali), Overlap (ove), Underlay (und_l, und_s), and Frechet Distance (FD). The validity is the ratio of valid elements greater than 0.1% of the canvas. We calculate all of the measures with valid elements. ali, ove, and und are assessed based on the methodology outlined in [14]. Additionally, we compute the Frechet distance (FD) [11] in the feature space pretrained by [21].

Content Measures. The aim of content measures is to evaluate the harmony between the generated layout elements and the canvas, following appropriate design rules. We employ content measures such as Occlusion (occ), and Readability score (rea) for our assessment. Occlusion measures the extent of the area occluded between the well-defined layout and salient objects, with the intuition that a well-defined layout minimizes the extent to which it obscures salient objects. The Readability score evaluates the clarity of text elements by assessing their gradient variation in the image space.

4.3 Comparison

Quantitative Result. In this section, we assess the performance of our PosterLlama in comparison to DS-GAN [14], LayoutPrompter [30], and RADM [25]—all of which are sophisticatedly designed methods for layout generation. The evaluation is conducted using eight different metrics. Due to the absence of text annotations in the PKU dataset, we exclusively compare RADM's performance on the CGL dataset. The quantitative results of the annotated test split without user constraints are summarized in Table 1. PosterLlama attains the highest scores across five metrics and the second-highest scores for FD, rea, and occ in the CGL dataset. Additionally, it achieves the highest score in all metrics except for FD in the PKU dataset. Further details on performance degradation will be discussed in the upcoming section addressing data leakage.

Data Leakage. We note that PosterLlama's performance falls behind in three specific metrics (FD, occ, rea) compared to RADM. These three metrics only show high performance when the generated layout overlaps with the actual layout without introducing any novel elements. This is because they solely focus on the overlap with the canvas saliency map and the feature distance to the real layout, without evaluating the quality of the layout itself. Upon examining the layouts produced by RADM for test samples, we find that they always overlap with elements from the real layout. We hypothesize that this phenomenon may be due to artifacts introduced during the inpainting of the training data. Unlike PosterLlama, which uses a frozen visual encoder, RADM trains its visual encoder during the learning process, potentially allowing it to detect these artifacts due to data leakage. To test this hypothesis, we inpaint parts of objects in the test dataset and examine whether RADM's generated layouts overlap with these inpainted areas, as illustrated in Fig. 3. Our results show that RADM places layout elements precisely at the inpainted areas (including the areas where original GT layouts were), suggesting it utilizes the inpainted artifacts to estimate the layout locations, unlike PosterLlama, which remain unaffected by the inpainted areas. Therefore, based on our analysis, we believe PosterLlama is more reliable than RADM despite its inferior performance in content metrics and FD.

Fig. 3. The data-leakage example of inpainted area. In addition to the original layout Red box area is deliberately erased and inpainted. Note that RADM puts the layouts precisely at the inpainted area. (Color figure online)

Qualitative Result. Here, we provide a qualitative comparison between our PosterLlama and the baseline method, as detailed in Table 1 and depicted in Fig. 6. DS-GAN tends to generate layouts that are misaligned and overlapped, frequently positioning them in the top-left corner. This tendency is attributed to the fixed number of elements and the placement of non-element layouts at the top-left (0, 0, 0, 0) position. Although Layout Prompter yields highly aligned layouts, it lacks content awareness, leading to significant occlusion. On the other hand, RADM exhibits structures that closely resemble real data across all samples. In contrast, our PosterLlama approach demonstrates the ability to generate well-fitted and sensible layouts without overfitting the real data.

Table 2. Model architecture comparison. Initially based on Llama2+Eva-vit [51], our approach incorporates Dinov2 for visual encoding and CodeLlama tailored for effective code language processing. We denote CodeLlama as CL.

PosterLlama	val↑	ove↓	ali↓	und_l↑	und_s↑	FD↓	rea↓	occ↓
Llama2+Eva-vit	0.9974	0.0040	0.0009	0.9898	0.9820	7.07	0.2169	0.1741
Llama2+DINOv2	**0.9992**	0.0042	0.0008	0.9682	0.9133	3.87	0.2111	0.1506
CL+DINOv2 (Ours)	0.9988	**0.0024**	**0.0003**	**0.9918**	**0.9905**	**2.08**	**0.2042**	**0.1418**

4.4 Analysis

Effect of Model Architecture. We delve into the architectural design choices for language models and vision transformers, as detailed in Table 2. Our initial approach leverages a language model for content-aware layout generation, primarily adopting the Llama2+Eva-vit configuration, as proposed by MiniGPT-v2 [5]. We subsequently integrate the sophisticated visual encoder DINOv2, as detailed in Sect. 3.2, along with CodeLlama, a variant of Llama2 fine-tuned for code language processing. The integration of DINOv2, as evidenced in the Llama2 results led to a significant improvement in the visual content awareness metric. Despite Llama2's adaptability to a broad spectrum of language formats, it is primarily not designed for code interpretation. Thus, switching to CodeLlama enables our model to achieve superior outcomes, demonstrating its efficacy in the context of content-aware layout generation. This adaptation underscores the importance of tailored model architecture in enhancing performance metrics, particularly in specialized applications such as visual content comprehension.

Effect of Augmentation. We assess the impact of our depth-guided augmentation technique, as discussed in Sect. 3.3. The detailed results are presented in Table 3. The table indicates an overall positive effect of our augmentation method. In the data-rich CGL-Dataset, most metrics show performance improvements, except for FD and und_s. The slight degradation in these metrics can be attributed to the increased diversity introduced by our augmentation method,

Table 3. Effect of depth-guided augmentation technique.

CGL-Dataset	val↑	ove↓	ali↓	und_l↑	und_s↑	FD↓	rea↓	occ↓
Ours w/o aug	0.9984	0.0029	0.0006	0.9903	**0.9987**	**1.43**	0.2058	0.1476
Ours	**0.9988**	**0.0024**	**0.0003**	0.9918	0.9905	2.08	**0.2042**	**0.1418**
PKU-Dataset	val↑	ove↓	ali↓	und_l↑	und_s↑	FD↓	rea↓	occ↓
Ours w/o aug	1.000	0.0087	0.0014	0.9911	0.9661	12.47	0.1882	0.2119
Ours	**1.000**	**0.0032**	**0.0009**	**0.9986**	**0.9910**	**12.44**	**0.1875**	**0.2087**

Table 4. Comparison on depth-guided augmented CGL-v2 test dataset.

Model	val↑	ove↓	ali↓	und_l↑	und_s↑	rea↓	occ↓
DS-GAN	0.8451	0.0336	0.0039	0.8848	0.5969	0.1169	0.0597
RADM	**1.0000**	0.0079	0.0026	0.9029	0.6817	**0.0973**	0.0528
RALF	**1.0000**	0.0156	0.0044	0.9820	0.9666	0.1126	0.0595
PosterLlama-T	**1.0000**	0.0009	0.0028	0.9909	0.9883	0.1116	0.0575
PosterLlama	**1.0000**	**0.0006**	**0.0001**	**1.0000**	**1.0000**	0.1142	**0.0513**

impacting strict underlay und_s and distribution FD. We additionally discuss this in supplementary material B. Remarkably, our augmentation method demonstrates significant performance improvements in the PKU-Dataset, characterized by a limited number of data samples. It shows performance gains across all metrics, highlighting the effectiveness of our augmentation in scenarios with sparse data. This emphasizes the utility of our augmentation approach in enhancing model performance, particularly in data-scarce environments (Table 4).

Evaluation on Augmented Samples. We endeavor to collect an artifacts-free clean dataset; however, even if we create the dataset ourselves, assessment is unfeasible as the specifically trained textual encoder of RADM is not publicly available. Therefore, for leakage-free evaluation, we evaluate models on CGL-v2 test samples using our augmentation technique, which effectively minimizes inpainting artifacts, as detailed in supplementary material E.3. Additionally, we evaluate the recent visual content-aware model, RALF [13] using the publically released model. To ensure a fair comparison, we also evaluate PosterLlama trained on the unaugmented dataset. The results demonstrate that the excessively high score of RADM at quantitative result decreases, indicating susceptibility to information leakage. In contrast, our model achieves the highest scores across most metrics, showcasing robustness to inpainting artifacts.

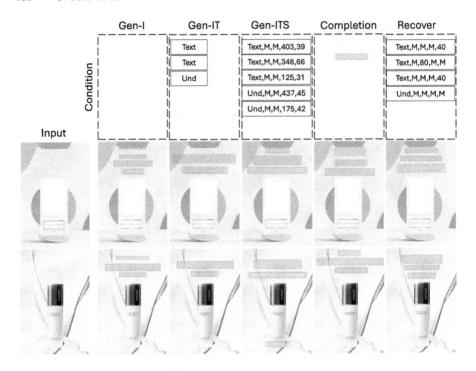

Fig. 4. Visualization of PosterLlama across five conditional generations, with the conditions for layout generation positioned at the top. The order of elements is (c, x, y, w, h).

Conditional Generation. Our model is capable of achieving all types of user-conditioned generation defining the condition as text format, which is also a key feature of our method. We conduct experiments with 5 conditions as shown in Fig. 4. The conditions are as follows: Gen-I involves image-conditioned layout generation, while Gen-IT and Gen-ITS additionally incorporate conditions based on category type and size. The completion task aims to generate a complete layout using partially placed elements. The recovery task involves restoring a randomly masked layout, with up to 80% of elements being masked. As observed, PosterLlama is adept at handling a variety of user constraints while producing high-quality layouts. We construct these above 5 conditions following [18] and additional conditioning results are detailed in the supplementary materials.

Poster Generation Pipeline. To extend the applicability of layout generation to real-world industries, we propose a one-click advertisement poster generation pipeline. We utilize the scene-text generation model, i.e., AnyText [43] to generate a completed advertisement poster from the generated layout and background image. We first generate text layout from layout generation conditioning on the user text. Then we build a mask from the text layout. Lastly, we generate text-rendered images through the scene-text-generation model. By utilizing the proposed pipeline, users can easily get advertisement posters in just one click. Figure 5 presents the outcomes of our poster generation pipelines. As demonstrated, our poster generation pipeline showcases applicability in creating plausible advertisement posters, indicating its usability in practical settings.

5 Limitation

We acknowledge two limitations as follows. Firstly, to accommodate the code language format suitable for CodeLlama, we translate the Chinese words provided in CGL-v2 into English for application in CodeLlama. Consequently, in our text-aware layout generation scheme, it becomes challenging to fully consider the text length in the generated layout. Furthermore, recent studies in scene text generation have highlighted issues with degraded inpainted text quality when there is a mismatch between the sizes of the text and its bounding box [31]. When we apply this scene text generation to our proposed poster generation pipeline, as shown in Fig. 5 right two samples, we observe occurrences where the quality of the generated poster is compromised due to discrepancies in text length. Secondly, PosterLlama's reliance on the LLM foundation model makes it less scalable compared to other models. This limitation poses challenges when attempting to deploy PosterLlama on edge devices.

Fig. 5. The example of proposed poster generation pipeline. The layout of scene text generation is generated by PosterLlama. Note that the right two samples exhibit unsatisfactory outcomes, attributed to inappropriate bounding box sizes.

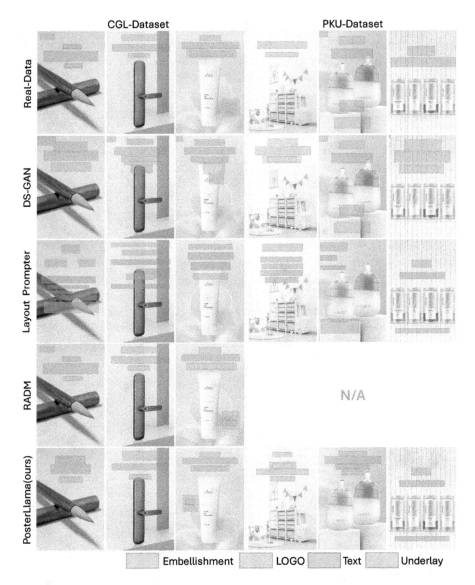

Fig. 6. Qualitative comparison of image conditioned generation with baseline models about PKU and CGL-Dataset. Each comparison is implemented on annotated test samples. PKU Dataset lacks text descriptions, hence RADM does not proceed.

6 Conclusion

In this paper, we present PosterLlama, both a visual and largely unexplored textual content-aware layout generation method utilizing LLM. For content-aware layout generation, we leverage an efficient Visual Question-answering training method to instill visual awareness into the LLM, handling the layout in a code format suitable for the language model. To overcome data scarcity, we propose depth-guided augmentation using off-the-shelf generative models, which also alleviates inpainting artifacts enabling a fair evaluation. Our extensive experiments demonstrate that PosterLlama outperforms existing approaches, achieving diverse conditional generation by handling conditions in text format and being robust against learning shortcuts caused by inpainting artifacts. Thanks to this robustness and our augmentation method, we demonstrate that PosterLlama is highly effective with small datasets and adaptable for real-world applications.

Acknowledgements. This work was supported by the National Research Foundation of Korea (NRF) grant funded by the Korea government (MSIT) (No.2022 R1C1C1008496), Institute of Information & communications Technology Planning & Evaluation (IITP) grant funded by the Korea government (MSIT)(No. RS-2020-II201336, Artificial Intelligence graduate school support (UNIST), RS-2022-II220959, (Part 2) Few-Shot Learning of Causal Inference in Vision and Language for Decision Making, RS-2021-II212068, Comprehensive Video Understanding and Generation with Knowledge-based Deep Logic Neural Network), and the Ministry of Science and ICT (MSIT, Korea) & Gwangju Metropolitan City (Artificial intelligence industrial convergence cluster development project). We also thank the supercomputing resources of the UNIST Supercomputing Center and Byung-hak Kim from the CJ AI Center for valuable discussions.

References

1. Arroyo, D.M., Postels, J., Tombari, F.: Variational transformer networks for layout generation. In: Proceedings of the IEEE/CVF Conference on Computer Vision and Pattern Recognition, pp. 13642–13652 (2021)
2. Cao, Y., Chan, A.B., Lau, R.W.: Automatic stylistic manga layout. ACM Trans. Graph. (TOG) **31**(6), 1–10 (2012)
3. Cao, Y., et al.: Geometry aligned variational transformer for image-conditioned layout generation. In: Proceedings of the 30th ACM International Conference on Multimedia, pp. 1561–1571 (2022)
4. Chai, S., Zhuang, L., Yan, F.: Layoutdm: transformer-based diffusion model for layout generation. In: Proceedings of the IEEE/CVF Conference on Computer Vision and Pattern Recognition, pp. 18349–18358 (2023)
5. Chen, J., et al.: Minigpt-v2: large language model as a unified interface for vision-language multi-task learning. arXiv preprint arXiv:2310.09478 (2023)
6. Dosovitskiy, A., et al.: An image is worth 16x16 words: transformers for image recognition at scale. arXiv preprint arXiv:2010.11929 (2020)
7. Feng, W., et al.: Layoutgpt: compositional visual planning and generation with large language models. In: Advances in Neural Information Processing Systems, vol. 36 (2023)
8. Fu, S., et al.: Dreamsim: learning new dimensions of human visual similarity using synthetic data. arXiv preprint arXiv:2306.09344 (2023)
9. Gandikota, R., Materzynska, J., Fiotto-Kaufman, J., Bau, D.: Erasing concepts from diffusion models. arXiv preprint arXiv:2303.07345 (2023)
10. Gupta, K., Lazarow, J., Achille, A., Davis, L.S., Mahadevan, V., Shrivastava, A.: Layouttransformer: layout generation and completion with self-attention. In: Proceedings of the IEEE/CVF International Conference on Computer Vision, pp. 1004–1014 (2021)

11. Heusel, M., Ramsauer, H., Unterthiner, T., Nessler, B., Hochreiter, S.: Gans trained by a two time-scale update rule converge to a local Nash equilibrium. In: Advances in Neural Information Processing Systems, vol. 30 (2017)
12. Holtzman, A., Buys, J., Du, L., Forbes, M., Choi, Y.: The curious case of neural text degeneration. arXiv preprint arXiv:1904.09751 (2019)
13. Horita, D., Inoue, N., Kikuchi, K., Yamaguchi, K., Aizawa, K.: Retrieval-augmented layout transformer for content-aware layout generation. In: CVPR (2024)
14. Hsu, H.Y., He, X., Peng, Y., Kong, H., Zhang, Q.: Posterlayout: a new benchmark and approach for content-aware visual-textual presentation layout. In: Proceedings of the IEEE/CVF Conference on Computer Vision and Pattern Recognition, pp. 6018–6026 (2023)
15. Hu, E.J., et al.: Lora: low-rank adaptation of large language models. arXiv preprint arXiv:2106.09685 (2021)
16. Hui, M., Zhang, Z., Zhang, X., Xie, W., Wang, Y., Lu, Y.: Unifying layout generation with a decoupled diffusion model. In: Proceedings of the IEEE/CVF Conference on Computer Vision and Pattern Recognition, pp. 1942–1951 (2023)
17. Inoue, N., Kikuchi, K., Simo-Serra, E., Otani, M., Yamaguchi, K.: Layoutdm: discrete diffusion model for controllable layout generation. In: Proceedings of the IEEE/CVF Conference on Computer Vision and Pattern Recognition, pp. 10167–10176 (2023)
18. Jiang, Z., et al.: Layoutformer++: conditional graphic layout generation via constraint serialization and decoding space restriction. In: Proceedings of the IEEE/CVF Conference on Computer Vision and Pattern Recognition, pp. 18403–18412 (2023)
19. Jiang, Z., Sun, S., Zhu, J., Lou, J.G., Zhang, D.: Coarse-to-fine generative modeling for graphic layouts. In: Proceedings of the AAAI Conference on Artificial Intelligence, vol. 36, pp. 1096–1103 (2022)
20. Jyothi, A.A., Durand, T., He, J., Sigal, L., Mori, G.: Layoutvae: stochastic scene layout generation from a label set. In: Proceedings of the IEEE/CVF International Conference on Computer Vision, pp. 9895–9904 (2019)
21. Kikuchi, K., Simo-Serra, E., Otani, M., Yamaguchi, K.: Constrained graphic layout generation via latent optimization. In: Proceedings of the 29th ACM International Conference on Multimedia, pp. 88–96 (2021)
22. Kong, X., et al.: BLT: bidirectional layout transformer for controllable layout generation. In: Avidan, S., Brostow, G., Cissé, M., Farinella, G.M., Hassner, T. (eds.) ECCV 2022. LNCS, vol. 13677, pp. 474–490. Springer, Cham (2022). https://doi.org/10.1007/978-3-031-19790-1_29
23. Kumar, R., Talton, J.O., Ahmad, S., Klemmer, S.R.: Bricolage: example-based retargeting for web design. In: Proceedings of the SIGCHI Conference on Human Factors in Computing Systems, pp. 2197–2206 (2011)
24. Lamy-Poirier, J.: Layered gradient accumulation and modular pipeline parallelism: fast and efficient training of large language models. arXiv preprint arXiv:2106.02679 (2021)
25. Li, F., et al.: Relation-aware diffusion model for controllable poster layout generation. In: Proceedings of the 32nd ACM International Conference on Information and Knowledge Management, pp. 1249–1258 (2023)
26. Li, J., Yang, J., Hertzmann, A., Zhang, J., Xu, T.: Layoutgan: generating graphic layouts with wireframe discriminators. arXiv preprint arXiv:1901.06767 (2019)

27. Li, J., Yang, J., Zhang, J., Liu, C., Wang, C., Xu, T.: Attribute-conditioned layout gan for automatic graphic design. IEEE Trans. Visual Comput. Graph. **27**(10), 4039–4048 (2020)
28. Li, Y., et al.: Gligen: open-set grounded text-to-image generation. In: Proceedings of the IEEE/CVF Conference on Computer Vision and Pattern Recognition, pp. 22511–22521 (2023)
29. Lian, L., Shi, B., Yala, A., Darrell, T., Li, B.: Llm-grounded video diffusion models. arXiv preprint arXiv:2309.17444 (2023)
30. Lin, J., Guo, J., Sun, S., Yang, Z., Lou, J.G., Zhang, D.: Layoutprompter: awaken the design ability of large language models. In: Advances in Neural Information Processing Systems, vol. 36 (2023)
31. Liu, R., et al.: Character-aware models improve visual text rendering. arXiv preprint arXiv:2212.10562 (2022)
32. Oquab, M., et al.: Dinov2: learning robust visual features without supervision. arXiv preprint arXiv:2304.07193 (2023)
33. Ordonez, V., Kulkarni, G., Berg, T.: Im2text: describing images using 1 million captioned photographs. Advances in neural information processing systems **24** (2011)
34. O'Donovan, P., Agarwala, A., Hertzmann, A.: Learning layouts for single-pagegraphic designs. IEEE Trans. Visual Comput. Graph. **20**(8), 1200–1213 (2014)
35. Patil, A.G., Ben-Eliezer, O., Perel, O., Averbuch-Elor, H.: Read: recursive autoencoders for document layout generation. In: Proceedings of the IEEE/CVF Conference on Computer Vision and Pattern Recognition Workshops, pp. 544–545 (2020)
36. Rasley, J., Rajbhandari, S., Ruwase, O., He, Y.: Deepspeed: system optimizations enable training deep learning models with over 100 billion parameters. In: Proceedings of the 26th ACM SIGKDD International Conference on Knowledge Discovery & Data Mining, pp. 3505–3506 (2020)
37. Roziere, B., et al.: Code llama: open foundation models for code. arXiv preprint arXiv:2308.12950 (2023)
38. Schrier, E., Dontcheva, M., Jacobs, C., Wade, G., Salesin, D.: Adaptive layout for dynamically aggregated documents. In: Proceedings of the 13th international conference on Intelligent user interfaces, pp. 99–108 (2008)
39. Schuhmann, C., et al.: Laion-5b: an open large-scale dataset for training next generation image-text models. Adv. Neural. Inf. Process. Syst. **35**, 25278–25294 (2022)
40. Sharma, P., Ding, N., Goodman, S., Soricut, R.: Conceptual captions: a cleaned, hypernymed, image alt-text dataset for automatic image captioning. In: Proceedings of the 56th Annual Meeting of the Association for Computational Linguistics (Volume 1: Long Papers), pp. 2556–2565 (2018)
41. Stein, G., et al.: Exposing flaws of generative model evaluation metrics and their unfair treatment of diffusion models. In: Advances in Neural Information Processing Systems, vol. 36 (2024)
42. Tang, Z., Wu, C., Li, J., Duan, N.: Layoutnuwa: revealing the hidden layout expertise of large language models. arXiv preprint arXiv:2309.09506 (2023)
43. Tuo, Y., Xiang, W., He, J.Y., Geng, Y., Xie, X.: Anytext: multilingual visual text generation and editing. arXiv preprint arXiv:2311.03054 (2023)
44. Yamaguchi, K.: Canvasvae: learning to generate vector graphic documents. In: Proceedings of the IEEE/CVF International Conference on Computer Vision, pp. 5481–5489 (2021)
45. Yang, Z., et al.: Reco: region-controlled text-to-image generation. In: Proceedings of the IEEE/CVF Conference on Computer Vision and Pattern Recognition, pp. 14246–14255 (2023)

46. Zhang, J., Guo, J., Sun, S., Lou, J.G., Zhang, D.: LayoutDiffusion: improving Graphic Layout Generation by Discrete Diffusion Probabilistic Models. In: Proceedings of the IEEE/CVF International Conference on Computer Vision (ICCV) (2023)
47. Zhang, L., Rao, A., Agrawala, M.: Adding conditional control to text-to-image diffusion models. In: Proceedings of the IEEE/CVF International Conference on Computer Vision, pp. 3836–3847 (2023)
48. Zhao, X., Pang, Y., Zhang, L., Lu, H.: Joint learning of salient object detection, depth estimation and contour extraction. IEEE Trans. Image Process. **31**, 7350–7362 (2022)
49. Zheng, X., Qiao, X., Cao, Y., Lau, R.W.: Content-aware generative modeling of graphic design layouts. ACM Trans. Graph. (TOG) **38**(4), 1–15 (2019)
50. Zhou, M., Xu, C., Ma, Y., Ge, T., Jiang, Y., Xu, W.: Composition-aware graphic layout GAN for visual-textual presentation designs. arXiv preprint arXiv:2205.00303 (2022)
51. Zhu, D., Chen, J., Shen, X., Li, X., Elhoseiny, M.: Minigpt-4: enhancing vision-language understanding with advanced large language models. arXiv preprint arXiv:2304.10592 (2023)

PreciseControl: Enhancing Text-to-Image Diffusion Models with Fine-Grained Attribute Control

Rishubh Parihar[1](✉), V. S. Sachidanand[1], Sabariswaran Mani[2], Tejan Karmali[1], and R. Venkatesh Babu[1]

[1] Vision and AI Lab, IISc Bangalore, Bengaluru, India
rishubhp@iisc.ac.in
[2] IIT Kharagpur, Kharagpur, India
https://rishubhpar.github.io/PreciseControl.home/

Abstract. Recently, we have seen a surge of personalization methods for text-to-image (T2I) diffusion models to learn a concept using a few images. Existing approaches, when used for face personalization, suffer to achieve convincing inversion with identity preservation and rely on semantic text-based editing of the generated face. However, a more fine-grained control is desired for facial attribute editing, which is challenging to achieve solely with text prompts. In contrast, StyleGAN models learn a rich face prior and enable smooth control towards fine-grained attribute editing by latent manipulation. This work uses the disentangled $\mathcal{W}+$ space of StyleGANs to condition the T2I model. This approach allows us to precisely manipulate facial attributes, such as smoothly introducing a smile, while preserving the existing coarse text-based control inherent in T2I models. To enable conditioning of the T2I model on the $\mathcal{W}+$ space, we train a latent mapper to translate latent codes from $\mathcal{W}+$ to the token embedding space of the T2I model. The proposed approach excels in the precise inversion of face images with attribute preservation and facilitates continuous control for fine-grained attribute editing. Furthermore, our approach can be readily extended to generate compositions involving multiple individuals. We perform extensive experiments to validate our method for face personalization and fine-grained attribute editing.

Keywords: Personalised Image Generation · Fine-grained editing

1 Introduction

Recent personalization methods [10,36] for large text-to-image (T2I) diffusion models [35,38] aim to learn a new concept (e.g., your pet) given a few input

---

R. Parihar and V. S. Sachidanand—Equal contribution
S. Mani—Work done during internship at VAL, IISc.

Supplementary Information The online version contains supplementary material available at https://doi.org/10.1007/978-3-031-73007-8_27.

Fig. 1. Given a single portrait image, we embed the subject into a text-to-image diffusion model for personalized image generation. The embedded subject can then be transformed or placed in a novel context using text conditioning. The proposed method can also compose multiple learned subjects with high fidelity and identity preservation. To obtain precise inversion of face, we condition the T2I model on the rich $\mathcal{W}+$ latent space of StyleGAN2. This enables our method to additionally perform fine-grained control over the generated face with continuous control over facial attributes such as age and beard.

images. The learned concept is then generated using text prompts in novel contexts (e.g. diverse backgrounds and poses) and styles, thus controlling coarse aspects of an image. Personalization of human portraits [50] is especially interesting due to the wide range of applications in entertainment and advertising. However, embedding faces into a generative model has its unique challenges, including faithful inversion of the subject's identity along with its fine facial features. More importantly, smooth control over facial attributes is crucial for precise editing of generated faces, which is challenging to achieve with only text (e.g., continuous increase in smile in Fig. 1).

Advancements in StyleGAN models [17,18] have enabled the generation of highly realistic face images by learning a rich prior over face images. Further, these models have semantically meaningful and disentangled $\mathcal{W}+$ latent space [40] that enable fine-grained attribute control in the generated images [1,13,30]. However, as these models are domain-specific and trained only on faces, they are limited to editing and generating cropped portrait images.

This raises the following question - *How can we combine the generalized knowledge from T2I models with the face-specific knowledge from StyleGAN models?* Such a framework will enjoy benefits from both groups, enabling coarse control with text and fine-grained attribute control through latent manipulation in the generation process. In this work, we propose a novel approach to combine these two categories of models by *conditioning the T2I model with $\mathcal{W}+$ space from StyleGAN2*. Conditioning on the $\mathcal{W}+$ provides a natural way for embedding faces in T2I model by projecting them into $\mathcal{W}+$ space using existing StyleGAN2 encoders [43]. The design of having $\mathcal{W}+$ as the inversion bottleneck has two major advantages: *1) excellent inversion of a face with precise reconstruction of attributes, and 2) explicit control over facial attributes for fine-grained attribute editing.* To the best of our knowledge, this is the first work to demonstrate the combination of two powerful generative models StyleGANs and T2I diffusion models for controlled generation.

To condition the T2I model on the $\mathcal{W}+$ space, we train a latent adaptor - a lightweight MLP, conditioned on the diffusion process's denoising timestep. It takes a latent code $w \in \mathcal{W}+$ of a face as input to predict a pair of time-dependent token embeddings that represent the input face and are used to condition the diffusion model cross-attention. We observe that having a different embedding for each timestep provides more expressivity to the inversion process. The latent adaptor is trained on a dataset of (image, w) latent pairs, guided by identity loss, a class regularization loss, and standard denoising loss. To further improve the inversion quality, we perform a few iterations of subject-specific U-Net tuning on the given input image using LoRA [15]. The embedded subject can then be edited in two ways: *i) coarse semantic edits using text (for e.g., changing the layout and background) and ii) fine-grained attribute edits by latent manipulation in $\mathcal{W}+$ (for e.g., smooth interpolation through a varying range of smiles, ages).* Some example edits are provided in Fig. 1. Our method *generalizes the fine-grained attribute edits from cropped faces (in StyleGANs) to in-the-wild and stylized face images* generated by T2I diffusion model.

The proposed method can be easily extended for multiple-person generation, which requires high fidelity identity of all the subjects (Fig. 1). We first predict separate token embeddings for each person and then perform subject-specific tuning to obtain personalized models. However, training a single personalized model for multiple subjects results in the problem of *attribute mixing* (Fig. 5) between faces, where attributes from one face are mixed with another. Instead, we learn separate subject-specific LoRA models, which are then jointly inferred with a chained diffusion process. The intermediate outputs for these processes are merged using an instance segmentation mask after each denoising step. This framework resolves *attribute-mixing* among subjects and preserves the identity from the fine-tuned models. The attributes of individual subjects can be edited in a fine-grained manner in $\mathcal{W}+$ while preserving the other subjects' attributes as shown in Fig. 1-(Attribute Editing).

We perform extensive experiments for embedding single and multiple subjects in StableDiffusion [35] model. Compared to the existing personalization method, the proposed method is extremely efficient and achieves a good tradeoff between identity preservation and text alignment. Next, we present results for fine-grained attribute editing with continuous control in the $\mathcal{W}+$ latent space and compare performance with existing editing methods. Finally, we compare the results for composing multiple subjects and attribute editing of individual subjects. In summary, our primary contributions are as follows:

- First approach to combine large text-to-image models with StlyeGAN2 by conditioning T2I on rich $\mathcal{W}+$ latent space
- Effective personalization method using a single portrait image enabling fine-grained attribute editing in $\mathcal{W}+$ space and coarse editing with text prompts.
- Novel approach to fuse multiple personalized models with chained diffusion processes for multi-person composition.

2 Related Work

Text-Based Image Generation. Large text-to-image diffusion models [32,35, 38] achieve excellent image generation performance when trained on internet scale captioned image datasets [39]. These models are scaled to high resolution by learning cascaded diffusion models [32,33,38] that generate low-resolution images followed by upsampling. Another promising approach is to train diffusion models in the compressed latent space of a pretrained autoencoder [35].

Personalization aims to embed a concept in T2I model, given a few input images. One group of methods optimize for object-specific token embeddings [2,10,50] via optimization. These approaches preserve text editability, however, they struggle to preserve identity. Another direction is based on fine-tuning diffusion model with strong regularization to avoid overfitting [19,21,36]. The third set of methods [11,37,44,47–49,52] learns a shared domain-specific encoder for faster inversion by leveraging the class-specific features.

Embedding Faces. Recently embedding human faces in T2I models has received a lot of attention [8,11,48,50,52] as the generic personalization methods [10,36] often fail to faithfully embed human faces. Celeb-basis [50] learns a basis of the celebrity names in the token embedding space. The weights of these basis vectors are then predicted by an encoder model applied to the input image. Profusion [52] proposes a regularization-free encoder-based approach. Photoverse [8] applies a dual branch conditioning in text and image domains for faster and more accurate inversion of faces. Although these methods are able to achieve good inversion, they do not allow for fine-grained attribute control. A concurrent work [23] aims to map the $\mathcal{W}+$ space to the T2I model, however, their method is limited in preserving identity.

Image Editing. Trained T2I models serve as strong image priors and enable various image editing and restoration applications [5,24,28,46]. For fine-grained image editing [14,27,28] localizes the object in the image space using attention masks and only allows editing in the specified region. However, these methods rely on text to change the localized object, which does not allow for fine-grained control. Promising approaches like [4,12] provide a finer control by interpolation in the noise space or training special sliders per attribute. Another set of works explores the intermediate feature space of unconditional diffusion models to obtain finer attribute control in generation [22,25]. However, they are limited to editing of generated images and not personalized subjects. We take inspiration from GAN models to attain fine-grained attribute control for real subjects by leveraging their disentangled and smooth latent spaces [16,40]. This enables precise attribute editing through latent manipulation [1,26,30,40] and we aim to embed these properties in pretrained T2I models.

3 Method

3.1 Preliminaries

Text-to-Image Diffusion Models. This work uses StableDiffusion-v2.1 [35] as a representative Text-to-image (T2I) diffusion model. Stable diffusion is based on the latent diffusion model, which applies the diffusion process in the latent space. Its training involves two stages: a) training a VAE or VQ-VAE autoencoder to map images to a compressed latent space, and b) training a diffusion model in its latent space conditioned on text for guiding the generation. This framework disentangles the learning of fine-grained details in the autoencoder and semantic features in the diffusion model, resulting in easier scaling.

Style-based GANs, [6,17,18] have been widely adapted to generate realistic object-specific images such as faces. Further, these models have disentangled latent space, which enables smooth interpolation between images and fine-grained attribute editing [30,40]. These properties are induced by mapping the Gaussian latent space to a learned latent space $\mathcal{W}/\mathcal{W}+$ with a mapper network. Further, GAN encoder models [34,43] can encode and edit real images that invert a given image into $\mathcal{W}+$ space, allowing for fine-grained editing of real images.

3.2 Overview

While the T2I models trade off the diversity in generation with an attribute-rich latent space, our goal is to condition the T2I model with an attribute-rich - $\mathcal{W}+$ space from StyleGAN2, that allows for disentangled and fine-grained control over the face attributes in the generated image. To condition the T2I model on $\mathcal{W}+$, we augment the T2I model with a learnable latent adaptor network $\mathcal{M}$ that projects a latent code $w \in \mathcal{W}+$ into the text embedding space. For embedding a new subject, we pass it through a pre-trained StyleGAN2 encoder $\mathcal{E}_{GAN}$ [43] to obtain w latent code, which is passed through $\mathcal{M}$ to obtain the corresponding text-embedding as shown in Fig. 2a). Conditioning on $\mathcal{W}+$ enables fine-grained attribute control in the generated image by latent manipulation. In the next sections, we discuss the details of the proposed latent adaptor, model training, and fine-grained attribute editing.

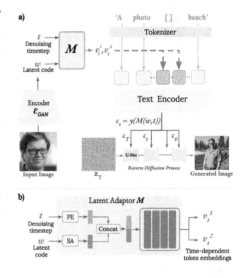

Fig. 2. Framework for personalization. Given a single portrait image, we extract its w latent representation from encoder $\mathcal{E}_{GAN}$. The latent w along with diffusion timestep t are passed through the latent adaptor $\mathcal{M}$ to generate a pair of time-dependent token embeddings (v_t^1, v_t^2) representing the input subject. Finally, the token embeddings are combined with arbitrary prompts to generate customized images.

3.3 Latent Adaptor $\mathcal{M}$

We implement the latent adaptor $\mathcal{M}$ as a shallow MLP network that maps the w latent code from StyleGAN to the token embedding space of the T2I model for any human face image as input. We learn two token embeddings (v^1, v^2) to represent a human subject as it is known to improve the embedding quality [50]. To extract the timestep-specific semantic information from the latent **w**, we condition $\mathcal{M}$ on the diffusion timestep t, as diffusion models represent the semantic hierarchy in a timestep-wise fashion [29]. The output of $\mathcal{M}$ is a set of pair of embedding vectors $\{(v_t^1, v_t^2)\}_{t=0}^{t=T}$, a pair for each timestep t. Time-dependent token embeddings allow for a richer representation space and improve identity preservation (shown in Fig. 12). The complete architecture of $\mathcal{M}$ is shown in Fig. 2b). The input t is first passed through positional encodings [42] and the flattened w latent code is passed through a self-attention layer to get relevant features. The encoded representations are then concatenated before passing through a set of linear layers. The obtained pair embedding (v_t^1, v_t^2) represents the person and is then passed at t^{th} denoising time-step in the U-Net for generation.

3.4 Training

We perform a two-stage training, where we first pretrain latent adaptor $\mathcal{M}$ on a face dataset, followed by an few iterations of subject-specific training of $\mathcal{M}$ and diffusion U-Net with low-rank updates for improving identity as detailed below.

Pretraining. The mapper $\mathcal{M}$ is pretrained with a paired dataset $\mathcal{D}_w$ consisting of (I, w) pairs, where I is a portrait face image and w is its corresponding latent code obtained as $\mathcal{E}_{GAN}(I)$. During training, we sample a pair (I, w) and a denoising timestep $t \in (1, T)$ which are passed through $\mathcal{M}$ to obtain the pair of token embeddings (v_t^1, v_t^2) corresponding to the input subject. We place the sampled tokens along with the neutral prompt - $y =$ 'A photo of a ... person' and pass through the text encoder to obtain the final text embeddings $c(y(\mathcal{M}(t,w)))$. We add the noise from the noise schedule at t to the image I and train diffusion loss along with additional regularization losses shown in Eq. 1 to train $\mathcal{M}$. In this stage of training, all the modules - $\mathcal{E}_{GAN}$, text-encoder, and U-Net are frozen, except for $\mathcal{M}$ which is a shallow MLP, making the training compute efficient.

Subject Specific Training. In the second stage, we fine-tune encoder $\mathcal{M}$ and U-Net for a few iterations with the single input image. Specifically, we perform low-rank weight updates (LoRA [15]) on the U-Net projection matrices and fine-tune

Fig. 3. Delayed identity injection results in better text editability.

$\mathcal{M}$, using the combined loss from Eq. 1. In LoRA training, the model weights are updated as $\mathbf{W} = \mathbf{W} + \alpha \Delta W$, where ΔW is the learned low-rank residual weights. The hyper-parameter α controls the extent of fine-tuning and allows for a trade-off between identity preservation and text editability. This second stage of low-rank tuning improves the subjects' identity without hurting the text editability (Fig. 12).

Loss Function. We train latent adaptor with a combination of denoising diffusion loss $\mathcal{L}_{Diffusion}$ and regularization loss $\mathcal{L}_{reg}$ following [11]. The diffusion loss enforces text-to-image consistency, and regularization loss ensures that the predicted token embedding is close to the token embedding of the superclass v_{cls}, such as *face*. Additionally, we add identity loss $\mathcal{L}_{ID}$ defined as the MSE between the face recognition embeddings from [9] to preserve the identity during inversion. The final loss is computed as a linear combination of these losses:

$$\begin{aligned}
\mathcal{L}_{Diffusion} &= E_{z,y,\epsilon,t}[||\epsilon - \epsilon_\theta(z, c(y(\mathcal{M}(t,w))))||_2^2] \\
\mathcal{L}_{reg} &= ||\mathcal{M}(t,w) - v_{cls}||_2^2 \\
\mathcal{L}_{ID} &= ||\mathcal{E}_{ID}(x_t) - \mathcal{E}_{ID}(I)||_2^2 \\
\mathcal{L} &= \mathcal{L}_{Diffusion} + \lambda_{reg}\mathcal{L}_{reg} + \lambda_{ID}\mathcal{L}_{ID}
\end{aligned} \quad (1)$$

where λ_{reg} and λ_{ID} are hyper-parameters and $\mathcal{E}_{ID}$ is pretrained face-recognition model [9]. To compute $\mathcal{L}_{id}$ at the intermediate denoising step, we use DDIM [41] approximation of the clean image $\hat{x}_0$ and pass it to the face detector (Fig. 4).

3.5 Inference

During inference, given a single image I, we obtain its token embedding as $(v_t^1, v_t^2) = \mathcal{M}(\mathcal{E}_{GAN}(I), t)$ for all the time steps $t \in (1, T)$. These embeddings can be added with text prompts to generate a novel composition of the learned subject. The image generation process in diffusion models follows a hierarchical structure, where the layout is formed in the first few steps, followed by the formation of object shape and appearance [28]. As our primary aim is to embed a subject's identity, we inject the obtained token embedding only after a time threshold $(t < \tau)$ to not hurt the layout generated during the initial timesteps.

For the initial denoising timesteps $(t > \tau)$, we use a celebrity name as a placeholder in the prompt, e.g. 'A photo of Brad Pitt as a star wars character' as the model generates improved image layouts when

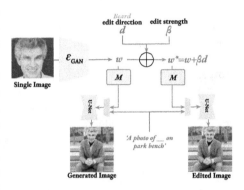

Fig. 4. Fine-grained attribute editing. We map the given input image into w latent code, which is shifted by a global linear attribute edit direction to obtain edited latent code $w*$. The edited latent code $w*$ is then passed through the T2I model to obtain fine-grained attribute edits. The scalar edit strength parameter β can be changed to obtain continuous attribute control.

prompted popular subjects. Empirically, we observe that the generations are not sensitive to the text celebrity name used, and it acts as a placeholder, so we fix a single celebrity name, and there is no overlap of the identity with the dataset used for evaluation. This delayed injection of the learned embedding improves text alignment, and passing the predicted token embeddings to all the timesteps results in poor compositions where the model output is a cropped face, as shown below in Fig. 3.

3.6 Fine-Grained Control over Face Attributes

Once trained, the latent adaptor $\mathcal{M}$ bridges between disentangled and smooth $\mathcal{W}+$ latent space and the text-conditioning of the diffusion model. This enables the *transfer* of latent attribute editing methods that function in the $\mathcal{W}+$ space of StyleGANs [13,30,40] to the diffusion model. Specifically, for a given source image I_s, we first obtain its corresponding w latent code with $\mathcal{E}_{GAN}$. Next, we edit the latent code w by adding a *global* linear attribute edit direction d with scalar weight β to obtain $\hat{w} = w + \beta d$. Note, the same global edit direction d generalizes for all the identities in the $\mathcal{W}+$ space [40]. The edited latent code $\hat{w}$ is then passed through $\mathcal{M}$ to obtain the edited token embedding $\hat{v}_t$. Notably, one can precisely control the strength of the attribute edit by changing the scalar β as shown in Fig. 9. To preserve the scene layout during editing, we use the same starting noise and copy the self-attention maps obtained during the generation with unedited w similar to [28]. Further, one can easily combine multiple edit directions by taking a weighted combination of individual attribute edits (Fig. 10), thanks to the linearity of the $\mathcal{W}+$ (Fig. 6).

Fig. 5. Composing multiple persons without finetuning results in identity distortion. Finetuning a single model for both the identities results in *attribute mixing*, the age and facial hairs from **v1** are transferred to **v2**. Combining outputs of individual finetuned models results in excellent identity preservation without *attribute mixing*.

3.7 Composing Multiple Persons

Our method can be extended to compose multiple subject identities in a single scene. Naively, embedding multiple token embeddings (one per subject) in the text prompt without subject-specific tuning results in identity distortion Fig. 5a). Jointly performing subject-specific tuning improves the identity but suffers from *attribute mixing*, where facial attributes from one subject are transferred to another, such as age and hairs in Fig. 5b). This is a well-known issue in T2I generation, where the model struggles with multiple objects in a scene and binds incorrect attributes [7]. We take an alternate approach inspired by MultiDiffusion [3], where we run multiple chained diffusion processes, one for each subject and one for the background. The outputs of these processes are combined at each denoising step using an instance segmentation mask. We run the diffusion process for each subject through its corresponding subject-specific finetuned model. This preserves the subjects' details learned by each finetuned model and enables high-fidelity composition of multiple persons without *attribute mixing*. To obtain an instance segmentation mask, we run a single diffusion process with a prompt containing two persons and apply the off-the-shelf segmentation model SAM [20]

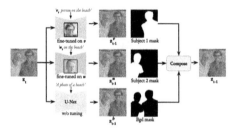

Fig. 6. Composing multiple subjects. We run multiple parallel diffusion processes, one per subject and one for the background, which are fused using instance masks at each denoising step. Importantly, the diffusion process for each subject is passed through its corresponding fine-tuned model, which results in excellent identity preservation.

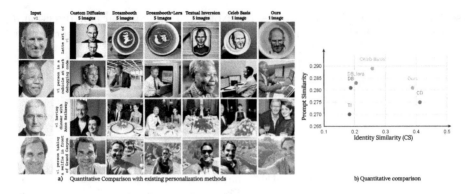

Fig. 7. Comparison for single subject personalization. Existing personalization methods designed for generic concepts either achieve good Identity similarity (Custom Diffusion) or Prompt similarity (Celeb Basis, Dreambooth, Dreambooth+Lora) but fail to achieve both simultaneously. Like ours, Celeb Basis is a single image, face-specific personalization method that achieves good prompt similarity. However, faces generated by Celeb Basis have a cartoonish look and lack realism. Our method strikes a perfect balance between Identity similarity and Prompt similarity, as shown in the plot, and generates highly photo realistic images following the text.

on the generated image. Further, we can perform fine-grained attribute edits on a single subject with latent manipulation in $\mathcal{W}+$ space while preserving other subjects, as shown in Fig. 1.

4 Experiments

We perform all our experiments on StableDiffusion-v2.1 [35] as a representative T2I model. For inversion, we use pre-trained StyleGAN2 e4e encoder [43] trained on the face dataset to map images in $\mathcal{W}+$. In the following sections, we first discuss the datasets and metrics (Sect. 4.1), followed by results on single-subject and multi-subject personalization (Sect. 4.2), fine-grained attribute editing (Sect. 4.3), and ablation studies (Sect. 4.4).

4.1 Dataset and Metrics

Dataset. The latent adaptor is trained with a combination of synthetic images generated from StyleGAN2 and real images from the FFHQ [17] dataset. The dataset contained $70K$ image and corresponding w latent codes obtained from e4e [43]. We collected a dataset of 30 subjects for evaluation, including scientists, celebrities, sports persons, and tech executives. We also evaluate on 'non-famous' identities and synthetic faces in supplementary. We use a set of 25 diverse text prompts, including texts for stylization, background change, and doing certain actions. Further details about the setup is provided in the supplementary.

Metrics. We evaluate the personalization performance using two widely used metrics for subject personalization: **Prompt similarity** - to measure the alignment of the prompt with the generated image using CLIP [31], and **Identity similarity(CS)** - to measure the identity similarity between the input image and the generated image using cosine similarity between face embeddings from [45]. To evaluate fine-grained attribute editing, we compute the change in Prompt similarity (Δ **CLIP**) with the attribute prompt (e.g., 'A $v1$ person smiling') before and after the edit. Additionally, we measure the change in the image during editing with **LPIPS** [51] and Identity similarity. For an ideal fine-grained attribute edit, a higher Δ CLIP indicates a meaningful edit and a lower LPIPS and higher ID-sim denotes the preservation of source identity.

4.2 Comparison with Personalization Methods

Single-Subject Personalization. We perform single-image personalization on the evaluation set with diverse text prompts in Fig. 7, 13 and S8. We compare with following fine-tuning-based personalization methods: Custom Diffusion [21], Dreambooth [36], Dreambooth+LoRA, which is Dreambooth with low-rank updates to avoid overfitting, Textual Inversion [10] and Celeb Basis [50]. All the methods are trained with 5 images per subject except Celeb-basis and ours, which operate on a single input image. Details about hyper-parameters for competitor methods are provided in the supplementary. Custom Diffusion embeds a subject while preserving its identity; however, it mostly generates closeup faces and does not follow the text prompts to stylize the subject or to have it perform an action. Dreambooth cannot embed the subject's identity faithfully, whereas with LoRA training, the identity is improved along with text alignment, which helps avoid overfitting.

Fig. 8. Comparison for multi-subject generation. Textual Inversion struggles to generate both the subjects and distorts their identities. Custom Diffusion and Celeb Basis can compose but suffer from *attribute mixing* - the same hairstyle and facial features are copied to both faces. This effect is more pronounced in Celeb Basis. Our method disentangles the attributes of the two subjects and generates identity-preserving composition.

Textual Inversion and Celeb Basis have poor identity preservation as they fine-tune only the token embedding and not the U-Net. Celeb Basis achieves the highest text alignment due to the strong regularization imposed by basis spanning across celebrity names. Our method strikes a perfect balance between text alignment and identity preservation. Note that ours and the Celeb Basis use only 1 input image, which slightly affects the identity, against Custom-diffusion that requires 5 images. We have provided an additional comparison with encoder-based models and the recent IP-adaptor [49] method in supplementary material.

Multi-subject Personalization. We present results for embedding multiple-person composition in Fig. 8, 14, S7. Specifically, we combine the intermediate outputs of subject-specific tuned models during generation. We compare against multi-concept personalization methods, Textual Inversion, Custom Diffusion, and Celeb Basis. For Textual Inversion and Celeb Basis, we learned two separate token embeddings, one for each subject. For Custom Diffusion, we jointly fine-tune the projection matrices on both subjects. Textual Inversion fails to generate both the subjects in the scene. Celeb Basis and Custom Diffusion generate both the subjects but suffer from *attribute mixing* (eyeglasses from $v4$ are transferred to $v3$). As noted earlier, Celeb Basis generates cartoonish faces in most cases. Our method resolves the attribute mixing by running multiple subject-specific diffusion processes and results in highly realistic compositions.

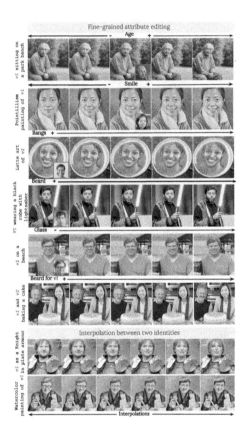

Fig. 9. **Attribute Control.** We perform continuous attribute editing by adding an attribute edit direction in $\mathcal{W}+$ and increasing its edit strength β. Our method performs disentangled edits for various attributes while preserving identity and generalizing to in-the-wild faces, styles, and multiple persons. **Identity Interpolation.** We can perform smooth interpolation between identities by interpolating between the corresponding w codes.

4.3 Fine-Grained Control by Latent Manipulation

The proposed method matches the disentangled $\mathcal{W}+$ latent space of StyleGANs to the token embedding space of T2I models, allowing for continuous control over the image attributes by latent space manipulation. We present two important image editing applications fueled by the disentangled latent space of StyleGANs: 1) fine-grained attribute editing and 2) smooth identity interpolation. Addi-

Table 1. Comparison of fine-grained attribute editing

	Imagic			InterfaceGAN			Ours + Prompt			Ours + $\mathcal{W}+$		
Attribute	Δ CLIP ↑	LPIPS ↓	CS ↑	Δ CLIP ↑	LPIPS ↓	CS ↑	Δ CLIP ↑	LPIPS ↓	CS ↑	Δ CLIP ↑	LPIPS ↓	CS ↑
Beard	4.184	0.533	0.492	2.644	0.221	0.732	0.986	0.232	0.877	2.473	0.185	0.731
Age	4.432	0.571	0.474	1.359	0.220	0.744	0.429	0.202	0.929	1.777	0.209	0.698
Smile	1.945	0.499	0.666	1.800	0.188	0.804	0.784	0.317	0.813	1.104	0.190	0.779
Asian	5.908	0.585	0.478	4.198	0.139	0.708	1.008	0.330	0.804	4.917	0.165	0.625
Black	5.806	0.546	0.347	1.748	0.147	0.825	0.897	0.222	0.898	1.877	0.133	0.834

tionally, our model can restore corrupted face images such as low resolution or inpainting masked facial features in supplementary.

Fine-Grained Attribute Editing. We perform attribute editing by adding a global latent edit direction in $\mathcal{W}+$ to w encoding of the input image. To have a unified method for all the attributes, we take a simplified approach to obtain edit directions, gathering a small set ($<$ 20) of paired portrait images before and after the attribute edit (generated using an off-the-shelf attribute editing method). Next, we take a difference between the corresponding paired w latent and average them to obtain a global edit direction. We found global edit direction for smile, age, beard, gender, race, and eyeglasses. We also show edits with directions obtained using InterfaceGAN [40] in Fig. S6 in supplementary. The results for fine-grained control editing are provided in Fig. 9 and 10, where we show disentangled continuous control for var-

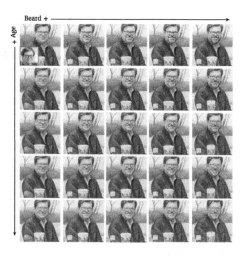

Fig. 10. Multi-attribute-control. We can perform continuous edits for two attributes simultaneously by taking a linear combination of attribute edit directions. Observe the smooth and disentangled edit transformations for age and beard attributes while preserving identity.

ious attributes by changing β (ref. Section 3.6) while preserving the identity. Our method *generalizes the edit directions in $\mathcal{W}+$, originally defined for portrait faces, to in-the-wild and stylized face images.* We evaluate attribute editing performance against **1)** StyleGAN-based global editing method InterfaceGAN [40], after encoding the image using e4e, **2)** Prompt-based editing of the learned subject (by giving prompts like 'A photo of $v1$ smiling'), **3)** Text-based editing method Imagic [19] build on single image personalization. The quantitative results are present in Table 1, and qualitative results are in Fig. 11. Our method achieves the lowest LPIPS scores with high Δ CLIP, indicating highly disentangled attribute editing. Both text-based editing methods fail to preserve the image regions (higher LPIPS). We achieve high CS scores during edits with higher Δ

CLIP, indicating identity-preserving attribute edits. Ours prompt-based editing achieves a superior CS because the edit is not performed in many cases indicated by lower LPIPS [51]. Like ours, InterfaceGAN works in $\mathcal{W}+$ latent space and performs similarly in preserving the image content, the identity of the subject, and editability. However, it is limited to the editing of portrait faces generated by StyleGANs and loses fine-facial features, whereas our method combines the best of both worlds, allowing for fine-grained latent editing with semantic editing in T2I models.

Identity Interpolation. $\mathcal{W}+$ space also allows for smooth interpolation between two identities. Given two input images, we obtain their corresponding w latent codes and perform linear interpolation to obtain the intermediate latent codes. When used as conditioning through the latent adaptor, these latents result in realistic face interpolations with

Fig. 11. Comparison for attribute editing. Using images for text-based attribute edits results in identity distortion and lacks realism after edit. As both InterfaceGAN and our method leverage the same disentangled latent space, they generate high-quality edits. However, InterfaceGAN is limited to cropped faces while we can edit in-the-wild images too.

smooth changes between the two faces, preserving background, as shown in Fig. 9-Bottom.

4.4 Ablations

We ablate over the design choices made in the proposed approach for personalization in Fig. 12. The Identity loss and regularization loss have a similar effect in pushing the token embeddings close to an embedding region for faces. Time-dependent token embedding is crucial to preserving subjects' identity as it provides a more expressive space to represent the face. Finally, subject-specific tuning with combined loss improves the Identity similarity as well as the Prompt similarity as the predicted token embeddings are pushed closer to the editable region with $\mathcal{L}_{reg}$ and $\mathcal{L}_{ID}$.

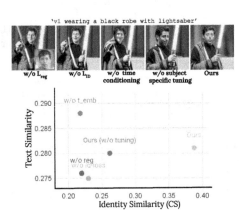

Fig. 12. Ablation study

5 Discussion

Conclusion. We present a novel framework to condition T2I diffusion models on $\mathcal{W}+$ space of StyleGAN2 models for fine-grained attribute control. Specifically, we learn a latent mapper that projects the latent codes from $\mathcal{W}+$ to the input token embedding space of the T2I model learned with denoising, regularization, and identity preservation losses. This framework provides a natural way to embed a real face image by obtaining its latent code using the GAN encoders model. The embedded face can then be edited in two manners - coarse text-based editing and fine-grained attribute editing by latent manipulation in $\mathcal{W}+$.

Limitations. The primary limitation is that the encoder-based inversion in $\mathcal{W}+$ is loose on some information hence we need to perform test time fine-tuning for a few iterations to recover identity similar to pivotal tuning. Additionally, the current approach utilizes multi-diffusion for composing multiple persons which requires multiple diffusion processes. Effectively composing more than two individuals with consistent identities proves challenging within the current method and is an interesting future direction to explore.

Fig. 13. Additional results for single subject personalization. Our method achieves excellent realism and text alignment with the prompts.

Fig. 14. Additional results for Multi-subject personalization. Our method preserves subject identity and follows the text prompt during generation.

Acknowledgements. We thank Aniket Dashpute, Ankit Dhiman and Abhijnya Bhat for reviewing the draft and providing helpful feedback. This work was partly supported by PMRF from Govt. of India (Rishubh Parihar) and Kotak IISc AI-ML Centre.

References

1. Abdal, R., Zhu, P., Mitra, N.J., Wonka, P.: Styleflow: attribute-conditioned exploration of stylegan-generated images using conditional continuous normalizing flows. ACM Trans. Graph. (TOG) **40**(3), 1–21 (2021)
2. Alaluf, Y., Richardson, E., Metzer, G., Cohen-Or, D.: A neural space-time representation for text-to-image personalization. ACM Trans. Graph. (TOG) **42**(6), 1–10 (2023)
3. Bar-Tal, O., Yariv, L., Lipman, Y., Dekel, T.: Multidiffusion: fusing diffusion paths for controlled image generation (2023)
4. Brack, M., Friedrich, F., Hintersdorf, D., Struppek, L., Schramowski, P., Kersting, K.: Sega: Instructing text-to-image models using semantic guidance. In: Thirty-seventh Conference on Neural Information Processing Systems (2023)
5. Brooks, T., Holynski, A., Efros, A.A.: Instructpix2pix: learning to follow image editing instructions. In: Proceedings of the IEEE/CVF Conference on Computer Vision and Pattern Recognition, pp. 18392–18402 (2023)
6. Chan, E.R., et al.: Efficient geometry-aware 3d generative adversarial networks. In: Proceedings of the IEEE/CVF Conference on Computer Vision and Pattern Recognition, pp. 16123–16133 (2022)
7. Chefer, H., Alaluf, Y., Vinker, Y., Wolf, L., Cohen-Or, D.: Attend-and-excite: attention-based semantic guidance for text-to-image diffusion models. ACM Trans. Graph.s (TOG) **42**(4), 1–10 (2023)

8. Chen, L., et al.: Photoverse: tuning-free image customization with text-to-image diffusion models. arXiv preprint arXiv:2309.05793 (2023)
9. Esler, T.: Github - face recognition using pytorch (2021). https://github.com/timesler/facenet-pytorch
10. Gal, R., et al.: An image is worth one word: personalizing text-to-image generation using textual inversion. arXiv preprint arXiv:2208.01618 (2022)
11. Gal, R., Arar, M., Atzmon, Y., Bermano, A.H., Chechik, G., Cohen-Or, D.: Encoder-based domain tuning for fast personalization of text-to-image models. ACM Trans. Graph. (TOG) **42**(4), 1–13 (2023)
12. Gandikota, R., Materzynska, J., Zhou, T., Torralba, A., Bau, D.: Concept sliders: lora adaptors for precise control in diffusion models. arXiv preprint arXiv:2311.12092 (2023)
13. Härkönen, E., Hertzmann, A., Lehtinen, J., Paris, S.: Ganspace: discovering interpretable gan controls. Adv. Neural. Inf. Process. Syst. **33**, 9841–9850 (2020)
14. Hertz, A., Mokady, R., Tenenbaum, J., Aberman, K., Pritch, Y., Cohen-or, D.: Prompt-to-prompt image editing with cross-attention control. In: The Eleventh International Conference on Learning Representations (2022)
15. Hu, E.J., et al.: Lora: low-rank adaptation of large language models. arXiv preprint arXiv:2106.09685 (2021)
16. Karmali, T., et al.: Hierarchical semantic regularization of latent spaces in StyleGANs. In: Avidan, S., Brostow, G., Cissé, M., Farinella, G.M., Hassner, T. (eds.) ECCV 2022. LNCS, vol. 13675, pp. 443–459. Springer, Cham (2022). https://doi.org/10.1007/978-3-031-19784-0_26
17. Karras, T., Laine, S., Aila, T.: A style-based generator architecture for generative adversarial networks. In: Proceedings of the IEEE/CVF Conference on Computer Vision and Pattern Recognition, pp. 4401–4410 (2019)
18. Karras, T., Laine, S., Aittala, M., Hellsten, J., Lehtinen, J., Aila, T.: Analyzing and improving the image quality of StyleGAN. In: Proceedings of the IEEE/CVF Conference on Computer Vision and Pattern Recognition, pp. 8110–8119 (2020)
19. Kawar, B., et al.: Imagic: text-based real image editing with diffusion models. In: Proceedings of the IEEE/CVF Conference on Computer Vision and Pattern Recognition, pp. 6007–6017 (2023)
20. Kirillov, A., et al.: Segment anything. arXiv preprint arXiv:2304.02643 (2023)
21. Kumari, N., Zhang, B., Zhang, R., Shechtman, E., Zhu, J.Y.: Multi-concept customization of text-to-image diffusion. In: Proceedings of the IEEE/CVF Conference on Computer Vision and Pattern Recognition, pp. 1931–1941 (2023)
22. Kwon, M., Jeong, J., Uh, Y.: Diffusion models already have a semantic latent space. arXiv preprint arXiv:2210.10960 (2022)
23. Li, X., Hou, X., Loy, C.C.: When StyleGAN meets stable diffusion: a w+ adapter for personalized image generation. arXiv preprint arXiv:2311.17461 (2023)
24. Meng, C., et al.: Sdedit: guided image synthesis and editing with stochastic differential equations. arXiv preprint arXiv:2108.01073 (2021)
25. Parihar, R., Bhat, A., Basu, A., Mallick, S., Kundu, J.N., Babu, R.V.: Balancing act: distribution-guided debiasing in diffusion models. In: Proceedings of the IEEE/CVF Conference on Computer Vision and Pattern Recognition, pp. 6668–6678 (2024)
26. Parihar, R., Dhiman, A., Karmali, T., Babu, R.V.: Everything is there in latent space: attribute editing and attribute style manipulation by StyleGAN latent space exploration. In: Proceedings of the 30th ACM International Conference on Multimedia, pp. 1828–1836 (2022)

27. Parmar, G., Kumar Singh, K., Zhang, R., Li, Y., Lu, J., Zhu, J.Y.: Zero-shot image-to-image translation. In: ACM SIGGRAPH 2023 Conference Proceedings, pp. 1–11 (2023)
28. Patashnik, O., Garibi, D., Azuri, I., Averbuch-Elor, H., Cohen-Or, D.: Localizing object-level shape variations with text-to-image diffusion models. arXiv preprint arXiv:2303.11306 (2023)
29. Patashnik, O., Garibi, D., Azuri, I., Averbuch-Elor, H., Cohen-Or, D.: Localizing object-level shape variations with text-to-image diffusion models. In: Proceedings of the IEEE/CVF International Conference on Computer Vision (ICCV) (2023)
30. Patashnik, O., Wu, Z., Shechtman, E., Cohen-Or, D., Lischinski, D.: Styleclip: text-driven manipulation of stylegan imagery. In: Proceedings of the IEEE/CVF International Conference on Computer Vision, pp. 2085–2094 (2021)
31. Radford, A., et al.: Learning transferable visual models from natural language supervision. In: International Conference on Machine Learning, pp. 8748–8763. PMLR (2021)
32. Ramesh, A., Dhariwal, P., Nichol, A., Chu, C., Chen, M.: Hierarchical text-conditional image generation with clip latents. arXiv preprint arXiv:2204.06125 **1**(2), 3 (2022)
33. Ramesh, A., et al.: Zero-shot text-to-image generation. In: International Conference on Machine Learning, pp. 8821–8831. PMLR (2021)
34. Richardson, E., et al.: Encoding in style: a StyleGAN encoder for image-to-image translation. In: Proceedings of the IEEE/CVF Conference on Computer Vision and Pattern Recognition, pp. 2287–2296 (2021)
35. Rombach, R., Blattmann, A., Lorenz, D., Esser, P., Ommer, B.: High-resolution image synthesis with latent diffusion models. In: Proceedings of the IEEE/CVF Conference on Computer Vision and Pattern Recognition, pp. 10684–10695 (2022)
36. Ruiz, N., Li, Y., Jampani, V., Pritch, Y., Rubinstein, M., Aberman, K.: Dreambooth: fine tuning text-to-image diffusion models for subject-driven generation. In: Proceedings of the IEEE/CVF Conference on Computer Vision and Pattern Recognition, pp. 22500–22510 (2023)
37. Ruiz, N., et al.: Hyperdreambooth: hypernetworks for fast personalization of text-to-image models. arXiv preprint arXiv:2307.06949 (2023)
38. Saharia, C., et al.: Photorealistic text-to-image diffusion models with deep language understanding. Adv. Neural. Inf. Process. Syst. **35**, 36479–36494 (2022)
39. Schuhmann, C., et al.: Laion-5b: an open large-scale dataset for training next generation image-text models. Adv. Neural. Inf. Process. Syst. **35**, 25278–25294 (2022)
40. Shen, Y., Gu, J., Tang, X., Zhou, B.: Interpreting the latent space of GANs for semantic face editing. In: Proceedings of the IEEE/CVF Conference on Computer Vision and Pattern Recognition, pp. 9243–9252 (2020)
41. Song, J., Meng, C., Ermon, S.: Denoising diffusion implicit models. arXiv preprint arXiv:2010.02502 (2020)
42. Tancik, M., et al.: Fourier features let networks learn high frequency functions in low dimensional domains. Adv. Neural. Inf. Process. Syst. **33**, 7537–7547 (2020)
43. Tov, O., Alaluf, Y., Nitzan, Y., Patashnik, O., Cohen-Or, D.: Designing an encoder for StyleGAN image manipulation. ACM Trans. Graph. (TOG) **40**(4), 1–14 (2021)
44. Valevski, D., Lumen, D., Matias, Y., Leviathan, Y.: Face0: instantaneously conditioning a text-to-image model on a face. In: SIGGRAPH Asia 2023 Conference Papers, pp. 1–10 (2023)

45. Wang, H., et al.: Cosface: large margin cosine loss for deep face recognition. In: Proceedings of the IEEE Conference on Computer Vision and Pattern Recognition, pp. 5265–5274 (2018)
46. Wang, J., Yue, Z., Zhou, S., Chan, K.C., Loy, C.C.: Exploiting diffusion prior for real-world image super-resolution. arXiv preprint arXiv:2305.07015 (2023)
47. Wang, Q., Bai, X., Wang, H., Qin, Z., Chen, A.: Instantid: zero-shot identity-preserving generation in seconds. arXiv preprint arXiv:2401.07519 (2024)
48. Xiao, G., Yin, T., Freeman, W.T., Durand, F., Han, S.: Fastcomposer: tuning-free multi-subject image generation with localized attention. arXiv preprint arXiv:2305.10431 (2023)
49. Ye, H., Zhang, J., Liu, S., Han, X., Yang, W.: Ip-adapter: text compatible image prompt adapter for text-to-image diffusion models. arXiv preprint arXiv:2308.06721 (2023)
50. Yuan, G., et al.: Inserting anybody in diffusion models via celeb basis. arXiv preprint arXiv:2306.00926 (2023)
51. Zhang, R., Isola, P., Efros, A.A., Shechtman, E., Wang, O.: The unreasonable effectiveness of deep features as a perceptual metric. In: Proceedings of the IEEE Conference on Computer Vision and Pattern Recognition, pp. 586–595 (2018)
52. Zhou, Y., Zhang, R., Sun, T., Xu, J.: Enhancing detail preservation for customized text-to-image generation: a regularization-free approach. arXiv preprint arXiv:2305.13579 (2023)

Author Index

A
Aich, Shubhra 57

B
Bao, Wentao 193
Bazin, Jean-Charles 57

C
Cai, Haoming 122
Cai, Xinhao 20
Chen, Badong 105
Chen, Qingchao 20
Chen, Yun-Hsuan 228
Chen, Yuxiao 193
Cho, Woojin 284
Choi, Shinkook 90
Chow, Ka-Ho 434

D
Dai, Yong 415
Dao, Trung 176
De la Torre, Fernando 57

F
Feng, Brandon Y. 122

G
Gupta, Saurabh 360

H
Ha, Taewook 284
Haene, Christian 57
Hayes, Wayne 304
Hema, Vishnu Mani 57
Hilliges, Otmar 140
Hong, Xiaopeng 415
Hu, Kun 159
Hu, Sihao 434
Huang, Daoji 140

Huang, Jia-Bin 122
Huang, Jiaxing 322
Huang, Tiansheng 434
Huang, Yi 265

J
Jiang, Kai 322

K
Kamal, Uday 74
Karmali, Tejan 469
Kim, Donghwan 284
Kim, Jeongsol 398
Kim, Minje 284
Kim, Sanghwan 140
Kim, Seojun 451
Kim, Tae-Kyun 284
Kong, Adams 265
Kong, Yu 193

L
Le, Thanh 176
Lee, Hyokeun 284
Lee, Jihyun 284
Lee, Wonjun 211
Lei, Jie 322
Lei, Ting 1
Lerman, Gilad 211
Li, Kai 193
Li, Yunsong 322
Liu, Ling 434
Liu, Qiang 342
Liu, Shaowei 360
Liu, Xiaoming 38
Liu, Xingchao 342
Liu, Yang 1, 20
Long, Chengjiang 159
Lu, Shijian 322
Lyu, Mingzhi 265

© The Editor(s) (if applicable) and The Author(s), under exclusive license to Springer Nature Switzerland AG 2025
A. Leonardis et al. (Eds.): ECCV 2024, LNCS 15140, pp. 489–490, 2025.
https://doi.org/10.1007/978-3-031-73007-8

M

Ma, Yaohui 415
Ma, Zhiheng 415
Mani, Sabariswaran 469
Manzur, Saad 304
Metaxas, Dimitris N. 193
Metzler, Christopher A. 122
Min, Martin Renqiang 193
Ming, Wenjie 228
Mo, Clinton 159
Mukhopadhyay, Saibal 74

N

Nguyen, Khoi 176
Nguyen, Thuan Hoang 176

P

Parihar, Rishubh 469
Park, Geon Yeong 398
Patel, Deep 193
Peng, Yuxin 1, 20
Pham, Cuong 176

Q

Quan, Ruijie 379

R

Ren, Zhongzheng 360
Ryu, Je-Hwan 284

S

Sachidanand, V. S. 469
Sawan, Mohamad 228
Seol, Jaejung 451
Shah, Sachin 122
Shao, Ling 322
Shim, Kyunghwan 90
Styborski, Jeremy 265
Sun, Leyuan 246

T

Tran, Anh 176

V

Van Gool, Luc 140
Venkatesh Babu, R. 469
Vu, Duc 176

W

Wang, Chao 379
Wang, Junzhe 228
Wang, Lei 105
Wang, Shenlong 360
Wang, Shuang 228
Wang, Xi 140
Wang, Yaowei 415
Wang, Zhiyong 159
Woo, Taeyun 284
Woo, Woontack 284

X

Xian, Yongqin 140
Xie, Mingyang 122
Xie, Weiying 322
Xu, Yankun 228
Xu, Yiran 122

Y

Yang, Jie 228
Yang, Yi 379
Yang, Yifei 211
Ye, Jong Chul 398
Yi, Minjae 284
Yin, Shaofeng 1
Yoo, Jaejun 451
Yoshiyasu, Yusuke 246
Yuan, Dong 159
Yuan, Zejian 105
Yun, Jaewoong 90

Z

Zheng, Minghang 20
Zheng, Zhedong 379
Zhu, Shengjie 38
Zhu, Yuanzhi 342
Zhu, Zhilin 415
Zhuang, Weijun 415
Zou, Dongmian 211

Printed in the USA
CPSIA information can be obtained
at www.ICGtesting.com
CBHW050744111024
15650CB00013B/44